Electromagnetic Metamaterials and Metasurfaces: From Theory To Applications

Long Li · Yan Shi · Tie Jun Cui
Editors

Electromagnetic Metamaterials and Metasurfaces: From Theory To Applications

西安电子科技大学出版社
http://www.xduph.com

Springer

Editors
Long Li
Key Laboratory of High-Speed Circuit Design and EMC
School of Electronic Engineering
Xidian University
Xi'an, Shaanxi, China

Yan Shi
School of Electronic Engineering
Xidian University
Xi'an, Shaanxi, China

Tie Jun Cui
State Key Laboratory of Millimeter Waves
Southeast University
Nanjing, Jiangsu, China

ISBN 978-981-99-7913-4 ISBN 978-981-99-7914-1 (eBook)
https://doi.org/10.1007/978-981-99-7914-1

Jointly published with Xidian University Press
The print edition is not for sale in China (Mainland). Customers from China (Mainland) please order the print book from Xidian University Press.

This Springer imprint is published by the registered company Springer Nature Singapore Pte Ltd.
The registered company address is: 152 Beach Road, #21-01/04 Gateway East, Singapore 189721, Singapore

Preface

Energy, materials, and information are the three pillars of modern science and technology, among which materials as a fundamental technology play an essential role in the development of science and technology. Human beings have been pursuing the exploration of new materials. More than 3000 years ago, human ancestors invented alloys to enhance the mechanical properties of metals. The conductivity of silicon can be significantly improved through doping, thereby promoting the development of semiconductor devices and integrated circuit industries. The nanotechnology in the twenty-first century directly manipulates atoms and molecules at the nanoscale to promote fabrication innovation in new materials. All the changes in these materials have driven the progress of human civilization and technology.

Over the past decade, three-dimensional bulk metamaterials and their two-dimensional counterparts metasurfaces have provided a new design concept to tailor the characteristics of materials according to people's needs. Metamaterials/metasurfaces generally are composed of periodic or non-periodic subwavelength "meta-atom", and thus, the resulting physical properties are derived not from the properties of the materials, but from their newly designed structures. By reasonably designing the meta-atoms as required and flexibly arranging them, metamaterials/metasurfaces can mimic arbitrary material property not found in nature. Therefore, people can achieve arbitrary control of electromagnetic fields and waves. Metamaterial/metasurface is an interdisciplinary research field covering electromagnetism, optics, acoustics, mechanics, thermotics, and many other fields. Metamaterials/metasurfaces-based groundbreaking works have been selected as one of the top ten breakthroughs by *Science Magazine* for many times, and they have been developed into a novel paradigm of modern science and technology.

The authors' research group called electromagnetic metamaterial innovation team has been engaged in the research of metamaterial since 2003. The research related to metamaterial/metasurface covers electromagnetic bandgap structures, left-handed media, reflectarray/transmitarray, frequency selective surface, coding and programmable metasurfaces, etc., and the metamaterial/metasurface-based designs have been applied to many engineering fields including wireless communication, electromagnetic compatibility, stealth, target recognition, tracking and positioning, etc. Authors feel that the book is necessary to summarize the metamaterial/metasurface-related works developed by the authors' research group. The book contains a collection of 12 chapters and is organized into two parts. One is the basic concepts, theoretical models, analysis, and modeling methods of metamaterial/metasurface, and the second part focuses on metamaterial/metasurface designs in the engineering application. The fundamental part includes wave manipulation of left-handed media, generalized Snell's law, Huygens' metasurface, digital coding metasurface, symmetry properties and group theory of the metasurface, local cavity model of electromagnetic bandgap structure, equivalent parameter extraction method of metamaterial, equivalent circuit analysis method of metasurface, fast full-wave simulation algorithm. The application section includes antenna designs, high-speed circuit noise suppression, absorber designs, vortex wave generator carrying orbital angular momentum, invisibility and radar cross section reduction, wireless power transfer and wireless energy harvesting, and metasurface design evolving from coding metasurface to programmable metasurface and finally to smart metasurface. This book does not pretend to be complete, as metamaterial/metasurface is a fast-developing subject. However, we hope that recent cutting-edge contributions presented in this book will give appropriate guidance to scientists, engineers, and graduate students in this attractive and promising field.

Finally, the authors invite the readers to point out any errors that come to their attention. They also welcome any comments and suggestions.

Xi'an, China — Long Li
Xi'an, China — Yan Shi
Nanjing, China — Tie Jun Cui

Acknowledgments

First, we are indebted to the collaborators of ten chapters and their contributions define this book: Xiangyu Cao and Sijia Li from Air Force Engineering University for Chap. 8, Zhang Jie Luo from Southeast University for Chap. 11, Shixing Yu from Guizhou University for Chaps. 7 and 9, Na Kou from Guizhou University for Chaps. 3 and 7, Pei Zhang from the 28th Research Institute of China Electronics Technology Group Corporation for Chaps. 9 and 10, Zhao Wu from Yulin Normal University for Chap. 4, Xuanming Zhang from Xi'an University of Posts and Telecommunications for Chap. 10, Haixia Liu from Xidian University for Chaps. 2, 3 and 4, Bian Wu and Yutong Zhao from Xidian University for Chap. 5, Linfen Shi from Xidian University for Chap. 4, Zhiwei Cui from Xidian University for Chap. 7, Jiaqi Han and Qiang Feng from Xidian University for Chap. 6, Hao Xue from Xidian University for Chap. 9, Xiaojie Dang from Xidian University for Chap. 2, Xi Chen from Xidian University for Chap. 3, Guangyao Liu from City University of Hong Kong for Chap. 6.

We also owe much to many graduate students who have worked together to make this work possible. In particular, we are also grateful to our Ph.D. supervisor Prof. Changhong Liang, who taught us the scientific spirit of pursuing truth. Long Li appreciates collaborative supervisors Prof. Chi Hou Chan from City University of Hong Kong, Prof. Kunio Sawaya from Tohoku University, and Prof. Raj Mittra from Pennsylvania State University, who provided excellent academic exchange and collaboration opportunities. Yan Shi also appreciates collaborative supervisors Prof. Chi Hou Chan from City University of Hong Kong and Prof. Jian-Ming Jin from the University of Illinois at Urbana Champaign, who provided him with opportunities to learn fast full-wave computational methods in time and frequency domains.

Financial support from the following organizations in the course of our research is gratefully acknowledged:

National Natural Science Foundation of China (NSFC) for Information Metamaterials Basic Science Center (No. 62288101);
National Key Research and Development Program of China (No. 2021YFA1401001, No. 2017YFA0700201, No. 2017YFA0700202, No. 2017YFA0700203);

National Natural Science Foundation of China (No. 62371355);
Key Research and Development Program of Shaanxi Province (No. 2021TD-07);
Outstanding Youth Science Foundation of Shaanxi Province (No. 2019JC-15);
Chang Jiang Scholars Program of the Ministry of Education of China.

Many thanks to who painstakingly read all chapters and checked for editorial corrections. Last but not least, we are grateful to our families for their supports and understanding.

Long Li
Yan Shi
Tie Jun Cui

Introduction

This book documents basic concepts and principles, design methods, and engineering applications of electromagnetic metamaterials and metasurfaces, covering equivalent parameter extraction method of metamaterial, equivalent circuit analysis method of metasurfaces, fast full-wave simulation algorithm of periodic structures, local cavity model of electromagnetic bandgap structures, planar reflectarray and transmitarray design method, reconfigurable metasurfaces, digital coding metasurfaces, programmable metasurfaces, information metasurfaces, and their engineering applications in the electromagnetic wave manipulation fields of antenna, scattering, electromagnetic compatibility, vortex wave carrying orbital angular momentum, wireless power transfer and wireless energy harvesting, new communication system, etc.

This book can be used as a textbook for undergraduate and graduate students with some preliminary background in electronic information and communications, as well as a reference book for one who intends to perform research in these areas. The variety of the topics covered is sufficient to nourish many different research directions in the metamaterials and metasurfaces fields.

Contents

Contributors

Xiangyu Cao Air Force Engineering University, Xi'an, Shaanxi, China

Xi Chen School of Electronic Engineering, Xidian University, Xi'an, Shaanxi, China

Tie Jun Cui State Key Laboratory of Millimeter Waves, Southeast University, Nanjing, Jiangsu, China

Zhiwei Cui School of Physics, Xidian University, Xi'an, Shannxi, China

Xiaojie Dang School of Electronic Engineering, Xidian University, Xi'an, Shaanxi, China

Qiang Feng School of Electronic Engineering, Xidian University, Xi'an, Shannxi, China

Jiaqi Han School of Electronic Engineering, Xidian University, Xi'an, Shannxi, China

Na Kou College of Big Data and Information Engineering, Guizhou University, Guiyang, China

Long Li School of Electronic Engineering, Xidian University, Xi'an, Shaanxi, China

Sijia Li Air Force Engineering University, Xi'an, Shaanxi, China

Guangyao Liu Department of Electrical Engineering, City University of Hong Kong, Hong Kong, SAR, China

Haixia Liu School of Electronic Engineering, Xidian University, Xi'an, Shaanxi, China

Zhang Jie Luo State Key Laboratory of Millimeter Waves, Southeast University, Nanjing, China

Linfeng Shi School of Electronic Engineering, Xidian University, Xi'an, Shaanxi, China

Yan Shi School of Electronic Engineering, Xidian University, Xi'an, Shaanxi, China

Bian Wu School of Electronic Engineering, Xidian University, Xi'an, Shaanxi, China

Zhao Wu School of Physics and Telecommunication Engineering, Yulin Normal University, Yulin, China

Hao Xue School of Electronic Engineering, Xidian University, Xi'an, Shaanxi, China

Shixing Yu College of Big Data and Information Engineering, Guizhou University, Guiyang, China

Pei Zhang The 28th Research Institute of China Electronics Technology Group Corporation, Nanjing, China

Xuanming Zhang School of Electronic Engineering, Xi'an University of Posts and Telecommunications, Xi'an, China

Yutong Zhao School of Electronic Engineering, Xidian University, Xi'an, Shaanxi, China

Chapter 1
Introduction to Electromagnetic Metamaterials and Metasurfaces

Long Li, Yan Shi, and Tie Jun Cui

Abstract Electromagnetic metamaterials/metasurfaces have long captivated the increasing interest and popularity in flexibly manipulating electromagnetic waves and gradually developed into a novel paradigm of modern science and technology. Over the past decade, there have been a great number of new discoveries and results reported in this exciting area, and metamaterials/metasurfaces have led to a myriad of new engineering applications. This chapter introduces the subject of this book, reviews the general description of the metamaterial/metasurface, and provides a historical perspective on the origin, concept, and milestone of electromagnetic metamaterials/metasurfaces.

Keywords Electromagnetic metamaterials · Electromagnetic metasurfaces · The origin · The concept · The milestone

1.1 Introduction

As well known, almost all of the electromagnetic phenomena are attributed to the interaction between electromagnetic waves and materials. Therefore, electromagnetic functionalities can be achieved by manipulating the behaviors of the electromagnetic waves in the materials. We all know that natural materials are composed of lots of atoms or molecules. Materials whose atoms are arranged in a regular periodical pattern are called crystals, while noncrystalline materials consist of atoms arranged in a random manner. When a beam with a fixed wavelength strikes a crystal

L. Li (✉) · Y. Shi
School of Electronic Engineering, Xidian University, Xi'an, 710071, Shaanxi, China
e-mail: lilong@mail.xidian.edu.cn

Y. Shi
e-mail: shiyan@mail.xidian.edu.cn

T. J. Cui
State Key Laboratory of Millimeter Waves, Southeast University, Nanjing 210096, Jiangsu, China
e-mail: tjcui@seu.edu.cn

L. Li et al. (eds.), *Electromagnetic Metamaterials and Metasurfaces: From Theory To Applications*, https://doi.org/10.1007/978-981-99-7914-1_1

at a certain angle, the reflection is reinforced by constructive interference, which is called Bragg reflection. This novel effect is at the basis of X-ray diffraction in crystals, which won the Braggs the Nobel Prize in 1915.

Metamaterials are artificial composite materials with electromagnetic properties beyond those found in nature. The term of metamaterials was introduced by R. M. Walser. In analogy with the natural materials, metamaterials generally are composed of periodic or non-periodic subwavelength "meta-atom". By reasonably designing the meta-atoms as required and flexibly arranging them, the bulky metamaterials can not only mimic known material responses, but tailor new, physically realizable responses, for example, negative permittivity and permeability, zero refractive index, etc. Metasurfaces as the two-dimensional counterparts of the volumetric metamaterials offer some advantages including less loss and easier-to-fabricate, etc., and can be engineered to control electromagnetic fields including wavefront transform, polarization conversion, etc. Metamaterials, including metasurfaces, are expected to achieve significant technological breakthroughs in the areas of new microwave/optical device design, broadband/miniaturized antenna design, high-resolution imaging, novel sensors, switches and modulators, military cloaks and absorbing materials. In the increasingly complex electromagnetic world, digital coding and programmable metamaterials and metasurfaces have been enabling commercial opportunities with broad impact on wireless communications. By controlling the electromagnetic amplitude, phase, polarization, spectrum, and their interactions with digital information, the incoming signals can be directly modulated by programmable metamaterials or metasurfaces. Further applications based on information metamaterials include novel microwave components, 5G/6G communication systems, reconfigurable intelligence surfaces (RIS), low-cost phased arrays, artificial-intelligence-driven designs, computational imaging, wireless power transfer and wireless energy harvesting, and microwave sensing and recognizing, and so on.

Since the concept of the metamaterial/metasurface was proposed, metamaterials/ metasurfacse have been applied into various fields, including electromagnetics, acoustics, optics, mechanics, thermotics, etc. In addition, electromagnetic metamaterial/metasurface can be categorized in terms of operating frequencies including microwave, THz, optics, etc. The new material construction paradigm brought by metamaterial/metasurface has broken the performance limits of conventional materials in nature, and its various exotic functions have derived many disruptive technologies. For this reason, metamaterial/metasurface has been selected as one of the top 10 breakthroughs by *Science Magazine* for many times. Since the subversive effect of metamaterials/metasurfaces has extended into various fields, it is impossible to cover all excellent advances on metamaterial/metasurface in this book. Facing the practical requirements in the fields of electronics, wireless communication and radar, etc., this book summarizes a part of the recent progress of the metamaterial/metasurface conducted by authors. Specifically, the book covers 12 chapters. This chapter is an introduction of the metamaterial/metasurface including basic concept and origin and progress. Chapter 2 discusses fundamental theoretical models of the metamaterial/

metasurface including wave manipulation by negative material parameters, generalized Snell's law, Huygens' metasurface, digital coding metasurface, symmetry properties and group theory of the metasurface. Chapter 3 gives the analysis and simulation methods of the metamaterial/metasurface, including the local resonant cavity model, equivalent parameter extraction methods, equivalent circuit analysis, and fast full-wave simulation method. Chapter 4 focuses on designs of electromagnetic bandgap high impedance surface and its applications including antenna designs and high-speed circuit noise suppression. Chapter 5 presents the graphene-based metasurface absorber and radiator designs. Chapter 6 investigates digital reconfigurable metasurfaces in frequency and spatial domains. Chapter 7 studies the reflection and transmission metasurface designs, especially for orbital angular momentum application. Chapter 8 discusses invisibility and radar cross section (RCS) reduction of the metamaterial/metasurface. Chapters 9 and 10 give the application of the metasurface into wireless power transfer and wireless energy harvesting, respectively. Chapter 11 provides the new concept and design of information metasurfaces that bridge the physical world and digital world. Chapter 12 is a summary of the electromagnetic metamaterials/metasurfaces.

1.2 Features and Classifications of Electromagnetic Metamaterials/Metasurfaces

Metamaterial is an artificial composite structure made of a periodic/quasi-periodic arrangement of so-called "meta-atoms" with the electrical size much smaller than the wavelength. The electromagnetic behaviour of the bulky metamaterial can be characterized by equivalent permittivity ε and permeability μ/refractive index n and impedance η, similar to the natural materials, as shown in Fig. 1.1. The materials including the natural materials and metamaterials can be categorized according to the equivalent permittivity and permeability in material parameter domain, as shown in Fig. 1.2. In the material parameter domain, the permittivity and permeability of the air/free space is ε_0 and μ_0, respectively. The permittivity and permeability of the natural materials, for example printed circuit board material including FR4, F4B, etc., are located in the first quadrant of the parameter domain, i.e., $\varepsilon > 0$ and $\mu > 0$. Moreover, most of natural materials are nonmagnetic, and thus their material parameters lie on the line of $\mu = \mu_0$. In the second quadrant of $\varepsilon < 0$ and $\mu > 0$, there are electric plasma, thus resulting in evanescent waves. Similarly, the fourth quadrant of $\varepsilon > 0$ and $\mu < 0$ denotes magnetic plasma, giving rise to evanescent waves. In the third quadrant, there are the left-handed materials (LHMs) with $\varepsilon < 0$ and $\mu < 0$, which supports the backward waves and the negative refraction. At the vicinity of $\mu = 0/\varepsilon = 0$, the materials are called μ/ε near zero material. Especially, at the origin, the materials have $\mu = 0$ and $\varepsilon = 0$ simultaneously, thus achieving perfect tunneling effect. With the design and arrangement of the meta-atoms, the metamaterial can theoretically mimic arbitrary material property in the parameter domain.

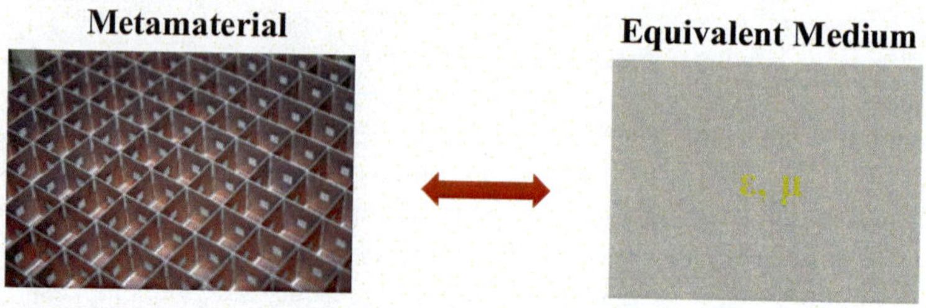

Fig. 1.1 Metamaterial characterized by equivalent medium

Fig. 1.2 Classification of all kinds of homogeneous materials in the ε–μ domain

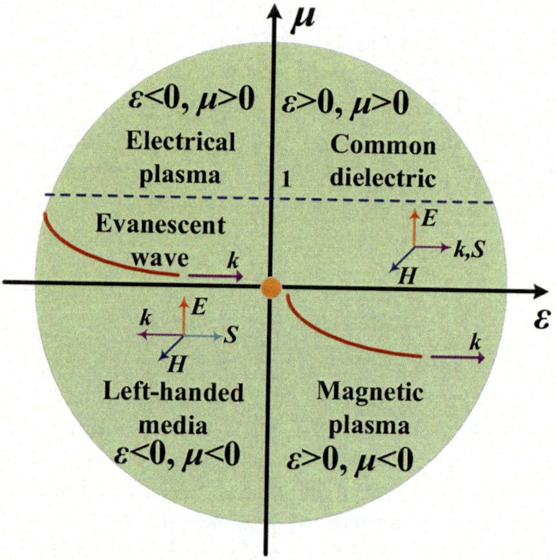

Metasurface as a surface version of the metamaterial is engineered by arranging a set of meta-atom into a two-dimensional (2D) pattern. Compared with the bulky metamaterials, the metasurfaces are of lower profile and less losses due to the subwave length thickness. Naturally, the metasurfaces can be considered as a homogenous thin slab characterized by effective material parameters. Considering the electrically thin thickness, the metasurface can be approximately regarded as a zero-thickness sheet for simplification. In this scenario, the interaction between the electromagnetic wave with the metasurface can be modelled by sheet boundary condition, as shown in Fig. 1.3. Specifically, the sheet boundary condition can model thin material layer as

$$
\begin{aligned}
Z\mathbf{J} &= \mathbf{n} \times \mathbf{E}_{av} \\
Y\mathbf{K} &= \mathbf{n} \times \mathbf{H}_{av}
\end{aligned}
\tag{1.1}
$$

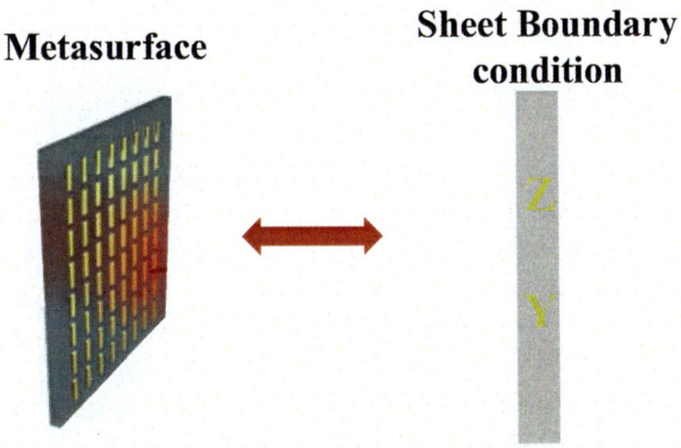

Fig. 1.3 Metasurface characterized by sheet boundary condition

where Z and Y are the equivalent electric impedance and the equivalent magnetic admittance, respectively, $\mathbf{J}$ and $\mathbf{K}$ are equivalent surface electric and magnetic current densities, respectively. Here $\mathbf{E}_{av}$ and $\mathbf{H}_{av}$ denote the average electric field and magnetic field at the thin sheet, respectively. Assume that the thickness of the thin slab is d and the corresponding wavenumber is k. The $\mathbf{E}_{av}$ and $\mathbf{H}_{av}$ can be approximately expressed in terms of tangential components of the electric and magnetic fields before and after the slab

$$\begin{aligned} \mathbf{E}_{av} &\approx \frac{\mathbf{E}_t^+ + \mathbf{E}_t^-}{kd} \tan\left(\frac{kd}{2}\right) \\ \mathbf{H}_{av} &\approx \frac{\mathbf{H}_t^+ + \mathbf{H}_t^-}{kd} \tan\left(\frac{kd}{2}\right) \end{aligned} \tag{1.2}$$

in which superscripts "+" and "−" represent two sides of the thin slab, respectively. When kd tends to be zero, Eq. (1.2) can be reduced to

$$\begin{aligned} \mathbf{E}_{av} &\approx \frac{\mathbf{E}_t^+ + \mathbf{E}_t^-}{2} \\ \mathbf{H}_{av} &\approx \frac{\mathbf{H}_t^+ + \mathbf{H}_t^-}{2} \end{aligned} \tag{1.3}$$

According to (1.1), we can know that in modelling the metasurface, the metasurface is approximately considered as a zero-thickness surface. The equivalent electric impedance and magnetic admittance relate the discontinuities of the electric and magnetic fields caused by the metasurface, respectively. The sheet boundary condition (1.1) corresponds to the isotropic metasurface. A general case, for example bianisotropic metasurface, can be straightforwardly generalized from (1.1) by introducing the equivalent electric impedance tensor, the equivalent magnetic admittance tensor, the equivalent electric-to-magnetic coupling tensor and the equivalent magnetic-to-electric coupling tensor.

1.3 Brief History of Electromagnetic Metamaterials/ Metasurfaces

In recent years, metamaterial/metasurface has become one of the fastest developing fields in electromagnetism and optics due to the abilities to manipulate electromagnetic waves. Up to now, there are countless examples of metamaterial/metasurface concepts and applications that have been reported in the literatures. But in fact, before the term 'metamaterial/metasurface' appeared, researchers have already carried out the work based on artificial periodic structures to control the frequency, amplitude, phase and polarization of electromagnetic waves. This regulation was achieved by employing the unit cells whose dimensions are in the order of operating wavelength, such as frequency selective surface (FSS), electromagnetic band gap (EBG) structures [1–5] and transmitarray (TA)/reflectarray (RA), etc. Therefore, for the research history of metamaterial/metasurface, we can establish two different timelines. Specifically, one is for the development of the concept of the metamaterial/metasurface, and the other covers the study of artificial electromagnetic media before the introduction of metamaterial/metasurface related concepts. It would be impossible to provide a complete review about the metamaterials/metasurfaces. Instead, a non-exhaustive review of metamaterials/metasurfaces is given. More details about the progress of the metamaterial/metasurface can be referred to [6–11].

First let us focus on the efforts of researchers on how to regulate electromagnetic waves through artificial media before the concept of the metamaterial/metasurface was born. More than a hundred years ago, researchers studied dense arrays of wires (so-called inductive mesh) [12, 13] in order to design light-weight reflectors and other RF devices, and thereby the earliest material homogenization theory was introduced. One of classical planar periodic arrays is FSS with spatial filtering characteristics [14–16]. Historically, physical insight of the FSS has evolved from the investigation of diffraction grating in optics. With the grating composed of the equally spaced hairs, write light is decomposed into some monochromatic lights [17, 18]. According to the simple frequency filtering procedure, various FSSs have been proposed and widely used in microwave, infrared, and even visible light bands. Through sophisticated design of shapes and sizes of the unit cells which have self-resonant effects, the resulting periodic array can exhibit a band-pass or band-stop filtering performance. The widely used self-resonant unit cells include 'Jerusalem cross' of the dipole-type [19] and 'slots' of the aperture-type [20], and so on. Since the conventional FSS is designed based on the half-wavelength size and the interaction between the electromagnetic wave and the FSS depends on Floquet-Bloch modes, it is difficult to be classified into the metasurface which are characterized by the sub-wavelength structure in the narrow sense. However, as the research progresses, some scholars have proposed the realization of FSS based on the sub-wavelength unit [21–25]. Moreover, response of the periodic array depends on not only the array layout but also the resonant characteristic of the unit cell [26]. Thus it also promotes the integration of FSS into the field of generalized metasurfaces.

In addition to the frequency filtering characterized by the FSS, photonic bandgap (PBG) structure realizes the localization and trapping of light in a periodically arranging bulk media with different refractive indices over a band of frequencies. L. Rayleigh studied one-dimensional PBG in 1887, showing a spectral range of large reflectivity known as stop-band [27]. Yablonovith [5] and John [28] published two milestone papers on PBG in 1987. Since then, the term "PBG" has emerged. In fact, there are many natural PBGs in nature, such as butterfly wings, peacock feathers, beetle shells, gem opal, etc., whose structural colors are produced by the selective reflection of the special periodic nanostructures of the PBGs to specific wavelengths. Many unusual physical phenomena of the PBG have been studied. However, it is difficult to fabricate the optical PBG, and thus early research focused more on microwave photonic crystals of centimeter size, i.e., EBG. In 1991, the first three-dimensional PBG in the microwave regime has been designed by Yablonovith [4], as shown in Fig. 1.4. The designed PBG is composed of a slab of material covered by a mask consisting of a triangular array of holes. Each hole is drilled through 3 times, at an angle of 35.26° away from normal, and spread out 120° on the azimuth. Another representative PBG is a three-dimensional stacked periodic dielectric structure constructed with layers of dielectric rods [29], as shown in Fig. 1.5.

According to the different structural arrangements, the PBG can be divided into one-dimensional, two-dimensional, and three-dimensional photonic crystals. The fabrication process of three-dimensional photonic crystals is relatively complex,

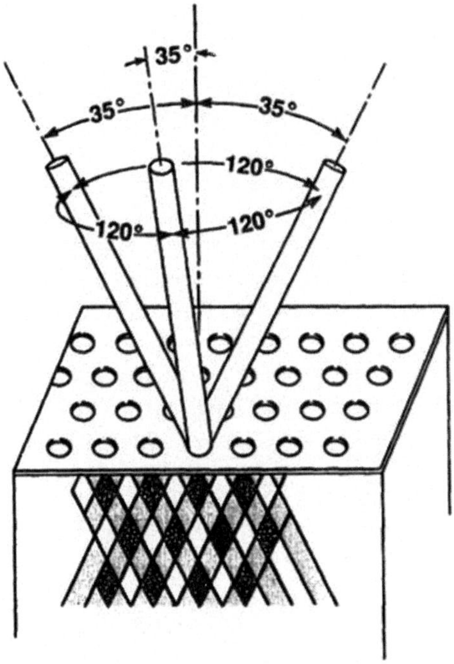

Fig. 1.4 Three-dimensional PBG proposed by Yablonovith [4]

Fig. 1.5 Three-dimensional stacked PBG proposed by Ho [29]

which limits their applications. On the other hand, two-dimensional photonic crystals are easier to fabricate. Especially with the mature printed circuit board (PCB) processing technology, various two-dimensional EBG structures have been designed and widely used in the design of structures such as antennas and microwave circuits. In 1999, D. Sievenpiper proposed a two-dimensional EBG structure consisting of an array of metal protrusions on a flat metal sheet [30], as shown in Fig. 1.6. They are arranged in a two-dimensional lattice, and can be visualized as mushrooms or thumbtacks protruding from the surface. The two-dimensional EBG exhibits high impedance and in-phase reflection in a certain frequency band [30–36], and thus it is also called high impedance surface (HIS) and artificial magnetic conductor (AMC). Based on the bandgap characteristic, the EBG can be used to suppress high speed circuit noise. When the EBG is used as the ground of the antenna, the antenna profile can be reduced owing to the in-phase reflection response. In addition, the EBG can also be employed for the RCS reduction. In 1999, the PBG has been selected as one of the top 10 breakthroughs by *Science Magazine.*

Except frequency manipulation of electromagnetic wave, the wave front can be reshaped by adjusting the phase of the electromagnetic wave, such as reflection, refraction and focusing [37]. Generally speaking, the spatial optical path difference gives rise to the phase difference. The key to control the beam is to make compensation for the phase delay. Although people have mastered the application of optical convergence and divergence as early as hundreds of years ago, it is difficult to achieve miniaturization and planarization of the bulk device. Prior to the birth of the concept of metasurface, planar RA [38–40]/TA [41, 42] based on half-wavelength resonant

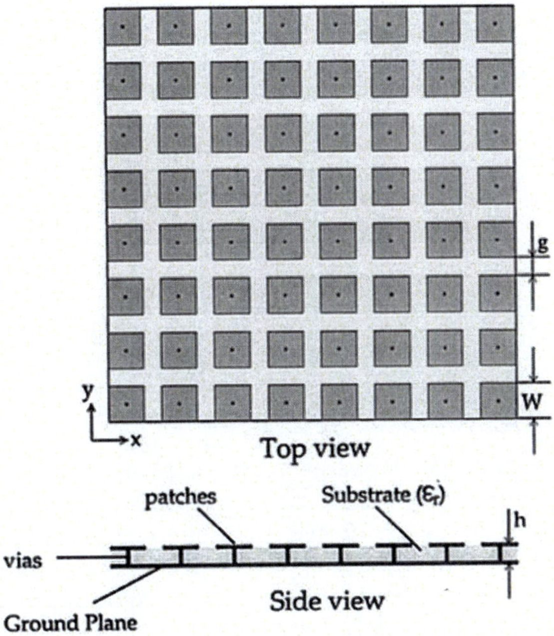

Fig. 1.6 Two-dimensional mushroom EBG proposed by Sievenpiper [30]

units have been gradually implemented with the development of PCB technology. By reasonably designing the unit size, extension line length, rotation angle, and other parameters, each unit can reflect or transmit electromagnetic waves after being irradiated by the incident wave to form the phase delay required by a certain wave front. When the structure thickness and unit size reach the sub-wavelength level, it can be regarded as non-uniform electric and magnetic current sheets, so that the theoretical characterization of metasurface can be obtained [43]. From single-beam to multi-beam [44, 45], single-frequency to dual-frequency/multi-frequency [46, 47], conventional beam to near-field focusing beam [48, 49], quasi-non-diffractive beam (such as Bessel beam) [50–52], passive to active reconfigurable [53–55], etc., various TA/RA have been thoroughly studied. In a broad sense, RA/TA can also be classified into the metasurface category based on the unit's self-resonance characteristics to achieve electromagnetic wave regulation.

Generally, metamaterials are considered to be originated from the Soviet physicist Veselago in the 1960s [56], who comprehensively analyzed that when electromagnetic waves propagate in a medium with the negative permittivity and negative permeability, the electric field **E**, the magnetic field **H**, the wave vector **k** follow the left-handed rule, in opposite to the right-handed rule in natural materials, and thus is called LHM or double negative medium (DNG). With the left-handed rules, the phase velocity is opposite to the group velocity in the LHM. In fact, as early as a hundred years ago, some scientific researchers have proposed the concept of backward-wave in the field of mechanical waves [57, 58]. However, since there is no such material in reality, the related research has been stuck at the theoretical level.

Fig. 1.7 The implementation of left-handed materials [61]

At the end of the twentieth century, Pendry et al. proposed that the electric negative properties [59] and the magnetic negative properties [60] could be realized, respectively, by the periodic metal wire structure and periodic split resonance ring (SRR) structure. At the beginning of the twenty-first century, D. R. Smith's team from Duke University realized the negative refraction characteristic at a certain frequency for the first time through combining the metal wire structures and SRR structures [61, 62], which was named one of the top ten scientific breakthroughs by *Science* in 2003, as shown in Fig. 1.7. The research on metamaterials has not been smooth sailing and has experienced great controversy. The reason is that some researchers believe that the properties of the LHM violate the condition that group velocity cannot exceed the speed of light and the law of causality. In 2004, T. J. Cui and J. A. Kong theoretically proved the energy conversation and causality of the electromagnetic wave hold in the LHM, thus clarifying controversy of the LHM [63, 64].

The terminology "metamaterial" was coined by R M. Walser from University of Texas at Austin in 1999, which is used to characterize the man-made three-dimensional periodic composite structure not found in nature. Hitherto, there is not yet a strict and authoritative definition of the metamaterial, and in the published literatures the definitions of the metamaterial are also different. The description about the metamaterial from Wikipedia is as follows [65].

A **metamaterial** (from the Greek word μετά *meta*, meaning "beyond" or "after", and the Latin word *materia*, meaning "matter" or "material") is any material engineered to have a property that is not found in naturally occurring materials. They are made from assemblies of multiple elements fashioned from composite materials such as metals and plastics. The materials are usually arranged in repeating patterns, at scales that are smaller than the wavelengths of the phenomena they influence. Metamaterials derive their properties not from the properties of the base materials, but from their newly designed structures. Therefore, LHM, EBG and FSS can be regarded as metamaterials.

Although the implementation of LHM is exciting, its inherent shortcomings of narrow-band and high-loss make the application path very tortuous [66]. Around 2005, the bending of electromagnetic waves by using gradient refractive index

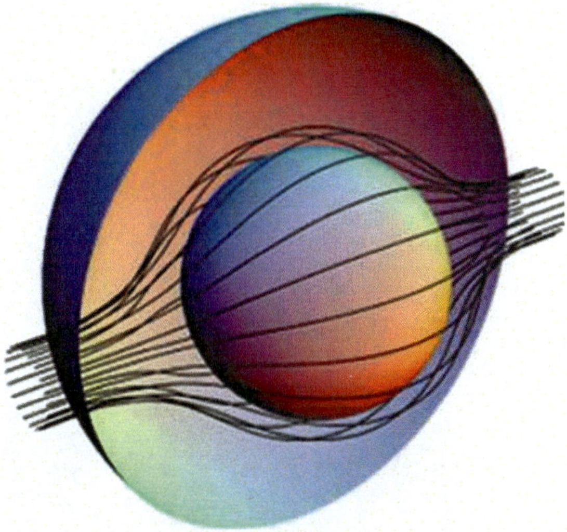

Fig. 1.8 Spherical invisible cloak [69]

medium set off a new wave of metamaterial research [67]. Subsequently, in 2006, two landmark papers were published consecutively in the same volume of *Science Magazine*, i.e., U. Leonhardt's paper entitled by "Optical conformal mapping" [68] and J. B. Pendry's paper entitled by "Controlling electromagnetic fields" [69], as shown in Fig. 1.8. The two works systematically reveal cloak design method through optical conformal transformation and coordinate transformation, respectively. Metamaterial electromagnetic cloak consisting of SRR structures at microwave frequencies was realized by Schurig et al. [70]. Thus metamaterials were once again named as one of the top ten scientific advances of the year by *Science Magazine* in 2006. Transformation optics has attracted widespread attention due to its ability to freely manipulate electromagnetic fields, and various cloaks [71–73] and functional devices [74–78] based on transformation optics have also been developed (Fig. 1.9).

The concept for space electromagnetic wave manipulation can be extended to guided wave, thus establishing composite right-left handed (CRLH) theory [79], as shown in Fig. 1.10. In terms of transmission line-based analysis and design, CRLH transmission line exhibits backward wave, anti-parallel phase and group velocities nonlinear responses, etc., thus resulting in various broadband, compact size device designs [80–82]. The introduction of metamaterial has provided new solutions for the design of new antennas and microwave devices [83], radar illusion [84], super-resolution imaging [85], and other fields. With the continuous improvement of application requirements, electronic equipment and products are developing towards the trends of miniaturization, low profile, and integration, and therefore researchers began to focus on the use of planar artificial media instead of the traditional three-dimensional metamaterials due to the limitations of bulk size.

The concept of metasurface is derived from the work published in *Science Magazine* by Capasso et al. [86] from Harvard University in 2011, who proposed the generalized Snell's law including the abnormal refraction and reflection, as shown in

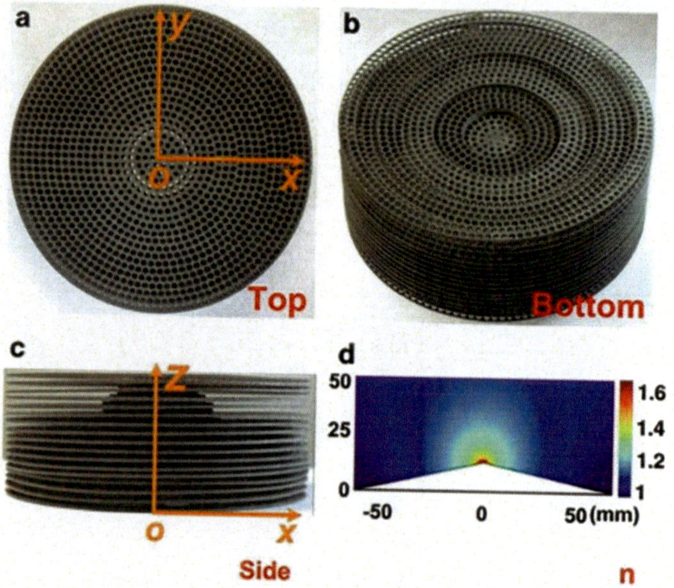

Fig. 1.9 Three-dimensional broadband ground-plane cloak [72]

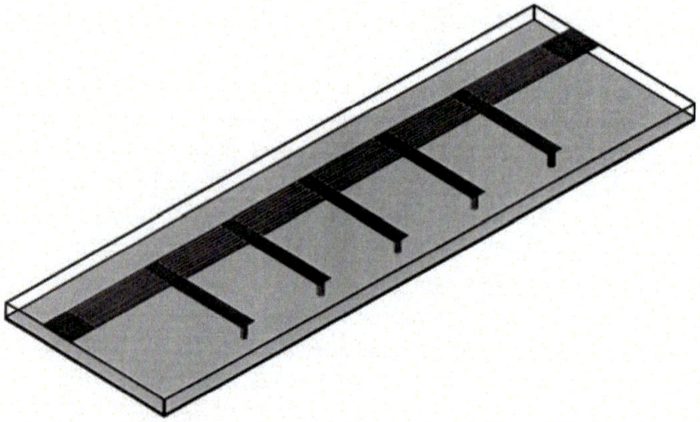

Fig. 1.10 Composite right-left handed structure [79]

Fig. 1.11. With the generalized Snell's law, researchers can design planar periodic or aperiodic arrays according to their needs to achieve the flexible control of the reflected and refracted waves, and achieve arbitrary phase gradient and wave front. A series of fascinating designs such as orbital angular momentum vortex wave [87], conversion from propagating waves to surface waves [88], polarization conversion [89, 90], holographic imaging [91, 92], perfect control of reflection and refraction [93, 94], wireless power transfer and wireless energy harvesting [95, 96], RCS reduction [97, 98], etc.

In response to the trend of high integration and miniaturization of electromagnetic systems, multiple electromagnetic functions are integrated into a sole metasurface, including multiple beams [99, 100], multiple polarizations [101, 102], and multiple frequencies [103, 104], and so on. In recent years, in order to achieve a higher degree of freedom regulation of electromagnetic waves, active devices, such as PIN diodes and varactors, or materials, for instance, liquid crystal, VO_2, and graphene, have been introduced into metasurfaces to achieve the design of active metasurfaces [105–109]. The concept of digital coding and programmable metasurfaces was proposed to digitally control EM waves by T. J. Cui from Southeast University in 2014, as shown in Fig. 1.12 [110]. An analogous concept applied to bulk metamaterials was developed by N. Engheta from the University of Pennsylvania [111]. Coding metasurfaces could simplify the design and optimization procedures owing to the digitalization of meta-atoms. Most importantly, coding metasurfaces have built up a bridge between the physical and digital worlds, making it possible to revisit metamaterials from the perspective of information science and be combined with algorithms in digital signal processing, which further promotes the trend of metamaterial/metasurface towards informatization [112] and intellectualization [113, 114].

Metamaterials/metasurfaces, have been developing faster than anyone could have imagined, rapidly becoming one of the most important branches in the field of electromagnetism and optics in the last twenty years. Its development extends from the original artificial electromagnetic media such as EBG and FSS to LHM, and now to various non-linear metasurface [8], holographic metasurface [9, 116], and reconfigurable metasurface [117–120], from theoretical hypotheses to experimental verification, then to characterization models and electromagnetic calculations. From

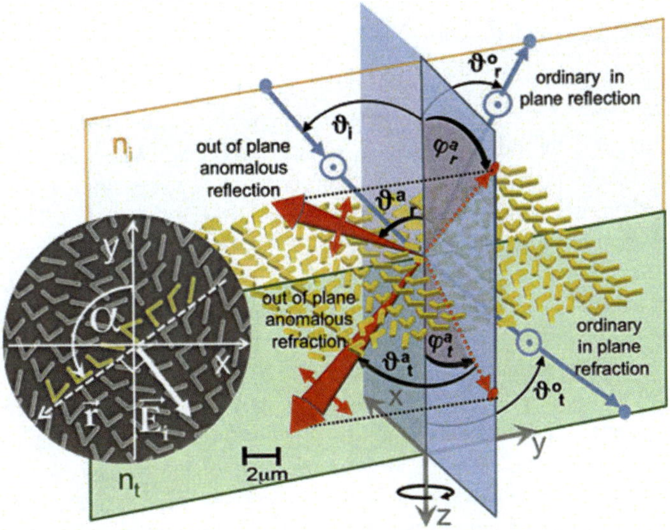

Fig. 1.11 Schematic diagram of metasurfaces based on generalized Snell's law [115]

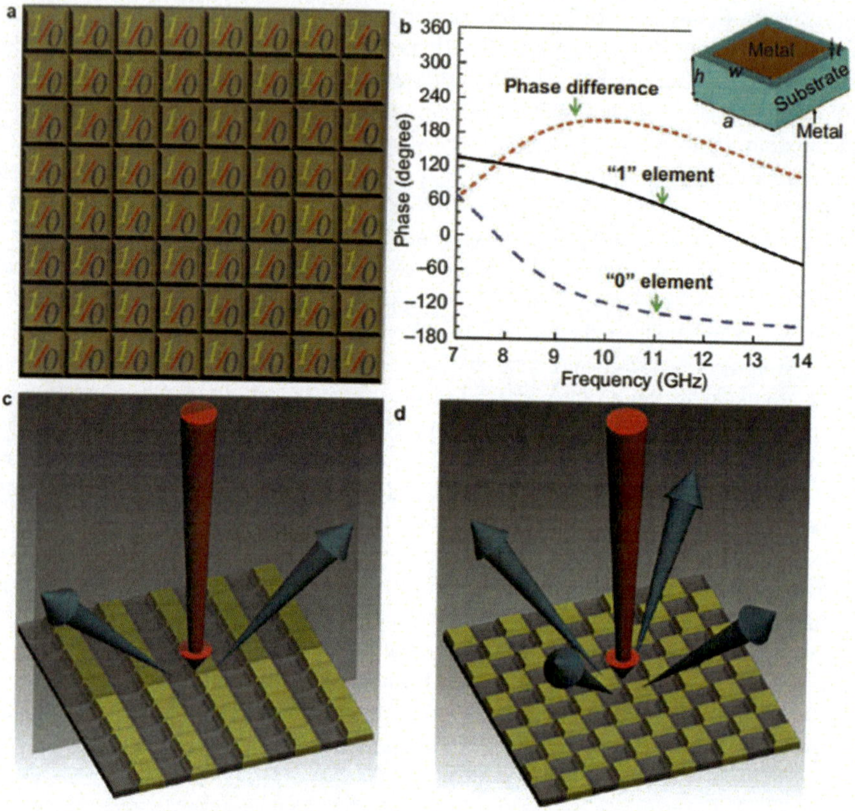

Fig. 1.12 1-bit digital metasurface and coding metasurface [110]

the aspect of composition of metasurface, more and more novel materials have been introduced to improve the performance, such as graphene [121–124], VO_2, GaAs, etc. Relying on 3D printing and micro-nano processing technologies such as laser etching, electron beam/ion beam exposure, higher frequency bands (i.e. THz metasurface [125–127], near-infrared metasurface [128] and visible light metasurface [115, 129, 130], etc.), more complicated structure and more small-scale of metasurfaces are moving from theory to reality. In the future, metasurface will also make rapid progress in the direction of multi-disciplinary systematization, diversified structure and functions, and practical products.

The subject of this book is the fast-developing area of research known as electromagnetic metamaterials/metasurfaces and their some engineering applications. In this books, the state-of-the arts of metamaterials/metasurfaces from theory to applications are comprehensively introduced. The theoretical side involves electrodynamics of left handed medium, generalized Snell's law, digital coding metamaterials/metasurfaces, group theory of metamaterials, information metamaterials and metasurfaces, etc. On the application side, a broad range of design examples including metamaterial antennas, electromagnetic interference suppression, frequency selective surface, wireless power transfer and energy harvesting, cloak and RCS reduction, orbital angular momentum, wireless communication, imaging, etc., have been discussed. The book can provide researchers, engineers, and graduate students a great number of new discoveries, results, information, and knowledge in the field of metamaterials and metasurfaces.

References

1. D. Maystre, Electromagnetic study of photonic band gaps. Pure Appl. Opt. J. Eur. Opt. Soc. Part A **3**, 975–993 (1994)
2. E. Yablonovitch, Photonic band-gap crystals. J. Phys. Condens. Matter **5**, 2443–2460 (1993)
3. C.M. Soukoulis, *Photonic Band Gap Materials* (Kluwer Academic Publishers, Dordrecht, The Netherlands, 1996)
4. E. Yablonovitch, T.J. Gmitter, K.M. Leung, Photonic band structure: the face-centered-cubic case employing nonspherical atoms. Phys. Rev. Lett. **67**, 2295–2298 (1991)
5. E. Yablonovitch, Inhibited spontaneous emission in solid-state physics and electronics. Phys. Rev. Lett. **58**, 2059–2062 (1987)
6. C.L. Holloway, E.F. Kuester, J.A. Gordon, J.O. Hara, J. Booth, D.R. Smith, An overview of the theory and applications of metasurfaces: the two-dimensional equivalents of metamaterials. IEEE Antennas Propag. Mag. **54**, 10–35 (2012)
7. S. Walia, C.M. Shah, P. Gutruf, H. Nili, D.R. Chowdhury, W. Withayachumnankul, M. Bhaskaran, S. Sriram, Flexible metasurfaces and metamaterials: a review of materials and fabrication processes at micro- and nano-scales. Appl. Phys. Rev. **2**, 011303 (2015)
8. A.E. Minovich, A.E. Miroshnichenko, A.Y. Bykov, T.V. Murzina, D.N. Neshev, Y.S. Kivshar, Functional and nonlinear optical metasurfaces. Laser Photonics Rev. **9**, 195–213 (2015)
9. P. Genevet, F. Capasso, Holographic optical metasurfaces: a review of current progress. Rep. Progr. Phys. Phys. Soc. **78**, 024401 (2015)
10. A.V. Kildishev, A. Boltasseva, V.M. Shalaev, Planar photonics with metasurfaces. Science **339**, 1232009 (2013)
11. N. Yu, F. Capasso, Flat optics with designer metasurfaces. Nat. Mater. **13**, 139–150 (2014)
12. G.G. Macfarlane, Surface impedance of an infinite parallel-wire grid at oblique angles of incidence. J. Inst. Electr. Eng. Part IIIA Radiolocation **93**(10), 1523–1527 (1946)
13. H. Lamb, On the reflection and transmission of electric waves by a metallic grating. Proc. London Math. Soc. **s1–29**, 523–546 (1897)
14. R. Mittra, C.H. Chan, T. Cwik, Techniques for analyzing frequency selective surfaces-a review. Proc. IEEE **76**, 1593–1615 (1988)
15. B.A. Munk, *Frequency Selective Surface and Grid Array* (Wiley, New York, 1995)
16. B.A. Munk, *Frequency Selective Surfaces. Theory and Design* (Wiley, New York, 2000)
17. D. Rittenhouse, An optical problem, proposed by Mr. Hopkinson, and solved by Mr. Rittenhouse. Trans. Am. Philos. Soc. **2**, 201–206 (1786)
18. G. Marconi, C.S. Franklin, Reflector for Use in Wireless Telegraphy and Telephony. US Patent, 1301473 A, 1919
19. I. Anderson, On the theory of self-resonant grids. Bell Syst. Tech. J. **54**, 1725–1731 (1975)
20. M. Al-Joumayly, N. Behdad, A new technique for design of low-profile, second-order, bandpass frequency selective surfaces. IEEE Trans. Antennas Propag. **57**, 452–459 (2009)
21. M.A. Al-Joumayly, N. Behdad, A generalized method for synthesizing low-profile, band-pass frequency selective surfaces with non-resonant constituting elements. IEEE Trans. Antennas Propag. **58**, 4033–4041 (2010)

22. Q.H. Sun, Q. Cheng, H.S. Xu, B. Zhou, T.J. Cui, A new type of band-pass FSS based on metamaterial structures, in *2008 International Workshop on Metamaterials*, pp. 267–269 (2008)
23. F. Bayatpur, K. Sarabandi, Single-layer high-order miniaturized-element frequency-selective surfaces. IEEE Trans. Microw. Theory Tech. **56**, 774–781 (2008)
24. K. Sarabandi, N. Behdad, A frequency selective surface with miniaturized elements. IEEE Trans. Antennas Propag. **55**, 1239–1245 (2007)
25. Y. Shi, P.P. Chu, Z.K. Meng, Ultra-wideband hybrid polarization conversion-absorption metasurface with a transmission window and narrow transition bands. J. Phys. D Appl. Phys. **56**, 095102 (2023)
26. X. Wu, Z.B. Pei, S.B. Qu, Z. Xu, P. Bai, J.F. Wang, X.H. Wang, H. Zhou, Design of metamaterial frequency selective surface with polarization selectivity. Acta. Phys. Sin. **60**, 271–275 (2011)
27. J.W.S. Rayleigh, On the remarkable phenomenon of crystalline reflexion described by Prof. Stokes. Phil. Mag. **26**, 256–265 (1888)
28. S. John, Strong localization of photons in certain disordered dielectric super lattices. Phys. Rev. Lett. **58**, 2486–2489 (1987)
29. K.M. Ho, C.T. Chan, C.M. Soukoulis, R. Biswas, M. Sigalas, Photonic band gaps in three dimensions: new layer-by-layer periodic structures. Solid State Commun. **89**, 413–416 (1994)
30. D. Sievenpiper, L. Zhang, R.F.J. Broas, N.G. Alexopolous, E. Yablonovitch, High-impedance electromagnetic surfaces with a forbidden frequency band. IEEE Trans. Microw. Theory Tech. **47**, 2059–2074 (1999)
31. A. Foroozesh, L. Shafai, Application of combined electric- and magnetic-conductor ground planes for antenna performance enhancement. Can. J. Electr. Comput. Eng. **33**, 87–98 (2008)
32. E. Rajo-Iglesias, Ó. Quevedo-Teruel, L. Inclan-Sanchez, Mutual coupling reduction in patch antenna arrays by using a planar EBG structure and a multilayer dielectric substrate. IEEE Trans. Antennas Propag. **56**, 1648–1655 (2008)
33. Y. Fan, Y. Rahmat-Samii, Microstrip antennas integrated with electromagnetic band-gap (EBG) structures: a low mutual coupling design for array applications. IEEE Trans. Antennas Propag. **51**, 2936–2946 (2003)
34. M.A. Hiranandani, A.B. Yakovlev, A.A. Kishk, Artificial magnetic conductors realised by frequency-selective surfaces on a grounded dielectric slab for antenna applications. IEE Proc. Microw. Antennas Propag. **153**, 487–493 (2006)
35. A.P. Feresidis, G. Goussetis, S. Wang, J.C. Vardaxoglou, Artificial magnetic conductor surfaces and their application to low-profile high-gain planar antennas. IEEE Trans. Antennas Propag. **53**, 209–215 (2005)
36. F. Costa, S. Genovesi, A. Monorchio, On the bandwidth of high-impedance frequency selective surfaces. IEEE Antennas Wirel. Propag. Lett. **8**, 1341–1344 (2010)
37. J. Sherman, Properties of focused apertures in the Fresnel region. IEEE Trans. Antennas Propag. **10**, 399–408 (1962)
38. J. Huang, J.A. Encinar, *Reflectarray Antenna* (Wiley, Piscataway, NJ, 2005)
39. D. Berry, R. Malech, W. Kennedy, The reflectarray antenna. IEEE Trans. Antennas Propag. **11**, 645–651 (1963)
40. R.D. Javor, X.-D. Wu, K. Chang, Design and performance of a microstrip reflectarray antenna. IEEE Trans. Antennas Propag. **43**, 932–939 (1995)
41. A.H. Abdelrahman, A.Z. Elsherbeni, F. Yang, Transmission phase limit of multilayer frequency-selective surfaces for transmitarray designs. IEEE Trans. Antennas Propag. **62**, 690–697 (2014)
42. A.H. Abdelrahman, Y. Fan, A.Z. Elsherbeni, P. Nayeri, Analysis and design of transmitarray antennas. Synth. Lect. Antennas **6**, 1–175 (2017)
43. S.B. Glybovski, S.A. Tretyakov, P.A. Belov, Y.S. Kivshar, C.R. Simovski, Metasurfaces: from microwaves to visible. Phys. Rep. **634**, 1–72 (2016)
44. P. Nayeri, F. Yang, A.Z. Elsherbeni, Design and experiment of a single-feed quad-beam reflectarray antenna. IEEE Trans. Antennas Propag. **60**, 1166–1171 (2012)

45. A.H. Abdelrahman, P. Nayeri, A.Z. Elsherbeni, Y. Fan, Design of single-feed multi-beam transmitarray antennas, in *2014 IEEE International Symposium on Antennas and Propagation & USNC/URSI National Radio Science Meeting* (2014)
46. B. Rahmati, M.O. Bagheri, H. Hassani, Dual-band, dual-polarized metallic slot transmitarray antenna. IET Microw. Antennas Propag. **11**, 402–409 (2016)
47. C. Han, J. Huang, K. Chang, A high efficiency offset-fed X/Ka-dual-band reflectarray using thin membranes. IEEE Trans. Antennas Propag. **53**, 2792–2798 (2005)
48. P. Zhang, L. Li, X. Zhang, H. Liu, Y. Shi, Design, measurement and analysis of near-field focusing reflective metasurface for dual-polarization and multi-focus wireless power transfer. IEEE Access **7**, 110387–110399 (2019)
49. S. Yu, H. Liu, L. Li, Design of near-field focused metasurface for high-efficient wireless power transfer with multifocus characteristics. IEEE Trans. Industr. Electron. **66**, 3993–4002 (2019)
50. S. Yu, L. Li, G. Shi, C. Zhu, X. Zhou, Y. Shi, Design, fabrication, and measurement of reflective metasurface for orbital angular momentum vortex wave in radio frequency domain. Appl. Phys. Lett. **108**, 121903 (2016)
51. S. Yu, L. Li, G. Shi, C. Zhu, Y. Shi, Generating multiple orbital angular momentum vortex beams using a metasurface in radio frequency domain. Appl. Phys. Lett. **108**, 241901 (2016)
52. N. Kou, S. Yu, L. Li, Generation of high-order Bessel vortex beam carrying orbital angular momentum using multilayer amplitude-phase-modulated surfaces in radiofrequency domain. Appl. Phys. Express **10**, 016701 (2017)
53. H. Kamoda, T. Iwasaki, T. Kuki, 60-GHz electrically reconfigurable reflectarray using p-i-n diode, in *2009 IEEE MTT-S International Microwave Symposium Digest*, pp. 1177–1180 (2009)
54. J.P. Gianvittorio, Y. Rahmat-Samii, Reconfigurable patch antennas for steerable reflectarray applications. IEEE Trans. Antennas Propag. **54**, 1388–1392 (2006)
55. J.Y. Lau, S.V. Hum, Reconfigurable transmitarray design approaches for beamforming applications. IEEE Trans. Antennas Propag. **60**, 5679–5689 (2012)
56. V.G. Veselago, The Electrodynamics of substances with simultaneously negative values of ε and μ. Phys. Usp. **10**, 509–514 (1968)
57. L. Horace, On group velocity, Proc. London Math. Soc. (1904)
58. G.D. Malyuzhinets, A note on the radiation principle. Zh. Tekh. Fiz. **21**, 940–942 (1951)
59. J.B. Pendry, A.J. Holden, W.J. Stewart, I. Youngs, Extremely low frequency plasmons in metallic mesostructures. Phys. Rev. Lett. **76**, 4773–4776 (1996)
60. J.B. Pendry, A.J. Holden, D.J. Robbins, W.J. Stewart, Magnetism from conductors and enhanced nonlinear phenomena. IEEE Trans. Microw. Theory Tech. **47**, 2075–2084 (1999)
61. R.A. Shelby, D.R. Smith, S. Schultz, Experimental verification of a negative index of refraction. Science **292**, 77–79 (2001)
62. D.R. Smith, N. Kroll, Negative refractive index in left-handed materials. Phys. Rev. Lett. **85**, 2933–2936 (2000)
63. T.J. Cui, J.A. Kong, Time-domain electromagnetic energy in a frequency-dispersive left-handed medium. Phys. Rev. B **70**, 205106 (2004)
64. T.J. Cui, J.A. Kong, Causality in the propagation of transient electromagnetic waves in a left-handed medium. Phys. Rev. B **70**, 165113 (2004)
65. https://en.wikipedia.org/wiki/Metamaterial
66. D.R. Smith, D. Schurig, M. Rosenbluth, S. Schultz, J.B. Pendry, Limitations on subdiffraction imaging with a negative refractive index slab. Appl. Phys. Lett. **82**, 1506–1508 (2003)
67. D.R. Smith, J.J. Mock, A.F. Starr, D. Schurig, Gradient index metamaterials. Phys. Rev. E, 036609 (2005)
68. U. Leonhardt, Optical conformal mapping. Science **312**, 1777–1780 (2006)
69. J.B. Pendry, Controlling electromagnetic fields. Science **312**, 1780–1782 (2006)
70. D. Schurig, J.J. Mock, B.J. Justice, S.A. Cummer, J.B. Pendry, A.F. Starr, D.R. Smith, Metamaterial electromagnetic cloak at microwave frequencies. Science **314**, 977–980 (2006)
71. W.X. Jiang, T.J. Cui, H.F. Ma, X.Y. Zhou, Q. Cheng, Cylindrical-to-plane-wave conversion via embedded optical transformation. Appl. Phys. Lett. **92**, 261903 (2008)

72. W.X. Jiang, T.J. Cui, Q. Cheng, J.Y. Chin, X.M. Yang, R. Liu, D.R. Smith, Design of arbitrarily shaped concentrators based on conformally optical transformation of nonuniform rational B-spline surfaces. Appl. Phys. Lett. **92**, 264101 (2008)
73. X.J. Wei, T.J. Cui, Y.Z. Xiao, M.Y. Xin, C. Qiang, Arbitrary bending of electromagnetic waves using realizable inhomogeneous and anisotropic materials. Phys. Rev. E **78**, 066607 (2009)
74. M. Rahm, D. Schurig, D.A. Roberts, S.A. Cummer, D.R. Smith, J.B. Pendry, Design of electromagnetic cloaks and concentrators using form-invariant coordinate transformations of Maxwell's equations. Photonics Nanostruct. Fundam. Appl. **6**, 87–95 (2008)
75. R. Liu, C. Ji, J.J. Mock, J.Y. Chin, T.J. Cui, D.R. Smith, Broadband ground-plane cloak. Science **323**, 366–369 (2009)
76. H.F. Ma, T.J. Cui, Three-dimensional broadband ground-plane cloak made of metamaterials. Nat. Commun. **1**, 21 (2010)
77. T. Ergin, N. Stenger, P. Brenner, J.B. Pendry, M. Wegener, Three-dimensional invisibility cloak at optical wavelengths. Science **328**, 337 (2010)
78. Q. Cheng, T.J. Cui, W.X. Jiang, B.G. Cai, An electromangetic black hole made of metamaterials. New J. Phys. **12**, 063006 (2010)
79. A. Lai, T. Itoh, C. Caloz, Composite right/left-handed transmission line metamaterials. IEEE Microw. Mag. **5**, 34–50 (2004)
80. C. Caloz, T. Itoh, A. Rennings, CRLH metamaterial leaky-wave and resonant antennas. IEEE Antennas Propag. Mag. **50**(5), 25–39 (2008)
81. C. Caloz, T. Itoh, A novel mixed conventional microstrip and composite right/left-handed backward-wave directional coupler with broadband and tight coupling characteristics. IEEE Microw. Wirel. Compon. Lett. **14**(1), 31–33 (2004)
82. S. Lim, C. Caloz, T. Itoh, Electronically scanned composite right/left handed microstrip leaky-wave antenna. IEEE Microwave Wirel. Compon. Lett. **14**(6), 277–279 (2004)
83. L. Li, Z. Jia, F. Huo, W. Han, A novel compact multiband antenna employing dual-band CRLH-TL for smart mobile phone application. IEEE Antennas Wirel. Propag. Lett. **12**, 1688–1691 (2013)
84. W.X. Jiang, T.J. Cui, Radar illusion via metamaterials. Phys. Rev. E **83**, 026601 (2011)
85. B.D.F. Casse, W.T. Lu, Y.J. Huang, E. Gultepe, L. Menon, S. Sridhar, Super-resolution imaging using a three-dimensional metamaterials nanolens. Appl. Phys. Lett. **96**, 023114 (2010)
86. N. Yu, P. Genevet, M.A. Kats, F. Aieta, J.-P. Tetienne, F. Capasso, Z. Gaburro, Light propagation with phase discontinuities: generalized laws of reflection and refraction. Science **334**, 333–337 (2011)
87. Q. Ma, C.B. Shi, G.D. Bai, T.Y. Chen, A. Noor, T.J. Cui, Beam-editing coding metasurfaces based on polarization bit and orbital-angular-momentum-mode bit. Adv. Opt. Mater. **5**, 1700548 (2017)
88. S. Sun, Q. He, S. Xiao, Q. Xu, X. Li, L. Zhou, Gradient-index meta-surfaces as a bridge linking propagating waves and surface waves. Nat. Mater. **11**, 426–431 (2012)
89. H.L. Zhu, S.W. Cheung, K.L. Chung, T.I. Yuk, Linear-to-circular polarization conversion using metasurface. IEEE Trans. Antennas Propag. **61**, 4615–4623 (2013)
90. P. Xu, G.C. Wang, X. Cai, H.Y. Shen, W.X. Jiang, Design and optimization of high-efficiency meta-devices based on the equivalent circuit model and theory of electromagnetic power energy storage. J. Phys. D Appl. Phys. **55**, 195303 (2022)
91. Z. Wang, X. Ding, K. Zhang, B. Ratni, S.N. Burokur, X. Gu, Q. Wu, Huygens metasurface holograms with the modulation of focal energy distribution. Adv. Opt. Mater. **6**, 1800121 (2018)
92. G. Zheng, H. Muhlenbernd, M. Kenney, G. Li, T. Zentgraf, S. Zhang, Metasurface holograms reaching 80% efficiency. Nat. Nanotechnol. **10**, 308–312 (2015)
93. N.M. Estakhrin, A. Alu, Wave-front transformation with gradient metasurfaces. Phys. Rev. X **6**, 041008 (2016)
94. V.S. Asadchy, M. Albooyeh, S.N. Tcvetkova, A. Diaz-Rubio, Y. Radi, S.A. Tretyakov, Perfect control of reflection and refraction using spatially dispersive metasurfaces. Phys. Rev. B **94**, 075142 (2016)

95. J. Zhou, P. Zhang, J. Han, L. Li, Y. Huang, Metamaterials and metasurfaces for wireless power transfer and energy harvesting. Proc. IEEE **110**, 31–55 (2022)
96. L. Li, X. Zhang, C. Song, W. Zhang, Y. Huang, Compact dual-band, wide-angle, polarization-angle-independent rectifying metasurface for ambient energy harvesting and wireless power transfer. IEEE Trans. Microw. Theory Tech. **69**, 1518–1528 (2021)
97. Y. Shi, H.X. Meng, H.J. Wang, Polarization conversion metasurface design based on characteristic mode rotation and its application into wideband and miniature antenna with low radar cross section. Opt. Express **29**(5), 6794–6809 (2021)
98. Y. Shi, Z.K. Meng, W.Y. Wei, W. Zheng, L. Li, Characteristic mode cancellation method and its application for antenna RCS reduction. IEEE Antennas Wirel. Propag. Lett. **18**(9), 1784–1788 (2019)
99. X. Zhang, Q. Li, F. Liu, Controlling angular dispersions in optical metasurfaces. Light Sci. Appl. **9**, 1286–1297 (2020)
100. M. Li, L. Shen, L. Jing, Origami metawall: mechanically controlled absorption and deflection of light. Adv. Sci. **6**, 1901434 (2019)
101. Q. Sun, Z. Zhang, Y. Huang, Asymmetric transmission and wavefront manipulation toward dual-frequency meta-hologram. ACS Photonics **6**, 1541–1546 (2019)
102. X.G. Zhang, Q. Yu, W.X. Jiang, Polarization-controlled dual-programmable metasurfaces. Adv. Sci. **7**, 1903328 (2020)
103. O. Avayu, E. Almeida, Y. Prior, T. Ellenbogen, Composite functional metasurfaces for multispectral achromatic optics. Nature Commun. **8**, 14992 (2017)
104. G.D. Bai, Q. Ma, S. Iqbal, L. Bao, H.B. Jing, L. Zhang, H.T. Wu, R.Y. Wu, H.C. Zhang, C. Yang, T.J. Cui, Multitasking shared aperture enabled with multiband digital coding metasurface. Adv. Opt. Mater. **6**, 18800657 (2018)
105. M.R.M. Hashemi, S.-H. Yang, T. Wang, N. Sepúlveda, M. Jarrahi, Electronically-controlled beam-steering through vanadium dioxide metasurfaces. Sci. Rep. **6**, 35439 (2016)
106. B. Scherger, M. Reuter, M. Scheller, K. Altmann, N. Vieweg, R. Dabrowski, J.A. Deibel, M. Koch, Discrete terahertz beam steering with an electrically controlled liquid crystal device. J. Infrared Millm. Terahertz Waves **33**, 1117–1122 (2012)
107. B.C. Smith, J.F. Whitaker, S.C. Rand, Steerable THz pulses from thin emitters via optical pulse-front tilt. Opt. Express **24**, 20755–20762 (2016)
108. Y. Monnai, K. Altmann, C. Jansen, H. Hillmer, M. Koch, H. Shinoda, Terahertz beam steering and variable focusing using programmable diffraction gratings. Opt. Express **21**, 2347–2354 (2013)
109. D. Shrekenhamer, J. Montoya, S. Krishna, W.J. Padilla, Four-color metamaterial absorber THz spatial light modulator. Adv. Opt. Mater. **1**, 905–909 (2013)
110. G.C. Della, N. Engheta, Digital metamaterials. Nat. Mater. **13**, 1115–1121 (2014)
111. T.J. Cui, M. Qi, X. Wan, J. Zhao, Q. Cheng, Coding metamaterials, digital metamaterials and programmable metamaterials. Light Sci. Appl. **3**(10), e218 (2014)
112. T.J. Cui, S. Liu, L. Zhang, Information metamaterials and metasurfaces. J. Mater. Chem. C **5**, 3644–3668 (2017)
113. Q. Ma, G. Bai, H. Jing, C. Yang, L. Li, T.J. Cui, Smart metasurface with self-adaptively reprogrammable functions. Light Sci. Appl. **8**, 98 (2019)
114. X. Zhang, Y. Sun, Q. Yu, Q. Cheng, W.X. Jiang, C.-W. Qiu, T.J. Cui, Smart doppler cloak operating in broad band and full polarizations. Adv. Mater. **33**, 2007966 (2021)
115. F. Aieta, P. Genevet, N. Yu, M.A. Kats, Z. Gaburro, F. Capasso, Out-of-plane reflection and refraction of light by anisotropic optical antenna metasurfaces with phase discontinuities. Nano Lett. **12**, 1702–1706 (2012)
116. X. Ni, A.V. Kildishev, V.M. Shalaev, Metasurface holograms for visible light. Nat. Commun. **4**, 2807 (2013)
117. K. Chen, Y. Feng, F. Monticone, J. Zhao, B. Zhu, T. Jiang, L. Zhang, Y. Kim, X. Ding, S. Zhang, A. Alu, C. Qiu, A reconfigurable active huygens' metalens. Adv. Mater. **29**, 1606422 (2017)

118. H. Wu, S. Liu, X. Wan, L. Zhang, D. Wang, L. Li, T.J. Cui, Controlling energy radiations of electromagnetic waves via frequency coding metamaterials. Adv. Sci. **4**, 1700098 (2017)
119. H.T. Wu, D. Wang, X. Fu, S. Liu, L. Zhang, G.D. Bai, C. Zhang, W.X. Jiang, H.L. Jiang, R.Y. Wu, Space-frequency-domain gradient metamaterials. Adv. Opt. Mater. **6**, 1801086 (2018)
120. L. Zhang, X.Q. Chen, S. Liu, Q. Zhang, J. Zhao, J.Y. Dai, G.D. Bai, X. Wan, Q. Cheng, G. Castaldi, V. Galdi, T.J. Cui, Space-time-coding digital metasurfaces. Nat. Commun. **9**, 4334 (2018)
121. J.Y. Lau, S.V. Hum, A wideband reconfigurable transmitarray element. IEEE Trans. Antennas Propag. **60**, 1303–1311 (2012)
122. P.Y. Chen, J. Soric, Y.R. Padooru, H.M. Bernety, A.B. Yakovlev, A. Alù, Nanostructured graphene metasurface for tunable terahertz cloaking. New J. Phys. **15**, 123029 (2013)
123. Y.R. Padooru, A.B. Yakovlev, C.S.R. Kaipa, G.W. Hanson, F. Medina, F. Mesa, Dual capacitive-inductive nature of graphene metasurface: transmission characteristics at low-terahertz frequencies, in *IEEE Antennas Propagation Society International Symposium*, pp. 1598–1599 (2013)
124. E. Carrasco, J. Perruisseau-Carrier, Reflectarray antenna at terahertz using graphene. IEEE Antennas Wirel. Propag. Lett. **12**, 253–256 (2013)
125. W.J. Padilla, M.T. Aronsson, C. Highstrete, M. Lee, A.J. Taylor, R.D. Averitt, Electrically resonant terahertz metamaterials: theoretical and experimental investigations. Phys. Rev. B **75**, 041102 (2007)
126. H.-T. Chen, J.F. O'Hara, A.J. Taylor, R.D. Averitt, C. Highstrete, M. Lee, W.J. Padilla, Complementary planar terahertz metamaterials. Opt. Express **15**, 1084–1095 (2007)
127. C. Jansen, I.A.I. Al-Naib, N. Born, M. Koch, Terahertz metasurfaces with high Q-factors. Appl. Phys. Lett. **98**, 051109 (2011)
128. M. Farmahini-Farahani, H. Mosallaei, Birefringent reflectarray metasurface for beam engineering in infrared. Opt. Lett. **38**, 462–464 (2013)
129. F. Qin, L. Ding, L. Zhang, F. Monticone, C.C. Chum, J. Deng, S. Mei, Y. Li, J. Teng, M. Hong, S. Zhang, A. Slu, C.W. Qiu, Hybrid bilayer plasmonic metasurface efficiently manipulates visible light. Sci. Adv. **2**, e1501168 (2016)
130. D. Sell, J. Yang, S. Doshay, K. Zhang, J.A. Fan, Visible light metasurfaces based on single crystal silicon. ACS Photonics **3**(10), 1919–1925 (2016)

Chapter 2
Theory Models of Electromagnetic Metamaterials and Metasurfaces

Long Li, Yan Shi, Haixia Liu, and Xiaojie Dang

Abstract Metamaterials are artificially engineered composite materials that exhibit special properties that are not found in nature and not observed in the constituent materials, including negative refraction, backward propagation, etc. With the subwavelength meta-atoms, the metamaterials can be characterized by the equivalent bulk parameters including permittivity and permeability. Metasurfaces can be considered as the two-dimensional (2D) equivalent of bulk metamaterials. The effective permittivity, permeability and refractive index are of less interest in metasurfaces. In contrast, of significant importance are the surface or interface reflection and transmission resulting from the tailored surface impedance, including their amplitude, phase, and polarization states. With the abrupt phase discontinuous, the anomalous reflection and transmission are achieved. In this chapter, the electromagnetic properties for metamaterials with double negative bulk parameters are first demonstrated. Then the generalized Snell's law for metasurfaces is elaborated. Furthermore, the digital coding method for the metamaterials/metasurfaces are discussed. Finally, the group theory is introduced to analyse the symmetry of the metamaterials/metasurfaces.

Keywords Metamaterials · Metasurfaces · Generalized Snell's law · Digital coding · Group theory

---

L. Li (✉) · Y. Shi · H. Liu · X. Dang
School of Electronic Engineering, Xidian University, Xi'an 710071, Shaanxi, China
e-mail: lilong@mail.xidian.edu.cn

Y. Shi
e-mail: shiyan@mail.xidian.edu.cn

H. Liu
e-mail: hxliu@xidian.edu.cn

X. Dang
e-mail: xjdang@mail.xidian.edu.cn

L. Li et al. (eds.), *Electromagnetic Metamaterials and Metasurfaces: From Theory To Applications*, https://doi.org/10.1007/978-981-99-7914-1_2

2.1 Introduction

In recent year, much attentions have been devoted to metamaterials due to novel electromagnetic (EM) properties [1–6]. One class of representative metamaterials is left-handed materials (LHM), also called double negative materials. The simultaneously negative permittivity and permeability gives rise to unique phenomena including negative refraction, backward wave propagation, etc. The discussions about the interesting behaviour of backward wave can date back to Lamb's work [7]. In that paper, he studied mechanical systems instead of electromagnetic wave and suggested the existence of the backward wave. In 1904, Schuster noted Lamb's work and gave a speculative discussion about the optical backward wave [8]. This question was considered in more detail by Mandelshtam [9]. Sivukhin [10] and Pafomov [11] described the backward wave governing by Maxwell's equation with the simultaneous permittivity and permeability. Veslago elaborated the electrodynamics properties of the materials with the simultaneous permittivity and permeability, and introduced the concept of negative refractive index [12]. The materials with negative permittivity and negative permeability were realized by Pendry [13, 14]. Later Smith et al. experimentally verified the LHM [15]. The term of "Metamaterials" is first proposed by R. M. Wasler at the University of Texas at Austin in 1999. Metamaterials are artificially engineered composite materials that exhibit special properties that are not found in nature and not observed in the constituent materials. Stimulated by the theoretical and experimental progress of the metamaterials, it has been extensively used in various fields, including invisible cloak [16–26], antenna designs [27–34], radar cross section (RCS) reduction [35–44], etc.

Metasurface as the two-dimensional (2D) metamaterial has been proposed to demonstrate the eminent ability in manipulating electromagnetic wave. The effective permittivity, permeability and refractive index are of less interest in metasurfaces. In contrast, of significant importance are the surface or interface reflection and transmission resulting from the tailored surface impedance, including their amplitude, phase, and polarization states. The introduction of gradient phase results in the anomalous reflection and transmission governing by generalized Snell's law [45–54]. Due to the negligible thickness as compared with the wavelength, metasurface has less loss, lower profile, and lower cost, and thus it has been widely developed in various applications, for instance, absorbers [55–61], orbital angular momentum (OAM) [62–68], wireless power transfer [69–76], etc.

In recent year, digital coding concept proposed by Cui et al. has been introduced into the metamaterial/metasurface [77, 78]. The digital coding metamaterial/metasurface designs are different from the conventional metamaterial/metasurface designs. In the former, the reflection or transmission phases are quantized, and the resultant quantizing elements are sequenced as required. By comparison, the latter is based on the continuous equivalent parameters used to characterize the metamaterials/metasurfaces. Therefore, the former is called digital metamaterials/metasurfaces, while the latter is called analog ones. The introduction of the digital coding method provides a feasible method to manipulate electromagnetic waves in a real-time way.

Most of the metamaterials/metasurfaces are of symmetry. The group theory is an effective mathematical method to characterize symmetries. The classical application fields of group theory include abstract mathematics and quantum mechanics. Recently, electronic engineers are slowly beginning to employ the group theory in their research, for example the symmetry of electric and magnetic fields, the antenna synthesis and the Babinet principle. Unlike quantum mechanics, the investigation of symmetry in electromagnetic metamaterials is not only in the field of geometry but also in combination with the symmetrical inherent of Maxwell's equation.

In this chapter, the electrodynamics properties of the LHM are first reviewed. And then, the generalized Snell's law is elaborated and the ideal transmission and reflection are discussed, followed by the digital coding method. Finally, the fundamentals of the group theory are introduced.

2.2 The Electrodynamics of Double Negative Metamaterials

As is well known, the materials in nature are made up of numerous small elements like atoms and molecules. Some of the materials are amorphous, and others are crystalline. The small elements in the former are heaped upon each other in a random manner, and those in the latter are arranged into some regular periodical pattern.

Consider the interaction between an electromagnetic wave and an artificial crystalline material in the framework of classic physics. Assume that the artificial crystalline material is composed of some small particles arranged in a regular lattice. The periodicity of the lattice is d. A plane wave with a wavelength of λ illuminates the crystalline material. When the d is proportional to the λ, i.e., $d \sim \lambda$, the plane wave interacts with the particles of the lattice and will be scattered, which is Bragg's effect. If the d is far smaller than the λ, i.e., $d < < \lambda$, the scattering from the particles becomes trivial. The wave is considerably affected when it enters the material. Therefore, the lattice-based artificial material can be considered as a homogeneous and continuous material.

Metamaterials are broadly defined as artificial structures which are engineered to obtain unusual properties not readily found in nature. With the subwavelength meta-atom, the metamaterials can be represented by permittivity and permeability. By tailoring the meta-atom as required, the permittivity and permeability of the metamaterial can acquire the arbitrary values. One of most fascinating features of metamaterials is simultaneously negative permittivity and permeability, which is also called LHM.

To investigate the electromagnetic properties of such a medium, it was first studied how the electromagnetic wave behaves when $\varepsilon < 0$ and $\mu < 0$. The Maxwell's equations can be written in differential form as follows.

$$\nabla \times \vec{H} = \vec{J} + \frac{\partial \vec{D}}{\partial t}$$
$$\nabla \times \vec{E} = -\frac{\partial \vec{B}}{\partial t}$$
$$\nabla \cdot \vec{D} = \rho$$
$$\nabla \cdot \vec{B} = 0 \tag{2.1}$$

where $\vec{E}$ is the electric field intensity, $\vec{B}$ is the magnetic flux density, $\vec{H}$ is the magnetic field intensity, and $\vec{D}$ is the electric flux density. The electric current density $\vec{J}$ and electric charge density ρ are the sources of the electromagnetic fields. For linear and isotropic media, $\vec{E}$ and $\vec{D}$ and $\vec{H}$ and $\vec{B}$ are related by the constitutive relations

$$\begin{aligned} \vec{B} &= \mu \vec{H} = \mu_0 \mu_r \vec{H} \\ \vec{D} &= \varepsilon \vec{E} = \varepsilon_0 \varepsilon_r \vec{E} \end{aligned} \tag{2.2}$$

where ε_0, μ_0, ε_r, and μ_r are the free-space permittivity, free-space permeability, relative permittivity, and relative permeability, respectively.

Let's consider the source-free Maxwell equations with the time-harmonic $e^{j\omega t}$. When a plane wave is propagating in the metamaterial, Maxwell's equations (2.1) can be reduced as

$$\vec{k} \times \vec{E} = \mu\omega \vec{H}, \ \vec{k} \times \vec{H} = -\varepsilon\omega \vec{E} \tag{2.3}$$

where ε and μ are equivalent permittivity and permeability of the metamaterial. Here $\vec{k} = k\hat{k}$ is wave propagating vector with the wavenumber k and the propagating direction $\hat{k}$. For simplicity, the lossless medium is considered. If $\varepsilon < 0$ and $\mu < 0$, it can be found from (2.3) that the $\vec{E}$, $\vec{H}$, and $\vec{k}$ form a left-hand triplet. This is the reason that the medium is named as the LHM. The phase velocity becomes

$$\vec{v}_p = \frac{\omega}{k}\hat{k} \tag{2.4}$$

On the other hand, the time average of the Ponyting vector is

$$\vec{S}_{av} = \frac{1}{2}\mathrm{Re}(\vec{E} \times \vec{H}^*) \tag{2.5}$$

Thus, the direction of the $\vec{v}_p$ is in the opposite to that of the $\vec{S}_{av}$, i.e., the direction of energy velocity, meaning that the wave in the LHM propagates backward to the source. Furthermore, the time-averaged density of energy in the dispersive medium is

$$W = \frac{1}{4}\left\{\frac{\partial(\omega\varepsilon)}{\omega}\left|\vec{E}\right|^2 + \frac{\partial(\omega\mu)}{\omega}\left|\vec{H}\right|^2\right\} \tag{2.6}$$

Note that any physical media other than vacuum must be dispersive. The positive W implies

$$\frac{\partial(\omega\varepsilon)}{\omega} > 0, \frac{\partial(\omega\mu)}{\omega} > 0 \tag{2.7}$$

which means that ε and μ are dispersive. In addition, the group velocity v_g is

$$v_g = \frac{\partial\omega}{\partial k} \tag{2.8}$$

Considering the relation between the phase velocity and the group velocity, i.e.,

$$\frac{\partial k^2}{\partial\omega} = \frac{2}{v_p v_g} \tag{2.9}$$

and

$$\frac{\partial k^2}{\partial\omega} = \frac{\partial(\omega\varepsilon\cdot\omega\mu)}{\partial\omega} = \omega\varepsilon\frac{\partial(\omega\mu)}{\partial\omega} + \omega\mu\frac{\partial(\omega\varepsilon)}{\partial\omega} < 0 \tag{2.10}$$

the phase velocity is in a reverse direction to the group velocity.

In addition, in the homogeneous and isotropic LHM medium, the wave number k can be expressed as in terms of the refractive index n

$$k = n\frac{\omega}{c} \tag{2.11}$$

The negative k implies the negative n, i.e.,

$$n = -\sqrt{\varepsilon_r\mu_r} \tag{2.12}$$

Therefore, on the interface between the LHM and the right-hand medium, for instance LHM-air interface, there will be negative refraction. The double negative metamaterials have unique electromagnetic properties such as the reversal Snell's law, the reversal Doppler effect, and the anomalous Vavilov-Cherenkov effect (radiation produced by a fast-moving particle as it travels through a medium). Equation (2.7) clearly indicates that ε and μ can be simultaneously negative, given the medium is frequency dispersive. Therefore, the LHM must be dispersive. Moreover, a medium with frequency dispersion is always dissipative. Following the causality principle, the real and imaginary parts of the permittivity are related by the Kramers–Kronig relations [79]. Since the imaginary parts of the permittivity, the permeability and the refractive index always coexist with the real parts in dispersive media, the left-handed material must be dissipative.

2.3 Generalized Snell's Law

Metasurfaces can be considered as the two-dimensional (2D) equivalent of bulk metamaterials. The effective permittivity, permeability and refractive index are of less interest in metasurfaces. In contrast, of significant importance are the surface or interface reflection and transmission resulting from the tailored surface impedance, including their amplitude, phase, and polarization states. A plane wave is obliquely incident on an interface ($z = 0$) between the medium 1 (ε_1, μ_1) and the medium 2 (ε_2, μ_2). A phase screen, i.e., a metasurface, which provides the additional phase shifts of $\Phi_r(x,y)$ and $\Phi_t(x,y)$ for the waves reflecting from and passing through it, respectively, is located on the interface. Note that the phase shifts are position dependent. Without loss of generality, consider a plane wave with a transverse electric field $\vec{E_i}$ and an incident angle θ_i, as shown in Fig. 2.1. Let us assume the reflected wave with a transverse electric field $\vec{E_r}$ and a reflection angle θ_r and the transmission wave with a transverse electric field $\vec{E_t}$ and a transmission angle θ_t. Furthermore, assume that the incident wave, the reflection wave and the transmission wave are located in the xoz plane. Therefore the incident electric and magnetic fields can be expressed as

$$\vec{E_i} = \hat{y} E_i e^{-jk_1(\sin\theta_i x + \cos\theta_i z)} \tag{2.13}$$

$$\begin{aligned}\vec{H_i} &= \frac{(\sin\theta_i \hat{x} + \cos\theta_i \hat{z})}{\eta_1} \times \left[\hat{y} E_i e^{-jk_1(\sin\theta_i x + \cos\theta_i z)}\right] \\ &= (\hat{z}\sin\theta_i - \hat{x}\cos\theta_i)\frac{E_i}{\eta_1} e^{-jk_1(\sin\theta_i x + \cos\theta_i z)}\end{aligned} \tag{2.14}$$

in which $k_1 = \omega\sqrt{\varepsilon_1\mu_1}$ and $\eta_1 = \sqrt{\mu_1/\varepsilon_1}$. The resulting electric and magnetic fields of the reflection wave and the transmission wave are

$$\vec{E_r} = \hat{y} E_r e^{-jk_1(\sin\theta_r x - \cos\theta_r z) + j\Phi_r} \tag{2.15}$$

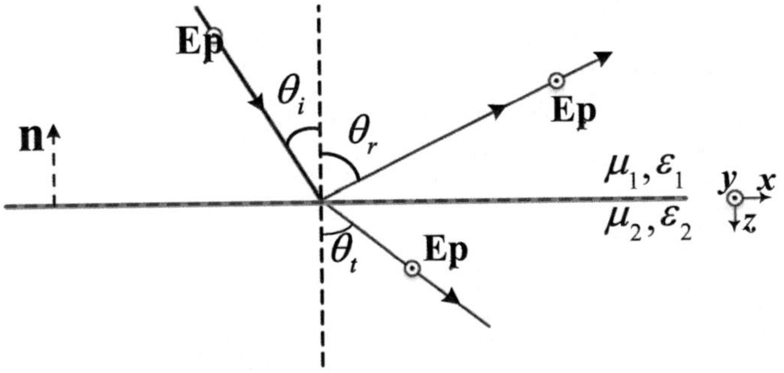

Fig. 2.1 The transverse electric (TE) case for the generalized Snell's law

$$\begin{aligned}\vec{H}_r &= \frac{(\sin\theta_r\hat{x} - \cos\theta_r\hat{z})}{\eta_1} \times \left[\hat{y}E_r e^{-jk_1(\sin\theta_r x - \cos\theta_r z)+j\Phi_r}\right] \\ &= (\hat{z}\sin\theta_r + \hat{x}\cos\theta_r)\frac{E_r}{\eta_1}e^{-jk_1(\sin\theta_r x - \cos\theta_r z)+j\Phi_r}\end{aligned} \tag{2.16}$$

$$\vec{E}_t = \hat{y}E_t e^{-jk_2(\sin\theta_t x + \cos\theta_t z)+j\Phi_t} \tag{2.17}$$

$$\begin{aligned}\vec{H}_t &= \frac{(\sin\theta_t\hat{x} + \cos\theta_t\hat{z})}{\eta_2} \times \left[\hat{y}E_t e^{-jk_2(\sin\theta_t x + \cos\theta_t z)+j\Phi_t}\right] \\ &= (\hat{z}\sin\theta_t - \hat{x}\cos\theta_t)\frac{E_t}{\eta_2}e^{-jk_2(\sin\theta_t x + \cos\theta_t z)+j\Phi_t}\end{aligned} \tag{2.18}$$

Here $k_2 = \omega\sqrt{\varepsilon_2\mu_2}$ and $\eta_2 = \sqrt{\mu_2/\varepsilon_2}$.

According to the continuity of the tangential components of the electromagnetic fields, we have

$$E_i e^{-jk_1\sin\theta_i x} + E_r e^{-jk_1\sin\theta_r x + j\Phi_r} = E_t e^{-jk_2\sin\theta_t x + j\Phi_t} \tag{2.19}$$

$$-\cos\theta_i\frac{E_i}{\eta_1}e^{-jk_1\sin\theta_i x} + \cos\theta_r\frac{E_r}{\eta_1}e^{-jk_1\sin\theta_r x + j\Phi_r} = -\cos\theta_t\frac{E_t}{\eta_2}e^{-jk_2\sin\theta_t x + j\Phi_t} \tag{2.20}$$

In order to guarantee (2.19) and (2.20) true, each exponential term in (2.19) and (2.20) should be same. Therefore, we have

$$k_1\sin\theta_r x - k_1\sin\theta_i x = \Phi_r \tag{2.21}$$

$$k_1\sin\theta_i x - k_2\sin\theta_t x = \Phi_t \tag{2.22}$$

It can be observed that Φ_r or Φ_t is not uniform on the interface, as long as $\sin\theta_r \neq \sin\theta_i$ or $k_1 \sin\theta_i \neq k_2 \sin\theta_t$.

Taking the derivatives of the (2.21) and (2.22) with respect to x, respectively, one gets

$$k_1 \sin\theta_r - k_1 \sin\theta_i = \frac{\partial \Phi_r}{\partial x} \tag{2.23}$$

$$k_1 \sin\theta_i - k_2 \sin\theta_t = \frac{\partial \Phi_t}{\partial x} \tag{2.24}$$

Furthermore, the reflection and transmission angles can be solved as

$$\theta_r = \sin^{-1}\left[\sin\theta_i + \frac{\lambda_1}{2\pi}\frac{\partial \Phi_r}{\partial x}\right] \tag{2.25}$$

$$\theta_t = \sin^{-1}\left[\frac{\lambda_2}{\lambda_1}\sin\theta_i - \frac{\lambda_2}{2\pi}\frac{\partial \Phi_t}{\partial x}\right] \tag{2.26}$$

in which $\lambda_1 = 2\pi / k_1$ and $\lambda_2 = 2\pi / k_2$. Equations (2.25) and (2.26) are called generalized Snell's reflection and transmission laws. With the phase gradient on the interface, the anomalous reflected and transmitted waves propagating in the arbitrary direction become possible. It can be found that if there is no phase gradient on the interface, (2.25) and (2.26) are reduced to the traditional Snell's law, i.e.,

$$\theta_r = \theta_i \tag{2.27}$$

$$\theta_t = \sin^{-1}\left[\frac{\lambda_2}{\lambda_1}\sin\theta_i\right] \tag{2.28}$$

The above discussion is based on the transverse electric (TE) scenario. The similar conclusion holds for the transverse magnetic (TM) case.

On the other hand, by substituting (2.21) and (2.22) into (2.19) and (2.20), we have

$$E_i + E_r = E_t \tag{2.29}$$

$$-\cos\theta_i \frac{E_i}{\eta_1} + \cos\theta_r \frac{E_r}{\eta_1} = -\cos\theta_t \frac{E_t}{\eta_2} \tag{2.30}$$

Therefore, the magnitudes of the reflection and transmission waves are expressed in terms of that of the incident wave

$$E_r = \frac{\eta_2 \cos\theta_i - \eta_1 \cos\theta_t}{\eta_2 \cos\theta_r + \eta_1 \cos\theta_t} E_i \tag{2.31}$$

$$E_t = \frac{\eta_2 \cos\theta_i + \eta_2 \cos\theta_r}{\eta_2 \cos\theta_r + \eta_1 \cos\theta_t} E_i \tag{2.32}$$

Let us look into (2.21) and (2.22) in more details. Assume the metasurface on the interface is a periodic array with the period of D. Therefore, the phase shifts satisfy

$$\Phi_r(x + D) = \Phi_r(x) \pm 2\pi \tag{2.33}$$

$$\Phi_t(x + D) = \Phi_t(x) \pm 2\pi \tag{2.34}$$

Substituting (2.33) into (2.21), we have

$$D = \frac{\lambda_1}{|\sin\theta_r - \sin\theta_i|} \tag{2.35}$$

Similarly, inserting (2.34) into (2.22), we obtain

$$D = \frac{1}{\left|\frac{\sin\theta_i}{\lambda_1} - \frac{\sin\theta_t}{\lambda_2}\right|} \tag{2.36}$$

The period of the metasurface is tightly related to anomalous reflection and transmission directions.

The above analyses are based on the assumption that the incident wave, the reflection wave and the transmission wave lie in the *xoz* plane. For an impinging wave with the incident direction (θ_i, φ_i), the generalized Snell's reflection and transmission laws can be expressed as

$$\begin{cases} \sin\theta_r \cos\varphi_r - \sin\theta_i \cos\varphi_i = \dfrac{\lambda_1}{2\pi} \dfrac{\partial \Phi_r}{\partial x} \\ \sin\theta_r \sin\varphi_r - \sin\theta_i \sin\varphi_i = \dfrac{\lambda_1}{2\pi} \dfrac{\partial \Phi_r}{\partial y} \end{cases} \tag{2.37}$$

$$\begin{cases} \dfrac{2\pi}{\lambda_1} \sin\theta_i \cos\varphi_i - \dfrac{2\pi}{\lambda_2} \sin\theta_t \cos\varphi_t = \dfrac{\partial \Phi_t}{\partial x} \\ \dfrac{2\pi}{\lambda_1} \sin\theta_i \sin\varphi_i - \dfrac{2\pi}{\lambda_2} \sin\theta_t \sin\varphi_t = \dfrac{\partial \Phi_t}{\partial y} \end{cases} \tag{2.38}$$

where the directions of the reflection wave and transmission wave are (θ_r, φ_r) and (θ_t, φ_t), respectively. With the above generalized Snell's laws, we highlight two special cases, i.e., the ideal transmission with zero reflection called Huygens' surface and the ideal reflection with zero transmission called meta-mirrors.

Case 1 Ideal transmission with zero reflection

Assume that the metasurface on the interface is lossless. The ideal transmission with zero reflection means full power transmission through the metasurface. In this scenario, according to (2.31), we have

$$\eta_2 \cos\theta_i = \eta_1 \cos\theta_t \tag{2.39}$$

which is in contradiction with that given by (2.24). In other words, the only transmission phase based metasurface design fails to generate a plane wave tilted toward an arbitrary direction with unitary efficiency. This is because it does not consider the impedance matching of the incident and transmitted waves in different directions, and thus some parasitic reflections are inevitably produced. This problem can be solved by designing the impedance of the metasurface to compensate for the field discontinuity across the interface. The tangential electric- and magnetic-field components at the two sides of the metasurface are expressed as in terms of an impedance matrix

$$\begin{bmatrix} \vec{E}_1^t \\ \vec{E}_2^t \end{bmatrix} = \begin{bmatrix} Z_{11} & Z_{12} \\ Z_{21} & Z_{22} \end{bmatrix} \begin{bmatrix} \hat{n} \times \vec{H}_1^t \\ -\hat{n} \times \vec{H}_2^t \end{bmatrix} \tag{2.40}$$

Here superscript "t" represents the tangential component and the subscript i ($i = 1$, 2) represents the medium i. The impedance parameters Z_{ij} depends on the topologies of the metasurface unit cell. Substituting (2.13), (2.14), (2.17) and (2.18) into (2.40), we can obtain

$$e^{-jk_1 \sin\theta_i x} = Z_{11} \frac{\cos\theta_i}{\eta_1} e^{-jk_1 \sin\theta_i x} - Z_{12} \frac{\cos\theta_t}{\eta_2} \frac{E_t}{E_i} e^{-jk_2 \sin\theta_t x + j\Phi_t} \tag{2.41}$$

$$e^{-jk_2 \sin\theta_t x + j\Phi_t} = Z_{21} \frac{\cos\theta_i}{\eta_1} \frac{E_i}{E_t} e^{-jk_1 \sin\theta_i x} - Z_{22} \frac{\cos\theta_t}{\eta_2} e^{-jk_2 \sin\theta_t x + j\Phi_t} \tag{2.42}$$

On the other hand, for the ideal transmission, we have

$$\mathrm{Re}\left[\frac{1}{2} \vec{E}_1^t \times \vec{H}_1^{t*}\right] = \mathrm{Re}\left[\frac{1}{2} \vec{E}_2^t \times \vec{H}_2^{t*}\right] \tag{2.43}$$

Therefore, we have

$$E_t = E_i \sqrt{\frac{\cos\theta_i}{\cos\theta_t}} \sqrt{\frac{\eta_2}{\eta_1}} \tag{2.44}$$

Inserting (2.44) into (2.41) and (2.42), one gets

$$e^{-jk_1 \sin\theta_i x} = Z_{11}\frac{\cos\theta_i}{\eta_1}e^{-jk_1 \sin\theta_i x} - Z_{12}\frac{\sqrt{\cos\theta_t \cos\theta_i}}{\sqrt{\eta_1\eta_2}}e^{-jk_2 \sin\theta_t x + j\Phi_t} \tag{2.45}$$

$$e^{-jk_2 \sin\theta_t x + j\Phi_t} = Z_{21}\frac{\sqrt{\cos\theta_i \cos\theta_t}}{\sqrt{\eta_1\eta_2}}e^{-jk_1 \sin\theta_i x} - Z_{22}\frac{\cos\theta_t}{\eta_2}e^{-jk_2 \sin\theta_t x + j\Phi_t} \tag{2.46}$$

According to (2.45) and (2.46), there are many solutions for the unknown Z-parameters. Considering the lossless metasurface, all the Z-parameters are purely imaginary, i.e., $Z_{ij} = jX_{ij}$. Therefore, (2.45) and (2.46) can be rewritten as

$$e^{-jk_1 \sin\theta_i x} = jX_{11}\frac{\cos\theta_i}{\eta_1}e^{-jk_1 \sin\theta_i x} - jX_{12}\frac{\sqrt{\cos\theta_t \cos\theta_i}}{\sqrt{\eta_1\eta_2}}e^{-jk_2 \sin\theta_t x + j\Phi_t} \tag{2.47}$$

$$e^{-jk_2 \sin\theta_t x + j\Phi_t} = jX_{21}\frac{\sqrt{\cos\theta_i \cos\theta_t}}{\sqrt{\eta_1\eta_2}}e^{-jk_1 \sin\theta_i x} - jX_{22}\frac{\cos\theta_t}{\eta_2}e^{-jk_2 \sin\theta_t x + j\Phi_t} \tag{2.48}$$

The solutions of (2.47) and (2.48) for X_{ij} are

$$\begin{cases} X_{11} = \dfrac{\eta_1}{\cos\theta_i}\cot\Psi_t \\ X_{22} = \dfrac{\eta_2}{\cos\theta_t}\cot\Psi_t \\ X_{12} = X_{21} = \dfrac{\sqrt{\eta_1\eta_2}}{\sqrt{\cos\theta_i \cos\theta_t}}\dfrac{1}{\sin\Psi_t} \end{cases} \tag{2.49}$$

in which

$$\Psi_t = -k_2 \sin\theta_t x + \Phi_t + k_1 \sin\theta_i x \tag{2.50}$$

Due to $X_{12} = X_{21}$, the metasurface is reciprocal. With $X_{11} \neq X_{22}$, it is an asymmetric metasurface. The possible realizations include split ring array, double patch array, and Ω-shaped inclusion array, etc.

Case 2 Ideal reflection with zero transmission

Similar to the ideal transmission, the normal component of the Poynting vector on the interface reads

$$P_n = \frac{1}{2}\mathrm{Re}(\vec{E}_1^t \times \vec{H}_1^{t*}) \cdot \hat{n} \tag{2.51}$$

in which $\hat{n}$ is unit normal vector of the interface.

Note that the field distribution on the interface is the superposition of the incident and reflection fields. In the case of the lossless metasurface, the ideal reflection gives rise to the periodical flow of power into the metasurface and back into space. When inserting (2.13)–(2.16) into (2.51), we have

$$E_r = E_i\sqrt{\frac{\cos\theta_i}{\cos\theta_r}} \tag{2.52}$$

With the introduction of input impedance, the relation between the tangential electric field and the tangential magnetic field is expressed as

$$\vec{E}_1^t = Z_{11}\hat{n}\times\vec{H}_1^t \tag{2.53}$$

Here the input impedance is

$$Z_{11} = \frac{\eta_1}{\sqrt{\cos\theta_i\cos\theta_r}}\frac{\sqrt{\cos\theta_r}+\sqrt{\cos\theta_i}e^{j\Psi_r}}{\sqrt{\cos\theta_i}-\sqrt{\cos\theta_r}e^{j\Psi_r}} \tag{2.54}$$

in which

$$\Psi_r = -k_1\sin\theta_r x + \Phi_r + k_1\sin\theta_i x \tag{2.55}$$

The real part of the input impedance in (2.54) periodically takes positive (loss) and negative (active) values. It implies that the power passing through the surface in the "lossy" regions is not absorbed but it is reradiated from the "active" regions. The average net power emerging from the surface is equivalently zero.

2.4 Digital Coding Metamaterials/Metasurfaces

Metamaterials/metasurfaces are artificial structures that are engineered to achieve the EM properties not found in nature, such as invisible cloak, super lens, negative refraction, etc. One of the important abilities of the metamaterials and metasurfaces is to tailor the equivalent parameters (e.g., permittivity, permittivity, surface impedance) for flexibly manipulating the electromagnetic waves. Depending on the continuity or discretization of the equivalent parameters, the metamaterials/metasurfaces are classified into an analog class and a digital class. Compared with the analog metamaterial/metasurface, the use of the digitizing procedure provides the possibility to achieve real-time manipulation of electromagnetic waves.

In order to better understand the digital coding procedure for the metamaterial/metasurface, a scattering problem from the metasurface is considered. Without loss of generality, the metasurface structure is composed of N-by-N equal-sized lattices with the dimension D. Here each lattice consists of a M-by-M meta-atom subarray such that the resulting phase response of the meta-atom is similar to that obtained in the infinite periodic scenario. The scattering phase of the (m,n)th lattice is φ_{mn} ($m = 1, \cdot\cdot\cdot, N, n = 1, \cdot\cdot\cdot, N$), as shown in Fig. 2.2. For the normal illumination

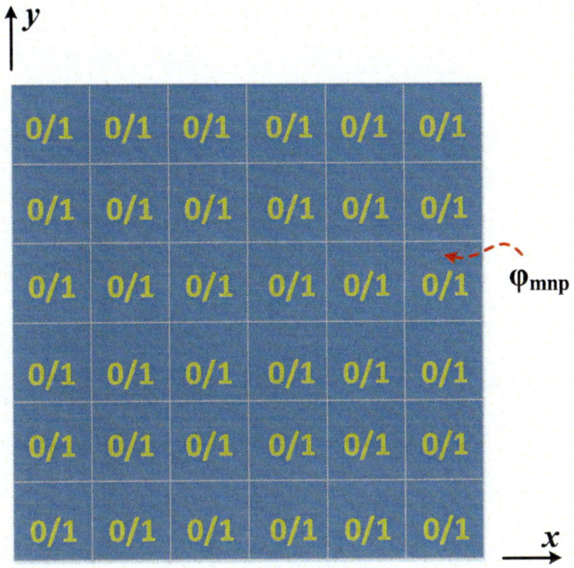

Fig. 2.2 1-bit metasurface composed of N-by-N lattices

of the plane wave, the scattering field can be solved as

$$f(\theta,\varphi) = f_e(\theta,\varphi)\cdot \sum_{m=1}^{N}\sum_{n=1}^{N}\exp\{-j[\varphi_{mn} + k(m-0.5)D\sin\theta\cos\varphi + k(n-0.5)D\sin\theta\sin\varphi]\} \tag{2.56}$$

in which k is the wave number in free space and $f_e(\theta,\varphi)$ denotes the pattern function of lattices. Here the assumption that each lattice has the same scattering pattern except a scattering phase.

Case 1 1-bit coding metamaterial/metasurface

The 1-bit coding metasurface is composed of two kinds of meta-atoms with an approximate phase difference of 180°. Therefore, the meta-atom with a 0 phase response is called the "0" element, while the meta-atom with a π phase response is defined as the "1" element. A simple example for the "0" and "1" elements is a perfectly magnetic conductor (PMC) and a perfectly electric conductor (PEC). The coding metasurface manipulates the electromagnetic wave by different coding sequences of "0" and "1" elements. With the 1-bit coding scheme, the scattering phase of the (m,n)th lattice is

$$\varphi_{mn} = \begin{cases} 0 & 0 \text{ element} \\ \pi & 1 \text{ element} \end{cases} \tag{2.57}$$

Without loss of generality, the metasurface is composed of 2-by-2 lattices, ignoring the pattern function of the lattice. Further assume that the dimension of the lattice is less than a half wavelength, i.e., $D \leq \lambda/2$. Thus, (2.56) is rewritten as

$$\begin{aligned} f(\theta,\varphi) = & \exp\{-j[\varphi_{11} + k0.5D\sin\theta\cos\varphi + k0.5D\sin\theta\sin\varphi]\} \\ & + \exp\{-j[\varphi_{12} + k0.5D\sin\theta\cos\varphi + k1.5D\sin\theta\sin\varphi]\} \\ & + \exp\{-j[\varphi_{21} + k1.5D\sin\theta\cos\varphi + k0.5D\sin\theta\sin\varphi]\} \\ & + \exp\{-j[\varphi_{22} + k1.5D\sin\theta\cos\varphi + k1.5D\sin\theta\sin\varphi]\} \end{aligned} \tag{2.58}$$

When all meta-atoms are chosen as the "0" or "1" elements, (2.58) is reduced as

$$|f(\theta,\varphi)| = 4 \cdot |\cos\Psi_1 \cos\Psi_2| \tag{2.59}$$

in which

$$\Psi_1 = \frac{kD\sin\theta\cos\varphi}{2} \tag{2.60}$$

$$\Psi_2 = \frac{kD\sin\theta\sin\varphi}{2} \tag{2.61}$$

It can be observed from (2.59) that when $\theta = 0°$ in the upper half space, the $|f(\theta,\varphi)| = 4$ is maximal.

When the meta-atom sequence is chosen as the "0101", (2.58) is reduced as

$$|f(\theta,\varphi)| = 4 \cdot |\cos\Psi_1 \sin\Psi_2| \tag{2.62}$$

In this scenario, two maximal scattering beams occur along the directions of $(\theta = \sin^{-1}(\pi/kD), \varphi = \pm\pi/2)$ in the upper half space.

When the meta-atom sequence is chosen as the "0110", (2.58) is reduced as

$$|f(\theta,\varphi)| = 4 \cdot |\sin\Psi_1 \sin\Psi_2| \tag{2.63}$$

There are four maximal scattering fields along the directions of $(\theta = \sin^{-1}(\sqrt{2}\pi/kD), \varphi = \pm\pi/4)$ and $(\theta = \sin^{-1}(\sqrt{2}\pi/kD), \varphi = \pm 3\pi/4)$ in the upper half space.

Therefore, we can find that for different coding sequences, the maximal scattering direction can be flexibly controlled.

Case 2 *n*-bit coding metamaterial/metasurface

The *n*-bit coding concept can be obtained by extending the 1-bit coding. In the case of *n*-bit coding, there are 2^n meta-atoms with the phase interval $2\pi/2^n$. For instance, 2-bit meta-atoms have the phase responses of 0, $\pi/2$, π, $3\pi/2$ which correspond to the "00", "01", "10", "11" elements, respectively. Compared with 1-bit coding, more sequences can be achieved by using the 2-bit coding, thus providing greater freedom to design metasurface structures, and furthermore more effectively manipulating the electromagnetic wave. It is worthwhile pointing out that with the increase of the bits, the metasurface design procedure becomes increasingly complicated.

2.5 Group Theory of Metamaterials

Metamaterials generally consist of periodic structures with sub-wavelength unit cells. Therefore, metamaterial has a certain symmetry. The group theory can be used to study the symmetry of the metamaterial.

2.5.1 *Periodicity of 2-D Photonic Crystals*

From the point of view of symmetry, any photonic crystal is a periodic structure. Specifically, it possesses a discrete translational symmetry. That is, they are invariant under translations of distances that are an integer multiple of the fixed step size. Besides, one can also consider the geometrical symmetry of dielectric elements, the symmetry of their material (for example, anisotropy), geometrical symmetry of the crystal unit cells.

Photonic crystals have distinct structures made up of a lattice and a basis. The group of atoms or molecules that repeats itself is called a basis or a unit cell, and the smallest possible unit cell is called a primitive unit cell. The vectors that define a primitive unit cell are called primitive translation vectors. The array of points generated by the primitive translation vectors is called a lattice.

$$\vec{r'} = \vec{r} + m_1\vec{a}_1 + m_2\vec{a}_2 = \vec{r} + \vec{R} \tag{2.64}$$

where $\vec{a_1}$ and $\vec{a_2}$ are the primitive translation vectors, m_1 and m_2 are integers, $\vec{r}$ is any vector in the lattice and $\vec{R}$ is lattice vector.

The choice of a unit cell is not unique. Wigner–Seitz cell is one of the most common cells in constructing a unit cell. The construction proceeds as follows: Firstly, a lattice point is chosen as the origin. Then, the straight lines are drawn to connect the origin and its nearest lattice points. Next, from the midpoints of these straight lines, new perpendicular straight lines are drawn. The region enclosed by the intersecting perpendicular lines is called the Wigner–Seitz cell, as shown in Fig. 2.3.

Wigner–Seitz cells share the symmetry of the lattice and enclose the region containing all points that are closer to the origin lattice point than to any other lattice point. In the following section of reciprocal space, the Wigner–Seitz cells have special significance.

Fig. 2.3 Illustration of the procedure for constructing the Wigner–Seitz cell

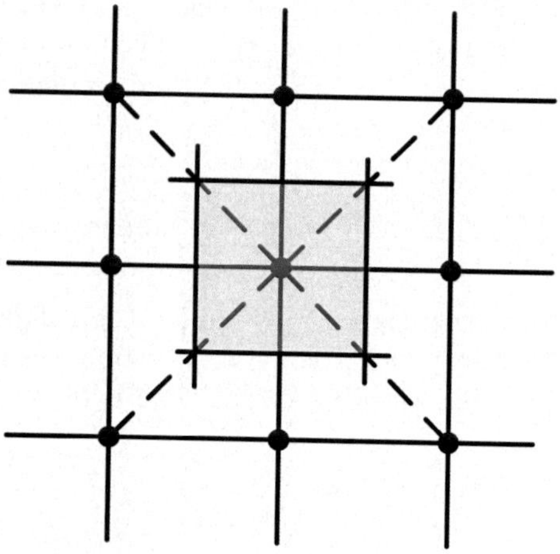

2.5.2 *Reciprocal Space*

A direct lattice describes the way the physical elements are arranged. The concept of reciprocal lattice is found in many solid-state physics textbooks. In terms of the periodicity of crystals, we can transform the crystal structures into the Fourier domain. The Fourier decomposition of the period function that describes a crystal lattice can be expressed as

$$f(\vec{r}) = \sum_{\vec{G}} f(\vec{G}) e^{j\vec{G}\cdot\vec{r}} \tag{2.65}$$

where $\vec{G}$ is an infinite set of vectors in k-space that define the spatial frequencies. Thus the Fourier decomposition is an expansion in a plane-wave basis. According to the periodicity of the crystal, the expression must satisfy

$$f(\vec{r} + \vec{R}) = \sum_{\vec{G}} f(\vec{G}) e^{j\vec{G}\cdot\vec{r}} e^{j\vec{G}\cdot\vec{R}} = f(\vec{r}) \tag{2.66}$$

To meet the above equation, one has $e^{j\vec{G}\cdot\vec{R}} = 1$ and $\vec{G} \cdot \vec{R} = 2m\pi$, where m is an integer. This is done by defining the vectors $\vec{b_1}$ and $\vec{b_2}$ with respect to the primitive translation vectors

$$\vec{G} = n_1 \vec{b_1} + n_2 \vec{b_2} \tag{2.67}$$

$$\vec{b_i} \cdot \vec{a_j} = 2\pi \delta_{ij} \tag{2.68}$$

where $i, j = 1, 2$, and δ_{ij} is the Kronecker delta function which can be defined as

$$\delta_{ij} = \begin{cases} 1 & i = j \\ 0 & i \neq j \end{cases} \tag{2.69}$$

The k-space can be called the reciprocal space. We consider the set $\vec{b}$ as the fundamental vectors of a lattice in k-space. Every lattice in the real space has a corresponding reciprocal lattice in the reciprocal space. One can construct the Wigner–Seitz cell that is called the first Brillouin zone in the reciprocal zone. The introduction of the first Brillouin zone can simplify the analysis of photonic crystals.

2.5.3 *Brillouin Zone*

As we know, each of the Bravais lattices in real space has its corresponding reciprocal lattice in the reciprocal space. The unit cells in reciprocal space can be formed by the same method as the one of real space. These unit cells are called Brillouin zones. For details, the Brillouin zones are constructed by drawing the vectors $\vec{k}$ defining the reciprocal lattice and then bisecting each of these with lines perpendicular to $\vec{k}$. The shape enclosed by these planes is the first Brillouin zone. This zone is repeated throughout the reciprocal lattice by translating it along reciprocal lattice vectors. Any point in a given Brillouin zone is equivalent to the corresponding point defined in the first Brillouin zone.

In the view of Bloch's theorem, the mode frequencies are also periodic in k-space, which simplifies the concern of $\vec{k}$ to the limited range. The range of nonredundant value is also called the first Brillouin zone. The Wigner–Seitz cell in the reciprocal space is the same as the nonredundant region of $\vec{k}$ in the Bloch's theorem.

2.5.4 *Symmetry of 2-D Photonic Crystals*

Any operation performed on a crystal that carries the crystal structure into itself is part of the symmetry group for that crystal. This may include translations, reflections through planes, rotations about axes, inversion through a point, and combinations of these operations. The fundamental types of crystal lattices are defined by their symmetry operations. These include translation group, point group, and space group symmetries. For a crystal, the translation group can be also considered as the periodicity of a crystal, as discussed in the above section. The point group consists of reflections, rotations and combinations of these operations. For point groups, there is at least one point remains unchanged when the operations above are applied. Here we will concentrate on the point group of 2-D crystals.

Case 1 Rotation symmetry

The square lattice has the rotation symmetry C_4. We can employ a matrix to describe the rotation symmetry.

$$\mathbf{R}_{C_4} = \begin{bmatrix} 0 & -1 \\ 1 & 0 \end{bmatrix} \tag{2.70}$$

where the rotation symmetry C_4 is defined as

$$\begin{aligned} C_N &= \{(C_N)_l | l = 1, 2, \cdots, N\} \\ (C_N)_l &\equiv \text{rotation by } \frac{2\pi l}{N} \end{aligned} \tag{2.71}$$

For the hexagonal lattice, it has the rotation symmetry C_6, which can also be expressed by the matrix

$$\mathbf{R}_{C_6} = \begin{bmatrix} \frac{1}{2} & -\frac{\sqrt{3}}{2} \\ \frac{\sqrt{3}}{2} & \frac{1}{2} \end{bmatrix} \tag{2.72}$$

Case 2 Mirror symmetry

There are two kinds of mirror symmetries in the square lattice, as shown in Fig. 2.4. σ_x and σ_y are the mirror symmetries through the x and y axes respectively. And σ_d' and σ_d'' are the mirror symmetries through the two diagonals.

$$\mathbf{R}_{\sigma_x} = \begin{bmatrix} -1 & 0 \\ 0 & -1 \end{bmatrix} \mathbf{R}_{\sigma_y} = \begin{bmatrix} 1 & 0 \\ 0 & -1 \end{bmatrix} \tag{2.73}$$

$$\mathbf{R}_{\sigma_d'} = \begin{bmatrix} 0 & 1 \\ 1 & 0 \end{bmatrix} \mathbf{R}_{\sigma_d''} = \begin{bmatrix} 0 & -1 \\ -1 & 0 \end{bmatrix} \tag{2.74}$$

According to the rotation and mirror symmetries, the resulting group of symmetry of the square lattice is C_{4v} point group which contains the following eight elements:

$$C_{4v} = \left\{E, C_4, C_4^{-1}, C_2, \sigma_x, \sigma_y, \sigma_d', \sigma_d''\right\} \tag{2.75}$$

The characteristics of a group can be described by a group multiplication table. Consider a group consisting of six elements represented by the letters A, B, C, D, E, and F that obey the multiplication table shown in Table 2.1. The elements in the table are the product of the element designating its column and the element designating its row. Following this convention, the table shows that the identity element is a member

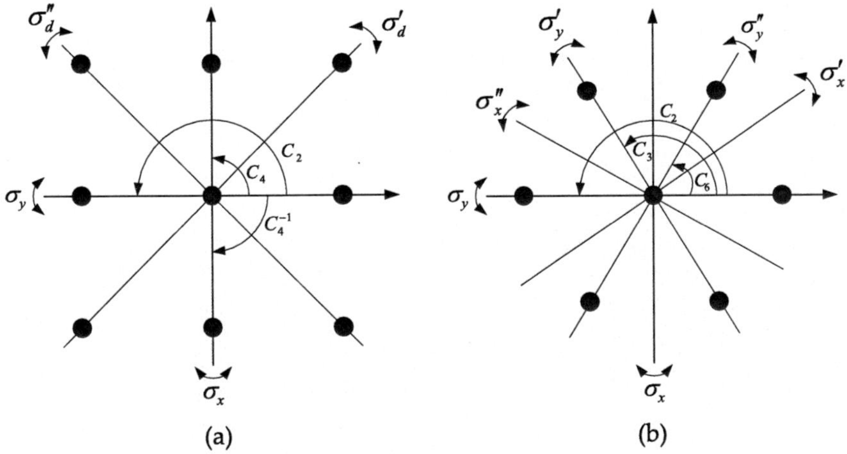

Fig. 2.4 Symmetry of **a** square lattice and **b** hexagonal lattice

of the group, the product of any two elements is an element of the group and each element appears only once in any given row or column.

The multiplication rules of C_{4v} point group are shown in Table 2.2, where E is the identity element, C_2 is a rotation by π around the *z*-axis, C_4 and C_4^{-1} are rotations by π/4 and −π/4, respectively.

And the resulting group of symmetry of the square lattice is C_{6v} point group which contains the following twelve elements:

$$C_{6v} = \{E, C_6, C_6^{-1}, C_3, C_3^{-1}, C_2, \sigma_x, \sigma_x', \sigma_x'', \sigma_y, \sigma_y', \sigma_y''\} \tag{2.76}$$

The multiplication rules of C_{6v} point group are shown in Table 2.3, where E is the identity element, C_6 and C_6^{-1} are rotations by π/6 and −π/6, respectively. There are other distinct types of 2-D lattices in the photonic crystals, in accord with different point groups. For instance, the rectangular, centered rectangular, and oblique lattices.

Table 2.1 Group multiplication

	E	A	B	C	D	F
E	E	A	B	C	D	F
A	A	E	D	F	B	C
B	B	F	E	D	C	A
C	C	D	F	E	A	B
D	D	C	A	B	F	E
F	F	B	C	A	E	D

Table 2.2 Multiplication rules of C_{4v}

C_{4v}	E	C_4	C_4^{-1}	C_2	σ_x	σ_y	σ_d'	σ_d''
E	E	C_4	C_4^{-1}	C_2	σ_x	σ_y	σ_d'	σ_d''
C_4	C_4	C_2	E	C_4^{-1}	σ_d''	σ_d'	σ_x	σ_y
C_4^{-1}	C_4^{-1}	E	C_2	C_4	σ_d'	σ_d''	σ_y	σ_x
C_2	C_2	C_4^{-1}	C_4	E	σ_y	σ_x	σ_d''	σ_d'
σ_x	σ_x	σ_d'	σ_d''	σ_y	E	C_2	C_4	C_4^{-1}
σ_y	σ_y	σ_d''	σ_d'	σ_x	C_2	E	C_4^{-1}	C_4
σ_d'	σ_d'	σ_y	σ_x	σ_d''	C_4^{-1}	C_4	E	C_2
σ_d''	σ_d''	σ_x	σ_y	σ_d'	C_4	C_4^{-1}	C_2	E

Table 2.3 Multiplication rules of C_{6v}

C_{6v}	E	C_6	C_6^{-1}	C_3	C_3^{-1}	C_2	σ_x	σ_x'	σ_x''	σ_y	σ_y'	σ_y''
E	E	C_6	C_6^{-1}	C_3	C_3^{-1}	C_2	σ_x	σ_x'	σ_x''	σ_y	σ_y'	σ_y''
C_6	C_6	C_3	E	C_2	C_6^{-1}	C_3^{-1}	σ_y''	σ_y	σ_y'	σ_x''	σ_x	σ_x'
C_6^{-1}	C_6^{-1}	E	C_3^{-1}	C_6	C_2	C_3	σ_y'	σ_y''	σ_y	σ_x'	σ_x''	σ_x
C_3	C_3	C_2	C_6	C_3^{-1}	E	C_6^{-1}	σ_x'	σ_x''	σ_x	σ_y'	σ_y''	σ_y
C_3^{-1}	C_3^{-1}	C_6^{-1}	C_2	E	C_3	C_6	σ_x''	σ_x	σ_x'	σ_y''	σ_y	σ_y'
C_2	C_2	C_3^{-1}	C_3	C_6^{-1}	C_6	E	σ_y	σ_y'	σ_y''	σ_x	σ_x'	σ_x''
σ_x	σ_x	σ_y'	σ_y''	σ_x''	σ_x'	σ_y	E	C_3^{-1}	C_3	C_2	C_6	C_6^{-1}
σ_x'	σ_x'	σ_y''	σ_y	σ_x	σ_x''	σ_y'	C_3	E	C_3^{-1}	C_6^{-1}	C_2	C_6
σ_x''	σ_x''	σ_y	σ_y'	σ_x'	σ_x	σ_y''	C_3^{-1}	C_3	E	C_6	C_6^{-1}	C_2
σ_y	σ_y	σ_x'	σ_x''	σ_y''	σ_y'	σ_x	C_2	C_6	C_6^{-1}	E	C_3^{-1}	C_3
σ_y'	σ_y'	σ_x''	σ_x	σ_y	σ_y''	σ_x'	C_6^{-1}	C_2	C_6	C_3	E	C_3^{-1}
σ_y''	σ_y''	σ_x	σ_x'	σ_y'	σ_y	σ_x''	C_6	C_6^{-1}	C_2	C_3^{-1}	C_3	E

Since a unit cell is not unique, we choose the basis vectors of a unit cell as the primitive translation vectors for the rectangular and centered rectangular lattices, as shown in Fig. 2.5.

Figure 2.5a shows a rectangular lattice, where $\left|\vec{a_1}\right| \neq \left|\vec{a_2}\right|$, and $\vec{a_1} \perp \vec{a_2}$. The resulting group of symmetry of the rectangular lattice is C_{2v} point group which contains the following four elements:

$$C_{2v} = \{E, C_2, \sigma_x, \sigma_y\} \tag{2.77}$$

The multiplication rules of C_{2v} point group are shown in Table 2.4.

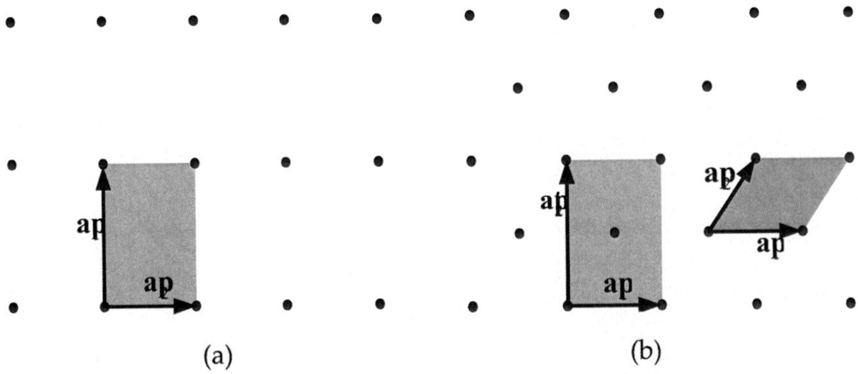

Fig. 2.5 Rectangular lattice, **a** rectangular lattice, **b** centered rectangular lattice

Table 2.4 Multiplication rules of C_{2v}

C_{2v}	E	C_2	σ_x	σ_y
E	E	C_2	σ_x	σ_y
C_2	C_2	E	σ_y	σ_x
σ_x	σ_x	σ_y	E	C_2
σ_y	σ_y	σ_x	C_2	E

Next, consider the centered rectangular lattice depicted in Fig. 2.5b. For the centred rectangular lattice, $\left|\vec{a_1}\right| \neq \left|\vec{a_2}\right|$ and $\left|\vec{a_1} \cdot \vec{a_2}\right| = \left|\vec{a_1}\right|^2 \Big/ 2$. In order to better describe the symmetry of this kind of lattice, we choose another primitive translation vectors $\vec{a_1}$ and $\vec{a_2'}$. Thus, the resulting group of symmetry is C_{2v}.

It can be seen from the example of the centered rectangular lattice that sometimes the unit cell, that is, the smallest periodic unit, can not reflect the symmetry of the lattice. Based on this, crystallography defines the concept of a conventional cell. In order to reflect the symmetry of the crystal, a larger cell may be chosen as the conventional cell. A conventional cell is, in some cases, a unit cell, and in some cases, not a unit cell. In crystallography, there are uniform rules about how to select conventional cells of different types of Bravais lattices. For example, the vectors $\vec{a_1}$ and $\vec{a_2'}$ of the centered rectangular lattice are chosen in accordance with the rules so as to reflect the symmetry of the lattice.

2.5.5 *Irreducible Brillouin Zone*

Due to the symmetry and periodicity of the first Brillouin zone, we can present the smallest region of the first Brillouin zone which is called the irreducible Brillouin zone. All points within the first Brillouin zone can be mapped onto points within the irreducible Brillouin zone through symmetry operations. So, we do not need to consider every k point in the first Brillouin zone. The irreducible Brillouin zone is an effective method to simplify the analysis of crystals. It is obvious that the irreducible Brillouin zone is closely related to the symmetry of crystals. The relationship can be expressed as

$$S_{\mathrm{IBZ}} = \frac{4\pi^2}{S|\mathrm{P}|} \tag{2.78}$$

where S is the area of the unit cell within real space, S_{IBZ} is the area of the irreducible Brillouin zone, and |P| is the rank of the point group P.

Figure 2.6 depicts the real lattice, the reciprocal lattice and the Brillouin zone of a square lattice. Within the Brillouin zone, the shaded region is the irreducible Brillouin zone of the square lattice, which is determined by (2.78). Due to the C_{4v} symmetry of a square lattice, |P| is 8. So, the area of the irreducible Brillouin zone is 1/8 of the area of the first Brillouin zone. Note that there are three highly symmetry points Γ, X and M within the irreducible Brillouin zone. They denote the points

$$\begin{aligned} \Gamma &: \frac{2\pi}{a}[0, 0] \\ \mathrm{X} &: \frac{2\pi}{a}\left[\frac{1}{2}, 0\right] \\ \mathrm{M} &: \frac{2\pi}{a}\left[\frac{1}{2}, \frac{1}{2}\right] \end{aligned} \tag{2.79}$$

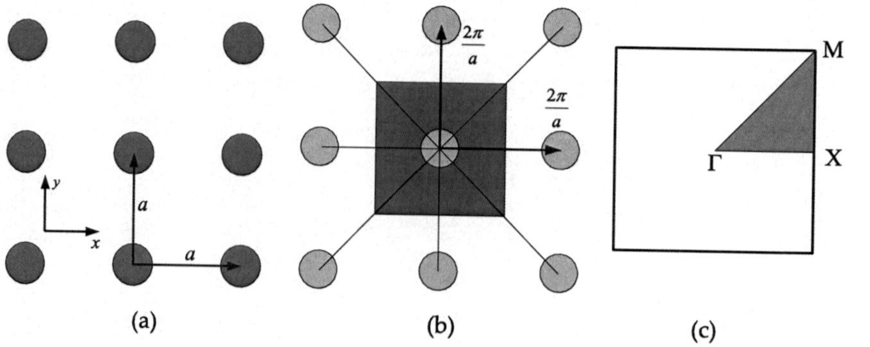

Fig. 2.6 Square lattice, **a** real lattice, **b** reciprocal lattice, **c** Brillouin zone

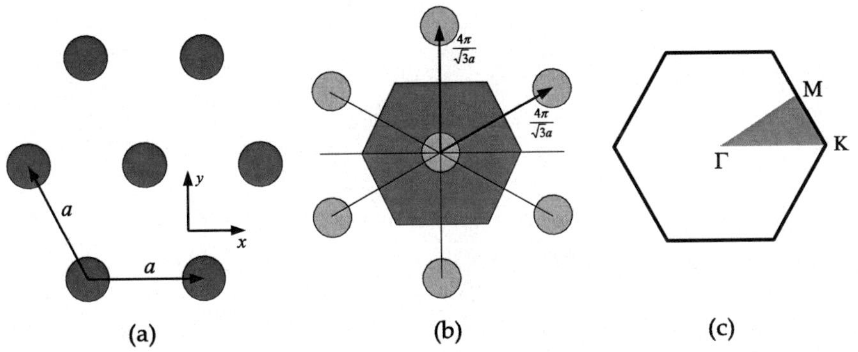

Fig. 2.7 Hexagonal lattice, **a** real lattice, **b** reciprocal lattice, **c** Brillouin zone

Figure 2.7 depicts the real lattice, the reciprocal lattice and the Brillouin zone of a hexagonal lattice. The shaded region is the irreducible Brillouin zone of the hexagonal lattice. Due to the C_{6v} symmetry of a hexagonal lattice, the rank of P is 12. So, the area of the irreducible Brillouin zone is 1/12 of the area of the first Brillouin zone. Different from a square lattice, there are three different highly symmetry points Γ, *K* and M within the irreducible Brillouin zone. They denote the points

$$\begin{aligned} \Gamma &: \quad \frac{2\pi}{a}[0,0] \\ \mathrm{K} &: \quad \frac{2\pi}{a}\left[\frac{2}{3},0\right] \\ \mathrm{M} &: \quad \frac{2\pi}{a}\left[\frac{1}{2},\frac{\sqrt{3}}{6}\right] \end{aligned} \tag{2.80}$$

Similarly, Fig. 2.8 shows the real lattice, the reciprocal lattice and the Brillouin zone of a rectangular lattice. The irreducible Brillouin zone is depicted by the shaded region. Due to the C_{2v} symmetry of a square lattice, the rank of P is 4. So, the area of the irreducible Brillouin zone is 1/4 of the area of the first Brillouin zone. Thus, there are four different highly symmetry points Γ, X, M and Y within the irreducible Brillouin zone. They denote the points

$$\begin{aligned} \Gamma &: \quad [0,0] \\ \mathrm{X} &: \quad \left[\frac{\pi}{a_1},0\right] \\ \mathrm{M} &: \quad \left[\frac{\pi}{a_1},\frac{\pi}{a_2}\right] \\ \mathrm{Y} &: \quad \left[0,\frac{\pi}{a_2}\right] \end{aligned} \tag{2.81}$$

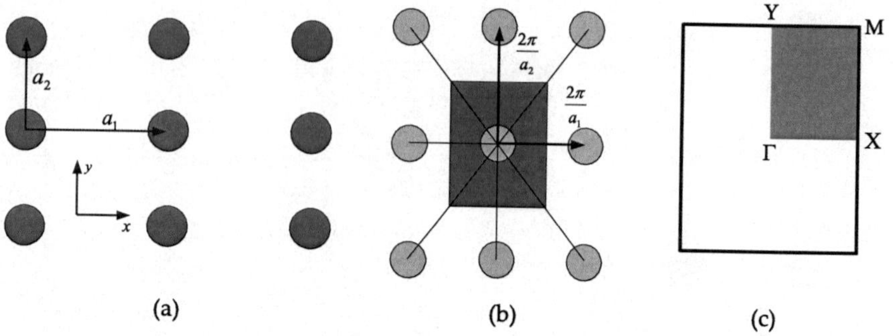

Fig. 2.8 Rectangular lattice, **a** real lattice, **b** reciprocal lattice, **c** Brillouin zone

To fully describe a crystal, we must learn both the period and symmetry properties and the contents of the unit cells. The period and symmetry properties can be reflected by the lattice of a crystal, which has been discussed above. The contents of the unit cell are often called basis. The discussion of the symmetry of a crystal must involve the symmetry of the lattice and basis. And the existence of basis may lead to a decrease of symmetry. Figure 2.9a depicts an electromagnetic bandgap (EBG) structure within which the bases are arranged by the square lattice. As we know, the symmetry of a square lattice is C_{4v}, but the structure of basis has the symmetry of C_{2v}. So, the entire description of symmetry is reduced to C_{2v}. Figure 2.9b shows the first Brillouin zone of the EBG in the reciprocal lattice, and the shaded region is the irreducible Brillouin zone. The C_{2v} symmetry makes the irreducible Brillouin zone account for 1/4 area of the first Brillouin zone.

Most of the EBG metamaterials are designed to have the same symmetry as the lattices, which makes the entire structure more symmetrical. But, under certain conditions, one intentionally design the bases with the less symmetry than the lattices. The breaking of symmetry can bring some new properties, one of which is polarization dependency. Taking the EBG shown in Fig. 2.9 as an example, we get different reflection phase of the EBG when the incident plane wave has different polarization states. Using this property, we can use the EBG as a polarization converter. When the incident wave is circularly polarized, the reflection phase difference between two principal polarizations determines the polarization of the reflected wave. The scattering wave can maintain the circular polarization, or be converted to the linear polarization depending on the reflection phase difference. Similarly, for the polarization dependent EBG, the surface wave bandgap can only exist in the certain direction of $\boldsymbol{k}$ rather than in the entire first Brillouin zone.

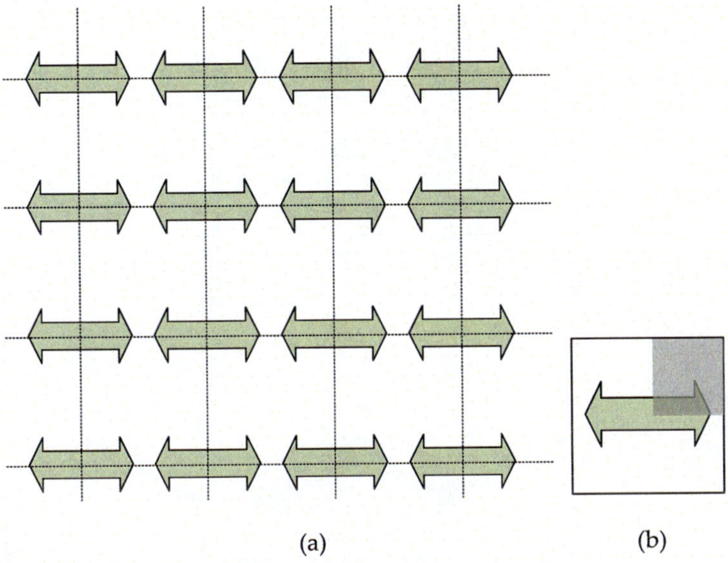

Fig. 2.9 Square lattice with basis **a** real space, **b** Brillouin zone

2.5.6 *Some Irreducible Brillouin Zones of EBG Metamaterials*

EBG metamaterials, also known as photonic crystals or photonic bandgap (PBG) materials are a novel class of artificially fabricated structures which have the ability to control the propagation of electromagnetic waves. Properly designed PCs can prohibit the propagation of light/ electromagnetic waves, or allow the propagation along only certain directions, or localize light in specified areas. With the new property, the EBG has a variety of applications. They can be constructed in one, two, and three dimensions (1D, 2D, and 3D) with either dielectric or/and metallic materials.

The following are some 2D EBG structures which are commonly used in the fields of metamaterial antennas, filters, etc. Figure 2.10 depicts a so-called mushroom type EBG structure. Metallic patches are fabricated on a dielectric substrate backed by a metallic ground. The metallic patches are connected to the ground with vertical metallic posts or vias.

Obviously, the periodic lattice of this EBG structure is a square lattice (C_{4v} symmetry), and the basis also has the symmetry of C_{4v}. The total structure satisfies the symmetry C_{4v}. In reciprocal space, the irreducible Brillouin zone is the shaded region shown in Fig. 2.10b. Another kind of mushroom type EBG structures is shown in Fig. 2.11. Its periodic lattice is a hexagonal lattice (C_{6v} symmetry). Also, the basis displays the C_{6v} symmetry. As shown in Fig. 2.11b, the irreducible Brillouin zone accounts for 1/12 area of the first Brillouin zone.

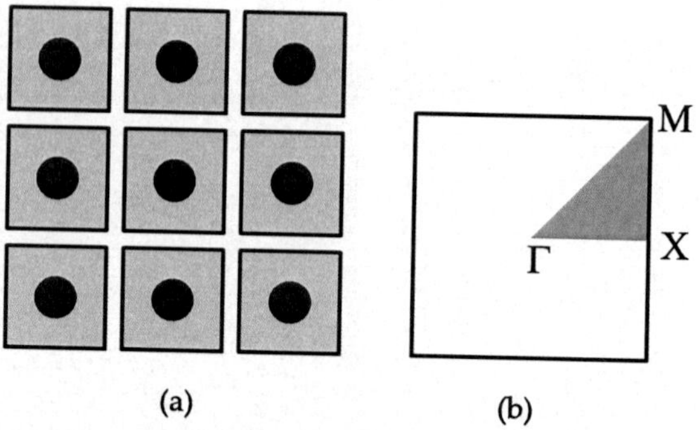

Fig. 2.10 **a** Square mushroom EBG structure, **b** Brillouin zone

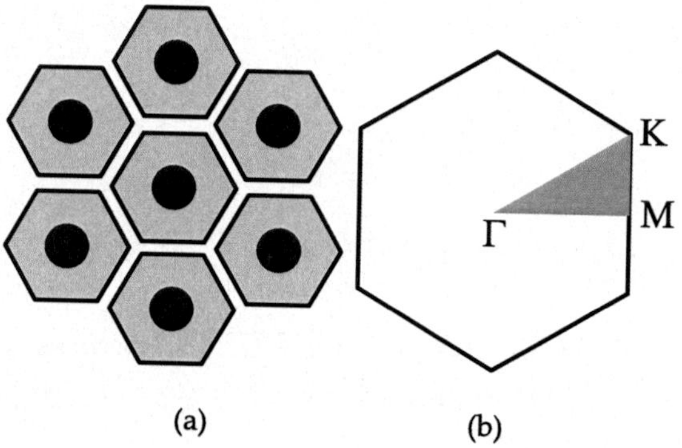

Fig. 2.11 **a** Hexagonal mushroom EBG structures, **b** Brillouin zone

Grounding vias make the fabrication of mushroom type EBGs complicated. In order to simplify the fabrication, a new kind of EBG structures has been presented by removing vias, as shown in Fig. 2.12. Since this structure exits in a complete plane, it is called uniplanar EBG (UC-EBG). According to the above discussion, the unit cell of this structure is arranged by a square lattice in the real space. Both the lattice and the basis satisfy the symmetry C_{4v}. The resulting irreducible Brillouin zone (shaded region) is shown in Fig. 2.12b.

Next, consider the EBG structure depicted in Fig. 2.13. This structure is also a kind of UC-EBG. The lattice is a square lattice, but the symmetry of the basis is merely C_{2v}. The entire symmetry of this structure is C_{2v}, and so the irreducible Brillouin zone (shaded region in Fig. 2.13b) is 1/4 rather than 1/8 area of the first Brillouin zone.

2.5.7 *Surface Wave and Symmetry*

Traditionally, the term 'surface waves' was used to describe waves bounded at the interface, with field strength that decays exponentially away from it. These waves are

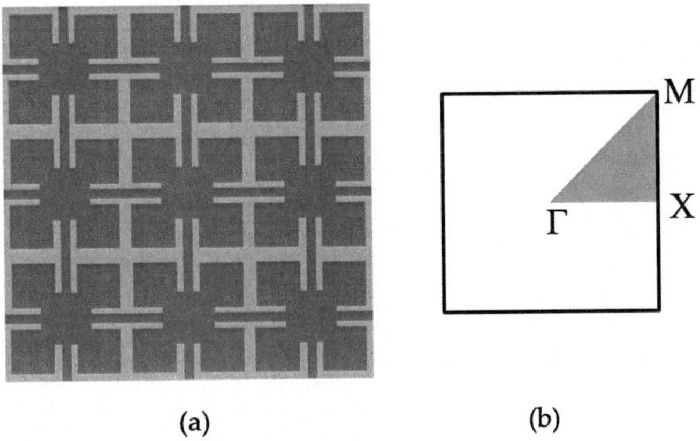

Fig. 2.12 **a** Uniplanar EBG (UC-EBG) structure with the symmetry C_{4v}, **b** Brillouin zone

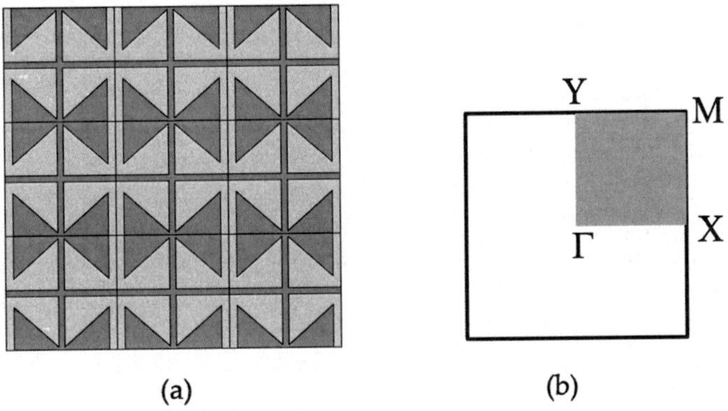

Fig. 2.13 **a** Uniplanar EBG (UC-EBG) structure with the symmetry C_{2v}, **b** Brillouin zone

characterized by tangential wavenumbers larger than that of free-space plane waves and are therefore termed as slow waves, since their phase velocity is less than the speed of light. When discussing the surface waves along an interface and the EBG, a dispersion diagram is an effective tool to present the existence of the surface waves. A dispersion diagram is the graphical representation of the wavenumber variation with frequency.

In the dispersion, there is usually a straight line called light line, which represents the relationships

$$k_0 = \frac{2\pi f}{c} \tag{2.82}$$

where f is the frequency and c is the light speed. The light line separates the dispersion diagram into two parts, slow and fast waves parts.

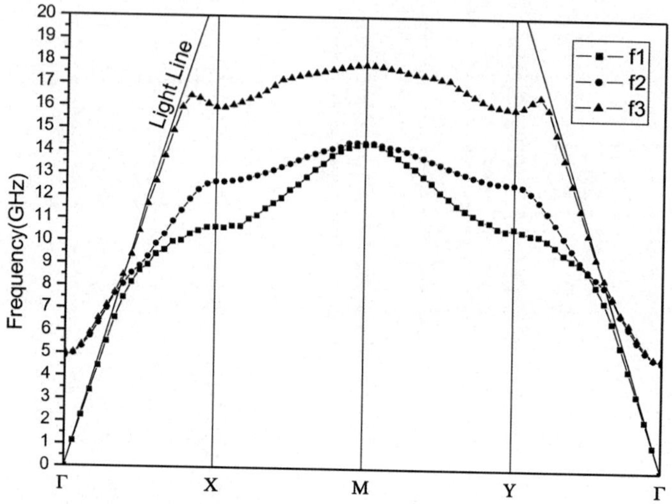

Fig. 2.14 Dispersion diagram

The surface waves of EBG structures are usually represented in dispersion diagrams. For such structures, the irreducible Brillouin zone defines the range of wavenumbers that must be covered in the computation of the surface wave. A complete characterization of the surface waves along the surface of an EBG structure requires the mapping of each point of the irreducible Brillouin zone to the first Brillouin zone by the symmetric operations. Bandgaps are identified as those frequency bands that do not correspond to any real wavenumber solution.

Although a full dispersion characterization requires the mapping of all wavenumbers in the irreducible Brillouin zone, we only need to show the dispersion around the contour of the irreducible Brillouin zone, as shown in Fig. 2.14. The contour is formed by the straight lines connecting the highly symmetry points.

As mentioned above, for infinite periodic structures, all possible wavenumbers can be reduced to values within the irreducible Brillouin zone. By definition, any periodic structure consists of an infinite arrangement of a minimum unit cell. A basic theory underlying the propagation of waves in periodic structures is known as Floquet's theorem. This theorem suggests the fact that electromagnetic fields are repeated at unit cell edges except for a propagation factor $e^{-j\beta d}$, where d is the length of a unit cell. Thus, the relationship of two adjacent unit cells can be represented as

$$\vec{E}(x, y, z+d) = e^{-j\beta d}\,\vec{E}(x, y, z) \tag{2.83}$$

The field in a periodic structure is described by a solution of the form

$$\vec{E}(x, y, z) = e^{-j\beta z}\,\vec{E}_p(x, y, z) \tag{2.84}$$

where $\vec{E}_p$ is periodic function of z with period d, i.e.,

$$\vec{E}_p(x, y, z + nd) = \vec{E}_p(x, y, z) \tag{2.85}$$

where n is any integer. According to Fourier series, we can obtain

$$\vec{E}_p(x, y, z) = \sum_{n=-\infty}^{\infty} \vec{E}_{pn}(x, y)e^{-j2n\pi z/d} \tag{2.86}$$

$$\begin{aligned}\vec{E}(x, y, z) &= e^{-j\beta z}\,\vec{E}_p(x, y, z)\\ &= \sum_{n=-\infty}^{\infty} \vec{E}_{pn}(x, y)e^{-j(\beta+2n\pi/d)z} = \sum_{n=-\infty}^{\infty} \vec{E}_{pn}(x, y)e^{-j\beta_n z}\end{aligned} \tag{2.87}$$

For the 1D case employed here as an example, this range represents the Brillouin zone corresponding to the periodicity d, which yields all possible mathematically different values of the wavenumber corresponding to the periodic structure. Hence all other wavenumber values relevant to the propagation within a periodic structure can be reduced to values within the Brillouin zone. In other words, the Brillouin zone contains all physically distinct wavenumbers for an infinite periodic structure.

The symmetry of the structure under consideration with respect to the $\pm z$-axis suggests a further reduction of the Brillouin zone while still maintaining all physically useful information. Waves propagating in the positive or negative z-direction (with wavenumbers $\pm\beta$), apart from the different directions, share identical characteristics. Hence all physical information included in the Brillouin zone $[-\pi/dz, \pi/dz]$ can be summarized in the range $\beta \in [0, \pi/dz]$. The irreducible Brillouin zone is defined as the Brillouin zone reduced by all possible symmetries.

As we know, in crystallography, the electromagnetic waves can be divided into two types: TE wave and TM wave due to the symmetry of two-dimensional photonic crystal. Similarly, we introduce the symmetry to the analysis of surface waves in 2D EBG structures. Taking the EBG structure shown in Fig. 2.15 as an example, we analyze the relationship between the surface wave and the symmetry of the structure. Because the EBG structure is not infinite in z-dimension, it is no longer satisfies the symmetry of the 2D photonic crystal, that is, the mirror symmetry of the plane xoy. But, for the wave vector $\vec{k}$ in the direction of Γ-X (y direction in Fig. 2.15), the EBG structure satisfies the mirror symmetry of plane yoz. We can obtain

$$\begin{aligned}&\mathcal{M}_x\tilde{\boldsymbol{H}}(r) = \tilde{\boldsymbol{H}}(r) \quad \text{even mode}\\ &\mathcal{M}_x\tilde{\boldsymbol{H}}(r) = -\tilde{\boldsymbol{H}}(r) \quad \text{odd mode}\end{aligned} \tag{2.88}$$

where $\mathcal{M}_\mathrm{x}$ is a mirror operator and the subscript x indicates that the mirror operation changes $\hat{x}$ to $-\hat{x}$ and keeps $\hat{y}$ and $\hat{z}$ unchanged. Thus, the surface wave is divided into two types: TE and TM surface waves. Only E_x, H_y, and H_z exist in

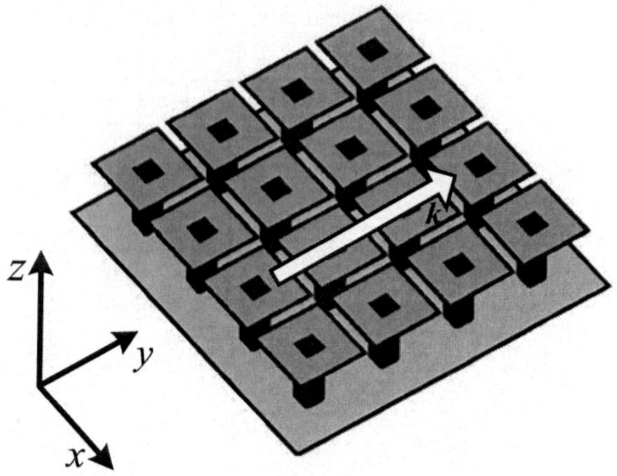

Fig. 2.15 EBG and wave vector of its surface wave

TE surface waves, whose field component E_x is perpendicular to the wave vector $\vec{k}$. Similarly, only H_x, E_y, and E_z exist in TM surface waves, whose field components H_x is perpendicular to the wave vector $\vec{k}$. Obviously, this definition is a common way to define the transmission systems such as waveguides. From the above analysis, it can be seen that the two kinds of surface waves, are essentially generated by the symmetry of the structure. This shows the importance of symmetry in the analysis of EBG metamaterials.

2.6 Conclusion

In this chapter, we review the development of metamaterials and metasurfaces. For the metamaterials, the electrodynamics properties of the left-handed metamaterials including backward wave propagation and negative refractive index are discussed. The generalized Snell's law for the metasurface is elaborated. Meanwhile, the digital coding method for the metamaterials/metasurfaces is discussed. Finally, the group theory is introduced to analyze the symmetry of the metamaterials/metasurfaces. The relationship between the irreducible Brillouin zone and the point group is presented, which regulates the irreducible Brillouin zone from the perspective of group theory. Some typical Brillouin zones of electromagnetic band gap metamaterials are given.

References

1. G.V. Eleftheriades, K.G. Balmain, *Negative-Refraction Metamaterials Fundamental Principles and Applications* (IEEE Press, New Jersey, 2005)
2. C. Caloz, T. Itoh, *Electromagnetic Metamaterials: Transmission Line Theory and Microwave Applications The Engineering Approach* (Wiley, New Jersey, 2006)
3. R. Marques, F. Martin, M. Sorolla, *Metamaterials with Negative Parameters Theory, Design and Microwave Applications* (Wiley, New Jersey, 2008)
4. L. Solymar, E. Shamonina, *Waves in Metamaterials* (Oxford University, New York, 2009)
5. T.J. Cui, R. Liu, D.R. Smith, *Metamaterials Theory, Design and Applications* (Springer, New York, 2009)
6. T.J. Cui, W.X. Tang, X.M. Yang, Z.L. Mei, W.X. Jiang, *Metamaterials Beyond Crystals, Noncrystals, and Quasicrystals* (CRC Press, New York, 2016)
7. H. Lamb, On group-velocity. Proc. London Math. Soc, Ser. 2, **1**, 473–479 (1904)
8. A. Schuster, *An Introduction to the Theory of Optics* (Edward Arnold, London, 1904)
9. L.I. Mandelshtam, Group velocity in crystalline arrays. Zh. Eksp. Teor. Fiz. **15**, 475–478 (1945). (in Russian)
10. D.V. Sivukhin, The energy of electromagnetic fields in dispersive media. Opt. Spektrosk. **3**, 308–312 (1957). (in Russian)
11. V.E. Pafomov, On transition radiation and the Vavilov-Cherenkov radiation. Zh. Eksp. Teor. Fiz. **36**, 1853–1858 (1959). (in Russian)
12. V.G. Veselago, The electrodynamics of substances with simultaneously negative values of ε and μ. Usp. Fiz. Nauk. **92**, 517–526 (1967). (in Russian)
13. J.B. Pendry, A.J. Holden, D.J. Robbins, W.J. Stewart, Low frequency plasmons in thin-wire structures. J. Phys. Condens. Matter. **10**, 4785–4809 (1998)
14. J.B. Pendry, A.J. Holden, D.J. Robbins, W.J. Stewart, Magnetism from conductors and enhanced nonlinear phenomena. IEEE Trans. Microw. Theory Tech. **47**(11), 2075–2084 (1999)
15. R.A. Shelby, D.R. Smith, S. Schultz, Experimental verification of a negative index of refraction. Science **292**, 77–79 (2001)
16. J.B. Pendry, D. Schurig, D.R. Smith, Controlling electromagnetic fields. Science **312**, 1780–1782 (2006)
17. U. Leonhardt, Optical conformal mapping. Science **312**, 1777–1780 (2006)
18. D. Schurig, J.J. Mock, B.J. Justice, S.A. Cummer, J.B. Pendry, A.F. Starr, D.R. Smith, Metamaterial electromagnetic cloak at microwave frequencies. Science **314**, 977–979 (2006)
19. R. Liu, C. Ji, J.J. Mock, J.Y. Chin, T.J. Cui, D.R. Smith, Broadband ground-plane cloak. Science **323**, 366–369 (2009)
20. L. Zhang, Y. Shi, C.H. Liang, Achieving illusion and invisibility of inhomogeneous cylinders and spheres. J. Optics **18**, 085101 (2016)
21. L. Zhang, Y. Shi, C.H. Liang, Optimal illusion and invisibility of multilayered anisotropic cylinders and spheres. Opt. Express **24**, 23333–23352 (2016)
22. Y. Shi, W. Tang, L. Li, C.H. Liang, Three-dimensional complementary invisibility cloak with arbitrary shapes. IEEE Antennas Wirel. Propag. Lett. **14**, 1550–1553 (2015)
23. Y. Shi, L. Zhang, W. Tang, L. Li, C.H. Liang, Design of a minimized complementary illusion cloak with arbitrary position. Int. J. Antennas Propag., pp. 932495 (2015)
24. Y. Shi, W. Tang, C.H. Liang, A minimized invisibility complementary cloak with a composite shape. IEEE Antennas Wirel. Propag. Lett. **13**, 1800–1803 (2014)
25. L. Zhang, Y. Shi, Bifunctional arbitrarily-shaped cloak for thermal and electric manipulations. Opt. Mater. Express **8**, 2600–2613 (2018)
26. Y. Shi, L. Zhang, Cloaking design for arbitrarily shape objects based on characteristic mode method. Opt. Express **25**, 32263–32279 (2017)
27. L. Li, X.J. Dang, B. Li, C.H. Liang, Analysis and design of waveguide slot antenna array integrated with electromagnetic band-gap structures. IEEE Antennas Wirel. Propag. Lett. **5**, 111–115 (2006)

28. L. Li, Z. Jia, F.F. Huo, W.Q. Han, A novel compact multiband antenna employing dual-band CRLH-TL for smart mobile phone application. IEEE Antennas Wirel. Propag. Lett. **12**, 1688–1691 (2013)
29. L. Li, F.F. Huo, Z. Jia, W.Q. Han, Dual zeroth-order resonance antennas with low mutual coupling for MIMO communications. IEEE Antennas Wirel. Propag. Lett. **12**, 1692–1695 (2013)
30. J. Xing, L. Li, L. Zhang, H. Sun, Y. Shi, Compact multiband antenna with CRLH-TL ZOR for wireless USB dongle applications. Microw. Opt. Technol. Lett. **56**, 1133–1138 (2014)
31. Y. Shi, L. Zhang, C.H. Liang, Dual zeroth-order resonant USB dongle antennas for 4G MIMO wireless communications. Int. J. Antennas Propag. Article ID 816051 (2015)
32. G.H. Li, H.Q. Zhai, L. Li, C.H. Liang, R.D. Yu, S. Liu, AMC-loaded wideband base station antenna for indoor access point in MIMO system. IEEE Trans. Antennas Propag. **63**(2), 525–532 (2015)
33. Y. Shi, K. Li, J. Wang, L. Li, C.H. Liang, An etched planar metasurface half maxwell fish-eye lens antenna. IEEE Trans. Antennas Propag. **63**, 3742–3747 (2015)
34. Y. Shi, Y.C. Li, Design of broadband leaky-wave antenna based on permeability-negative transmission line. Microw. Opt. Technol. Lett. **60**, 699–704 (2018)
35. M. Paquay, J.C. Iriarte, I. Ederra, Thin AMC structure for radar cross-section reduction. IEEE Trans. Antennas Propag. **55**, 3630–3638 (2007)
36. W. Chen, C.A. Balanis, C.R. Birtcher, Checkerboard EBG surfaces for wideband radar cross section reduction. IEEE Trans. Antennas Propag. **63**, 2636–2645 (2015)
37. Y. Shi, Z.K. Meng, W.Y. Wei, W. Zheng, L. Li, Characteristic mode cancellation method and its application for antenna RCS reduction. IEEE Antennas Wireless Propag. Lett. **18**, 1784–1788 (2019)
38. H. Zhang, Y. Lu, J. Su, Z. Li, J. Liu, Y. Yang, Coding diffusion metasurfaces for ultra-wideband RCS reduction. Electron. Lett. **53**, 187–189 (2017)
39. J.F. Han, X.Y. Cao, J. Gao, J. Wei, Y. Zhao, S. Li, Z. Zhang, Broadband radar cross section reduction using dual-circular polarization diffusion metasurface. IEEE Antennas Wirel. Propag. Lett. **17**, 969–973 (2018)
40. R. Zaker, A. Sadeghzadeh, A low-profile design of polarization rotation reflective surface for wideband RCS reduction. IEEE Antennas Wirel. Propag. Lett. **18**, 1794–1798 (2019)
41. Y. Lu, J.X. Su, J.B. Liu, Q.X. Guo, H.C. Yin, Z.R. Li, J.M. Song, Ultrawideband monostatic and bistatic RCS reductions for both copolarization and cross polarization based on polarization conversion and destructive interference. IEEE Trans. Antennas Propag. **67**, 4936–4941 (2019)
42. Y. Shi, X.F. Zhang, Z.K. Meng, L. Li, Design of low-RCS antenna using antenna array. IEEE Trans. Antennas Propag. **67**, 6484–6493 (2019)
43. Z.K. Meng, Y. Shi, W.Y. Wei, X.F. Zhang, Multifunctional scattering antenna array design for orbital angular momentum vortex waves and RCS reduction. IEEE Access **8**, 109289–109296 (2020)
44. Y. Shi, H.X. Meng, H.J. Wang, Polarization conversion metasurface design based on characteristic mode rotation and its application into wideband and miniature antenna with low radar cross section. Opt. Express **29**(5), 6794–6809 (2021)
45. N. Yu, P. Genevet, M.A. Kats, F. Aieta, J.-P. Tetienne, F. Capasso, Z. Gaburro, Light propagation with phase discontinuities: generalized laws of reflection and refraction. Science **334**, 333–337 (2011)
46. C.L. Holloway, E.F. Kuester, J.A. Gordon, J. O'Hara, J. Booth, D.R. Smith, An overview of the theory and applications of metasurfaces: the two-dimensional equivalents of metamaterials. IEEE Antennas Propag. Mag. **54**(2), 10–35 (2012)
47. S. Sun, K.-Y. Yang, C.-M. Wang, T.-K. Juan, W.T. Chen, C.Y. Liao, Q. He, S. Xiao, W.-T. Kung, G.-Y. Guo, L. Zhou, D.P. Tsai, High-efficiency broadband anomalous reflection by gradient meta-surfaces. Nano Lett. **12**, 6223–6229 (2012)
48. A. Pors, M.G. Nielsen, R.L. Eriksen, S.I. Bozhevolnyi, Broadband focusing flat mirrors based on plasmonic gradient metasurfaces. Nano Lett. **13**, 829–834 (2013)

49. A. Pors, S.I. Bozhevolnyi, Plasmonic metasurfaces for efficient phase control in reflection. Opt. Express **21**, 27438–27451 (2013)
50. M. Farmahini-Farahani, H. Mosallaei, Birefringent reflectarray metasurface for beam engineering in infrared. Opt. Lett. **38**, 462–464 (2013)
51. M. Kim, A.M.H. Wong, G.V. Eleftheriades, Optical Huygens' metasurfaces with independent control of the magnitude and phase of the local reflection coefficients. Phys. Rev. X **4**, 041042 (2014)
52. N.M. Estakhri, A. Alu, Wave-front transformation with gradient metasurface. Phys. Rev. X **6**, 041008 (2016)
53. V.S. Asadchy, M. Albooyeh, S.N. Tcvetkova, A. Diaz-Rubio, Y. Radi, S.A. Tretyakov, Perfect control of reflection and refraction using spatially dispersive metasurface. Phys. Rev. B **94**, 075142 (2016)
54. A. Diaz-Rubio, V.S. Asadchy, A. Elsakka, S.A. Tretyakov, From the generalized reflection law to the realization of perfect anomalous reflectors. Sci. Adv. **3**, e1602714 (2017)
55. N.I. Landy, S. Sajuyigbe, J.J. Mock, D.R. Smith, W.J. Padilla, Perfect metamaterial absorber. Phys. Rev. Lett. **100**, 207402 (2008)
56. L. Li, Y. Yang, C.H. Liang, A wide-angle polarization-insensitive ultra-thin metamaterial absorber with three resonant modes. J. Appl. Phys. **110**, 063702 (2011)
57. Y. Shi, J. Yang, H. Shen, Z. K. Meng, T. Hao, Design of broadband metamaterial-based ferromagnetic absorber. Mater. Sci., Adv. Compos. Mater. **2**, 1–7 (2018)
58. Y. Zhang, Y. Shi, C.H. Liang, A broadband tunable graphene-based metamaterial absorber. Opt. Mater. Express **6**, 3036–3044 (2016)
59. Y. Shi, Y.C. Li, T. Hao, L. Li, C.H. Liang, A design of ultra-broadband metamaterial absorber. Waves Random Complex **27**, 381–391 (2017)
60. L. Li, R. Xi, H. Liu, Z. Lv, Broadband polarization-independent and low-profiles optically transparent metamateial absorber. Appl. Phys. Express **11**, 052001 (2018)
61. L. Zhang, Y. Shi, J.X. Yang, X. Zhang, L. Li, Broadband transparent absorber based on indium tin oxide-polyethylene terephthalate film. IEEE Access **7**, 137848–137855 (2019)
62. S. Yu, L. Li, G. Shi, C. Zhu, X. Zhou, Y. Shi, Design, fabrication, and measurement of reflective metasurface for orbital angular momentum vortex wave in radio frequency domain. Appl. Phys. Lett. **108**, 121904 (2016)
63. S. Yu, L. Li, G. Shi, C. Zhu, Y. Shi, Generating multiple orbital angular momentum vortex beams using a metasurface in radio frequency domain. Appl. Phys. Lett. **108**, 241901 (2016)
64. S. Yu, L. Li, G. Shi, Dual-polarization and dual-mode orbital angular momentum radio vortex beam generated by using reflective metasurface. Appl. Phys. Express **9**, 082202 (2016)
65. S. Yu, L. Li, N. Kou, Generation, reception and separation of mixed-state orbital angular momentum vortex beams using metasurfaces. Opt. Mater. Express **7**, 3312–3321 (2017)
66. N. Kou, S. Yu, L. Li, Generation of high-order Bessel vortex beam carrying orbital angular momentum using multilayer amplitude-phase-modulated surfaces in radiofrequency domain. Appl. Phys. Express **10**, 016701 (2017)
67. Y. Shi, Y. Zhang, Generation of wideband tunable orbital angular momentum vortex waves using graphene metamaterial reflectarray. IEEE Access **6**, 5341–5347 (2018)
68. Z.K. Meng, Y. Shi, W.Y. Wei, Y. Zhang, L. Li, Graphene-based metamaterial transmitarray antenna design for the generation of tunable orbital angular momentum vortex electromagnetic waves. Opt. Mater. Express **9**, 3709–3716 (2019)
69. B. Wang, K.H. Teo, T. Nishino, W. Yerazunis, J. Barnwell, J. Zhang, Experiments on wireless power transfer with metamaterials. Appl. Phys. Lett. **98**, 254101 (2011)
70. L. Li, Y.C. Fan, S.X. Yu, C. Zhu, C.H. Liang, Design, fabrication and measurement of highly sub-wavelength double negative metamaterials at HF frequencies. J. Appl. Phys. **113**, 271372 (2013)
71. A. Rajagopalan, A.K. Ramrakhyani, D. Schurig, G. Lazzi, Improving power transfer efficiency of a short-range telemetry system using compact metamaterials. IEEE Trans. Microw. Theory Techn. **62**(4), 947–955 (2014)

72. G. Lipworth, J. Ensworth, K. Seetharam, D. Huang, J.S. Lee, P. Schmalenberg, T. Nomura, M.S. Reynolds, D.R. Smith, Y. Urzhumov, Magnetic metamaterial superlens for increased range wireless power transfer. Sci. Rep. **4**, 3642 (2015)
73. Y. Cho, J.J. Kim, D.-H. Kim, S. Lee, H. Kim, C. Song, S. Kong, H. Kim, C. Seo, S. Ahn, J. Kim, Thin PCB-type metamaterials for improved efficiency and reduced EMF leakage in wireless power transfer systems. IEEE Trans. Microw. Theory Techn. **64**(2), 353–364 (2016)
74. L. Li, H. Liu, H. Zhang, W. Xue, Efficiency wireless power transfer system integrating with metasurface for biological applications. IEEE Trans. Ind. Electron. **65**, 3230–3239 (2018)
75. S. Yu, H. Liu, L. Li, Design of near-field focused metasurface for high efficient wireless power transfer with multi-focus characteristics. IEEE Trans. Industrial Electron. **66**, 3993–4002 (2019)
76. W. Lee, Y.-K. Yoon, Wireless power transfer systems using metamaterials: a review. IEEE Access **8**, 147930–147947 (2020)
77. T.J. Cui, M.Q. Qi, X. Wan, J. Zhao, Q. Cheng, Coding metamaterials, digital metamaterials and programmable metamaterials. Light Sci. Appl. **3**, e218 (2014)
78. S. Liu, T.J. Cui, Concept, working principles, and applications of coding and programmable metamaterials. Adv. Opt. Mater. **5**, 1700624 (2017)
79. V. Lucarini, J.J. Saarinen, K.-E. Peiponen, E.M. Vartiainen, *Kramers-Kronig Relations in Optical Materials Research* (Springer, Berlin Heidelberg, New York, 2005)

Chapter 3
Analysis and Design Methods of Metamaterials and Metasurfaces

Yan Shi, Xi Chen, Na Kou, Haixia Liu, and Long Li

Abstract Metamaterial, a material engineered to have a property that is not found in nature, is composed of periodic or non-periodic structure with subwavelength structural unit cell. With diversities of the unit cell topologies, the metamaterials exhibit various unique properties. In order to effectively analyse and design the metamaterials, four classes of analysis methods have been developed in this chapter. Firstly, the physics-based local resonant cavity cell (LRCC) model has been developed to study the electromagnetic band gap structure. Secondly, three effective medium theory based retrieval methods have been developed to extract the effective material parameters of the metamaterials. Thirdly, the equivalent circuit based model has been developed to characterize the metasurfaces. Finally, the full-wave simulation technique, called multilevel Green's function interpolation method (MLGFIM), has been developed to analyse the periodic metamaterial array.

Keywords Metamaterial · Local resonant cavity cell (LRCC) · Effective medium · Equivalent circuit · Multilevel Green's function interpolation method (MLGFIM)

3.1 Introduction

Since theoretical speculation by Viktor Veselago on the electromagnetics of substrates with simultaneously negative permittivity and permeability in 1968 [1], electromagnetic metamaterials have attracted tremendous attention due to their

Y. Shi · X. Chen · H. Liu · L. Li (✉)
School of Electronic Engineering, Xidian University, Xi'an 710071, Shaanxi, China
e-mail: lilong@mail.xidian.edu.cn

Y. Shi
e-mail: shiyan@mail.xidian.edu.cn

X. Chen
e-mail: chenxi@mail.xidian.edu.cn

N. Kou
College of Big Data and Information Engineering, Guizhou University, Guiyang 550025, China

L. Li et al. (eds.), *Electromagnetic Metamaterials and Metasurfaces: From Theory To Applications*, https://doi.org/10.1007/978-981-99-7914-1_3

unusual electromagnetic properties. The concept of metamaterials [2–6] has a much broader scope than that of well-known double negative medium (DNM) or left-handed material (LHM). Electromagnetic metamaterials are broadly defined as artificially effective homogeneous electromagnetic structures with unusual properties not readily available in nature. Generally, a metamaterial is a periodic or non-periodic composite structured material with the subwavelength structural unit cell.

With the diverse unit cell topologies, various kinds of metamaterials have been presented. Due to the surface-wave suppression features and the in-phase reflection phase properties, electromagnetic bandgap (EBG) structures or high-impedance electromagnetic surfaces (HIS) have attracted increasing attention [7–10]. Within the surface-wave frequency bandgap, HIS can be used to block the propagation of the surface wave, enhance antenna gain, decrease backward radiation, and reduce mutual coupling between antennas. The HIS is regarded as the perfect magnetic surface owing to its in-phase reflection phase properties, which can be used as ground plane for a low profile antenna to get good return loss. It can be used to enhance antenna input impedance bandwidth. The novel ultra-thin radar absorbing material (RAM) can be realized by using the high-impedance ground plane [11, 12].

In the long wavelength limit, the metamaterial can be characterized by an effective medium. Various levels of equivalence can be introduced to address levels of sophistication of the metamaterial model [13], for instance, external equivalence, dispersion equivalence, single-mode equivalence, double-mode equivalence, etc. Among these equivalences, one primary concern is to consider the external scattering response by the metamaterials, and thus the effective constitutive parameters involving the wave impedance, index of refraction, permittivity, and permeability can be retrieved based on either simulated or measured scattering parameters. Many techniques [14–32] have been developed to extract the effective electromagnetic properties of various metamaterial structures, including isotropic characteristics, bianisotropic characteristics, chiral characteristics, homogeneous structure, inhomogeneous structure, electrically small size, and moderate size, etc. By utilizing the effective medium parameters, a bridge between the metamaterial structure and the device design can be set up, thus giving rise to novel device functionalities, such as perfect lens [33], invisible cloaking [34–42], absorber [43–47], antennas [48–53], etc.

The surface version of the metamaterial called metasurface has caused wide attention in recent years. With the ultrathin dimension, the metasurfaces have less loss as compared with the metamaterials and therefore have enabled plenty of emerging functions and applications, including ultrathin planar lenses, vortex beam generators, metasurface holograms, etc. The most common way to analyze electromagnetic responses of the metasurface is the full-wave numerical methods, for instance, the method of moments (MoM), finite element method (FEM), and finite-difference time-domain (FDTD) with periodic boundary conditions. In spite of their powerful ability in the model of arbitrary metasurfaces, the conventional full-wave methods consume many computational resources, including computational memory and CPU time. In order to solve it, a fast full-wave algorithm, i.e., multilevel Green's function interpolation method (MLGFIM) has been presented to solve the periodic structure. As an alternative to the full-wave numerical algorithms, the equivalent circuit network based method has been given to achieve a very efficient analysis and design procedure of the frequency-selective metasurface.

In this chapter, four kinds of the analysis approaches including local resonant cavity cell (LRCC) model, effective medium parameter retrieval methods, equivalent circuit method and full-wave simulation method, have been presented to model the metamaterials and metasurface.

3.2 Local Resonant Cavity Cell Model for EBG Metamaterials

The metamaterials are macroscopic man-made composites designed to produce the properties not available in nature. With the unique properties, metamaterials have a wide range of potential applications at frequencies from the microwave to the optics. In optics, a class of well-known metamaterials is photonic crystal [54]. The photonic crystals are periodically engineered structures which exhibit a photonic band gap, i.e., the existence of a stop band where waves cannot propagate through it. In the microwave and millimeter wave region, the metamaterials with similar characteristics are called electromagnetic band-gap (EBG) structures [8–12] or high-impedance electromagnetic surfaces (HIS) [7]. This section focuses on the local resonance behavior of EBG structures, and a simply locally resonant cavity cell (LRCC) model [55] for mushroom-like EBG structures is presented to gain insight to the physical mechanism of the EBG structures and the interaction of electromagnetic waves with EBG structures. The LRCC model proposes two kinds of main resonance modes, i.e., mono-polarized (MP) mode and cross-coupling polarized (CCP) mode. Based on the quasi-static assumption in principle, the mono-polairzed mode is equivalent to the LC parallel resonance, and thus predicts the position of the surface wave suppression bandgap. The cross-coupling polarized mode exhibits some novel resonant phenomena, which has not been found previously. Some parametric studies of the mushroom EBG structure have been comprehensively implemented by using the LRCC model, including the radius effect of the metal-plated vias. Good agreement between the simulated and measured results of EBG structures, such as square, triangle, and hexagon lattices, are observed to verify the effectiveness and correctness of the LRCC model.

3.2.1 Local Resonant Cavity Cell Model

In this section, the mushroom-like EBG structure is analysed, as shown in Fig. 3.1. The EBG structure is composed of an array of square metal patches in a two-dimension square lattice. The top square patch and the metallic ground are fabricated on top and bottom surfaces of the substrate, a metallic vias is used to connect the patch and the ground. The unit cell is a electrical small structure in terms of the operating wavelength. In [7], D. Sievenpiper has demonstrated the parallel LC resonant circuit analysis of the EBG structure including capacitors and inductors. The equivalent LC circuit as a two-dimensional electric filter blocks the current flowing along the sheet.

When an electromagnetic wave illuminates the EBG structure, currents are induced on the top patches. A voltage applied parallel to the top surface results in charge accumulation at the ends of the plates, which can be regarded as a capacitance. As the charges slosh back and forth, they flow around a long path through the vias and bottom plate. The resulting currents are associated with magnetic field, thus resulting in an inductance. The origin of the capacitance and inductance are depicted in Fig. 3.2a. The response of the structure can characterized by a parallel resonant LC circuit, as shown in Fig. 3.2b. It is worthwhile pointing out that the capacitance and inductance of the model are actually the sheet capacitance and sheet inductance which depend on the value of each element as well as their arrangement, instead of the capacitance and inductance of the individual elements. The values of the inductor L and the capacitor C are solved by the following formula [7]:

$$C = \frac{\varepsilon_0(1+\varepsilon_r)W}{\pi}\cosh^{-1}\left(\frac{a}{g}\right) \tag{3.1}$$

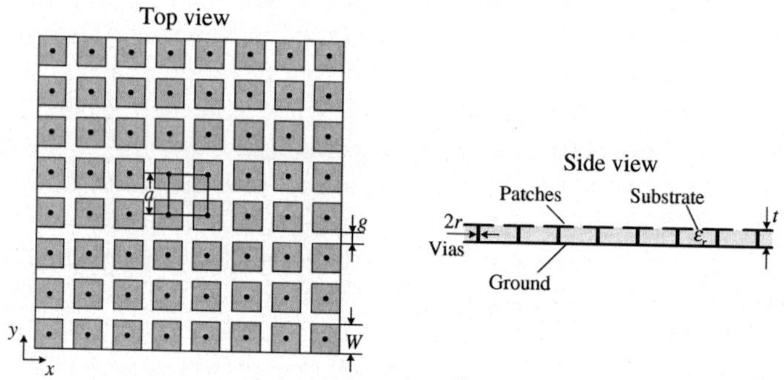

Fig. 3.1 Geometry of a mushroom-like EBG structure [55]

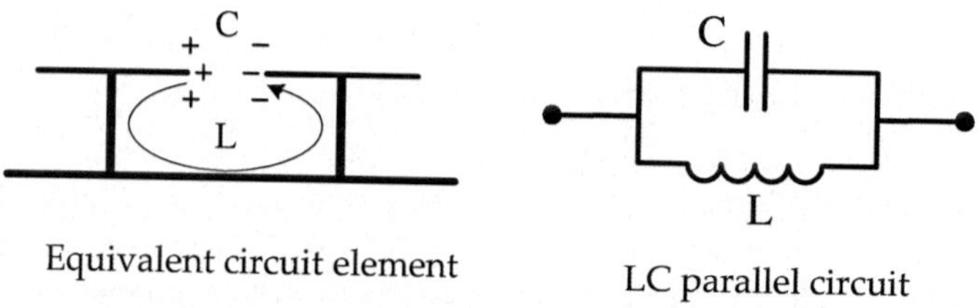

Fig. 3.2 Equivalent LC parallel resonant circuit for EBG structure [55]

$$L = \mu_0 t \tag{3.2}$$

where a is a period of the EBG structure, i.e., $a = W + g$. μ_0 and ε_0 are the permeability and permittivity in free space, respectively. The resonant frequency and the effective surface impedance can be calculated in terms of the capacitance and inductance as

$$\omega_0 = \frac{1}{\sqrt{LC}} \tag{3.3}$$

$$Z = \frac{j\omega L}{1 - \omega^2 LC} \tag{3.4}$$

It can be observed according to (3.4) that the surface impedance is inductive at low frequencies and TM surface waves are supported, while at high frequencies the capacitive surface impedance is achieved and TE surface waves are supported. Near the resonance frequency, the surface impedance becomes very high. Within this frequency region, each row of metal patches has opposite charge, and thus TM waves become the standing waves. On the other hand, TE waves which are not bound to the surface radiate readily into the surrounding space as leaky waves. In this scenario, there are no surface waves flowing along the EBG structure in the frequency region, and the surface-wave suppression bandgap is established. The resonant frequency ω_0 determines the center position of the whole surface wave suppression bandgap. Although (3.1)–(3.4) give simply and qualitatively descriptions of the EBG local resonant mechanism, the obtained results are not very accurate under the quasi-static assumption. More importantly, the radius effect of vias cannot be involved in (3.1) and (3.2). It will be shown by the following experiments and numerical simulations that the vias radius has an important effect on the bandgap responses.

According to EBG local resonance response, a LRCC model is established. For the mushroom-like EBG structure shown in Fig. 3.1, its LRCC model is depicted in Fig. 3.3.

For simplification, the ideal LRCC model without dielectric and conductor losses is considered. As shown in Fig. 3.3, the top and bottom walls of the equivalent cavity are regarded as the perfectly electric conductors (PEC) except a cross slot on the top wall used to provide electromagnetic coupling with outer space. The side walls excluding four corners are equivalent to be the perfectly magnetic conductors (PMC) where the tangential component of magnetic field equals zero. The each corner of the equivalent cavity consists of one-quarter metal post or via. The cavity is filled by the medium with permittivity ε_r. It is worthwhile noticing that the LRCC is a slot-cavity-posts resonance model with all characteristic parameters of the EBG structures, including the patch width W, gap width g, substrate thickness t and permittivity ε_r, and the radius of metal plated vias r. Here the PMC boundary at the four side walls is introduced with the following assumptions: (1) The induced currents on the top patches flow around a long path through the vias and on bottom plate, and the current lines are parallel to the four side walls of the cavity. Thus the normal

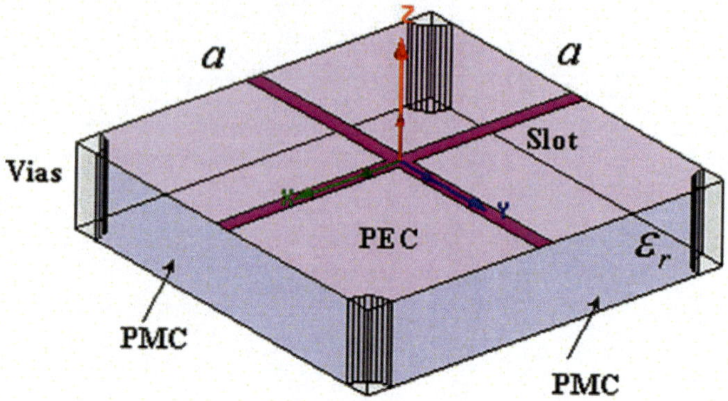

Fig. 3.3 A LRCC model for the square EBG structure [55]

currents and tangential magnetic fields cannot be induced on the side walls. (2) The thickness of EBG structures is much smaller compared to the operating wavelength ($t << \lambda_0$), and thus the electromagnetic fields can be assumed to be constant along z-direction, which is similar to microstrip circuits and patch antennas. (3) With the image symmetries of EBG structures, the tangential electric fields in the cross slot should be continuous.

A rigorous eigenmode analysis of the LRCC model is implemented to explore the relation of the inherent modes of the LRCC with the electromagnetic repsonses of EBG structures, i.e., the surface-wave suppression bandgap and the reflection phase. Different from the quasi-static LC resonant circuit, the LRCC model may have infinite free resonant modes with different resonance frequencies theoretically, that is, the multi-resonance property of the microwave cavity. With the analysis of eigenmodes of the LRCC model, two kinds of main resonance modes of EBG structures are revealed, i.e., MP mode and CCP mode. The former is equivalent to the LC parallel resonant circuit proposed by Sievenpiper in principles. The resonant frequency of the MP mode is more accurate than the LC resonance frequency ($\omega_0 = 1\Big/\sqrt{LC}$). The latter exhibits some special and interesting resonant phenomena. Due to the complexity of the mode field and the particularity of the mode, the CCP mode has not caused wide attention previously. Here, two resonant modes can be denoted according to the E-field vector distributions on the top wall (including the metal patches and the cross slot) of the LRCC model, as illustrated in Fig. 3.4.

It is observed that mode 2 and mode 3 are degenerate, with the same resonant frequency. As an electromagnetic wave illuminates the EBG structures, different resonance modes of EBG structures are excited, giving rise to various electromagnetic properties. Figure 3.5 demonstrates the E-field distributions of the CCP mode and the MP mode on the surface of the EBG structures, respectively. Note that the E-field is solved by applying the PMC image symmetry of the LRCC model. For the CCP mode, there are the slot E-field couplings along x and y directions of the

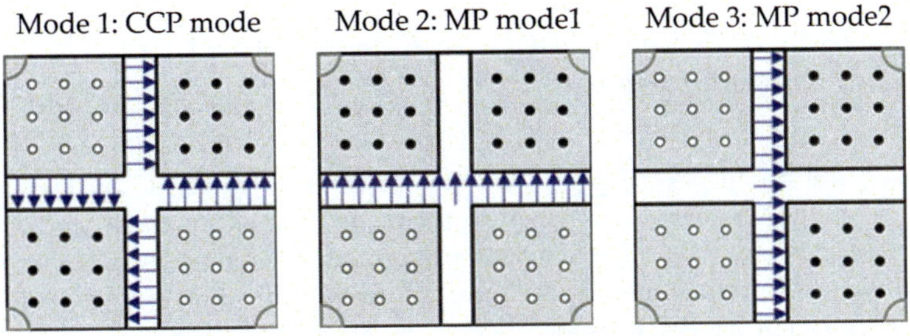

Fig. 3.4 Mode patterns for two kinds of resonance modes of the LRCC model. ($\vec{E}$: →, $\vec{E}$ into paper o o o, $\vec{E}$ out of paper • • •) [55]

EBG surface simultaneously. That's why the mode is called the cross-coupling polarization mode. On the other hand, for the MP mode, the E-field distributes along a certain direction, with a similar characteristic to the *LC* parallel resonant circuit. At the mode resonant frequency, opposite charges or electric fields are generated at each row of metal patches, and the surface waves along the EBG surface are trapped in the cavity to achieve the standing waves, leading to the surface waves suppression band gap. However, the active power will be coupled out from the slot, which results in the leaky waves radiation.

The MP and CCP modes are the dominant modes of the LRCC model, owing to the lowest resonance frequencies. On the other hand, the eigenmode fields of these two modes mainly distribute at the cross-slot cut on the top surface of EBG structures. It is worthwhile pointing out that the LRCC model has high-order resonant modes because of its multi-resonance property. However, the fields of high-order modes distribute between the top and bottom walls of the LRCC model, and their resonant frequencies are out of the surface wave bandgap, which are difficult to be excited in practical applications.

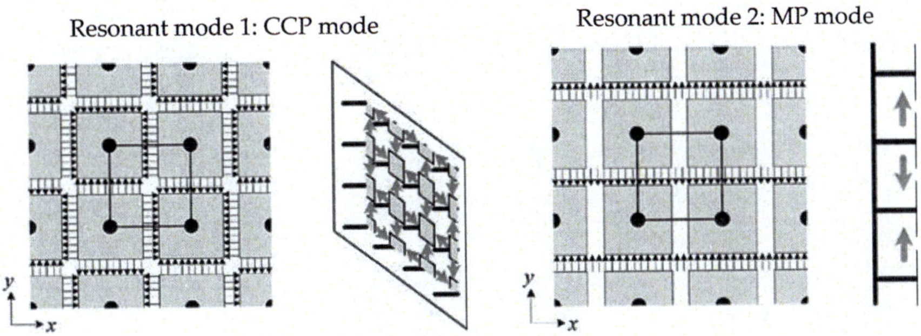

Fig. 3.5 E-field distribution characteristics of two kinds of resonant modes of a mushroom-like EBG structure [55]

3.2.2 Numerical Simulations and Experiments of Two Kinds of Resonant Modes

In the following, numerical simulations and experiments are provided to demonstrate the local resonant behavior of EBG structures and the relationship between two kinds of resonant modes and the surface wave bandgap and reflection phase feature of EBG structures by using the proposed LRCC model. As shown in Fig. 3.1, the dimensions of the mushroom-like EBG structure are $W = 7.0$ mm, $g = 0.35$ mm, $h = 2.5$ mm, $\varepsilon_r = 2.65$, $r = d/2 = 0.5$ mm. A square LRCC model for the EBG structure is constructed, as shown in Fig. 3.3. The lower four eigenmodes are calculated by using the Ansys-HFSS Eigenmode Solver, with the convergence precision of 0.1%. Figure 3.6 depicts the discretized mesh graph of the LRCC model based on the finite element method (FEM), and the normalized electric field amplitudes at the cavity surface, and the vector electric fields on the cross-slot of the four eigenmodes. The first eigenmode is a CCP mode with the lowest eigenfrequency of 4.412 GHz. The second and third eigenmodes which have the same eigenfrequency of 5.205 GHz are MP mode and its degenerate mode. The fourth eigenmode is a high-order mode with the eigenfrequency of 8.220 GHz.

In order to clearly understand the characteristics of these eigenmodes, Fig. 3.7 gives the normalized three-dimension E-field distributions of the four eigenmodes in the cavity. The E-fields of the modes 1–3 distribute on the cross-slot region, while the E-field of the mode 4 distributes between the top wall and the bottom wall of the

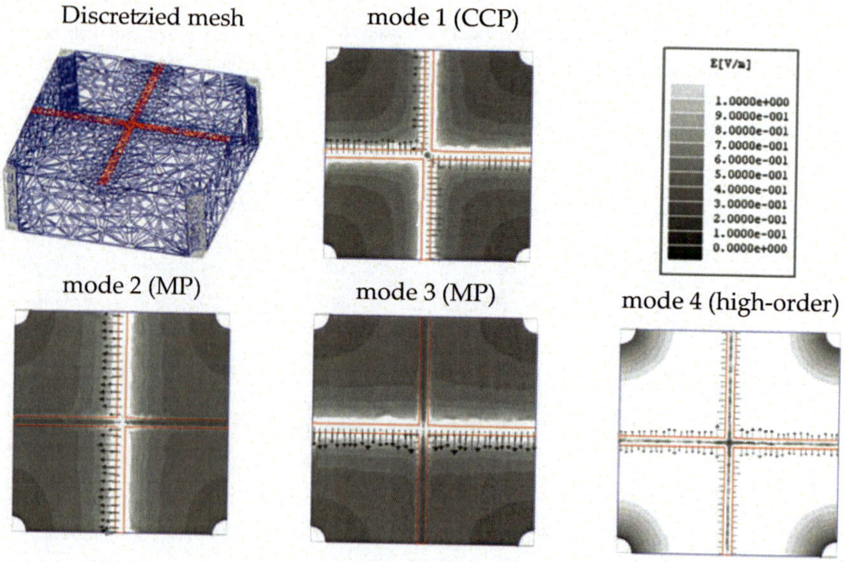

Fig. 3.6 Eigenmodes of LRCC model for the EBG structure with square patch and square array [55]

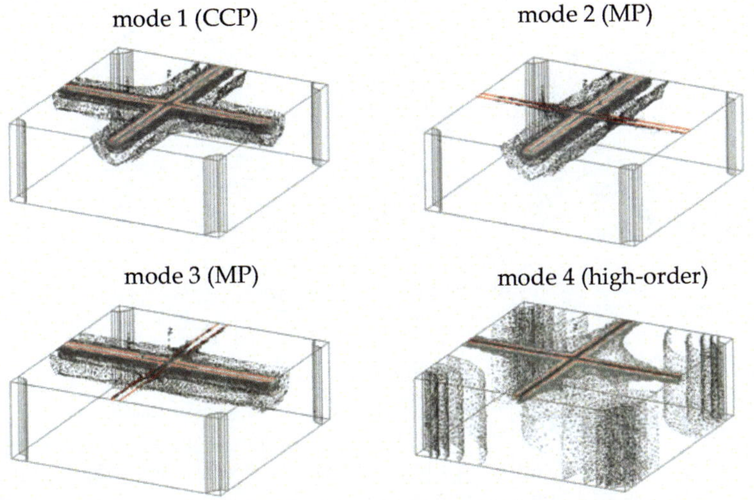

Fig. 3.7 The distributions of E-field for various modes in LRCC mode [55]

cavity. It is worthwhile pointing out that the higher order modes including the mode 4 are difficult to be excited in practice application, and thus their contribution to the electromagnetic bandgap is negligible. Here, we study the resonant behaviors of two dominant modes (CCP and MP modes) related to the surface waves suppression bandgap and the in-phase reflection characteristics. Two kinds of different experiments are conducted i.e., a waveguide probe based measurement of reflection characteristics S_{11} and the other is a coaxial probe based measurement of surface wave coupling S_{21}.

3.2.2.1 Waveguide Probe Experiments

In the waveguide probe experiment, the appropriate waveguide (C-band WR187 rectangular waveguide with $wa = 47.55$ mm, $wb = 22.149$ mm, and $wt = 1.5$ mm for this test) and calibration kit are required. Two EBG structures with 20×20 and 3×6 lattices were fabricated by using the Teflon-based fiberglass laminates with dielectric constant 2.65, and two kinds of measurement schemes are used to observe the electromagnetic properties of the EBG structure, as shown in Fig. 3.8. In the measurement scheme 1, the whole EBG structure with 20×20 lattices is tightly placed at the waveguide probe termination as a load. By contrast, the measurement scheme 2 is to embed an EBG structure (3×6 lattices) with the same size as the waveguide aperture into the waveguide. With the two schemes, the CCP mode and the MP mode can be effectively excited.

The 1-port reflection S_{11} of the test samples is measured by an Agilent 8719ES vector network analyzer. Figure 3.9 shows the measured reflection coefficient magnitude ($|S_{11}|$) and reflection phase in two schemes. The mode resonant frequencies of

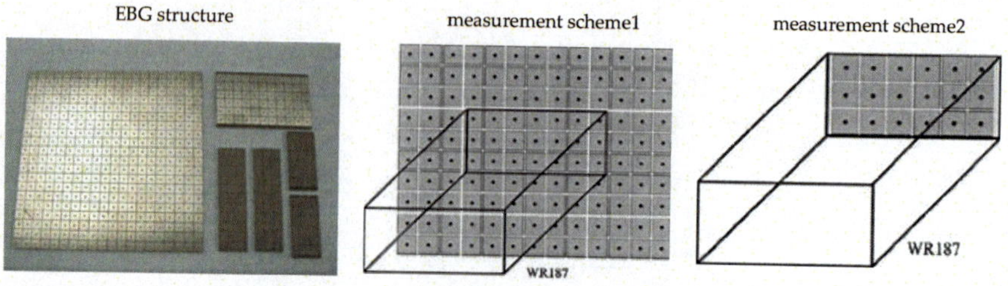

Fig. 3.8 The waveguide probe measurements of the EBG structure [55]

LRCC are indicated by triangles on the curves. It can be observed that the EBG structure exhibits different resonant phenomena in two schemes. In the experiment scheme 1, the CCP mode resonating at 4.412 GHz is effectively excited, with a narrower bandwidth. A good return loss can be achieved at the resonant frequency of the CCP mode, meaning that the incident wave is strongly coupled into the inner of the EBG structure and propagates along all directions. According to the reflection phase curve, in-phase reflection can be obtained nearby the resonant frequency of the CCP mode in spite of the narrow band. In the experiment scheme 2, the MP mode with an effective bandwidth much wider than that of the CCP mode is excited. It can be seen from Fig. 3.9 that there is the little return loss caused by the dielectric loss, conductor loss, and the imperfect enclosure of the waveguide aperture in the measurement. The reflection phase changes continuously from $+180°$ to $-180°$ with the increase of the frequency, and the in-phase reflection region (+90° to $-90°$) spans a considerable frequency band. Note that the angle and polarization of incident plane waves generally affect the reflection phase of the EBG structure. The reflection field in the waveguide probe experiments is obtained under TE wave oblique incidence. Therefore, the measured reflection phase at the resonant frequencies is not exactly zero. Due to the dominant mode limit of the WR187 rectangular waveguide, the experiment is conducted in the frequency band of 4.0–6.0 GHz.

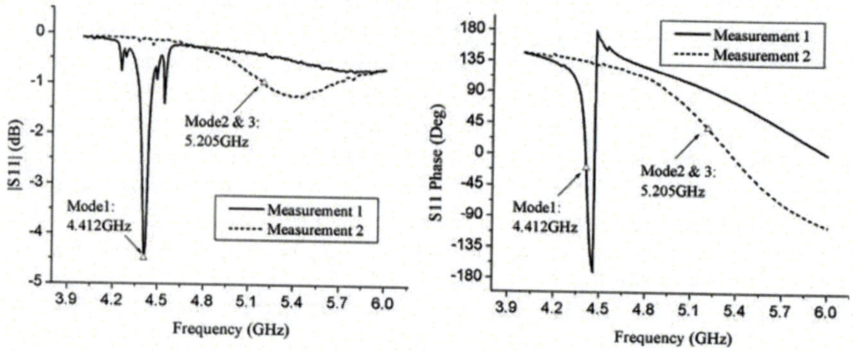

Fig. 3.9 Measured reflection coefficient magnitude and reflection phase results of the EBG structure in two measurement schemes [55]

In order to elaborate physical mechanism of two kinds of resonant modes, the E-field characteristics on the interface between the waveguide port and the EBG structure surface are analyzed, as shown in Figs. 3.10 and 3.11. In the measurement scheme 1, the electric field distribution of the CCP mode on the interface is very complicated. By contrast, in the measurement scheme 2, the E-field distribution of the MP mode nearby the interface is always preponderant in the *y*-direction.

In order to further demonstrate the resonant behavior of the CCP mode, two same waveguide probes are placed in a distance of 60.0 mm. The EBG structure with 20×20 lattices shown in Fig. 3.8 and a metal plate tightly cover two waveguide probes, respectively, as shown in Fig. 3.12. The ports 1 and 2 are the exciting and receiving ports, respectively. The measured transmission characteristic S_{21} versus frequency is given in Fig. 3.13. When the metallic plate covers two waveguide probes, the signal transmission level between two ports is about −95 dB, achieving the isolation between two ports. The tiny signal level is due to noise field. On the other hand, as two waveguide probes are covered by the EBG structure, there is an obvious signal transmission at the resonant frequency of 4.412 GHz. The transmission signal level is about −70 dB, and thus it is not noise effect but the CCP mode resonant phenomenon. At the frequencies different from the resonant frequency of CCP mode,

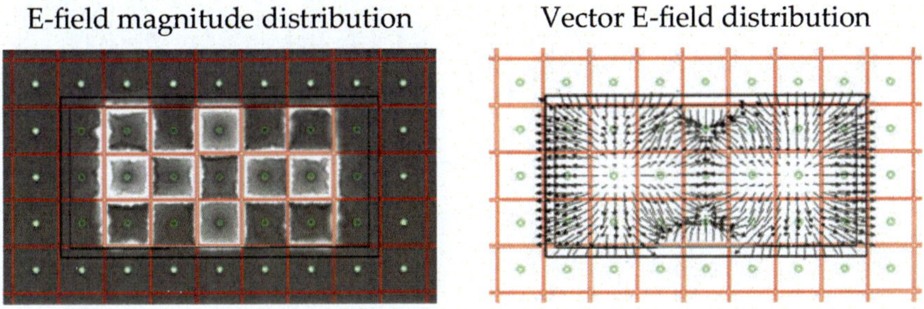

Fig. 3.10 Simulated E-field distribution on the interface between the waveguide aperture and the EBG structure surface at 4.412 GHz in the measurement scheme 1 [55]

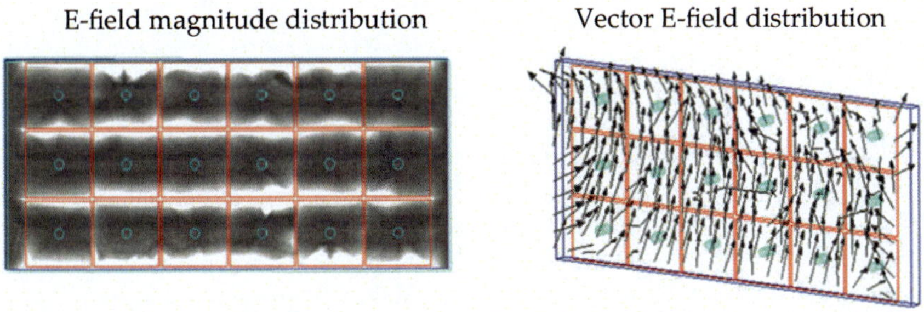

Fig. 3.11 Simulated E-field distribution on the interface between the waveguide aperture and the EBG structure surface at 5.205 GHz in the measurement scheme 2 [55]

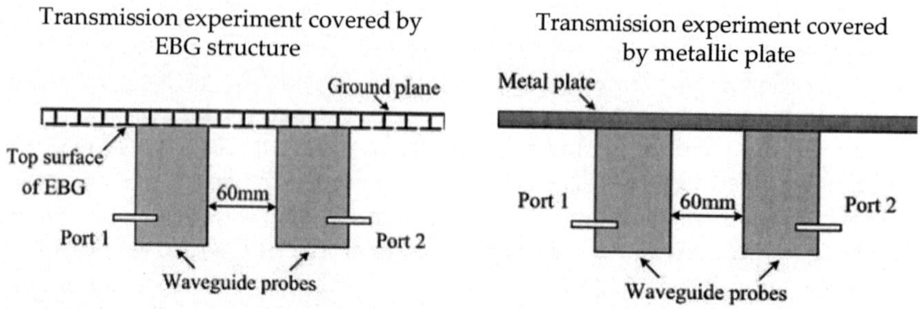

Fig. 3.12 The experimental models for waveguide probes transmission covered by the EBG structure and the metal plate [55]

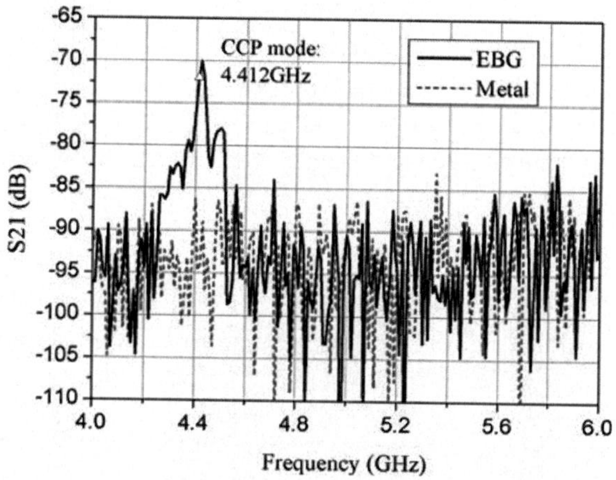

Fig. 3.13 Comparison of transmission characteristics of two waveguide probes covered by the EBG structure and the metal plate [55]

the transmission response of the EBG structure is similar to that of the metal plate, including the MP mode resonant frequency. This is because the mode does not be excited in this case. It is worthwhile noting that the reflection characteristic of port 1 is the same as the results of measurement 1 shown in Fig. 3.9.

3.2.2.2 Coaxial Probes Experiments

Similar to the method used by Sievenpiper [7], the surface wave suppression bandgap is obtained by measuring the transmission coefficient S_{21} with a pair of coaxial probes normal or parallel to the surface, which is used to couple the TM and TE surface waves, respectively. Figure 3.14 gives the numerically simulated k-β dispersion diagram of the surface modes propagating in the EBG structure. The first (dominant) surface mode is TM_0, without cutoff frequency, and the second surface wave mode is TE_1. A complete stopband between the first mode TM_0 and the second mode TE_1 in the frequency band 4.64–6.18 GHz is observed. In this figure, Γ, X and M

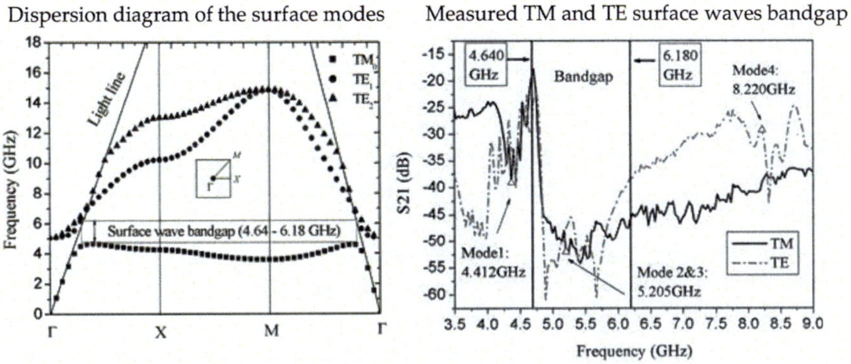

Fig. 3.14 Measurement and simulation of surface wave bandgap of the EBG structure with the given geometry parameters [55]

represent symmetric points in the irreducible Brillouin zone. The Γ-X branch represents $\beta_x a/\pi$ with $\beta_y = 0$. The X-M branch represents $\beta_y a/\pi$ with $\beta_x = a/\pi$ and M-Γ branch represents $\beta_x a/\pi$ with $\beta_x = \beta_y$.

Figure 3.14 also demonstrates the transmission coefficient S_{21} measured by an Agilent network analyzer 8719ES versus the frequency. Good agreement between the measured and simulated surface-wave bandgaps is observed. In Fig. 3.14, the resonant frequencies of the CCP and MP modes are indicated by triangles on the curves. It is interesting to note that the MP resonant modes of the LRCC model, i.e., the degenerate modes, are very close to the center of the bandgap, in accordance with the previous theoretical analysis. By analyzing the eigenmodes of the LRCC model, the location of the surface-wave suppression bandgap of the EBG structures can be effectively predicted. The transmission drop can be observed nearby the CCP mode frequency. A number of experiments including different EBG structure sizes and different measurement positions demonstrate a little transmission drop in the neighbor of the surface wave bandgap, which is caused by the CCP mode resonance, instead of the multipath interference.

3.2.2.3 Low Profile Inverse-L Antenna

Considering the special reflection phase feature, the EBG structure can be employed as a ground plane of low-profile antennas [7, 8]. The low profile design usually refers to the overall height of antenna structure less than one tenth of the wavelength at the operating frequency. When the PEC plate is used as the ground plane, the impedance matching of the inverse-L antenna is very poor, since the PEC surface with the reflection phase of 180° induces the direction of the image current opposite to that of the original antenna. This problem is often addressed by placing the radiating element above the PEC (or metal) ground plane in a quarter-wavelength distance. But the resulting structure requires a thickness of about λ/4.

When the EBG surface is used as the ground plane, the low-profile antennas with the good impedance matching characteristic can be obtained in the surface wave frequency bandgap where the reflection phase convering $-180°$~$180°$ changes with the frequency. However, the resonant behavior caused by mode 1 (CCP mode) has not been caused much attention. It will be demonstrated in the following that when the mode 1 is excited, the good S_{11} can also be obtained. Figure 3.15 depicts an inverse-L antenna located above a mushroom-like EBG structure with a distance of 1.5 mm. The geometric parameters of the EBG structure are $W = 6.0$ mm, $g = 1.0$ mm, $t =$ 2.0 mm, $r = 0.75$ mm, $\varepsilon_r = 2.2$. The energy is radiated along all directions transverse to the antenna, and the phase response for each direction is centered about the phase of the specular reflection (reflection at normal incidence). The resonant frequency of the CCP mode for the EBG structure is 8.292 GHz, and the resonant frequency of the degenerate MP mode is 9.190 GHz, and the high-order mode resonant frequency is 11.58 GHz. The measured surface wave bandgap of the EBG surface is given in Fig. 3.16. The variation of S_{11} with the inverse-L antenna length is shown in Fig. 3.17. For comparison, the S_{11} of the reference antenna without EBG structure is also given. When the antenna length varies from $0.33\lambda_{8GHz}$ to $1.0\lambda_{8GHz}$, the good S_{11} can be obtained in both the CCP and MP mode regions.

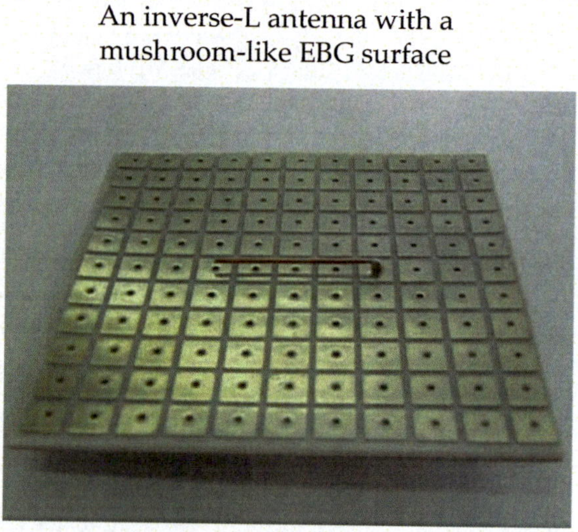

Fig. 3.15 A low profile inverse-L antenna over a mushroom-like EBG surface [55]

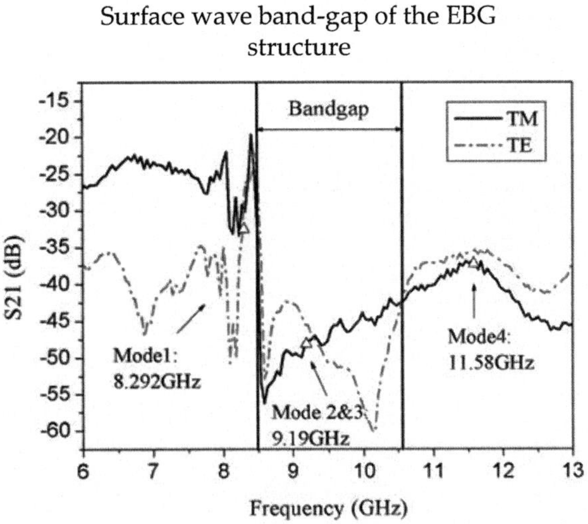

Fig. 3.16 Measured surface wave band-gap of the EBG structure [55]

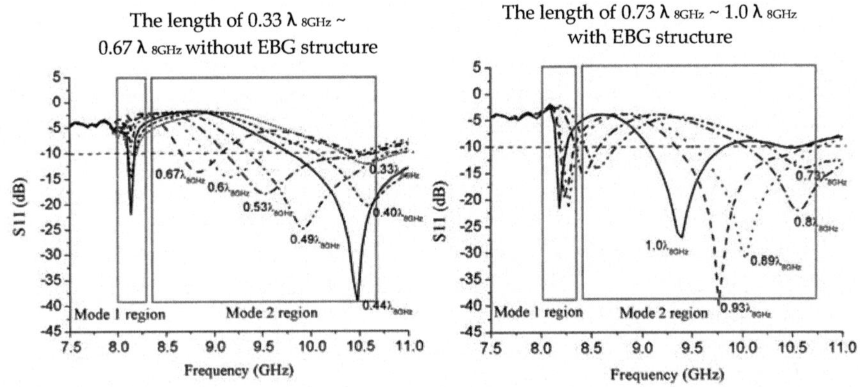

Fig. 3.17 Measured S_{11} of the inverse-L antenna over a mushroom-like EBG surface with its length variations [55]

3.2.3 LRCC Analyses of Various EBG Structures

3.2.3.1 Parameters Analysis of the LRCC Model

The surface wave bandgap and reflection phase are determined by five parameters of the mushroom-like EBG structure shown in Fig. 3.1, i.e., patch width (W), gap width (g), substrate thickness (t), substrate permittivity (ε_r), and radius of vias (r). we will discuss the effects of these parameters one by one based on the LRCC model, thus providing some engineering guidelines for EBG structure designs. It is worthwhile pointing out that according to the band structure of surface-wave modes

Fig. 3.18 Effect of parameters of the mushroom-like EBG structure on the position of the surface wave bandgap: patch width ($0.08–0.20\lambda_{6GHz}$), gap width ($0.002–0.026\lambda_{6GHz}$), substrate thickness ($0.01–0.07\lambda_{6GHz}$) and relative permittivity (1.0–10.2) [55]

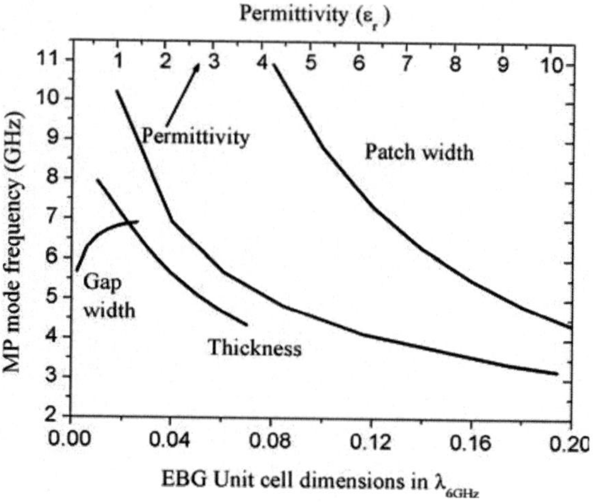

and experiments, the vias radius has an important effect on the bandgap of surface wave suppression, which is neglected in available models [8, 56, 57]. In the LRCC model, the effect of the vias radius on the eigenmodes can be involved accurately, and therefore the analysis results based on the LRCC model are in good agreement with the simulated and measured results. Here the initial design parameters of the EBG structure are $W = 0.14\lambda_{6GHz}$, $g = 0.006\lambda_{6GHz}$, $t = 0.03\lambda_{6GHz}$, $r = 0.01\lambda_{6GHz}$, and $\varepsilon_r = 2.65$, where λ_{6GHz} is the wavelength in free space at 6 GHz. Figure 3.18 shows the effect of parameters on the resonant frequency of the degenerate MP mode which is associated with the position of the surface wave bandgap, i.e., patch width of $0.08–0.20\lambda_{6GHz}$, gap width of $0.002–0.026\lambda_{6GHz}$, substrate thickness of $0.01–0.07\lambda_{6GHz}$, and relative permittivity of 1.0–10.2.

As shown in Fig. 3.18, as the patch width is increased from $0.08\lambda_{6GHz}$ to $0.20\lambda_{6GHz}$ with other parameters kept unchanged, the frequency band position decreases. The variation in the gap width has the opposite effect to the variation in the patch width. Specifically, with the increase of the gap width, the frequency band position increases accordingly. Relative permittivity (ε_r) is another effective parameter for controlling the frequency band. The widely used commercial materials such as RT/Duroid substrates, TMM substrates as well as air are investigated. It can be observed from Fig. 3.18 that the frequency band decreases as the permittivity is increased. With the air substrate, the EBG structure has the widest frequency band. Generally, the substrate thickness is small owing to a low-profile structure. With the increase of the substrate thickness, the frequency band decreases, which is similar to the effect of patch width and permittivity. Note that the obtained results based on the LRCC model are consistent with those of Yang [8] computed by FDTD models for the reflection phase of plane waves.

The radius of vias also plays an important role in determining the frequency band. Table 3.1 gives the mode resonant frequency when the vias radius changes from $0.002\lambda_{6GHz}$ to $0.03\lambda_{6GHz}$, with other parameters kept unchanged. Especially, when the vias radius is zero, i.e., without vias, the resonant frequencies of the LRCC model do not locate at the position of the surface wave suppression bandgap, meaning that the structure without vias is not the electromagnetic bandgap structure any more.

In addition, it can be seen from Table 3.1 that with the increase of the vias radius, the frequency band position increases, which is opposite to the effect of the substrate thickness. For verification, we fabricated three EBG surfaces with radii of $r = 0.005\lambda_{6GHz}$, $0.010\lambda_{6GHz}$, and $0.015\lambda_{6GHz}$, and measured their surface-wave bandgaps for TE case, as illustrated in Fig. 3.19. Good agreement between the simulated and measured results is obtained, and the resonant behavior of mode 1 can also be observed for different radii of vias.

Table 3.1 Vias radius effect of LRCC model [55]

Radius (λ_{6GHz})	Resonant frequency of CCP mode	Resonant frequency of degenerate MP mode	Resonant frequency of higher order mode
0	18.24	18.51	20.96
0.002	3.97	4.557	5.816
0.005	4.654	5.344	6.903
0.01	5.452	6.272	8.302
0.015	6.158	7.082	9.594
0.02	6.81	7.828	10.92
0.025	7.64	8.727	12.36
0.03	8.462	9.606	13.96

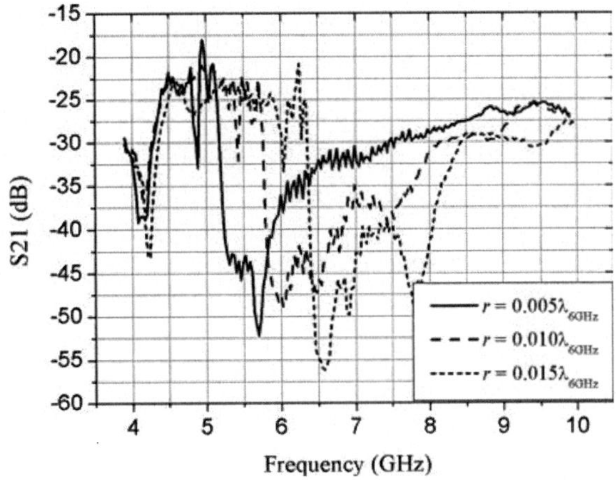

Fig. 3.19 Measured surface wave bandgaps of the mushroom-like EBG structures with different radii of vias [55]

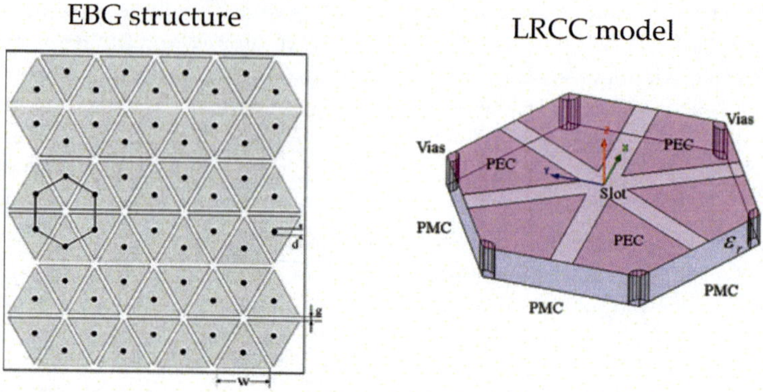

Fig. 3.20 EBG structure with triangle patch and hexagon array and its LRCC mode [55]

3.2.3.2 EBG Structures with Triangle Patch and Hexagon Array

As shown in Fig. 3.20, the geometry parameters of the mushroom-like EBG structure are as follows: $W = 11.0$ mm, $g = 0.9$ mm, $t = 1.5$ mm, $\varepsilon_r = 2.65$, and $r = d/2 = 0.5$ mm. Considering its periodicity layout, the hexagonal LRCC model illustrated in Fig. 3.20 can be established. The lower five eigenmodes of the LRCC model is solved by the HFSS Eigenmode Solver. Here the obtained eigenstates are different from those of the square cavity. Owing to the geometric symmetry, both modes 2 & 3 and modes 4 & 5 are degenerate. The electric field amplitude distributions on the cavity surface and the vector electric field of each mode in the cross-slot are shown in Fig. 3.21. The transmission coefficient S_{21} measured by a pair of monopole probes connected to an Agilent 8719ES vector network analyser is shown in Fig. 3.22. The measured surface-wave bandgap is from 5.94 to 7.40 GHz. The mode resonant frequencies of the LRCC model are also indicated by triangles on the curves in Fig. 3.22. It can be seen that the second degenerate mode of the LRCC model is very close to the center of the bandgap. Furthermore, it is shown that based on the LRCC model, the position of the surface-wave suppression bandgap of the EBG structures can be predicted quickly and accurately by the resonant frequencies of the degenerate eigenmodes.

3.2.3.3 EBG Structures with Hexagon Patch and Triangle Array

Figure 3.23 shows the EBG structure with hexagon patch and triangle array, with the following geometric parameters: $W = 3.7$ mm, $g = 1.0$ mm, $t = 1.5$ mm, $\varepsilon_r = 2.65$, and $r = d/2 = 0.5$ mm. A triangle LRCC model shown in Fig. 3.23 is constructed. With the eigenmode analysis of the LRCC model, the first three resonant modes can be solved, as shown in Fig. 3.24. It can be found that the lowest mode of the LRCC model is degenerate, indicating the center of the frequency band, as illustrated in

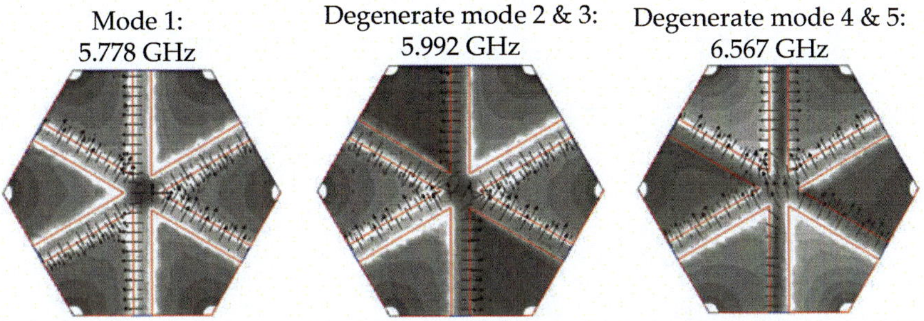

Fig. 3.21 Eigenmodes of LRCC model for the EBG structure with triangle patch and hexagon array [55]

Fig. 3.22 Measured surface wave band-gap of the EBG structure with triangle patch and hexagon array [55]

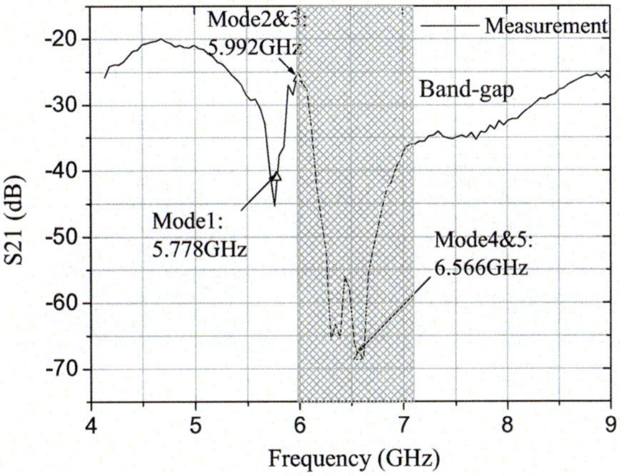

Fig. 3.25. This observation further validates the local resonant behaviors of the EBG structures and the close relationship between the eigenmodes of the LRCC model and the surface wave bandgap and reflection phase characteristics.

An analysis of a LRCC model for the mushroom-like EBG structures provides a physical insight into EBG structures bandgap properties and the local resonance behaviors. It has been shown that the surface wave bandgap and the reflection resonant behaviors of the EBG structures can be characterized by the eigenmodes of the LRCC models. The degenerate MP mode determines the center position of the complete band-gap of surface wave suppression. Its resonant behavior is equivalent to that of the *LC* parallel resonant circuit based on quasi-static assumption. The CCP mode exhibits significant resonant phenomena. The LRCC model revealing the intrinsic properties of EBG structures is independent of the complex excitation and polarization of incident waves. In practical applications, such as low-profile EBG antennas or EBG microwave devices, it is important to determine which mode can be excited when electromagnetic waves interact with EBG structures. It should be pointed out that the LRCC model can accurately determine the bandgap position,

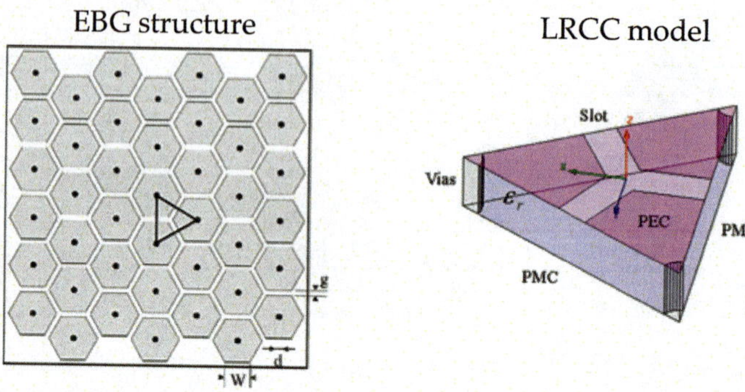

Fig. 3.23 EBG structure with hexagon patch and triangle array and its LRCC model [55]

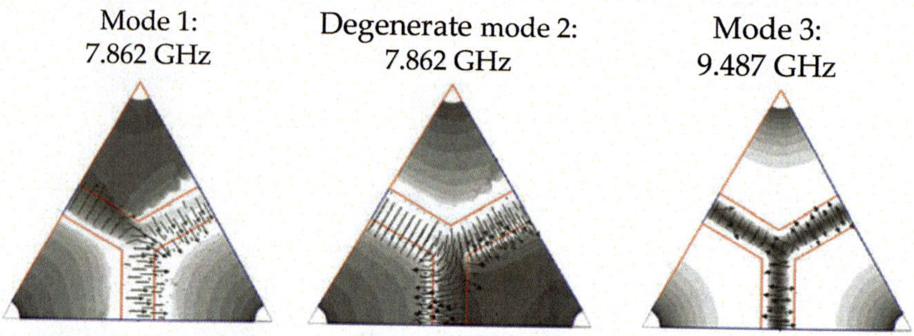

Fig. 3.24 Eigenmodes of LRCC model for the EBG structure with triangle patch and hexagon array [55]

Fig. 3.25 Measured surface wave band-gap of the EBG structure with hexagon patch and triangle array [55]

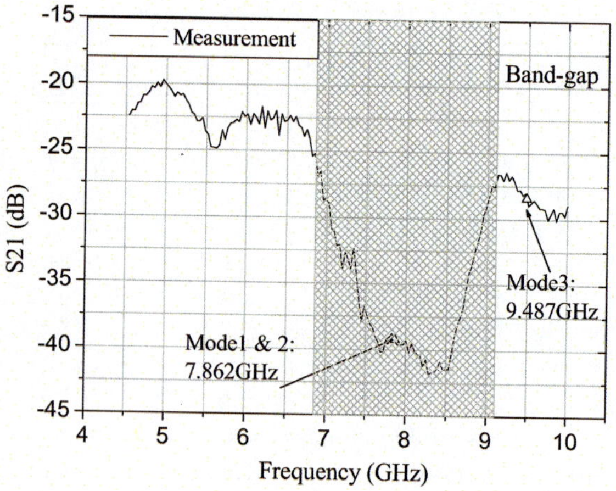

but it is incapable of estimating the bandgap width owing to its cell model. If the complex frequency theory might be introduced into the EBG structures, the complex natural resonant modes of LRCC models will include its resonant frequencies and leaky wave radiation Q factor associated with the bandgap width.

3.3 Equivalent Medium Theory for Metamaterials

The macroscopic electromagnetic parameters can be theoretically derived from atomic or molecular structure. The metamaterials use subwavelength meta-atoms to mimic the natural material. Therefore, we resort to effective medium theory to characterize the metamaterials. Specifically, the responses of the metamaterials to the electromagnetic wave are equivalent to that of a hypothetical continuous material. Different levels of equivalence corresponding to different responses can be considered to address increasing levels of sophistication of the metamaterial model. The widely used equivalence is external equivalence, which extracts the effective constitutive parameters, including permittivity, permeability, refractive index, and impedance to provide the correct scattering parameters. In this section, several extraction techniques of the effective medium parameters for the metamaterials are given.

3.3.1 Nicolson–Ross–Weir (NRW)-Based Retrieval Approach

Metamaterial is composed of periodic or non-periodic structures with sub-wavelength unit cells. According to the effective medium theory, the responses of the metamaterial to electromagnetic waves can be equivalent to the electromagnetic responses of a hypothetical continuous material, and thus the metamaterial properties are characterized by the effective medium parameters, including relative permittivity ε_r, relative permeability μ_r, normalized wave impedance η and refractive index n, which is similar to the conventional material.

To extract the effective material parameters of the metamaterial structure, consider a plane wave incident on a symmetry metamaterial structure with a thickness of d, as shown in Fig. 3.26. Here time dependence $e^{j\omega t}$ is used. In this scenario, the complex S parameters of the metamaterial structure at N distinct frequency points can be obtained by full-wave simulation or measurement. Due to the symmetry structure, we have $S_{11} = S_{22}$. In addition, assuming reciprocal materials in the metamaterial structure, the condition of $S_{21} = S_{12}$ is satisfied.

According to electromagnetic field theory, the relationship between the scattering parameters and the material parameters for a homogenous material slab can be written as [14, 15]

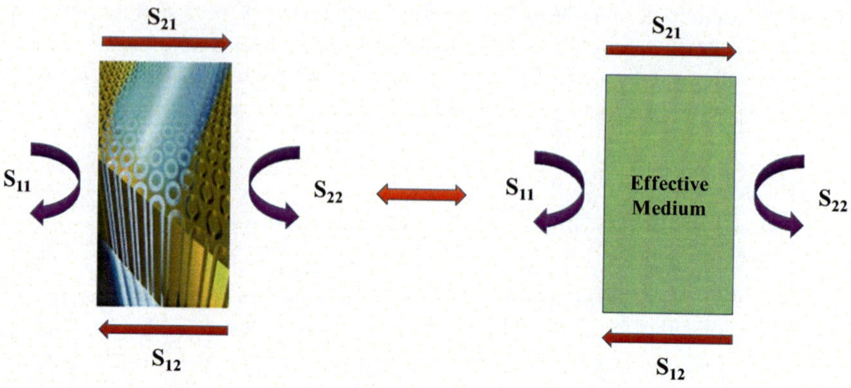

Fig. 3.26 Metamaterial and its effective medium

$$S_{11} = \frac{(\eta^2 - 1)(1 - Z^2)}{(\eta + 1)^2 - (\eta - 1)^2 Z^2}, \tag{3.5}$$

$$S_{21} = \frac{4\eta Z}{(\eta + 1)^2 - (\eta - 1)^2 Z^2}, \tag{3.6}$$

where

$$Z = e^{-jk_0 nd}. \tag{3.7}$$

Here k_0 is wave number in free space. Introducing three intermediate variables, i.e.,

$$V_1 = S_{21} + S_{11}, \tag{3.8}$$

$$V_2 = S_{21} - S_{11}, \tag{3.9}$$

$$\Gamma = \frac{\eta - 1}{\eta + 1}, \tag{3.10}$$

Equations (3.5) and (3.6) can be written as

$$Z = \frac{1 + V_1 V_2}{V_1 + V_2} \pm \sqrt{\left(\frac{1 + V_1 V_2}{V_1 + V_2}\right)^2 - 1}, \tag{3.11}$$

$$\Gamma = \frac{1 - V_1 V_2}{V_1 - V_2} \pm \sqrt{\left(\frac{1 - V_1 V_2}{V_1 - V_2}\right)^2 - 1}. \tag{3.12}$$

The signs in Eqs. (3.11) and (3.12) can be determined to maintain $|Z| \leq 1$ and $|\Gamma| \leq 1$. Once Γ is solved, the wave impedance η can be determined according to

(3.10). To solve the refractive index n according to Z, (3.7) is rewritten as

$$Z = \left[e^{-j\frac{k_0 nd + 2m\pi}{N}}\right]^N \quad (m = 0, \pm 1, \pm 2 \ldots), \tag{3.13}$$

in which N is an integer number. Note that in (3.13), the integer m denotes the branch index caused by the complex exponential function. For the electrically small thickness of the metamaterial, one has $m = 0$ in (3.13). If N is chosen to be large enough, (3.10) can be well approximated according to Taylor's series as

$$Z = \left(1 - j\frac{k_0 nd}{N}\right)^N. \tag{3.14}$$

Solving Nth root in (3.14) and taking real and imaginary parts into consideration, we can obtain

$$n' = -\frac{N \cdot \mathrm{Im}(Z^{\frac{1}{N}})}{k_0 d}, \tag{3.15}$$

$$n'' = N \cdot \frac{\mathrm{Re}(Z^{\frac{1}{N}}) - 1}{k_0 d}, \tag{3.16}$$

in which n' and n'' are real and imaginary parts of the refractive index, respectively. The Nth root of Z is solved as

$$Z^{\frac{1}{N}} = |Z|^{\frac{1}{N}} e^{j\frac{\phi + 2(l-1)\pi}{N}} \quad (l = 1, \ldots, N), \tag{3.17}$$

where ϕ is the principal value of the argument of Z. It is worthwhile noticing that the Nth root of Z has N solutions. According to (3.14), the integer l should be chosen to maintain the root of Z closest to the real axis in the first or fourth quadrants of the complex plane.

With the obtained n and η, the equivalent permittivity and permeability can be calculated as

$$\varepsilon = n/\eta, \tag{3.18}$$

$$\mu = n\eta. \tag{3.19}$$

Figure 3.27 shows a fishnet metamaterial unit cell. Here two thicknesses of 0.1 mm and 0.2 mm for the unit cell are considered, respectively. Owing to electrically small thickness in the concerned frequency band from 10 to 18 GHz, the branch index m in (3.13) is equal to zero. The NRW method is used to extract the effective parameters of the fishnet structure. The retrieval refractive index is shown in Fig. 3.27 and compared with Smith's retrieval method in [19]. It can be seen that when N is chosen as a large

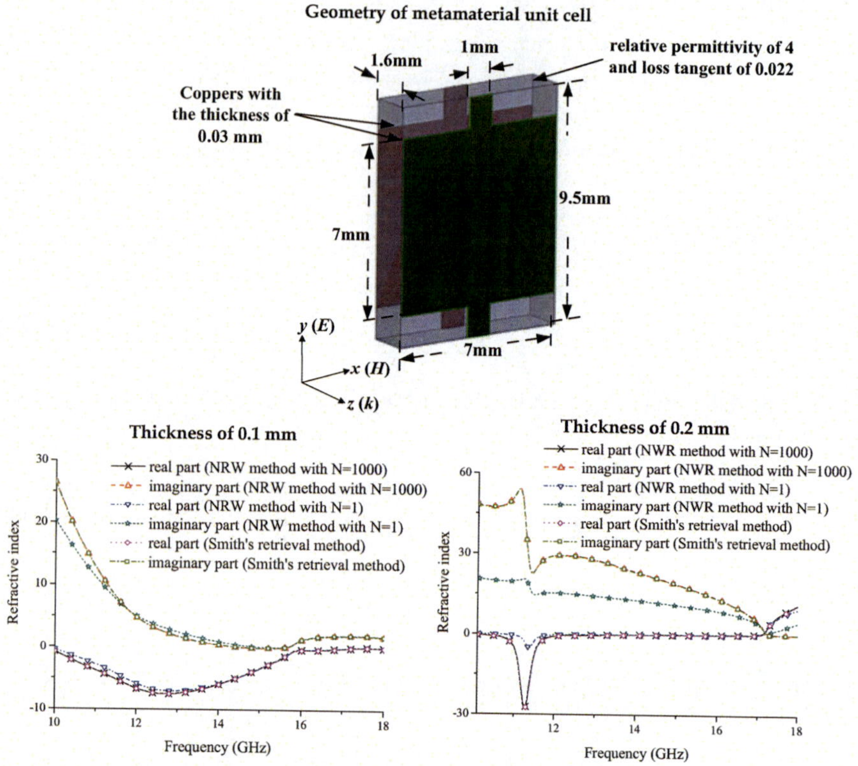

Fig. 3.27 The retrieval refractive index for fishnet metamaterial slab [26]

number, i.e., $N = 1000$, the resultant retrieval result agrees well with that obtained by the method in [19]. However, the accuracy of the retrieval result for $N = 1$ is worse than that for $N = 1000$. This is because the accuracy of Taylor's series expansion becomes better with the increase of N.

3.3.2 *Phase Unwrapping-Based Extraction Approach*

According to the previous discussion, we know that the branch index m equals zero when the thickness of the homogeneous metamaterial structure is electrically small. However, with the increase of the thickness, the branch index becomes nonzero. Therefore, we need to determine the branch index m for the metamaterial structure with moderate thickness. Resolving Eqs. (3.5) and (3.6), one has

$$Z = e^{-jnk_0 d} = \frac{S_{21}}{1 - S_{11}\Gamma}. \tag{3.20}$$

Considering the nonzero branch index m, Z in (3.20) can be rewritten as

$$e^{-jnk_0d} = Ae^{j(\phi+2m\pi)}, \tag{3.21}$$

in which

$$A = \left|\frac{S_{21}}{1 - S_{11}\Gamma}\right|, \tag{3.22}$$

$$\phi = \arg\left(\frac{S_{21}}{1 - S_{11}\Gamma}\right), \tag{3.23}$$

where arg(·) denotes the principal value of the argument of the complex function. In this scenario, we have $\phi \in (-\pi, \pi]$. Solving (3.21) and taking real and imaginary parts into account, one has

$$n'' = \frac{\ln A}{k_0d}, \tag{3.24}$$

$$n' = -\frac{\phi + 2m\pi}{k_0d}. \tag{3.25}$$

According to (3.24), the imaginary part of the refractive index is determined uniquely, while the branch index related to its real part needs to be solved. On the other hand, the analytical continuation of the refractive index n in the upper half of a complex angular frequency plane is satisfied [58]. Hence, the phase $\phi + 2m\pi$ in (3.25) is continuous. Here a phase unwrapping technique can be used to determine the branch index m. The unwrapping procedure is to connect the discontinuous $-\pi$-to-π segments into a continuous curve. When the frequency approaches zero, the thickness of the metamaterial structure becomes electrically small, and thus one has $m = 0$. ϕ discontinuously varies from π to $-\pi$ as the frequency increases. In this scenario, the branch index m is set as 1 to maintain the continuous phase of the refractive index. On the other hand, if ϕ discontinuously changes from $-\pi$ to π, $m = -1$ is chosen. This unwrapping procedure is iteratively used in the entire operating frequency band.

Figure 3.28 demonstrates a metamaterial unit cell consisting of two split ring resonators (SRRs) and four capacitively loaded strips (CLSs), which are printed on both sides of the nonmagnetic substrate with relative permittivity of 2.2 and a thickness of 93 mils. The refractive index is extracted by using the phase unwrapping technique and compared with the result obtained by the Kramers–Kronig (K-K) relationship-based retrieval method [22]. It can be seen that with the increase of the upper limit of the K-K integral, the refractive indexes solved by the phase unwrapping method and the K-K method are in increasing agreement with each other. To solve the refractive index at 12 GHz, the K-K-based method uses the S parameters at the frequencies up to 16 GHz, while the phase unwrapping method only employs the S parameter at the frequency of 12 GHz. The S parameter comparison between

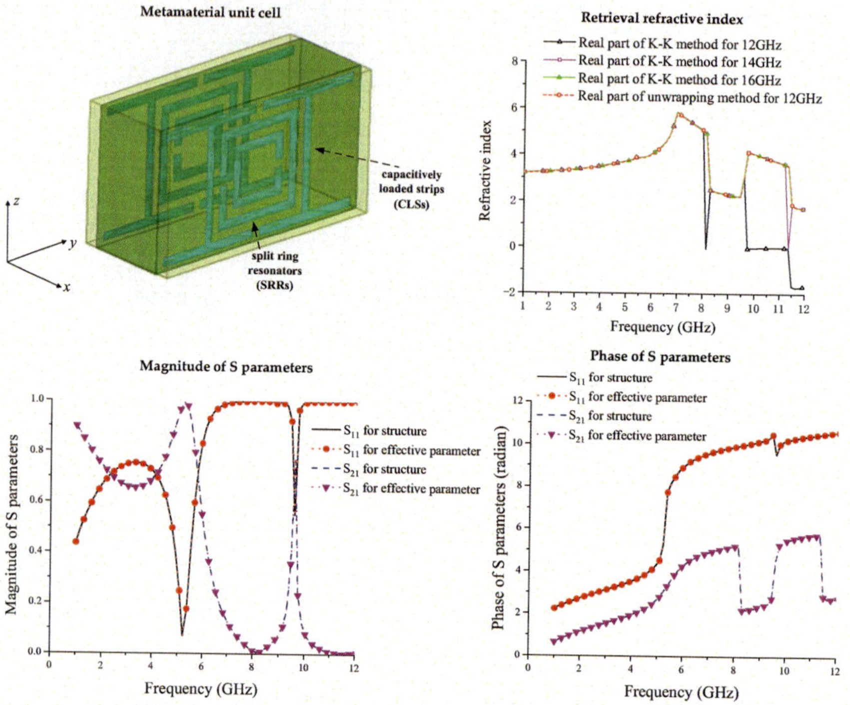

Fig. 3.28 Effective parameter extraction based on phase unwrapping technique [27]

the structure and the retrieval effective parameters are given in Fig. 3.28, in good agreement.

3.3.3 *Extraction Approach for Inhomogeneous Metamaterials*

The previous retrieval approaches are valid under the condition of $S_{11} = S_{22}$ due to the symmetry property of the homogeneous metamaterial structure. However, when the metamaterial structure is inhomogeneous, S_{11} is unequal to S_{22}. To achieve the extraction of the effective parameters, the inhomogeneous metamaterial structure is divided into many thin layers of piecewise homogeneous layers, as shown in Fig. 3.29. Assume that the thickness of each layer is denoted by d_i ($i = 1, 2, \ldots, N$), and the effective refractive index and normalized wave impedance are n_i and η_i, respectively. Four S-parameters of the inhomogeneous metamaterial structure can be obtained by full-wave simulation or measurement, i.e., S_{11}, S_{12}, S_{21}, and S_{22}. Considering the reciprocity property, we have $S_{12} = S_{21}$. To solve the scattering parameters of the layered media, an electromagnetic field-based recursive method has been used [32].

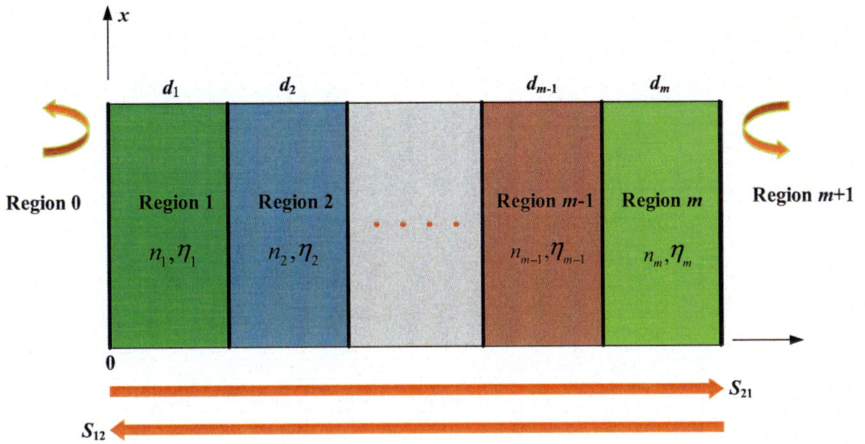

Fig. 3.29 A m-layer inhomogeneous structure

Consider an incident wave with an electric field parallel to $+x$ axis normally incident on the m-layer structure along $+z$ axis. Assume that complex magnitudes of the incident and reflected waves at the left side of the interface between regions l and $l+1$ are E^i_{l-1} and E^r_{l-1}, respectively, as shown in Fig. 3.30. According to the boundary conditions at the interface between regions l and $l+1$, we can obtain

$$\begin{cases} 2E^i_l \cdot e^{-in_l k_0 d_l} = \left[1 + \dfrac{\eta_l}{\eta_{l-1}}\right] \cdot E^i_{l-1} + \left[1 - \dfrac{\eta_l}{\eta_{l-1}}\right] \cdot E^r_{l-1} \\ 2E^r_l \cdot e^{in_l k_0 d_l} = \left[1 - \dfrac{\eta_l}{\eta_{l-1}}\right] \cdot E^i_{l-1} + \left[1 + \dfrac{\eta_l}{\eta_{l-1}}\right] \cdot E^r_{l-1} \end{cases}. \tag{3.26}$$

Considering that there is no reflected wave in region $m+1$, we have at the interface between regions m and $m+1$

$$\begin{cases} 2E^i_m = \left[1 + \dfrac{\eta_m}{\eta_0}\right] \cdot E^i_{m+1} \\ 2E^r_m = \left[1 - \dfrac{\eta_m}{\eta_0}\right] \cdot E^i_{m+1} \end{cases}. \tag{3.27}$$

In this scenario, we can denote the reflection coefficient $S^{(l-1)}_{11}$ at the interface between regions $l-1$ and l and the transmission coefficient $S^{(l-1)}_{21}$ from region l-1 to region $m+1$ respectively as follows:

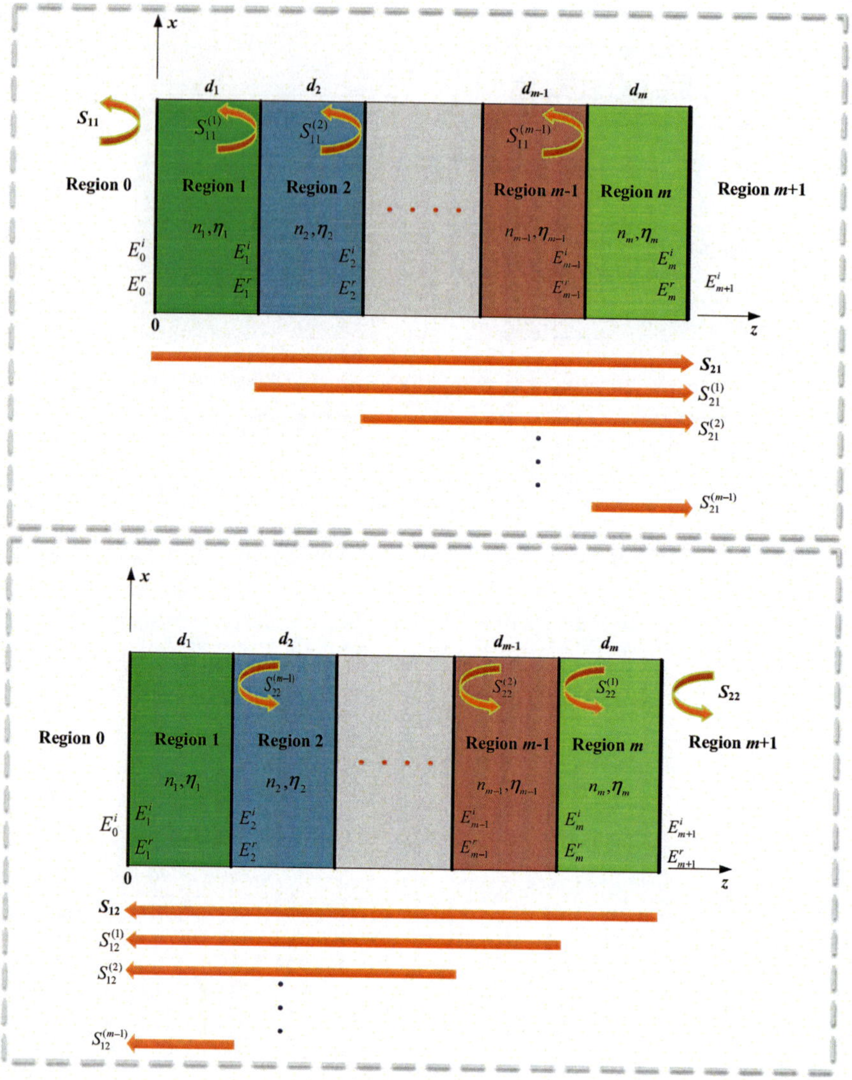

Fig. 3.30 *S*-parameter recursion representation [32]

$$\begin{cases} S_{11}^{(l-1)} = \dfrac{E_{l-1}^r}{E_{l-1}^i} \\ S_{21}^{(l-1)} = \dfrac{E_{m+1}^i}{E_{l-1}^i} \end{cases}. \tag{3.28}$$

Combining Eqs. (3.26)–(3.28), we can obtain the following recursive relationships

$$\begin{cases} S_{11}^{(1)} = \dfrac{\eta_1 - \eta_0 - S_{11}(\eta_1 + \eta_0)}{S_{11}(\eta_1 - \eta_0) - \eta_1 - \eta_0} e^{-i2n_1k_0d_1} \\ S_{21}^{(1)} = \dfrac{2\eta_0 S_{21}}{\eta_0 + \eta_1 + S_{11}(\eta_0 - \eta_1)} e^{-in_1k_0d_1} \end{cases}, \tag{3.29}$$

$$\begin{cases} S_{11}^{(l)} = \dfrac{\eta_l - \eta_{l-1} - S_{11}^{(l-1)}(\eta_l + \eta_{l-1})}{S_{11}^{(l-1)}(\eta_l - \eta_{l-1}) - \eta_l - \eta_{l-1}} e^{-i2n_lk_0d_l} \\ S_{21}^{(l)} = \dfrac{2\eta_{l-1} S_{21}^{(l-1)}}{\eta_{l-1} + \eta_l + S_{11}^{(l-1)}(\eta_{l-1} - \eta_l)} e^{-in_lk_0d_l} \end{cases} (l = 2, \ldots, m-1). \tag{3.30}$$

Therefore, the effective electromagnetic parameters in region m can be solved according to $S_{11}^{(m-1)}$ and $S_{21}^{(m-1)}$ as

$$\begin{cases} \eta_m = \eta_0 \eta_{m-1} \sqrt{\dfrac{(1 + S_{11}^{(m-1)})^2 - (S_{21}^{(m-1)})^2}{\left[\eta_0(1 - S_{11}^{(m-1)})\right]^2 - (\eta_{m-1} S_{21}^{(m-1)})^2}} \\ e^{-in_mk_0d_m} = \dfrac{\eta_0}{\eta_{m-1}} \dfrac{S_{11}^{(m-1)}(\eta_{m-1} - \eta_m) + \eta_{m-1} + \eta_m}{S_{21}^{(m-1)}(\eta_m + \eta_0)} \end{cases}. \tag{3.31}$$

Similarly, in the case of an incident wave propagating along -z axis, the scattering parameter recursive relationships can be derived as

$$\begin{cases} S_{22}^{(1)} = \dfrac{\eta_m - \eta_0 - S_{22}(\eta_m + \eta_0)}{S_{22}(\eta_m - \eta_0) - \eta_m - \eta_0} e^{-i2n_mk_0d_m} \\ S_{12}^{(1)} = \dfrac{2\eta_0 S_{12}}{\eta_0 + \eta_m + S_{22}(\eta_0 - \eta_m)} e^{-in_mk_0d_m} \end{cases}, \tag{3.32}$$

$$\begin{cases} S_{22}^{(l)} = \dfrac{\eta_{m+1-l} - \eta_{m+2-l} - S_{22}^{(l-1)}(\eta_{m+1-l} + \eta_{m+2-l})}{S_{22}^{(l-1)}(\eta_{m+1-l} - \eta_{m+2-l}) - \eta_{m+1-l} - \eta_{m+2-l}} e^{-i2n_lk_0d_l} \\ S_{12}^{(l)} = \dfrac{2\eta_{m+2-l} S_{12}^{(l-1)}}{\eta_{m+2-l} + \eta_{m+1-l} + S_{22}^{(l-1)}(\eta_{m+2-l} - \eta_{m+1-l})} e^{-in_lk_0d_l} \end{cases} (l = 2, \ldots, m-1). \tag{3.33}$$

Therefore, the effective electromagnetic parameters in region 1 can be solved according to $S_{22}^{(m-1)}$ and $S_{12}^{(m-1)}$ as

$$\begin{cases} \eta_1 = \eta_0 \eta_2 \sqrt{\dfrac{(1 + S_{22}^{(m-1)})^2 - (S_{12}^{(m-1)})^2}{\left[\eta_0(1 - S_{22}^{(m-1)})\right]^2 - (\eta_2 S_{12}^{(m-1)})^2}} \\ e^{-in_1k_0d_1} = \dfrac{\eta_0}{\eta_2} \dfrac{S_{22}^{(m-1)}(\eta_2 - \eta_1) + \eta_2 + \eta_1}{S_{12}^{(m-1)}(\eta_1 + \eta_0)} \end{cases}. \tag{3.34}$$

On the other hand, only given four S parameters of the inhomogeneous reciprocal metamaterial structure, i.e., S_{11}, S_{12}, S_{21}, and S_{22}, the effective material parameters of *m*-layer structure cannot be uniquely determined. Therefore, two kinds of solution procedures, i.e., iterative solution and direct solution, are used.

3.3.3.1 Iterative Solution Procedure

Step 1 The effective material parameters $n_i^{(0)}$ and $\eta_i^{(0)}$ ($i = 1, \ldots, m-1$) are first determined. Specifically, *S*-parameters of the metamaterial structure in the region *i* (*i* = 1, …, $m-1$) are solely simulated or measured. Considering the homogeneity, the retrieval methods given in previous sections can be used to extract the corresponding effective material parameters.

Step 2 Inserting $n_i^{(0)}$ and $\eta_i^{(0)}$ ($i = 1, \ldots, m-1$) into Eqs. (3.29)–(3.30), and solve $n_m^{(0)}$ and $\eta_m^{(0)}$ from Eq. (3.31). In this case, if we replace the inhomogeneous metamaterial structure by the *m*-layer structure with $n_i^{(0)}$ and $\eta_i^{(0)}$ ($i = 1, \ldots, m$), S_{21} and S_{11} of the resulting effective materials can coincide with those of the inhomogeneous metamaterial structure. However, S_{22} of the resulting effective materials is not equal to that of the metamaterial structure.

Step 3 In order to satisfy S_{22} of the metamaterial structure, the obtained $n_i^{(0)}$ and $\eta_i^{(0)}$ ($i = 2, \ldots, m$) are substituted into Eqs. (3.32)–(3.34) for the solution of the effective material parameters of the first layer, i.e., $n_1^{(1)}$ and $\eta_1^{(1)}$. Hence the calculated $n_1^{(1)}$, $\eta_1^{(1)}$, $n_i^{(0)}$ and $\eta_i^{(0)}$ ($i = 2, \ldots, m$) can lead to the same S_{22} and S_{12} as the metamaterial structure. But the resulting S_{11} does not coincide with that of the metamaterial structure.

Step 4 Repeat *Steps* 2 and 3, until the solved effective electromagnetic parameters can give rise to the same four *S*-parameters as the metamaterial structure.

3.3.3.2 Direct Solution Procedure

Step 1 Implement *Steps* 1–3 in the iterative solution procedure for the solution of the effective material parameters $n_i^{(0)}$ and $\eta_i^{(0)}$ ($i = 1, \ldots, m$).

Step 2 The middle *m*-2 slabs with the effective material parameters of $n_i^{(0)}$ and $\eta_i^{(0)}$ ($i = 2, \ldots, m-1$) are regarded as the known, and the effective material parameters of the first and the last slabs are solved from nonlinear equation systems of Eqs. (3.31) and (3.34) with the initial solution of $n_1^{(0)}$, $\eta_1^{(0)}$, $n_m^{(0)}$ and $\eta_m^{(0)}$.

Figure 3.31a demonstrates a five-layer structure consisting of four slabs of materials and an SRR-Wire structure. Four slabs of nonmagnetic materials are mica material with relative permittivity of 5.7, marble material with relative permittivity of 8.3, glass material with relative permittivity of 5.5, and FR4 material with relative permittivity of 4.4. In the SRR-Wire structure, inner and outer rings are fabricated on the same side of the substrate, while the wire is fabricated on the other side of the substrate. The unit cell of the SRR-Wire structure is cubic, with a cell dimension of 2.5 mm. A 0.25 mm thick substrate of FR4 is used in the SRR-Wire structure. The width of the wire is 0.14 mm, and it runs the length of the unit cell. The outer ring length of SRR is 2.2 mm, and both rings have a linewidth of 0.2 mm. The gap in each ring is 0.3 mm, and the gap between the inner and outer rings is 0.15 mm. The constitutive parameters of four slabs of the materials are separately extracted, and the effective parameters of the SRR-Wire structure are retrieved by using the S-parameter recursion procedure with an iterative solution method, as shown in Fig. 3.31a. It can be observed that the retrieved effective parameters by using the recursion procedure are different from those solely obtained by the traditional NRW method. Figures 3.31b, c show the comparison of S-parameters between the inhomogeneous structure model and the corresponding effective parameter model, in good agreement with each other. By comparison, the S-parameters according to the effective parameters solely extracted by the traditional NRW method do not coincide with those of the material-structure model. It is worthwhile pointing out that in this example, the separately extracted constitutive parameters of four slabs of the materials are entirely accurate.

Figure 3.32a shows a three-layer metamaterial structure. The first layer consists of two rings fabricated on two sides of the substrate, and the third one is composed of two patches fabricated on two sides of the substrate. The second layer is an FR4 material. The dimensions of the unit cell for each layer are 2.5 mm × 2.5 mm × 2.5 mm. The material parameters in the second layer are solely extracted, and a direct solution procedure is used to extract the effective electromagnetic parameters in the first and third layers. The resultant refractive indexes and relative impedances are given in Fig. 3.32a. Figure 3.32b gives the S-parameter comparison between the inhomogeneous structure model and the corresponding effective parameter model, in good agreement.

3.4 Equivalent Circuit Model for Metasurfaces

Metasurfaces have emerged as the 2D form of the metamaterials to flexibly manipulate the electromagnetic wave. Metasurface is an ultrathin planar array made of subwavelength elements. For the metasurface composed of a periodic array, its analysis can be classified into three kinds. At the low frequency where the periodicity of the array is far less than the wavelength, the homogenization theory can be used to analyse the periodic array. In the resonant region, where the periodicity of the array is smaller than the wavelength but the meta-atom becomes resonant, the empirical close-form expressions for them are unavailable. The equivalent circuit theory can be retrieved to model the periodic array. In the Floquet-Bloch region where the periodicity of the array is comparable with the wavelength, the multimode network is required to represent the periodic array. In this section, a new equivalent circuit model has been presented to analyse and design the frequency selective metasurface.

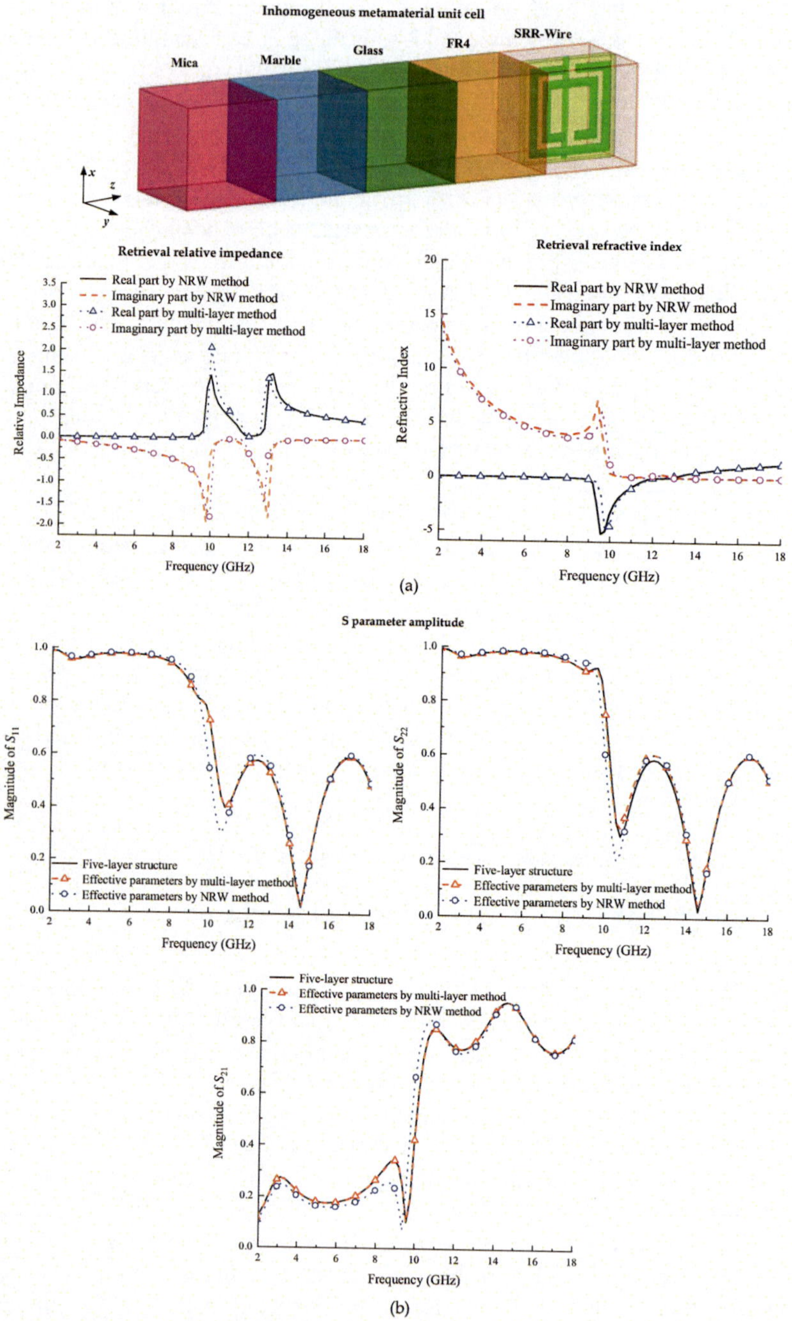

Fig. 3.31 A five-slab structure consisting of FR4, marble, mica, glass and SRR-Wire structure: **a** geometry and retrieval equivalent parameters; **b** S parameter magnitude; **c** S parameter phase [32]

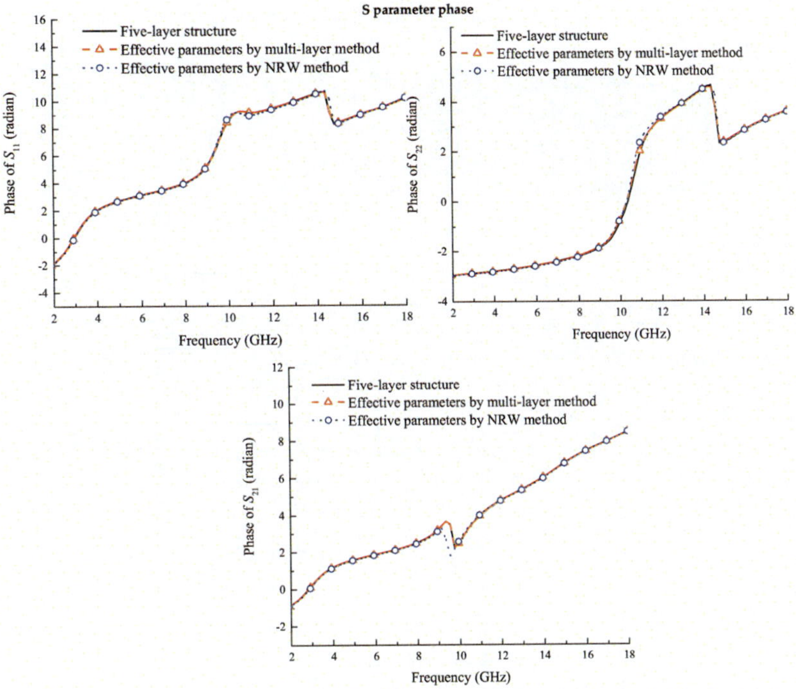

(c)

Fig. 3.31 (continued)

3.4.1 Equivalent Circuit Analytical Method for Frequency Selective Metasurfaces

Typically, frequency selective surfaces (FSS) could be divided into two categories - band stop FSS and bandpass FSS, which are patch type and aperture type, respectively, as shown in Figs. 3.33 and 3.34. Hence, the equivalent circuit of FSS could be extracted by using LC band stop and band pass networks.

3.4.2 Dual-Resonant Band Pass and Band Stop LC Circuit Models

By transforming the low-pass LC network via the frequency transformation technique [59], the corresponding equivalent circuit models can be categorized into three types: an asymmetrical network, a symmetrical T-network and a symmetrical π-network, as shown in Figs. 3.35 and 3.36. By using the S-parameters of the FSS solved by the full-wave simulation at few key frequency points distributing in the frequency

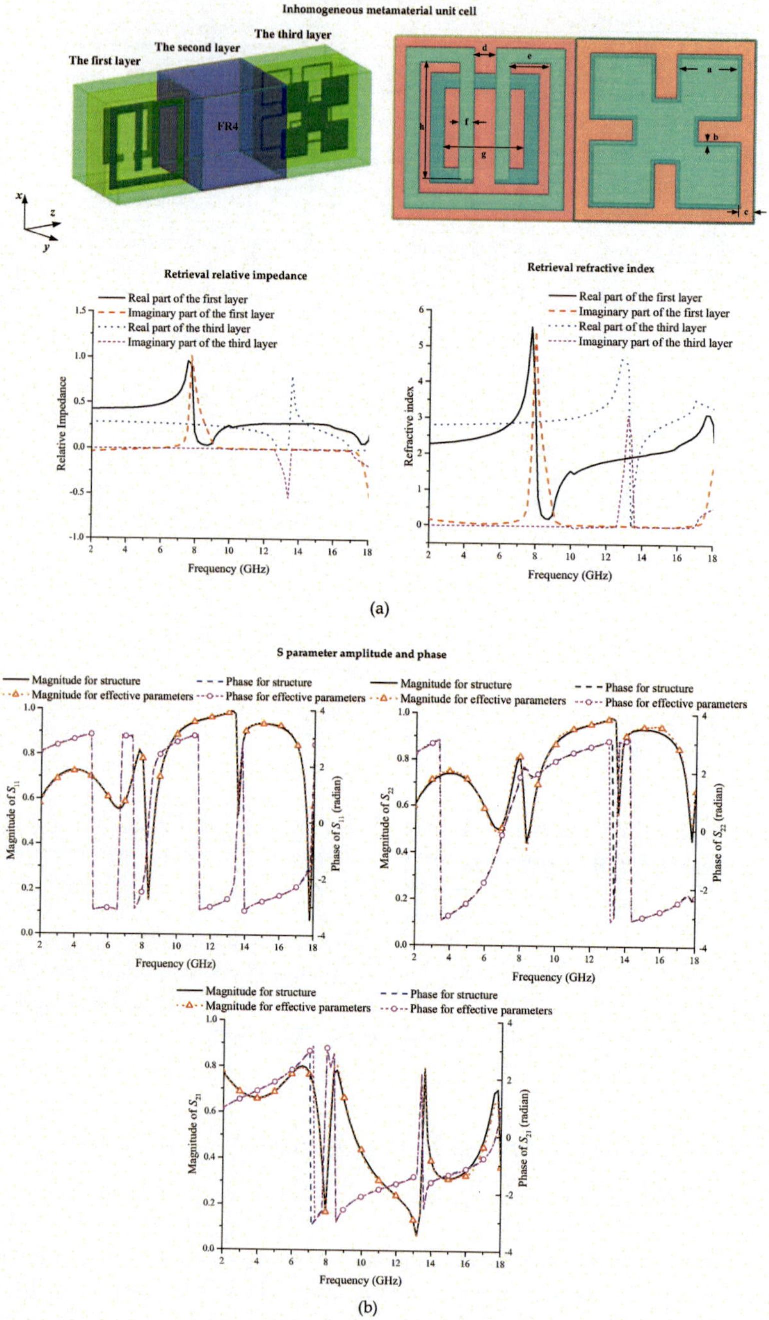

Fig. 3.32 A three-slab structure: **a** geometry and retrieval equivalent parameters; **b** S parameter amplitude and phase. All dimensions are in millimeters: a = 0.8, b = 0.05, c = 0.25, d = 0.3, e = 0.55, f = 0.2, g = 1.1, h = 1.8 [32]

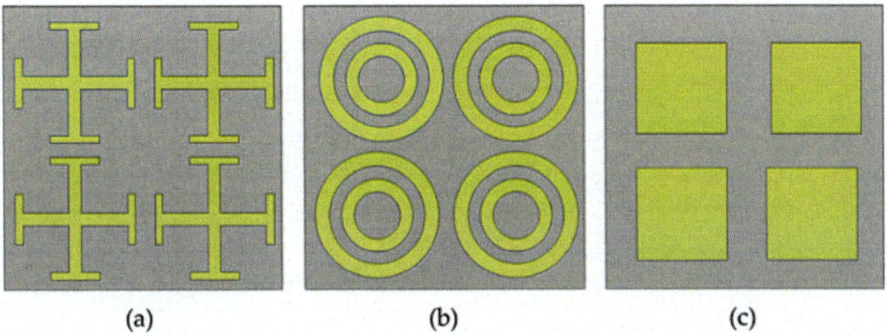

Fig. 3.33 Typical band stop FSS with patch type structure. **a** Jerusalem cross patch. **b** Dual-ring patch. **c** Square patch

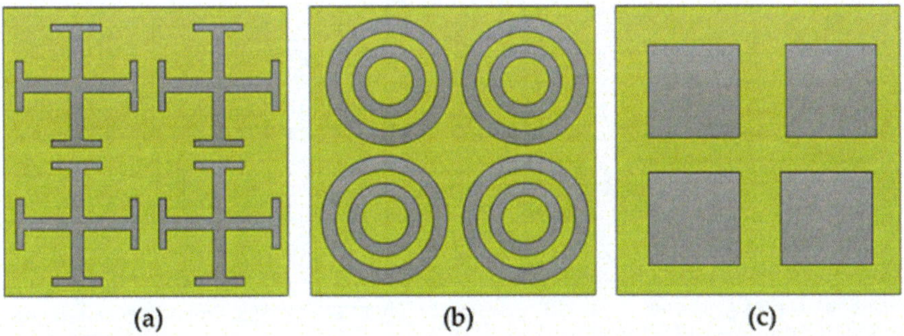

Fig. 3.34 Typical band pass FSS with aperture type structure. **a** Jerusalem cross slot. **b** Dual-ring slot. **c** Wire grid patch

range on average, ABCD parameters are first obtained. With ABCD parameters and three types of dual-pass band circuit prototypes, the lumped parameters in the circuit model can be solved by an optimized curve fitting technique.

3.4.2.1 Extraction of Lumped Elements in Dual-Band Pass LC Equivalent Circuit

According to the equivalent circuit topology, the ABCD parameters can be expressed by impedance Z and admittance Y. Specifically, for the asymmetrical network, we have:

$$\begin{bmatrix} A & B \\ C & D \end{bmatrix} = \begin{bmatrix} 1 + ZY & Z \\ Y & 1 \end{bmatrix} \tag{3.35}$$

Full-wave simulation can solve the complex S parameters at N distinct frequency points for a plane wave with normal incidence on a general FSS structure with a dual

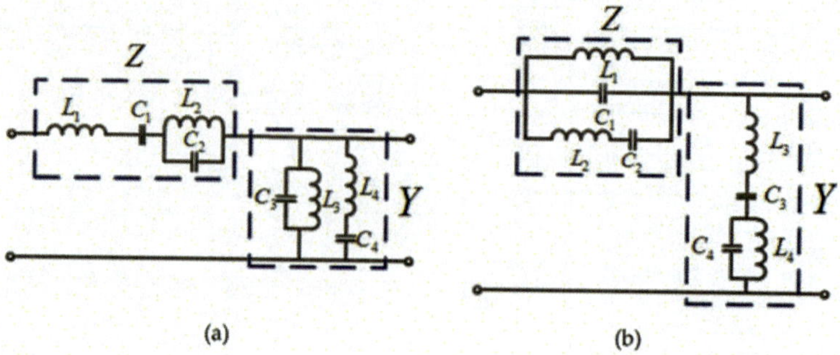

Fig. 3.35 The dual-resonant asymmetrical equivalent LC circuits: **a** band pass equivalent LC network, **b** band stop equivalent LC network

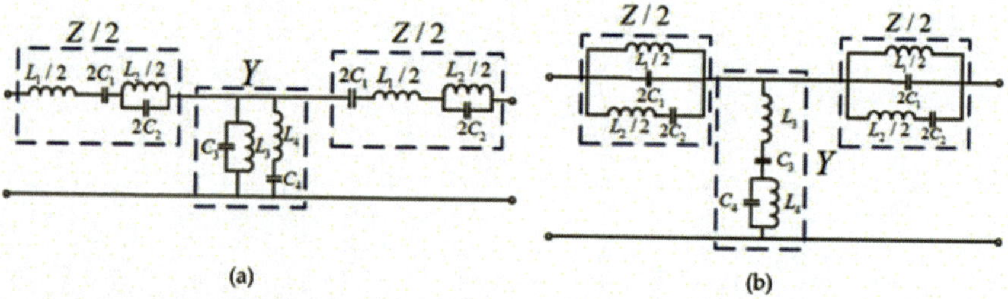

Fig. 3.36 The dual-resonant T-network equivalent LC circuits: **a** band pass equivalent LC network, **b** band stop equivalent LC network

pass-band response. Considering the relationship between the S parameters and the transmission matrix-ABCD, i.e.,

$$A = \frac{(1 + S_{11})(1 - S_{22}) + S_{12}S_{21}}{2S_{21}} \tag{3.36}$$

$$B = Z_0\frac{(1 + S_{11})(1 + S_{22}) - S_{12}S_{21}}{2S_{21}} \tag{3.37}$$

$$C = \frac{(1 - S_{11})(1 - S_{22}) - S_{12}S_{21}}{2Z_0S_{21}} \tag{3.38}$$

$$D = \frac{(1 - S_{11})(1 + S_{22}) + S_{12}S_{21}}{2S_{21}} \tag{3.39}$$

The S-parameters can be expressed in terms of the impedance Z and admittance Y:

$$S_{11} = \frac{ZY + Z + Y}{2 + ZY + Z + Y} \tag{3.40}$$

$$S_{12} = S_{21} = \frac{2}{2 + ZY + Z + Y} \tag{3.41}$$

$$S_{22} = \frac{Z - Y - ZY}{2 + ZY + Z + Y} \tag{3.42}$$

For the symmetrical T-network, the ABCD parameters can be obtained as

$$\begin{bmatrix} A & B \\ C & D \end{bmatrix} = \begin{bmatrix} 1 + \frac{ZY}{2} & Z(1 + \frac{ZY}{4}) \\ Y & 1 + \frac{ZY}{2} \end{bmatrix} \tag{3.43}$$

Substituting (3.43) into (3.36)–(3.39), the S-parameters can be rewritten as

$$S_{11} = S_{22} = \frac{Z + (Z^2Y/4) - Y}{2+ZY + Z + Y + (Z^2Y/4)} \tag{3.44}$$

$$S_{12} = S_{21} = \frac{2}{2+ZY + Z + Y + (Z^2Y/4)} \tag{3.45}$$

Similarly, for the symmetrical π-network, we have the ABCD parameters as follows:

$$\begin{bmatrix} A & B \\ C & D \end{bmatrix} = \begin{bmatrix} 1 + \frac{ZY}{2} & Z \\ Y(1 + \frac{ZY}{4}) & 1 + \frac{ZY}{2} \end{bmatrix} \tag{3.46}$$

Inserting (3.46) into (3.36)–(3.39), we can get the S-parameters of the symmetrical π-network as follows

$$S_{11} = S_{22} = \frac{Z - (ZY^2/4) - Y}{2+ZY + Z + Y + (ZY^2/4)} \tag{3.47}$$

$$S_{12} = S_{21} = \frac{2}{2+ZY + Z + Y + (ZY^2/4)} \tag{3.48}$$

Note that the relationships between S-parameters and the impedance Z and admittance Y are different for the three types of equivalent circuits. Then the following procedure is to determine the specific lumped element values for bandpass LC networks. For three types of equivalent circuits, Im(Z) and Im(Y) (Im represents the imaginary part of the complex) can be expressed in terms of the capacitances and inductances as follows:

$$\mathrm{Im}[Z] = \omega L_1 - \frac{1}{\omega C_1} + \frac{1}{(1/\omega L_2) - \omega C_2} \tag{3.49}$$

$$\mathrm{Im}(Y) = \omega C_3 - \frac{1}{\omega L_3} + \frac{1}{(1/\omega C_4) - \omega L_4} \tag{3.50}$$

where ω is the operating angular frequency. To obtain the values of L_k and C_k ($k = 1, 2, 3, 4$), the nonlinear Eqs. (3.49) and (3.50) can be solved by the curve fitting technique based on the bisquare method. The purpose of the curve fitting is to find a function $f(\omega)$ in a function class Φ for the data (ω_i, y_i) where $i = 1, 2, \ldots, N - 1$. The function $f(\omega)$ minimizes the residual under the weight w. The residual is the distance between the data samples and $f(\omega)$. A smaller residual means a better fit. In geometry, curve fitting is a curve $y = f(\omega)$ that fits the data (ω_i, y_i) where $i = 1, 2, \ldots, N - 1$. Here the frequency range is from 0.1 GHz to 30 GHz, namely ω is from $2\pi \times 0.1 \times 10^9$ rad/s to $2\pi \times 30 \times 10^9$ rad/s.

Here we choose several key frequency points distributing in the frequency range on average to determine the lumped elements. In other words, the full-wave simulation could be carried on with much fewer frequency points (for example, from 0.1 GHz to 30 GHz, around 8 frequency points is enough to extract the equivalent circuit parameters). The bisquare method iteratively searches for the optimal lumped element parameters to minimize the following residuals:

$$\frac{1}{N}\sum_{i=1}^{N} w_i \left| \omega_i L_1 - \frac{1}{\omega_i C_1} + \frac{1}{(1/\omega_i L_2) - \omega_i C_2} - \mathrm{Im}[Z(\omega_i)] \right| \tag{3.51}$$

$$\frac{1}{N}\sum_{i=1}^{N} w_i \left| \omega_i C_3 - \frac{1}{\omega_i L_3} + \frac{1}{(1/\omega_i C_4) - \omega_i L_4} - \mathrm{Im}[Y(\omega_i)] \right| \tag{3.52}$$

in which w_i is defined as the weights. The details of the bisquare method can refer to [60]. Figure 3.38 shows the comparison of the impedance Z and admittance Y obtained by the full-wave simulation and the fitting results solved by the bisquare method. The black dots show the sampling points that are used to calculate the equivalent circuit parameters. The impedance Z and admittance Y are obtained from the S-parameters of the infinite periodical structure with first-order multi-layer FSS shown in Fig. 3.39. In the ultra-wide frequency band, the fitting results are in good agreement with the simulated results.

3.4.2.2 Some Examples of Frequency Selective Metasurfaces

Here some dual band-pass frequency selective metasurfaces are presented to show the validity and the efficiency of the proposed method. As the first example, consider a low-profile, band-pass FSS composed of non-resonant constituting elements proposed in [61]. The FSS is a first-order multi-layer structure, where three metallic layers are separated from one another by very thin dielectric substrates with the permittivity of 2.65, as shown in Fig. 3.39a. Each metallic layer is in the form of a two dimensional periodic arrangement of sub-wavelength capacitive patches or a 2D wire grid with sub-wavelength periodicity, as shown in Fig. 3.39b. In the frequency range of 0.1–30 GHz, the equivalent circuit of the FSS structure is extracted with 8 frequency points (e.g., 0.1, 5, 12, 15, 20, 25, 28 and 30 GHz). Note that in the concerned frequency band, the FSS has two pass bands, and the second one is a harmonic band, as shown in Fig. 3.40a. Considering the symmetrical FSS structure and the relationship between S-parameters and the impedance Z and admittance Y, the symmetrical π-network is chosen, as shown in Fig. 3.37a. Following the proposed method for the solution of the lumped elements, we can obtain all parameters as follows: $L_1 = 0.08018$ nH, $L_2 = 0.117$ nH, $L_3 = 0.3057$ nH, $L_4 = 0.2597$ nH, $C_1 = 100$ pF, $C_2 = 0.7458$ pF, $C_3 = 1.256$ pF, and $C_4 = 0.1472$ pF. Figure 3.40 shows the S-parameter comparison between the full-wave simulation and the equivalent circuit. It can be seen from Fig. 3.40 that the equivalent dual band-pass circuit of the multi-layer FSS can simultaneously match S_{21} in the magnitude and phase during a wide range of frequency band.

The second example considers a dual-loop structure with a dual band-pass characteristic. It is the complementary structure of the concentric rings [61] printed on the dielectric substrates with a permittivity of 2.65, as shown in Fig. 3.41. The two loops generate the two passbands, respectively. The asymmetrical network as the equivalent circuit of this structure is chosen, as shown in Fig. 3.35a, because of its

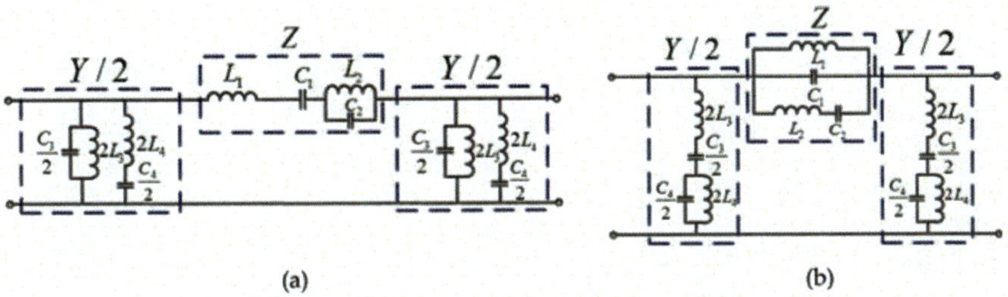

Fig. 3.37 The dual-resonant π-network equivalent LC circuits: **a** band pass equivalent LC network, **b** band stop equivalent LC network

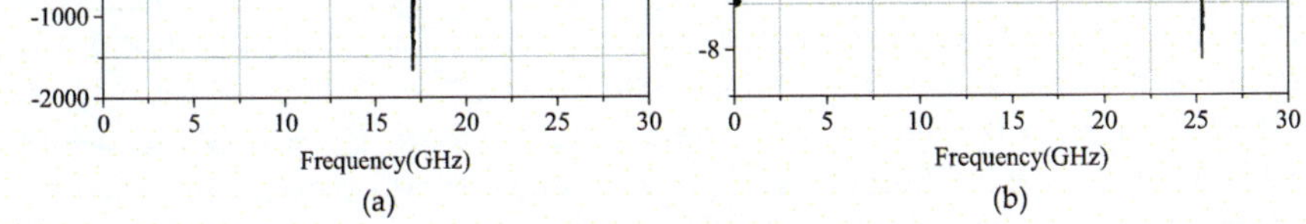

Fig. 3.38 Comparison of impedance and admittance among the original data, fitting data and the full-wave simulation: **a** comparison of imaginary part of impedance, **b** comparison of imaginary part of admittance

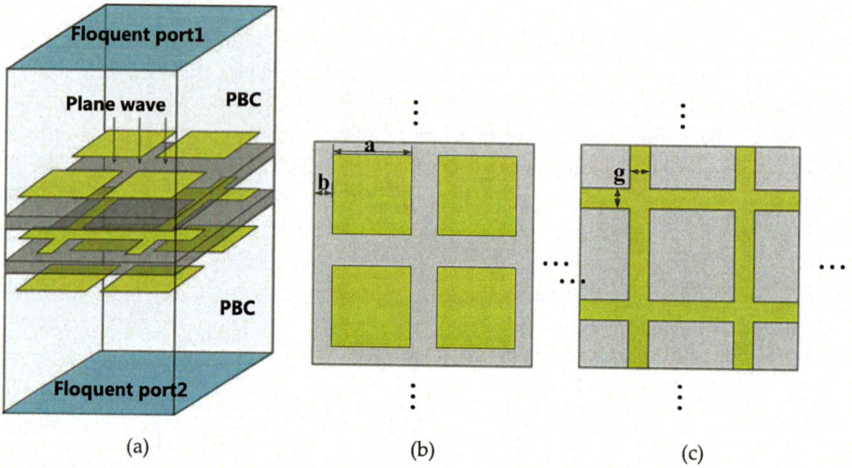

Fig. 3.39 Multi-layer FSS structure, **a** 3-D topology, **b** top views and structure parameters of the FSS

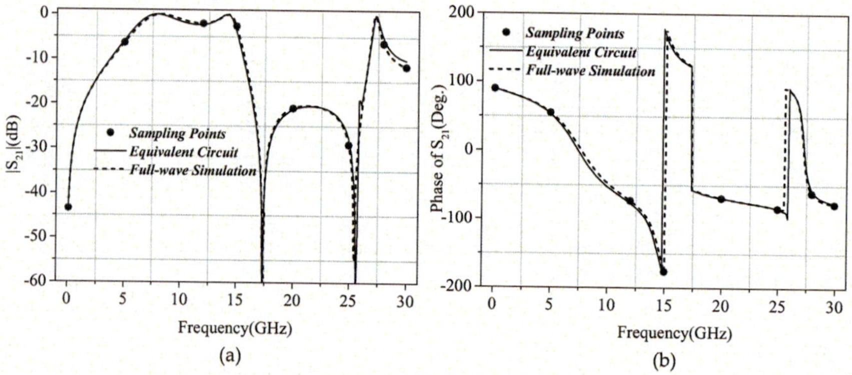

Fig. 3.40 S-parameter comparison of the full-wave simulation and the equivalent circuit of multi-layer FSS, **a** magnitude of S_{21} and **b** phase of S_{21}

asymmetrical structure. With the proposed procedure of the parameter extraction, in the frequency range of 0.1–20 GHz, the equivalent circuit of the FSS structure is extracted with 7 frequency points (i.e., 0.1, 5, 9, 10, 12, 15, and 20 GHz). The lumped element parameters are obtained as follows: $L_1 = 0.1738$ nH, $L_2 = 0$ nH, $L_3 = 0.216$ 9nH, $L_4 = 0.7176$ nH, $C_1 = 100$ pF, $C_2 = 500$ pF, $C_3 = 1.339$ pF, and $C_4 = 0.3768$ pF. Figure 3.42 shows the S-parameter comparison between the full-wave simulation and the equivalent circuit. Good agreement between the two results can be observed. It can be seen from the above numerical examples that the accurate equivalent circuit of the FSS can be obtained by using the proposed method. With

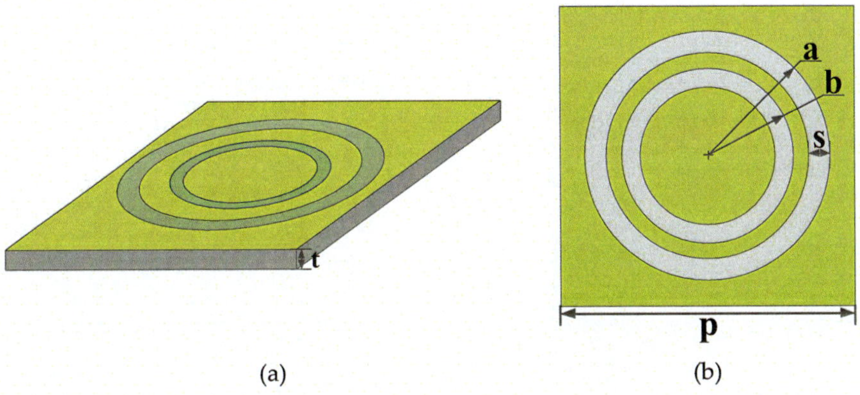

Fig. 3.41 A dual ring FSS, **a** 3-D topology, **b** top views and structure parameters of the FSS

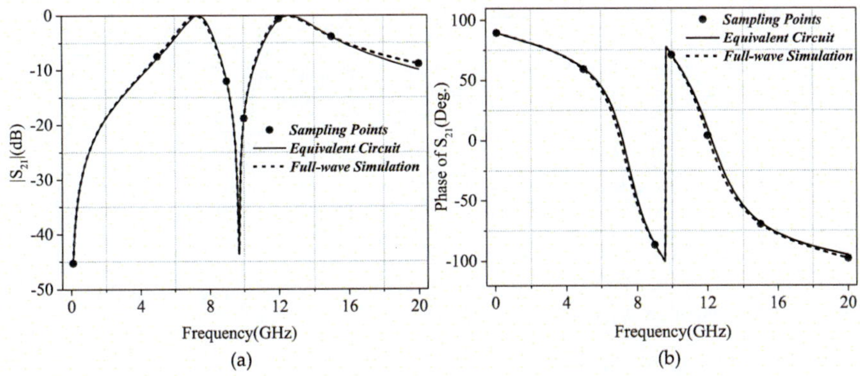

Fig. 3.42 S-parameter comparison of the full-wave simulation and equivalent circuit of the dual ring FSS, **a** magnitude of S_{21} and **b** phase of S_{21}

the retrieval equivalent circuit, the design of the frequency selective metasurface can be easily realized.

3.4.3 Triple-Resonant Band Pass and Band Stop LC Circuit Models

Similarly, the corresponding triple band-pass and band-stop asymmetric equivalent circuit model can be obtained by transforming the low-pass LC network via the frequency transformation technique, as shown in Figs. 3.43, 3.44 and 3.45, respectively.

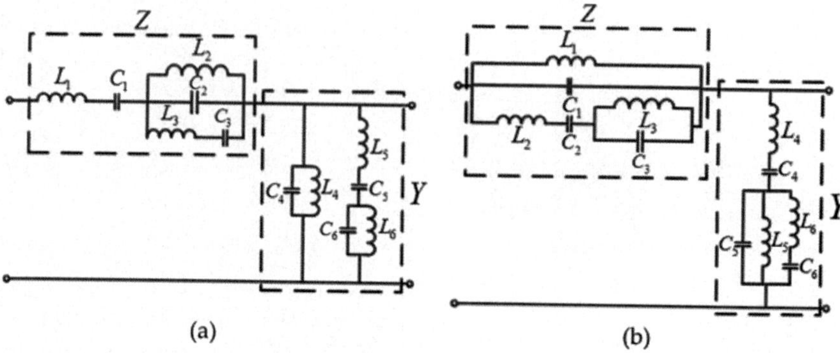

Fig. 3.43 The triple-resonant asymmetrical equivalent LC circuits: **a** band pass equivalent LC network, **b** band stop equivalent LC network

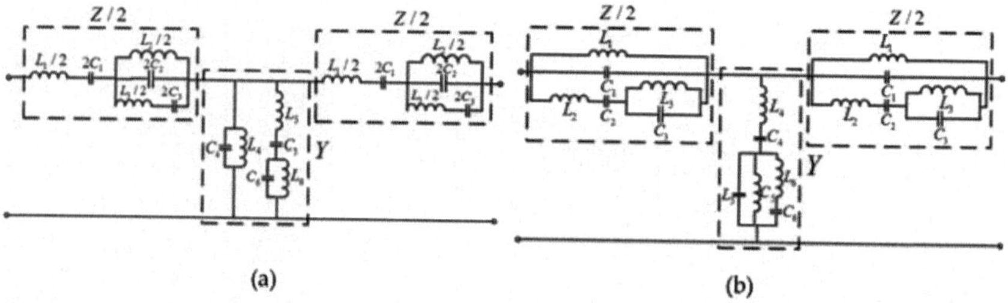

Fig. 3.44 The triple-resonant T-network equivalent LC circuits: **a** bandpass equivalent LC network, **b** band stop equivalent LC network

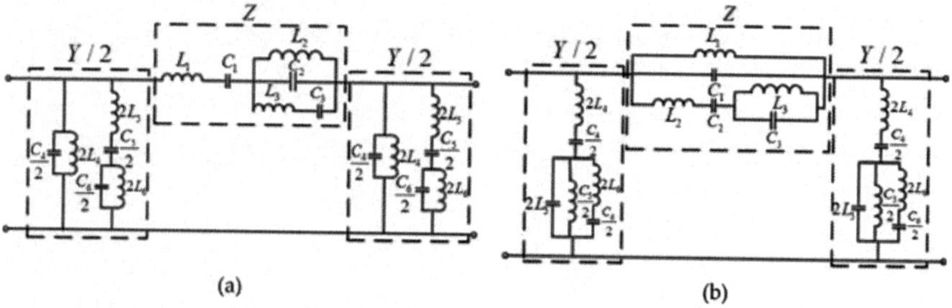

Fig. 3.45 The triple-resonant π-network equivalent LC circuits: **a** band pass equivalent LC network, **b** band stop equivalent LC network

3.4.3.1 Extraction of Lumped Elements in Triple-Band Stop LC Equivalent Circuit

The extraction procedure of lumped elements in LC equivalent circuit is the same as in Sect. 3.4.2.1. However, to determine the specific lumped element values for band stop triple-resonant LC networks, the capacitances and inductances in Figs. 3.43, 3.44 and 3.45 can be expressed as follows:

$$\mathrm{Im}\left[\frac{1}{Z}\right] = -\omega C_1 + \frac{1}{\omega L_1} - \frac{1}{(1/\omega C_2) - \omega L_2 + 1/(\omega C_3 - (1/\omega L_3))} \tag{3.53}$$

$$\mathrm{Im}\left[\frac{1}{Y}\right] = -\omega L_4 + \frac{1}{\omega C_4} - \frac{1}{(1/\omega L_5) - \omega C_5 + 1/(\omega L_6 - (1/\omega C_6))} \tag{3.54}$$

We also choose several key frequency points uniformly distributed in the frequency range to determine the lumped elements. The bisquare method iteratively searches for the optimal lumped element parameters to minimize the following residuals:

$$\frac{1}{N}\sum_{i=1}^{N} w_i \left| \omega_i C_1 - \frac{1}{\omega_i L_1} + \frac{1}{(1/\omega_i C_2) - \omega_i L_2 + 1/(\omega_i C_3 - (1/\omega_i L_3))} + \mathrm{Im}\left[\frac{1}{Z(\omega_i)}\right]\right| \tag{3.55}$$

$$\frac{1}{N}\sum_{i=1}^{N} w_i \left| \omega_i L_4 - \frac{1}{\omega_i C_4} + \frac{1}{(1/\omega_i L_5) - \omega_i C_5 + 1/(\omega_i L_6 - (1/\omega_i C_6))} + \mathrm{Im}\left[\frac{1}{Y(\omega_i)}\right]\right| \tag{3.56}$$

From (3.55) and (3.56) we can see that there are six unknowns to be solved. As a result, at least six sampling points are needed to solve the problem. Namely, six or more sampling points are enough to extract the lumped elements of the equivalent circuit. Moreover, when more than three resonances of an FSS exist in frequency response, higher-order resonant equivalent circuit could be utilized for further analysis with the same procedures.

3.5 Fast Full-Wave Algorithm Simulation for Periodic Array

In addition to the equivalent medium theory and equivalent circuit method, the full-wave simulation method is a general tool to analyze and design the metamaterials. The metamaterial generally is a periodic array composed of metallic and dielectric materials. A fast algorithm is a good candidate to solve the electromagnetic responses of the metamaterials because their complicated structures easily result in huge unknowns. In this section, a kernel-independent fast algorithm called MLGFIM [62–68] is used to fast analyze the periodic array.

3.5.1 Volume-Surface Integral Equation for Periodic Array

An incoming plane wave with an incident angle (θ^i, ϕ^i) is incident on the periodic structure shown in Fig. 3.46. Without loss of generality, the array is assumed to be periodic in *xoy* plane and has a general skewed lattice defined by the primitive vector $\vec{\rho}_a$ and $\vec{\rho}_b$. The (m, n) cell of the array is obtained by shifting the $(0, 0)$ cell through the relation

$$\vec{\rho}_{mn} = m\vec{\rho}_a + n\vec{\rho}_b \tag{3.57}$$

According to the equivalent principle, the total field $\vec{E}^s$ scattering from an arbitrarily infinite periodic structure can be formulated in terms of the fields $\vec{E}_S^s$ radiated by the equivalent electric current on the surface of conducting body and the fields $\vec{E}_V^s$ radiated by equivalent volume current in the dielectric. Specifically, we have

$$\vec{E}^s\left(\vec{r}\right) = \vec{E}_S^s\left(\vec{r}\right) + \vec{E}_V^s\left(\vec{r}\right) \tag{3.58}$$

in which

$$\vec{E}_\alpha^s\left(\vec{r}\right) = -j\omega\vec{A}_\alpha\left(\vec{r}\right) - \nabla\phi_\alpha\left(\vec{r}\right) \quad (\alpha = S, V) \tag{3.59}$$

where the magnetic vector potential and electric scalar potential are

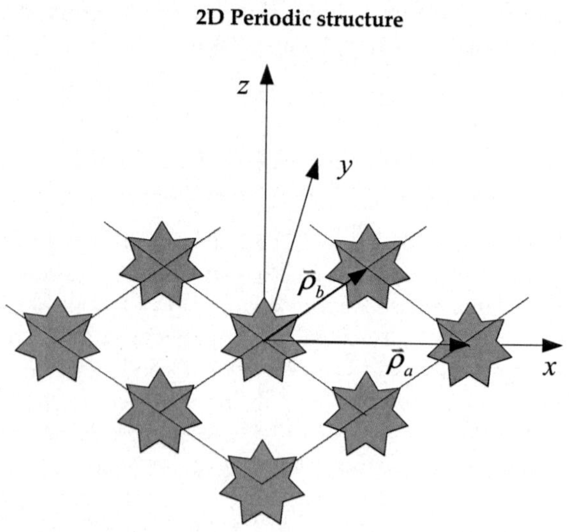

Fig. 3.46 Infinite periodic structures [68]

$$\vec{A}_\alpha\left(\vec{r}\right) = \frac{\mu_0}{4\pi}\int_\alpha G\left(\vec{r},\vec{r}'\right)\vec{J}_\alpha\left(\vec{r}'\right)d\alpha' \tag{3.60}$$

$$\phi_\alpha\left(\vec{r}\right) = -\frac{1}{j\omega 4\pi\varepsilon_0}\int_\alpha G\left(\vec{r},\vec{r}'\right)\nabla'\cdot\vec{J}_\alpha\left(\vec{r}'\right)d\alpha' \tag{3.61}$$

Here the Green's function in free space is

$$G\left(\vec{r},\vec{r}'\right) = \frac{e^{-jk_0\left|\vec{r}-\vec{r}'\right|}}{\left|\vec{r}-\vec{r}'\right|} \quad \left(k_0 = \omega\sqrt{\mu_0\varepsilon_0}\right) \tag{3.62}$$

In the dielectric region and on the conducting surface, the electric field integral equation (EFIE) can be formulated, respectively, as

$$\frac{\vec{J}_V}{j\omega\varepsilon_0(\tilde{\varepsilon}_r-1)} - \left\{\frac{\nabla\nabla\cdot\vec{A}_S + k^2\vec{A}_S}{j\omega\mu_0\varepsilon_0} + \frac{\nabla\nabla\cdot\vec{A}_V + k^2\vec{A}_V}{j\omega\mu_0\varepsilon_0}\right\} = \vec{E}^i \quad \left(\vec{r}\in V\right) \tag{3.63}$$

$$\hat{n}\times\vec{E}^i = -\hat{n}\times\left\{\frac{\nabla\nabla\cdot\vec{A}_S + k^2\vec{A}_S}{j\omega\mu_0\varepsilon_0} + \frac{\nabla\nabla\cdot\vec{A}_V + k^2\vec{A}_V}{j\omega\mu_0\varepsilon_0}\right\} \quad \left(\vec{r}\in S\right) \tag{3.64}$$

where

$$\vec{J}_V\left(\vec{r}\right) = x\left(\vec{r}\right)\vec{C}\left(\vec{r}\right) \tag{3.65}$$

in which

$$x\left(\vec{r}\right) = \frac{\tilde{\varepsilon}\left(\vec{r}\right) - \varepsilon_0}{\tilde{\varepsilon}\left(\vec{r}\right)} \tag{3.66}$$

Here $\tilde{\varepsilon}(\vec{r})$ is the complex permittivity of the dielectric, V is the volume of the dielectric, and S is the surface of the conductor.

The Floquet's theorem implies the following relations, i.e.,

$$\vec{J}_s\left(\vec{r}+\vec{\rho}_{mn}\right) = \vec{J}_s\left(\vec{r}\right)e^{-j\vec{k}_{t00}\cdot\vec{\rho}_{mn}} \tag{3.67}$$

$$\vec{C}\left(\vec{r}+\vec{\rho}_{mn}\right) = \vec{C}\left(\vec{r}\right)e^{-j\vec{k}_{t00}\cdot\vec{\rho}_{mn}} \tag{3.68}$$

in which

$$\vec{k}_{t00} = k_0 \hat{k}_{t00} \tag{3.69}$$

$$\hat{k}_{t00} = -\sin\theta^i \cos\phi^i \hat{x} - \sin\theta^i \sin\phi^i \hat{y} \tag{3.70}$$

Equations (3.67) and (3.68) tell us that the analysis of the periodic array can be implemented by solving the volume/surface integral equation in a single unit cell only if the Green's function in free space is replaced by the periodic Green's function. According to Floquet's theorem, the periodic Green's function $G_p\left(\vec{r}, \vec{r}'\right)$ can be expressed as

$$G_p\left(\vec{r}, \vec{r}'\right) = \sum_{m=-\infty}^{\infty} \sum_{n=-\infty}^{\infty} e^{-j\vec{k}_{t00}\cdot\vec{\rho}_{mn}} \frac{e^{-jk_0 R_{mn}}}{4\pi R_{mn}} \tag{3.71}$$

where

$$R_{mn} = \left|\vec{r} - \vec{r}' - \vec{\rho}_{mn}\right| \tag{3.72}$$

3.5.2 Periodic Boundary Condition

In implementing the MoM-based fast algorithm to solve the periodic array, the array is discretized in terms of the tetrahedral elements for the dielectric body and triangle patches for PEC surface, where the Schaubert–Wilton–Glisson (SWG) basis function and the Rao–Wilton–Glisson (RWG) basis function are defined, respectively. Considering that the solution procedure is implemented in a single unit cell of the periodic structure, the half-basis functions are introduced at the sidewall of the unit cell. Figure 3.47 shows two half basis functions of equivalent volume currents, i.e., $\vec{f}_f$ and $\vec{f}_b$. According to periodic boundary conditions (3.67) and (3.68), the relation between the $\vec{f}_f$ and the $\vec{f}_b$ becomes

$$\vec{f}_f = \vec{f}_b e^{-j\vec{k}_{t00}\cdot\vec{\rho}_a} \tag{3.73}$$

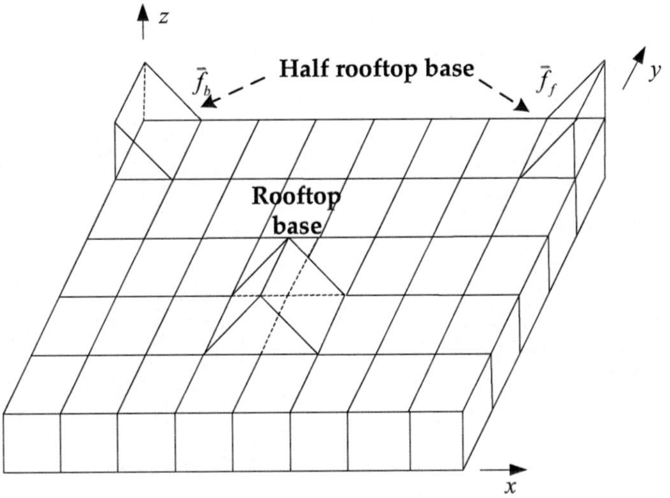

Fig. 3.47 Hexahedral elements and quadrilateral elements [68]

3.5.3 Periodic Octary-Cube-Tree

The MLGFIM method is a multilevel algorithm, and in order to implement it an octary-cube tree based data structure is built. Specifically, the entire unit cell is enclosed in a large cube, which is divided into eight smaller cubes. Each subcube is recursively subdivided into smaller cubes until the finest cubes satisfy the termination criterion. According to the octary-cube tree structure, the cubes who is adjacent to a cube i and whose parent cube is adjacent to the parent cube of the cube i is called the neighbours of the cube i in the certain level. The cubes which are not the neighbours of the cube i and whose parent cube is adjacent to the parent cube of the cube i called the interaction lists of the cube i in this level. In the MLGFIM, the interaction between the cube i and its neighbours is directly calculated, while that between the cube i and its interaction list is fast and approximately solved. According to the periodic boundary condition, we can know that there are relationships between the cubes on the sidewall of the unit cell. Hence definitions of the neighbor and interaction list should be redefined. Here the image cube is introduced, as shown in Fig. 3.48. The cube A is defined as the neighbor/interaction list cube of the cube B when the cube A is near/well separated from either the cube B or the image cubes of the cube B, and the parent cube of the cube A is the neighbor of the parent cube of either the cube B or the image cubes of the cube B.

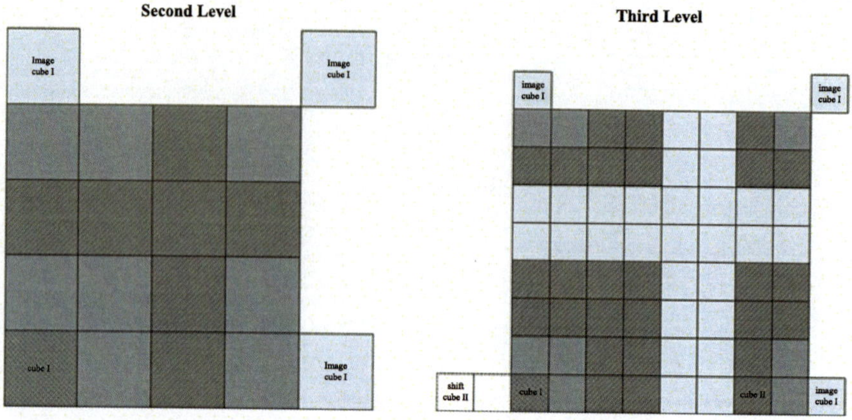

Fig. 3.48 Two dimensional pictorial representation of periodic octary-cube-tree [68]

3.5.4 Multilevel Green's Function Interpolation Method

In the MLGFIM method, the near-field interactions are directly calculated, and the far-field interactions are approximately computed using Green's function interpolation method. The matrix–vector multiplication can be divided into two parts, i.e.,

$$\overline{AJ} = \overline{A}_{near}\overline{J} + \overline{A}_{far}\overline{J} = I + II \tag{3.74}$$

Take the impedance matrix related to the magnetic vector potential of the conductor as an example. Specifically, with the Galerkin testing procedure, we have

$$\begin{aligned} Z_{mn} &= \frac{k_0\eta_0}{j4\pi}\int_{S_m} ds \int_{S_n} ds' G\left(\vec{r}, \vec{r'}\right) \vec{f}_m(\vec{r}) \cdot \vec{f}_n(\vec{r'}) \\ &= \sum_{\xi=x,y,z} \frac{k_0\eta_0}{j4\pi}\int_{S_m} ds \int_{S_n} ds' G\left(\vec{r}, \vec{r'}\right) f_{m\xi}(\vec{r}) \cdot f_{n\xi}(\vec{r'}) \end{aligned} \tag{3.75}$$

Here $\vec{f}_m$ and $\vec{f}_n$ are testing and basis functions defined on the surface patches S_m and S_n, respectively, and $f_{m\xi}$ is the ξ-component of the $\vec{f}_m$. The testing function is located in the field cube G_u and the basis function is located in the source cube G_v, and the field cube is the interaction list of the source cube. In this scenario, a suitable interpolation function is chosen to approximate $G\left(\vec{r}, \vec{r'}\right)$ as

$$G(\vec{r}, \vec{r'}) = \sum_{p=1}^{N}\sum_{q=1}^{N} w_{m,p}(\vec{r}) w_{n,q}(\vec{r'}) G(\vec{r}_p, \vec{r'}_q) \tag{3.76}$$

Here the interpolation functions $w_{m,p}(\vec{r})$ and $w_{n,q}(\vec{r'})$ are defined in the field cube and source cube, respectively, and $\vec{r}_p$ and $\vec{r'_q}$ are the corresponding interpolation points. Inserting (3.76) into (3.75), we have

$$Z_{mn}^{(\xi)} = \left[\int\limits_{S_m} ds f_{m\xi}\left(\vec{r}\right) \overline{w}_u^T\left(\vec{r}\right) \right] \overline{\overline{G}} \left[\int\limits_{S_n} ds' f_{n\xi}\left(\vec{r'}\right) \overline{w}_v\left(\vec{r'}\right) \right] = \overline{v}_{u,m}^T \cdot \overline{\overline{G}} \cdot \overline{v}_{v,n} \tag{3.77}$$

in which $\overline{w}_v\left(\vec{r'}\right)$ is the interpolation function matrix consisting of $w_{n,q}(\vec{r'})$ and $\overline{\overline{G}}$ denotes the Green's function matrix consisting of $G(\vec{r}_p, \vec{r'}_q)$. Therefore, the impedance submatrix A_{uv} between field cube G_u and source cube G_v can be expressed as

$$A_{uv} = \begin{bmatrix} [\overline{v}_{u,1}]^T \\ [\overline{v}_{u,2}]^T \\ \vdots \\ [\overline{v}_{u,M_m}]^T \end{bmatrix} \cdot \overline{\overline{G}} \cdot \left[\overline{v}_{v,1} \; \overline{v}_{v,2} \cdots \overline{v}_{v,N_n} \right] = \left[\overline{W}(v)_u \right]^T \cdot \overline{\overline{G}} \cdot \overline{W}(v)_v \tag{3.78}$$

Here M_m and N_n is the number of unknowns in G_u and G_v. The multiplication given by (3.78) between the matrix and the vector in the MLGFIM includes three procedures: 1. Upward pass from the finest level L to the top level l_t using the $l + 1$th-to-lth level interpolation matrix $\overline{C}_{n_l,l;n_{l+1},l+1}$; 2. Translation at the top level l_t using Green's function matrix $\overline{G}_{m_{l_t},l_t;n_{l_t},l_t}$ between the cube $G_{m_{l_t},l_t}$ and its interaction list $G_{n_{l_t},l_t}$; 3. Downward pass from the top level l_t to the finest level L using Green's function matrix and the lth-to-$l + 1$th level interpolation matrix. Detailed steps for the calculation of II in (3.74) are given in Fig. 3.49.

3.5.5 Numerical Examples

The dielectric slab with the embedded periodic material blocks is considered, as shown in Fig. 3.50. The dielectric slab is illuminated by a TM wave under the incident angles of $\theta = 0^0$, $\varphi = 0^0$. The specular reflection coefficient is calculated by MLGFIM and compared with the results in [69] and [70], as shown in Fig. 3.50. The result is in good agreement with the FEM-BI result [70], and there is a slightly shift to a lower frequency between our result and the result of VIE in the spectral domain [69].

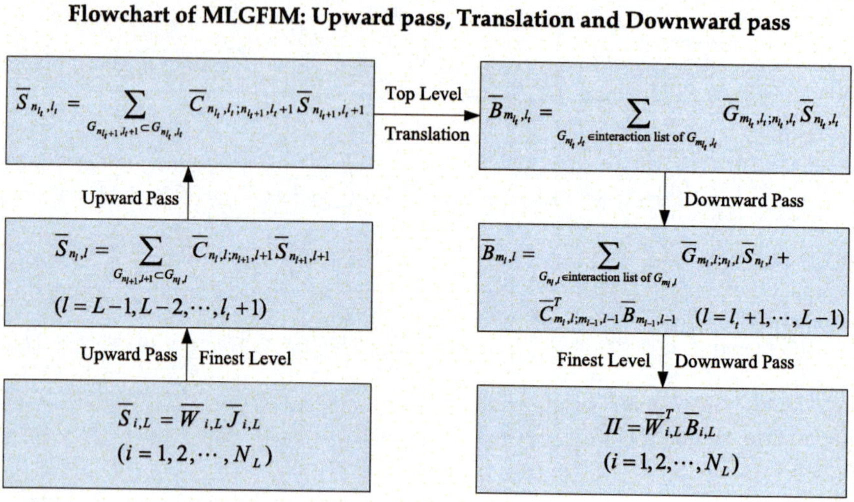

Fig. 3.49 The far-field interactions of MLGFIM [68]

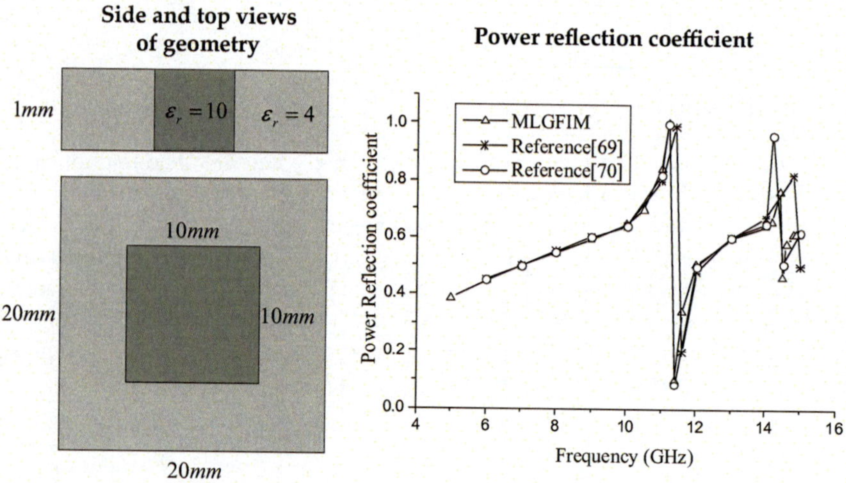

Fig. 3.50 A dielectric slab with planarly embedded periodic material block [68]

A 2D lattice of the conducting SRRs [71] with and without backing dielectric is considered, as shown in Fig. 3.51a. In this structure, the conducting wire width, gap and length are 0.33 mm, 0.33 mm, and 3 mm, respectively. The thickness of dielectric material is 0.33 mm and the permittivity is 4. Both of the periods of the structure in the *x*-direction and *y*-direction are 3.66 mm. Here we consider three cases: the first is the single conducting SRRs parallel to *xoy* plane; the second is the double conducting SRRs which are parallel to *xoy* plane and *xoz* plane, respectively;

the third is the triple conducting SRRs which are parallel to *xoy* plane, *xoz* plane and *yoz* plane, respectively, as shown in Fig. 3.51a. A uniform plane wave with the electric field parallel to the *x*-axis is incident along the normal direction, and the TM reflection coefficient calculated by MLGFIM is shown in Fig. 3.51b. It can be seen from Fig. 3.51b that when the conducting SRRs is normal to the magnetic field, there are the stop band and pass band for the cases of the conducting SRRs without and with backing dielectric, respectively. In this example, the four-level MLGFIM is used, and CPU time and memory approximately obey $O(N)$, as shown in Fig. 3.51c.

3.6 Conclusion

In this chapter, four analysis methods have been presented for the metamaterials and metasurfaces. The local resonant cavity cell model is first proposed to analyze the electromagnetic bandgap structures, and the relationship between the surface-wave suppression and the in-phase reflection phase is elaborated. According to the effective medium theory, the NRW-based retrieval method and the phase unwrapping technique have been developed to model the electrically thin and thick metamaterials, respectively. Moreover, for the inhomogeneous metamaterial structure, the planar multi-layer model with piecewise homogeneous isotropic media is introduced to extract the effective material parameters by using iterative solutions or direct solution procedures. In addition, an analytical method based on the multiple-resonant equivalent circuit for the metasurfaces has been introduced in this chapter. The semi-analytical method incorporates a filter circuit model and full-wave simulation to fully analyze the periodic array in a relatively wide frequency band range, and the resulting theoretical results agree well with the full-wave simulated ones. Finally, a full-wave simulation method, i.e., multilevel Green's function interpolation method, has been developed to fast solve the periodic array. The volume/surface integral equation is formulated, and the periodic tree data structure is developed. Numerical examples are given to demonstrate the good accuracy and low computational complexity of the fast algorithm.

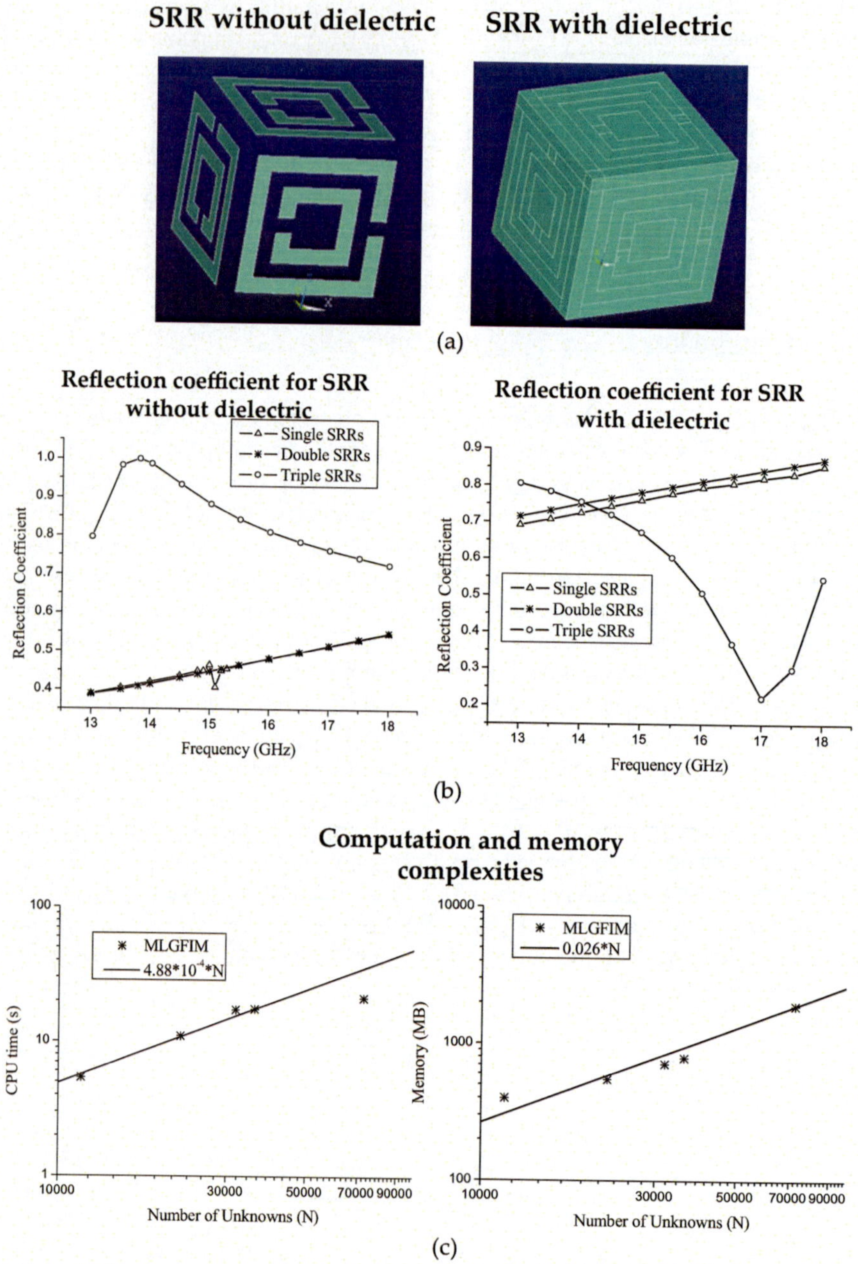

Fig. 3.51 The three-dimensional split ring resonators. **a** geometry. **b** reflection coefficient, **c** computational and memory complexities [68]

References

1. V. Veselago, The electrodynamics of substances with simultaneously negative values of ε and μ. Sov. Phys. Usp. **10**, 509–514 (1968)
2. N. Engheta, R.W. Ziolkowski, *Metamaterials physics and engineering explorations* (IEEE Press, New Jersey, 2006)
3. C. Caloz, T. Itoh, *Electromagnetic Metamaterials: Transmission Line Theory and Microwave Applications* (Wiley and IEEE Press, New Jersey, 2005)
4. T.J. Cui, D.R. Smith, R.P. Liu, *Metamaterials Theory, Design, and Applications* (Springer, New York, 2010)
5. D.H. Werner, D. Kwon, *Transformation Electromagnetics and Metamaterials* (Springer, London, 2014)
6. D.H. Werner, *Broadband Metamaterials in Electromagnetics* (Pan Stanford Publishing, Singapore, 2017)
7. D. Sievenpiper, L. Zhang, R.F.J. Broas, N.G. Alexopolus, E. Yablonovitch, High-impedance electromagnetic surfaces with a forbidden frequency band. IEEE Trans. Microw. Theory Tech. **47**, 2059–2074 (1999)
8. F. Yang, Y. Rahmat-Sammi, Reflection phase characterizations of the EBG ground plane for low profile wire antenna application. IEEE Trans. Antennas Propag. **51**(10), 2691–2703 (2003)
9. F. Yang, Y. Rahmat-Samii, Microstrip antennas integrated with electromagnetic band-gap (EBG) structures: a low mutual coupling design for array applications. IEEE Trans. Antennas Propag. **51**(10), 2939–2949 (2003)
10. L. Li, Q. Chen, Q. Yuan, C.H. Liang, K. Sawaya, Surface-wave suppression band gap and plane-wave reflection phase band of mushroomlike photonic band gap structures. J. Appl. Phys. **103**, 023513 (2008)
11. Y.Q. Li, H. Zhang, Y.Q. Fu, N.C. Yuan, RCS reduction of ridged waveguide slot antenna array using EBG radar absorbing material. IEEE Antennas Wirel. Propag. Lett. **7**, 473–476 (2008)
12. X. Chen, L. Li, C.-H Liang, Z.-J Su, Locally resonant cavity cell model for meandering slotted electromagnetic band gap structure. IEEE Antennas Wirel. Propag. Lett. **9**, 3–7 (2010)
13. E. Martini, G.M. Sardi, S. Maci, Homogenization processes and retrieval of equivalent constitutive parameters for multisurface-metamaterials. IEEE Trans. Antennas Propag. **62**(4), 2081–2092 (2014)
14. A.M. Nicolson, G. Ross, Measurement of the intrinsic properties of materials by time–domain techniques. IEEE Trans. Instrum. Meas. **19**(4), 377-382 (1970)
15. W.B. Weir, Automatic measurement of complex dielectric constant and permeability at microwave frequencies. Proc. IEEE **62**(1), 33-36 (1974)
16. J. Baker-Jarvis, E.J. Vanzura, W.A. Kissick, Improved technique for determining complex permittivity with the transmission/reflection method. IEEE Trans. Microw. Theory Techn. **38**(8), 1096-1103 (1990)
17. A.H. Boughriet, C. Legrand, A. Chapoton, Noniterative stable transmission/reflection method for low-loss material complex permittivity determination. IEEE Trans. Microw. Theory Techn. **45**(1), 52–57 (1997)
18. X. Chen, T.M. Grzegorczyk, B.-I. Wu, J. Pacheco, Jr., J.A. Kong, Robust method to retrieve the constitutive effective parameters of metamaterials. Phys. Rev. E **70**, 016608 (2004)
19. D.R. Smith, D.C. Vier, T. Koschny, C.M. Soukoulis, Electromagnetic parameter retrieval from inhomogeneous metamaterials. Phys. Rev. E **71**, 036617 (2005)
20. V.V. Varadan, R. Ro, Unique retrieval of complex permittivity and permeability of dispersive materials from reflection and transmitted fields by enforcing causality. IEEE Trans. Microw. Theory Techn. **55**(10), 2224–2230 (2007)
21. K. Chalapat, K. Sarvala, J. Li, G.S. Paraoanu, Wideband reference plane invariant method for measuring electromagnetic parameters of materials. IEEE Trans. Microw. Theory Techn. **57**(9), 2257–2267 (2009)
22. S. Zsolt, G.-H. Park, H. Ravi, E.-P. Li, A unique extraction of metamaterial parameters based on Kramers-Kronig relationship. IEEE Trans. Microw. Theory Techn. **58**(10), 2646–2653 (2010)

23. O. Luukkonen, S.I. Maslovski, S.A. Tretyakov, A stepwise Nicolson–Ross–Weir-based material parameter extraction method. IEEE Antennas Wirel. Propag. Lett. **10**, 1295–1298 (2011)
24. J.J. Barroso, U.C. Hasar, Resolving phase ambiguity in the inverse problem of transmission/reflection measurement methods. J. Infrared Milli. Terahertz Waves **32**(6), 857–866 (2011)
25. J.J. Barroso, U.C. Hasar, Constitutive parameters of a metamaterial slab retrieved by the phase unwrapping method. J. Infrared Milli. Terahertz Waves **33**(2), 237–244 (2012)
26. Y. Shi, T. Hao, L. Li, C.-H. Liang, An improved NRW method to extract electromagnetic parameters of metamaterials. Microw. Opt. Technol. Lett. **58**(3), 647–652 (2016)
27. Y. Shi, Z.Y. Li, L. Li, C.H. Liang, An electromagnetic parameters extraction method for metamaterials based on phase unwrapping technique. Wave Random Complex **26**(4), 417–433 (2016)
28. Z. Li, K. Aydin, E. Ozbay, Determination of the effective constitutive parameters of bianisotropic metamaterials from reflection and transmission coefficients. Phys. Rev. E **79**, 026610 (2009)
29. Z. Jiang, J.A. Bossard, X. Wang, D.H. Werner, Synthesizing metamaterials with an angularly independent effective medium properties based on an anisotropic parameter retrieval technique coupled with a genetic algorithm. J. Appl. Phys. **109**, 013515 (2011)
30. Z. Li, R. Zhao, T. Koschny, M. Kafesaki, K.B. Alici, E. Colak, H. Caglayan, E. Ozbay, C.M. Soukoulis, Chiral metamaterials with negative refractive index based on four "u" split ring resonators. Appl. Phys. Lett. **97**(8), 081901 (2010)
31. M.E. Baginski, D.L. Faircloth, M.D. Deshpande, Comparison of two optimization techniques for the estimation of complex permittivities of multilayered structures using waveguide measurements. IEEE Trans. Microw. Theory Techn. **53**(10), 3251–3259 (2005)
32. Y. Shi, Z.Y. Li, K. Li, L. Li, C.H. Liang, A retrieval method of effective electromagnetic parameters for inhomogeneous metamaterials. IEEE Trans. Microw. Theory Techn. **65**(4), 1160–1178 (2017)
33. J.B. Pendry, Negative refraction makes a perfect lens. Phys. Rev. Lett. **85**(18), 3966–3969 (2000)
34. Y. Lai, H.Y. Chen, Z.Q. Zhang, C.T. Chan, Complementary media invisibility cloak that cloaks objects at a distance outside the cloaking shell. Phys. Rev. Lett. **102**, 093901 (2009)
35. W.X. Jiang, T.J. Cui, Radar illusion via metamaterials. Phys. Rev. E **83**, 026601 (2011)
36. Y. Shi, W. Tang, C.H. Liang, A minimized invisibility complementary cloak with a composite shape. IEEE Antennas Wirel. Propag. Lett. **13**, 1800–1803 (2014)
37. F.F. Huo, L. Li, T. Li, Y.M. Zhang, C.H. Liang, External invisibility cloak for multi-objects with arbitrary geometries. IEEE Antennas Wirel. Propag. Lett. **13**, 273–276 (2014)
38. Y. Shi, W. Tang, L. Li, C.H. Liang, Three-dimensional complementary invisibility cloak with arbitrary shapes. IEEE Antennas Wirel. Propag. Lett. **14**, 1550–1553 (2015)
39. Y. Shi, L. Zhang, W. Tang, L. Li, C.H. Liang, Design of a minimized complementary illusion cloak with arbitrary position. Int. J. Antenna Propag., Article ID 932495 (2015)
40. L. Zhang, Y. Shi, C.H. Liang, Achieving illusion and invisibility of inhomogeneous cylinders and spheres. J. Optics **18**, 085101 (2016)
41. L. Zhang, Y. Shi, C.H. Liang, Optimal illusion and invisibility of multilayered anisotropic cylinders and spheres. Opt. Express **24**, 23333–23352 (2016)
42. Y. Shi, L. Zhang, Cloaking design for arbitrarily shape objects based on characteristic mode method. Opt. Express **25**, 32263–32279 (2017)
43. N.I. Landy, S. Sajuyigbe, J.J. Mock, D.R. Smith, W.J. Padilla, Perfect metamaterial absorber. Phys. Rev. Lett. **100**(20), 207402 (2008)
44. L. Li, Y. Yang, C.H. Liang, A wide-angle polarization-insensitive ultra-thin metamaterial absorber with three resonant modes. J. Appl. Phys. **110**, 063702 (2011)
45. Y. Zhang, Y. Shi, C.H. Liang, A broadband tunable graphene-based metamaterial absorber. Opt. Mater. Express **6**(9), 3036–3044 (2016)
46. Y. Shi, Y.C. Li, T. Hao, L. Li, C.H. Liang, A design of ultra-broadband metamaterial absorber. Wave Random Complex **27**, 381–391 (2017)

47. L. Zhang, Y. Shi, J.X. Yang, X. Zhang, L. Li, Broadband transparent absorber based on Indium Tin Oxide-Polyethylene Terephthalate film. IEEE Access **7**(1), 137848–137855 (2019)
48. Y. Shi, K. Li, J. Wang, L. Li, C.H. Liang, An etched planar metasurface half maxwell fish-eye lens antenna. IEEE Trans. Antennas Propag. **63**, 3742–3747 (2015)
49. S. Yu, L. Li, G. Shi, C. Zhu, X. Zhou, Y. Shi, Design, fabrication, and measurement of reflective metasurface for orbital angular momentum vortex wave in radio frequency domain. Appl. Phys. Lett. **108**(12), 121904 (2016)
50. S. Yu, L. Li, G. Shi, C. Zhu, Y. Shi, Generating multiple orbital angular momentum vortex beams using a metasurface in radio frequency domain. Appl. Phys. Lett. **108**(24), 241901 (2016)
51. Y. Shi, Y. Zhang, Generation of wideband tunable orbital angular momentum vortex waves using graphene metamaterial reflectarray. IEEE Access **6**(1), 5341–5347 (2018)
52. Y. Shi, Y.C. Li, Design of broadband leaky-wave antenna based on permeability-negative transmission line. Microw. Opt. Technol. Lett. **60**(3), 699–704 (2018)
53. Z.K. Meng, Y. Shi, W.Y. Wei, X.F. Zhang, Multifunctional scattering antenna array design for orbital angular momentum vortex wave and RCS reduction. IEEE Access **8**, 109289–109296 (2020)
54. J.D. Joannopoulos, R.D. Meade, J.N. Winn, *Photonic Crystals* (Princeton Univ. Press, Princeton, NJ, 1995)
55. L. Li, B. Li, H.X. Liu, C.H. Liang, Locally resonant cavity cell model for electromagnetic bandgap structure. IEEE Trans. Antennas Propag. **54**, 90–100 (2006)
56. D. Sievenpiper, *High-Impedance Electromagnetic Surfaces*. Ph.D. dissertation, Dept. Elect. Eng. Univ. California at Los Angeles, Los Angeles, CA, 1999.
57. R.E. Diaz, J.T. Aberle, W.E. Mckinzie, TM mode analysis of a Sievenpiper high-impedance reactive surface, in *Proceedings of Antennas and Propagation Society International Symposium*, vol. 1, (2000), pp. 327–330
58. V. Lucarini, J.J. Saarinen, K.E. Peiponen, E.M. Variainen, *Kramers-Kronig Relations in Optical Materials Research* (Springer, Berlin, 2005)
59. D.M. Pozar, *Microwave Engineering*, 3rd edn. (Wiley & Sons, New York, 2004)
60. http://www.ni.com/white-paper/6954/en/
61. E.A. Parker, S.M.A. Hamdy, R.J. Langley, Arrays of concentric rings as frequency-selective surfaces. Electron. Lett. **17**, 880–881 (1981)
62. H.G. Wang, C.H. Chan, L. Tsang, A new multilevel Green's function interpolation method for large-scale low-frequency EM simulations. IEEE Trans. Comput. Aided Des. Integr. Circuits Syst. **24**, 1427–1443 (2005)
63. H.G. Wang, C.H. Chan, The implementation of multilevel Green's function interpolation method for full-wave electromagnetic problems. IEEE Trans. Antennas Propag. **55**, 1348–1358 (2007)
64. L. Li, H.G. Wang, C.H. Chan, An improved multilevel Green's function interpolation method with adaptive phase compensation. IEEE Trans. Antennas Propag. **56**(5), 1381–1393 (2008)
65. Y. Shi, H.G. Wang, L. Li, C.H. Chan, Multilevel Green's function interpolation method for scattering from composite metallic and dielectric objects. J. Opt. Soc. Am. A **25**(10), 2535–2548 (2008)
66. Y. Shi, C.H. Chan, An OpenMP parallelized multilevel Green's function interpolation method accelerated by fast Fourier transform technique. IEEE Trans. Antennas Propag. **60**, 3305–3313 (2012)
67. Y. Shi, C.H. Chan, Comparison of interpolating functions and interpolating points in full-wave multilevel Green's function interpolation method. IEEE Trans. Antennas Propag. **58**(8), 2691–2699 (2010)
68. Y. Shi, C.-H. Chan, Multilevel Green's function interpolation method for analysis of 3-D frequency selective structures using volume/surface integral equation. J. Opt. Soc. Am. A **27**, 308–318 (2010)
69. H.Y.D. Yang, R. Diaz, N.G. Alexopoulos, Reflection and transmission of waves from multilayer structures with planar implanted periodic material blocks. J. Opt. Soc. Am. B **14**, 2513–2521 (1997)

70. T.F. Eibert, J.L. Volakis, D.R. Wilton, D.R. Jackson, Hybrid FE/BI modeling of 3-D doubly periodic structures utilizing triangular prismatic elements and an MPIE formulation accelerated by the Ewald transformation. IEEE Trans. Antennas Propag. **47**, 843–850 (1999)
71. T. Koschny, P. Markos, D.R. Smith, C.M. Soukoulis, Resonant and antiresonant frequency dependence of the effective parameters of metamaterials. Phys. Rev. E **68**, 065602 (2003)

Chapter 4
Analysis and Applications of Electromagnetic Bandgap Metasurfaces

Linfeng Shi, Zhao Wu, Haixia Liu, and Long Li

Abstract Artificial electromagnetic materials, including electromagnetic band gap (EBG) and left-handed material (LHM), exhibit unique electromagnetic properties and thus have attracted much attention in recent years. In this chapter, two bandgap characteristics of the EBG, i.e., surface-wave suppression and in-phase reflectivity have been theoretically elaborated, and the relationship between them is clarified. With these attractive properties, the application of the EBG into various practical engineering problems, including high-speed circuit noise suppression and antenna substrate and superstrate designs have been presented.

Keywords Electromagnetic band gap (EBG) · Surface-wave suppression · In-phase reflectivity · high-speed circuit noise suppression · antenna substrate and superstrate designs

4.1 Introduction

Artificial electromagnetic materials, such as photonic band gap (PBG) structures/ electromagnetic bandgap (EBG) [1, 2], frequency selective surfaces [3], and left-handed materials [4–6], are broadly classified as metamaterials, which are typically fabricated using two- or three-dimensional periodic metallic and dielectric structures [7–9]. They have attracted significant research interests in recent years due to their special electromagnetic properties applicable to a wide range of electromagnetic devices.

L. Shi · H. Liu · L. Li (✉)
School of Electronic Engineering, Xidian University, Xi'an 710071, Shaanxi, China
e-mail: lilong@mail.xidian.edu.cn

Z. Wu
School of Physics and Telecommunication Engineering, Yulin Normal University, Yulin 537000, China

L. Li et al. (eds.), *Electromagnetic Metamaterials and Metasurfaces: From Theory To Applications*, https://doi.org/10.1007/978-981-99-7914-1_4

The EBG structures have two novel electromagnetic features, i.e., surface wave band gap and in-phase reflection phase band gap, and thus have been used in high-speed circuits, antennas, microwaves, etc. The surface wave band gap can be used to reduce power noise propagation, improve signal integrity, and enhance antenna and microwave device performance, and the in-phase reflection phase band gap can be used to design low profile antenna, antenna stealth, and other aspects. In this chapter, the EBG theory has been reviewed and the relationship between surface wave band gap and in-phase reflection phase band gap has been elaborated. In addition, some simulation and measurement models for the surface wave band gap and in-phase reflection phase band gap have been discussed. Finally, the applications of the EBG in high-speed circuits and antenna designs have been given.

4.2 Electromagnetic Bandgap and High Impedance Theory

Mushroom-like EBG structures first proposed by D. Sievenpiper have the attractive features of compactness, planarity, low loss and broad stopband [8, 9], which are important in communication antenna and array applications. The traditional two-dimensional mushroom-like EBG structure is composed of planar periodic square patches fabricated on a dielectric substrate and grounded by vertical posts or metal-plated vias, as shown in Fig. 4.1. It is worthwhile pointing out that in the mushroom-like EBG structure, the metal patches can have arbitrary shapes and the corresponding array arrangements also are arbitrary, such as rectangular arrangement, hexagonal arrangement, triangular arrangement, etc. The physical mechanism of the mushroom-like EBG structure can be simply explained from the viewpoint of an LC parallel resonant circuit.

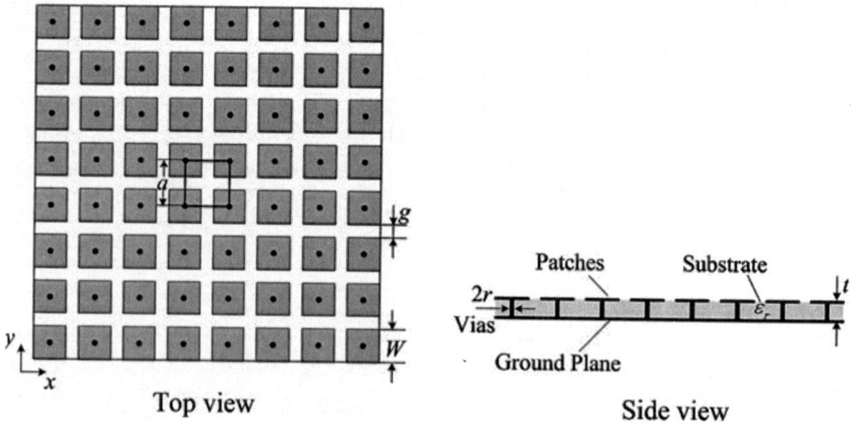

Fig. 4.1 Schematic diagram of traditional mushroom-like EBG structure

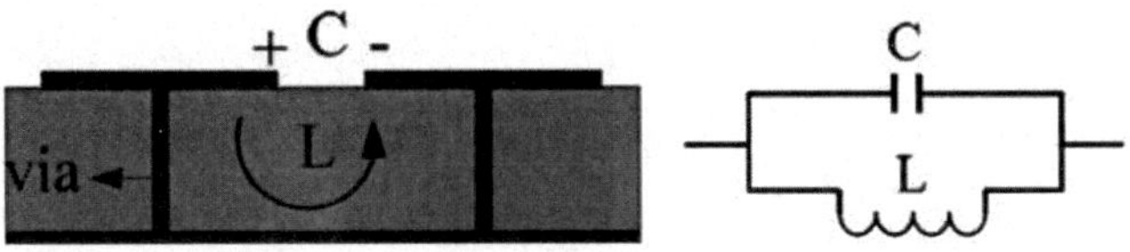

Fig. 4.2 Equivalent circuit structure model

As can be seen from Fig. 4.2, when the electromagnetic wave illuminates the EBG structure, two kinds of induced charges accumulate between adjacent metal patches and thus result in the potential difference between them, which is equivalent to the capacitance *C*. On the other hand, the resulting potential difference forms a current loop by the vias, which is equivalent to the inductance *L*. The calculation formula of capacitance *C* and inductance *L* can be obtained by the following formula [10]:

$$C = \frac{\varepsilon_0(1+\varepsilon_r)W}{\pi}\cosh^{-1}\left(\frac{a}{g}\right) \tag{4.1}$$

$$L = \mu_0 t \tag{4.2}$$

where *W* is the side length of a square metal patch, the slot width between the patches is *g*, *t* represents the substrate thickness, and *a* represents the unit period. In addition to the quasi-static approximation method, the capacitance *C* can also be calculated by the low-frequency transmission line method as [11]

$$C = \frac{\varepsilon_0(1+\varepsilon_r)a}{\pi}\log\left(\frac{2a}{\pi g}\right) \tag{4.3}$$

The resonant frequency and bandgap width of the equivalent *LC* parallel resonant circuit can be calculated as follows:

$$\omega_0 = \frac{1}{\sqrt{LC}} \tag{4.4}$$

$$BW = \frac{\Delta\omega}{\omega_0} = \frac{1}{\eta}\sqrt{\frac{L}{C}} \tag{4.5}$$

where η is the wave impedance of free space, i.e. $\eta = 120\pi$. The impedance can be solved in terms of the inductance *L* and the capacitance *C* as

$$Z_S = \frac{j\omega L}{1-\omega^2 LC} \tag{4.6}$$

At the resonant frequency, the impedance of the *LC* resonant circuit tends to be infinite. Figure 4.3 shows the variation of the impedance with the frequency. It can be observed that the EBG structure exhibits the high impedance characteristic

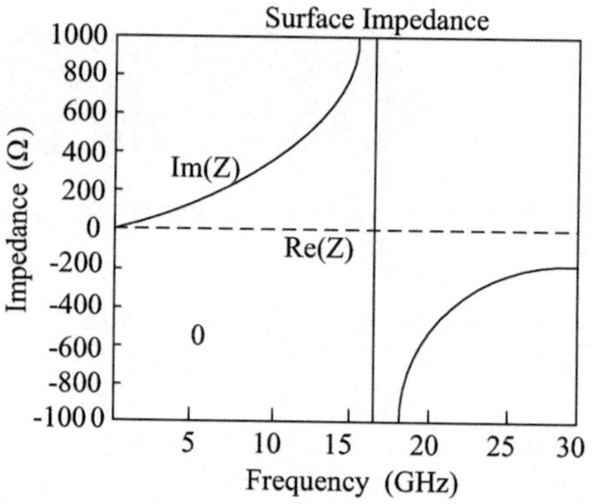

Fig. 4.3 Impedance variation curve with frequency in the equivalent circuit

near the resonant frequency. The equivalent resonant circuit is inductive below the resonant frequencies and capacitive above the resonant frequencies. Therefore, in the band near the resonant frequency, the resonant circuit is equivalent to an open circuit, suppressing the propagation of surface wave and generating a surface wave suppression band gap. Meanwhile, the in-phase reflected phase band gap is achieved because of the band's high surface impedance characteristic. Therefore, the reflected phase of the incoming wave remains unchanged [9].

4.2.1 *Surface Wave Suppression Band Gap*

Surface wave propagation along the surface of a medium is usually characterized by surface current. The EBG structure can suppress the surface wave propagation. In order to elaborate this phenomenon, the surface impedance and dispersion formulas for transverse magnetic (TM) and transverse electric (TE) waves are given, respectively, as

$$Z_S(TM) = \frac{j\alpha}{\omega\varepsilon} \tag{4.7}$$

$$k_{TM} = \frac{\omega}{c}\sqrt{1 - \frac{Z_{TM}^2}{\eta^2}} \tag{4.8}$$

$$Z_S(TE) = \frac{-j\omega\mu}{\alpha} \tag{4.9}$$

$$k_{TE} = \frac{\omega}{c}\sqrt{1 - \frac{\eta^2}{Z_{TE}^2}} \tag{4.10}$$

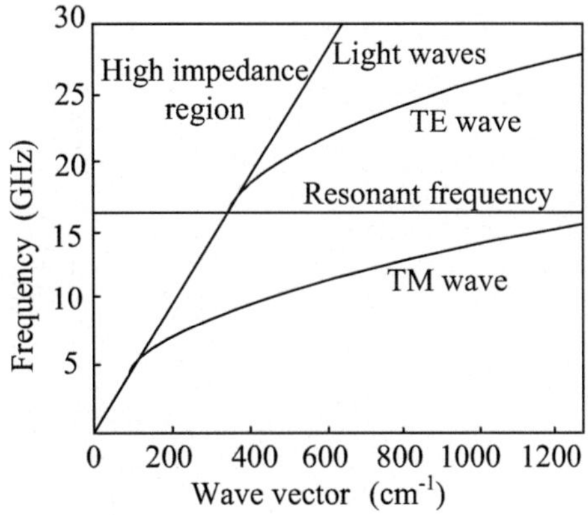

Fig. 4.4 Schematic diagram of surface wave dispersion

By substituting the above formula Eq. (4.6) into Eqs. (4.8) and (4.10), the surface wave dispersion curve can be drawn, as shown in Fig. 4.4. It can be seen from that the TM wave and the TE wave overlap in the low-frequency region, but they cover different frequency band ranges, respectively. Among them, the TM wave mainly propagates in the frequency band less than the resonant frequency, while the TE wave mainly propagates in the frequency band greater than the resonant frequency. In the vicinity of the resonant frequency, neither of the two waves can propagate, thus forming the suppression effect of the surface wave, which is called the surface wave suppression band gap.

4.2.2 EBG Reflective Phase Band Gap

In addition to surface bang gap property, the in-phase reflected phase band gap of the EBG structure is an attractive characteristic. In general, when a wave with the electric field E_{in} and the magnetic field H_{in} normally illuminates on a surface located at $x = 0$, a reflected wave with the electric field E_{re} and the magnetic field H_{re} is generated, as shown in Fig. 4.5. The total field in the incident region with the wave impedance η can be expressed as

$$E(x) = E_{in}e^{jkx} + E_{re}e^{-jkx} \tag{4.11}$$

$$H(x) = H_{in}e^{jkx} + H_{re}e^{-jkx} \tag{4.12}$$

The impedance at the impedance surface $x = 0$ can be expressed as:

$$Z_S = \frac{E(x=0)}{H(x=0)} \tag{4.13}$$

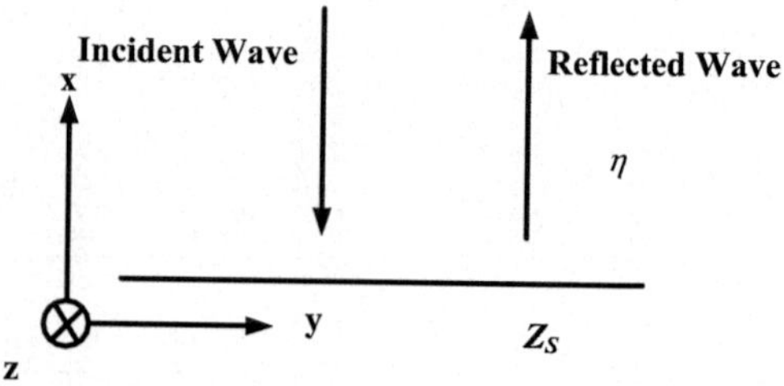

Fig. 4.5 Schematic diagram of a plane wave normal incidence on a plane

The reflection coefficient of the impedance plane is:

$$\Gamma = \frac{Z_S - \eta}{Z_S + \eta} \tag{4.14}$$

Furthermore, the reflection phase can be deduced as follows:

$$\Phi = \mathrm{Im}\left\{\ln\left(\frac{Z_S - \eta}{Z_S + \eta}\right)\right\} \tag{4.15}$$

Equation (4.15) gives the relationship between impedance and reflection phase [10]. It can be seen that when the impedance is infinite, which corresponds to the perfectly magnetic conductor, the reflection phase is 0 and the reflection coefficient is 1. When the impedance is 0, which corresponds to the perfectly electric conductor, the reflection phase is $\pm\ \pi$ and the reflection coefficient is -1. Substituting the impedance of the EBG given by the equivalent circuit (4.6) into (4.15), the variation of the reflection phase with the frequency can be shown in Fig. 4.6. It can be seen that the reflection phase of the EBG changes periodically in the range of $-180° \sim 180°$. An outstanding characteristic of the EBG is the reflection phase of 0°, compared with the metallic reflection phase of 180°. The in-phase reflection phase band gap is defined as a frequency band covering a phase range of $-90° \sim 90°$ [9, 10].

4.2.3 Relationship Between Surface-Wave Suppression Bandgap and Plane-Wave In-Phase Reflection Phase Band

The surface-wave band gap and in-phase reflection phase bands of the mushroom-like EBG structure are mainly determined by geometric parameters. An infinite periodic structure, as shown in Fig. 4.7, was used to analyze the effects of the geometric parameters on the two bands respectively [12]. Consider a mushroom-like EBG structure with the geometric parameters: $W = 0.14\lambda_{6\mathrm{GHz}}$, $g = 0.006\lambda_{6\mathrm{GHz}}$, $t = 0.03\lambda_{6\mathrm{GHz}}$, $r = 0.01\lambda_{6\mathrm{GHz}}$, $\varepsilon_r = 2.65$. Figure 4.8a shows the simulated $k-\beta$ dispersion diagram of the surface modes propagating in the EBG structure. Here $k = \omega/c$ is

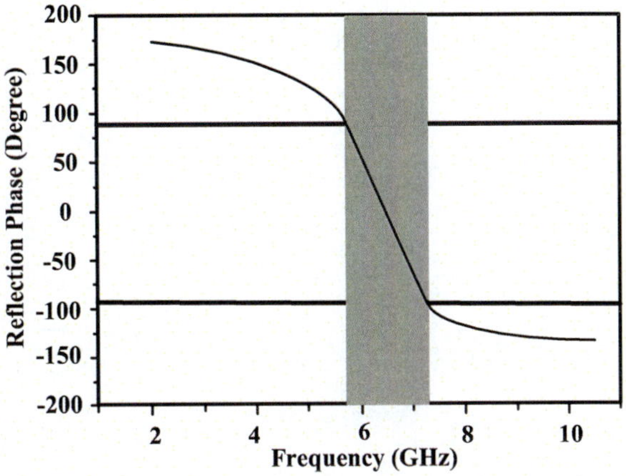

Fig. 4.6 Variation of reflection phase with frequency

the free space wave number, and β is the surface-wave propagation constant. The first (dominant) surface wave mode is the TM mode, which has no cutoff frequency; the second is the TE mode. There was a complete stop-band between the first mode, TM_0, and the second mode, TE_1, in the frequency band 5.771–8.45 GHz. In this figure, Γ, X and M represent the symmetric points of the irreducible Brillouin zone. The $\Gamma - X$ branch represents $\beta_x a/\pi$ with $\beta_y = 0$, the $X - M$ branch represents $\beta_y a/\pi$ with $\beta_x = \pi/a$, and the $M - \Gamma$ branch represents $\beta_x a/\pi$ with $\beta_x = \beta_y$. In this way, we spanned all possible wave vectors in the x and 45° directions, and we also spanned two special points (the minimum and maximum wave vectors) in the other directions. Calculation inside this special Brillouin zone provides sufficient surface-wave bandgap information. Figure 4.8b shows the frequency response of transmission coefficient S_{21}, where both TM and TE surface waves were measured by using a pair of small monopole antennas oriented vertically (TM mode) to the PBG surface and a pair of small shielded loop antennas oriented parallel (TE mode) to the PBG surface, respectively. The simulated surface-wave bandgap was in good agreement with the measured result [13].

The reflection phase band generally depends not only on the EBG structure itself but also on the incident angle and polarization of a plane wave. For normal incidence, the reflection phases are independent of the polarization, which are stable and can be used to identify the surface-wave band gap features. In the model used for calculating the reflection phase of plane waves at normal incidence shown in Fig. 4.7b, the observational plane was set in the far-field zone of the EBG surface. A perfectly matched layer (PML) was positioned to absorb the reflection waves. The scattered E-field on the observational plane was recorded, and the reflection phase at this plane was calculated using

$$\varphi'_{EBG} = \frac{\int_S \text{Phase}(E^S_{EBG})ds}{\int_S \hat{s} \cdot d\vec{s}}, \tag{4.16}$$

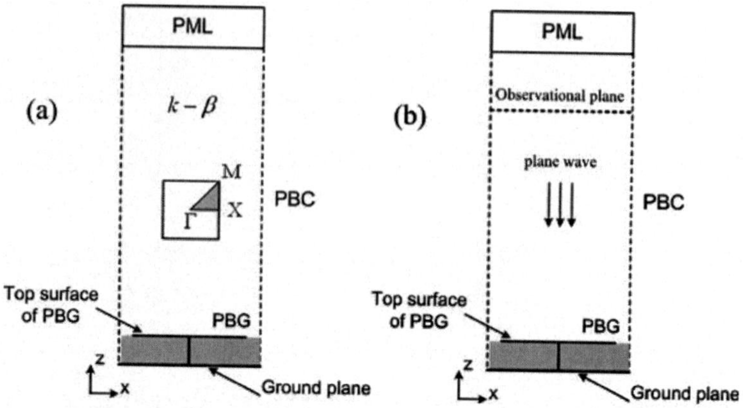

Fig. 4.7 Infinite periodic models of the mushroomlike EBG structure. **a** Model for dispersion diagram in irreducible Brillouin zone. **b** Model for reflection phase of plane wave at normal incidence [13]

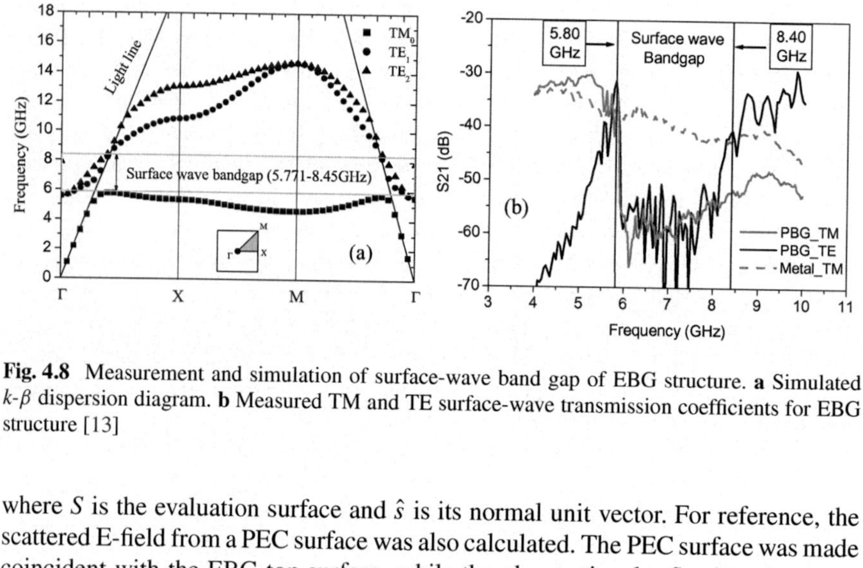

Fig. 4.8 Measurement and simulation of surface-wave band gap of EBG structure. **a** Simulated *k*-β dispersion diagram. **b** Measured TM and TE surface-wave transmission coefficients for EBG structure [13]

where S is the evaluation surface and $\hat{s}$ is its normal unit vector. For reference, the scattered E-field from a PEC surface was also calculated. The PEC surface was made coincident with the EBG top surface, while the observational reflection plane was left the same. A factor of π was added to the phase to account for the reference of the PEC surface, which is known to have a phase-shift of π radians. Specifically,

$$\varphi'_{PEC} = \frac{\int_S \text{Phase}(E^S_{PEC})ds}{\int_S \hat{s} \cdot d\vec{s}} - \pi \tag{4.17}$$

According to (4.16) and (4.17), the reflection phase of the EBG can be obtained as

$$\varphi_{EBG} = \varphi'_{EBG} - \varphi'_{PEC} \tag{4.18}$$

Parametric studies were performed to identify the effects of the EBG structure parameters on the two bands and on their relationship. Figure 4.9a–e show the surface-wave bandgap and the reflection phase of a plane wave normally incident on the EBG surface for different patch widths, gap widths, substrate thicknesses, permittivities, and via radii, respectively. The frequency range between two circles on each curve in Fig. 4.9 represents the position of the surface-wave suppression bandgap of the corresponding EBG structure, which was calculated by simulating dispersion diagrams. The electromagnetic properties and the relationship between the two bands were obtained from these results [13].

(1) The surface-wave band gap was not necessarily related to the in-phase reflection band. The corresponding relationship between the two bands varied with the parameter values, especially for a significant change in the TM surface-wave band edge.
(2) The variations in the characteristics of the two bands with the parameter values were almost the same, meaning that we can predict the effects of EBG parameters on the position and bandwidth of the surface-wave suppression band gap by using their effects on the position and slope of the in-phase reflection band.
(3) The metal vias play an important role in determining the surface-wave suppression bandgap. As shown in Fig. 4.9e and f, although there was no bandgap in the surface-wave suppression for the via-less structure, there was still an in-phase reflection band. The reflection phase characteristics were barely changed when the via radius was much less than the patch width. However, when the radius was increased, the reflection phase band increased and its bandwidth decreased due to the coupling between the metal vias and the gap between adjacent patches. Moreover, the edge of the TM surface-wave band gradually approached the frequency point of 90° reflection phase with an increase in via radius.
(4) The correspondence between the two bands can be effectively adjusted by changing the ratio of the period to the thickness a/t, assume r is kept small. Numerical tests showed that, when $a/t < 2$, the surface-wave band gap was between the ±90° reflection phase frequencies for normal incidence. In practice, it is highly desirable to find a material that simultaneously shows in-phase reflection and surface-wave bandgap in a certain frequency band.

The interaction between the metal patch and the ground plane can be viewed as far-zone interaction when substrate thickness t is large compared to the period a of an EBG structure, even if t and a are small compared to the operating wavelength [11]. This means that we need to consider only the fundamental-mode plane wave between the array and the ground, which leads to the transmission-line formula for the equivalent surface impedance of the EBG structure. However, if t is smaller than a, we need to take into account the effect of higher-order evanescent modes reflected by the ground plane. The effect factor of the evanescent modes can be expressed simply as [14].

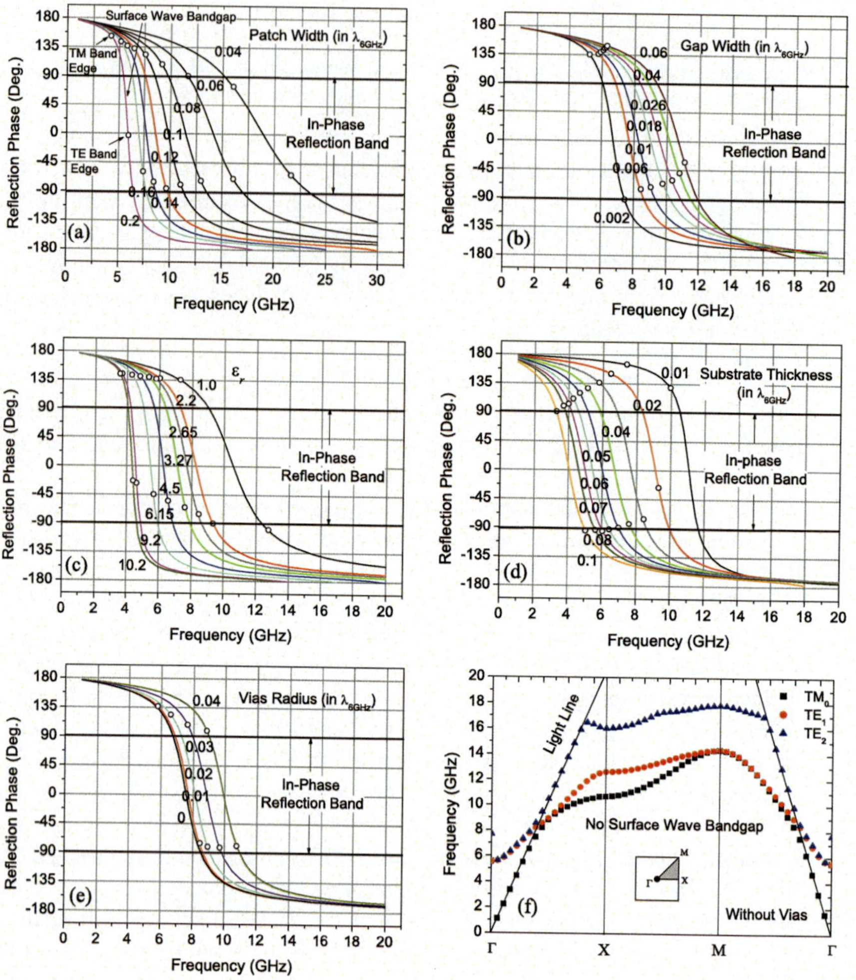

Fig. 4.9 Parametric analyses and corresponding relationships between two bands of EBG structures: **a** effect of changing patch width from $0.04\lambda_{6GHz}$ to $0.20\lambda_{6GHz}$, **b** gap width from $0.002\lambda_{6GHz}$ to $0.06\lambda_{6GHz}$, **c** permittivity from 1.0 to 10.2, **d** substrate thickness from $0.01\lambda_{6GHz}$ to $0.1\lambda_{6GHz}$, and **e** via radius from 0 to $0.04\lambda_{6GHz}$, while keeping other parameters unchanged, **f** dispersion diagram of PBG structure without vias, which indicates that there is no surface-wave suppression band gap in the structure. Note that the frequency region between two circles on each curve represents the position of surface-wave suppression band gap of the corresponding PBG structure [12]

$$\gamma = \frac{2a}{\lambda}\log\left(1 - e^{-\frac{4\pi t}{a}}\right) < 0, (a << \lambda). \tag{4.19}$$

When $a << \lambda$ is not satisfied, the effect factor can be expressed using a convergent series [15]:

$$\gamma = \frac{2a}{\lambda}\sum_{m=1}^{\infty}\left(\frac{1 - e^{-4\pi\frac{t}{a}\sqrt{m^2-(a/\lambda)^2}}}{\sqrt{m^2 - (a/\lambda)^2}} - \frac{1}{m}\right). \tag{4.20}$$

It can be seen that the higher-order evanescent mode influence is negligible if $t \approx a/2$, which is consistent with our numerical results. When $a/t < 2$, the plane-wave in-phase reflection band generally corresponds to the surface-wave suppression band gap. Reviewing the EBG structure consisting of a triangular array of hexagonal patches, as proposed by Sievenpiper [8], with $a/t = 2.54/1.55 \approx 1.639$, we see that the surface-wave band gap edges occur where the reflection phase is equal to ±90°. If a/t is increased (>2), the band edge of a TM surface wave gradually deviates from the frequency point of the +90° reflection phase and enters the out-of-phase region, as shown in Fig. 4.9a and d. The square EBG structure of Yang [16] has $a/t = 0.14/0.04 = 3.5$, and the TM surface-wave band edge is approximately located at the frequency point of the 138.1° reflection phase. The results show that the results of previous investigations are reconcilable and that their primary differences are due to the different corresponding relationships for the two bands of the mushroom-like EBG structures.

(5) The relative permittivity of the substrate used to fabricate the EBG structure plays a less important role in adjusting the correspondence of the two bands, as shown in Fig. 4.9c.

4.2.4 Simulation and Measurement for Surface Wave Suppression

In order to achieve the above analytical analysis, simulation verification is required. The following two simulation methods for surface wave suppression band gap are mainly introduced.

4.2.4.1 Waveguide Transmission Method

The waveguide transmission method uses a rectangular waveguide cavity to calculate the transmission between two ports of the waveguide to characterize the surface wave suppression of EBG structures [17]. As shown in Fig. 4.10, the corresponding simulation model is built in the rectangular waveguide air cavity with the PEC boundary conditions at the upper and bottom planes and the PMC boundary conditions at the left and right planes. The wave ports are placed at the front and back planes. The EBG structure with more than four unit cells is placed in the waveguide. Note that the EBG structure should be located at least λ/8 apart from the waveguide port to ensure the continuity of the field. The S_{21} reflecting the transmission between two ports is investigated, and the frequency band range for $S_{21} \leq -10$ dB corresponds to the surface wave band gap.

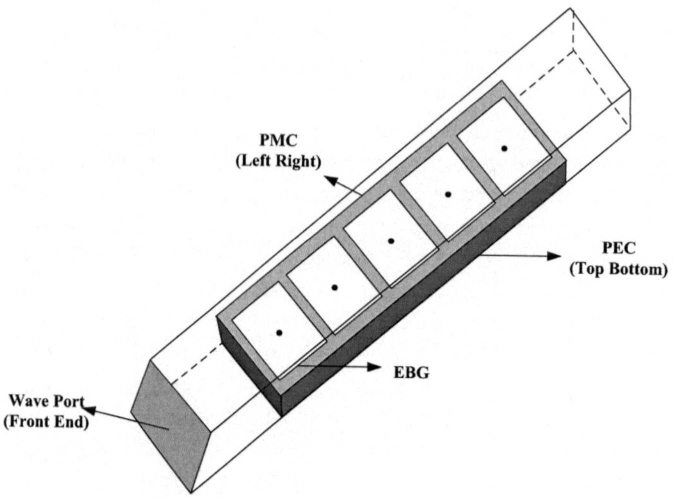

Fig. 4.10 Waveguide based simulation model

4.2.4.2 Suspension Microstrip Method

Figure 4.11 shows the simulation model of the suspended microstrip line. In the model, the substrate of the suspended microstrip line is the same as that of the EBG, and the microstrip line is located above the EBG [18]. The lumped port excitation is implemented at both ends of the microstrip line to simulate surface wave propagation. The S_{21} is simulated, and the frequency band of S_{21} less than −10 dB corresponds to the surface wave band gap.

The above two methods are often used in the simulation of EBG surface wave band gap, but they have their own advantages and disadvantages. The waveguide transmission method is simple and easy-to-design, and thus the less simulation time is required. The suspended microstrip line method is convenient and usually used in the actual measurement and processing. However, both methods cannot accurately simulate the surface wave transmission between the two ports. To obtain accurate results, the dispersion mode method can be used.

Fig. 4.11 Suspended microstrip based simulation model

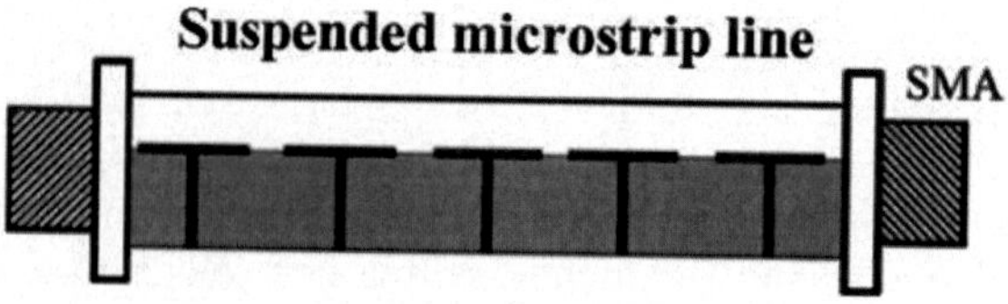

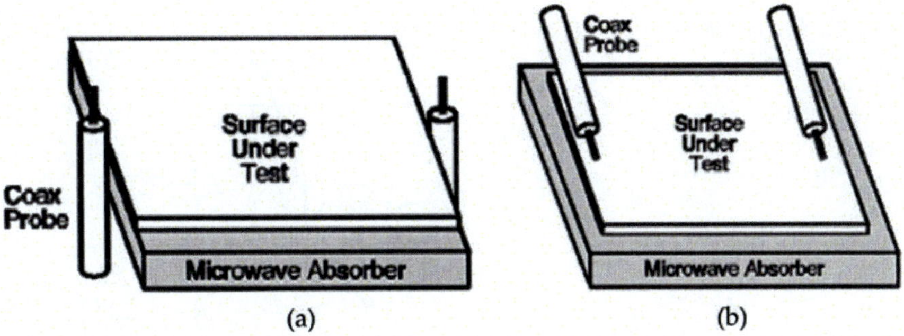

Fig. 4.12 Diagram of actual measuring device for surface wave band gap. **a** TM surface wave. **b** TE surface wave [9]

4.2.4.3 Measurement of Surface Wave Band Gap

As shown in Fig. 4.12, two coaxial probes as the receiver and the transmitter are placed on both sides of the EBG structure, respectively, to investigate the transmission S_{21} between the two coaxial probes under the influence of the EBG structure. It should be noted that two kinds of modes, i.e., TM wave and TE wave, are excited to measure the surface wave band gap. As shown in Fig. 4.12, the excitation mode of the coaxial probe placed vertically is TM mode, while the excitation mode of the coaxial probe placed horizontally corresponds to TE mode [9].

4.2.5 Simulation and Measurement for Reflective Phase Characteristics

4.2.5.1 Infinite Period Model

The widely used simulation model for the reflection phase gap of the EBG is an infinite period model using the Floquet port, as shown in Fig. 4.13. The master–slave boundaries are set to enclose the unit cell of the EBG. The S_{11} is calculated, and the band for S_{11} phase covering -90° ~ +90° is defined as the in-phase reflected phase band gap [13].

4.2.5.2 Measurement of Reflection Phase Characteristics

As shown in Fig. 4.14, the measurement schematic diagram of the reflected band gap is shown. The whole test environment is carried out in the microwave anechoic chamber to avoid the influence of electromagnetic scattering. At one side of the chamber, two horn antennas are used as transceivers and a wave-absorbing material

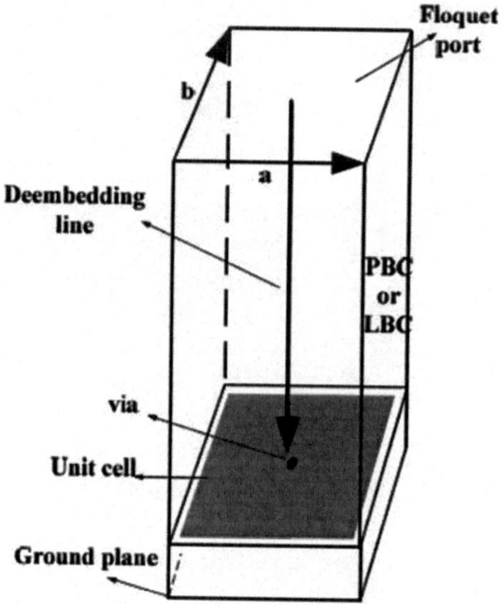

Fig. 4.13 Simulation model of reflective phase characteristics

is placed between the antennas to reduce the mutual coupling between them. At the other side, a sample to be measured is placed. In order to ensure the plane wave incidence, the distance between the transceivers and the EBG meets the far field condition of the antenna. The whole test needs to be completed in two steps. First, a metallic plate with the same size as the EBG is chosen as the reference to measure the phase received by the receiver. Second, the phase of the EBG is measured. The reflection phase of the EBG can be obtained by subtracting the phase obtained in step 1 from the phase obtained in step 2 and adding the 180° caused by the reference metallic plate to the resultant phase [9], as shown in Eq. (4.18).

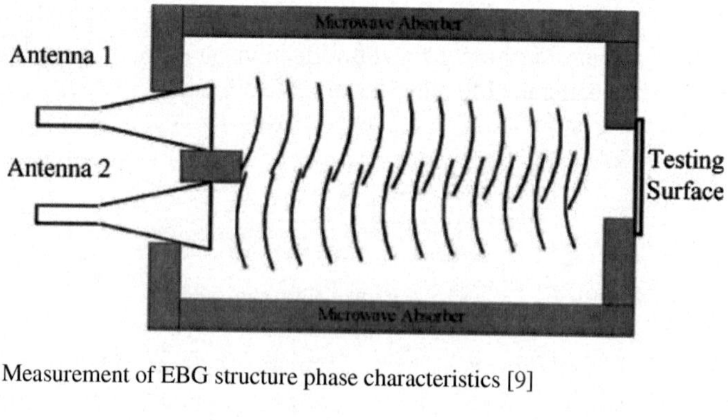

Fig. 4.14 Measurement of EBG structure phase characteristics [9]

4.3 EBG Power-Ground Plane for High-Speed Circuit Noise Suppression

With the trends of mix-signal system integration, high-speed microprocessors, radio frequency (RF) circuits, memory, sensors, and optical devices, etc., are required to be integrated into a tight module, which is known as "system on package". The ground bounce noise (GBN) [19], also known as simultaneous switching noise (SSN), on the power/ground planes is becoming one of the major bottlenecks for designing high-speed circuits in multilayer PCB. As the systems operate toward a higher frequency range, the GBN will excite the resonance modes of the parallel-plate waveguide structure between power and ground planes, which behave as radial waves and travel outwards from the noise source to the edges of the circuit board. Some portion of the energy is reflected inwards affecting the signal integrity (SI), and some part is radiated into the free space causing electromagnetic interference (EMI) problems [20]. The EBG structures [21–30] have been applied to mitigate GBN in power/ground plane pairs. The introduction of effective bridges or fractal high-impedance topologies into the EBG structures can give rise to the suppression of the ultra-wideband (UWB) GBN.

4.3.1 L-Shaped Bridges and Slits (LBS)

As shown in Fig. 4.15a, the unit cell of the EBG power plane consists of one square patch with four narrow slits inserted at the boundary of the patch, and four L-shaped bridges on each side of the patch. The unit cell of the LBS-EBG and its corresponding notations of parameters are shown in Fig. 4.15b. The proposed L-shaped bridges not only improve the inductance between two neighboring patches so that they can suppress the noise at lower frequencies but also keep signal quality acceptably well [31]. The narrow slits inserted at the boundary of the patch change the flow paths of currents. Due to the slits perturbation, the resonant mode of the square patch will be split so that they can suppress the noise at higher frequencies. As a result, the geometry of the slit and bridge can be designed properly based on the requirement of suppression bandwidth.

To demonstrate the effectiveness of the LBS-EBG power plane, we consider a two-layer PCB with the dimension 90 × 150 mm consisting of 3 × 5 unit cells. Figure 4.16 shows the measured and simulated $|S_{21}|$ for the designed LBS-EBG power/ground planes. The insertion loss of the reference board with both power and ground planes being solid is also presented in this figure for comparison. An ultra-wideband suppression for $|S_{21}| \leq -40$ dB covering 432 MHz–15 GHz is observed.

Furthermore, a four-layer PCB structure with two transmission lines passing from the top layer (SG1) to the bottom layer (SG2), with two via transitions between the power plane (PWR) and solid ground plane (GND) is considered, as shown in Fig. 4.17. A scheme of locally-embedded EBG elements is adopted to improve SI. The

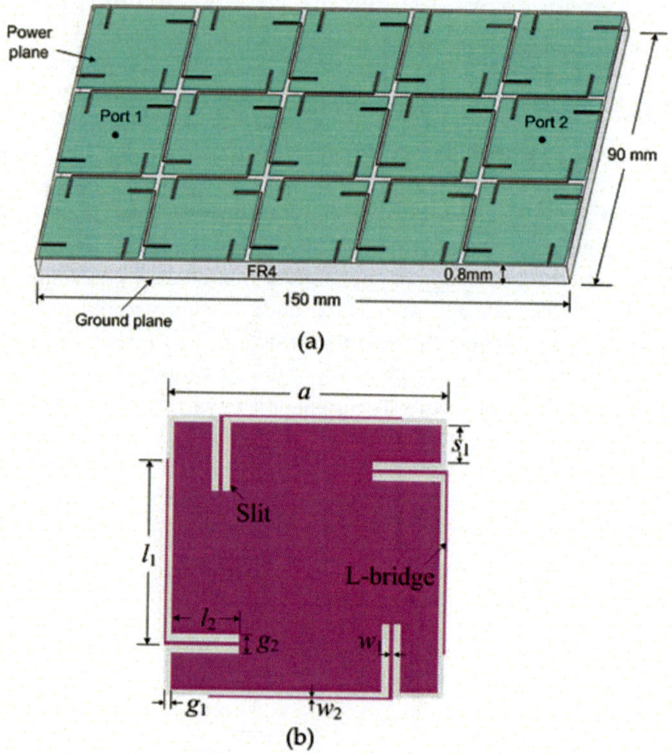

Fig. 4.15 The proposed EBG power plane. **a** 3D view. **b** Unit cell with the L-bridges with slits and dimensions [31]

Fig. 4.16 Comparison of $|S_{21}|$ between the LBS-EBG board and the reference board by the numerical simulation and the measurement [31]

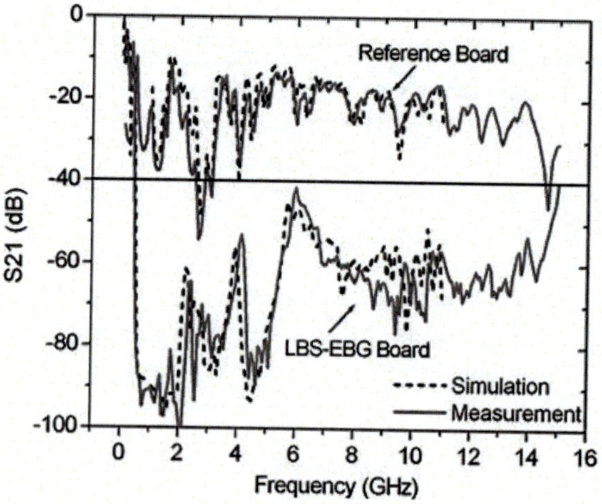

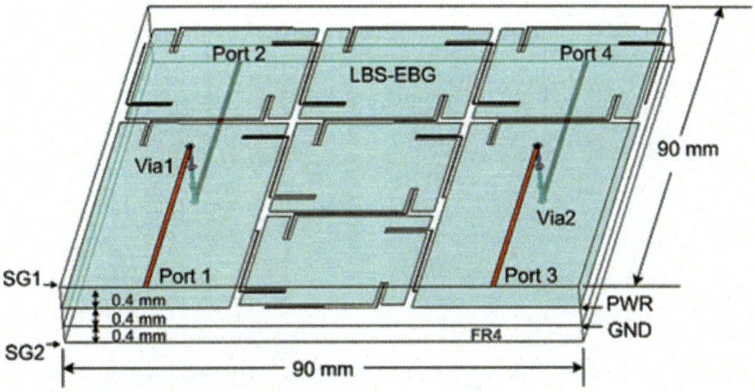

Fig. 4.17 Four-layer PCB structure with transmission lines transition between the locally embedded LBS-EBG power plane and solid ground plane [31]

local regions of power plane corresponding to signal traces are still kept continuous, and other regions are replaced with LBS-EBG elements. All local regions can be connected by L-bridges. The traces are designed as 50 Ω for the transmission signal. The prototype of the corresponding multilayer structure is fabricated, as shown in Fig. 4.18. The four-port S-parameters for the two kinds of boards are simulated and measured. Figure 4.19a and b show the comparison of measured S-parameters for the reference board and the locally-embedded LBS-EBG board. The signal trace which passes through via will couple energy to the power bus and propagate between the power and ground planes with multi-reflections. While GBN occurs and propagates outward along the power bus, some power energy is reflected, and some power will be coupled to the signal trace by through-hole-via. In Fig. 4.19b, S_{31}, which is almost the same as S_{41}, shows that the GBN coupling can be effectively suppressed by using LBS-EBG locally embedded in the power plane. The characteristics of S_{11} and S_{21} show that the energy suppressed is mainly transported from port 1 to port 2, as expected. Figures 4.20 and 4.21 show the normalized E-field distribution on each layer of the reference and LBS-EBG boards, respectively. It can be seen from Fig. 4.20 that the parallel-plate waveguide structure has strong resonance modes between power and ground planes excited by GBN in the reference board. But only a local resonance mode exists in the LBS-EBG board due to a locally-embedded scheme. The GBN is hardly coupled to signal traces in other blocks constituting system, as shown in Fig. 4.21.

4.3.2 Fractal Topologies

As is well known, the principle for a coplanar EBG structure to effectively suppress the noise when applied in the power/ground plane pairs is that it can offer a low-impedance path for noise, acting as an LC low pass filter. The capacitance of the unit cell depends on the topology of the metal patch and the gap between two neighboring

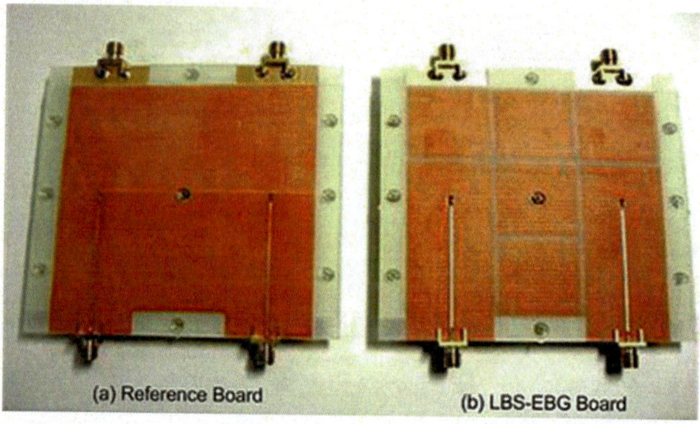

Fig. 4.18 Fabricated multilayer PCBs. **a** Reference board. **b** LBS-EBG board [31]

unit cells, while the inductance is due to the bridges connecting the adjacent unit cells. By using the fractal topology, the capacitance component is fixed, and the lower bandgap onset frequency mainly depends on the inductance component. In this scenario, the meander lines and S-bridges can be selected as inductance components because they can increase the effective inductance of EBG efficiently. Based on the fractal topology concepts, two novel designs are proposed here in Fig. 4.22 [32].

The first design is composed of a fractal high-impedance topology with one iteration and meander lines. The simulated and measured $|S_{21}|$ for the first design are shown in Fig. 4.23a. The measured −49 dB suppression bandwidth between port 1 and port 2 ranges from 202 MHz to 20 GHz. Figure 4.23b also gives other measured S-parameters. Port 1 is chosen as the receiving port, and the noise excitation is located at other different test ports. An apparent stopband can be observed from 194 MHz to 20 GHz between port 1 and port 3 with the insertion loss lower than −49 dB. The measured −40 dB suppression bandwidth between port 1 and port 4 ranges from 246 MHz to 20 GHz, and between port 1 and port 5 is from 215 MHz to 20 GHz.

The second design is proposed using a fractal high-impedance topology with two iterations and slots and S-bridges. The simulated and measured S-parameters for the second design and reference board between port 1 and port 2 are plotted in Fig. 4.24a. It can be seen that the second design also can mitigate the noise effectively. The measured −38 dB suppression bandwidth between port 1 and port 2 is from 249 MHz to 20 GHz. The lower edges of −40 dB suppression bandwidth for the first and second designs are 190 and 251 MHz, respectively, as shown in Fig. 4.24b.

A four-layer structure with single-ended traces is given to evaluate the impact of the power plane with the first design on the SI, as shown in Fig. 4.25. The 60 mm trace passes from the first layer to the fourth layer and back to the first layer, and the trace is designed to have 50 Ω characteristic impedance. There are two via transitions

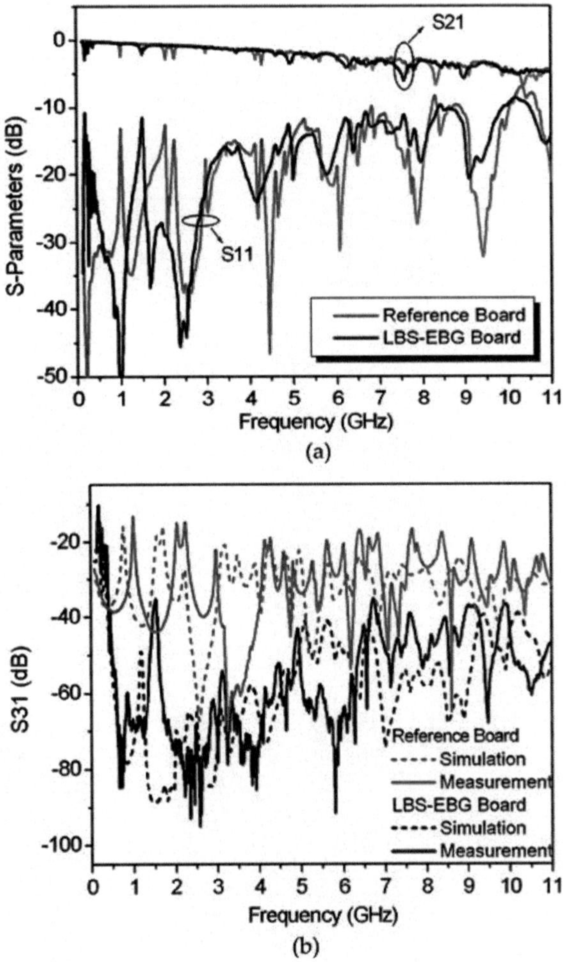

Fig. 4.19 Comparison of S-parameters of multilayer reference board and LBS-EBG board. **a** Measured S_{11} and S_{21}. **b** Simulated and measured S_{31} [31]

along the signal path. The second layer is the 9-cell power plane with the first design; the third layer is the solid ground plane. The S-parameters are simulated by the full-wave electromagnetic simulator, and the eye diagrams are finally generated in the ADS environment by launching a pattern source of random bit flows, nonreturn to zero (NRZ), coded at 1.95 GHz. The bit-sequence swing is 1 V and the nominal rise/fall time is 120 ps. The eye diagrams for the reference board and the board with first design as the power plane are investigated and shown in Fig. 4.26. It can be observed that for the reference board, MEO = 0.464 V and MEW = 509.4 ps; and for the 9-cell board with the first design, MEO = 0.331 V and MEW = 484.4 ps. Compared with the reference board, the degradation of the MEO and MEW for the 9-cell board with the first design is 28.7 and 4.9%.

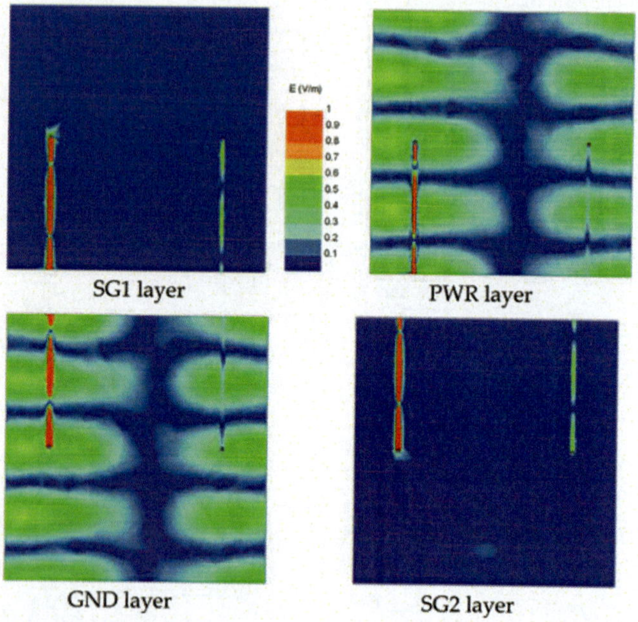

Fig. 4.20 Normalized E-field distribution on each layer of reference board at 3.3 GHz [31]

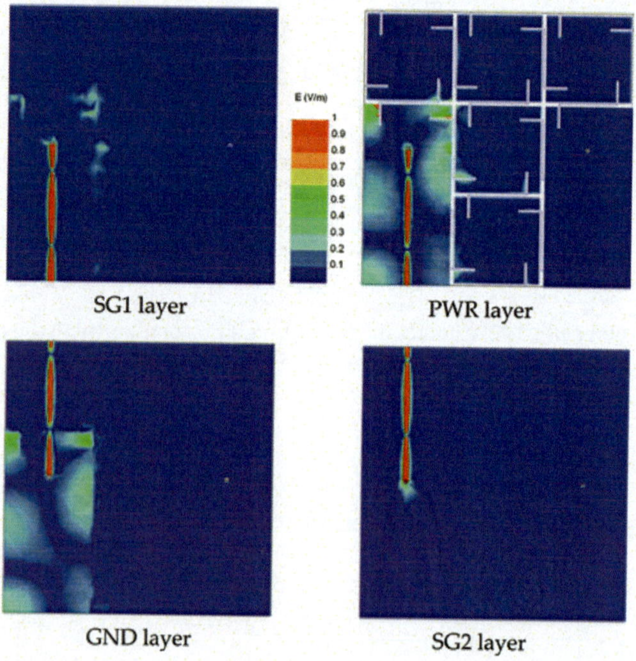

Fig. 4.21 Normalized E-field distribution on each layer of LBS-EBG board at 3.3 GHz [31]

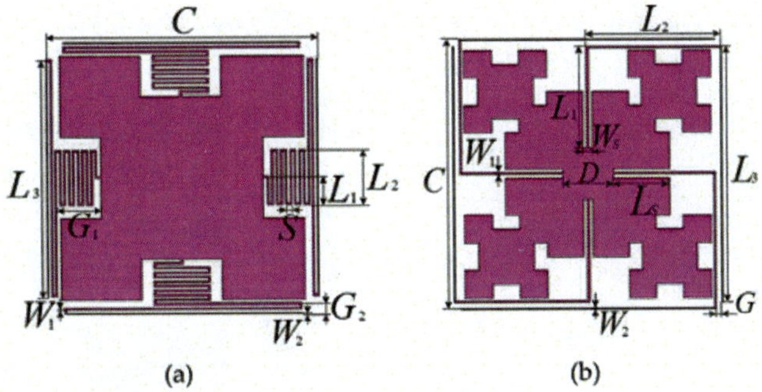

Fig. 4.22 Schematic of the proposed designs. **a** First design. **b** Second design [32]

Fig. 4.23 GBN suppression behavior of the first design for the noise excitation located at different test ports. **a** Simulated and measured $|S_{21}|$. **b** Other measured S parameters [32]

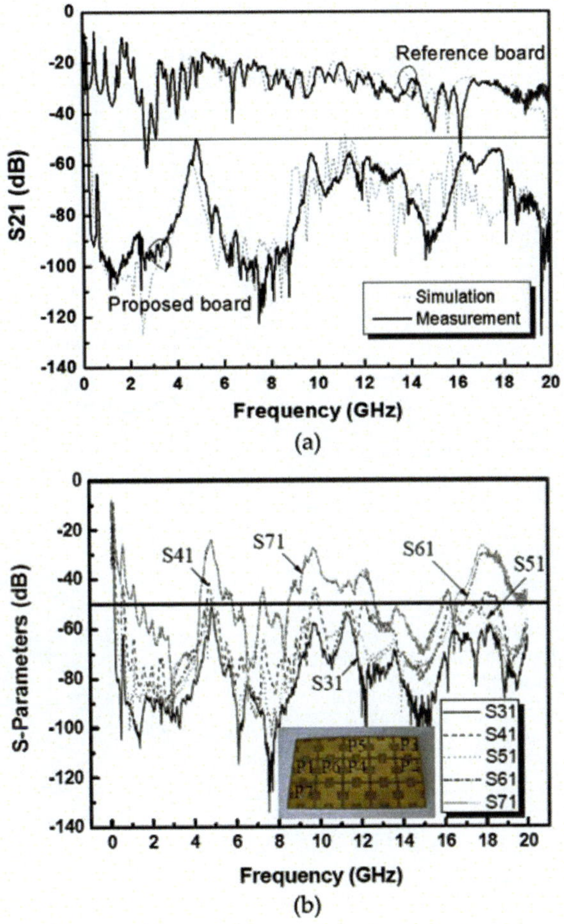

Fig. 4.24 Measured GBN suppression behaviors. **a** Measured and simulated $|S_{21}|$ of the second design. **b** Measured lower frequencies characteristics of $|S_{21}|$ for the reference board and two design boards [32]

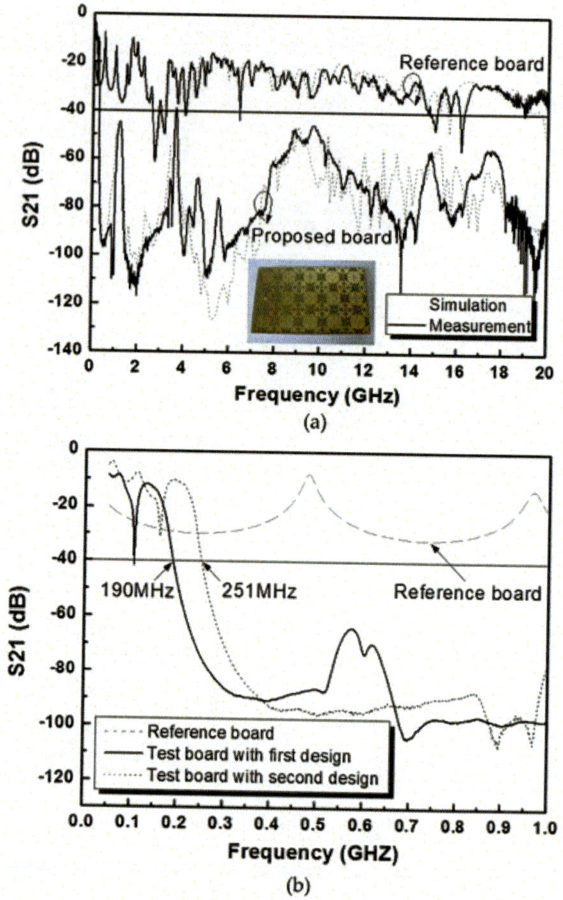

Fig. 4.25 Four-layer structure with transmission line transient between the 9-cell power plane with first design and the solid ground plane [32]

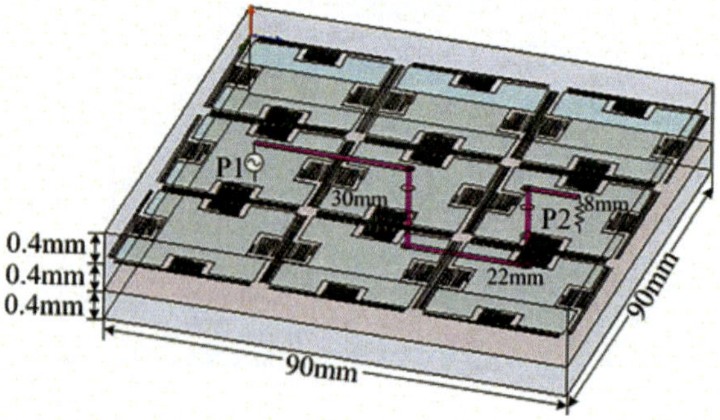

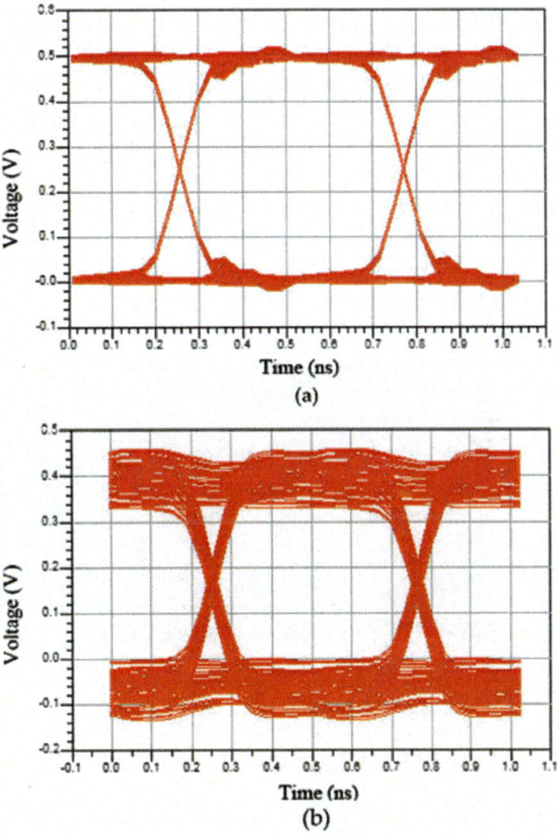

Fig. 4.26 Simulated eye diagrams. **a** Reference board with single-end trace. **b** 9-cell first design board with single-end trace [32]

Except the fractal structures with the meander lines and S-bridges designed above, the Sierpinski curve based fracture can be used in the EBG structure to achieve the ultra-wideband SSN suppression [33]. As shown in Fig. 4.27, the proposed design consists of a Sierpinski curve with one iteration and meander lines. The test board with the proposed design has been fabricated and measured simultaneously with a reference board whose power and ground planes are solid shown in Fig. 4.28. Figure 4.29 shows the $|S_{21}|$ comparison between the reference board and the designed board. An apparent stopband between port 1 and port 2 is observed within the frequency range of 263 MHz–19 GHz. Compared with the reference board, about −30 dB enhancement of the S_{21} coupling level can be achieved for the EBG power plane with the first design. Other measured S-parameters, including $|S_{11}|$ and $|S_{14}|$ are plotted in Fig. 4.30. The −39 dB suppression bandwidth between port 1 and port 4 is from 316 MHz to 19 GHz. The −48 dB suppression bandwidth between port 1 and port 3 is from 250 MHz to 19 GHz.

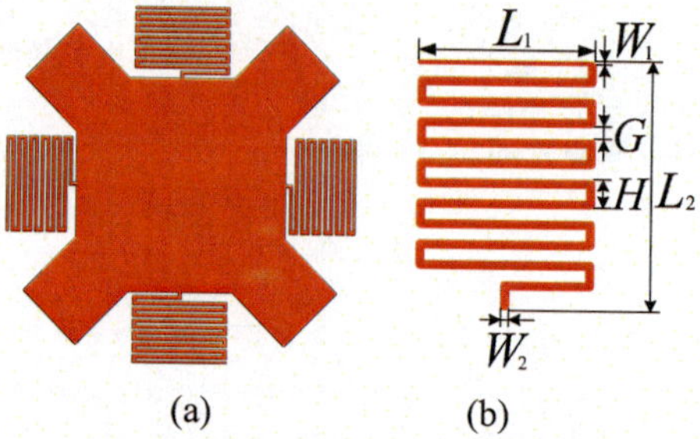

Fig. 4.27 Design using Sierpinski curve with one iteration and meander lines. **a** Topology. **b** Parameters of meander lines [33]

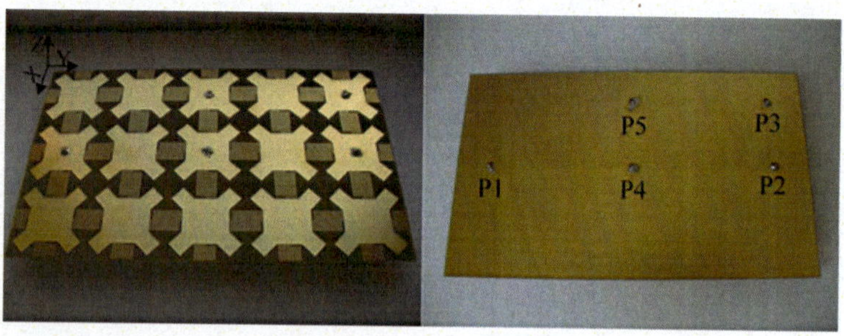

Fig. 4.28 Prototype of the test board with the proposed design and the reference board [33]

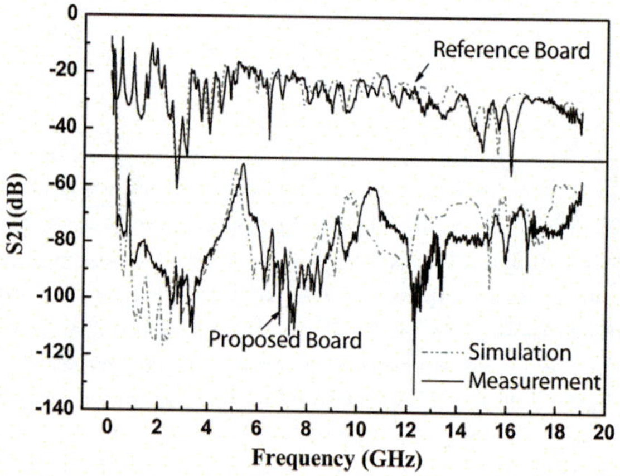

Fig. 4.29 Comparison of $|S_{21}|$ between the test board with the proposed design and the reference board obtained by simulation and measurement [33]

4.4 EBG Substrates and Superstrates for Antennas Designs

The EBG structures have been proven to be an effective measure to reduce the surface wave coupling and found wide applications in the antenna communities. The EBG structures have been widely used as a reflector, because the two aspects of its band gap characteristics are surface wave suppression characteristic and in-phase reflection

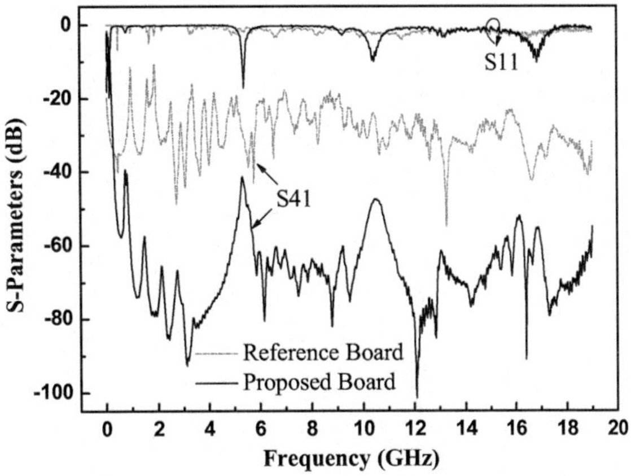

Fig. 4.30 Comparison of measured S-parameters including $|S_{11}|$ and $|S_{14}|$ between the test board with the proposed design and the reference board [33]

characteristic. Surface wave suppression characteristics can improve antenna gain and reduce the back lobe, the radar scattering cross section, and the mutual coupling between the array antenna elements, etc. In-phase reflection characteristic is used in the design of artificial magnetic conductors to reduce the profile of the antenna. In this section, some EBG structures are designed for antenna applications [34–37].

4.4.1 Application of EBG into Waveguide End-Slot Phased Array (WESPA) Antenna

The WESPA is a phased array with rectangular waveguides loaded by radiating slots on an infinite conducting ground plane [37], which is depicted in Fig. 4.31. The elements are spaced in a rectangular grid and each slot is assumed to be fed by the dominant TE_{10} mode phased accordingly with the desired scanning angle. The distance between two adjacent elements is dx in the x-direction and dy in the y-direction. There are $M \times N$ elements assumed in the arrays. The size of the waveguide is denoted by $a \times b$. The slot is L in length, W in width, and t in depth, as shown in Fig. 4.32. The $(m, n)^{\text{th}}$ element of the array is phased for the scanning angle (θ, φ) by $\exp\left[j(n-1)(2\pi/\lambda)dx\sin\theta\cos\phi + j(m-1)(2\pi/\lambda)dy\sin\theta\sin\phi\right]$. Scanning is controlled in the H-plane where $\varphi = 0$ or $\varphi = \pi$ and in the E-plane where $\varphi = \pm\pi/2$. The geometry of the problem can be divided into three regions: waveguide region a, cavity region b, and half-free space region c, as shown in Fig. 4.32. Applying continuity conditions to the tangential components of the magnetic field on the slot surfaces, we obtain a set of integral equations

$$\frac{1}{j\omega\mu}\left(k^2+\frac{d^2}{dx^2}\right)\left[\iint\limits_{s_i^1}(-M_i^1)G_a^{xx}(\vec{r},\vec{r}\,\prime)ds+\iint\limits_{s_i^1}(-M_i^1)G_b^{xx}(\vec{r},\vec{r}\,\prime)ds+\iint\limits_{s_i^2}M_i^2G_b^{xx}(\vec{r},\vec{r}\,\prime)ds\right]=H_{it}^{in}(\vec{r}) \quad (4.21)$$

$$\frac{1}{j\omega\mu}\left(k^2+\frac{d^2}{dx^2}\right)\left[\iint\limits_{s_i^1}M_i^1G_b^{xx}(\vec{r},\vec{r}\,\prime)ds+\iint\limits_{s_i^2}(-M_i^2)G_b^{xx}(\vec{r},\vec{r}\,\prime)ds+\sum_{j=1}^{M\times N}\iint\limits_{s_j^2}(-M_j^2)G_c^{xx}(\vec{r},\vec{r}\,\prime)ds\right]=0 \quad (4.22)$$

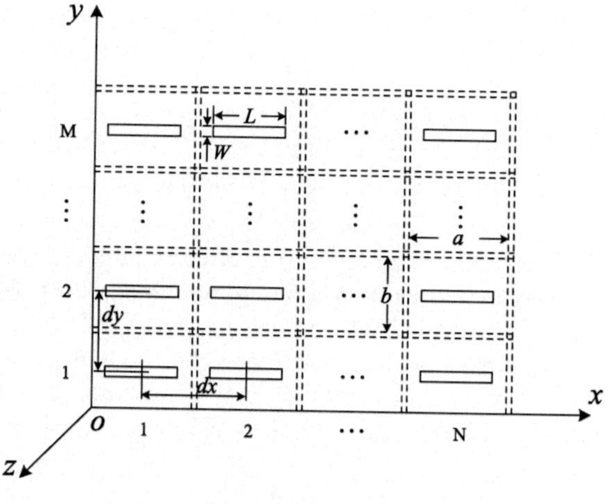

Fig. 4.31 Phased array of rectangular waveguides loaded by radiating slots (waveguide end-slot) of $M \times N$ elements [37]

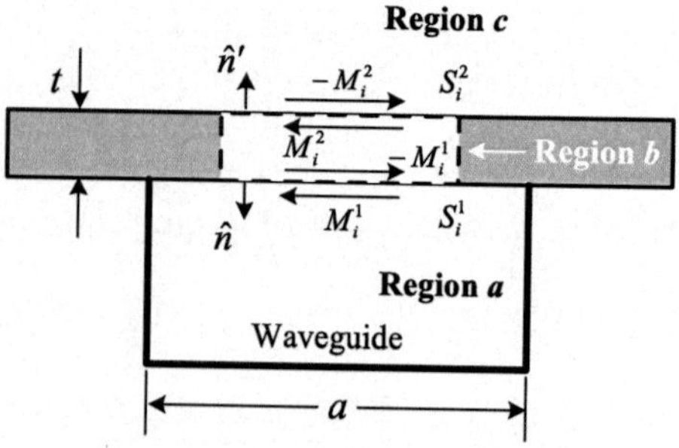

Fig. 4.32 Equivalent magnetic current distribution on a slot [37]

Here M_i^1 and M_i^2 (where $i = 1, 2, \cdots M \times N$) are the equivalent magnetic currents on the surfaces of S_i^1 and S_i^2, G_a^{xx}, G_b^{xx}, and G_c^{xx} represent the xx-components of the dyadic Green's functions in regions a, b, and c, respectively. $H_{it}^{in}(\overrightarrow{r})$ is the x-component of the domain mode TE_{10} on S_i^1. The unknown surface magnetic currents can be solved by using the method of moments (MoM). Figure 4.33a and b show the active reflection coefficient of the WESPA center element versus the scan angle in the E-plane and H-plane, respectively. It can be observed that the slot array for the uniform case has a large value for $|S_{11}|$, a total reflection (scan blindness) for an infinite slot array, near 29° in the E-plane.

In order to overcome the "blind" scan directions of the WESPA, a mushroom-like EBG shown in Fig. 4.34a is introduced. Assume the center operating frequency of the WESPA antenna is 9.375 GHz. Figure 4.34b shows the numerically simulated $k - \beta$ dispersion diagram of the surface modes propagating in the EBG structure. The first (dominant) surface wave mode is TM_0 which has no cutoff frequency, and the second

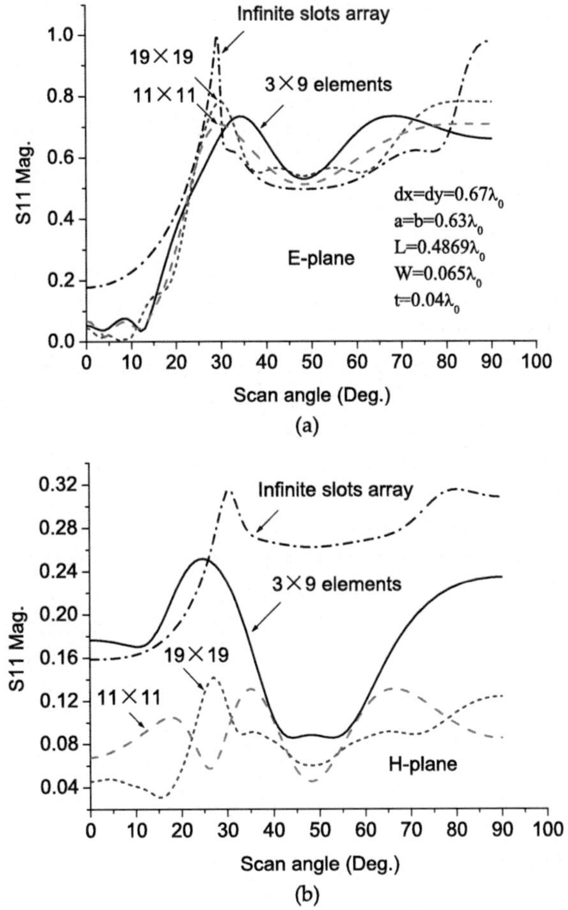

Fig. 4.33 Reflection coefficients of the WESPA center element versus scan angle in **a** E-plane, and **b** H-plane [37]

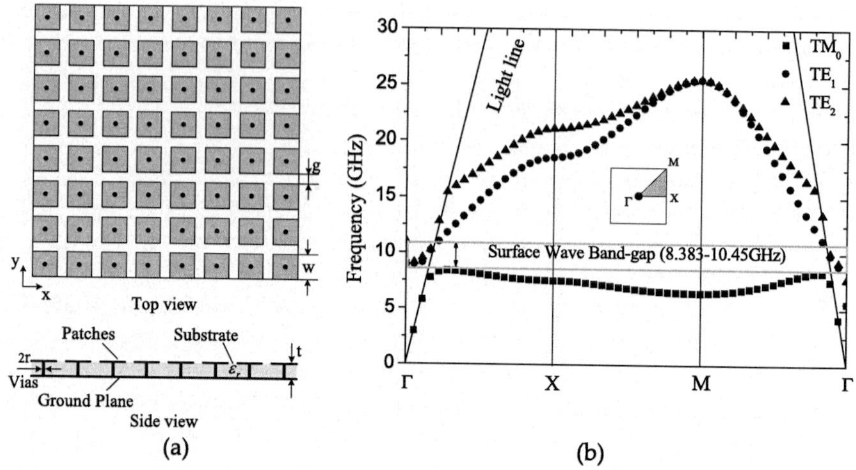

Fig. 4.34 **a** Geometry of a mushroom-like EBG structure. **b** Simulated surface wave band structure [37]

surface wave mode is TE_1. One can observe a complete stopband between the first mode TM_0 and the second mode TE_1 within the frequency band 8.383–10.45 GHz. In this figure, Γ, X and M stand for symmetric points in the irreducible Brillouin zone. The $\Gamma - X$ branch represents $\beta_x a/\pi$ with $\beta_y = 0$, the $X - M$ branch represents $\beta_y a/\pi$ with $\beta_x = \pi/a$, and $M - \Gamma$ branch represents $\beta_x a/\pi$ with $\beta_x = \beta_y$.

The EBG structure is integrated with the waveguide end-slot phased array, as shown in Fig. 4.35. Note that an important tradeoff is the space *D* between the EBG element and the slot element. The coupling of EBG and the slot decreases with the increase of space *D*. On the other hand, there should be sufficient EBG elements (at least two periods are needed to achieve the surface wave suppression phenomenon) when the array space is determined (the grating lobe is suppressed in the visible region). Here space *D* is reasonably chosen to locate four EBG elements between the slots in the E-plane.

The variation of the reflection coefficient S_{11} of the infinite array with EBG (4 unit cells) versus the scan angle is demonstrated in Fig. 4.36. The total reflection of the blindness in the original array is removed, which makes the array work near the blindness angle. Thus the scan ability of the array can be improved.

Figure 4.37a shows the normalized distribution of the surface electric field on the metal ground of the original array. And the case of the array with EBG is shown in Fig. 4.37b when the excitation and the normalized standard are the same. A comparison of two cases illustrates that EBG structures in the E-plane obviously reduce the coupling of the slots.

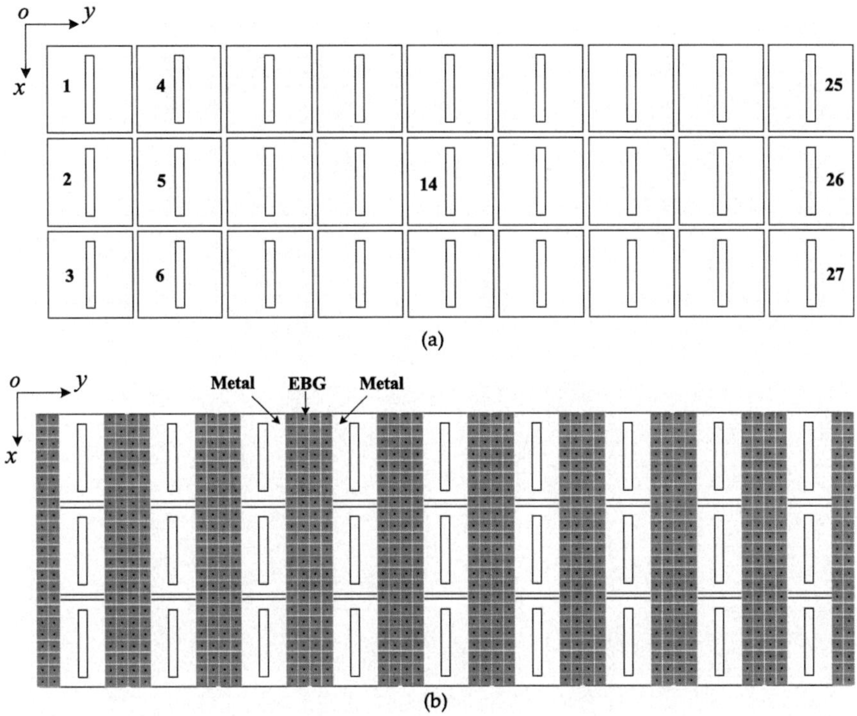

Fig. 4.35 Top view of 3 × 9 waveguide end slots phased array, **a** original array (without EBG), and **b** EBG array integrated with 4 unit cells [37]

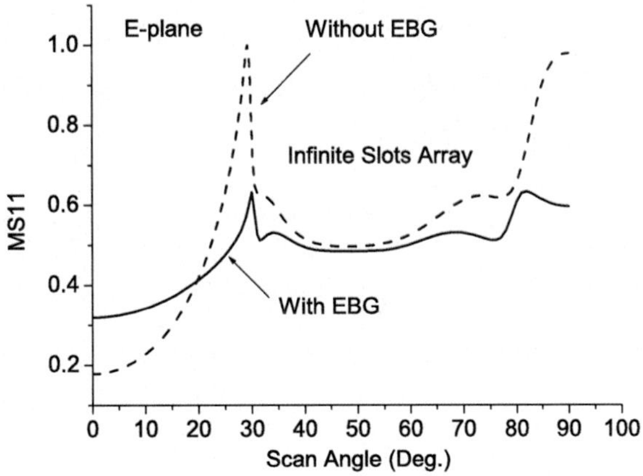

Fig. 4.36 A comparison of reflection coefficient of element in the infinite phased array with and without EBG [37]

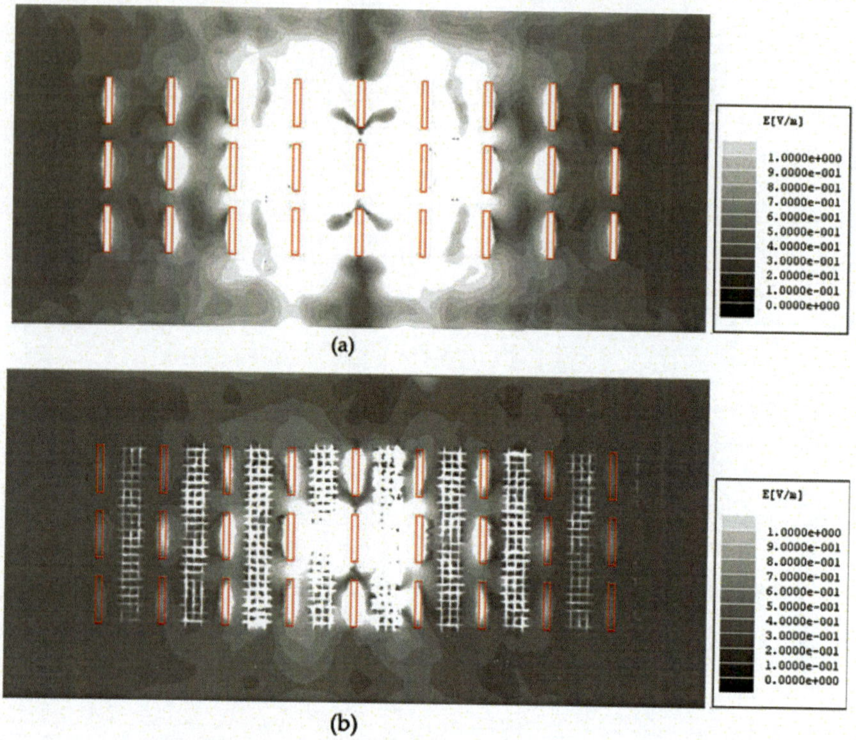

Fig. 4.37 Distribution of the surface electric field on the array ground, **a** without EBG, and **b** with EBG integrated with 4 unit cells [37]

4.4.2 *EBG as a Superstrate for Phased Array Antenna [38–40]*

In addition to the isolation improvement, the EBG as a superstrate can be used to enhance the radiation. As shown in Fig. 4.38, the EBG array unit consists of two parts: a rectangular mushroom structure array composed of 12 unit cells arranged in a 3×4 layout printed on a dielectric substrate, and a slot antenna with the slot on the ground plane orthogonal to the coplanar waveguide (CPW) feed-line [40]. The array antenna is composed of eight antenna units arranged in a straight line to form a 1×8 linear array, as shown in Fig. 4.39.

There are two operating frequencies of the proposed array. The lower frequency band is from 3.1 to 3.7 GHz with -10 dB bandwidth of 17.1%, while the upper frequency band is from 5.2 to 6.0 GHz with -10 dB bandwidth of 15.8%, as shown in Fig. 4.40. The isolation between two adjacent ports in the two working frequency bands are below -10 and -15 dB, respectively, as shown in Fig. 4.41.

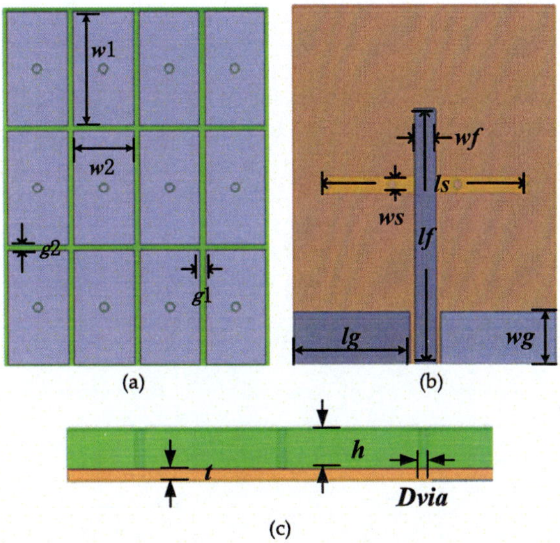

Fig. 4.38 Geometry of the proposed dual-band array unit, **a** front view, **b** rear view, and **c** side view [40]

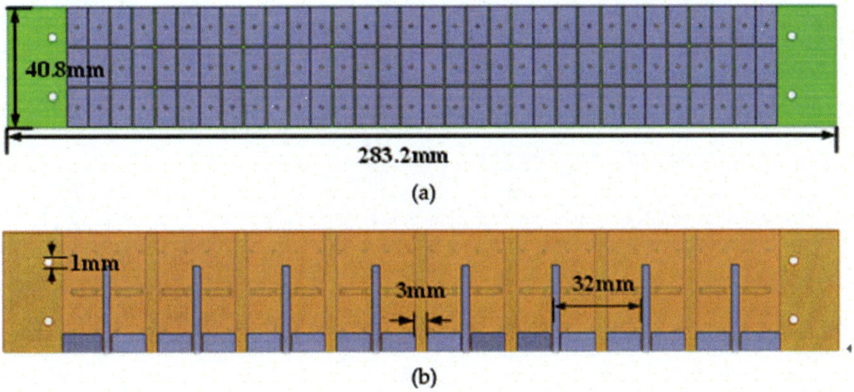

Fig. 4.39 Geometry of the proposed dual-band metamaterial-based phased array antenna. **a** Front view. **b** Rear view

In order to elaborate the dual-band characteristic, a rectangular mushroom-like superstrate antenna unit is studied. Figure 4.42 shows the simulated impedance characteristics for the proposed antenna and other cases without top patches or vias, respectively. As illustrated in Fig. 4.42, the antenna unit can work in two WLAN bands. Without the vias, the antenna can only cover the 5 GHz WLAN band and cannot work in the 2.4 GHz band. When the rectangular patches are removed, the antenna only operates in a narrow band of around 5.7 GHz. It is worthwhile to point out that the rectangular mushroom structure with vias is critical to generate the 2.4 GHz operational band. The dispersion diagrams of the rectangular mushroom unit cell for two directions in the cases with/without vias are analyzed and

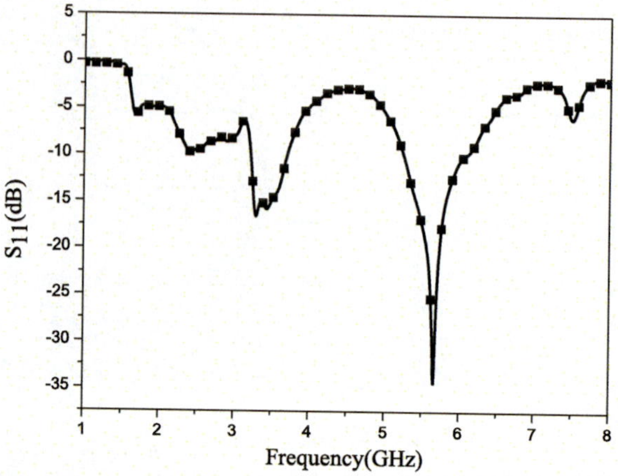

Fig. 4.40 Simulated reflection coefficient of the proposed dual-band metamaterial-based phased array antenna

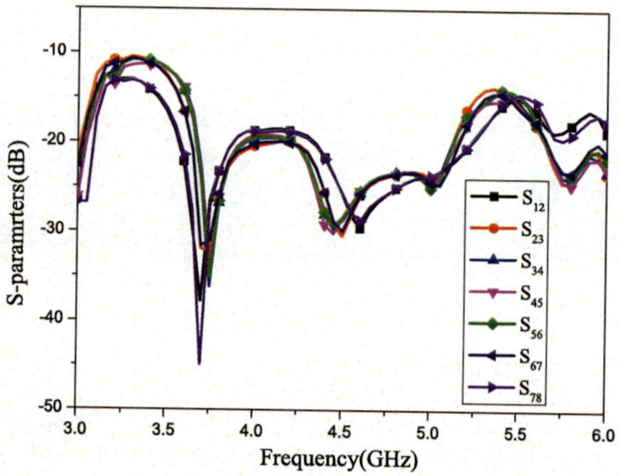

Fig. 4.41 Simulated isolation of the proposed dual-band metamaterial-based phased array antenna

shown in Fig. 4.43. It can be seen that there is a zeroth-order resonance (ZOR) mode in *x*- and *y*-directions with vias, respectively, but the ZOR modes will disappear if without vias. The utilization of rectangular patches can improve the bandwidth of the 2.4 GHz band by overlapping the two ZOR modes. In addition, the TM_{10} and antiphase TM_{20} resonance modes of the CRLH mushroom structure in the high-frequency band (5 GHz band) can be simultaneously excited for wideband characteristics, whether the metallic vias exist or not [41].

The prototype of the proposed array antenna is fabricated, and three microstrip phase shifters with phase differences of 0°, 90° and 130° are designed to connect with the antenna prototype, as shown in Fig. 4.44. As shown in Fig. 4.45, the impedance bands of the antenna connected to the three kinds of phase shifters for $|S_{11}| \leq -10$ dB are from 3.1 to 3.68 GHz and 5.3–6.0 GHz, meaning that the proposed array antenna operates at the bands of 3.4 GHz and 5.6 GHz simultaneously. Figure 4.46 shows the measurement and simulation patterns of the array antenna at two operating

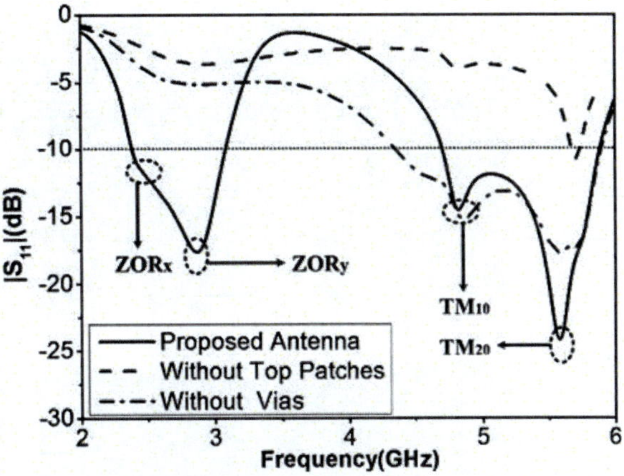

Fig. 4.42 Comparison of $|S_{11}|$ curves of the EBG superstrate antenna unit in different cases [40]

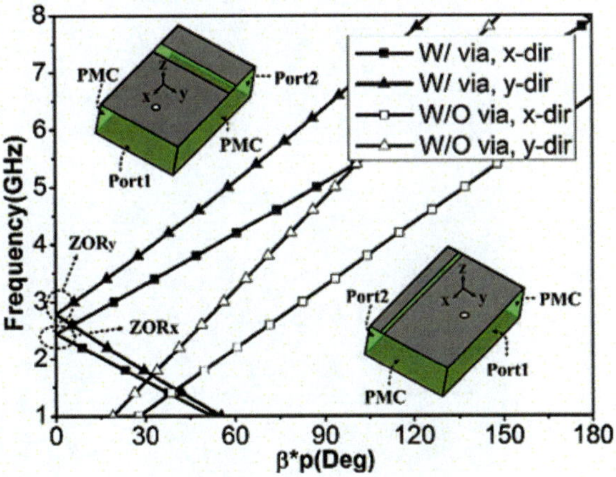

Fig. 4.43 Dispersion diagrams of the mushroom unit cell for $x-$ and $y-$ polarization directions in the cases with/without vias [40]

frequencies of 3.4 GHz and 5.6 GHz. As can be seen from the figure, in the lower band the proposed phase array can achieve the radiation from the boresight direction to the endfire direction. In the higher band, the scanning range of the proposed phase array covers $-36° \sim +36°$. It is noted that with the increase of scanning angle, the beamwidth and the sidelobe level increases and the gain decreases.

4.4.3 EBG as a Superstrate for Circular Polarization Antenna

Figure 4.47 depicts a low-profile circularly polarized structure, which is composed of two parts: one is the 3 × 4 rectangular patches metasurface array and the other is a planar slot coupling antenna with a slanted slot embedded on the ground plane [42]. To integrate the EBG superstrate with the slot coupling antenna, the coplanar

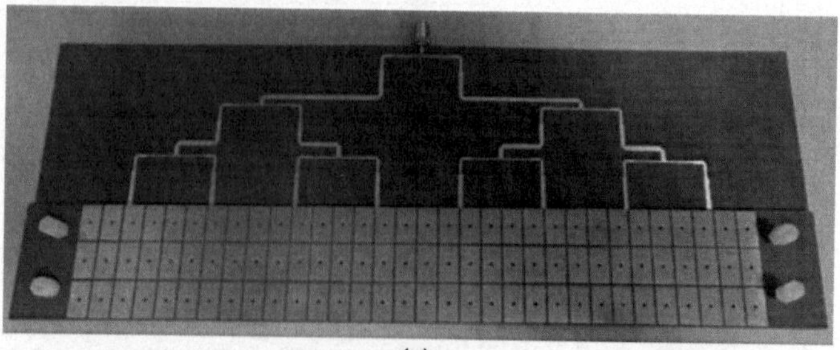

(a)

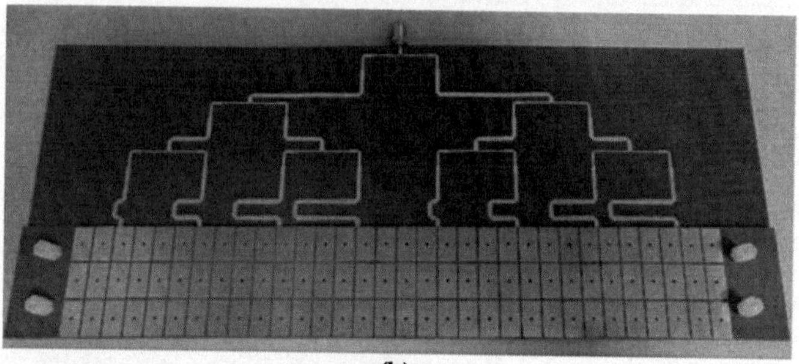

(b)

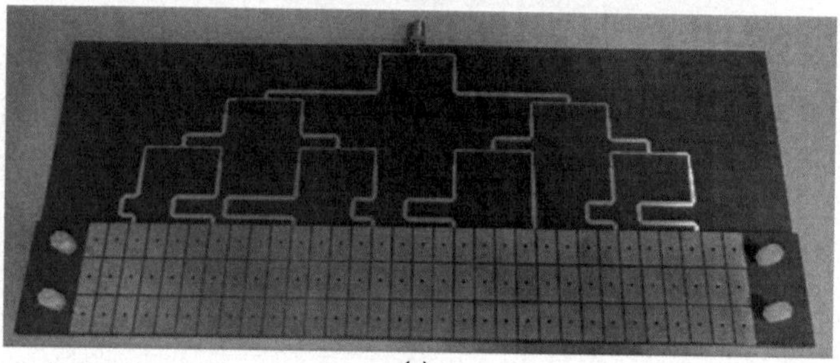

(c)

Fig. 4.44 Fabricated prototype of the proposed dual-band array antenna with EBG superstrate. **a** Phase shifters with phase differences of 0°. **b** Phase shifters with phase differences of 90°. **c** Phase shifters with phase differences of 130°

waveguide (CPW) feed-line was adopted in this design. The polarization-dependent rectangular patches and inclined slot coupling are designed to produce the circularly polarized wave.

To illustrate why circular polarization radiation can be generated by the proposed antenna, we examine how the surface current distribution on the EBG varies with time. Figure 4.48 shows surface current distributions of the proposed antenna at 5.2 GHz for three different time phases (ωt), from 0° to 90°, with an interval of 45°. At ωt = 0°, the dominant surface current flows in $-y$ axis direction. At ωt = 45°, the dominant surface current flows in the diagonal direction of $-x$ and $-y$ axis. When ωt = 90°, the dominant $-x$ directed current is observed. As the time changes, the

Fig. 4.45 Measured reflection coefficient of the proposed dual-band superstrate array antenna

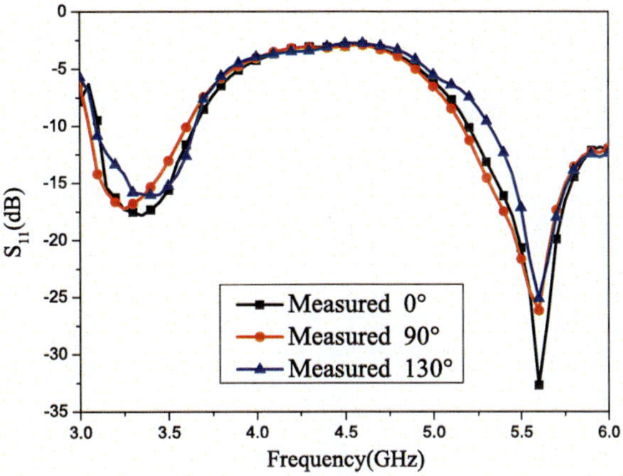

surface currents turn in a clockwise direction. Hence the polarization characteristic is the left-hand circular polarization (LHCP) in the $+ z$ direction.

The proposed antenna is fabricated, as shown in Fig. 4.49, and then measured by an Agilent N9918A vector network analyzer. Measured and simulated S_{11} are shown in Fig. 4.50a. The measured impedance bandwidth ($|S_{11}| < -10$ dB) is 1.7 GHz, about 33.7% from 4.2 to 5.9 GHz. Figure 4.50b shows the measured axial ratio (AR) and realized gain in the boresight direction of the proposed antenna. The 3 dB AR bandwidth is wide, from 4.9 to 5.9 GHz, about 18.5%. The average gain level of 5.8 dBic can be obtained.

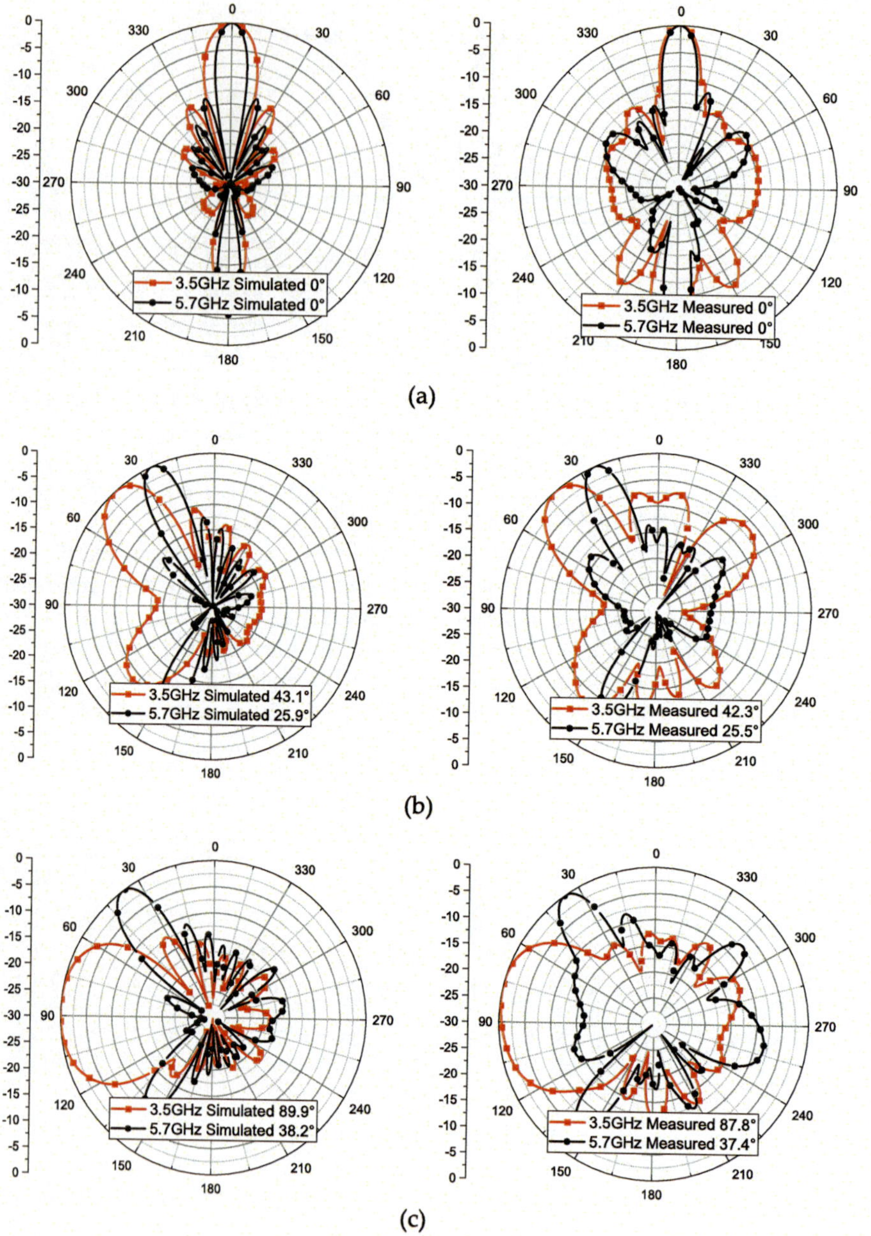

Fig. 4.46 Simulated and measured normalized radiation patterns at 3.5 and 5.7 GHz at different phases

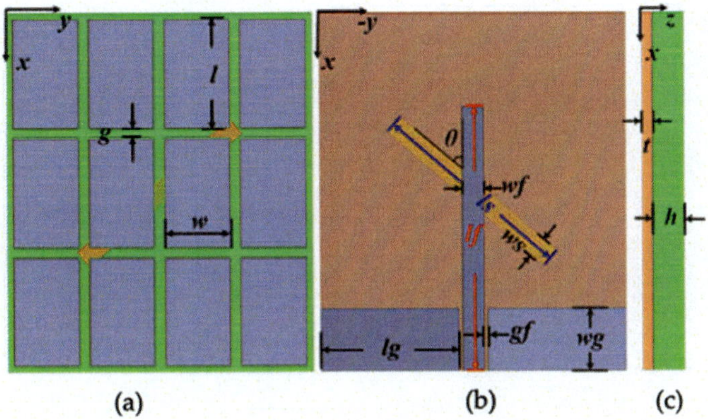

Fig. 4.47 Geometry of the proposed antenna, **a** top view, **b** rear view, and **c** side view [42]

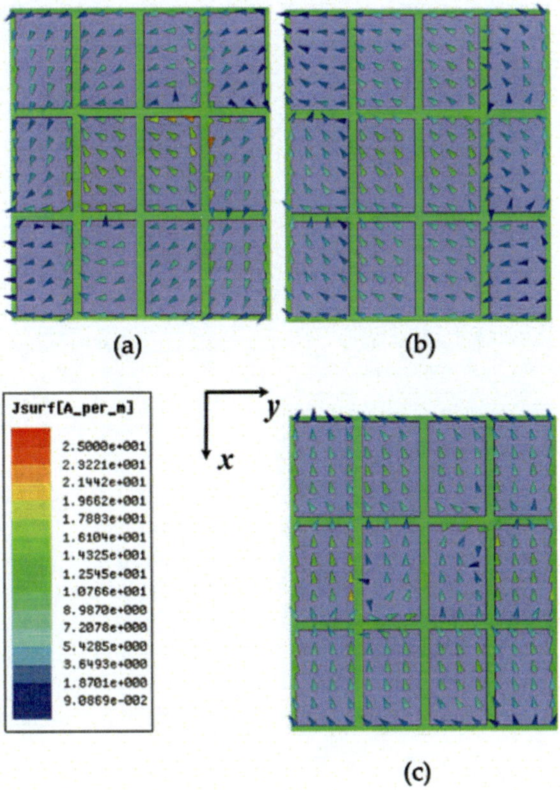

Fig. 4.48 Simulated surface current distributions of the proposed antenna at 5.2 GHz for different time instants, **a** ωt = 0°, **b** ωt = 45°, and **c** ωt = 90° [42]

Fig. 4.49 Photograph of the fabricated EBG superstrate antenna [42]

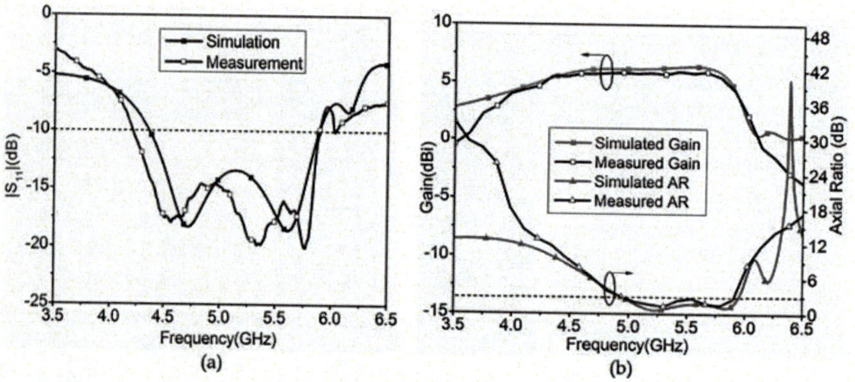

Fig. 4.50 Simulated and measured characteristics of the proposed antenna, **a** $|S_{11}|$, and **b** axial ratio and gain characteristics [42]

4.5 Conclusion

In this chapter, the theoretical aspects of the electromagnetic bandgap, including surface-wave suppression and the in-phase reflection phase, have been elaborated. With the $k - \beta$ dispersion diagram and in-phase reflection phase curve, the relationship between the surface-wave suppression and the in-phase reflection phase has been clarified. Some simulation and measurement models for the two characteristics have been presented. The EBG-based noise suppression in high-speed circuits and the antenna designs with the EBG as the substrate and superstrate have been given.

References

1. E. Yablonovitch, Inhibited spontaneous emission in solid state physic and electronics. Phys. Rev. Lett. **58**, 2059–2062 (1987)
2. S. John, Strong localization of photons in certain disordered dielectric superlattices. Phys. Rev. Lett. **58**, 2486–2489 (1987)
3. B.A. Munk, *Frequency Selective Surface and Grid Array* (Wiley, New York, 1995)
4. V. Veselago, The electrodynamics of substances with simultaneously negative values of ε and μ. Sov. Phys. Usp. **10**, 509–514 (1968)
5. J.B. Pendry, Negative refraction makes a perfect lens. Phys. Rev. Lett. **85**(18), 3966–3969 (2000)
6. R.A. Shelby, D.R. Smith, S. Schultz, Experimental verification of a negative refractive index of refraction. Science **292**, 77–79 (2002)
7. F.R. Yang, K.P. Ma, Y. Qian, T. Itoh, A uniplanar compact photonic-bandgap (UC-PBG) structure and its applications for microwave circuit. IEEE Trans. Microw. Theor. Tech. **47**, 1509–1514 (1999)
8. D. Sievenpiper, L. Zhang, R.F.J. Broas, N.G. Alexopolus, E. Yablonovitch, High-impedance electromagnetic surfaces with a forbidden frequency band. IEEE Trans. Microw. Theor. Tech. **47**, 2059–2074 (1999)
9. D. Sievenpiper, *High-impedance electromagnetic surfaces*. Ph. D. Dissertation, Dep. Elect. Eng. Univ. California at Los Angeles, Los Angeles, CA, (1999)
10. N. Engheta, R.W. Ziolkowski, *Metamaterials Physics and Engineering Explorations* (IEEE Press, New Jersey, 2006)
11. S.A. Tretyakov, C.R. Simovski, Dynamic model of artificial reactive impedance surfaces. J. Electromagn. Waves Appl. **17**(1), 131–145 (2003)
12. R. Remski, Analysis of PBG surfaces using Ansoft HFSS. Microw. J. **43**(9), 190–198 (2000)
13. L. Li, Q. Chen, Q. Yuan, C. Liang, K. Sawaya, Surface-wave suppression band gap and plane-wave reflection phase band of mushroomlike photonic band gap structures. J. Appl. Phys. **103**, 023513 (2008)
14. S.A. Tretyakov, in *Analytical Modeling in Applied Electromagnetics* (Artech House, Norwood, MA, 2003)
15. J.R. Wait, On the theory of an antenna with an infinite corner reflector. Can. J. Phys. **32**(6), 365–371 (1954)
16. F. Yang, Y. Rahmat-Samii, Reflection phase characterizations of the EBG ground plane for low profile wire antenna applications. IEEE Trans. Antennas Propagat. **51**, 2691–2703 (2003)
17. F. Yang, Y. Rahmat-Samii, Polarization-dependent electromagnetic band gap (PDEBG) structures: designs and applications. Microw. Opt. Technol. Lett. **41**(6), 1153–1158 (2004)
18. Y. Horii, M. Tsutsumi, Harmonic control by photonic bandgap on microstrip patch antenna. IEEE Microw. Guided Wave Lett. **9**(l), 13–15 (1999)
19. S. Van Den Berghe, F. Olyslager, D. De Zutter, J. De Moerloose, W. Temmerman, Study of the ground bounce caused by power plane resonances. IEEE Trans. Electromagn. Compat. **40**(2), 111–119 (1998)
20. T.L. Wu, S.T. Chen, J.N. Huang, Y.H. Lin, Numerical and experimental investigation of radiation caused by the switching noise on the partitioned dc reference planes of high speed digital PCB. IEEE Trans. Electromagn. Compat. **46**(1), 33–45 (2004)
21. T. Kamgaing, O.M. Ramahi, A novel power plane with integrated simultaneous switching noise mitigation capability using high impedance surface. IEEE Microw. Wireless Compon. Lett. **13**, 21–23 (2003)
22. T.L. Wu, Y.Y. Lin, C.C. Wang, S.T. Chen, Electromagnetic bandgap power/ground planes for wideband suppression of ground bounce noise and radiated emission in high-speed circuits. IEEE Trans. Microw. Theory Tech. **53**(9), 2935–2942 (2005)
23. T.L. Wu, C.C. Wang, Y.H. Lin, T.K. Wang, G. Chang, A novel power plane with super-wideband elimination of ground bounce noise on high speed circuits. IEEE Microw. Wireless Compon. Lett. **15**(3), 174–176 (2005)

24. X.H. Wang, B.Z. Wang, Y.H. Bi, W. Shao, A novel uniplanar compact photonic bandgap power plane with ultra-broadband supprsssion of ground bounce noise. IEEE Microw. Wireless Compon. Lett. **16**(5), 267–268 (2006)
25. J. Qin, O.M. Ramahi, Ultra-wideband mitigation of simultaneous switching noise using novel planar electromagnetic bandgap structures. IEEE Microw. Wireless Compon. Lett. **16**(9), 487–489 (2006)
26. M.S. Zhang, Y.S. Li, C. Jia, L.P. Li, J. Pan, A double-surface electromagnetic bandgap structure with one surface embedded in power plane for ultra-wideband SSN suppression. IEEE Microw. Wireless Compon. Lett. **17**(10), 706–708 (2007)
27. S.H. Joo, D.Y. Kim, H.Y. Lee, A S-bridged inductive electromagnetic bandgap power plane for suppression of ground bounce noise. IEEE Microw. Wireless Compon. Lett. **17**(10), 709–711 (2007)
28. L.F. Shi, G. Zhang, M.-M. Jin, S. Chen, X.J. Hu, Novel subregional embedded electromagnetic bandgap structure for SSN suppression. IEEE Trans. Compon. Pack. Manufact. Technol. **6**(4), 613–621 (2016)
29. L.F. Shi, Z. Wei, C.R. Wang, EBG combined isolation slots with a bridge on the ground for noise suppression. Int. J. Electron. **103**(10), 1726–17355 (2016)
30. L.F. Shi, D.L. Zhou, Selectively embedded electromagnetic bandgap structure for suppression of simultaneous switching noise. IEEE Trans. Electromagn. Compat. **56**(6), 1370–1376 (2014)
31. L. Li, Q. Chen, Q. Yuan, K. Sawaya, Ultrawideband suppression of ground bounce noise in multilayer PCB using locally embedded planar electromagnetic bandgap. IEEE Antennas Wirel. Propag. Lett. **8**, 740–743 (2009)
32. Y. He, L. Li, C.H. Liang, EBG structures with fractal topologies for ultra-wideband ground bounce noise suppression. J. Electromagn. Waves Appl. **24**(10), 1365–1374 (2010)
33. Y. He, L. Li, H. Zhai, X. Dang, C.H. Liang, Q.H. Liu, Sierpinski space-filling curves and their application in high-speed circuits for ultrawideband SSN suppression. IEEE Antennas Wirel. Propag. Lett. **9**, 568–571 (2010)
34. L. Zhang, J.A. Castaneda, N.G. Alexopoulos, Scan blindness free phased array design using PBG materials. IEEE Trans. Antennas Propaga. **52**(8), 2000–2007 (2004)
35. Z. Iluz, R. Shavit, R. Bauer, Microstrip antenna phased array with electromagnetic bandgap substrate. IEEE Trans. Antennas Propaga. **52**(6), 1446–1453 (2004)
36. Y.Q. Fu, N.C. Yuan, Elimination of scan blindness in phased array of microstrip patches using electromagnetic bandgap materials. IEEE Antennas Wirel. Propaga. Lett. **3**, 63–65 (2004)
37. L. Li, C.H. Liang, C.H. Chan, Waveguide end-slot phased array antenna integrated with electromagnetic bandgap structure. J. Electromagn. Waves Appl. **21**(2), 161–174 (2007)
38. W. Liu, Z.N. Chen, X.M. Qing, Metamaterial-based low-profile broadband mushroom antenna. IEEE Trans. Antennas Propag. **62**(3), 1165–1172 (2014)
39. W. Liu, Z.N. Chen, X.M. Qing, 60-GHz thin broadband high-gain LTCC metamaterial-mushroom antenna array. IEEE Trans. Antennas Propag. **62**(9), 4592–4601 (2014)
40. Z. Wu, L. Li, X. Chen, K. Li, Dual-band antenna integrating with rectangular mushroom-like superstrate for WLAN applications. IEEE Antennas Wirel. Propag. Lett. **15**, 1269–1272 (2016)
41. C. Caloz, T. Itoh, *Electromagnetic Metamaterials: Transmission Line Theory and Microwave Applications* (Wiley and IEEE Press, New Jersey, 2005)
42. Z. Wu, L. Li, Y. Li, X. Chen, Metasurface superstrate antenna with wideband circular polarization for satellite communication application. IEEE Antennas Wirel. Propag. Lett. **15**, 374–377 (2016)

Chapter 5
Graphene-Based Metamaterial Absorbers

Bian Wu and Yutong Zhao

Abstract This chapter introduces the basic concept and the main research advance of graphene-based metamaterial absorbers. First of all, we review the development of graphene and summarize the modeling methods of graphene-based Salisbury screen, Jaumann absorber, and frequency selective surface (FSS) absorber. Then we demonstrate both the microwave absorption and near-field radiation behavior of large-area multilayer chemical vapor deposition (CVD) graphene. In detail, we present several graphene-based metamaterial absorbers, including graphene-based transparent shielding enclosure, which can suppress unwanted cavity resonant modes, graphene-based quasi-transverse electromagnetic (TEM) wave microstrip absorber that prevents the transmission line main mode, graphene-based microwave (MW) FSS absorber that has a low profile and flexible or tunable wideband absorption, graphene-based millimeter wave (MMW) optically transparent absorber which may absorb plane wave energy with broad bandwidth, and graphene-based switchable THz absorber that enables fast electrical controlling of terahertz absorption and reflection. The various potential applications of graphene in the field of electromagnetic wave absorption are revealed and validated.

Keywords Multilayer graphene · Metamaterial absorber · Optically transparent absorber · Microwave FSS absorber · Switchable THz absorber

5.1 Introduction

Graphene, a flat monolayer consisting of carbon atoms arranged in a two-dimensional honeycomb lattice (Fig. 5.1), serves as a fundamental building block for graphitic materials in various dimensions. It can be transformed into zero-dimensional

B. Wu (✉) · Y. Zhao
School of Electronic Engineering, Xidian University, Xi'an 710071, Shaanxi, China
e-mail: bwu@mail.xidian.edu.cn

Y. Zhao
e-mail: bwu@mail.xidian.edu.cn

L. Li et al. (eds.), *Electromagnetic Metamaterials and Metasurfaces: From Theory To Applications*, https://doi.org/10.1007/978-981-99-7914-1_5

fullerenes, rolled into one-dimensional nanotubes, or stacked into three-dimensional graphite. The study of graphene, also known as '2D graphite,' has a history of over sixty years [1–3] and is widely utilized to describe the properties of diverse carbon-based materials. During the 1980s, researchers realized that graphene also serves as a valuable condensed-matter model for (2 + 1)-dimensional quantum electrodynamics [4–6], which significantly enhanced its status as a theoretical toy model. Conversely, although graphene was known as an integral component of three-dimensional materials, it was previously assumed to be nonexistent in a free state. It was considered an 'academic' material and believed to be unstable, giving rise to curved structures like soot, fullerenes, and nanotubes. However, the vintage model turned into reality when researchers unexpectedly discovered freestanding graphene [7, 8]. Subsequent experiments [9, 10] further confirmed that the charge carriers in graphene were indeed massless Dirac fermions. As a result, the graphene 'gold rush' commenced [11].

Graphene, a remarkable nanomaterial, is a monolayer of graphite with an incredibly thin thickness of only 0.34 nm. Its structure arises from carbon atoms in an sp2 hybridization state, forming a planar lattice where each carbon atom is covalently bonded to three others. This lattice takes the form of a honeycomb pattern, with two interpenetrating triangular sublattices. Graphene's physical properties are truly astonishing, earning it the title of "the wonder material".

In various devices, the measured carrier mobility of graphene at room temperature ranges from 8000 to 10,000, but in suspended graphene, it can reach as high as

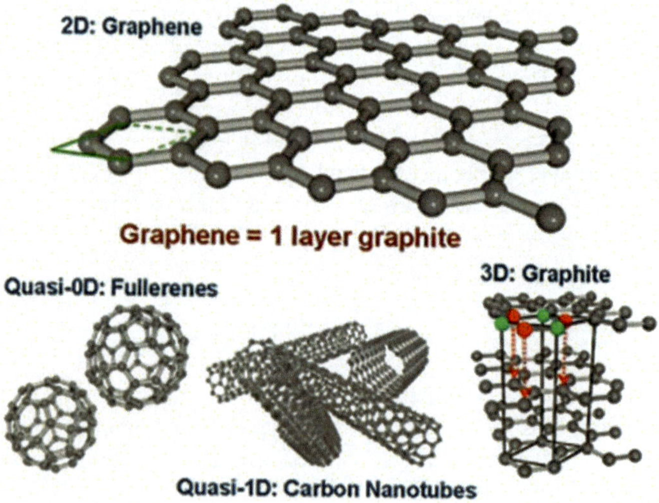

Fig. 5.1 Graphene, the ultimate graphitic form, serves as a foundational 2D building material for carbon structures of all dimensions. It can be transformed into zero-dimensional (0D) fullerenes, one-dimensional (1D) nanotubes, or stacked to form three-dimensional (3D) graphite

200,000. The mean-free path for ballistic carrier transport in graphene is approximately 300–500 nm at room temperature, while the resistivity of graphene depends significantly on the gate voltage applied, as shown in Table 5.1.

The dispersion relation in graphene exhibits a linear behavior for both electrons and holes at low energies E. This relationship is described by a simple equation where the Fermi velocity V_F and the wavenumber k of the charge carriers play crucial roles. The linear dispersion consists of two intersecting lines, forming what is known as the Dirac point. In regions far from the Dirac point, carrier transport in graphene is unipolar, while in the vicinity of the Dirac point, it becomes ambipolar, giving rise to significant recombination processes [12–15] .

From the initial method of graphene mechanical exfoliation using adhesive tape, the growth techniques for synthesizing graphene on metal or semiconductor substrates have rapidly advanced in just a few years, particularly with the development of advanced chemical vapor deposition (CVD) methods [16]. Currently, there are four primary methods employed for graphene production, as shown in Table 5.2.

The first method involves mechanical exfoliation using adhesive tape applied to highly ordered pyrolytic graphite (HOPG). Although this method yields low quantities, it offers the highest quality graphene achieved thus far. The second method is epitaxial growth, where graphene is grown on a silicon carbide (SiC) substrate, requiring heating to temperatures exceeding 1000 °C [17]. The third method involves the use of graphene oxide (GO), which is dispersed in hydrazine and deposited onto various substrates as a uniform film containing single- or few-layer graphene [18]. The final method, which shows great promise in terms of yield and reproducibility, is the use of CVD techniques to produce graphene. This technique has enabled the fabrication and testing of a graphene-based microelectrical mechanical system (MEMS) switch [19]. Furthermore, CVD has been successfully employed in the roll-to-roll production of 76.2 cm (30 in) graphene films on flexible copper substrates, finding applications in flexible electronics such as touch screens where graphene films serve as transparent electrodes [20] .

Graphene has gained significant attention in recent years due to its extraordinary properties as a two-dimensional monolayer of carbon atoms arranged in a

Table 5.1 Graphene's main properties

Parameter	Value and units	Observations
Thermal conductivity	5000 W/mK	Better thermal conductivity than in most crystals
Young modulus	1.5 TPa	Ten times greater than in steel
Mobility	40,000 $cm^2 V^{-1} s^{-1}$	At room temperature (intrinsic mobility), maximum mobility: 200,000 $cm^2 V^{-1} s^{-1}$, on suspended graphene
Mean free path (ballistic transport)	$\approx$ 400 nm	At room temperature
Fermi velocity	c/300 = 1,000,000 m/s	At room temperature

Table 5.2 Methods to produce graphene

Starting material	Brief description of the method	Yield	Quality	Area
Highly ordered pyrolytic graphite	Repetitive peeling highly ordered pyrolytic graphite	Low	Very high	Small
Silicon carbide	Reduction of silicon atoms at high temperature	Low	Medium	Large (3–4 inches wafers)
Graphene oxide	Graphene oxide dispersion into hydrazine	High	Medium	Large
Gas mixture	Chemical vapor deposition	Very high	High	Very large (30 inches)

honeycomb lattice. It possesses unique characteristics, including high optical transparency, exceptional flexibility, high electron mobility, and adjustable conductivity. Monolayer graphene films exhibit optical transparency levels ranging from 97 to 98% and undoped sheet resistances varying between 6000 and 125 Ω^{-2}. As a result, these films have found diverse applications in various fields that require low sheet resistance and high optical transparency. Some notable applications include transparent electrodes, optical modulators, polarizers, plasmonic devices, photodetectors, hyperlenses, cloaks, and absorbers [21–30].

However, it was only with the development of reproducible large-scale synthesis techniques such as chemical vapor deposition (CVD) that the production of graphene on a large scale with improved conductivity became possible [31–34]. Despite these advancements, challenges still remain in terms of synthesizing and transferring high-quality, large-scale, and highly conductive monolayer CVD graphene films reliably.

Due to its special physical properties, graphene will be widely used in many different areas. Suspended graphene has been theoretically characterized by Hanson [35], and experimental characterizations of graphene in the microwave region have been carried out using coplanar waveguide (CPW) [36], waveguide [37], or cylinder resonant cavity [38]. Active graphene FET mixers [39] have been fully studied, and various graphene-based metasurfaces [40] have been proposed to design dynamically controllable radio-frequency (RF) devices. Due to its relatively high sheet resistance, graphene is found to possess good absorption properties at microwaves [41, 42]. In previous works, we have experimentally demonstrated a transparent millimeter-wave absorber with 28% fractional bandwidth at 140 GHz [43] and also studied the microwave absorption and near-field radiation behavior of monolayer and few-layer CVD graphene in the C and X bands [44]. Typical applications of graphene in the electronic engineering area are shown in Fig. 5.2.

As operating frequencies and device integration continue to increase, electromagnetic interference (EMI) poses challenges for communication systems [45]. To address this issue, electromagnetic shielding is crucial. Over the past two decades, efforts have been made to identify effective shielding materials, including metals, metallic mesh [46], doped carbon nanotubes [47], composite carbon nanotube flexible films [48], and conducting polymers [49]. However, these materials have limitations, making the emergence of graphene highly significant in this context.

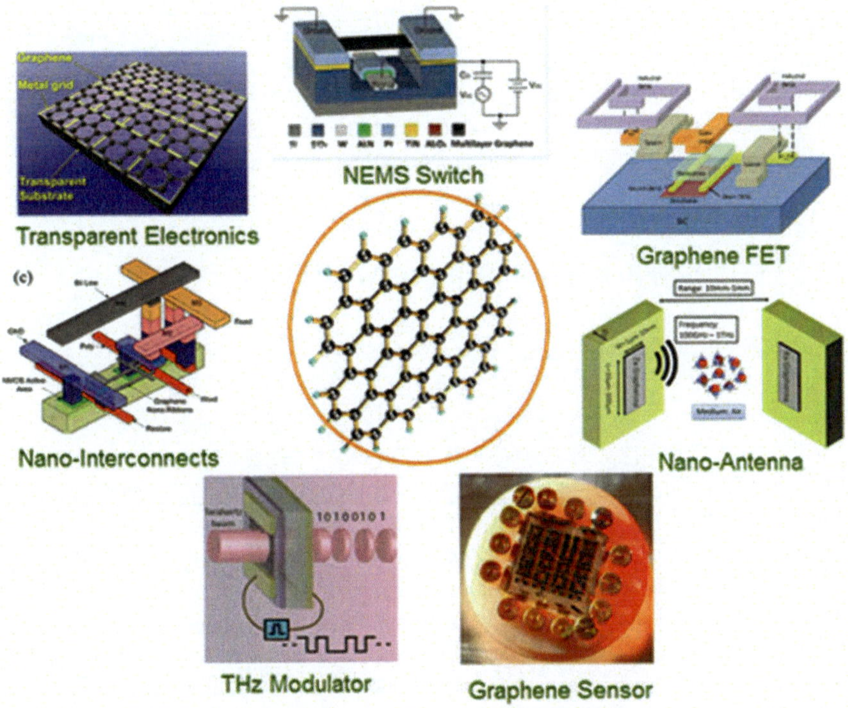

Fig. 5.2 Typical applications of graphene in the electronic engineering area

Graphene offers great potential in the development of radar-absorbing materials (RAMs), which aim to reduce an object's observable radar cross-section (RCS) within specific bandwidths [50]. Traditional absorbing screens suffer from various disadvantages, such as angular sensitivity, narrow bandwidth, and large size [51]. Hence, there is a clear demand for materials that can minimize angular sensitivity, reduce the thickness of broadband absorbers, and broaden their bandwidth. Graphene emerges as a leading candidate for fulfilling these requirements.

Metamaterial absorbers (MAs) have undergone extensive exploration since their initial theoretical and experimental demonstration by Landy et al. in [52]. Since then, a diverse range of MA designs has been showcased across various technologically relevant spectral ranges, encompassing optical [53], near-infrared (IR) [54], mid-IR [55], terahertz (THz) [56–59], and millimeter-wave bands [60]. The significance of MAs is particularly pronounced in the realm of THz frequencies, where the quest for strong and frequency-selective terahertz absorbers proves challenging [61]. THz near-perfect MAs boast crucial technological applications, including sensors [62], thermal emitters [63], and imaging devices [64].

Switchable MAs exemplify metamaterial structures capable of modifying or tuning their absorptivity through external stimuli [65]. This versatility renders switchable MAs highly valuable in applications such as beam steering antennas, hyperspectral imaging, and dynamic frequency-selective spatial light modulators. Given graphene's finite conductivity, flexible structure, and optical transparency, it holds significant potential for diverse applications within the metamaterial absorber domain. This chapter serves to introduce the theoretical characterization, structural design, and experimental verification of graphene-based metamaterial absorbers.

5.2 Formulation of the Graphene Model

5.2.1 *Electronic Model of Graphene*

Figure 5.3 illustrates an arrangement of laterally infinite graphene positioned in the x–z plane at the interface between two distinct mediums. The material parameters for these mediums are represented by (μ_1, ε_1) for medium 1 and (μ_2, ε_2) for medium 2. It is important to note that all material parameters, including those for conductivity, may possess complex values. In this depiction, graphene is represented as an infinitesimally thin, local two-sided surface, characterized by a surface conductivity $\sigma(\omega, \mu_c, \Gamma, T)$, where ω is radian frequency, μ_c is chemical potential, Γ is a phenomenological scattering rate that is assumed to be independent of energy E, and T is temperature.

The conductivity of graphene has been extensively studied in various recent works [66–73]. In this context, we utilize the expression obtained from the Kubo formula [74] to describe the conductivity of graphene.

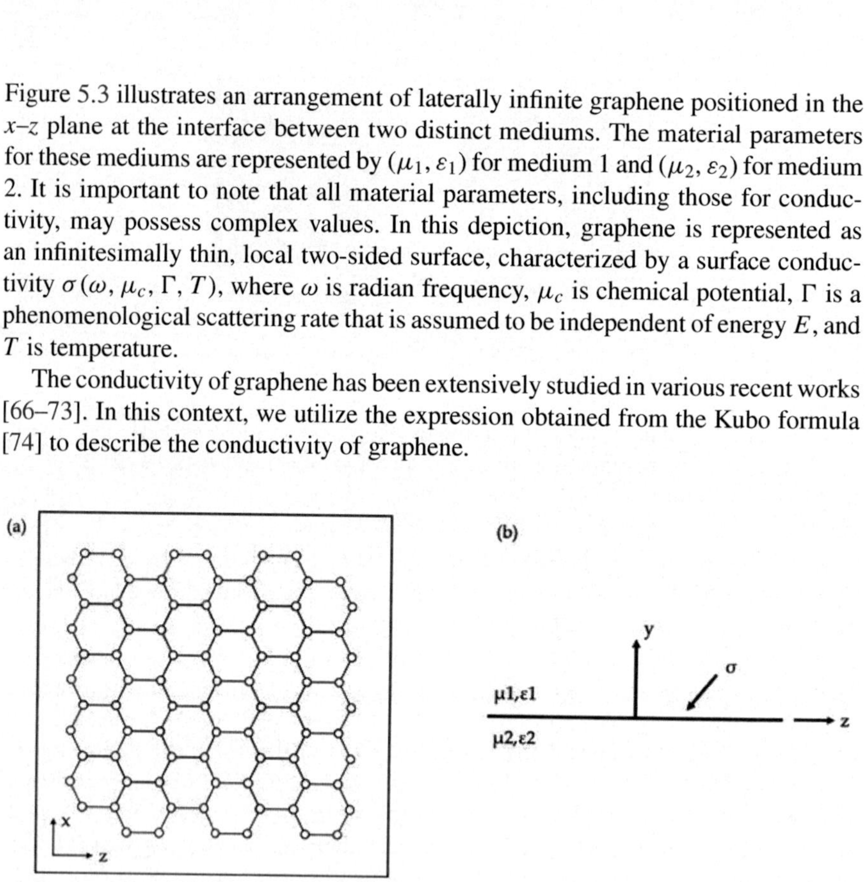

Fig. 5.3 **a** The diagram shows a top view depiction of graphene, where the small circles represent carbon atoms. **b** The diagram illustrates the side view of graphene characterized by conductance at the interface between two dielectrics [35]

$$\sigma(\omega, \mu_c, \Gamma, T) = \frac{je^2(\omega - j2\Gamma)}{\pi \hbar^2}\left[\frac{1}{(\omega - j2\Gamma)^2}\int_0^\infty \varepsilon\left(\frac{\partial f_d(\varepsilon)}{\partial \varepsilon} - \frac{\partial f_d(-\varepsilon)}{\partial \varepsilon}\right)d\varepsilon - \int_0^\infty \frac{f_d(-\varepsilon) - f_d(\varepsilon)}{(\omega - j2\Gamma)^2 - 4(\varepsilon/\hbar)^2}d\varepsilon\right] \tag{5.1}$$

where e is the charge of an electron, $\hbar = h/2\pi$ is the reduced Planck's constant, $f_d(\varepsilon) = (e^{(\varepsilon-\mu_c)/k_B T} + 1)^{-1}$ is the Fermi–Dirac distribution, and k_B is Boltzmann's constant. In the absence of an external magnetic field, we assume that the local conductivity of graphene is isotropic, meaning there is no Hall conductivity. The first term in Eq. (5.1) accounts for intraband contributions to conductivity, while the second term represents interband contributions. This applies to an isolated graphene sheet, the chemical potential μ_c is determined by the carrier density n_s, i.e.,

$$n_s = \frac{2}{\pi \hbar^2 v_F^2}\int_0^\infty \varepsilon[f_d(\varepsilon) - f_d(\varepsilon + 2\mu_c)]d\varepsilon \tag{5.2}$$

where $v_F \cong 9.5 \times 10^5$m/s is the Fermi velocity. The carrier density can be controlled by the application of a gate voltage and/or chemical doping. For the undoped, ungated case at $T = 0$ K, n_s=μ_c= 0. The intraband term in Eq. (5.1) can be evaluated as

$$\sigma_{intra}(\omega, \mu_c, \Gamma, T) = -j\frac{e^2 k_B T}{\pi \hbar^2(\omega - j2\Gamma)}\left(\frac{\mu_c}{k_B T} + 2\ln(e^{-\mu_c/k_B T} + 1)\right) \tag{5.3}$$

For the case of $\mu_c = 0$, Eq. (5.3) was first derived in [75] for graphite (with the addition of a factor to account for the interlayer separation between graphene planes) and corresponds to the intraband conductivity of a single-wall carbon nanotube in the limit of infinite radius [76]. With $\sigma = \sigma' + j\sigma''$, it can be seen that $\sigma'_{\text{intra}} \geq 0$ and $\sigma''_{\text{intra}} < 0$. As we will delve into further in the discussion, the imaginary part of conductivity plays a crucial role in the propagation of surface waves that are guided by the graphene sheet [67]. The interband conductivity can be approximated for $k_B T << |\mu_c|$, as [77]

$$\sigma_{\text{inter}}(\omega, \mu_c\Gamma, 0) \approx \frac{-je^2}{4\pi\hbar}\ln\left(\frac{2|\mu_c| - (\omega - j2\Gamma)\hbar}{2|\mu_c| + (\omega - j2\Gamma)\hbar}\right) \tag{5.4}$$

For $\Gamma = 0$ and $2|\mu_c| > \hbar\omega$, $\sigma_{\text{inter}} = j\sigma''_{\text{inter}}$ with $\sigma''_{\text{inter}} > 0$, and for $\Gamma = 0$ and $2|\mu_c| < \hbar\omega$, σ_{inter} is complex-valued, i.e.,

$$\sigma'_{inter} = \frac{\pi e^2}{2h} = \sigma_{\min} = 6.085 \times 10^{-5}(S) \tag{5.5}$$

and $\sigma''_{inter} > 0$ for $\mu_c \neq 0$.

5.2.2 Plane Wave Reflection and Transmission Coefficients

For a planarly layered, piecewise medium shown in Fig. 5.3b, assume that the current is located in the upper half-space (region 1), i.e.,

$$\vec{J}^{(1)}(r) = \hat{\alpha}\frac{j4\pi r_0}{\omega\mu_1}\delta(r - r_0) \tag{5.6}$$

where $\hat{\alpha} = \hat{x}$ or $\hat{y}$, $r_0 = \hat{y}y_0$, and $y_0 >> 0$. The electric and magnetic fields in region n ($n = 1, 2$) can be obtained in terms of Hertzian potential $\prod^{(n)}(\vec{r})$ as

$$\begin{aligned}\vec{E}^{(n)}\left(\vec{r}\right) &= \left(k_n^2 + \nabla\nabla\right)\Pi^{(n)}\left(\vec{r}\right)\\ \vec{H}^{(n)}\left(\vec{r}\right) &= j\omega\varepsilon_n\nabla\times\Pi^{(n)}\left(\vec{r}\right)\end{aligned} \tag{5.7}$$

where

$$\begin{aligned}\Pi^{(1)}\left(\vec{r}\right) &= \Pi_1^p\left(\vec{r}\right) + \Pi_1^s\left(\vec{r}\right) = \int_\Omega\left[\overline{\overline{g_1^p}}\left(\vec{r},\vec{r}'\right) + \overline{\overline{g_1^s}}\left(\vec{r},\vec{r}'\right)\right]\frac{\vec{J}^{(1)}\left(\vec{r}'\right)}{j\omega\varepsilon_1}d\vec{r}'\\ \Pi^{(1)}\left(\vec{r}\right) &= \Pi_2^s\left(\vec{r}\right) = \int_\Omega\left[\overline{\overline{g_2^s}}\left(\vec{r},\vec{r}'\right)\right]\frac{\vec{J}^{(1)}\left(\vec{r}'\right)}{j\omega\varepsilon_1}d\vec{r}'\end{aligned} \tag{5.8}$$

and $k_n = \omega\sqrt{\varepsilon_n\mu_n}$. The principle and scattered Green's dyadic are

$$\overline{\overline{g_1^p}}(\vec{r},\vec{r}') = \overline{\overline{I}}\frac{e^{-jk_1R}}{4\pi R} = \overline{\overline{I}}\frac{1}{2\pi}\int_{-\infty}^{+\infty} e^{-p_1|y-y'|}\frac{H_0^{(2)}(k_\rho\rho)}{4p_1}k_\rho dk_\rho \tag{5.9}$$

$$\overline{\overline{g_1^s}}(\vec{r},\vec{r}') = \hat{y}\hat{y}g_{n1}^s(\vec{r},\vec{r}') + \left(\hat{y}\hat{x}\frac{\partial}{\partial x} + \hat{y}\hat{z}\frac{\partial}{\partial z}\right)g_{c1}^s(\vec{r},\vec{r}') + \left(\hat{x}\hat{x} + \hat{z}\hat{z}\right)g_{t1}^s(\vec{r},\vec{r}') \tag{5.10}$$

$$\overline{\overline{g_2^s}}(\vec{r},\vec{r}') = \hat{y}\hat{y}g_{n2}^s(\vec{r},\vec{r}') + \left(\hat{y}\hat{x}\frac{\partial}{\partial x} + \hat{y}\hat{z}\frac{\partial}{\partial z}\right)g_{c2}^s(\vec{r},\vec{r}') + \left(\hat{x}\hat{x} + \hat{z}\hat{z}\right)g_{t2}^s(\vec{r},\vec{r}') \tag{5.11}$$

$$\overline{\overline{g_{\beta1}^s}}(\vec{r},\vec{r}') = \frac{1}{2\pi}\int_{-\infty}^{+\infty} R_\beta\frac{H_0^{(2)}(k_\rho\rho)e^{-p_1(y+y')}}{4p_1}k_\rho dk_\rho \quad (\beta = t, n, c) \tag{5.12}$$

$$\overline{\overline{g_{\beta2}^s}}(\vec{r},\vec{r}') = \frac{1}{2\pi}\int_{-\infty}^{+\infty} T_\beta\frac{H_0^{(2)}(k_\rho\rho)e^{(p_2y-p_1y')}}{4p_1}k_\rho dk_\rho \quad (\beta = t, n, c) \tag{5.13}$$

in which $p_n^2 = k_\rho^2 - k_n^2$, $\rho = \sqrt{(x-x')^2 + (y-y')^2}$, $R = |\vec{r} - \vec{r}'| = \sqrt{(y-y')^2 + \rho^2}$, $\overline{\overline{I}}$ is the unit dyadic,

$$R_t = \frac{M^2 p_1 - p_2 - j\sigma\omega\mu_2}{M^2 p_1 + p_2 + j\sigma\omega\mu_2} = \frac{N^H(k_\rho, \omega)}{Z^H(k_\rho, \omega)} \tag{5.14}$$

$$R_n = \frac{N^2 p_1 - p_2 + \frac{\sigma p_1 p_2}{j\omega\varepsilon_1}}{N^2 p_1 + p_2 + \frac{\sigma p_1 p_2}{j\omega\varepsilon_1}} = \frac{N^E(k_\rho, \omega)}{Z^E(k_\rho, \omega)} \tag{5.15}$$

$$R_c = \frac{2p_1\left[\left(N^2M^2 - 1\right) + \frac{\sigma p_2 M^2}{j\omega\varepsilon_1}\right]}{Z^H Z^E} \tag{5.16}$$

$$T_t = \frac{(1 + R_t)}{N^2 M^2} = \frac{2p_1}{N^2 Z^H} \tag{5.17}$$

$$T_n = \frac{p_1(1 - R_n)}{p_2} = \frac{2p_1}{Z^E} \tag{5.18}$$

$$T_c = \frac{2p_1\left[\left(N^2M^2 - 1\right) + \frac{\sigma p_1}{j\omega\varepsilon_1}\right]}{N^2 Z^H Z^E} \tag{5.19}$$

and $N^2 = \varepsilon_2/\varepsilon_1$, $M^2 = \mu_2/\mu_1$. When a unit-amplitude, α-polarized, transverse electromagnetic (TEM) plane wave is normally incident on the interface, it generates a current. The reflection and transmission coefficients for this scenario can be obtained by solving the equations provided in reference [35] as

$$\mathrm{R} = \frac{\eta_2 - \eta_1 - \sigma\eta_1\eta_2}{\eta_2 + \eta_1 + \sigma\eta_1\eta_2} \tag{5.20}$$

$$\mathrm{T} = \frac{2\eta_2}{(\eta_2 + \eta_1 + \sigma\eta_1\eta_2)} \tag{5.21}$$

where $\eta_n = \sqrt{\mu_n/\varepsilon_n}$ is the wave impedance in region n ($n = 1, 2$). The plane wave reflection and transmission coefficients obviously reduce to the correct results for $\sigma = 0$, and in the limit $\sigma \to \infty$, $\mathrm{R} \to -1$ and $\mathrm{T} \to 0$ as expected. In the special case $\varepsilon_1 = \varepsilon_2 = \varepsilon_0$ and $\mu_1 = \mu_2 = \mu_0$,

$$\mathrm{R} = -\frac{\frac{\sigma\eta_0}{2}}{1 + \frac{\sigma\eta_0}{2}}, \quad \mathrm{T} = \frac{1}{\left(1 + \frac{\sigma\eta_0}{2}\right)}, \tag{5.22}$$

where $\eta_0 = \sqrt{\mu_n/\varepsilon_n} \approx 377\Omega$. The reflection coefficient agrees with the result presented in [67] for normal incidence.

5.2.3 Surface Waves Guided by Graphene

Pole singularities present in the Sommerfeld integrals indicate the existence of discrete surface waves that are guided by the medium [78–80]. Based on Eqs. (5.14)–(5.19), the dispersion equation for transverse-electric (TE) surface waves, also known as *H*-waves, can be expressed as

$$Z^H(k_\rho, \omega) = M^2 p_1 + p_2 + j\sigma\omega\mu_2 = 0 \tag{5.23}$$

whereas for transverse-magnetic (TM) waves (*E*-waves),

$$Z^E(k^\rho, \omega) = N^2 p_1 + p_2 + \frac{\sigma p_1 p_2}{j\omega\varepsilon_1} = 0 \tag{5.24}$$

In the limit that $\varepsilon_1 = \varepsilon_2 = \varepsilon_0$, $Z^{H,E}$ agree with TE and TM dispersion equations in [67]. The surface wave field can be obtained from the residue contribution of the Sommerfeld integrals. For example, the electric field in region 1 associated with the surface wave excited by a Hertzian dipole current $\vec{J}(r) = \hat{y}A_0\delta(x)\delta(y)\delta(z)$ is

$$\vec{E}^{(1)}(\rho_0) = \frac{A_0 k_\rho^2 R_n'}{4\omega\varepsilon_1} e^{-p_1 y}\left[\left(\hat{x}\frac{x}{\rho_0} + \hat{z}\frac{z}{\rho_0}\right) H_0^{(2)'}(k_\rho\rho_0) - \hat{y}\frac{(k_1^2 + p_1^2)H_0^{(2)}(k_\rho\rho_0)}{k_\rho\sqrt{k_1^2 - k_\rho^2}}\right], \tag{5.25}$$

where $R_n' = N^E/(\partial Z^E/\partial k_\rho), H_0^{(2)'}(\alpha) = \partial H_0^{(2)}(\alpha)/\partial\alpha$, and $\rho_0 = \sqrt{x^2 + z^2}$. The term $e^{-p_1 y}$ leads to exponential decay away from the graphene surface on the proper sheet [$\mathrm{Re}(p_n) > 0, n = 1, 2$]. The surface wave mode may or may not lie on the proper Riemann sheet, depending on the value of surface conductivity. In general, only modes on the proper sheet directly result in physical wave phenomena, although leaky modes on the improper sheet can be used to approximate parts of the spectrum in restricted spatial regions and to explain certain radiation phenomena [80].

5.3 Modeling of Graphene-Based Absorber

5.3.1 Modeling of Graphene-Based Salisbury Screen Absorber

The schematic layout of the graphene-based Salisbury screen absorber is depicted in Fig. 5.4a. It consists of a large-area resistive graphene sheet and a metallic ground plane separated by a transparent substrate slab with a relative permittivity of ε_r and a thickness of h. The monolayer graphene film can be modeled as an infinitesimally thin, two-sided surface characterized by surface conductivity, as shown in Eq. (5.3).

In terms of TEM transmission line theory, the substrate layer can be modeled as a short transmission line with propagation constant β_d and characteristic impedance Z_d, and the resistive graphene sheet can be represented by the surface impedance $Z_S = 1/\sigma_s = R_s + jX_s$, where R_s is sheet resistance and X_s is sheet reactance (Fig. 5.4b). The propagation constant of free space is denoted by β_0 and the characteristic impedance by Z_0. From Fig. 5.5a, R_s remains frequency independent while X_s slowly rising up with frequency. At room temperature $T = 300$ K and $\mu_c = 0$ eV, when Γ increases from 0.1 to 1.1 meV, R_s increases from 72 to 792 Ω^{-2}, while X_s remains the same. If Γ is fixed to 1 meV and μ_c varies from 0 to 1 eV, both R_s and X_s will be reduced significantly, as shown in Fig. 5.5b. At microwaves, X_s is much smaller than R_s and can be ignored, so R_s dominates the resistivity of the graphene sheet. In practice, R_s can be controlled by multilayer structure or field bias [43].

The general analytical expressions for TE and TM polarized waves with incident angle θ can be expressed as [81].

$$k_0 = \frac{\omega}{c}, \beta_0 = k_0 \cos\theta, \beta_d = k_0\sqrt{\varepsilon_r - \sin^2\theta} \tag{5.26}$$

$$Y_0^{TE} = \frac{\cos\theta}{\eta_0}, Y_0^{TM} = \frac{1}{\eta_0 \cos\theta} \tag{5.27}$$

$$Y_d^{TE} = \frac{\sqrt{\varepsilon_r - \sin^2\theta}}{\eta_0}, Y_d^{TM} = \frac{\varepsilon_r}{\eta_0\sqrt{\varepsilon_r - \sin^2\theta}} \tag{5.28}$$

where c is the speed of light in vacuum, ω is the angular frequency, and $\eta_0 = \sqrt{\mu_0/\varepsilon_0}$ is the intrinsic free space wave impedance. For the graphene-based Salisbury screen absorber in Fig. 5.4, the input impedance is given by

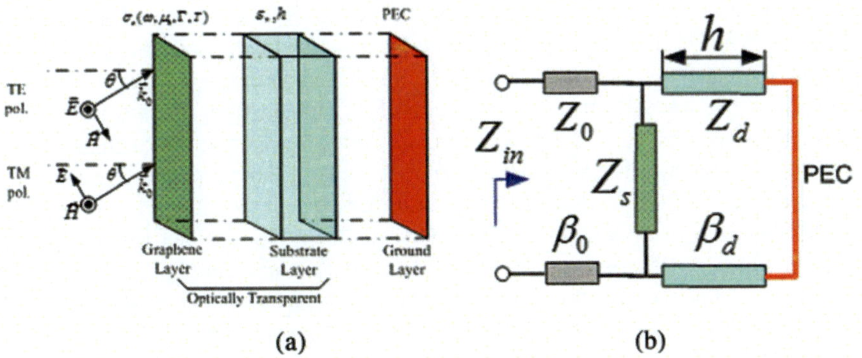

Fig. 5.4 Schematic of the graphene-based transparent Salisbury screen absorber. **a** Structure layout. **b** Equivalent circuit

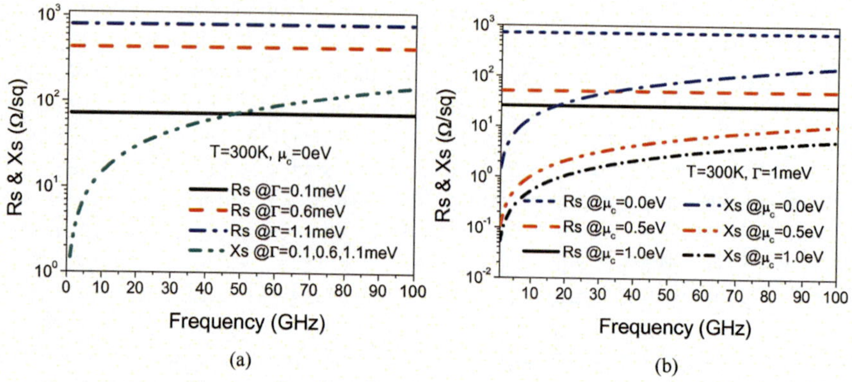

Fig. 5.5 Sheet resistance and sheet reactance of graphene at microwave frequency and room temperature **a** with different scattering rates Γ, **b** with different chemical potential μ_c

$$Z_{in}^{TE,TM} = \frac{Z_s Z_d^{TE,TM}}{Z_d^{TE,TM} - jZ_s \cot(\beta_d h)} \tag{5.29}$$

Considering the total reflection on the ground surface, the reflection coefficient S_{11} and absorption coefficient A of the graphene-based absorber can be obtained as

$$S_{11}^{TE,TM} = \frac{Z_{in}^{TE,TM} - Z_0^{TE,TM}}{Z_{in}^{TE,TM} + Z_0^{TE,TM}} \tag{5.30}$$

$$A^{TE,TM} = 1 - \left|S_{11}^{TE,TM}\right|^2 \tag{5.31}$$

The highest absorption is achieved when $Z_{in} = Z_0$, and the absorption peaks correspond to reflection zeros. As an example, we assume the graphene film has a chemical potential of 0.2 eV and a scattering rate of 5 meV, which corresponds to a sheet resistance of 645 Ω^{-2} since this value is very close to the extracted sheet resistance of 0.6 kΩ^{-2} for a 3-layer graphene sample from the microwave cavity measurement [44]. Moreover, quartz with $\varepsilon_r = 3.8$ is chosen as the transparent substrate, which is usually used as the dielectric support for graphene films. Figure 5.6 depicts the absorption spectra of the graphene-based microwave absorber with different thicknesses of quartz. The first absorption peak appears at the fundamental resonant frequency $f_0 = c/4h\sqrt{\varepsilon_r}$, and the periodic absorption peaks locate at $f_i = (2i-1)f_0$ (where $i = 1, 2\ldots$). As the thickness h increases, both the resonant frequency and the bandwidth of the absorption band will be decreased accordingly.

In Fig. 5.7, the absorption spectra of the graphene-based absorber are presented for the oblique incidence case, with a quartz thickness of $h = 1.3$ mm. Different incident angles are considered, and the results are shown. As the incident angle increases from 0° to 60°, the absorption peaks and resonant frequencies gradually increase for TE polarization. However, for TM polarization, the absorption peaks tend to decrease.

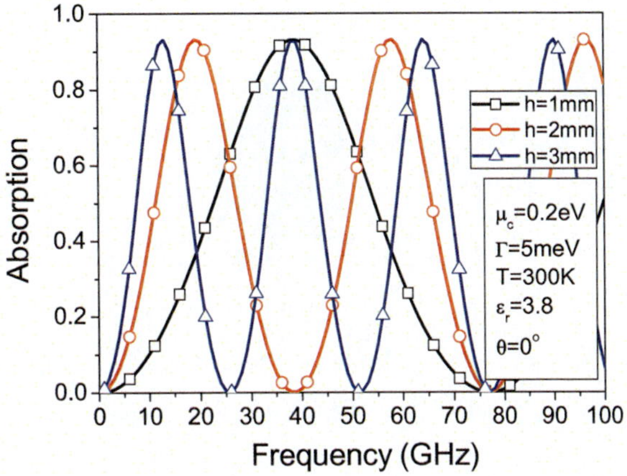

Fig. 5.6 Absorption spectra of the graphene-based Salisbury screen absorber with different thicknesses of transparent quartz slab

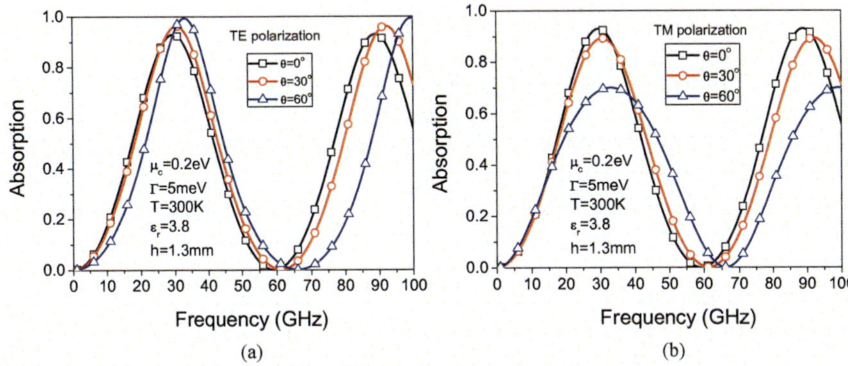

Fig. 5.7 Absorption spectra of the graphene-based Salisbury screen absorber with different incident angles. **a** TE polarization. **b** TM polarization

Nevertheless, despite these variations, the proposed structure remains effective as an absorber for most incident wave cases.

5.3.2 *Modeling of Graphene-Based Jaumann Absorber*

To achieve wide-frequency-range absorption of a normally incident plane wave, a multilayer Jaumann absorber based on graphene is proposed, as illustrated in Fig. 5.8a. This absorber configuration comprises a metallic ground and multiple graphene-dielectric stacks. The equivalent circuit, representing the absorber within the framework of transmission line theory, is depicted in Fig. 5.8b.

To calculate the input impedance from the input port of the absorber for both TE and TM polarizations, an iterative process is employed. The specific steps of this

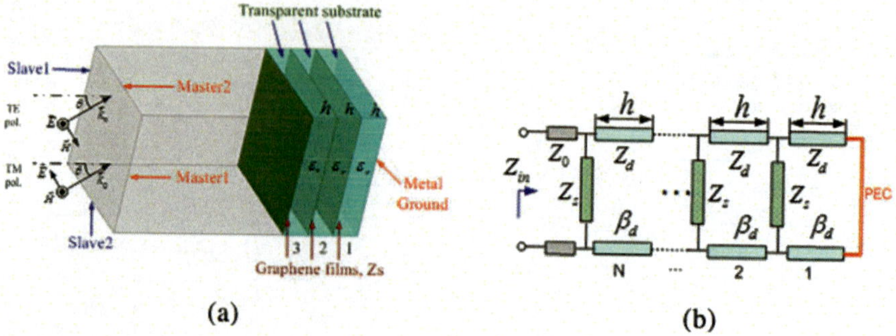

Fig. 5.8 **a** Simulation model and **b** equivalent circuit of the graphene-based wideband Jaumann absorber

iteration process are as follows:

$$Z_{i,in} = \frac{Z_s Z_d[Z_{i-1,in} + jZ_d \tan(\beta_d h)]}{Z_d[Z_{i-1,in} + Z_s] + j\tan(\beta_d h)[Z_d^2 + Z_{i-1,in} Z_s]} (i = 1, 2 \ldots, N) \quad (5.32)$$

The reflection and absorption coefficients can then be calculated according to Eqs. (5.30) and (5.31).

In Fig. 5.9, the absorption responses achieved by both circuit model calculations and HFSS EM simulations are compared, showing good agreement. The graphene films in the simulations are assumed to possess a scattering rate and a chemical potential, corresponding to a sheet resistance of 645 Ω^{-2} achievable through multilayer CVD technology. The substrate utilized has a relative permittivity of 1.8 and a height of 1 mm, resembling a transparent polytetrafluoroethylene (PTFE) thin film.

When $N = 1$ (referring to the number of graphene layers), the absorber exhibits a single absorption peak of 93% centered around 56 GHz. On the other hand, for N

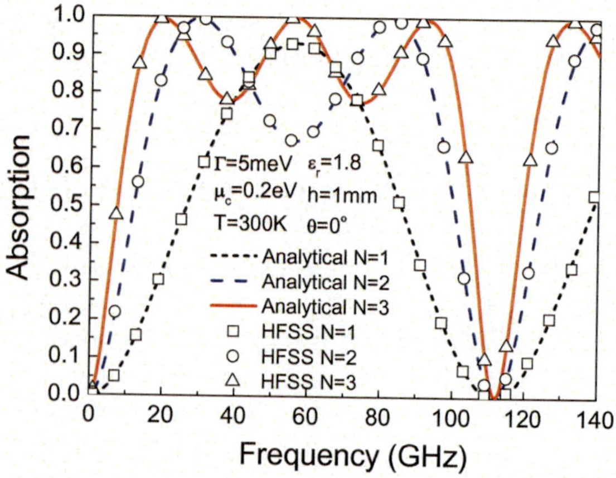

Fig. 5.9 Absorption spectra of $N = 1, 2, 3$ graphene-based wideband Jaumann absorbers achieved by circuit model calculation (lines) and HFSS simulation (symbols)

= 2 or $N = 3$, the graphene absorbers demonstrate multiple absorption peaks and a wide absorption band spanning from 20 to 100 GHz.

5.3.3 Modeling of Graphene-Based FSS Absorber

Figure 5.10 shows the side view of the graphene-based FSS absorber. An FSS pattern is positioned on the top surface of the substrate to induce resonant effects. Additionally, a conductor plate is placed on the bottom of the substrate to aid in blocking the transmitted electromagnetic (EM) waves. This arrangement enhances the trapping and absorption of the waves, leading to improved wave absorption characteristics. The transmission line model of the absorbing structure is established, and Z_0 is the characteristic impedance of the air. The patterned graphene sheet is treated as a conductive film with the sheet impedance $Z_s = R_s + jX_s$ due to the ultrathin thickness compared with the operating wavelength, and the imaginary part comes from the resonance between graphene micropatterns.

The substrate layer is modeled as a transmission line, where ε_r and d are the relative permittivity and the thickness of the substrate layer. When a plane wave is incident on the graphene-based FSS screen, the input impedance Z_{in} of the absorber is derived as

$$Z_{\text{in}} = Z_s//jZ_c\tan(k_z d) = \frac{Z_s \times jZ_c\tan(k_z d)}{Z_s + jZ_c\tan(k_z d)} \tag{5.33}$$

where $Z_c = \omega\mu_0/k_z$ is the characteristic impedance of the substrate, $k_z = \sqrt{k^2 - k_0^2\sin^2\theta}$ is the propagation constant along the z-direction, $k = \omega\sqrt{\mu_0\varepsilon_0\varepsilon_r}$ is the wave number in the substrate, and θ indicates the incident angle of a plane wave. For simplicity, here we consider the condition of normal incidence and set θ=0°.

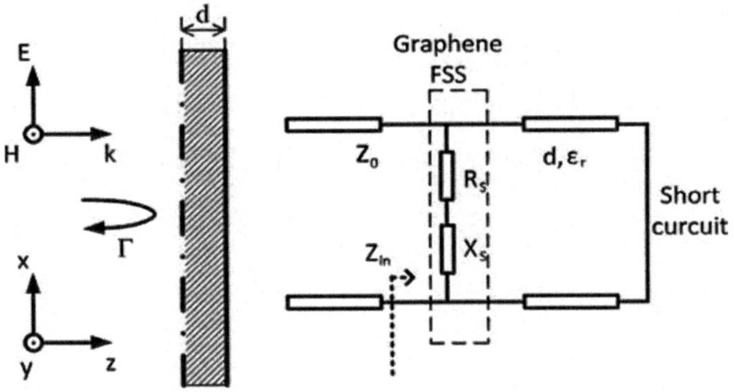

Fig. 5.10 Conductor-backed graphene-based FSS absorber and its equivalent transmission line model

The reflection coefficient and the absorptivity of the absorber can be calculated by Eqs. (5.30) and (5.31), respectively.

5.4 Microwave Absorption and Radiation of Multilayer Graphene

5.4.1 Sample Preparation and Characterization

Monolayer graphene was synthesized using the thermal CVD method. Samples were obtained by dicing from 4 in graphene/Cu (1 μm)/SiO2 (200 nm)/Si wafers and transferred onto fused silica quartz substrates. The transfer process involved utilizing a 200 nm thick layer of polymethyl methacrylate (PMMA) as a supportive polymer. The stacking technique comprised iteratively transferring the PMMA-graphene films onto diced substrates composed of graphene/Cu/SiO2/Si, followed by etching the underlying Cu catalyst in a 2.2% w/v ammonium persulfate solution for 12 h. Subsequently, the graphene/PMMA films were transferred onto the desired quartz substrates (refer to Fig. 5.11a). This method was deliberately selected to prevent the accumulation of PMMA residues between the layers of graphene, as removing them entirely through thermal annealing and acetone baths can be quite challenging. Through this process, graphene samples consisting of one to five layers (5 L) with dimensions of 17 × 8.5 mm were successfully prepared.

The opacity of suspended monolayer graphene was determined to be 2.3 ± 0.1% [15]. The number of stacked graphene samples was confirmed by UV-Vis optical transmittance measurements (Fig. 5.11b), which correlated with the number of transfer steps performed. These results were consistent with the in-situ Raman data, indicating the presence of monolayer as-grown graphene. The Spectrosil fused

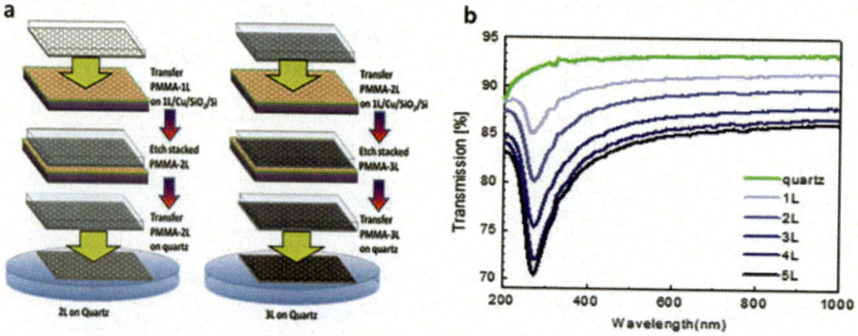

Fig. 5.11 **a** The diagram illustrates a schematic representation of the multi-step transfer-and-etch cycle for the fabrication of 2 and 3 L stacked graphene. **b** UV-Vis spectroscopy optical transmittance measurements are shown for bare quartz as well as monolayer and 2–5 L stacked graphene on quartz [44]

silica quartz substrate used had over 90% transmission throughout the optical window of interest (200–1000 nm). Monolayer and few-layer stacked samples exhibited characteristic absorption in the range of 250–300 nm, with relatively uniform absorption from 600 to 1000 nm, which increased with the number of layers. The 5 L sample showed only a 0.7% higher absorption compared to the 4 L sample, suggesting that it may be a partial 4 L sample with some regions containing 5 L graphene.

5.4.2 *Microwave Resonant Cavity Absorption of Multilayer Graphene*

Microwave cavity absorption techniques offer a valuable approach for the evaluation of graphene's resistivity. This method shares a fundamental principle with the millimeter-wave technique utilized for extracting sheet resistance, where power transmission through thin films is measured by coupling the source and power meter via dielectric waveguides. In this investigation, a straightforward microwave cylindrical cavity method, well-suited for large-area samples, is employed, as depicted in Fig. 5.12a.

In this method, the graphene-quartz samples are directly placed at the base of the cylindrical cavity. The cavity is characterized by its height H and radius R, and it is excited by a pair of probes positioned at $H/4$ and $3H/4$ to ensure weak coupling. At the cavity base, the graphene-bearing quartz substrate acts as a Salisbury screen-like absorber, resulting in a reduction in both the transmission coefficient and the quality factor. This effect is further amplified as the conductivity of the graphene improves with an increasing number of layers.

For monolayer graphene on quartz, the transmission responses show a suppressed resonance peak and an increased 3 dB-bandwidth $\mathrm{BW}_{3dB} = f_R - f_L$, while the total quality factor of the cavity represented by $Q_t = f_0/\mathrm{BW}_{3\mathrm{dB}}$ is reduced (Fig. 5.12b). The optical transmittance measurements reveal that the loading of 2 L graphene on quartz results in a remarkable peak suppression of approximately 10 dB. Subsequently, from 3 to 4 L, the suppression gradually increases. Surprisingly, the 5 L sample does not exhibit any further enhancement in suppression and behaves similarly to the 4 L sample. In the microwave range, the complex resistivity of graphene displays a constant real term (resistance) and negligible reactance, allowing the sheet resistance obtained at a single frequency using our microwave cavity approach to represent a wideband resistivity [35, 83]. The relationship between the quality factor or resonance peak of the graphene-loaded cavity and the sheet resistance is determined through Ansys HFSS full-wave electromagnetic simulations, as depicted in Fig. 5.12c. The total quality factor of the graphene-loaded cavity can also be defined as $Q_t = 1/(1/Q_c + 1/Q_q + 1/Q_g)$, where Q_c, Q_q, Q_g represent the quality factors of the cavity, quartz, and graphene, respectively. The minimum quality factor (or the minimum resonance peak) corresponds to the maximum absorption point at R_{s0}. For $R_s < R_{s0}$, graphene behaves as a quasi-metallic material with its quality factor

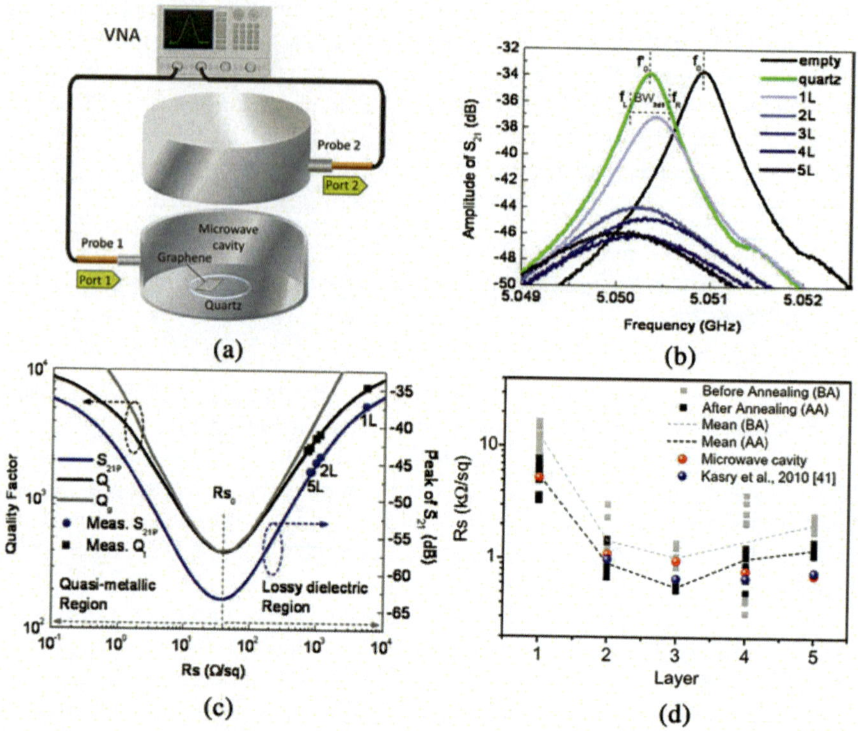

Fig. 5.12 **a** The figure illustrates the setup for microwave cavity measurements. **b** The graph shows the measured transmission spectra of a microwave cavity loaded with graphene. **c** The plot displays the relationship between the predicted and measured quality factor and resonance peak with sheet resistance. R_{s0} represents the maximum absorption point, serving as a boundary for two operational regions. **d** A comparison is made between the DC sheet resistance (before and after annealing) and microwave cavity measurements in this study, along with the DC sheet resistance of multilayer graphene before doping from the work by Kasry et al. [44, 82]

inversely proportional to the sheet resistance, whilst for $R_s > R_{s0}$, graphene exhibits characteristics of a lossy dielectric, with the quality factor being directly proportional to the sheet resistance. We classify the two regions as 'quasi-metallic' and 'lossy dielectric,' respectively. In the lossy dielectric region, the measured resonance peaks and quality factors decrease as the number of graphene layers increases.

The sheet resistance of multilayer graphene is determined by fitting the measured and simulated transmission responses. In Fig. 5.12d, it can be observed that the extracted sheet resistance obtained from the microwave cavity measurements exhibits a noticeable decrease with an increasing number of graphene layers. For monolayer graphene, the extracted sheet resistance is $5.26\ \Omega^{-2}$. However, for bilayer graphene (2 L), the sheet resistance sharply drops to $1.1\ \mathrm{k}\Omega^{-2}$. Subsequently, for 3 L graphene, the sheet resistance decreases incrementally to $0.96\ \mathrm{k}\Omega^{-2}$, and further reduces to 0.72 $\mathrm{k}\Omega^{-2}$ for 5 L graphene. This trend demonstrates a monotonically decreasing sheet resistance as the number of graphene layers increases. The most significant drop in sheet resistance occurs from 1 to 2 L, while the subsequent layers exhibit a relatively smaller reduction.

5.4.3 Microwave Near-Field Radiation of Multilayer Graphene

To assess the influence of the number of graphene layers on microwave near-field radiation, we utilized monolayer and multilayer graphene samples placed on a 50 Ω microstrip line, treated as a capacitively coupled patch antenna. The microstrip line was integrated onto a 1.57 mm thick Duroid substrate measuring 60 × 40 mm, with a relative permittivity of 2.2. The monopole probe was positioned in the XOY plane, running parallel to the microstrip feed line, as illustrated in Fig. 5.13a.

The reflection coefficients of the microstrip-coupled graphene patch were measured and compared to the case of a bare quartz substrate, as shown in Fig. 5.13b. It was observed that the monolayer graphene exhibited a slightly suppressed reflection compared to the bare quartz substrate due to its relatively high sheet resistance (5.3 Ω^{-2}). Consequently, the effects of absorption and radiation were found to be negligible in this case.

Similarly to the microwave cavity measurement, the reflection from 2 L graphene exhibited a significant suppression from 2 to 14 GHz, while the reduction became less pronounced for 3–5 L graphene samples. However, full-wave simulations indicate that a resonance effect occurs as a patch antenna only in the quasi-metallic region. Achieving microwave resonance requires a low sheet resistance (typically < 0.01 Ω^{-2}), indicating the need for further improvement in the conductivity of the graphene film to realize a functional microwave graphene patch antenna.

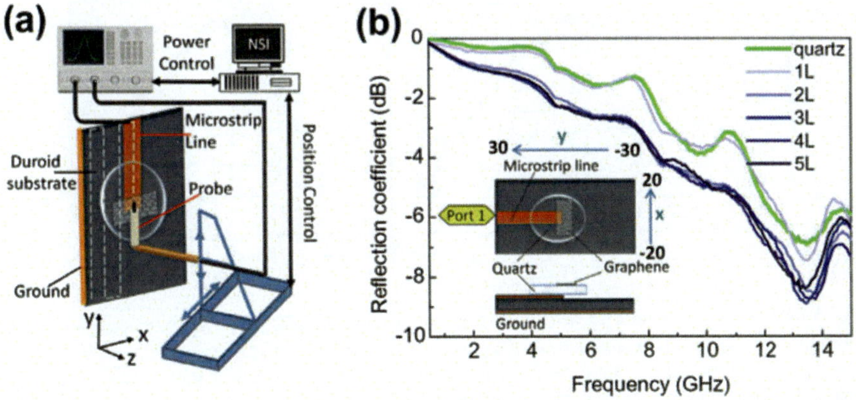

Fig. 5.13 **a** Schematic of near-field radiation measurement set-up. **b** Measured reflection coefficients of multilayer graphene patches (1–5 L) on quartz fed by a 50 Ω microstrip line, showing decreased reflection from 1 to 5 L and no distinct resonance [44]

The absorptive behavior of the lossy graphene patch results in a reduction of the radiation efficiency compared to a copper (Cu) patch antenna in the far field. To study the radiation properties of multilayer graphene, near-field measurements were conducted. Figure 5.14 presents the normalized radiation amplitudes of monolayer and multilayer graphene at the XOZ and YOZ cross-section planes, specifically at 5 and 8 GHz, with bare quartz serving as the reference.

Figure 5.15 depicts the measured near-field radiation patterns for 1, 3, and 5 L graphene on quartz, as well as bare quartz, at 5 and 8 GHz. At 5 GHz (Fig. 5.15a), the graphene patches with increasing numbers of layers exhibit reduced radiation compared to bare quartz. At 8 GHz, the coupling effect of the microstrip line weakens, causing the peak near-field radiation position for both quartz and multilayer graphene to shift from $Y = 15$ mm to $Y = 0$ mm (Fig. 5.15b). Additionally, the radiation amplitude noticeably decreases for a greater number of graphene layers compared to bare quartz. The lossy nature of graphene at microwave frequencies significantly affects the radiation properties of the patch.

The full-wave simulations of the far-field radiation reveal that the peak gain of the graphene patch in the lossy dielectric region is significantly lower than −5 dBi, a stark contrast to the 6 dBi achieved with a copper (Cu) patch of the same geometry

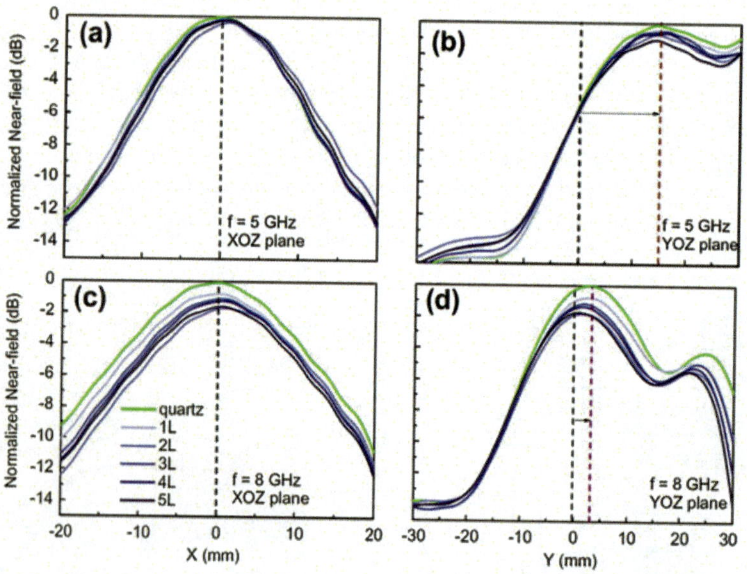

Fig. 5.14 **a, b** The normalized near-field amplitudes of multilayer graphene patches (1–5 L) on quartz in the XOZ and YOZ planes at 5 GHz demonstrate that the graphene patch emits slightly less radiation compared to bare quartz. Furthermore, it can be observed that as the number of layers increases, the radiation decreases, particularly when $Y > 0$ mm. The normalized near-field amplitudes in the XOZ and YOZ planes at 8 GHz exhibit a smaller shift of the radiation peak from the center and a more pronounced reduction in radiation for a greater number of layers. These observations indicate that the graphene patch attenuates the radiation, and the effect becomes more significant with an increasing number of layers [44]

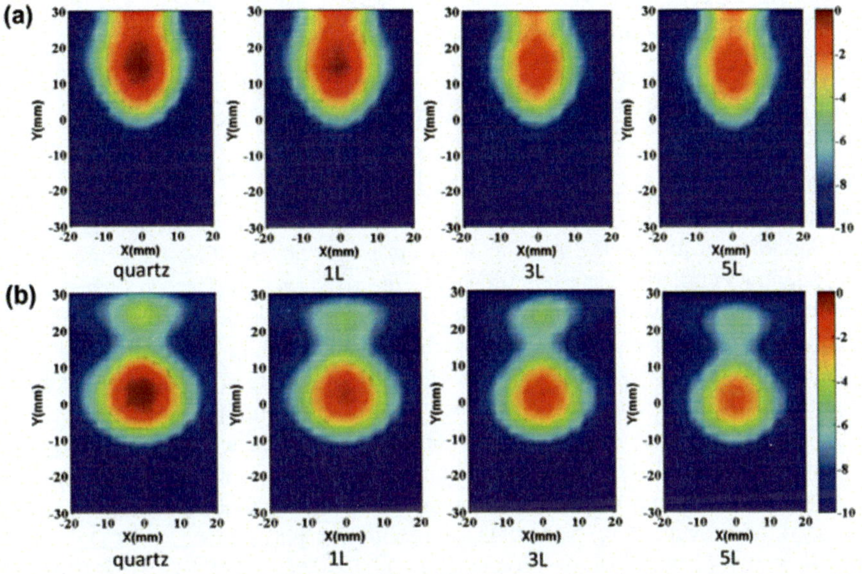

Fig. 5.15 **a** The near-field radiation patterns of multilayer graphene patches on quartz at 5 GHz were measured. It can be observed that as the number of layers increases, the radiation decreases, and even becomes less than that of bare quartz. This suggests that the graphene patch attenuates the radiation, with a more pronounced effect as more layers are added. **b** The near-field radiation patterns at 8 GHz were also measured. The radiation peaks are located at the center of the patch, and the intensity of radiation decreases as the number of layers increases. This further confirms the trend observed at 5 GHz, showing that the graphene patch exhibits reduced radiation with an increasing number of layers [44]

and slightly lower than the bare quartz case. These results strongly suggest that the graphene patch, due to its high sheet resistance, behaves more like an absorber than an antenna, which is consistent with the observations from the previous near-field measurements.

However, for lower sheet resistance values ($R_s < 10\ \Omega^{-2}$) in the quasi-metallic region, the peak gain of the graphene patch rises to a satisfactory level (> 0 dBi). This enhancement can be achieved through further doping or electrical biasing. In such scenarios, the graphene patch may find valuable applications in transparent and flexible radiating antennas, offering improved radiation efficiency compared to existing technologies.

The near-field measurements clearly indicate that as the number of graphene layers increases in the lossy dielectric region, the radiation performance does not improve; instead, the radiation amplitude decreases due to heightened absorption. To achieve enhanced radiation performance of the graphene patch as an antenna, it is essential to use lower sheet resistivity graphene in the quasi-metallic region.

5.5 Graphene-Based Metamaterial Absorbers

5.5.1 Graphene-Based Transparent Shielding Enclosure

With ever-increasing operating frequencies and higher degrees of integration of electronic devices, electromagnetic interference in communication systems has become an important problem [45]. Adding an electromagnetic shielding enclosure acts as an essential countermeasure. As shown in Fig. 5.16a, two adverse problems would be led to by the cavity resonant modes in the metal shielding enclosure. Firstly, due to the occurrence of cavity modes, the mutual coupling between the different RF modules increases significantly; secondly, the shielding effect (SE) of the metal enclosure will be deteriorated, on account of its strong coupling with EMI [84].

To solve the above two problems simultaneously, cavity resonant modes of shielding enclosure should be controllable. In addition, to maintain good optical transparency, transparent films such as graphene and indium tin oxide (ITO) can be used, as shown in Fig. 5.16b. ITO film with a sheet resistance of 9 Ω^{-2} (sheet conductivity of 0.11 S^{-2}) in the proposed structure could shield the external EMI and increase the SE. However, other needless cavity resonant modes would be excited by the highly reflective cavities which is made of ITO or other conductive materials. The graphene exhibits impedance characteristics in the microwave frequency band, and the needless electromagnetic resonances can be suppressed by a hybrid structure consisting of ITO and graphene. When the graphene film is attached at the bottom of quartz, surface currents would be induced at the graphene layer, and the current is absorbed by the lossy graphene, which explains the decrease of resonant intensity [85].

As shown in Fig. 5.17, an aluminum enclosure coated with PET films (Fig. 5.17a) is fabricated as a reference design, whose dimension corresponds to the schematic diagram in Fig. 5.16b. The proposed shielding enclosure of the same dimensions as the reference is shown in Fig. 5.17b, which is composed of ITO film with a sheet resistance of 9 Ω^{-2}. To ensure a good electric connection between the cavity walls, aluminum foil tapes are only used at the joints. Graphene on quartz substrates is set in

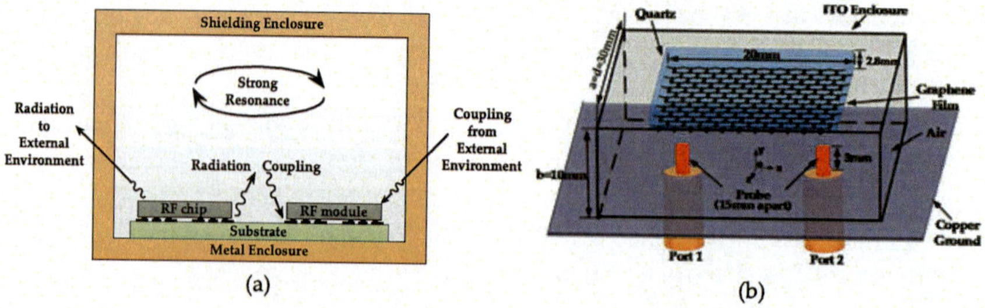

Fig. 5.16 **a** Schematic diagram of EMI phenomena of packaged microwave module at resonant frequencies. **b** Shielding enclosure model for simulation and experiment [85]

the inner walls of the enclosure as shown in Fig. 5.17b. Dimensions of the quartz and graphene-PET are 20 mm × 20 mm × 2.8 mm and 20 mm × 20 mm, respectively. A pair of 3 mm-long SMA connectors are mounted on a copper sheet of size 100 × 100 mm to excite and measure the resonant cavity modes as shown in Fig. 5.17d.

The monolayer graphene film used in the experiment is fabricated by CVD method before being transferred onto PET substrates. The sheet resistance of the monolayer graphene is measured 500 Ω^{-2} approximately, which can be modified by controlling transfer conditions and post-processing method [85–88].

The simulated and measured transmission coefficients (S_{21}) of the above two enclosures are presented in Fig. 5.18. It shows that the metal shielding enclosure exhibits two resonant modes (TE_{101} and TE_{201}) at 5.5 GHz and 11.5 GHz, respectively. And the transmission of the two resonant modes is normalized to 0 dB. For the graphene-based transparent shielding enclosure, TE_{101} and TE_{201} modes are suppressed by nearly 17 dB. The experiment was carried out using Agilent

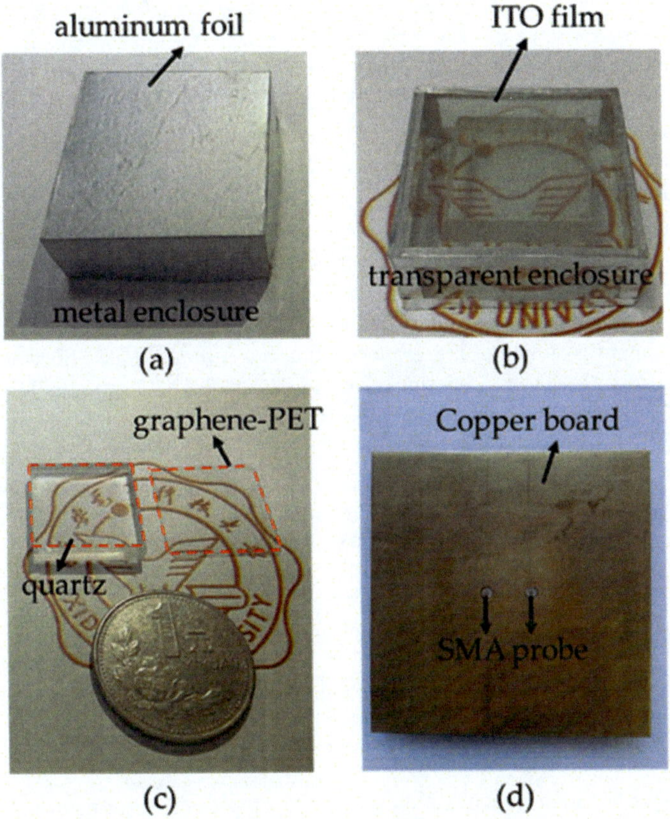

Fig. 5.17 **a** The metal shielding enclosure is made of aluminum foil tape. **b** The proposed transparent shielding enclosure is made of ITO films. **c** Quartz and CVD graphene (coated on PET) were used in the experiment. **d** SMA connectors excite and measure the resonant cavity modes [85]

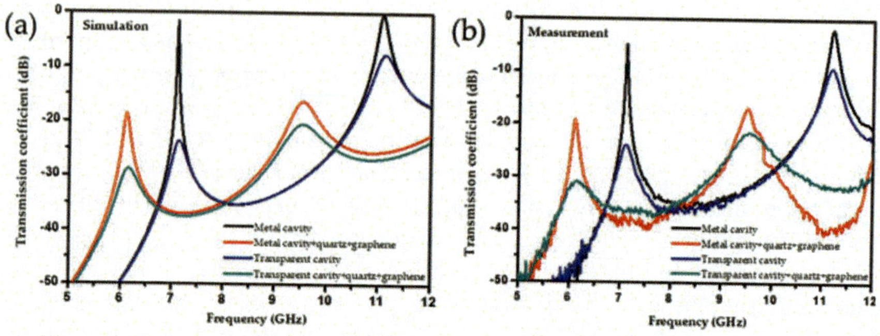

Fig. 5.18 Simulated and measured transmission coefficients of the metallic enclosure (ME) and ITO transparent enclosure (TE) in different imbedded statuses of graphene (Gr) film and quartz (Qz). **a** Simulation. **b** Measurement [85]

N9918A vector network analyzer, and the simulation is performed using the Ansoft HFSS EM software, which is in good agreement with the measurement. Compared with the metal shielding enclosure, the proposed shielding enclosure in Fig. 5.17b shows two major advantages. Firstly, it can be used in electronic systems, where optical transparency is an important application requirement. Secondly, the ability of graphene (with a sheet resistance of 500 Ω^{-2}) to absorb electromagnetic waves is very significant.

A closer investigation of the electric field distributions and intensity within the two shielding enclosures is discussed in Fig. 5.19. It shows that cavity resonant modes can be further suppressed by lowering the sheet resistance of graphene. Figure 5.19a and d show the electric field distributions of the ITO enclosure without graphene film. The sheet resistance of the graphene film has a remarkable effect on the field strength of resonant modes. As shown in Fig. 5.19b and e, TE_{101} and TE_{201} modes of the ITO enclosure will be suppressed by graphene film with a sheet resistance of 5000 Ω^{-2}. Besides, the electric field will be further suppressed when the sheet resistance is reduced to 500 Ω^{-2} as shown in Fig. 5.19c and f. It is worth mentioning that the suppression effect of the fundamental mode TE_{101} is greater than that of TE_{201} in the above analysis. This may be related to the thickness of the graphene/quartz substrate in the design.

5.5.2 *Graphene-Based Quasi-TEM Wave Microstrip Absorber*

The shielding absorber above is mainly used to suppress the resonant modes of the shielding enclosure, but usually, we need to suppress the main transmission mode, such as the quasi-TEM mode in the microstrip transmission line. A novel interdigital feed line is used to introduce the electromagnetic resonance and utilize a transparent graphene-quartz absorber on the top to couple with the interdigital line and absorb

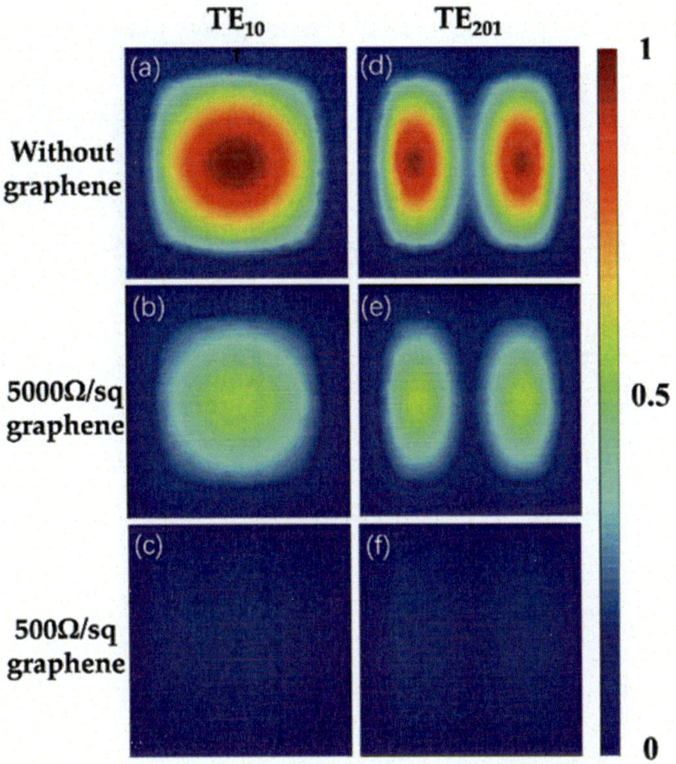

Fig. 5.19 Contrast contour plot of the normalized electric field distribution of the ITO enclosure (top view and $y = 6$ mm). TE_{101} mode (**a**, **b**, **c**) and TE_{201} mode (**d**, **e**, **f**) are normalized to 3000 V/m and 6000 V/m [85]

the energy at the resonance. The simulation model is illustrated in Fig. 5.20a and the fabricated sample is shown in Fig. 5.20b. The interdigital feed line has five open-ended stubs with a length of 7.5 mm and a width of 1 mm, which are combined by a microstrip section with a length of 17 mm and a width of 1 mm, then they are connected by a 50 Ω feed line. The Duroid substrate has a relative permittivity of 2.2 and a thickness of 1.57 mm. The graphene film with a size of 17 × 8.5 mm almost covers the interdigital feed line area and is separated by the cylindrical quartz slab.

The simulated reflection coefficients without and with graphene absorber are shown in Fig. 5.21a. The bare interdigital feed line possesses almost total reflection across the frequency range from 1 to 8 GHz. After loading the transparent graphene-quartz absorber on the top of the interdigital line, a reflection zero occurs at 4.4 GHz with a reflection coefficient of −25.3 dB, which is mainly due to the absorption effect since the simulated radiation gain is smaller than −5 dB at 4.4 GHz. In Fig. 5.21b, the measured reflection response also has a reflection zero at 4.6 GHz with −28.9 dB, which validates the simulation.

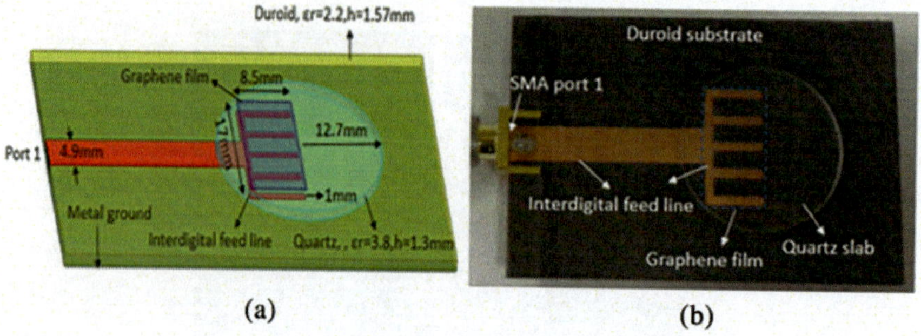

Fig. 5.20 The graphene-based transparent TEM wave absorber using an interdigital feed line. **a** Simulation model. **b** Fabricated sample

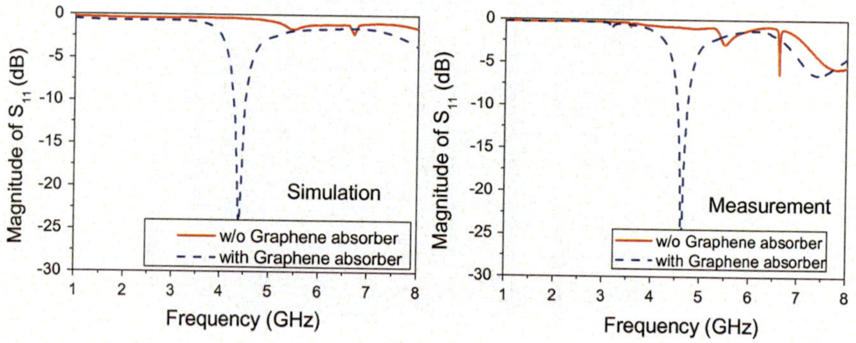

(a) (b)

Fig. 5.21 Reflection spectra of the microstrip interdigital feed lines with and without graphene-based TEM wave absorber. **a** Simulated reflection responses. **b** Measured reflection responses

5.5.3 *Graphene-Based Microwave FSS Absorber*

The wideband FSS absorber based on graphene and ITO adopts a three-layer structure, as shown in Fig. 5.22. The transparent substrate in the middle adopts polycarbonate with a low dielectric constant of 2.7. The bottom layer adopts the sheet resistance of 9 Ω^{-2} ITO conductive film, and the top layer is composed of a patch array. The top layer uses an array of $m \times n$ square periodic units and an array of $(m + 1) \times (n + 1)$ cross-shaped units to form a graphene patch array with 100 Ω^{-2} multilayer graphene. Compared with an array composed of a single array element, this structure can ensure impedance matching between the absorber and free space in a wide frequency band. Figure 5.22a is the schematic view of the structure of the FSS absorber with $m = n = 2$ and Fig. 5.22b gives the top view. Appropriate impedance matching to the air is the primary condition of a high-performance broadband absorbers. It can make the electromagnetic wave incident into the structure without reflection, so as to maximize the absorption of the incident waves. A periodic resonant structure is used to improve the performance of impedance matching [89].

To achieve the broadband absorption, the geometric parameters marked out in Fig. 5.22 are chosen first to obtain the desired two resonance frequencies. In order to provide broadband characteristics, these parameters need to be further optimized. The following paragraphs will take a microwave transparent electromagnetic absorber as an example to illustrate the design ideas.

The reflection and transmission coefficients of the transparent absorber are simulated by HFSS electromagnetic simulation software. As depicted in Fig. 5.23a, the reflection coefficient of the transparent absorber is less than −10 dB from 8.5 to 18 GHz. The transmission coefficient is less than −27 dB in 8–18 GHz and the transmittance approaches zero. Figure 5.23b shows the absorbance curves of the graphene composite patch array with different sheet resistances. With the increase of the sheet resistance, the absorption bandwidth is gradually narrowed and the absorption rate is increased. Here, the sheet resistance of the top and bottom layers are fixed to 100 Ω^{-2} and 9 Ω^{-2}, respectively. Figure 5.23c shows the absorbance variation with the substrate's thickness h. It shows that when the thickness h increases, the resonant frequency decreases. The final thickness of the substrate h was chosen to be 3 mm.

In reference [90], a broadband multilayer graphene FSS absorber is designed. The absorber is composed of interconnected Jerusalem cross patterns, and its surface and unit cell are depicted in Fig. 5.24a and b respectively. The geometric parameters used for the design are as follows: $D = 5$ mm, $d = 3.5$ mm, $l = 1.5$ mm, $t = 2$ mm, and $p = 13$ mm. The dielectric constant of the FR4 layer is 4.4. The simulated and calculated absorption coefficients are shown in Fig. 5.24c. When the sheet resistance of the multilayer graphene is low, the structure acts as a dual-band absorber with absorption peaks at 10.5 and 20.2 GHz. As the sheet resistance (R_s) increases, the absorptivity between the two peaks also increases while remaining almost unchanged

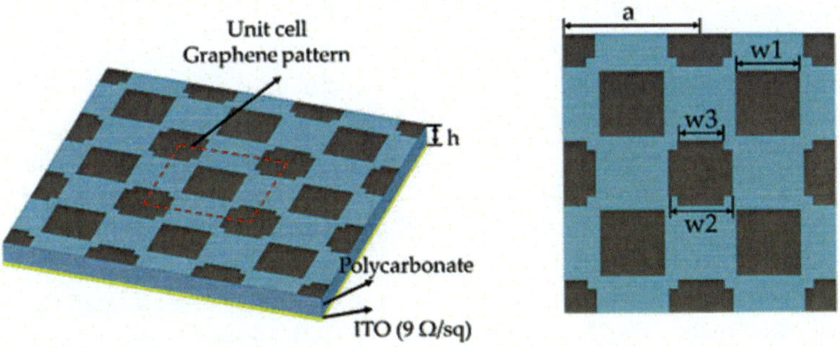

Fig. 5.22 Schematic of the proposed transparent wideband absorber: **a** perspective view and **b** top view of the absorber. (Design parameters of the microwave absorber: $w1 = 7$ mm, $w2 = 10$ mm, $w3 = 5$ mm, $a = 16$ mm, $h = 3$ mm; design parameters of the millimeter absorber: $w1 = 3.4$ mm, $w2 = 3.8$ mm, $w3 = 2$ mm, $a = 7$ mm, $h = 1.5$ mm) [89]

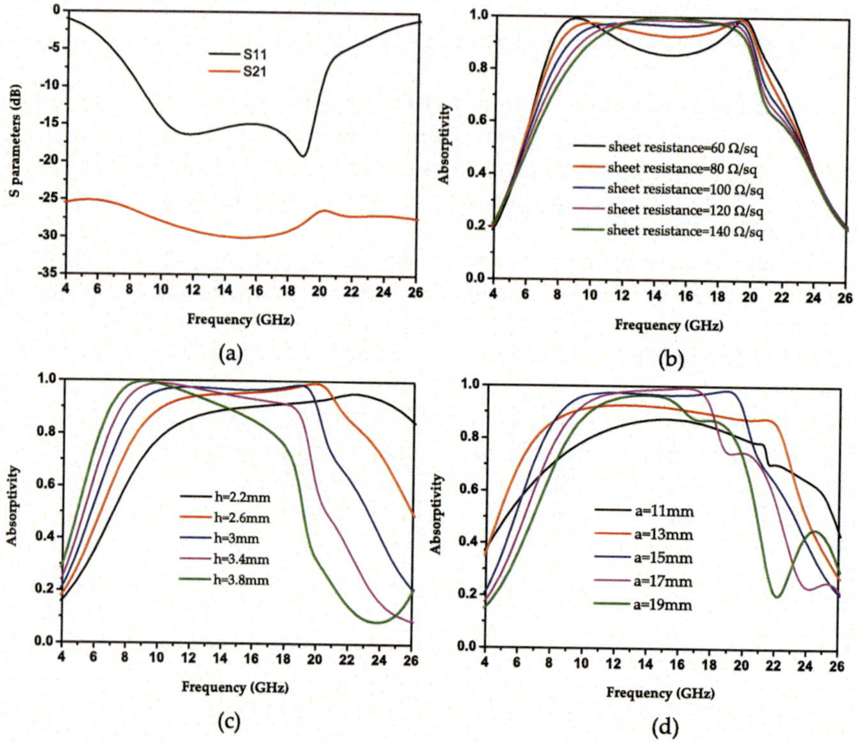

Fig. 5.23 **a** S parameters simulation curve. **b** Simulated absorption with different sheet resistances of the upper surface. (The 90% absorption bandwidth at 100 Ω^{-2}: 71.7%). **c** Simulated absorption with different thicknesses of the substrate. **d** Simulated absorption with different periods of the composite patch array (the 90% absorption bandwidth at 16 mm: 71.7%) [89]

at the two peaks. When R_s reaches 70 Ω^{-2}, the absorptivity between the two peaks exceeds 0.8, indicating broadband absorption properties.

Figure 5.25a and b illustrate the steps involved in preparing multilayer graphene (MLG) on polyvinyl chloride (PVC) substrates. MLG samples were synthesized using CVD on 25 μm thick nickel foils at various temperatures ranging from 800 to 1100 °C under ambient pressure. H_2, CH_4 and Ar gases were used during the growth process. CH_4 gas served as a carbon feedstock and was only sent during the growth. Flow rates of the H_2, CH_4, and Ar were set as 99 sccms, 42 sccms, 71 sccms, respectively. Growth times were 10 min. After synthesis, 75 μm thick PVC films were laminated onto the graphene-coated foils at 150 °C. Etching the foils with diluted nitric acid resulted in large-area graphene films on a transparent substrate.

In the microwave region, the resistance term of graphene impedance remains nearly constant, while the reactance term tends to zero and can be neglected. Thus, graphene can be approximated as a resistive sheet without dispersion properties at the microwave band.

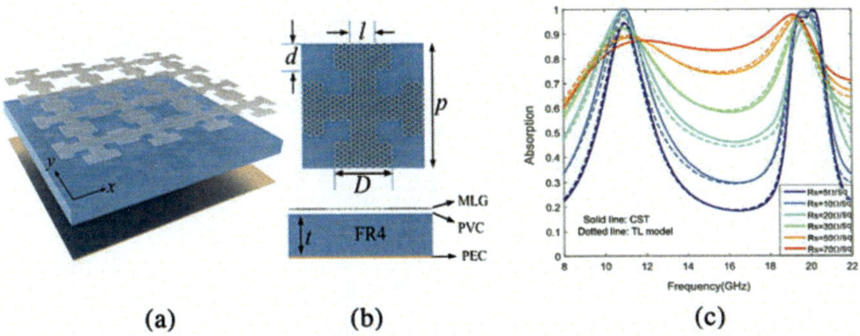

Fig. 5.24 **a** Schematic diagram of a broadband absorber composed of multiple graphene FSS layers. **b** The unit cell of the structure is depicted with the dimensions; top view of the unit cell (the top diagram) and side view of the unit cell (the bottom diagram). **c** Absorption coefficients of the proposed broadband tunable absorber versus frequency under different sheet resistances of multilayer graphene [90]

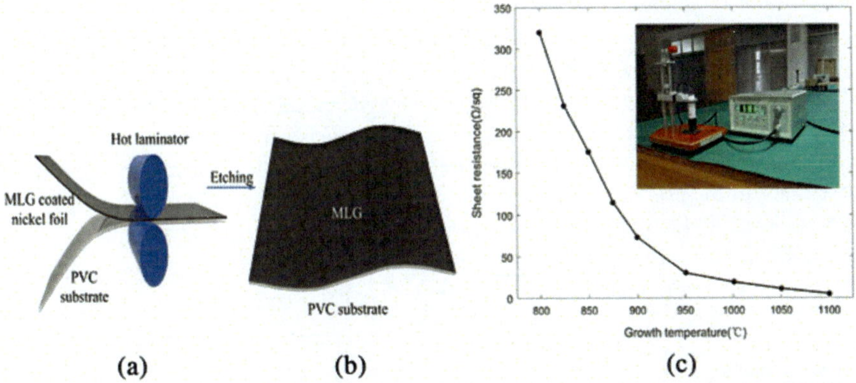

Fig. 5.25 **a** Large area MLG transfer printing was carried out on flexible PVC substrates by the laminating process. MLG was synthesized on nickel foil, and then 75 μm thick PVC film was formed on the graphene-coated nickel foil. **b** Etched metal foils yield flexible multilayer graphene electrodes on the PVC support. **c** The sheet resistance of MLG changes with the growth temperature in the range of 800–1100°C. The inset is the photograph of the four-point measurement system [90]

Figure 5.25c depicts the variation of the sheet resistance of the synthesized MLG at different growth temperatures, as measured using a four-wire system. The growth temperature influences the amount of carbon dissolved in the nickel foils, which in turn affects the number of graphene layers grown on the foils. As a result, the sheet resistance of MLG varies with the growth temperature. It is important to note that the relationship between MLG resistance and growth temperature is not constant and can also be influenced by factors such as the CVD furnace and the type of nickel foils used. The graph shows a decrease in sheet resistance from 325 to 5 Ω^{-2}.

Two multilayer graphene MLG FSS consisting of 14 × 14 unit cells (prototype 1) or 11 × 11 unit cells (prototype 2) were fabricated, as shown in Fig. 5.26a. The growth temperatures of two kinds of the MLG FSS layers are chosen to be (1100, 925, 800 °C) and (1100, 950, 900 °C), which correspond to the sheet resistances of (5 Ω^{-2}, 40 Ω^{-2}, 200 Ω^{-2}) and (5 Ω^{-2}, 20 Ω^{-2}, 70 Ω^{-2}), respectively. The ground plane is made of copper foil with a thickness of 25 μm.

Figure 5.26b depicts the measured absorptivity of the second type of absorber under different growth temperatures. The absorber acts as a dual-band absorber when the growth temperatures are 1100 and 950 °C. The measured absorptivity peaks are at 10.2 and 20.2 GHz, and the absorptivity between the two peaks increases with the decrease of growth temperature obviously. The absorber behaves as a broadband absorber when the growth temperature decreases to 900 °C. The absorptivity is above 0.8 from 10.3 to 20 GHz. Compared with the simulation results, the absorbing peak frequencies correspond well when the growth temperature varies from 1100 to 900 °C, while the slight amplitude difference is caused by the deviation in the synthesis process of the MLG layer and the manufacturing process of the absorber. Due to the limit of horn antennas, the measured results are combined of two bands (8–18 and 18–22 GHz), which leads to some fluctuation in the measured results at 18 GHz.

In [91], a flexible transparent microwave absorber (FTMA) with a wide bandwidth is introduced. The absorber utilizes a graphene FSS (GFSS) combined with an oxide-metal-oxide film as the metal ground. The GFSS consists of electrically disconnected unit cells made from monolayer graphene synthesized via the CVD method. The sheet resistance of the graphene is significantly reduced to around 105 Ω^{-2} by doping with HNO3. The FTMA offers wide bandwidth, low profile, high optical transparency, and good flexibility, making it suitable for applications in electromagnetic-compatible facilities and stealth technologies.

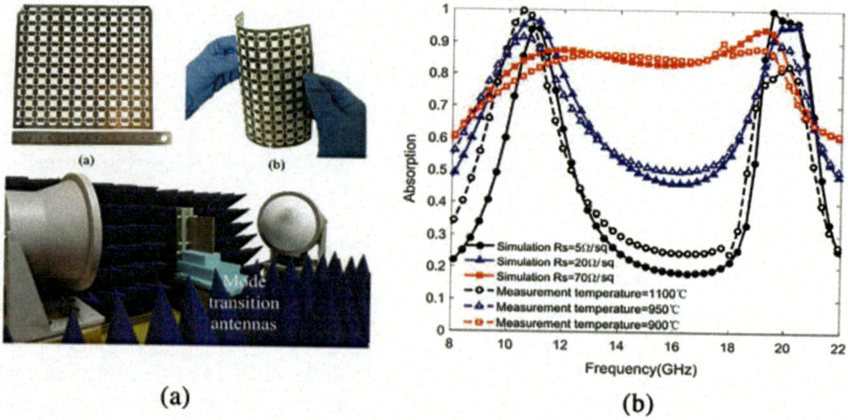

Fig. 5.26 **a** The fabricated broadband tunable absorber and the measured environment. **b** Measured and simulated absorption of the proposed absorber [90]

In [92], a dynamically tunable microwave absorber with a low profile is proposed. The absorber is composed of a graphene-based sandwiched structure and a high-impedance surface. It provides a dynamically tunable reflection range from greater than −3 dB to less than −30 dB at 11.2 GHz (center working frequency). The thickness of the proposed absorber is only 2.8 mm, which is approximately one-tenth of the working wavelength. The absorption mechanism is explained using a modified equivalent circuit model. This work serves as a reference for the design and fabrication of dynamically tunable microwave absorbers based on large-scale graphene and promotes the practical applications of graphene at microwave frequencies. A wideband microwave absorber with dynamically tunable absorption is proposed in [93], which is composed of a random metasurface layer and a few layers of large-area graphene. Due to the superimposition of a few layers of graphene, the tunable range of graphene sheet resistance is lowered to 80–380 Ω^{-2}, which is easier to match the impedance of free space as the resistance film for broadband microwave absorber. Besides, 12 proper elements of the metasurface are adopted and distributed randomly, and more resonance frequencies and phase responses can be achieved, which broaden the bandwidth of the absorber and reduce the profile at the same time. The sheet resistance of graphene could be manipulated by applying a bias voltage, and the absorptivity can be tuned from 80 to 50% from 5 to 31 GHz.

5.5.4 Graphene-Based Millimeter-Wave Wideband Absorber

CVD graphene films grown on four-inch Cu/SiO_2/Si wafers are researched, which are free of pin-holes by optical and electron microscopy. Raman spectroscopic mapping and optical microscopy confirmed that the samples were of high uniformity with > 90% monolayer coverage, [94]. Films were transferred to fused silica quartz substrates using spin-coated 200 nm thick PMMA as a supporting layer (Fig. 5.27a). By the multiple transfer-and-etch method, multilayer graphene samples were processed. This involves the repetitive transfer of the PMMA-graphene films onto diced graphene on Cu/SiO_2/Si substrates and etching them in an aqueous ammonium persulfate solution before finally transferring the released PMMA/graphene onto the quartz substrates. This method avoids significant PMMA residue accumulation between the graphene layer stacks and reduced the average sheet resistance of ~ 0.9 kΩ^{-2} for 2 L and ~ 0.6 kΩ^{-2} for 3 L. The number of graphene layers was confirmed via UV-Vis Spectro-photometer. Optical transmittances of 85–91% at 700 nm for quartz-supported 1–4 L graphene were noted (Fig. 5.27b).

In Fig. 5.27c, a broadband absorber is shown where multilayer graphene samples are stacked on top of a ground plate, supported by quartz. Figure 5.27d provides optical images of the quartz-supported 2 and 3 L graphene layers, each measuring 17 × 8.5 mm. The *N*-unit samples, with similar sheet resistances, are stacked onto the ground plate to create broadband absorbers.

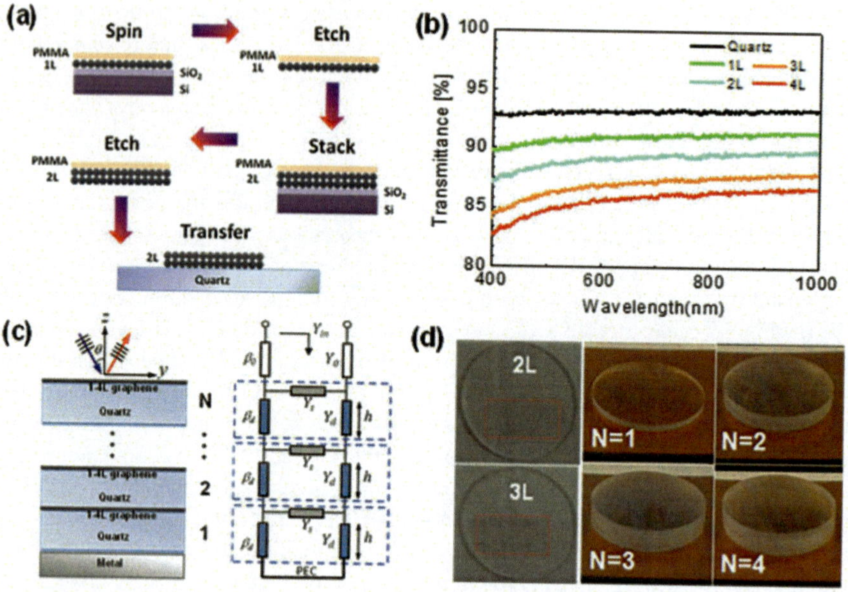

Fig. 5.27 Optical images and schematic of the multilayer graphene on stacked graphene-quartz and quartz absorbers. **a** Flow chart of the multiple transfer-etch processing for a 2 L device. **b** UV–Vis spectra of the 1.3 mm thick bare quartz and 1–4 L graphene samples. **c** Schematic of the *N*-unit stacked absorber and the equivalent circuit model. **d** Optical images of 2 and 3 L absorbers and *N* stacked graphene-quartz structures backed with a ground plate (*N* is the number of stacked graphene-quartz, $N = 1$–4) [43]

To study the nanostructured absorber, reflection spectra are measured using a free-space millimeter wave reflectometry technique. These spectra are then transformed into absorption spectra using Eq. (5.31). The experimental setup, depicted in Fig. 5.28a, utilizes an HP N5244A vector network analyzer with millimeter wave extension heads, operating in the frequency range of 110–170 GHz.

Firstly, single absorbers with 1–4 L graphene were tested. Subsequently, two graphene-quartz samples were stacked to create a 2-unit absorber ($N = 2$), as shown in Fig. 5.28b. Absorbers of up to $N = 5$ units were measured. The stacked graphene-quartz structures are supported by electrically conductive ground floors that are firmly fixed to metal brackets to ensure perpendicular to the incident waves. Just normal incidence was considered in these measurements. All graphene layers had dimensions of 8.5 × 17 mm to ensure that the beam width of the incident wave could be covered.

The measured and analytical calculations reflection and absorption results of the single graphene-quartz absorbers with 1–4 L graphene on quartz are shown in Fig. 5.29. The calculated results show the effect of the chemical potential on the absorption and reflection properties of the proposed absorber. When $\mu_c = 0$ eV and $\Gamma = 7$ meV, corresponding to a sheet resistance of 5044 Ω^{-2}, which makes it difficult to match the input impedance to free space. So, the peak absorption is lower than 40% indicating poor absorption. In steps $\mu_c = 0.1$ eV, the sheet resistance of

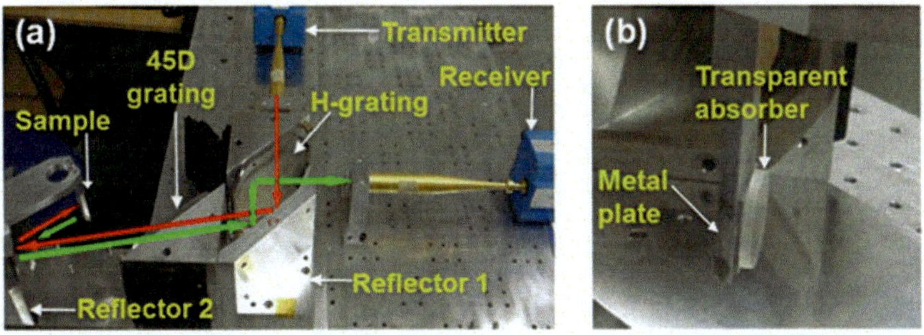

Fig. 5.28 Millimeter wave reflectometer measurement environment. **a** Experimental set-up. The incident wave from the transmitter to the sample (Red lines); the reflected wave from the sample to the receiver (Green lines). The H-grating transmits vertically polarized waves but reflects horizontally polarized waves. The 45D grating selects the E-field components with 45° rotation. **b** The proposed transparent absorber [43]

the graphene is reduced and the peak absorption improves as the chemical potential increases. When $\mu_c = 0.3$ eV and $\Gamma = 5$ meV, the corresponding sheet resistance is 430 Ω^{-2} tending to the free space impedance. Due to the good impedance matching, the peak absorption is near 100% at 148 GHz. The graphene surface resistance is mainly effected by the intraband contribution from microwaves to far infrared waves. Besides, the real part of the intraband conductivity increases linearly with the chemical potential at 10 GHz [35] which contributes to energy absorption or dissipation [29]. The calculations indicate that the same behavior applies at the operating frequency of 140 GHz. Similarly, an increase in surface conductivity can be achieved by artificially stacking multiple layers of graphene [82]. The measured improvement in absorption for multilayer graphene, as shown in Fig. 5.29d, aligns with the calculated absorption curves in Fig. 5.29c when varying the chemical potential from 0–300 meV. The chemical potential can be altered through chemical doping or by applying a bias voltage [35, 82, 95]. In this case, increasing the number of stacked layers achieves a similar enhancement in absorption capability.

First, the bare quartz substrate on the ground plate as a background reference was tested in the reflectometer. In addition to the inherent system noise at high frequencies (> 160 GHz), total reflection without absorption is observed throughout the frequency range. Then, the single-unit graphene-quartz absorbers with 1–4 L graphene were measured. A 1 L absorber shows a small absorption peak of around 30% at 148 GHz, which is similar to the case of $\mu_c = 0$ eV and $\Gamma = 7$ meV (Fig. 5.29c, d). In contrast, a 2 L absorber shows a peak absorption of around 95%, which is similar to the calculated case of $\mu_c = 0.2$ eV and $\Gamma = 5$ meV. The change in the thickness of the actual quartz plate (±2%) and the air gap between the quartz and the floor (~ 0.1 mm) cause a small frequency shift. The absorption peak increases marginally (+1.2%) for the 3 L case and falls slightly (–1%) for the 4 L case. Multilayer graphene can thusly be used to derive a turbostratic, stacked, artificial graphite-like material of sufficiently reduced sheet resistance capable of near matching the free space

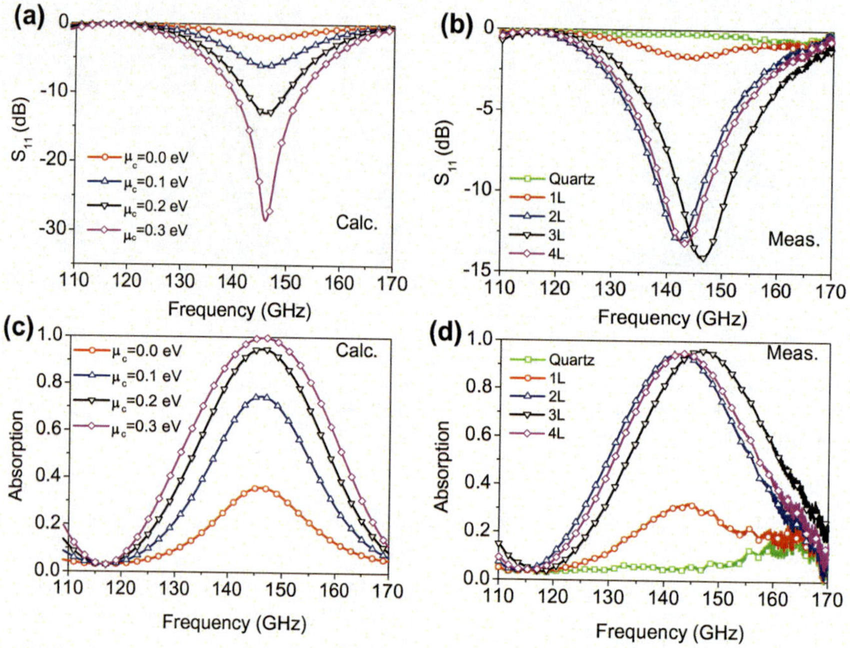

Fig. 5.29 Measured and calculated results of the proposed single graphene-quartz absorbers. **a**, **c** Calculated absorptivity. The scattering rates are chosen as $\Gamma = 7$ meV for $\mu_c = 0.0$ eV and $\Gamma = 5$ meV for the others, $T = 300$ K. **b**, **d** Measured reflection and absorption results of single ($N = 1$) graphene-quartz absorbers with 1–4 L graphene on quartz ($h = 1.3$ mm and $\varepsilon_r = 3.8$). The measurements show that the improvement in absorption is substantial from 1 to 2 L multilayer graphene while less significant after 2 L. The measured reflection and absorption spectra in **b** and **d** are similar to the calculated spectra in **a** and **c** [43]

impedance. However, it is challenging to improve the sheet resistance further for samples consisting of more than 3 layers, which could be due to water residue between the layers that prevent good contact between the interfacing layers in the present samples.

Figure 5.30a and c depict the calculated results of the stacked graphene-quartz absorber. Design for simplicity, assume that the graphene film has the same parameters as the initial calculation ($\Gamma = 5$ meV, $\mu_c = 0.15$ eV) and the corresponding sheet resistance is 859 Ω^{-2}. The calculated reflection spectrum in Fig. 5.30a has the same number of reflection zeros as the measured stacked units, which expands the absorption bandwidth while keeping the center frequency around 148 GHz. A similar phenomenon also exists in the absorption spectra in Fig. 5.30c, as the number of layers increases, the absorption peak increases and the absorption band widens. The mutual coupling of Fabry–Perot resonators helps to produce multiple absorption peaks in the band. The measured results in Fig. 5.30b and d are in good agreement with the calculated results, except for a small frequency shift of reflection zeros and an increased reflection in the band. The difference is possibly due to additional losses in the practical samples, parameter errors, and the small air gap between the adjacent units that induces multiple reflections. For the 5-unit stacked absorber, nearly 90% absorption can be achieved for 125–165 GHz, which shows that the practical millimeter wave absorber has a fractional absorption bandwidth of 28% and optical transparency.

To achieve optical transparency, a Jaumann absorber design is proposed using an ultra-thin and lossy resistive component separated by a transparent quartz dielectric filler, with a ground plate backing. The goal is to match the impedance to free space by manipulating the surface resistance properties of the multilayer graphene and leveraging the impedance transformation of the transparent dielectric filler. This design allows for the absorption of 90% of the incident millimeter-wave without the need to consider permeability.

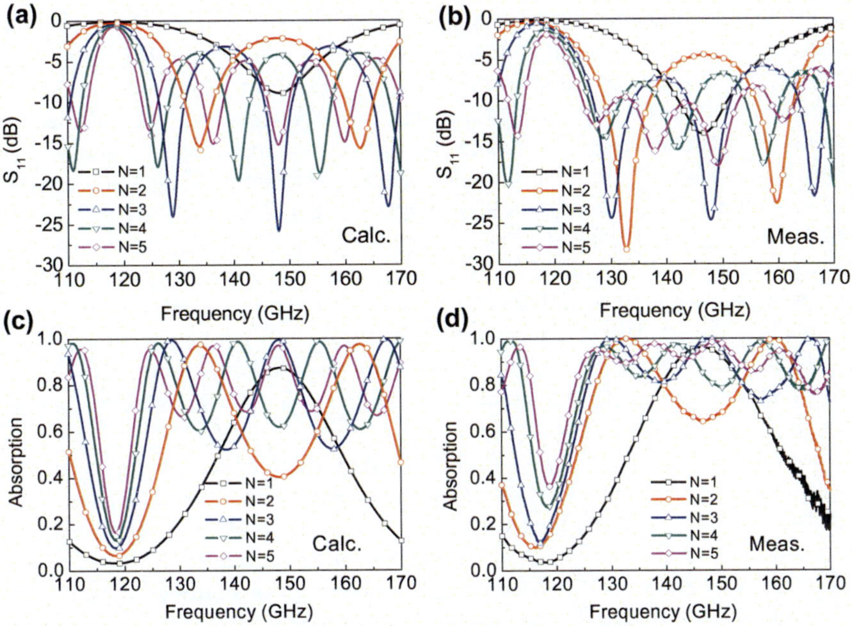

Fig. 5.30 Calculated and measured results of stacked graphene-quartz absorbers. **a, c** The calculated reflection and absorption results show that the central frequency is 148 GHz, the number of reflection zeros (or absorption peaks) increases, and the absorption bandwidth increases from $N =$ 1 to $N = 5$, (N represents the number of stacked graphene-quartz units). Assume that all graphene sheets have the same parameters ($\mu_c = 0.15$ eV, $\Gamma = 5$ meV) and are separated by a uniform quartz substrate $\varepsilon_r = 3.8, h = 1.3$ mm). **b, d** Measurement results show a similar response to the calculations [43]

5.5.5 Graphene-Based Switchable THz Absorber

Switchable MAs are a type of metamaterial structure that can modulate their absorption properties through external stimuli [65]. In this context, a broadband switchable terahertz absorber/reflector is introduced, utilizing the control of graphene's chemical potential to achieve high switching intensity (72%) while maintaining a broad bandwidth (0.53–1.05 THz). By applying a bias voltage between a gold electrode and p-type silicon, the conductivity of the graphene layer can be easily adjusted, enabling control over the absorption behavior. The proposed MA allows for switching between reflection (reflection > 82%) and absorption (absorption > 90%) in the low-terahertz spectrum. The absorption principle and switching mechanism of this MA will be discussed, and it demonstrates excellent performance for large incident angles with a thin thickness (approximately 1/6 wavelength for mid-frequency).

The unit structure of the proposed absorber/reflector is shown in Fig. 5.31a and b. It consists of six layers as follows: an aluminum layer on the bottom, a lossy polydimethylsiloxane (PDMS) layer, a p-type doped silicon layer, a silicon dioxide layer, a monolayer CVD graphene layer, and a gold pattern layer. The gold pattern, with six equilateral triangles arranged in a regular hexagon, is used to enhance the capacitance effect. The distance between different equilateral triangles and adjacent hexagons is $g_1 = 2\ \mu m$ and $g_2 = 90\ \mu m$, respectively. A graphene patch is placed between the silicon dioxide layer and gold pattern which is designed to be a regular hexagon with a side length of $d_2 = 60\ \mu m$, and. The combination of graphene and gold pattern can be regarded as a graphene-gold hybrid metasurface [96]. Gold (thickness $t_1 = 0.5\ \mu m$) and aluminum (thickness $t_5 = 1\ \mu m$) are modeled with a conductivity of 4×10^7 S/m and 3.8×10^7 S/m, respectively. A lossy PDMS layer with thickness $t_4 = 60\ \mu m$ and permittivity $\varepsilon = 2.35 - j0.047$ [97] is adopted in this design.

As shown in Fig. 5.31c, a gold feed line, with a width of $g_3 = 2\ \mu m$, is adapted to provide equivalent bias voltage for each unit cell. A small piece of rectangular graphene is used to connect the feed line and gold pattern, which could avoid unexpected resonance. By applying a gate voltage (a static electric field) on the electrode, the chemical potential or the conductivity of graphene can be controlled on purpose. A 3D simulation model based on the unit cell shown in Fig. 5.31a is adopted to investigate the characteristics of the switchable absorber/reflector in the low-terahertz band. Due to the symmetry of the structure, asymmetric and symmetric boundary conditions are applied in the x and y directions respectively, which could reduce the simulation time effectively [47].

Suppose that a terahertz plane wave is illuminated on the switchable absorber/reflector normally with polarization along the y direction. Based on the finite element method, the simulated absorption spectrum of the unit cell with periodic boundary is shown in Fig. 5.32. The absorption $A(\omega)$ is obtained by $A(\omega) = 1 - \mathrm{T}(\omega) - \mathrm{R}(\omega)$, where $\mathrm{T}(\omega)$ is transmission and equal to zero, due to a 1-μm-thick aluminum layer used as a ground plane. $\mathrm{R}(\omega)$ is the reflectivity of the absorber. For the undoped and

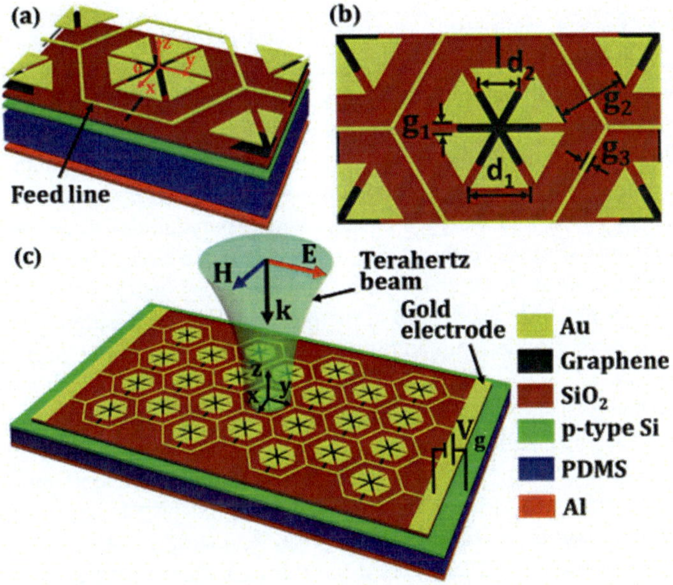

Fig. 5.31 Schematic diagram of the proposed absorber **a** perspective view, **b** top view. The geometrical dimensions are as follows: $d_1 = 70$ µm; $d_2 = 6$ µm; $g_1 = 2$ µm; $g_2 = 90$ µm; $g_3 = 2$ µm. The thickness of the Au, SiO2, *p*-type Si, PDMS, and Al layers are $t_1 = 0.5$ µm; $t_2 = 0.3$ µm; $t_3 = 0.5$ µm; $t_4 = 65$ µm and $t_5 = 1$ µm, respectively, **c** schematic diagram of the proposed switchable absorber/reflector. A DC bias voltage is applied between the gold electrode and p-type Si to control the sheet conductivity of graphene [96]

ungated case at $T = 0$ K, which means the chemical potential $\mu_c = 0$ eV, absorptivity over 90% is obtained from 0.53 to 1.05 THz, with a wide bandwidth of 65.8%.

Fig. 5.32 Absorptivity of the proposed absorber. The curve without graphene patch (w/o Graphene) or gold pattern (w/o Gold) show a weak absorption. The curve of lossy and lossless PDMS are nearly equal [96]

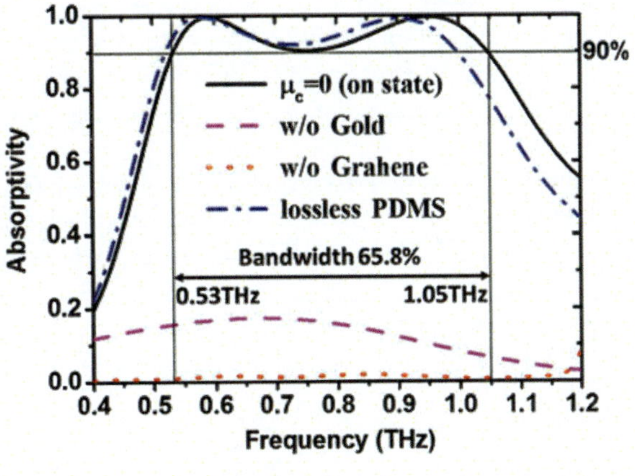

The investigation of the loss mechanism aims to understand the contributions of each component. It is found that the absorption spectra under a lossless PDMS substrate are similar to those under a lossy one, indicating that the incident electromagnetic energy is not dissipated in the dielectric. When considering the structure without the presence of the graphene layer or gold pattern, maximum absorption of 2 or 19% is achieved across the frequency range of interest, respectively. The slot area, formed by two adjacent gold patches with a perpendicular component to the electric field, introduces a certain amount of capacitance effect. The graphene patch can be treated as an impedance element, which reduces the Q factor and broadens the impedance-matching bandwidth in the presence of a gold pattern. The effective absorption is primarily determined by the combined effects of the gold pattern and graphene patch.

The thickness of the graphene film is only about 0.34 nm, but it can be found from Fig. 5.33a that with the increases of chemical potential μ_c (or Fermi energy E_F), the effect of the graphene patch on the absorptivity is remarkable. By using an electrostatic field to change the chemical potential of graphene, a method of adjusting graphene's electrical conductivity can be generated. When $T = 300$ K, $\Gamma = 1$ meV, $f = 0.8$ THz, and $\mu_c = 0$ eV, the sheet conductivity of graphene is 0.37-j0.61 mS, and the absorber is in the "on" state. When $\mu_c = 0.2$ eV, the graphene sheet conductivity is 2.1-j3.4 mS and the absorptivity is decreased to less than 30%. By adopting an electrode/graphene/SiO2/pSi structure, it is easy to change the graphene sheet conductivity 5 to 6 times. When the chemical potential increased to 0.3 eV, the graphene sheet conductivity is 3.1-j5.1 mS, which is around 8 times bigger than the conductivity of $\mu_c = 0$ eV. When the absorber is in the "off" state, the reflectivity is more than 82%. As a result, a high switching intensity (>72%) is obtained.

For the high switching intensity, the possible reasons can be derived from Fig. 5.33b. Since the electric field is strongly concentrated in the slot area, the graphene patch in the slot area has a significant effect on absorptivity. As the chemical potential increases from $\mu_c = 0$ eV to $\mu_c = 0.3$ eV, the sheet conductivity of graphene can be varied around 8 times. When $\mu_c = 0.3$ eV, the valuable interaction between the graphene patch and gold pattern which dominantly contributes to the absorption is noticeably damped, and the excellent impendence matching is destroyed. As a result, the absorption is diminished to 16%. Through biasing at different voltages to turn ON and OFF of the proposed absorber, we are able to switch the structure between absorption (absorption > 90%) and reflection (reflection > 82%) at a low-terahertz spectrum.

Figure 5.34a–d shows the central side view of the simulated electric field intensity distributions at 1 THz. The corresponding coordinate systems are shown in Fig. 5.34a. The electric field direction of the normal incident wave is along with the y-axis.

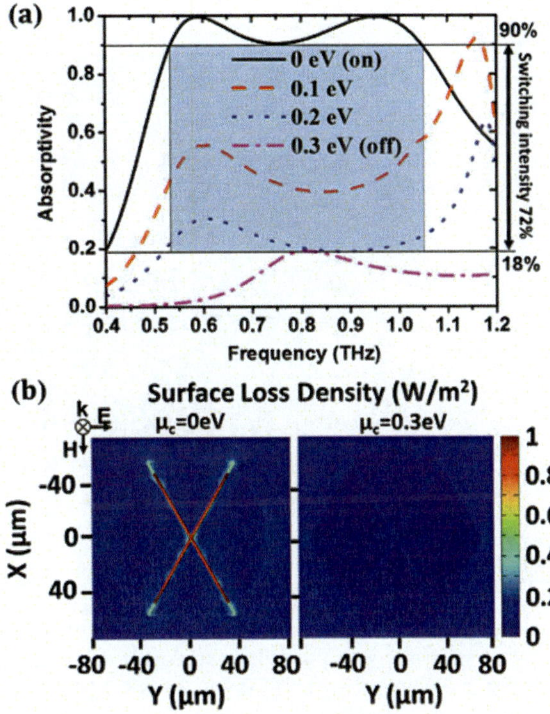

Fig. 5.33 **a** Absorptivity with different graphene chemical potential μ_c. The proposed absorber is in "on" state when $\mu_c = 0$ eV and in "off" state when $\mu_c = 0.3$ eV. **b** The surface loss density of graphene and gold at 0.8 THz, which is normalized to 1×10^9 W/m2 [96]

As shown in Fig. 5.34a and b ($\mu_c = 0$ eV), the incident electric field is concentrated on the slot area strongly formed by the neighboring gold triangle patches. Figure 5.34a also verifies the existence of the capacitance effect in the slot area. When the phase of the incident wave is increased from 0° to 90°, the wave peak of the electric field moves toward the absorber. Notably, a traveling wave exists above the absorber when $\mu_c =$ 0 eV. In Fig. 5.34c and d ($\mu_c = 0.3$ eV), a standing wave is formed and the absorber is switched into a reflector successfully for the fixed wave peak and the reduced field strength. Figure 5.35a shows the absorptivity of the proposed absorber at different polarization angles under normal incidence. The C6 symmetric unit cell structure of the MA is the inherent reason for the excellent polarization independence.

A Floquet port is set in the simulation and TE and TM polarizations could be simulated. Generally speaking, TE and TM polarizations are defined as follows: the wave vector k of the incident light is in the XOZ or YOZ plane (for TE or TM) and the electric field is in the x direction (TE) or in the YOZ plane (TM). The simulated absorptivity at different incident angles under TE and TM polarizations are plotted in Fig. 5.35b and c, respectively.

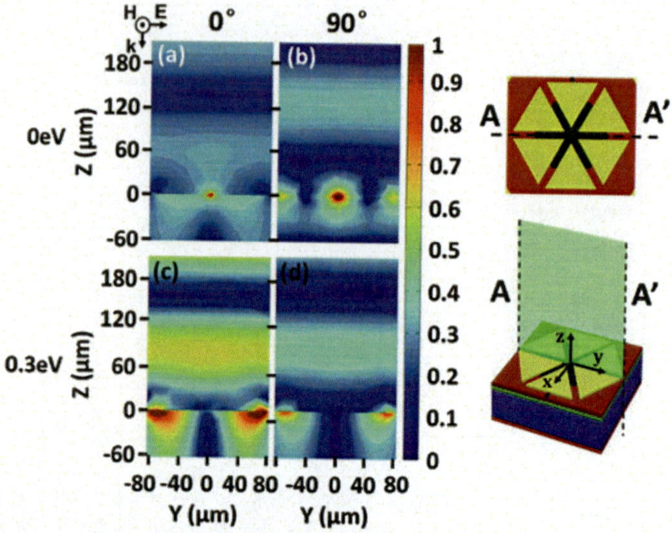

Fig. 5.34 Cross section of the normalized electric field intensity at 1 THz. For **a** and **b**, the incident wave is strongly concentrated in the graphene-gold interface when $\mu_c = 0$ eV and the traveling wave is observed. For **c** and **d**, the standing wave is observed due to the high reflectivity of the absorber when $\mu_c = 0.3$ eV and the maximum density occurs in a fixed position. All of the electric intensity distributions are normalized to 2.5×10^5 V/m [96]

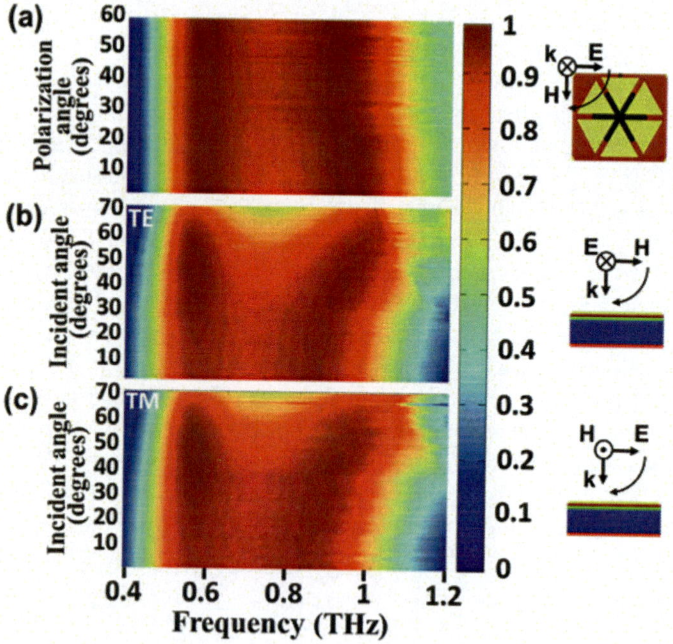

Fig. 5.35 **a** Absorptivity spectrum at different polarization angles. **b** Absorptivity at different incident angles under TE **c** TM polarizations [96]

5.6 Conclusion

Due to its finite conductivity, flexible structure, and optical transparency, graphene has a unique application potential in the metamaterial absorber area. In this chapter, we mainly introduce the theoretical characterization, structural design, and experimental verification of graphene-based metamaterial absorbers. At first, the modeling of the graphene-based Salisbury screen, Jaumann absorber, and FSS absorber are given. Then we demonstrated the microwave absorption and near-field radiation of multilayer large-area CVD graphene and presented several kinds of graphene-based metamaterial absorbers, including graphene-based transparent shielding enclosure, graphene-based quasi-TEM wave microstrip absorbers, graphene-based microwave FSS absorber, graphene-based MMW wideband transparent absorber, and graphene-based switchable THz absorber. We believe that more and more applications will be explored for graphene-based metamaterial absorbers.

References

1. P.R. Wallace, The band theory of graphite. Phys. Rev. **119**(2), 622–634 (1947)
2. J.W. McClure, Diamagnetism of graphite. Phys. Rev. **119**(2), 606–613 (1956)
3. J.C. Slonczewski, P.R. Weiss, Band structure of graphite. Phys. Rev. **109**(2), 272–279 (1958)
4. G.W. Semenoff, Condensed-matter simulation of a three-dimensional anomaly. Phys. Rev. Lett. **53**, 2449–2452 (1984)
5. E. Fradkin, Critical behavior of disordered degenerate semiconductors. Phys. Rev. B **33**, 3263–3268 (1986)
6. F.D.M. Haldane, Model for a quantum Hall effect without Landau levels: condensed-matter realization of the parity anomaly. Phys. Rev. Lett. **61**(18), 2015–2018 (1988)
7. K.S. Novoselov, A.K. Geim, S.V. Morozov, D. Jiang, Y. Zhang, S.V. Dubonos, I.V. Grigorieva, A.A. Firsov, Electric field effect in atomically thin carbon films. Science **306**(5696), 666–669 (2004)
8. K.S. Novoselov, D. Jiang, F. Schedin, T.J. Booth, V.V. Khotkevich, S.V. Morozov, A.K. Geim, Two-dimensional atomic crystals. Proc. Natl. Acad. Sci. U.S.A. **102**(30), 10451–10453 (2005)
9. K.S. Novoselov, A.K. Geim, S.V. Morozov, D. Jiang, M.I. Katsnelson, I.V. Grigorieva, S.V. Dubonos, A.A. Firsov, Two-dimensional gas of massless Dirac fermions in graphene. Nature **438**, 197–200 (2005)
10. Y. Zhang, Y.-W. Tan, H.L. Stormer, P. Kim, Experimental observation of the quantum Hall effect and Berry's phase in graphene. Nature **438**, 201–204 (2005)
11. A.K. Geim, K.S. Novoselov, The rise of graphene. Nat. Mater. **6**, 183–191 (2007)
12. A.K. Geim, Graphene: status and prospects. Science **324**(5934), 1530–1534 (2009)
13. K.S. Novoselov, V.I. Fal'ko, L. Colombo, P.R. Gellert, M.G. Schwab, K. Kim, A roadmap for graphene. Nature **490**, 192–200 (2012)
14. X. Du, I. Skachko, A. Barker, E.Y. Andrei, Approaching ballistic transport in suspended graphene. Nat. Nanotechnol. **3**, 491–495 (2008)
15. R.R. Nair, P. Blake, A.N. Grigorenko, K.S. Novoselov, T.J. Booth, T. Stauber, N.M.R. Peres, A.K. Geim, Fine structure constant defines visual transparency of graphene. Science **320**(5881), 1308–1308 (2008)
16. S. Caterina, A. Mahmood, E. Dujardin, Production, properties and potential of graphene. Carbon **48**(8), 2127–2150 (2010)

17. J. Hass, W.A. De Heer, E.H. Conrad, The growth and morphology of epitaxial multilayer graphene. J. Phys.: Condens. Matter **20**, 323202 (2008)
18. V.C. Tung, M.J. Allen, Y. Yang, R.B. Kaner, High-throughput solution processing of large-scale graphene. Nat. Nanotechnol. **4**, 25–29 (2009)
19. K.M. Milaninia, M.A. Baldo, A. Reina, J. Kong, All graphene electromechanical switch fabricated by chemical vapor deposition. Appl. Phys. Lett. **95**(18), 183105 (2009)
20. S. Bae, H. Kim, Y. Lee, X. Xu, J.-S. Park, Y. Zheng, J. Balakrishnan, T. Lei, H.R. Kim, Y.I. Song, Y.-J. Kim, K.S. Kim, B. Özyilmaz, J.-H. Ahn, B.H. Hong, S. Iijima, Roll-to-roll production of 30-inch graphene films for transparent electrodes. Nat. Nanotechnol. **5**(8), 574–578 (2010)
21. P. Blake, P.D. Brimicombe, R.R. Nair, T.J. Booth, D. Jiang, F. Schedin, L.A. Ponomarenko, S.V. Morozov, H.F. Gleeson, E.W. Hill, A.K. Geim, K.S. Novoselov, Graphene-based liquid crystal device. Nano Lett. **8**(6), 1704–1708 (2008)
22. M. Liu, X. Yin, E. Ulin-Avila, B. Geng, T. Zentgraf, L. Ju, F. Wang, X. Zhang, A graphene-based broadband optical modulator. Nature **474**, 64–67 (2011)
23. Q. Bao, H. Zhang, B. Wang, Z. Ni, C.H.Y.X. Lim, Y. Wang, D.Y. Tang, K.P. Loh, Broadband graphene polarizer. Nat. Photonics **5**, 411–415 (2011)
24. Z. Fei, A.S. Rodin, G.O. Andreev, W. Bao, A.S. McLeod, M. Wagner, L.M. Zhang, Z. Zhao, M. Thiemens, G. Dominguez, M.M. Fogler, A.H. Castro Neto, C.N. Lau, F. Keilmann, D.N. Basov, Gate-tuning of graphene plasmons revealed by infrared nano-imaging. Nature **487**, 82–85 (2012)
25. M. Tamagnone, J.S. Gómez-Díaz, J.R. Mosig, J. Perruisseau-Carrier, Reconfigurable terahertz plasmonic antenna concept using a graphene stack. Appl. Phys. Lett. **101**, 214102 (2012)
26. L. Ju, B. Geng, J. Horng, C. Girit, M. Martin, Z. Hao, H.A. Bechtel, X. Liang, A. Zettl, Y.R. Shen, F. Wang, Graphene plasmonics for tunable terahertz metamaterials. Nat. Nanotechnol. **6**, 630–634 (2011)
27. N.M. Gabor, J.C.W. Song, Q. Ma, N.L. Nair, T. Taychatanapat, K. Watanabe, T. Taniguchi, L.S. Levitov, P. Jarillo-Herrero, Hot carrier–assisted intrinsic photoresponse in graphene. Science **334**(6056), 648–652 (2011)
28. A. Andryieuski, A.V. Lavrinenko, D.N. Chigrin, Graphene hyperlens for terahertz radiation. Phys. Rev. B **86**(12), 121108 (2012)
29. P.-Y. Chen, A. Alù, Atomically thin surface cloak using graphene monolayers. ACS Nano **5**(7), 5855–5863 (2011)
30. A. Andryieuski, A.V. Lavrinenko, Graphene metamaterials based tunable terahertz absorber: effective surface conductivity approach. Opt. Express **21**(7), 9144–9155 (2013)
31. K.S. Kim, Y. Zhao, H. Jang, S.Y. Lee, J.M. Kim, K.S. Kim, J.-H. Ahn, P. Kim, J.-Y. Choi, B.H. Hong, Large-scale pattern growth of graphene films for stretchable transparent electrodes. Nature **457**, 706–710 (2009)
32. A. Reina, X. Jia, J. Ho, D. Nezich, H. Son, V. Bulovic, M.S. Dresselhaus, J. Kong, Large area, few-layer graphene films on arbitrary substrates by chemical vapor deposition. Nano Lett. **9**(1), 30–35 (2009)
33. X. Li, Y. Zhu, W. Cai, M. Borysiak, B. Han, D. Chen, R.D. Piner, L. Colombo, R.S. Ruoff, Transfer of large-area graphene films for high-performance transparent conductive electrodes. Nano Lett. **9**(12), 4359–4363 (2009)
34. X. Li, W. Cai, J. An, S. Kim, J. Nah, D. Yang, R. Piner, A. Velamakanni, I. Jung, E. Tutuc, S.K. Banerjee, L. Colombo, R.S. Ruoff, Large-area synthesis of high-quality and uniform graphene films on copper foils. Science **324**(5932), 1312–1314 (2009)
35. G.W. Hanson, Dyadic Green's functions and guided surface waves for a surface conductivity model of graphene. J. Appl. Phys. **103**, 064302 (2008)
36. M. Dragoman, D. Neculoiu, A. Cismaru, A.A. Muller, G. Deligeorgis, G. Konstantinidis, D. Dragoman, R. Plana, Coplanar waveguide on graphene in the range 40 MHz–110 GHz. Appl. Phys. Lett. **99**(3), 033112 (2011)
37. J.S. Gómez-Díaz, J. Perruisseau-Carrier, P. Sharma, A. Ionescu, Non-contact characterization of graphene surface impedance at micro and millimeter waves. J. Appl. Phys. **111**(11), 114908 (2012)

38. J. Krupka, W. Strupinski, Measurements of the sheet resistance and conductivity of thin epitaxial graphene and SiC films. Appl. Phys. Lett. **96**(8), 082101 (2010)
39. O. Habibpour, J. Vukusic, J. Stake, A 30-GHz integrated subharmonic mixer based on a multichannel graphene FET. IEEE Trans. Microw. Theory Tech. **61**(2), 841–847 (2013)
40. A. Fallahi, J. Perruisseau-Carrier, Design of tunable biperiodic graphene metasurfaces. Phys. Rev. B **86**(19), 195408 (2012)
41. C. Wang, X. Han, P. Xu, X. Zhang, Y. Du, S. Hu, J. Wang, X. Wang, The electromagnetic property of chemically reduced graphene oxide and its application as microwave absorbing material. Appl. Phys. Lett. **98**(7), 072906 (2011)
42. Z. Zheng, C. Zhao, S. Lu, Y. Chen, Y. Li, H. Zhang, S. Wen, Microwave and optical saturable absorption in graphene. Opt. Express **20**(21), 23201–23214 (2012)
43. B. Wu, H.M. Tuncer, M. Naeem, B. Yang, M.T. Cole, W.I. Milne, Y. Hao, Experimental demonstration of a transparent graphene millimetre wave absorber with 28% fractional bandwidth at 140 GHz. Sci. Rep. **4**, 4130 (2014)
44. B. Wu, H.M. Tuncer, A. Katsounaros, W. Wu, M.T. Cole, K. Ying, L. Zhang, W.I. Milne, Y. Hao, Microwave absorption and radiation from large-area multilayer CVD graphene. Carbon **77**, 814–822 (2014)
45. J. Sohn, S.H. Han, Electromagnetic noise suppression characteristics of a coplanar waveguide transmission line integrated with a magnetic thin film. J. Appl. Phys. **100**(12), 124510 (2006)
46. Z. Lu, H. Wang, J. Tana, S. Lin, Microwave shielding enhancement of high-transparency, double-layer, submillimeter-period metallic mesh. Appl. Phys. Lett. **105**(24), 241904 (2014)
47. Z. Wang, G. Wei, G.-L. Zhao, Enhanced electromagnetic wave shielding effectiveness of Fe doped carbon nanotubes/epoxy composites. Appl. Phys. Lett. **103**(18), 183109 (2013)
48. Z.P. Wu, D.M. Cheng, W.J. Ma, J.W. Hu, Y.H. Yin, Y.Y. Hu, Y.S. Li, J.G. Yang, Q.F. Xu, Electromagnetic interference shielding effectiveness of composite carbon nanotube macro-film at a high frequency range of 40–60 GHz. AIP Adv. **5**(6), 067130 (2015)
49. R. Kumaran, M. Alagar, S. Dinesh Kumar, V. Subramanian, K. Dinakaran, Ag induced electromagnetic interference shielding of Ag-graphite/PVDF flexible nanocomposites thin films. Appl. Phys. Lett. **107**(11), 113107 (2015)
50. U.F. D'Elia, G. Pelosi, S. Stefano, R. Taddei, A carbon-nanotube-based frequency-selective absorber. Int. J. Microw. Wirel. Technol. **2**(5), 479–485 (2010)
51. K. Naishadham, P.K. Kadaba, Measurement of the microwave conductivity of a polymeric material with potential applications in absorbers and shielding. IEEE Trans. Microw. Theory Tech. **39**(7), 1158–1164 (1991)
52. N.I. Landy, S. Sajuyigbe, J.J. Mock, D.R. Smith, W.J. Padilla, Perfect metamaterial absorber. Phys. Rev. Lett. **100**, 207402 (2008)
53. C.-H. Lin, R.-L. Chern, H.-Y. Lin, Polarization-independent broad-band nearly perfect absorbers in the visible regime. Opt. Express **19**(2), 415–424 (2011)
54. N. Liu, M. Mesch, T. Weiss, M. Hentschel, H. Giessen, Infrared perfect absorber and its application as plasmonic sensor. Nano Lett. **10**(7), 2342–2348 (2010)
55. N. Zhang, P. Zhou, D. Cheng, X. Weng, J. Xie, L. Deng, Dual-band absorption of mid-infrared metamaterial absorber based on distinct dielectric spacing layers. Opt. Lett. **38**(7), 1125–1127 (2013)
56. H. Li, D.R. Chowdhury, S. Ramani, M.T. Reiten, S.-N. Luo, A.J. Taylor, H.-T. Chen, Experimental demonstration of terahertz metamaterial absorbers with a broad and flat high absorption band. Opt. Lett. **37**(2), 154–156 (2012)
57. Y. Wen, W. Ma, J. Bailey, G. Matmon, X. Yu, G. Aeppli, Planar broadband and high absorption metamaterial using single nested resonator at terahertz frequencies. Opt. Lett. **39**(6), 1589–1592 (2014)
58. D.Y. Shchegolkov, A.K. Azad, J.F. O'Hara, E.I. Simakov, Perfect subwavelength fishnetlike metamaterial-based film terahertz absorbers. Phys. Rev. B **82**(20), 205117 (2010)
59. R. Yahiaoui, J.P. Guillet, F. de Miollis, P. Mounaix, Ultra-flexible multiband terahertz metamaterial absorber for conformal geometry applications. Opt. Lett. **38**(23), 4988–4990 (2013)

60. M. Gokkavas, K. Guven, I. Bulu, K. Aydin, R.S. Penciu, M. Kafesaki, C.M. Soukoulis, E. Ozbay, Experimental demonstration of a left-handed metamaterial operating at 100 GHz. Phys. Rev. B **73**(19), 193103 (2006)
61. Y. Lu, J. Li, S. Zhang, J. Sun, J.Q. Yao, Polarization insensitive, broadband terahertz metamaterial absorber. Opt. Lett. **36**(17), 3476–3478 (2011)
62. F.B.P. Niesler, J.K. Gansel, S. Fischbach, M. Wegener, Metamaterial metal-based bolometers. Appl. Phys. Lett. **100**(20), 203508 (2012)
63. F. Alves, B. Kearney, D. Grbovic, G. Karunasiri, Narrowband terahertz emitters using metamaterial films. Opt. Express **20**(19), 21025–21032 (2012)
64. F. Alves, D. Grbovic, B. Kearney, N.V. Lavrik, G. Karunasiri, Bi-material terahertz sensors using metamaterial structures. Opt. Express **21**(11), 13256–13271 (2013)
65. W. Xu, S. Sonkusale, Microwave diode switchable metamaterial reflector/absorber. Appl. Phys. Lett. **103**(3), 031902 (2013)
66. L.A. Falkovsk, S.S. Pershoguba, Optical far-infrared properties of a graphene monolayer and multilayer. Phys. Rev. B **76**(15), 153410 (2007)
67. S.A. Mikhailov, K. Ziegler, New electromagnetic mode in graphene. Phys. Rev. Lett. **99**(1), 016803 (2007)
68. V.P. Gusynin, S.G. Sharapov, Transport of Dirac quasiparticles in graphene: hall and optical conductivities. Phys. Rev. B **73**(24), 245411 (2006)
69. V.P. Gusynin, S.G. Sharapov, J.P. Carbotte, Unusual microwave response of Dirac quasiparticles in graphene. Phys. Rev. Lett. **96**(25), 256802 (2006)
70. N.M.R. Peres, F. Guinea, A.H. Castro Neto, Electronic properties of disordered two-dimensional carbon. Phys. Rev. B **73**(12), 125411 (2006)
71. N.M.R. Peres, A.H. Castro Neto, F. Guinea, Conductance quantization in mesoscopic graphene. Phys. Rev. B **73**(19), 195411 (2006)
72. K. Ziegler, Minimal conductivity of graphene: nonuniversal values from the Kubo formula. Phys. Rev. B **75**(23), 233407 (2007)
73. L.A. Falkovsky, A.A. Varlamov, Space-time dispersion of graphene conductivity. Euro. Phys. J. B **56**(4), 281–284 (2007)
74. V.P. Gusynin, S.G. Sharapov, J.P. Carbotte, Magneto-optical conductivity in graphene. J. Phys.: Condens. Matter **19**(2), 026222 (2006)
75. P.R. Wallace, The band theory of graphite. Phys. Rev. **71**(9), 622 (1947)
76. G.Y. Slepyan, S.A. Maksimenko, A. Lakhtakia, O. Yevtushenko, A.V. Gusakov, Electrodynamics of carbon nanotubes: dynamic conductivity, impedance boundary conditions, and surface wave propagation. Phys. Rev. B **60**(24), 17136 (1999)
77. V.P. Gusynin, S.G. Sharapov, J.P. Carbotte, Sum rules for the optical and hall conductivity in graphene. Phys. Rev. B **75**(16), 165407 (2007)
78. W.C. Chew, *Waves and Fields in Inhomogeneous Media* (IEEE press, New York, 1995)
79. A. Ishimaru, Electromagnetic Wave Propagation, Radiation, and Scattering. Prentice-Hall (1991)
80. T. Tamir, A.A. Oliner, Guided complex waves. Part 1: fields at an interface. Proc. IEEE **110**(2), 310–324 (1963)
81. Y.R. Padooru, A.B. Yakovlev, C.S.R. Kaipa, F. Medina, F. Mesa, Circuit modeling of multiband high-impedance surface absorbers in the microwave regime. Phys. Rev. B **84**(3), 035108 (2011)
82. A. Kasry, M.A. Kuroda, G.J. Martyna, G.S. Tulevski, A.A. Bol, Chemical doping of large-area stacked graphene films for use as transparent, conducting electrodes. ACS Nano **4**(7), 3839–3844 (2010)
83. G.W. Hanson, Dyadic Green's functions for an anisotropic, non-local model of biased graphene. IEEE Trans. Antennas Propag. **56**(3), 747–757 (2008)
84. C.-Q., Jiao, Z.H. Zhao, Resonance suppression and electromagnetic shielding effectiveness improvement of an apertured rectangular cavity by using wall losses. Chin. Phys. B **22**(8), 084101 (2013)
85. Y. Zhao, B. Wu, Y. Zhang, Y. Hao, Transparent electromagnetic shielding enclosure with CVD graphene. Appl. Phys. Lett. **109**(10), 124510 (2016)

86. O. Shaforost, K. Wang, S. Goniszewski, M. Adabi, Z. Guo, S. Hanham, J. Gallop, L. Hao, N. Klein, Contact-free sheet resistance determination of large area graphene layers by an open dielectric loaded microwave cavity. J. Appl. Phys. **117**(2), 024501 (2015)
87. J.A. Robinson, M. LaBella, M. Zhu, M. Hollander, R. Kasarda, Z. Hughes, K. Trumbul, R. Cavalero, D. Snyder, Contacting graphene. Appl. Phys. Lett. **98**(5), 053103 (2011)
88. W. Li, Y.R. Liang, D.M. Yu, L.M. Peng, K.P. Pernstich, T. Shen, A.R.H. Walker, G.J. Cheng, C.A. Hacker, C.A. Richter, Q.L. Li, D.J. Gundlach, X.L. Liang, Ultraviolet/ozone treatment to reduce metal-graphene contact resistance. Appl. Phys. Lett. **102**(18), 183110 (2013)
89. B. Xue, Y. Hu, B. Wu, L. Chen, W. Zhang, A wideband transparent absorber for microwave and millimeter wave application. APCAP 16–19 Oct 2017, Xi'an
90. H. Chen, W.-B. Lu, Z.-G. Liu, J. Zhang, A.Q. Zhang, B. Wu, Experimental demonstration of microwave absorber using large-area multilayer graphene-based frequency selective surface. IEEE Trans. Microw. Theory Tech. **66**(8), 3807–3816 (2018)
91. W.B. Lu, J.W. Wang, J. Zhang, Z.G. Liu, H. Chen, W.J. Song, Z.H. Jiang, Flexible and optically transparent microwave absorber with wide bandwidth based on graphene. Carbon **152**, 70–76 (2019)
92. J. Zhang, Z. Liu, W. Lu, H. Chen, B. Wu, Q. Liu, A low profile tunable microwave absorber based on graphene sandwich structure and high impedance surface. Int. J. RF Microw. Comput. Aided Eng. **30**(2), e22022 (2020)
93. B.B. Xing, Z.G. Liu, W.B. Lu, H. Chen, Q.D. Zhang, Wideband microwave absorber with dynamically tunable absorption based on graphene and random metasurface. IEEE Antennas Wirel. Propag. Lett. **18**(12), 2602–2606 (2020)
94. C. Li, M.T. Cole, W. Lei, K. Qu, K. Ying, Y. Zhang, A.R. Robertson, J.H. Warner, S. Ding, X. Zhang, B. Wang, W.I. Milne, Highly electron transparent graphene for field emission triode gates. Adv. Funct. Mater. **24**(9), 1218–1227 (2014)
95. K.K. Kim, A. Reina, Y. Shi, H. Park, L.J. Li, Y.H. Lee, J. Kong, Enhancing the conductivity of transparent graphene films via doping. Nanotechnology **21**(28), 285205 (2010)
96. Y. Zhao, B. Wu, B. Huang, Q. Cheng, Switchable broadband terahertz absorber/reflector enabled by hybrid graphene-gold metasurface. Opt. Express **25**(7), 7161–7169 (2017)
97. S. Walia, C.M. Shah, P. Gutruf, H. Nili, D.R. Chowdhury, W. Withayachumnankul, M. Bhaskaran, S. Sriram, Flexible metasurfaces and metamaterials: a review of materials and fabrication processes at micro-and nano-scales. Appl. Phys. Rev. **2**(1), 011303 (2015)

Chapter 6
Frequency-Domain and Space-Domain Reconfigurable Metasurfaces

Jiaqi Han, Guangyao Liu, Qiang Feng, and Long Li

Abstract Since its first appearance, digital reconfigurable metasurfaces have been developed rapidly in the past few years. This type of devices can dynamically modulate EM wave characteristic parameters (e.g., scattering amplitude, phase, and polarization), molding the wavefronts into shapes that can be arbitrarily designed, and greatly enhancing the nonlinear response of the system. With the deepening of research, reconfigurable metasurfaces have gradually transitioned from single-function devices to multi-functional modules and advanced information systems. In this chapter, the design methods and recent progress of exotic metasurfaces operating at frequency ranging from microwave to visible are reviewed first. Following the above basis, we emphatically discuss the concepts and the functions of the digital reconfigurable metasurfaces, including polarization conversion, beam-forming, and beam-shaping, orbital angular momentum (OAM) generation, etc.

Keywords Digital reconfigurable metasurface · Coding · Multi-functional · Polarization conversion · Beam-forming and beam-shaping · Orbital angular momentum (OAM) generation

6.1 Introduction

As a two-dimensional (2D) structure of metamaterials, metasurfaces have greatly expanded existing metamaterials' modulation capabilities and application fields. Metasurfaces are considered as one of artificially functional materials [1–4] due to their advantages of flexible phase control ability [5], optional conversion modes in wideband or multiple bands [6], ultra-small thickness, and other finely tailorable

J. Han · Q. Feng · L. Li (✉)
School of Electronic Engineering, Xidian University, Xi'an 710071, Shannxi, China
e-mail: lilong@mail.xidian.edu.cn

G. Liu
Department of Electrical Engineering, City University of Hong Kong, Hong Kong, SAR, China

L. Li et al. (eds.), *Electromagnetic Metamaterials and Metasurfaces: From Theory To Applications*, https://doi.org/10.1007/978-981-99-7914-1_6

characteristics [7–9]. There have been many metasurface devices applied to manipulate electromagnetic (EM) waves, such as reflective and transmissive polarization converters [6, 10], beam deflecting and shaping [11, 12], and carrying information for applications such as sensing, imaging and communication [13, 14], etc. The above-mentioned metasurfaces, which are designed using sheet transition condition or generalized Snell's law, have continuous characteristic parameters between adjacent elements, and such devices are called "analog metasurfaces". Correspondingly, a series of coding metasurfaces is designed and applied by researchers, who are inspired by the idea of binary coding. Della Giovampaola et al. proposed a method for constructing a "digital metamaterial" in the spatial domain through an appropriate combination of "digital metamaterial bits", in which the "digital metamaterial bits" are material particles (e.g., Ag and Si) with different properties [15]. At the same time, T. J. Cui et al. developed the first binary phase coding (digital states "0" or "1") metasurface in the microwave frequency, which successfully modulated electromagnetic waves by orderly arranging coding units in a 2D space [16]. The concept of the coding metasurface greatly simplifies the metasurface design and meanwhile builds a bridge between the physical world and the digital information world. Thus, coding metasurfaces have been increasingly popular.

Most previously mentioned devices have fixed structures and are generally difficult to adjust in real time, which greatly limits the scope of applications for metasurfaces. The recently proposed reconfigurable metasurfaces allow dynamically manipulating the EM waves. In general, the reconfigurability of the device can be achieved by using external electromagnetic modulation or controllable components of the element. On the one hand, the polarization multiplexing technology related to the element topology can reconstruct the characteristic of the metasurface according to the polarization direction of the incoming EM wave, thus giving rise to reconfigurable polarization converters (RPCs). In 2015, L. Li reported a polarization-reconfigurable converter using a multi-layer metasurface, which was proposed to convert the polarization states of the reflected wave by rotating the polarization of the incident wave [17]. In addition, the direction of re-radiated beams or the functions of the metasurface can be switched according to the polarization direction of the incident wave [18–20]. On the other hand, reconfigurability can also be achieved by using tunable components that tune the metasurface responses, namely, mechanical parts [17], PIN diodes or switches [21], varactors, or variable resistors [22], microelectromechanical systems (MEMS) [23, 24].

Because the "ON" and "OFF" states of the switches correspond to the "0" and "1" of the binary code, a programmable metasurface can be formed when the elements loaded with the active components are arranged periodically. With the deepening of research, the unit control method of programmable metasurface transitions from multi-unit synchronous control to discrete control. The central controller has also changed from a microcontroller (MCU) [25] to a field programmable gate array (FPGA) [26], which greatly enhances the real-time information processing capability of the metasurface. Recently, the integration of metasurfaces is becoming higher and higher. Some researchers have integrated sensors into the elements so that metasurfaces can sense environmental information [27]. According to the development vision of the digital metasurface (intelligent metasurface), this hybrid system can finally sense the changes of the external physics field (sound, light, electricity, etc.) in real-time, switch functions freely under the driving of self-adaptive algorithms and finally complete the information interaction with the environment.

Programmable metasurface has proven helpful in solving the challenge of packaging multiple functions for different scenarios in a single device. Some researchers have realized beam-forming and perfect absorption using programmable metasurfaces. Yang et al. has designed a reconfigurable metasurface with integrated functions including polarization conversion and wavefront modulation [28]. Cui et al. combined a polarized grid and PIN diodes to design programmable transmission-reflective metasurfaces (TRMs) [26]. In addition to continuously expanding the functions of the metasurface, Cui et al. further proposed space–time-coding digital metasurface, which greatly improved the flexibility of programmable metasurface [12]. Frequency domain and space domain programmable metasurface and multi-domain reconfigurable metasurface have also become new research directions. However, multi-functions of the metasurface cause some adverse problems, such as the integration of the complicated control circuits and the parasitical coupling suppression, etc. At the same time, the control system must be sufficiently powerful and extremely low-loss, which imposes strict restrictions on the actual design of circuits and systems, and therefore encourages highly simplified solutions at the circuit, system, and overall architecture levels.

Different from the modulation methods of the above-mentioned microwave programmable metasurfaces, the metasurfaces working in the terahertz (THz) frequency band and even the visible light frequency band are combined with advanced materials (such as silicon island [29] or liquid crystals [30]) to achieve reconfigurability. In particular, Shrekenhamer et al. reported a four-color metasurface absorber in which the reflection and absorption in each pixel are dynamically controlled by all-electronic means [31]. Such a reconfigurable type of device has been widely used in the fields including detection and compression imaging. In addition, the design strategies of reconfigurable metasurfaces have also been extended to other fields, of which the spoof surface plasmon polaritons (SPPs) are a very constructive achievement. Zhou and Xiao have proposed and experimentally demonstrated a band-notched surface plasmonic filter, which is mounted on active components across the slit cut in the C-shaped ring, and dynamic control of rejection of spoof SPPs can be accomplished [32]. Ma et al. presented an electronically controlled programmable SPPs waveguide, whose dispersions can be manipulated in real-time at a fast speed by programming the bias voltage. The device has three different modes in different frequency bands [33].

This chapter will review the working mechanism and design strategy of frequency-domain and space-domain reconfigurable metasurfaces and show the novel manipulation of EM waves using digital metasurfaces. And this chapter is organized as follows. Section 6.2 presents the basic design and working principles of reconfigurable metasurfaces. Section 6.3 introduces spatial domain reconfigurable metasurfaces and their typical applications, such as beam scanning, imaging, and OAM beam generation. In Sect. 6.4, multi-function and multi-domain reconfigurable metasurfaces are introduced. At the end of this chapter, the important developing directions of metasurfaces and metamaterials are summarized, as well as the discussion on future research spots and challenges.

6.2 Scattered Electric Field Modulation Method for Reconfigurable Metasurface

In the EM environment, a reconfigurable metasurface is made of groups of elements with different EM responses in frequency, time, and spatial domains. Here, a method to design a reconfigurable metasurface using scattering parameters that are related to frequency, time, and spatial azimuth is proposed. The scattering matrix **S** of an interface relates the incident and re-radiation electric fields on the same or different sides of the pth element

$$\begin{bmatrix} E_x^{(o)} \\ E_y^{(o)} \\ E_u^{(o)} \\ \vdots \end{bmatrix} = \boldsymbol{S} \begin{bmatrix} E_x^{(i)} \\ E_y^{(i)} \\ E_u^{(i)} \\ \vdots \end{bmatrix} \tag{6.1}$$

where the superscripts i and o indicate the incidence and re-radiation, and the subscripts x, y, and u indicate the polarization directions of the electric field, respectively. In order to achieve the unit-cell adjustable in the frequency, time, and spatial domains simultaneously according to the polarization direction of the incident wave, the scattering matrix of the pth element is given by

$$\boldsymbol{S}_p = \begin{bmatrix} S_{xp}(\theta, \varphi, f, t) & & & \\ & S_{yp}(\theta, \varphi, f, t) & & \\ & & S_{up}(\theta, \varphi, f, t) & \\ & & & \ddots \end{bmatrix} \tag{6.2}$$

where S_{xp}, S_{yp}, and S_{up} are the scattering parameters of the pth unit cell related to the space, time, and frequency domains, f is the working frequency, t is the instantaneous time, and θ and φ are the elevation and azimuth angles corresponding to the scattered beam of the element, respectively. The scattering parameter matrix is a Jordan standard type so that the frequency, time, and spatial domain functions of the unit cell do not interfere with each other. According to the modulated field $\boldsymbol{E}_p^{(o)}$ of each element, the re-radiation field of the metasurface can be expanded as

$$\boldsymbol{E} = \sum_{p=1} \boldsymbol{E}_p^{(o)} = \left(\sum_{p=1} \boldsymbol{S}_p \right) \boldsymbol{E}_p^{(i)} \tag{6.3}$$

The scattering parameters in different fields can be expressed as

$$S_p(f) = \sum_{n=0} \left[\frac{a_{pn}}{n!}(f - f_1)^n + \frac{b_{pn}}{n!}(f - f_2)^n + \frac{c_{pn}}{n!}(f - f_3)^n + \cdots \right] \tag{6.4}$$

$$S_p(t) = \sum_{n=0} \Gamma_{pn} \int_0^{n\tau} e^{-j2\pi m f t} dt \tag{6.5}$$

$$S_{sp}(\theta, \varphi) = \alpha_p^n \cdot \exp[jk(x \sin\theta \cos\varphi + y \sin\theta \sin\varphi)] \tag{6.6}$$

Scattering parameters are expanded in the frequency domain according to Taylor series, where $f_1, f_2, \ldots$ are different resonant frequencies, a_{pn}, b_{pn}, … are nth-order resonance coefficients at the corresponding resonant frequencies. Γ_{pn} is the time-domain modulation coefficient, α_p^n is a set of spatial modulation coefficients, and k is the wavenumber. When $n = 0$ (0 th-order), the element is just like a gradient metasurface with a fixed topology, in which Eq. (6.4) is resolved as

$$S_p(f) = a_{p0} + b_{p0} + c_{p0} + \cdots \tag{6.7}$$

It is difficult for the fixed structure to switch the states in different resonance modes. In this design method, we introduce frequency modulation technology to freely change the resonant frequency of the unit cell according to the Taylor series. The scattering parameters can also modulate EM waves in multiple directions, which will be analyzed in the later section.

6.3 Spatial Domain Reconfigurable Metasurfaces and Applications

6.3.1 *Reconfigurable Metasurface Designs and Beam-Scanning*

Since 2014, digital metamaterials have been proposed to extend the applications of metamaterials with "0" and "1" bits sequences [15]. By properly selecting two metamaterial bits with opposite relative permittivity, the digital metamaterial can be formed to achieve scattering, the convex lens, and the graded-index lens. At the same time, a coding metasurface [16] different from that in [15] has been proposed. In [15], the metamaterial is characterized by using discrete equivalent material parameters. By contrast, the coding concept has been ingeniously developed in [16], thus simplifying the metamaterial design and optimization. With the concept of coding metasurfaces, various devices to manipulate electromagnetic waves, such as the transmission or reflection beams forming [34, 35], phase gradient surface [36], mantle cloak and absorber [37], etc., have been presented. However, the building blocks of the conventional coding metasurfaces are passive, which cannot be altered once it is designed. As the active device and coding metasurface are combined, the novel programmable metasurfaces can be formed to control microwaves, terahertz waves, and holograms [30, 38, 39].

Generally, reconfigurable metasurfaces have very narrow bandwidths, for instance, microstrip elements with lumped components [40]. For conventional microstrip array antennas, the element bandwidth is the main factor limiting the gain bandwidth. Some improved methods have been proposed. By introducing two resonances in the element, a varactor diode-tuned element with improved bandwidth was proposed [41]. In this chapter, a wideband reconfigurable metasurface is designed using PIN diodes instead of the varactor. And topology of the proposed wideband 1-bit element is shown in Fig. 6.1. The element consists of three metallic layers and two dielectric substrate layers (FR4, $\varepsilon_r = 4.4$, tan$\delta = 0.02$). The top layer contains a simple slotted square patch, and the separated patches are connected with a PIN diode to evoke 180° phase shift. The middle layer is the metallic plane that serves as radio frequency (RF) and direct circuit (DC) ground. The DC loop is shown in Fig. 6.1c. Here Skyworks SMP1340-040LF PIN diodes are selected as the lumped components. Simulated results of the reflection phase and magnitude of the 1-bit reconfigurable element when the diode turns "ON" and "OFF" are shown in Fig. 6.1d. It can be seen from the figure that 180° ± 20° phase difference (dark blue with a triangle symbol curve) is achieved from 4.7 to 5.3 GHz. The reflection loss is less than 0.9 dB within the working frequency band. In this design, the proposed reconfigurable metasurface element realizes a maximum phase shift of 200° at 5.0 GHz, and two 180° phase shifts occur at 4.8 GHz and 5.2 GHz. This method leads to 12.0% fractional bandwidth at the central frequency of 5.0 GHz.

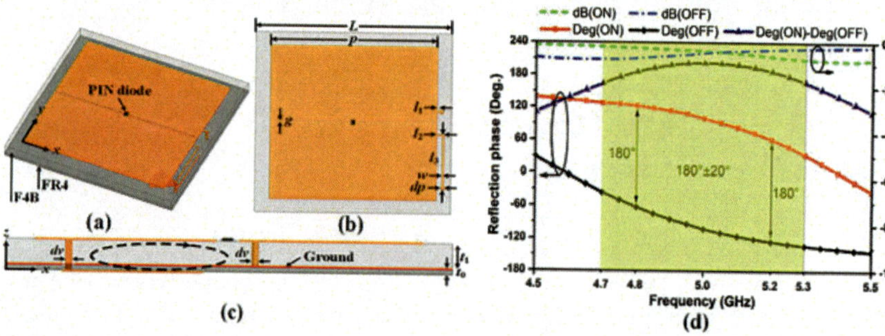

Fig. 6.1 Topology and characteristics of proposed 1-bit reconfigurable element: **a** perspective view, **b** top view, **c** side view with DC loop, and **d** simulated results of reflection phase (including phase difference of ON and OFF states), reflection magnitude, and operating band of the proposed element [25]

Based on the proposed wideband 1-bit active tuned element, a 12 × 12 reconfigurable metasurface is designed. In the design, the control board that can regulate 144 output channels independently is elaborately designed, which is shown in Fig. 6.2a. Therefore, it is flexible to steer the reflection waves to realize two-dimensional beam-scanning. The board is controlled by an 8-bit microcontroller unit (MCU). One hundred forty-four light-emitting diodes (LEDs) are in series with every PIN diode control channel to compose a display grid. It can be judged from the variation of the LEDs grid that the target phase shift distribution is changed, namely, to launch the desired beam.

Generally, metasurface antennas collimate the incoming waves fed by a horn antenna to generate target beams. Using lumped components, the beam-scannable metasurface can be realized by dynamically tuning the phase shifts of all reflection elements independently. The required phase shift of each element can be determined with the scattered electric field modulation method. In this design, inserting Eq. (6.6) into Eq. (6.1), one has

$$E_{yp}^{(o)} = \alpha_p^n \cdot \exp[jk(x \sin\theta \cos\varphi + y \sin\theta \sin\varphi)] \cdot E_{yp}^{(i)} \tag{6.8}$$

The re-radiated electric field from the reconfigurable metasurface in an arbitrary direction can be generally expressed as

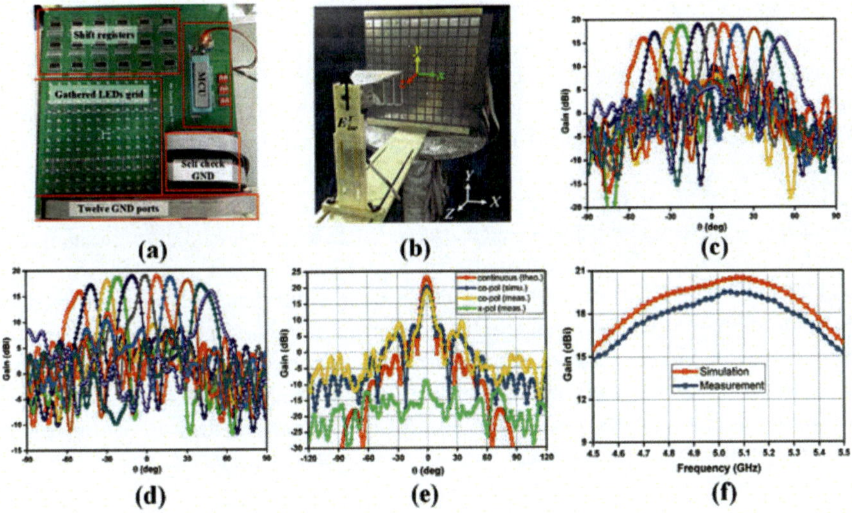

Fig. 6.2 Metasurface prototype system, test environment, and its radiation characteristics: **a** phase shift real-time displayed control board and its functional areas, **b** measurement system configurations of beam-scanning, **c** measured radiation patterns of beam-scanning within ± 50° in *xoz*-plane, **d** *yoz*-plane at 5.0 GHz, **e** comparisons of simulated and measured radiation patterns at 5.0 GHz with cross-polarization pattern and continuous phase shift pattern, and **f** comparison of simulated and measured gain bandwidth [25]

$$E(\hat{r}) = \sum_{m=1}^{M}\sum_{n=1}^{N} F_{mn}(\vec{r_{mn}} \cdot \vec{r_f}) \cdot A_{mn}(\vec{r_{mn}} \cdot \hat{r}_0) \cdot A_{mn}(\hat{r} \cdot \hat{r}_0) \cdot \exp\begin{bmatrix} jk_0(\alpha_{mn} + \vec{r_{mn}} \cdot \hat{r}) \\ +j\phi_{mn}^b \end{bmatrix} \tag{6.9}$$

where F_{mn} is the feed-to-element pattern function, A_{mn} is the reflective element pattern function, $\vec{r_{mn}}$ is the position vector of the element, $\hat{r}_0$ is the desired beam direction, and α_{mn} is the feed-to-element phase shift. Consider that we choose the 1-bit reconfigurable metasurface working with phase shift 200°, which is different from conventional 1-bit quantization. In this design, we quantized the phase shifts as

$$\phi_{mn}^b = \begin{cases} 0^\circ & 0 \le \phi_{mn} < 200^\circ \\ 200^\circ & 200^\circ \le \phi_{mn} < 360^\circ \end{cases} \tag{6.10}$$

where $\phi_{mn} = -k_0(\alpha_{mn} + \vec{r_{mn}} \cdot \hat{r})$ is the analog compensation phase.

The assembled prototype is shown in Fig. 6.2b to measure beam-scanning in *xoz*-plane. The focal diameter ratio of this reconfigurable metasurface is 0.9. Figure 6.2c and d report the measured results in *xoz*- and *yoz*-planes at 5.0 GHz, respectively. It can be clearly seen that good beam-scanning performance within two-dimensional ± 50° elevation angle is obtained. Average sidelobe levels (SLLs) are less than −10 dB. Simulated and measured boresight patterns at 5.0 GHz in *xoz*-plane are shown in Fig. 6.2e. The measured cross-polarization level is about 28 dB less than the co-polarization level. In addition, the pattern obtained by the continuous phase shift elements is also illustrated in Fig. 6.2e, and the quantization loss is about 2.5 dB. The measured maximum gain is 19.22 dBi, and the aperture efficiency is 15.26% at 5.0 GHz.

Moreover, the gain bandwidth of the reconfigurable metasurface in the boresight direction was investigated, and the measured results combined with the simulated one are shown in Fig. 6.2f. The measured −1 dB gain bandwidth is achieved from 4.85 to 5.275 GHz, giving rise to 8.4% fractional bandwidth that indicates a wideband reconfigurable metasurface compared to the conventional designs. A comparison of the proposed 1-bit reconfigurable metasurface with some previous designs verified that the choice of the working frequency in terms of 200° phase shift where the maximum phase difference occurs could effectively enhance the bandwidth of the 1-bit reconfigurable metasurface.

6.3.2 Reconfigurable Metasurfaces Designs and Imaging Applications

The topology of the proposed element in Sect. 6.3.1 can be transplanted to other frequency bands readily with good characteristics. Accordingly, a 1-bit coding element tuned by a PIN diode operating at 35 GHz is proposed. In particular, it should be noted that the PIN diode here uses MACOM MADP-000907. The dielectric substrate consists of FR4 and Rogers 4350 ($\varepsilon_r = 3.66$, $\tan\delta = 0.004$). The phase difference at 35 GHz is 180° when the PIN diode turns "ON" and "OFF", and the bandwidth of this element is 5.4%. The reflection loss within the working frequency band is less than 2.1 dB. A 1-bit reconfigurable metasurface (RM) with 20×20 cells is designed and fabricated based on the proposed element. The front view of the prototype is shown in Fig. 6.3a. Such integration gives a new opportunity for combining the electromagnetic (EM) field manipulation ability of the RM and board-level high-speed signal process unified.

As flexibly manipulating EM waves, the RM can be considered a low-cost, efficient millimeter-wave imaging solution. Applying the beams as temporal EM pulse to illuminate the object, we could receive the reflected waves for reconstructing targets. If the tangential electric field (E-field) is known, the far-field and near-field distributions can be calculated using Fourier optics. The spectrum functions can be expressed with Fourier transform as

$$F_{x/y}(\sin\theta\cos\varphi, \sin\theta\sin\varphi) = \frac{1}{\lambda^2}\iint\limits_{surf} E^t_{x/y}(x, y)\exp\left[jk_0\begin{pmatrix}\sin\theta\cos\varphi x\\ +\sin\theta\sin\varphi y\end{pmatrix}\right]\mathrm{d}x\mathrm{d}y \tag{6.11}$$

For far-field imaging, the beam-forming region should be determined by rapidly scanning the objects and denoted by ($\theta_{\min}$, $\theta_{\max}$) for the elevation angle and ($\varphi_{\min}$, $\varphi_{\max}$) for the azimuth angle. Using the scanning region combined with the high-gain beamwidth of the RM, beam pointing directions of temporal EM pulse should be calculated and denoted by (θ_q, φ_q), where $q = 1, 2, \ldots Q$. Then, the EM pulse signals are sent to the transmitter (Tx), and the receiver (Rx) records the scattered field information. Iterate on this loop, and finally, imaging of the objects could be reconstructed according to the echo signals. The original target and the reconstructed image are shown in Fig. 6.3b and c, respectively. The 1-bit RM is located in the *xoy*-plane, and the object is an aircraft model, which is parallel to the RM with a distance of 10 m. As can be seen, the outline of the aircraft model can be clearly

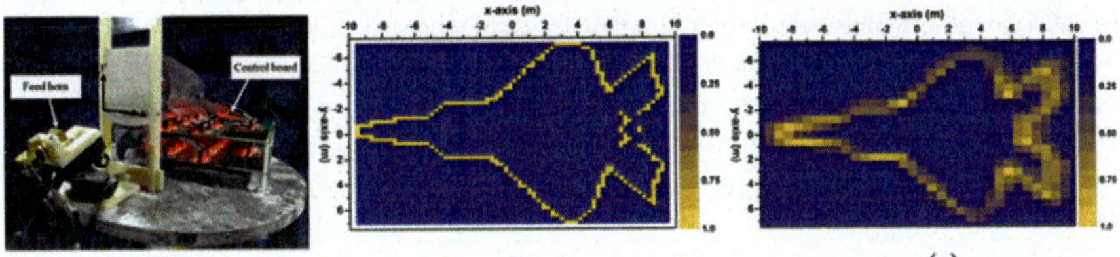

Fig. 6.3 **a** Sample design and actual system settings, **b** the original far-field millimeter-wave imaging for aircraft model, and **c** the reconstructed far-field millimeter-wave imaging using 1-bit RM for aircraft model [43]

recognized from the reconstructed image. All data in the image are normalized. It can be improved by using a larger aperture RM that has a narrower beamwidth and lower SLL. Compared to conventional aircraft imaging with radar adaptive beam-forming [42], the 1-bit RM is low-cost and easy to be realized.

The previous part shows the simulation model of far-field millimeter-wave imaging for large objects using 1-bit RM. However, for near-field millimeter-wave imaging, the high-gain beam is inappropriate since the beam pattern is unstable within the Fresnel region, and the patterns are related to the transmission distance. Thereby, a stochastic beam pattern is a promising solution for near-field millimeter-wave imaging. Using high-Q metasurface elements with random arrangements, we can realize the imaging restoration in the frequency domain [43, 44]. According to computational imaging, we can describe the measurement mathematically using the matrix $\mathbf{g} = \mathbf{Hf} + \mathbf{n}$, where $\mathbf{g}$ is the measurement, $\mathbf{H}$ is a matrix with I by J that each row represents a random pattern with J pixels, and $\mathbf{f}$ is the sampled representation vector of the object, and $\mathbf{n}$ is the noise coefficient in the measurement. To verify that randomly generated coding sequences possess low mutual coherence, we calculate 500 groups of near E-field patterns using RM parameters. Four of the 500 groups of E-field patterns are shown in Fig. 6.4a. The mutual coherence is only 0.015, indicating a good candidate for imaging restoration. In the actual measurement, the mutual coherence of the measured 64 patterns is 0.12. Such mutual coherence is good enough for imaging reconstruction. Four actual E-field patterns measured at 35 GHz are shown in Fig. 6.4b. As can be seen, within the scanning plane, the field patterns are varied dramatically in each measurement mode.

Here, a virtual object of the human body is built, and the sampled pixel $\mathbf{f}$ is shown in Fig. 6.4c. The reconstructed image is shown in Fig. 6.4d. As can be seen, the virtual body is properly reconstructed. Figure 6.4e illustrates the reconstructed image using 2000 groups of random patterns with a mutual coherence of 0.0133. From the results, it can be clearly seen that more details are restored as randomly distributed patterns increase. By adjusting the random pattern generation from the frequency domain to the time domain, more patterns with low mutual coherence can be formed. Without the requirement of a wide frequency band and high-Q metasurface element, complex image restoration could be realized.

Here, a virtual object with different letters is used as $\mathbf{f}$, as shown in Fig. 6.4f. The pixel density is 17×17 because the measured modes are only 64 groups. Nevertheless, for these simple objects, the measured data are sufficient. From Fig. 6.4g, the letters in the restoration images can be clearly recognized as their counterparts in the original images. Particularly, the non-uniformed amplitudes, which are intentionally designed, can also be recognized. Through imaging reconstruction, it is verified that the proposed RM is a promising technique for millimeter-wave imaging with excellent performance.

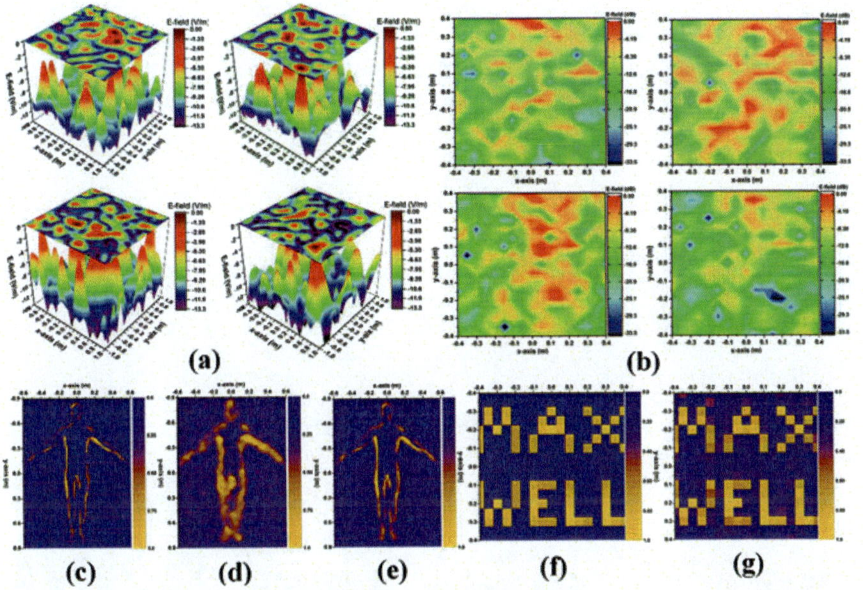

Fig. 6.4 Simulation and measurement results of millimeter-wave near-field imaging **a** four calculated near-field patterns using the random coding sequences, **b** four measured E-field patterns of total 64 patterns at 35 GHz, **c** a virtual human body imaging, **d** reconstructed image using 500 groups of random patterns, **e** reconstructed image using 2000 groups of random patterns, **f** the original letters "MAXWELL" with the non-uniformed amplitudes, and **g** the restoration letters "MAXWELL" [43]

6.3.3 Digital Orbital Angular Momentum Vortex Beam Generator

The interesting characteristic of orbital angular momentum (OAM) carrying beams is that they form a complete orthogonal modal basis set, thus establishing a new set of data carriers that do not depend on polarization or frequency. In the optical field, spiral phase plates, cylindrical lenses, and synthesized holograms are widely used to create phase singularity artificially [45–47]. In 2011, Tamburini et al. experimentally verified that non-integer OAM could be achieved by using radio techniques [48]. At RF frequencies, the researchers used circular arrays of antennas [49], spiral phase plates [50], circular and elliptical patch antennas [51], leaky wave antennas, and transmitarrays [52] to achieve such a concept. Recently, metasurfaces have been adopted to twist the wavefront phase of EM waves so that multiple OAM vortex beams can be generated [53–55].

Utilizing the 1-bit element proposed in Sect. 6.3.1, we designed a 12×12 digital OAM generator, as shown in Fig. 6.5a. To ensure the performance of the generator, we fabricated the prototype with the center operating frequency of 5 GHz. The substrate thickness of F4B and FR4 are 2 mm and 0.5 mm, respectively. For $l = 1$ and $l = 2$

OAM modes, the working status of all 1-bit elements can be visualized and verified using LEDs, as shown in Fig. 6.5a. The theoretical formula for the generation of the OAM vortex beam using a typical reflective metasurface defines the compensation phase as

$$\phi_{mn} = l\Phi_0 - jk_0(\alpha_{mn} + \vec{r_{mn}} \cdot \hat{r}),\ l = 0, \pm 1, \pm 2, ... \tag{6.12}$$

where Φ_0 is the azimuthal angle of the mn^{th} 1-bit element, l is the OAM mode number. Therefore, a digital compensation phase distribution is presented by binary coding sequences after phase fuzzification. As PIN diodes are integrated with the

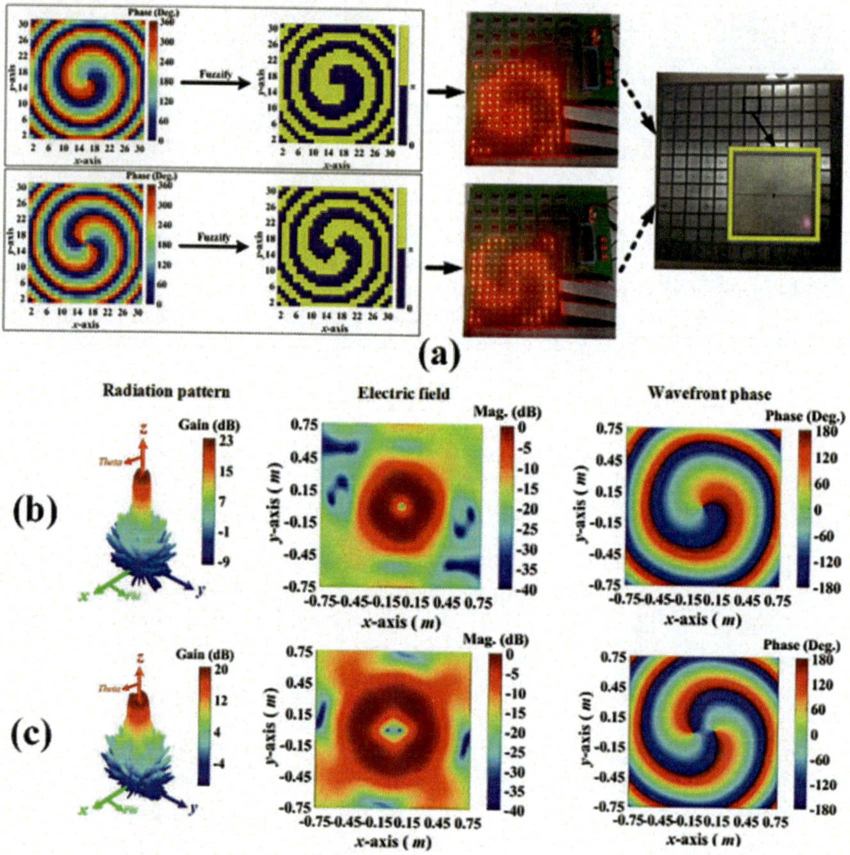

Fig. 6.5 Theoretical analysis results and experimental prototype of a 1-digit OAM generator: **a** 1-digit OAM generator control strategy, corresponding control board PIN diodes working state, experimental prototype and analog compensation phase and binary compensation phase distributions of different modes ($l = 1$ and $l = 2$), **b** theoretical analysis results of OAM generator with $l = 1$, and **c** $l = 2$ at 4.75 GHz [56]

binary elements, the digital metasurface can dynamically generate the "0" and "1" bit sequences so that the desired OAM vortex beam can be flexibly launched.

In order to validate the 1-bit digital OAM generator, a theoretical model of the coding reflective metasurface is constructed and analyzed. Different analog compensation phases and their corresponding binary compensation phases of the OAM modes are illustrated in Fig. 6.5a. It can be clearly seen that the binary compensation phase distributions also exhibit helical characteristics. Theoretical analysis results of two different OAM modes are shown in Fig. 6.5b and c. The whole vortex wavefront phases can be clearly recognized for two different OAM modes at 4.75 GHz. If a circular metasurface is adopted, the circularly shaped intensity nulls could be observed. Meanwhile, as the OAM mode number increases, the maximum gain of the radiation pattern decreases, and the OAM beam diffuses slightly at the sampling plane. Based on the above analysis, the 1-bit digital OAM generator based on a digital reconfigurable metasurface is feasible to generate different OAM vortex beams.

With the flexibility of analog metasurfaces, linear polarization multiple beams and dual-polarization dual-mode beams carrying OAM are experimentally demonstrated [54, 55]. However, limited by compensating for phase quantization errors, it is difficult for bit-type digital metasurfaces to generate multiple OAM beams, high-order OAM beams, and mixed OAM mode beams. Based on the previous section, a reconfigurable OAM vortex beam generator loaded with varactors is proposed. The proposed reconfigurable metasurface element is a three-layer structure, as shown in Fig. 6.6a. The unit cell is similar to that in Sect. 6.3.1, but the thicknesses of two substrates (F4B and FR4) in these three layers are 2.0 mm and 0.5 mm, respectively. In addition, a Skyworks SMV1405 varactor diode is loaded on the element. Figure 6.6b and c illustrate the amplitude and phase of reflection coefficient versus bias voltage by a full-wave simulator with periodic boundary conditions. It can be seen that the reconfigurable element can work in a broadband by adjusting the bias voltage for desired compensating phase. Here, the reflection phase of the element covers over 320°, and its reflection amplitude is less than 1.8 dB at 5 GHz. For the required compensation phase that exceeds 320°, we artificially assign the bias voltage of the element as 5 V, which corresponds to 320°. It is found that this processing method has little impact on vortex beam generation through numerical analyses.

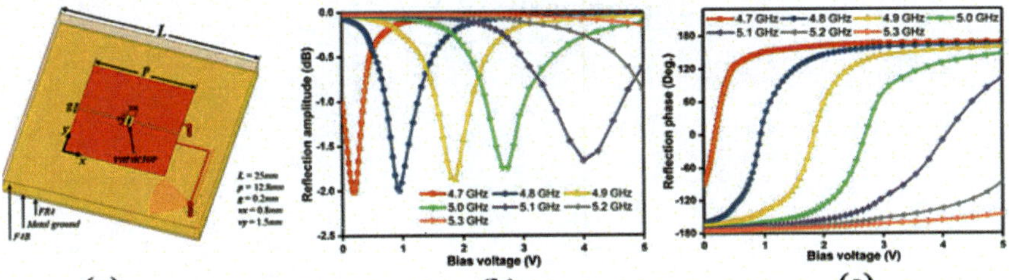

Fig. 6.6 **a** Geometry and biasing line layout of reconfigurable metasurface element, **b** reflection amplitude, and **c** reflection phase versus bias voltage ranging from 0 to 5 V [57]

In order to demonstrate the versatility of the proposed generator, a reconfigurable metasurface with 30 × 30 proposed elements is also theoretically analyzed. Figure 6.7a shows the generated two vortex beams with OAM mode numbers of l = 1 and l = 2 in two different propagation directions, which are $\theta = 20°$, $\varphi = 0°$ and $\theta = 20°$, $\varphi = 180°$ simultaneously. Electric field intensity and phase maps indicate that two helical beams are generated successfully. Figure 6.7b shows a vortex beam with high-order OAM mode number of l = 5. It can be seen that the OAM mode purity of l = 5 is up to 0.983. Figure 6.7c shows a mixed OAM vortex beam that combines the modes of l = 1 and l = 4. The weights of l = 1 and l = 4 are 0.411 and 0.412 respectively. The parasitic mode will disappear by using a larger metasurface to generate a mixed vortex beam combined with l = 1 and l = 4. These theoretical analysis results show that the proposed varactor-loaded OAM vortex beam generator has good flexibility and multi-functionality.

The prototype of the 16 × 16 reconfigurable reflective metasurface and control bias circuit board is fabricated and measured. To regulate the 256 varactor diodes independently, 64 chips of digital-to-analog converter (DAC), each of which has four voltage output channels, are selected to be the essential parts of the control board. The four OAM beams are generated dynamically by switching the four-state OAM switcher. Figure 6.8 shows the comparison of the theoretical analysis results and the measurement results. The OAM mode number weights of the four beams are 0.733 (OAM beam #1, l = 1), 0.673 (OAM beam #2, l = 2), 0.667 (OAM beam #3, l = 2), and 0.731 (OAM beam #4, l = 1), respectively. The measured results show that the generated OAM vortex beams by using the versatile generator have good performance. The fact that there is an upmost ± 11% variation in varactor diode junction capacitance [58]. And it leads to distortion of the reflection phase

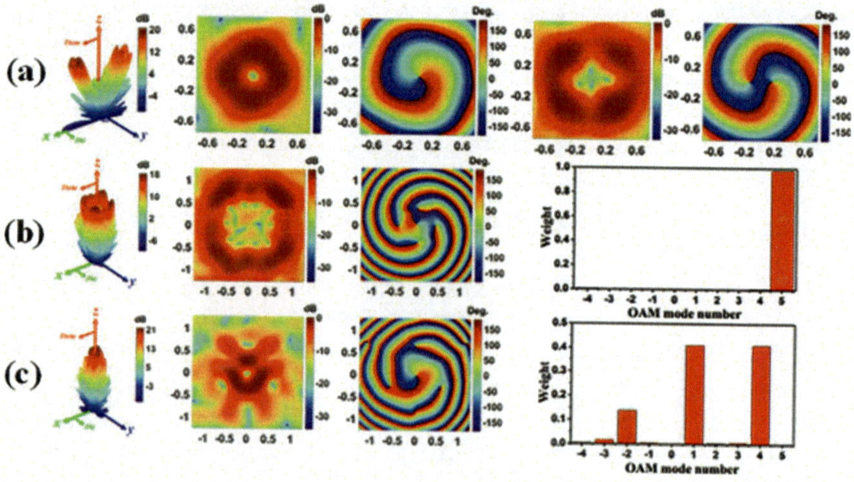

Fig. 6.7 Theoretical analysis results of 30 × 30 reconfigurable reflective metasurfaces **a** multiple OAM vortex beams, **b** higher-order OAM beam, and **c** mixed OAM modes beam [57]

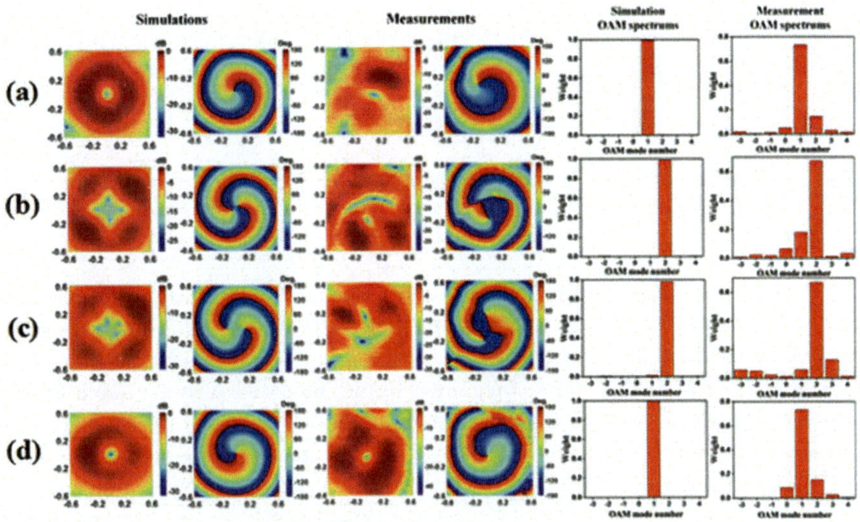

Fig. 6.8 Comparison of simulation results and measurement results of **a** OAM beam #1, **b** OAM beam #2, **c** OAM beam #3, and **d** OAM beam #4 [57]

versus reverse voltage, which has an impact on electric field intensity distributions. It might be possible to feedback on the voltage that is introduced by variation junction capacitance to suppress the intensity map deformation.

Digital reconfigurable metasurfaces could be further exploited to realize the high-level functionality of self-adaptive beam-forming. The proposed reconfigurable metasurfaces are very attractive for a variety of applications, such as controlling the radiation beams of antennas (similarly to phase-array antennas), reducing the scattering features of targets, and realizing other smart metamaterials. The above design concepts can be extended to millimeter-wave and terahertz frequencies.

6.4 Multifunction Reconfigurable Metasurfaces and Applications

With the demand for miniaturization and versatility, more and more devices with special functions have been developed, such as space-surface wave regulators [59], polarization and beam modulation metasurface [35], dual-mode and dual-band array antennas [60]. So far, the reconfigurable multifunctional metasurfaces are made in many forms. A metasurface proposed by Luo et al. can change its local phase distribution to achieve dynamical beam deflection and polarization transformation in the spatial domain [20]. A few achievements in realizing multiple functions on the same device in the frequency and spatial domains have been reported. But designing a real-time reconfigurable metasurface with polarization-independent manipulation of reflection and transmission wavefronts has still been a challenging problem.

6.4.1 Reflection and Transmission Reconfigurable Metasurface

Most of the existing passive transmission-reflective metasurfaces (TRMs) are composed of FSS structures, and their functions (transmission and reflection) are switched according to the polarization direction of the incident wave [18]. In addition, although EM wave control capabilities of current active reconfigurable TRMs have been greatly improved, there are some limitations of polarization modulation [26, 42, 61]. In this section, a new reconfigurable element with a 1-bit phase resolution for TRM is presented [62]. The element is composed of two layers of internally slotted patches, of which the active patch is additionally loaded with a microstrip line. PIN diodes are used to individually adjust the 1-bit reflection phase or transmission phase of the unit cell. In order to obtain good isolation between the two operating states, the transmission and reflection functions of the device are designed to work in two different frequency bands. This design exhibits low insertion loss and wide working bandwidth in both states.

The geometry of the reconfigurable metasurface unit cell is shown in Fig. 6.9. The element contains two different layers of metal patterns wherein the active layer patch is loaded with an O-slot. Two PIN diodes (#1 and #2) are soldered on the O-slot of the active patch along the x-axis. In addition, a PIN diode #3 is embedded in the middle of the microstrip line. The passive layer patch of the element is loaded with a U-slot, and both layers are connected by a metalized via at the center of the unit cell. Two layers of metal patches are printed on two identical dielectric substrates (Rogers RO4350B, $\varepsilon = 3.48$, $\tan\delta = 0.0037$, $H = 1.524$ mm), which are joined together by a bonding film (Rogers RO4450F, $\varepsilon = 3.52$, $\tan\delta = 0.004$, $H_b = 0.1$ mm). The back of the upper dielectric substrate is a metal ground (GND). And 82 nH inductors are soldered to suppress high-frequency signals from the patches to the bias lines. As shown in Fig. 6.9d, the minimum value of S_{11} is about −2 dB. From 6.85 to 7.39 GHz, the reflection phase difference is within $180° \pm 20°$ range. It can meet the requirement of 1-bit reflection phase regulation. In the transmission state, the S_{21} from 8.0 GHz to 9.0 GHz is greater than −3 dB, and the minimum insertion loss is about 0.8 dB. It is worth mentioning that the phase curves of the two transmission cases in Fig. 6.9e are basically parallel throughout the frequency band, and the phase difference is about 180°. Therefore, by adjusting the state of the PIN diodes in the time domain, different functions of the unit cell can be dynamically switched.

A TRM array consisting of 12×12 elements is designed and simulated. Based on Sect. 6.3.3, TRM is used to further study the OAM beams. The coding scheme corresponding to the TRM unit cells in the two states is shown in Fig. 6.10. In the reflection state and the transmission state, the number of modes of the OAM beams

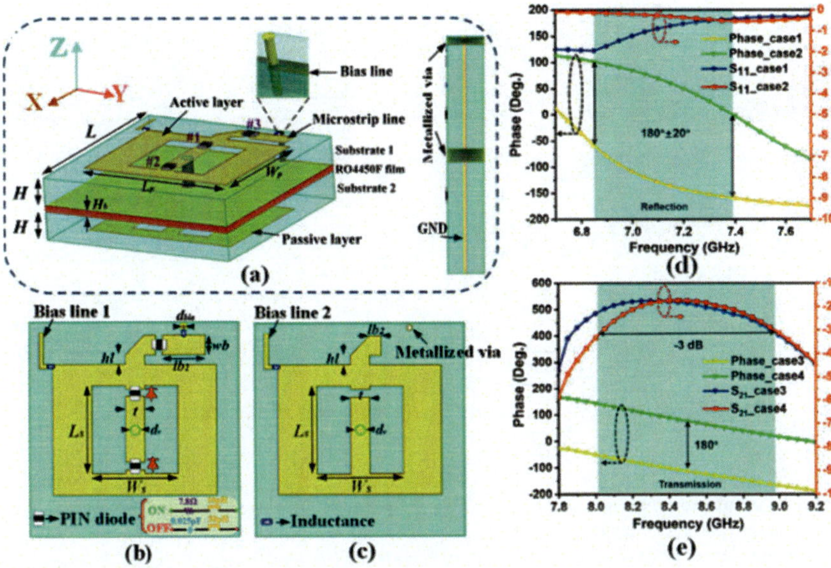

Fig. 6.9 Geometry and simulated results of the designed metasurface element: **a** schematic and side view, **b** active layer, **c** passive layer, **d** S-parameters and phases of reflection states (case1 and case2), and **e** S-parameters and phases of transmission states (case3 and case4) [62]

re-radiated by the metasurface is set to $l = 1$ and $l = 2$, respectively. The electric field amplitude (Mag_E) and electric field phase (Phase_E) on the observation window at 7.3 GHz (reflection state) and 8.5 GHz (transmission state) are shown in Fig. 6.10a and b. Since the size of the TRM array is relatively small, a few amounts of energy are radiated in other directions in both states. The waveguide method is used to measure the element characteristics. The element is enlarged to the same size as the waveguide port to facilitate connection to the waveguide, and the active and passive layers of the unit cell are shown in Fig. 6.10c. The other parameters of the element were not changed except for the sizes of the substrates and the GND. It can be seen from Fig. 6.10e that the magnitude of reflection coefficient S_{11} from 7.40 to 7.74 GHz is greater than −2 dB, while the reflection phases satisfy the 1-bit quantization requirement. Besides, the signal can be transmitted within the virtual waveguide from 8.33 to 9.25 GHz, which corresponds to −3 dB bandwidth of the transmission coefficient S_{21}, as shown in Fig. 6.10f.

Besides, the corresponding characteristics of the actual prototype have been measured between 7 and 10 GHz. A system used to test the reflection and transmission performances of the unit cell consists of two waveguides, two coaxial adapters, and the prototype placed between both waveguides (waveguide-sample-waveguide), as shown in Fig. 6.10d. The experimental results in the waveguide are shown in Fig. 6.10g and h. The reflection phase difference of the prototype satisfies the 1-bit quantization condition from 7.11 to 7.4 GHz. At the same time, the maximum loss of

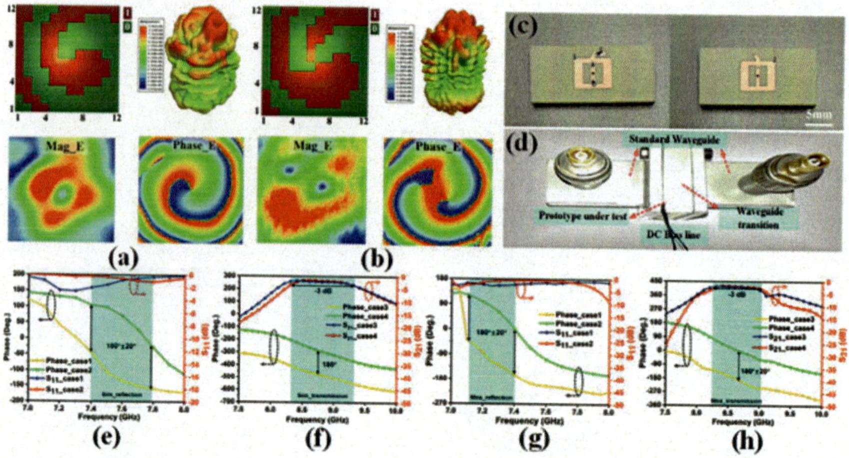

Fig. 6.10 Simulated and measured results of reconfigurable TRM: **a** reflection OAM beam with $l = 1$, **b** transmission OAM beam with $l = 2$, **c** photographs of the active and passive layers of the fabricated prototype, **d** the two-port waveguide experimental system setup, **e** simulated S-parameters and phases under different reflection cases and, **f** transmission cases, **g** measured S-parameters and phases under different reflection cases, and **h** transmission cases [62]

the element is about 2.8 dB in both reflection cases. In addition, the measured transmission coefficient is also slightly smaller than the simulation, with a 3 dB bandwidth from 8.25 to 9.05 GHz. And the actual transmission results show that the entire 3 dB bandwidth can achieve 1-bit transmission phase modulation, but the phase difference is not maintained at about 180°. As shown in Figs. 6.9 and 6.10, in both transmission and reflection states, the operating frequency bands of the unit cell in the waveguide are a little different from that in the periodic boundary. This phenomenon results from the changes in the size and length–width ratio of the enlarged unit-cell.

6.4.2 *Frequency-Spatial-Domain Reconfigurable Metasurface*

In order to further expand the capabilities of reconfigurable metasurfaces, a general frequency-spatial-domain reconfigurable metasurface (FSRM) is proposed [63]. The geometrical configuration and detailed dimensions of the FSRM element are shown in Fig. 6.11a. These metal layers are spaced by two F4B substrates, and the driven patch layer is a square patch, which is divided into three blocks with different sizes by the gaps parallel to the diagonal direction of the patch. It is worth mentioning that the size ratio and shapes of the three blocks of the driven layer enable the unit cell to evidence reconfigurable characteristics in the frequency domain, and the resonant frequencies are f_1, f_2, f_3, and f_4 in this configuration. The rectangular groove increases the path of the induced current movement so that the element can provide 1-bit phase modulation in the spatial domain and have a wide working frequency bandwidth. The rectangular groove also brings variable factors to the scattering matrix **S** of the FSRM element, which may cause the frequency-domain resonance of the unit cell to further increase ($f_5, f_6, \ldots$). Combining the conditions described above, the scattering matrix **S** of this metasurface unit-cell is related to the spatial and frequency domains, i.e.,

$$\begin{bmatrix} E_x^{(o)} \\ E_y^{(o)} \\ E_u^{(o)} \end{bmatrix} = \begin{bmatrix} S_{xp}(f) & & \\ & S_{yp}(f) & \\ & & S_{up}(\theta, \varphi) \end{bmatrix} \begin{bmatrix} E_x^{(i)} \\ E_y^{(i)} \\ E_u^{(i)} \end{bmatrix} \tag{6.13}$$

where S_{xp}, S_{yp} and S_{up} are the scattering parameters of the pth unit cell in the frequency domain and spatial domain, and f is the working frequency.

When the incident EM wave is LP with the electric field along the y-direction or x-direction, the metasurface can modulate the LP wave in the frequency domain. Accordingly, FSRM can be regulated as a multifunctional polarization converter (MPC). Four states of 2-bit ("00" "01" "10" "11") digital coding are used to describe the working states of the two PIN diodes in the unit cell, in which state "01" means that PIN diode #1 is "OFF," and PIN diode #2 is "ON". Switching the working states of the diodes, the interconnection of the patches of the driven layer is changed, and the resonant state (scattering parameter) of the driven layer is varied accordingly. As

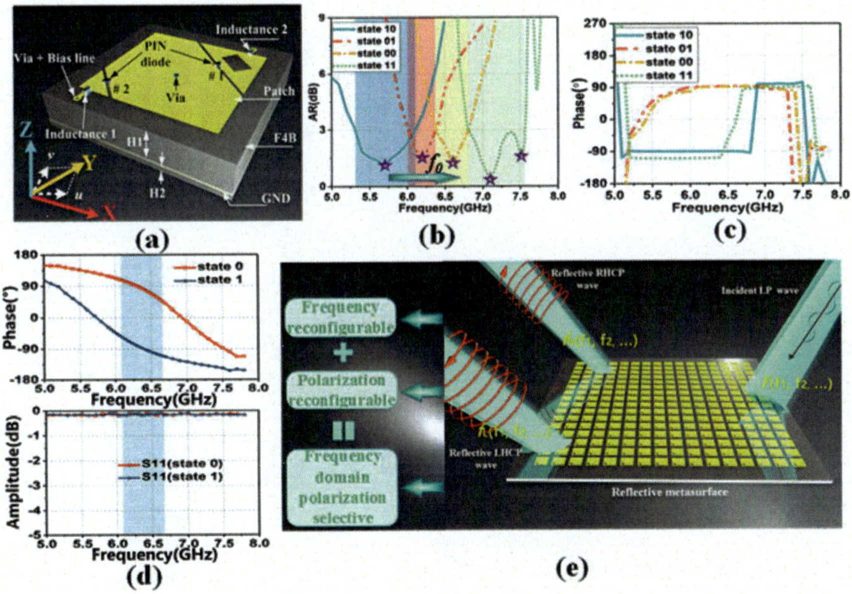

Fig. 6.11 Polarization working mechanism and polarization transition conditions of FSRM: **a** the 3D topology expanded view of the FSRM unit cell, **b** reflected wave axis ratio, **c** phase difference of the components (u and v directions) in the frequency domain, **d** 1-bit reflection characteristics of unit cell, and **e** polarization modulation mechanism of MPC [63]

shown in Fig. 6.11b, the 3-dB AR bandwidth and the corresponding center resonant frequency of the reflected waves are different along with the states of PIN diodes. Figure 6.11c shows the phase difference of components in u and v directions in different states. The left-hand circularly polarized (LHCP) and right-hand circularly polarized (RHCP) waves can be realized in different states when the excitation is along the y-polarization direction. Therefore, the MPC has the frequency-domain polarization selective function. More interestingly, the CP waves can be realized in different frequency bands under different states, and this MPC function is called frequency reconfigurable polarization conversion. The frequency reconfigurability allows the element to achieve different polarizations at the same frequency, and this feature is called polarization reconfigurable. The polarization modulation mechanism of the MPC is shown in Fig. 6.11e. Furthermore, when the excitation is along the x-polarized direction, the reflected wave's polarization direction (LHCP or RHCP) in different states will be reversed.

When the polarization direction of incident LP waves is 45° (counter-clockwise rotation) from the $+x$ axis, the FSRM can modulate the incident waves in the spatial domain based on the scattering parameters and convert itself into a 1-bit programmable metasurface. As the states of the two diodes are inverted simultaneously, the reflection phase of the unit cell is around 90° (digital state 0) in the same frequency band. In Fig. 6.11d, the amplitudes of S_{11} in two states are larger than −0.15 dB, indicating that the unit cells can achieve nearly perfect 1-bit reflection characteristics. The proposed element is periodically arranged to construct a programmable metasurface ($M \times Q$). The reflected E-field of the mq^{th} unit cell in the spatial domain is given by

$$\boldsymbol{E}_{mq}^{(r)} = \boldsymbol{E}_{mq}^{(i)} \cdot S_{smq}(\theta, \varphi) = \boldsymbol{E}_{mq}^{(i)} \cdot \alpha_{mq}^{n} \cdot \exp\left[-jk\begin{pmatrix} x \sin\theta \cos\varphi \\ +y \sin\theta \sin\varphi \end{pmatrix} + j\Phi_{mq}\right]. \tag{6.14}$$

The subscript mq is a 2D representation of the pth unit cell. Correspondingly, $\alpha_{mq}^{n} \approx \cos^{q_e}(\theta)$ is the spatial modulation parameter of the mq^{th} unit cell. As the reflected E-field is manipulated, the reflection phase of the metasurface element is compensated according to the continuous phase compensation formula. Then the continuous compensation phase ϕ_{mq} is discretized as

$$\phi_{cmq} = \begin{cases} -90^\circ & -180^\circ \le \phi_{mq} < 0^\circ \\ 90^\circ & 0^\circ \le \phi_{mq} < 180^\circ \end{cases} \tag{6.15}$$

The discrete phases in Eq. (6.15) are the compromise of the phase quantization error in the entire operating band. In the same situation, if the quantization phases of the unit cell are set to other values, the beam deflection error will increase, and the metasurface radiation efficiency will decrease. According to the discrete compensation phase, the working states (state 0 or state 1) of the elements are determined. The scattering far-field of the programmable metasurface at the n-order harmonic frequency is written as

$$E(\theta,\varphi)=\sum_{m=1}^{M}\sum_{q=1}^{Q}\left|\boldsymbol{E}_{mq}^{(r)}\right|\cdot\exp\left(j\phi_{cmq}\right) \tag{6.16}$$

Based on Eq. (6.16), several functions such as 2D beam scanning, vortex beam carrying OAM, and beam splitting of reflected beams can be implemented by encoding the metasurface at the zero-order harmonic frequency. The specific functions are shown in Fig. 6.12.

As shown in Fig. 6.13a, a metasurface sample with 12 × 12 elements was fabricated. For the beam scanning function, four typical scanning angles θ (0°, 15°, 30°, 45°) are verified in the measurement. It can be seen from the measured 2D scattering patterns at 6.3 GHz in Fig. 6.13b that the actual main beam pointing directions in the four states are 0°, 14.5°, 31.8°, and 45°, respectively. The steering angles are in good agreement with the theoretical predictions. Certainly, more beam-deflection angles are expected through the ingenious design of the programming state imparted on the metasurface.

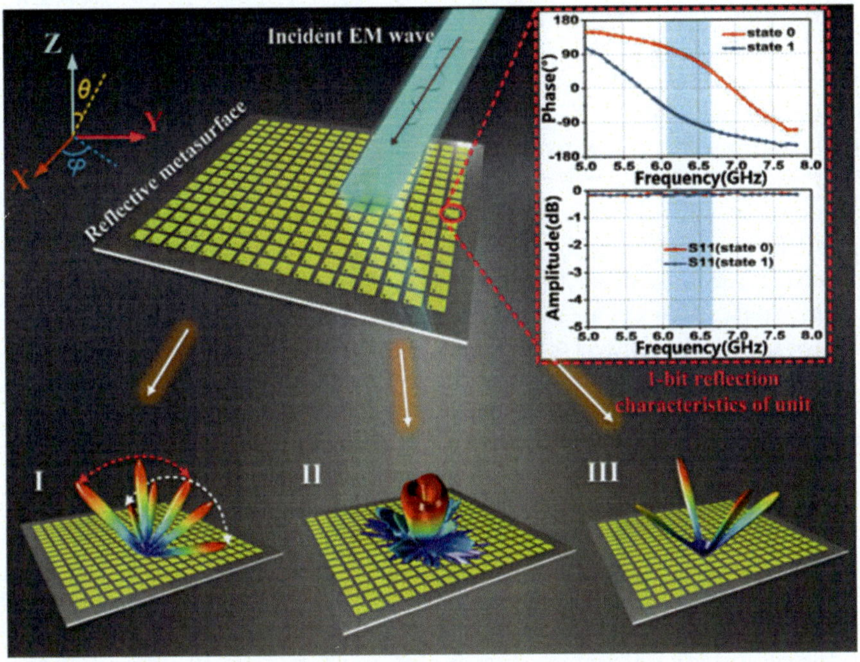

Fig. 6.12 Multiple functions of the 1-bit programmable reflective FSRM. Function I: 2D beam dynamic scanning. Function II: vortex-beam generation of OAM. Function III: specific splitting beams with desired transmission directions [63]

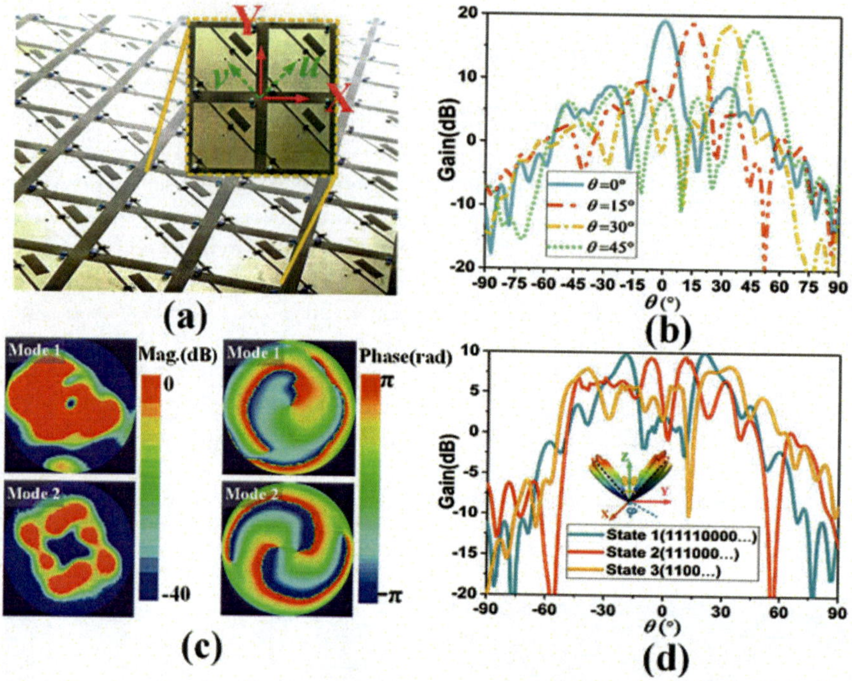

Fig. 6.13 Experimental verification to the 1-bit programmable performance of FSRM at 6.3 GHz: **a** fabricated metasurface, **b** 2D radiation patterns under four scanning angles, **c** near-field amplitude and phase distributions of the vortex beams at modes $l = 1$ and $l = 2$, and **d** 2D scattering patterns [63]

Next, the measured near-field amplitude and phase distributions of the vortex beams at different modes are shown in Fig. 6.13c, respectively. It was demonstrated that the continuous controls of the vortex beam with different transmission modes could be successfully realized using the metasurface. Meanwhile, the scattering patterns of the FSRM in the spatial domain were also verified. As shown in Fig. 6.13d, the two symmetrical scattered beams are modulated to direct in different deflection angles based on three different periodic coding states (11110000…, 111000…, 1010…). Since the incident wave is not a perfect plane wave and the size of this sample is small, the main lobes of the scattering patterns are wider, and the side lobes are larger. Accordingly, as the deflection angle increases, the symmetry of the main beams deteriorates. But these errors also provide useful help for the RCS reduction function.

The design strategies presented in this section can also allow the tailoring of EM waves in many other intriguing ways for further exploration. This work paves the road to realizing a programmable metasurface with multi-domain and multi-function radiation properties. The proposed metasurfaces can also be extended to the optical and acoustic regimes and provide possibilities to perfectly manipulate EM waves in other applications.

6.5 Conclusion

This chapter reviews the basic concept, working mechanism, and design strategy of reconfigurable metasurfaces. Some classical system-level works are presented with metasurfaces to demonstrate their powerful manipulations of EM waves. By controlling the response phase of the metasurface unit cell, the general cases of dynamical 2D beam scanning, vortex-beam generations with OAM modes, and specific beam transmission with pre-designed directions can be realized. Looking at the development trend of reconfigurable metamaterials and metasurfaces over the past decades, we can summarize the following main stages. The first milestone, the binary coding method, was successfully introduced into the design of metamaterials, and the digital world and traditional physical materials were gradually linked. In the second stage, techniques such as phase superposition, convolution operation, and scattering modulation were proposed based on the generalized Snell's theorem. The theoretical system has been greatly improved, which has laid the foundation for the emergence of dynamic digital metasurfaces. In addition, the recently reported perceptual and self-adaptive metasurfaces push reconfigurable metasurfaces into a new phase.

With the combination of sensors and reconfigurable metasurfaces, digital metasurfaces are gradually transitioning to smart metasurfaces. However, with the complication of device functions, the integration of control circuits and its own coupling suppression urgently needs to be solved together. The control system must be sufficiently powerful and extremely low-loss, which imposes strict restrictions on the actual design of circuits and systems. At the same time, advanced control components and control methods need to be developed. To realize the concept of the intelligent metasurface, not only hardware but software framework also need to be further improved. They combine existing technologies with advanced algorithms and platforms such as artificial intelligence, big data processing, and cloud computing to provide a powerful "brain" for metasurfaces.

In the future, we envision that metasurfaces and metamaterials should have the ability to perceive, learn, process, and remember information from the surrounding environment. Such materials can be called quasi-living metasurfaces. With the continuous development of high-speed communication networks, the application potential of metamaterials and metasurfaces will be fully explored in the context of the interconnection of everything. From smart dust to communication equipment, quasi-living metamaterials will be everywhere.

References

1. F. Falcone, T. Lopetegi, M.A.G. Laso, J.D. Baena, J. Bonache, M. Beruete, R. Marqués, F. Martín, M. Sorolla, Babinet principle applied to the design of metasurfaces and metamaterials. Phys. Rev. Lett. **93**(19), 197401 (2004)
2. A.A. High, R.C. Devlin, A. Dibos, M. Polking, D.S. Wild, J. Perczel, N.P. de Leon, M.D. Lukin, H. Park, Visible-frequency hyperbolic metasurface. Nature **522**, 192 (2015)

3. L. Huang, X. Chen, B. Bai, Q. Tan, G. Jin, T. Zentgraf, S. Zhang, Helicity dependent directional surface plasmon polariton excitation using a metasurface with interfacial phase discontinuity. Light: Sci. Appl. **2**(3), e70 (2013)
4. X. Ni, A.V. Kildishev, V.M. Shalaev, Metasurface holograms for visible light. Nat. Commun. **4**(1), 1–6 (2013)
5. S. Yu, H. Liu, L. Li, Design of near-field focused metasurface for high-efficient wireless power transfer with multifocus characteristics. IEEE Trans. Ind. Electron. **66**(5), 3993–4002 (2019)
6. P.C. Wu, W. Zhu, Z.X. Shen, P.H.J. Chong, W. Ser, D.P. Tsai, A.-Q. Liu, Broadband wide-angle multifunctional polarization converter via liquid-metal-based metasurface. Adv. Opt. Mater. **5**(7), 1600938 (2017)
7. X. Ding, F. Monticone, K. Zhang, L. Zhang, D. Gao, S.N. Burokur, A. de Lustrac, Q. Wu, C.-W. Qiu, A. Alù, Ultrathin Pancharatnam-Berry metasurface with maximal cross-polarization efficiency. Adv. Mater. **27**(7), 1195–1200 (2015)
8. L. Liu, X. Zhang, M. Kenney, X. Su, N. Xu, C. Ouyang, Y. Shi, J. Han, W. Zhang, S. Zhang, Broadband metasurfaces with simultaneous control of phase and amplitude. Adv. Mater. **26**(29), 5031–5036 (2014)
9. N. Yu, F. Capasso, Flat optics with designer metasurfaces. Nat. Mater. **13**(2), 139–150 (2014)
10. X. Ma, W. Pan, C. Huang, M. Pu, Y. Wang, B. Zhao, J. Cui, C. Wang, X. Luo, An active metamaterial for polarization manipulating. Adv. Opt. Mater. **2**(10), 945–949 (2014)
11. Z. Wei, Y. Cao, X. Su, Z. Gong, Y. Long, H. Li, Highly efficient beam steering with a transparent metasurface. Opt. Express **21**(9), 10739–10745 (2013)
12. L. Zhang, X. Q. Chen, S. Liu, Q. Zhang, J. Zhao, J. Y. Dai, G. D. Bai, X. Wan, Q. Cheng, G. Castaldi V. Galdi, T.J. Cui, Space-time-coding digital metasurfaces. Nat. Commun. **9**(1), 1–11 (2018)
13. J.N. Gollub, O. Yurduseven, K.P. Trofatter, D. Arnitz, M.F. Imani, T. Sleasman, M. Boyarsky, A. Rose, A. Pedross-Engel, H. Odabasi, T. Zvolensky, G. Lipworth, D. Brady, D.L. Marks, M.S. Reynolds, D.R. Smith, Large metasurface aperture for millimeter wave computational imaging at the human-scale. Sci. Rep. **7**, 42650 (2017)
14. J. Zhao, B. Sima, N. Jia, C. Wang, B. Zhu, T. Jiang, Y. Feng, Achieving flexible low-scattering metasurface based on randomly distribution of meta-elements. Opt. Express **24**(24), 27849–27857 (2016)
15. C. Della Giovampaola, N. Engheta, Digital metamaterials. Nat. Mater. **13**(12), 1115–1121 (2014)
16. T.J. Cui, M.Q. Qi, X. Wan, J. Zhao, Q. Cheng, Coding metamaterials, digital metamaterials and programmable metamaterials. Light: Sci. Appl. **3**(10), e218 (2014)
17. L. Li, Y. Li, Z. Wu, F. Huo, Y. Zhang, C. Zhao, Novel polarization-reconfigurable converter based on multilayer frequency-selective surfaces. Proc. IEEE **103**(7), 1057–1070 (2015)
18. T. Cai, G.-M. Wang, X.-L. Fu, J.-G. Liang, Y.-Q. Zhuang, High-efficiency metasurface with polarization-dependent transmission and reflection properties for both reflectarray and transmitarray. IEEE Trans. Antennas Propag. **66**(6), 3219–3224 (2018)
19. H. Yue, L. Chen, Y. Yang, L. He, X. Shi, Design and implementation of a dual frequency and bidirectional phase gradient metasurface for beam convergence. IEEE Antennas Wirel. Propag. Lett. **18**(1), 54–58 (2018)
20. C. Huang, C. Zhang, J. Yang, B. Sun, B. Zhao, X. Luo, Reconfigurable metasurface for multifunctional control of electromagnetic waves. Adv. Opt. Mater. **5**(22), 1700485 (2017)
21. H. Kamoda, T. Iwasaki, J. Tsumochi, T. Kuki, O. Hashimoto, 60-GHz electronically reconfigurable large reflectarray using single-bit phase shifters. IEEE Trans. Antennas Propag. **59**(7), 2524–2531 (2011)
22. B. Ratni, A. de Lustrac, G.P. Piau, S.N. Burokur, Electronic control of linear-to-circular polarization conversion using a reconfigurable metasurface. Appl. Phys. Lett. **111**(21), 214101 (2017)
23. E. Arbabi, A. Arbabi, S.M. Kamali, Y. Horie, M. Faraji-Dana, A. Faraon, MEMS-tunable dielectric metasurface lens. Nat. Commun. **9**(1), 812 (2018)

24. B. Schoenlinner, A. Abbaspour-Tamijani, L.C. Kempel, G.M. Rebeiz, Switchable low-loss RF MEMS Ka-band frequency-selective surface. IEEE Trans. Microw. Theory Tech. **52**(11), 2474–2481 (2004)
25. J. Han, L. Li, G. Liu, Z. Wu, Y. Shi, A wideband 1 bit 12× 12 reconfigurable beam-scanning reflectarray: design, fabrication, and measurement. IEEE Antennas Wirel. Propag. Lett. **18**(6), 1268–1272 (2019)
26. R.Y. Wu, L. Zhang, L. Bao, L.W. Wu, Q. Ma, G.D. Bai, H.T. Wu, T.J. Cui, Digital metasurface with phase code and reflection–transmission amplitude code for flexible full-space electromagnetic manipulations. Adv. Opt. Mater. **7**(8), 1801429 (2019)
27. X.G. Zhang, W.X. Tang, W.X. Jiang, G.D. Bai, J. Tang, L. Bai, C.W. Qiu, T.J. Cui, Light-controllable digital coding metasurfaces. Adv. Sci. **5**(11), 1801028 (2018)
28. H.H. Yang, X.Y. Cao, F. Yang, J. Gao, S.H. Xu, M.K. Li, X.B. Chen, Y. Zhao, Y.J. Zheng, S.J. Li, A programmable metasurface with dynamic polarization, scattering and focusing control. Sci. Rep. **6**, 35692 (2016)
29. J. Gu, R. Singh, X. Liu, X. Zhang, Y. Ma, S. Zhang, S.A. Maier, Z. Tian, A.K. Azad, H.-T. Chen, Active control of electromagnetically induced transparency analogue in terahertz metamaterials. Nat. Commun. **3**(1), 1–6 (2012)
30. J. Murray, D. Ma, J.N. Munday, Electrically controllable light trapping for self-powered switchable solar windows. ACS Photonics **4**(1), 1–7 (2017)
31. D. Shrekenhamer, J. Montoya, S. Krishna, W.J. Padilla, Four-color metamaterial absorber thz spatial light modulator. Adv. Opt. Mater. **1**(12), 905–909 (2013)
32. Y.J. Zhou, Q.X. Xiao, Electronically controlled rejections of spoof surface plasmons polaritons. J. Appl. Phys. **121**(12), 123109 (2017)
33. M. Wang, H.F. Ma, W.X. Tang, H.C. Zhang, Z.X. Wang, T.J. Cui, Programmable controls of multiple modes of spoof surface plasmon polaritons to reach reconfigurable plasmonic devices. Adv. Mater. Technol. **4**(3), 1800603 (2019)
34. Z. Shen, B. Jin, J. Zhao, Y. Feng, L. Kang, W. Xu, J. Chen, P. Wu, Design of transmission-type coding metasurface and its application of beam forming. Appl. Phys. Lett. **109**(12), 121103 (2016)
35. Z. Tao, X. Wan, B.C. Pan, T.J. Cui, Reconfigurable conversions of reflection, transmission, and polarization states using active metasurface. Appl. Phys. Lett. **110**(12), 121901 (2017)
36. Q. Zheng, Y. Li, J. Zhang, H. Ma, J. Wang, Y. Pang, Y. Han, S. Sui, Y. Shen, H. Chen, Wideband, wide-angle coding phase gradient metasurfaces based on Pancharatnam-Berry phase. Sci. Rep. **7**, 43543 (2017)
37. B. Zhu, Y. Feng, J. Zhao, C. Huang, T. Jiang, Switchable metamaterial reflector/absorber for different polarized electromagnetic waves. Appl. Phys. Lett. **97**(5), 051906 (2010)
38. K. Chen, L. Cui, Y. Feng, J. Zhao, T. Jiang, B. Zhu, Coding metasurface for broadband microwave scattering reduction with optical transparency. Opt. Express **25**(5), 5571–5579 (2017)
39. L. Li, T. J. Cui, W. Ji, S. Liu, J. Ding, X. Wan, Y. B. Li, M. Jiang, C.-W. Qiu, S. Zhang, Electromagnetic reprogrammable coding-metasurface holograms. Nat. Commun. **8**(1), 1–7 (2017)
40. D. Pozar, Bandwidth of reflectarrays. Electron. Lett. **39**(21), 1490–1491 (2003)
41. C. Liu, S.V. Hum, An electronically tunable single-layer reflectarray antenna element with improved bandwidth. IEEE Antennas Wirel. Propag. Lett. **9**, 1241–1244 (2010)
42. B.D. Steinberg, Microwave imaging of aircraft. Proc. IEEE **76**(12), 1578–1592 (1988)
43. J. Han, L. Li, S. Tian, G. Liu, H. Liu, Y. Shi, Millimeter-wave imaging using 1-bit programmable metasurface: simulation model, design and experiment. IEEE J. Emergi. Select. Topics Circ. Syst. **10**(1), 52–61 (2020)
44. J. Hunt, T. Driscoll, A. Mrozack, G. Lipworth, M. Reynolds, D. Brady, D.R. Smith, Metamaterial apertures for computational imaging. Science **339**(6117), 310–313 (2013)
45. L. Allen, M.W. Beijersbergen, R. Spreeuw, J. Woerdman, Orbital angular momentum of light and the transformation of Laguerre-Gaussian laser modes. Phys. Rev. A **45**(11), 8185–8189 (1992)

46. M. Uchida, A. Tonomura, Generation of electron beams carrying orbital angular momentum. Nature **464**(7289), 737–739 (2010)
47. A.M. Yao, M.J. Padgett, Orbital angular momentum: origins, behavior and applications. Adv. Opt. Photonics **3**(2), 161–204 (2011)
48. F. Tamburini, E. Mari, B. Thidé, C. Barbieri, F. Romanato, Experimental verification of photon angular momentum and vorticity with radio techniques. Appl. Phys. Lett. **99**(20), 204102 (2011)
49. O. Edfors, A.J. Johansson, Is orbital angular momentum (OAM) based radio communication an unexploited area? IEEE Trans. Antennas Propag. **60**(2), 1126–1131 (2011)
50. Y. Yan, G. Xie, M.P. Lavery, H. Huang, N. Ahmed, C. Bao, Y. Ren, Y. Cao, L. Li, Z. Zhao, High-capacity millimetre-wave communications with orbital angular momentum multiplexing. Nat. Commun. **5**(1), 1–9 (2014)
51. M. Barbuto, F. Trotta, F. Bilotti, A. Toscano, Circular polarized patch antenna generating orbital angular momentum. Prog. Electromagnet. Res. **148**, 23–30 (2014)
52. D. Zelenchuk, V. Fusco, Split-ring FSS spiral phase plate. IEEE Antennas Wirel. Propag. Lett. **12**, 284–287 (2013)
53. B. Xu, C. Wu, Z. Wei, Y. Fan, H. Li, Generating an orbital-angular-momentum beam with a metasurface of gradient reflective phase. Opt. Mater. Exp. **6**(12), 3940–3945 (2016)
54. S. Yu, L. Li, G. Shi, Z. Cheng, Y. Shi, Generating multiple orbital angular momentum vortex beams using a metasurface in radio frequency domain. Appl. Phys. Lett. **108**(24), 241901 (2016)
55. S. Yu, L. Li, G. Shi, Z. Cheng, X. Zhou, Y. Shi, Design, fabrication, and measurement of reflective metasurface for orbital angular momentum vortex wave in radio frequency domain. Appl. Phys. Lett. **108**(12), 121903 (2016)
56. J. Han, L. Li, H. Yi, Y. Shi, 1-bit digital orbital angular momentum vortex beam generator based on a coding reflective metasurface. Opt. Mater. Express **8**(11), 3470–3478 (2018)
57. J. Han, L. Li, H. Yi, W. Xue, Versatile orbital angular momentum vortex beam generator based on reconfigurable reflective metasurface. Jpn. J. Appl. Phys. **57**(12), 120303 (2018)
58. S.V. Hum, M. Okoniewski, R.J. Davies, Modeling and design of electronically tunable reflectarrays. IEEE Trans. Antennas Propag. **55**(8), 2200–2210 (2007)
59. S. Sun, Q. He, S. Xiao, Q. Xu, X. Li, L. Zhou, Gradient-index meta-surfaces as a bridge linking propagating waves and surface waves. Nat. Mater. **11**(5), 426–431 (2012)
60. C.-X. Mao, S. Gao, Y. Wang, Q. Luo, Q.-X. Chu, A shared-aperture dual-band dual-polarized filtering-antenna-array with improved frequency response. IEEE Trans. Antennas Propag. **65**(4), 1836–1844 (2017)
61. A. Clemente, L. Dussopt, R. Sauleau, P. Potier, P. Pouliguen, Wideband 400-element electronically reconfigurable transmitarray in X band. IEEE Trans. Antennas Propag. **61**(10), 5017–5027 (2013)
62. G. Liu, H. Liu, J. Han, Y. Mu, L. Li, Reconfigurable metasurface with polarization-independent manipulation for reflection and transmission wavefronts. J. Phys. D-Appl. Phys. **53**, 045107 (2020)
63. G. Liu, L. Li, J. Han, H. Liu, X. Gao, Y. Shi, T.J. Cui, Frequency-domain and spatial-domain reconfigurable metasurface. ACS Appl. Mater. Interfaces. **12**(20), 23554–23564 (2020)

Chapter 7
Reflective and Transmission Metasurfaces for Orbital Angular Momentum Vortex Waves Generation

Shixing Yu, Na Kou, Long Li, and Zhiwei Cui

Abstract Metasurfaces are a novel class of two-dimensional (2D) or quasi-two-dimensional (Q2D) artificial electromagnetic surfaces comprised of subwavelength microstructures. These structures allow for precise control over the propagation of electromagnetic waves. Reflective and transmissive metasurfaces, emerging as groundbreaking concepts and techniques, have experienced rapid development in recent years. Meanwhile, vortex waves with orbital angular momentum (OAM) have attracted considerable interest due to their distinctive helical phase distributions. Initially observed in optics, OAM vortex waves have since found applications in microwave and terahertz domains, offering promising opportunities for improving communication modes, expanding channel capacity, and enhancing spectral efficiency. This chapter aims to investigate the characteristics of electromagnetic vortex waves with OAM. Firstly, it offers a comprehensive review of OAM vortex waves in optics. Subsequently, it explores the generation of electromagnetic vortex waves with OAM in the radio frequency range using reflective and transmissive metasurfaces. The combination of metasurfaces and OAM vortex waves holds great potential for manipulating and controlling electromagnetic waves in radio frequency applications. This research contributes to furthering our understanding of these phenomena and opens up new possibilities for advanced communication systems.

Keywords Reflective metasurface · Transmissive metasurface · Orbital angular momentum (OAM) · Electromagnetic vortex wave

---

S. Yu · N. Kou
College of Big Data and Information Engineering, Guizhou University, Guiyang 550025, China
e-mail: sxyu1@gzu.edu.cn

L. Li (✉)
School of Electronic Engineering, Xidian University, Xi'an 710071, Shaanxi, China
e-mail: lilong@mail.xidian.edu.cn

Z. Cui
School of Physics, Xidian University, Xi'an 710071, Shannxi, China
e-mail: zwcui@mail.xidian.edu.cn

L. Li et al. (eds.), *Electromagnetic Metamaterials and Metasurfaces: From Theory To Applications*, https://doi.org/10.1007/978-981-99-7914-1_7

7.1 Introduction

The development of metamaterials has ushered in a new era in the field of modulating electromagnetic waves. Traditionally, achieving a variety of functions with flexibility required bulky and challenging-to-manufacture three-dimensional (3D) block structures. However, researchers have now proposed a groundbreaking solution—the use of simple planar two-dimensional (2D) structures as a replacement. This innovation has rapidly emerged as one of the most promising areas in modern theoretical and applied electromagnetics. With the integration of planar structures, the cumbersome nature of the traditional 3D approach can be overcome, while still achieving efficient modulation of electromagnetic waves [1]. The adoption of a planar structure, known as a metasurface, offers significant advantages over traditional materials due to its ability to leverage the mature printed circuit board (PCB) process for fabrication. Comprising sub-wavelength cells arranged in a periodic array, the metasurface enables easy manufacturing with a thickness much smaller than the working wavelength. Notably, microwave metasurfaces have emerged as a powerful tool in modern electromagnetics, allowing for flexible control of amplitude and phase distributions in both transmitted and reflected waves [2].

Prior to the proposal of metasurfaces, other similar structures with two-dimensional periodic arrangements, such as reflectarrays [3], transmitarrays [4], and polarization rotators [5], had already been employed in antenna design for beam deflection and modulation of the phase distribution of electromagnetic waves. With the introduction of metasurfaces, these traditional planar structures can be recognized as a subset of metasurfaces. Metasurfaces have gained significant attention worldwide due to their diverse applications and their ability to demonstrate novel phenomena and functions in various frequency ranges, including microwave [6], millimeter-wave [7], terahertz [8], infrared [9], and visible light [10] domains. As a result, metasurface research has opened up new possibilities for microwave device and antenna design, overcoming limitations in terms of volume, loss, and complexity associated with conventional metamaterials. The field of metasurfaces is rapidly advancing towards practical engineering and wider applications.

A groundbreaking paper by Capasso's group in 2011 [11] played a pivotal role in establishing the concept of metasurfaces. Their work focused on the generalized Snell's law and its application to metasurfaces for precise regulation of electromagnetic waves. Through experiments, they observed abnormal refraction and reflection effects. The generalized Snell's law mathematically describes the mechanism of phase gradient on electromagnetic waves using a differential equation, providing a solid theoretical foundation for metasurfaces. By leveraging metasurfaces based on the generalized Snell's law, researchers have achieved full control over the phases of reflection and transmission. Building upon the concept of phase gradient, researchers have subsequently designed ultrathin flat lenses [12], holographic metasurfaces [13], optical vortex phase plates [11], and even ultrathin invisibility skin cloaks [14]. These advancements demonstrate the versatility and potential of metasurfaces in various applications.

The advancement of metasurfaces has led to rapid developments in wireless communication technology. Today, high-speed and ultra-wideband devices are at the forefront of wireless communication technology. However, the limited spectrum resources present significant challenges to the rapid evolution of wireless communication. In recent years, the utilization of vortex waves with orbital angular momentum (OAM) [15] has garnered considerable attention across various fields, including optical manipulation [16–18], super-resolution imaging [19], and optical communication [20–24]. The OAM vortex wave is characterized by a phase term described by $\exp(jl\varphi) = 0$, where φ represents the azimuthal coordinate, l denotes the topological charge (TC) associated with the OAM of per photon, and corresponds to the Planck constant divided by 2π. l is an integer, indicating the number of intertwined helices, while the positive or negative sign of l denotes the clockwise or counterclockwise rotation direction of the phase fronts, respectively. These beams, also known as vortex beams, possess annular-shaped transverse profiles resulting from central phase singularities [15]. With a theoretically unlimited number of orthogonal modes in Hilbert space, the OAM vortex wave has emerged as a promising technique for wireless communication.

7.2 OAM Vortex Beams in Optics

7.2.1 Generation of Vortex Beams

Various methods have been proposed and demonstrated for the generation of optical vortex beams, as depicted in Fig. 7.1. These approaches can be mainly classified into two types: intra-cavity methods and mode conversion methods. In the intra-cavity method, vortex beams are directly obtained from the output of the laser cavity [25–27]. Conversely, the mode conversion method typically requires an external converter, such as a spiral phase plate (SPP) [28–30], computer-generated hologram (CGH) [31–33], liquid crystal spatial light modulator (LC-SLM) [34–37], q-plates [38], metamaterials [39, 40], or a digital micro-mirror device (DMD) [41]. Among these methods, SPP, CGH, and SLMs are the most commonly used experimental techniques for generating vortex beams.

7.2.2 Measurement OAM States

Various methods have been proposed for the detection of OAM or optical vortices. One commonly used approach is to observe the interference patterns resulting from the interference of an optical vortex beam with a plane wave [42], a spherical wave [43], or its conjugate beam, often employing a Mach-Zehnder interferometer [44–46]. The magnitude and sign of the topological charge can be determined using a

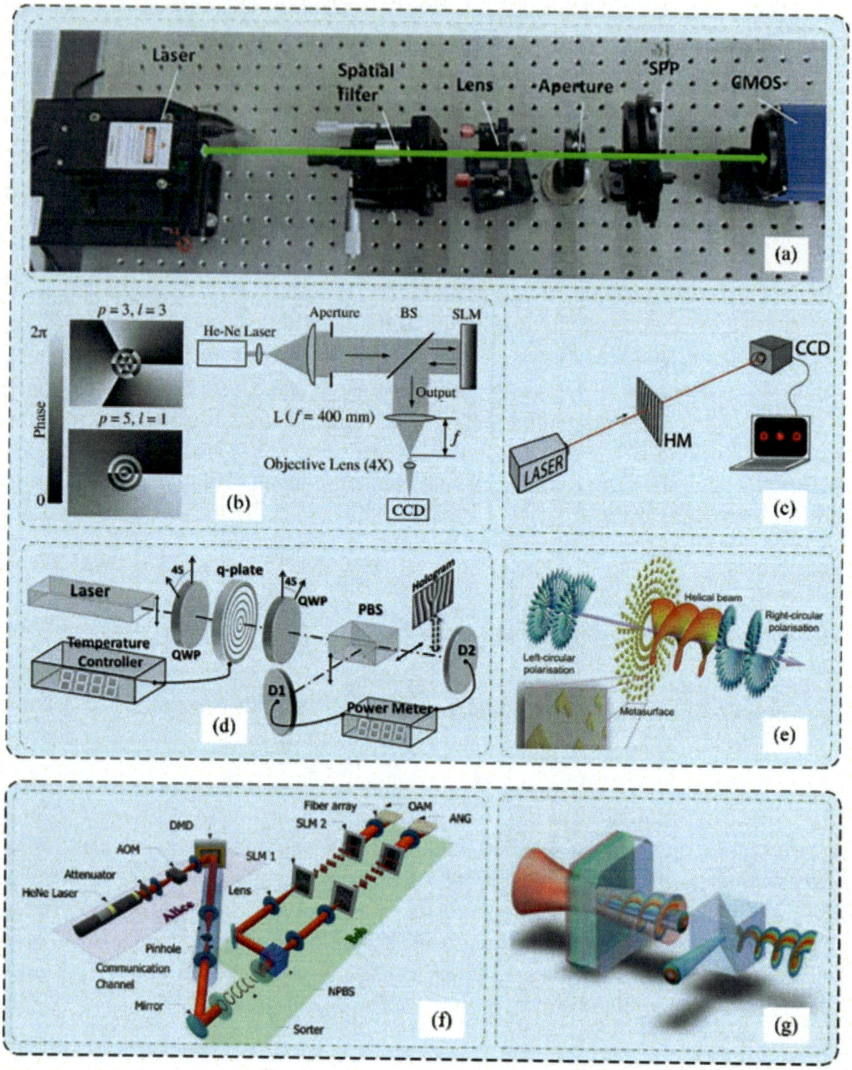

Fig. 7.1 Generation of OAM vortex beams. **a** SPP [29]. **b** SLM [35] **c** CGH [33]. **d** q-plates [38]. **e** metamaterials [39]. **f** DMD [41]. **g** Direct generation of vortex beam from a laser cavity [27]

double-slit [47] or a single-slit [48]. Furthermore, specially designed geometric apertures such as annular apertures [49], triangular apertures [50], and circular apertures [51] have been introduced to characterize the charge of the vortex by analyzing the far-field diffraction intensity pattern produced by these apertures. Another technique, presented by Berkhout et al. [52], involves sorting OAM states using Cartesian-to-log-polar coordinate transformation. Diffraction optical gratings, including amplitude gratings [53–56] and phase gratings [57–60], have also been utilized. Recently, a simple measurement scheme utilizing a spherical lens [61] or a single cylindrical lens [62, 63] was proposed and experimentally implemented.

Spiral Phase Grating

It is important to note that existing literature primarily focuses on diagnosing the azimuthal index l of Laguerre-Gaussian (LG) beams, assuming that the radial index $p = 0$ is not applicable for a priori unknown radial high-order LG beams. Therefore, the detection of the radial index associated with an incident vortex beam becomes a significant task in addition to detecting the azimuthal index. Moreover, utilizing both the radial and azimuthal index of LG beams for information transfer can enhance communication modes and increase system capacity. In this study, we propose a method for determining the mode indices of LG laser beams using a spiral phase grating. Initially, we generate optical vortices using CGH. Then, we let the LG beams illuminate the spiral phase grating displayed by a SLM. Subsequently, we observe bright spot array diffraction patterns on a white screen or capture them using a charge-coupled device (CCD) camera, which provide information about the mode indices. The feasibility and reliability of the measurement system are demonstrated through both simulations and experimental results.

We will now discuss the process of determining the mode indices of an optical vortex beam [64]. The schematic diagram illustrating the measurement of mode indices of vortex beams using a spiral phase grating is presented in Fig. 7.2. The transverse amplitude fields and corresponding helical wavefronts of the incident LG modes are displayed in Fig. 7.2a. The designed spiral phase grating is depicted in Fig. 7.2b, with the position of light illuminating the grating indicated by the red torus. When the LG beams pass through the grating, a certain distribution pattern of bright stripes will be observed on the screen, as shown in Fig. 7.2c. It is noteworthy that only the positive first-order diffraction will appear. The grating effectively directs energy into the desired diffraction order, thereby enhancing the sensitivity of the measurement system. It has been demonstrated that the number and orientation of the diffraction bright spots provide information about the radial and azimuthal mode indices of the LG beams.

A spiral phase grating is composed of two optical elements: the vortex phase plate and the holographic axicon. Considering the phase functions of these two elements are described as $q\phi$ and $2\pi r / D$, respectively, the transmission function of a spiral phase grating can be described as

$$t(r, \phi) = \exp\left[-j\left(q\phi + 2\pi r / D\right)\right] \tag{7.1}$$

where r and ϕ are the polar coordinates, q is the grating spoke number, and D is the radial period of the spiral phase grating. Let an LG beam be incident on a spiral phase grating. Then the electric field of the incident LG beams at the waist plane ($z = 0$) is written as [58, 60, 65]

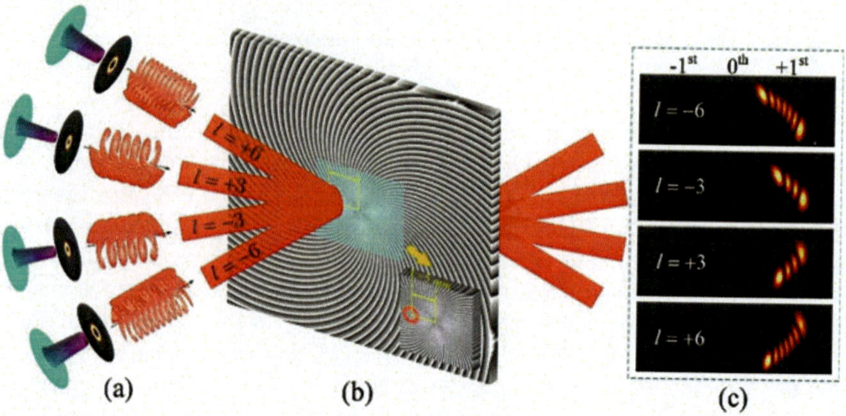

Fig. 7.2 Schematic diagram of measuring mode indices of LG beams with spiral phase grating. **a** incident LG laser beams, **b** spiral phase grating, and **c** far-field diffraction patterns [64]

$$u(r,\phi) = \sqrt{\frac{2p!}{\pi\omega_0^2(p+|l|)!}}\left[\frac{\sqrt{2}r}{\omega_0}\right]^{|l|} L_p^{|l|}\left(\frac{2r^2}{\omega_0^2}\right)\exp\left(-\frac{r^2}{\omega_0^2}\right)\exp(-jl\phi) \tag{7.2}$$

where ω_0 represents the beam waist radius of the fundamental mode; $L_p^{|l|}(\cdot)$ designates the associated Laguerre polynomial with p and l denoting the radial and azimuthal mode indices, respectively. Complex amplitude immediately after the grating will evolve to

$$U_0(r,\phi) = u(r,\phi)\exp[-j(q\phi + 2\pi r/D)] \tag{7.3}$$

Then the far-field diffraction intensity patterns at a distance z behind the grating can be obtained using Fresnel diffraction integral in the following form

$$U(\rho,\theta) = \frac{\exp(-jkz)}{-j\lambda z}\exp\left(\frac{-jk}{2z}\rho^2\right)\int_0^\infty\int_0^{2\pi} \begin{matrix} U_0(r,\phi)exp\left(\frac{-jk}{2z}r^2\right) \\ \exp\left\{\frac{jk}{z}\rho r\times\cos(\phi-\theta)\right\} \end{matrix} rdrd\phi \tag{7.4}$$

where (r, ϕ) is the polar coordinate at the diffraction plane, (ρ, θ) denotes the polar coordinate of the far-field, and $k = 2\pi/\lambda$ is the wavenumber with λ referring to a wavelength of the beam source.

Obtaining the complete analytic solution for Eq. (7.4) through complex integral calculations can be a daunting task. However, we can simplify the equation by leveraging Fourier transformation and perform numerical simulations of the far-field diffraction intensity distributions. As a result, Eq. (7.4) can be effectively expressed in the following manner:

$$I(\rho,\theta) = \left| \frac{\exp(-jkz)}{-j\lambda z} \exp\left(\frac{-jk}{2z}\rho^2\right) \times \mathcal{F}[u(r,\phi) \times t(r,\phi)] \right|^2 \tag{7.5}$$

where $\mathcal{F}$ represents the two-dimensional Fourier transform.

Based on the analytical formulas, we can generate numerical simulations for the far-field intensity distributions when the LG beams pass through the spiral phase grating. These simulations are depicted in Figs. 7.3 and 7.4. The simulation parameters used for these calculations are as follows: the wavelength of the incident beam is $\lambda = 632.8$ nm, the beam waist radius is $\omega_0 = 0.5$ mm, the spoke number of the spiral phase grating is $q = 80$, the radial period of the grating is $D = 0.03$ mm, and the center distance between the centers of the grating and incident vortex beams is $L = 3.5$ mm.

In Fig. 7.3, we observe that the annular-shaped LG beams transform into a ribbon-like diffraction intensity distribution, characterized by the presence of several bright spots. After carefully counting the number of these bright spots (N), we determine that the azimuthal index value is equal to $N - 1$. It is important to note that the direction of the bright spots is directly related to the sign of the azimuthal index l. For clarity in discussing the results, we mark the position of the intensity distribution with a yellow dashed line. The angle between this line and the positive direction of the horizontal axis is represented as α. The azimuthal index l is positive in the case of $\alpha < 90°$; otherwise, the sign of l is negative.

To study the relationship between the radial index values and the diffraction patterns, we also calculated the radial high-order condition ($p \neq 0$). In Fig. 7.4, we present the far-field diffraction patterns of a spiral phase grating illuminated by LG beams with $l = \pm 1, \pm 2, \pm 3$ and $p = 1, 2$. It is evident that the multiple-ring-shaped incident optical vortices transform into a bright spot array comprised of several ribbon-like diffraction intensities. Unlike the results depicted in Fig. 7.3, the

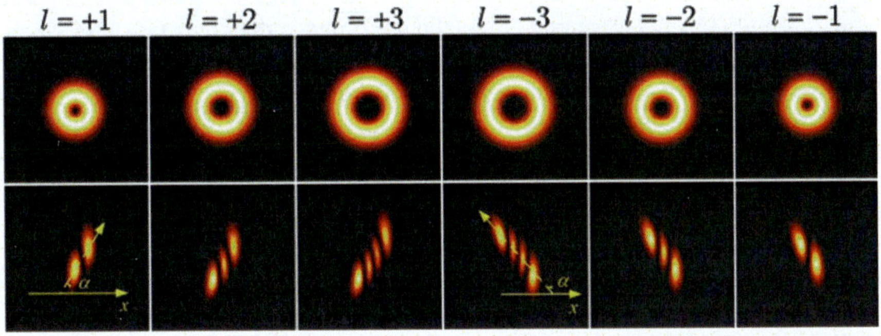

Fig. 7.3 Simulation results of incident LG beams and the corresponding far-field diffraction patterns when the fundamental radial mode ($p = 0$) vortex beams incident on the spiral grating [64]

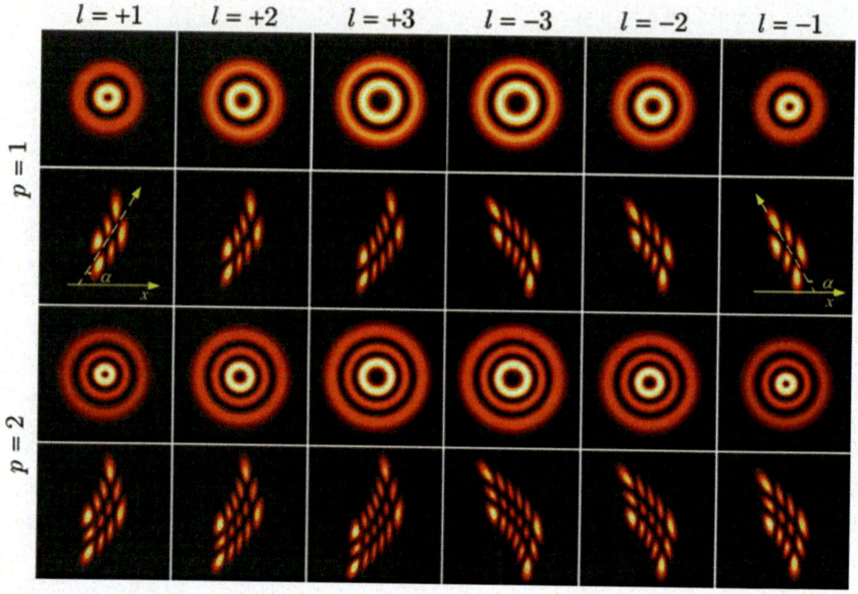

Fig. 7.4 Simulation results of incident LG beams and the corresponding far-field diffraction patterns when the radial high-order ($p \neq 0$) vortex beams incident on the spiral grating [64]

number of diffraction bright spots is no longer equal to the azimuthal index value plus one. Instead, it assumes a new value $(p + |l| + 1) \times (p + 1)$. To efficiently and quantitatively determine the mode indices information, we have designed a pattern recognition algorithm outlined in Fig. 7.5. In the first step, we analyze the received intensity pattern and quantify the number of bright spots ($a \times b$). The values of a and b represent the number of bright spots along the red and white line directions, respectively. Additionally, we observe the angle between the yellow dashed line and the positive direction of the x axis. The next step involves comparing the values of a and b, and subsequently assigning the smaller and larger values to two new variables m and n, respectively. We use a constant as a flag to identify the value of α.

Using the pattern recognition algorithm, we are able to automatically calculate and output the two-mode indices of the LG beams. Consequently, we obtain the values of the radial mode index and the azimuthal mode index, denoted as $p = m - 1$ and $l = (-1)^{\tau}(n - m)$, respectively. Furthermore, digital image processing technologies and artificial neural networks can also be employed to identify the transmitted mode intensity patterns captured at the receiver [66]. These techniques offer additional means for analyzing and interpreting the results.

The experiment setup to monitor the radial and azimuthal indices of LG vortex beams is sketched in Fig. 7.6. A 2-mW intensity stabilized He-Ne laser with 632.8 nm wavelength is used as the light source to emit a fundamental Gaussian beam. To avoid the exorbitant power causing damage to the optical elements, the optical power attenuation of the incident Gaussian beam is achieved by using a neutral density

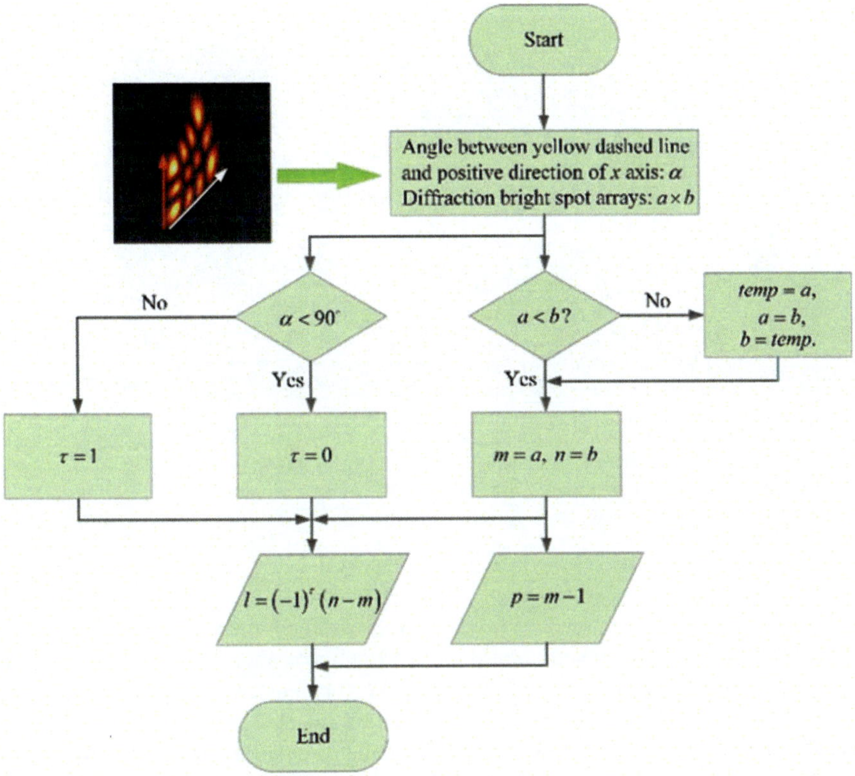

Fig. 7.5 The flow chart of the detection algorithm [64]

filter. Then the light is collimated and expanded by a beam expander (BE). After passing through the polarizer (P1), the horizontal linearly polarized beam is incident on a computer-controlled SLM (SLM1). Appropriate spot size is shaped to match the panel of SLM by inserting a pinhole (PH1) between BE and P1. A beam splitter (BS) is placed before the SLM1 to ensure that the light is incident perpendicularly on the SLM1. The CGH is uploaded on SLM1 to generate the input unknown LG beam. Subsequently, the modulated beam is split into two equal energy beams with the help of the BS. One of the beams through the transmission of another polarizer (P2) and pinhole (PH2) illuminates the second SLM (SLM2). The spiral phase grating is written into this SLM. The PH2 is needed to select the desired positive first diffraction order and block other unwanted orders. The optical field transformation is realized by placing lens L1 in a proper position. An image of the SLM2 plane is constructed in the plane of the last lens L3 via the convex lens L2. Finally, the far-field intensity pattern is captured by a CCD camera placed at the focal plane of L3 and is displayed on the monitor screen. The resolution of the SLMs (PLUTO-NIR-011, Holoeye) is 1920×1080 pixels, the pixel size is 8.0 μm, and the phase level is 8-bit depth.

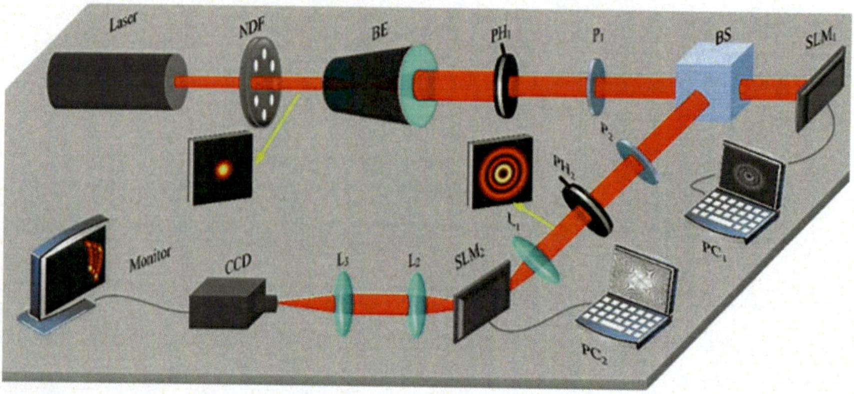

Fig. 7.6 Experiment setup for the generation of LG beams and detection of mode indices. Laser: He-Ne laser, 632.8 nm; NDF: neutral density filter; BE: beam expander; PH1, PH2: pinholes; P1, P2: polarizers; BS: beam splitter; SLM1, SLM2: spatial light modulators; PC1, PC2: personal computers; L1, L2, L3: convex lens, $f_1 = f_2 = 100$ mm, $f_3 = 120$ mm; CCD: charge-coupled device [64]

In the conducted experiment, the optical measurement setup depicted in Fig. 7.6 is utilized. The obtained results are partially presented in Figs. 7.7 and 7.8. To begin, the transverse intensity profiles of the incident LG beams are recorded by the CCD camera when SLM2 is not loaded with any hologram. This allows for the characterization of the incident LG beams. Subsequently, the vortex beams illuminate SLM2, which is loaded with the spiral phase grating. The far-field diffraction patterns resulting from this configuration are then measured. These patterns provide information about the effects of the spiral phase grating on the LG beams. The parameters used in the experimental setup are chosen to be consistent with the parameters mentioned in the previous simulation or theoretical analysis.

According to the experimental results displayed in Fig. 7.7, the measurements of the fundamental radial mode LG vortex beams with $l = \pm 1$, ± 2 and ± 3 specific

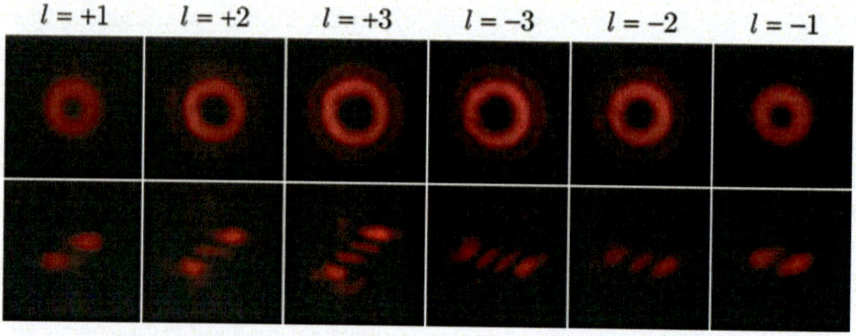

Fig. 7.7 Observed far-field diffraction patterns after LG beams with $p = 0$ and $l = \pm 1, \pm 2, \pm 3$ illuminating the spiral phase grating [64]

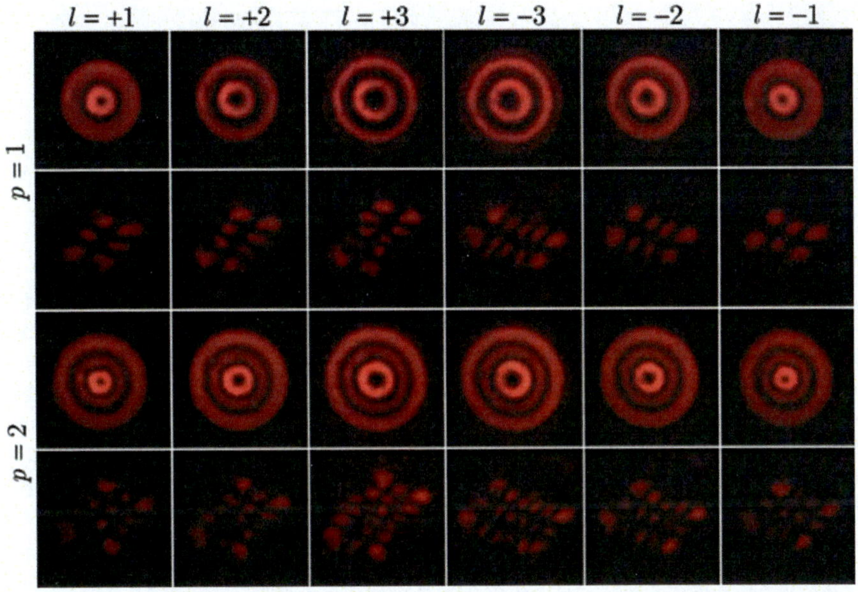

Fig. 7.8 Observed far-field diffraction patterns after LG beams with $p = 1, 2$ and $l = \pm1, \pm2, \pm3$ illuminating the spiral phase grating [64]

values of the radial and azimuthal indices demonstrate excellent agreement with the predictions presented in Fig. 7.3. The number of bright spots minus one corresponds to the magnitude of the azimuthal index, and the orientation of these spots can be used to determine the sign of l. In Fig. 7.8, it is observed that when the LG vortex beams illuminate a spiral phase grating with non-zero radial indices, simply counting the total number of spots would not provide the azimuthal state values. However, using the universal recognition algorithm illustrated in Fig. 7.4, the desired information can still be obtained. This algorithm allows for the determination of the mode indices based on the relationship between the mode indices and the far-field diffraction patterns. Therefore, this detection method can be effectively employed in practical measurements.

Indeed, it is worth noting that in the experimental setup described, the detection speed is primarily constrained by the low refresh rate of the SLM, which operates at 60 Hz, and the CCD camera with a frame rate of 90 frames per second (fps). As a result, the highest achievable detection rate for incident unknown LG beams would be limited to 60 states (p and l). However, this limitation can be addressed by developing an SLM with a higher frame rate. By utilizing an SLM device with an improved refresh rate, the detection speed can be significantly enhanced, allowing for faster measurements and characterization of the LG vortex beams. Additionally, an alternative to the CCD camera for capturing the far-field diffraction intensity patterns is to employ a photodiode. By using a photodiode instead of a CCD, the detection system can potentially achieve higher data acquisition rates, further improving the overall detection speed and efficiency of the measurement setup.

Gradually Changing-Period Spiral Spoke Grating

It appears that in order to expand the detectable range for OAM modes, a new scheme utilizing a gradually changing-period spiral spoke grating (GCPSSG) has been proposed [67]. The experimental results indicate that this scheme enables the detection of OAM states up to ±160, surpassing the previous limit of ±120 OAM modes that have been derived worldwide. One notable advantage of the GCPSSG scheme is its robustness and effectiveness in OAM diagnostics. It demonstrates good tolerance to environmental vibrations and beam misalignment, making it more reliable in practical applications. This further enhances its potential for accurately determining the OAM modes of structured beams.

The relevant phase elements and their superposition process are illustrated in Fig. 7.9. To produce the phase function of the desired grating, we first combine the holograms of an SPP and an axicon and then let the obtained hologram interfere with the gradually changing-period phase grating given in [60]. Considering that the phase profiles of an axicon and the gradually changing-period phase grating are described as $\exp(-j2\pi r/D)$ and $\exp(-jd\cos(\varsigma))$, respectively, the phase distribution of a GCPSSG can be expressed as

$$\phi(r,\phi) = 2 + 2\cos\left[m\phi + 2\pi r/D - d * \cos(\varsigma)\right] \tag{7.6}$$

where m is the grating spoke number, $d = -2000$ is an adjustable constant that determines the variation speed of the period of the grating, and $|\varsigma| \leq \mu$ with μ being the interval maximum. Based on Eq. (7.6), a GCPSSG is generated as the figure depicted in Fig. 7.9e. In the practical experiments, the phase value of this grating ranges from 0 to 2π when written into an SLM.

The LG beam is a typically and commonly used light beam containing a vortex phase. So, we choose the LG beam as the target beam. The electric field of the incident LG beam ($p = 0$) at the waist plane takes the form of

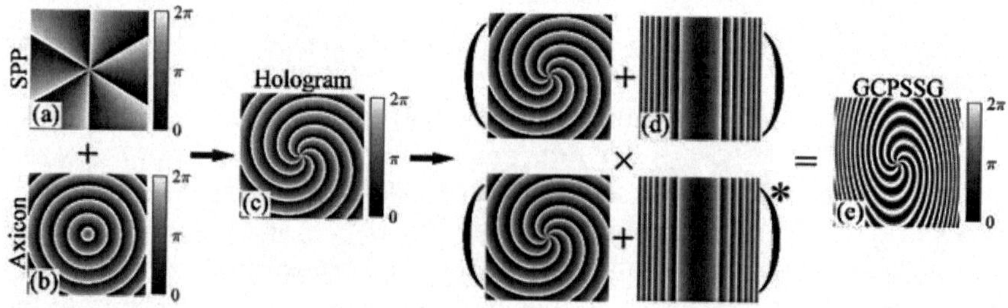

Fig. 7.9 Schematic diagrams of generating the diffractive optical mask loaded on the SLM. **a** spiral phase plate. **b** holographic axicon. **c** the phase hologram that results from combing **a** and **b**. **d** gradually changing-period phase grating. **e** gradually changing-period spiral spoke grating [67]

$$u(r, \phi, z = 0) = \left[\frac{\sqrt{2}r}{\omega_0} \right]^{|l|} L_p^{|l|}\left(\frac{2r^2}{\omega_0^2}\right) \exp\left(-\frac{r^2}{\omega_0^2}\right) \exp(-jl\phi) \quad (7.7)$$

To verify the practical validity of the proposed method, an experimental setup was devised for the generation and detection of vortex beams, as illustrated in Fig. 7.10. The setup utilizes a 632.8 nm He-Ne laser with an output power of 2 mW as the light source. The laser beam initially passes through a neutral density filter, which enables adjustment of the laser power. Subsequently, the beam is collimated and expanded using a beam expander, resulting in a wavefront resembling a plane wave. The expanded beam then proceeds through a polarizer and a BS1 before reaching the first SLM1. The placement of a polarizer before SLM1, which exclusively modulates horizontally linearly polarized beams, serves the purpose of effectively controlling the modulations performed by the SLM. SLM1, with a resolution of 1920 × 1080 pixels and an 8.0 μm pixel pitch, generates high-quality LG beams based on CGHs displayed on its surface. Following modulation, the vortex beam reflected by SLM1 is separated into two equal-energy beams with the assistance of the BS1. One of these beams passes through the transmission of a second BS2 and is then projected onto the second SLM2. SLM2, which is loaded with the calculated GCPSSG, contributes to further modulation. A circular aperture is introduced between SLM1 and SLM2 to selectively isolate the positive first-order vortex beam and eliminate any extraneous orders. Finally, the resulting diffracted light intensity pattern is captured by a monochrome CCD camera positioned at the image focal plane of a convex lens (L) with a focal length of $f = 120$ mm. This meticulously designed experimental setup allows for the generation and detection of vortex beams using a He–Ne laser and a series of optical components. By analyzing the captured diffracted light intensity pattern, the effectiveness of the proposed method can be examined.

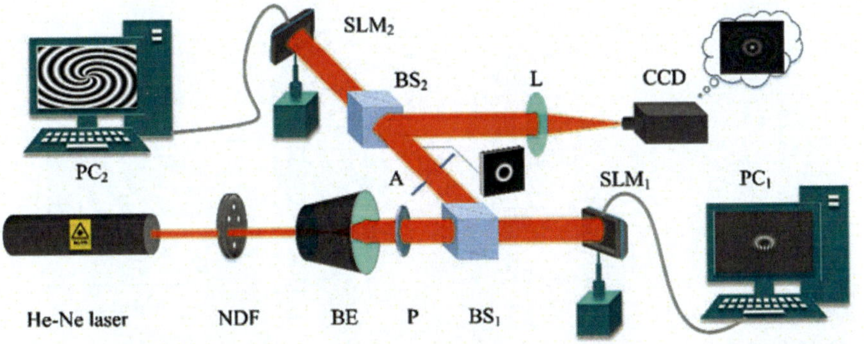

Fig. 7.10 Experiment setup for generating the optical vortices and measuring TC. He-Ne laser, 632.8 nm; NDF, neutral density filter; BE, beam expander; P, polarizer; BS1, BS2, beam splitters; A, aperture; SLM1, SLM2, spatial light modulators; PC1, PC2, personal computers; L, the convex lens with focus $f = 120$ mm; CCD, charge-coupled device [67]

In order to study the impact of phase distributions on the detection performance, four GCPSSGs were calculated and loaded onto the SLM. The parameters μ are set to $\mu = 0.8$, $\mu = 0.5$, $\mu = 0.1$, and $\mu = 0.01$ from left to right, separately. The grating used in the setup has a period of $D = 0.25$ mm, and the TC of the incident vortex beam is $l = +6$. The grating spoke number is $m = +6$. Figure 7.11 illustrates the GCPSSG graphs with different values of μ and their respective experimentally recorded diffraction patterns captured by the CCD. As depicted in the upper row figures, the visibility of the GCPSSG becomes increasingly evident as μ decreases. It is also observed that the transformed intensity patterns become more distinguishable when an appropriate value of μ is selected. The diffraction intensity patterns exhibit a prominent high-intensity spot at the center, indicating the conversion of the incident vortex beam into a Gaussian beam [20]. Additionally, a number of discontinuous spiral-shaped petals can be observed around the bright spot in the central region of the optical field. Importantly, the rotation direction of these diffraction petal fringes is consistently opposite to the uploaded gratings associated with the TC sign. Furthermore, the number of spiral fringes is twice the grating spoke number m, which allows for the verification of the TC measurement of vortex beams. Consequently, by utilizing the GCPSSG, it becomes straightforward to determine both the sign and magnitude information of the TC.

To further support our analyses, we conducted experimental investigations. Figure 7.12 presents the diffraction patterns obtained when LG beams with varied l illuminated different designed GCPSSGs with $m = \pm 4, \pm 6, \pm 8$. The left-most column in Fig. 7.12 displays the CGHs obtained using the method described in Ref. [68]. The second column, labeled as Fig. 7.12a$_2$, illustrates the resultant light beams generated with $l = +4, +6, +8, -4, -6$, and -8. The middle three columns, depicted as Figs. 7.12b$_1$–b$_3$, show the experimental results obtained after the vortex

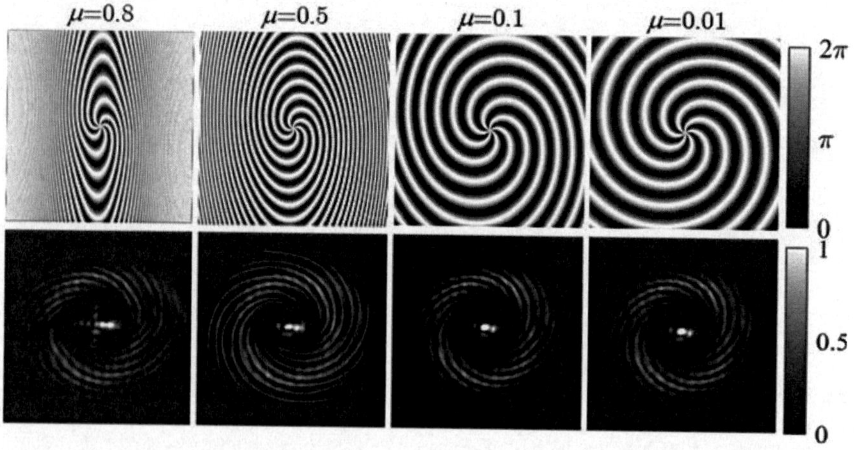

Fig. 7.11 Outline of the GCPSSG with different values of μ (upper row). Intensity distributions after the vortex beam with $l = +6$ illuminates the corresponding phase holograms of μ (lower row) [67]

beams in the second column pass through the GCPSSGs with $m = +4, +6, +8$, separately. Additionally, apart from discussing the diffraction patterns arising from GCPSSGs with positive grating spoke numbers $l = m$, we also evaluated the gratings with distinct negative spoke numbers, as showcased in Fig. 7.12c_1–c_3. Analyzing the recorded diffraction patterns in Fig. 7.12, we observe that a bright spot prominently appears in the central region for cases with $l \neq m$, while the intensity profile at the center maintains a doughnut structure for cases with $|l - m|$. The yellow dashed circle marks the position of the bright spot region. It is important to note that for cases where the value of $|l\text{-}m|$ is near zero, some on-axis split spots may be observed. However, the diffraction patterns remain easily distinguishable in such scenarios, as there is only a single central bright spot in the case of $l = m$ Therefore, the value of TC can be determined by observing the energy distribution of the central position.

Furthermore, upon examining Fig. 7.12, we observe that the radius of the dark center area becomes wider as the absolute value of the difference between l and m (i.e., $|l - m|$) increases. Moreover, for a positive (negative) value of m in the GCPSSG, the

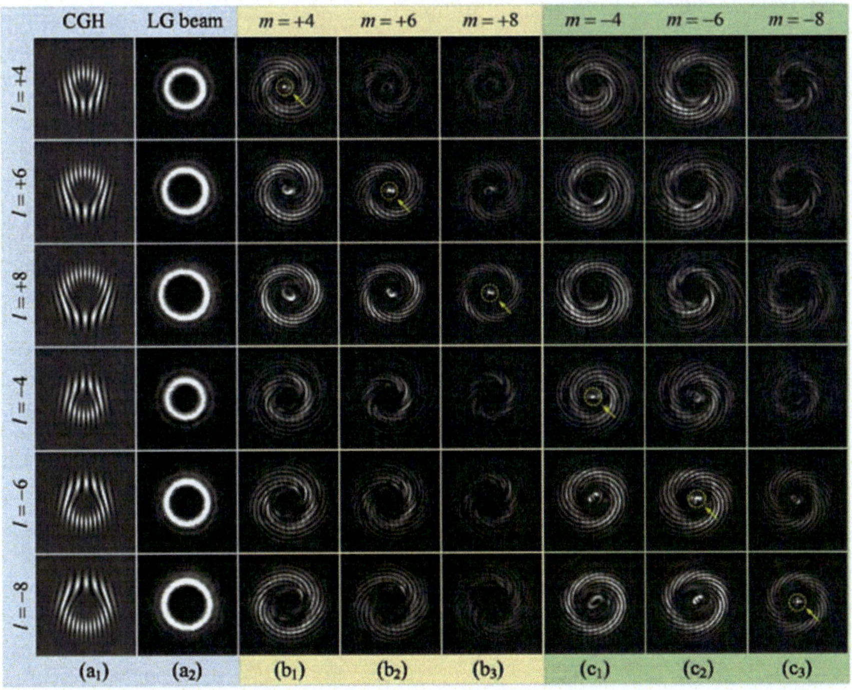

Fig. 7.12 Incident optical vortex beams and their diffraction patterns. $\mathbf{a_1}$ CGHs uploaded on SLM1. $\mathbf{a_2}$ experimentally generated optical vortices with topological charge $l = \pm 4, \pm 6, \pm 8$. $\mathbf{b_1}$–$\mathbf{b_3}$ diffraction patterns after vortex beams pass through the GCPSSGs with $m = +4, +6, +8$, separately. $\mathbf{c_1}$–$\mathbf{c_3}$ diffraction patterns after vortex beams pass through the GCPSSGs with $m = -4, -6, -8$, separately [67]

spiral petals rotate counterclockwise (clockwise). Although the number and orientation of the diffraction fringes do not change with the incident optical vortices, the rotation direction of the fringes provides valuable information for accurately measuring the sign of TC. Notably, the rotation direction of the GCPSSG and the corresponding intensity distributions exhibit an opposite relationship. This enables the determination of the sign of l by identifying the deflection direction of the spiral petals when the central intensity reaches a locally maximum value.

In order to further evaluate the reliability of the proposed scheme, we conducted additional experiments to demonstrate the behavior of unknown LG beams passing through the off-axis position of the GCPSSG. A concept diagram and partial experimental results are presented in Fig. 7.13. Figure 7.13a illustrates the incident vortex beams with different TC values. In Fig. 7.13b, the grating spoke number of the GCPSSG $m = -6$ is chosen, while the other parameters remain the same as those in Fig. 7.12. As indicated in Fig. 7.13d, the diffraction intensity patterns were measured when the center distance between the grating and the incident vortex beams was set to $S = 2$ mm. Comparing the results shown in Fig. 7.12 with those in Fig. 7.13, we observe that when the light beams illuminate the off-axis position of the GCPSSG, the diffraction intensity distributions exhibit Hermite-Gaussian-like modes instead of the spiral fringe petals observed in the previous case. It becomes evident that the intensity distributions are influenced by the sign and value of the incident beams. By carefully counting the number of bright stripes, it is observed that there will be $|l|+1$ bright stripes when an LG beam with the mode l passes through the grating. Furthermore, the direction of the stripes provides information about the sign of the TC. These observed patterns align well with the simulation results presented in Fig. 7.13c. This indicates that the incident light beams do not need to strictly illuminate the central position of the grating, highlighting the flexibility and robustness of the proposed scheme.

Furthermore, the experimental results of the far-field diffraction patterns obtained after the LG beams with high-order OAM modes $l = -30, \ -90, \ -120$ and -160 ($l = +30, \ +90, \ +120, \ +160$) pass through the GCPSSG are depicted in Fig. 7.14 (Fig. 7.15). It is evident that the fringe pattern becomes denser as the TC value increases, as there exists a relationship between the number of bright stripes and the TC value of the vortex beams. However, the high-order intensity patterns are not as clear as the results shown in Fig. 7.13 for low-order OAM states. The locally enlarged insets in Figs. 7.14 and 7.15 reveal that the diffraction bright stripes still appear ambiguous. The detection sensitivity has been enhanced by employing the concept of gradually changing period, as proposed in [57]. Importantly, our method, based on the GCPSSG, allows for the measurement of TCs as high as $\pm$ 160, with high diffraction efficiency compared to $\pm$100 in [60]. This indicates that the proposed detection method can be applicable in practical measurement scenarios.

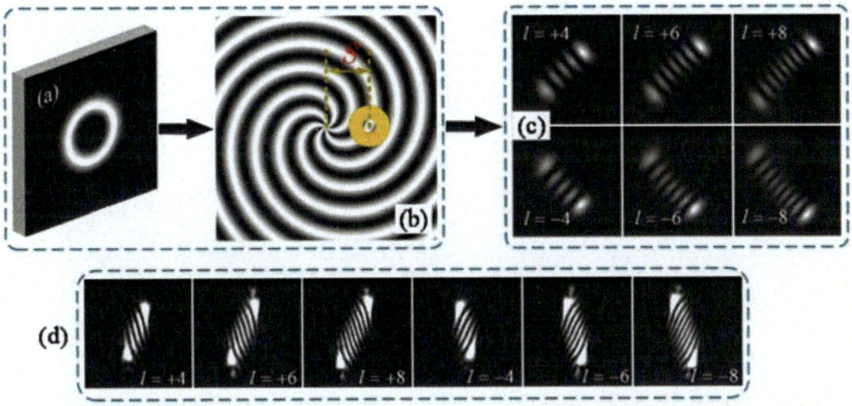

Fig. 7.13 Measurement results when vortex beams are incident under misalignment conditions. **a** incident vortex beams. **b** GCPSSG loaded on SLM2; S, the center distance between phase grating and input fields. **c** and **d** simulated and measured + 1st diffraction patterns behind the grating, respectively [67]

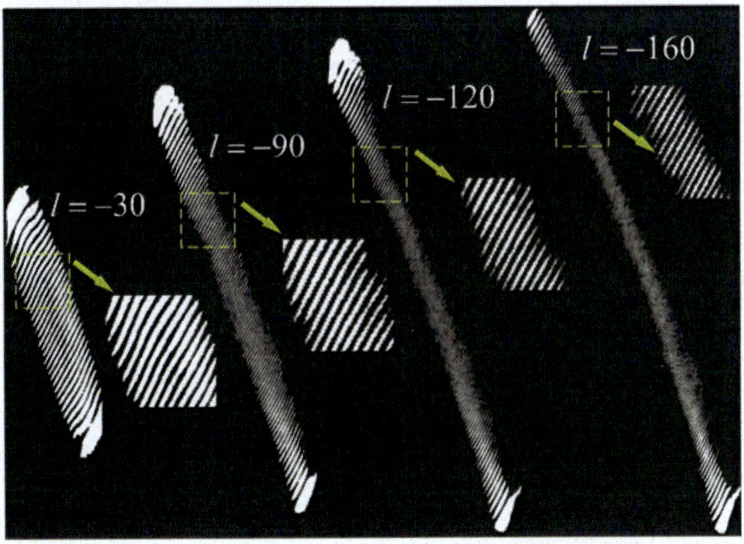

Fig. 7.14 Experimental results of the diffraction intensity distributions after vortex beams with larger TC values $l = -30, -90, -120$ and -160 pass through the phase grating [67]

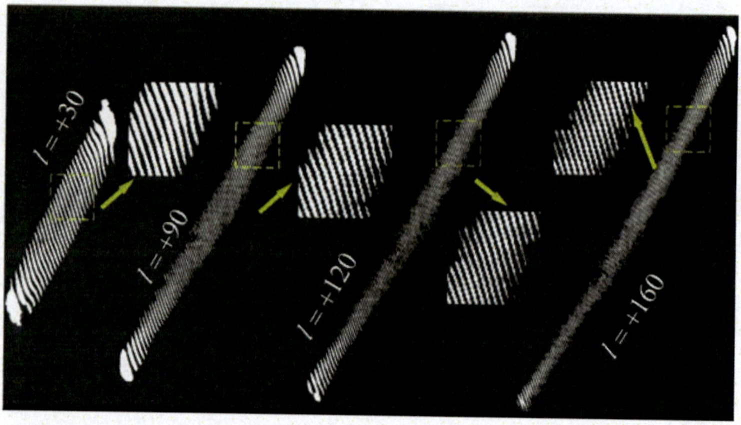

Fig. 7.15 Experimental results of the diffraction intensity distributions after vortex beams with larger TC values $l = +30, +90, +120$ and $+160$ pass through the phase grating

7.3 Design and Applications of Reflective Metasurface: Generating OAM Vortex Wave in Radio Frequency

7.3.1 Analysis and Design of Reflective Metasurface

A reflective metasurface typically comprises a metal ground plate that is printed on one side of a dielectric substrate. Unlike transmission metasurfaces, which require simultaneous consideration of both reflected and transmitted waves, reflective metasurfaces are designed solely to regulate the reflected beam. This makes the design process simpler, as illustrated in Fig. 7.16.

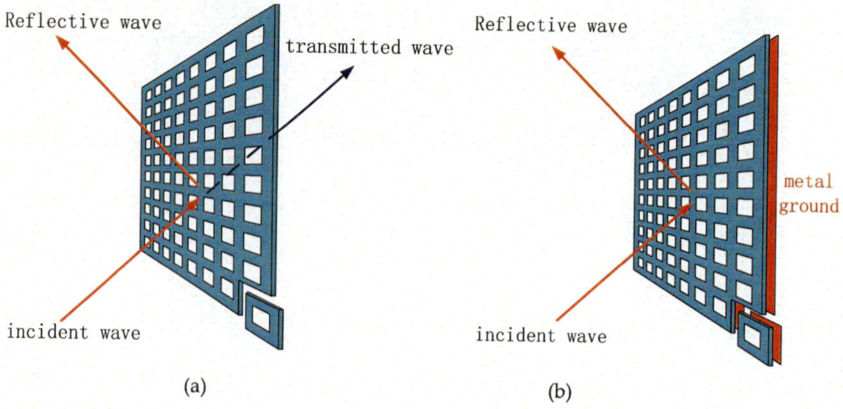

Fig. 7.16 Geometry of metasurfaces for electromagnetic waves regulation: **a** transmission metasurface, **b** reflective metasurface

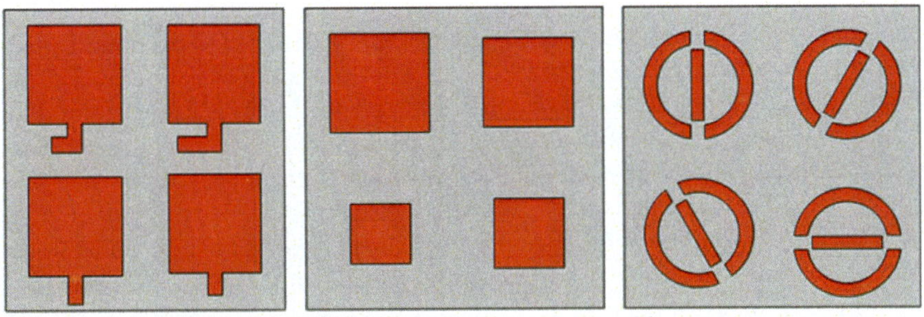

Fig. 7.17 **a** elements with phase/time delay lines, **b** elements with variable sizes, **c** elements with variable rotation angles

The selection of reflective metasurface elements is crucial for designing metasurfaces with optimal performance. A key consideration is to minimize the slope of the reflection phase curve to reduce sensitivity to element dimensions. A steep curve can lead to problems with operational bandwidth and fabrication tolerances. It is also important to ensure that the metasurface element has enough variation in dimensions to cover a phase range of 360°. In addition, achieving a smooth and linear phase variation is desirable [69]. Currently, there are three main types of units for tuning the phase of the reflective metasurface, as shown in Fig. 7.17, i.e., (1) elements with phase/time delay lines, (2) elements with variable sizes, (3) elements with variable rotation angles.

In the case of elements with phase/time delay lines, these elements are connected to transmission lines of varying lengths. The element receives the incoming electromagnetic wave from the feed and converts it into a guided wave propagating along the transmission line. The different lengths of the transmission lines introduce controlled time delays, thereby achieving the desired phase shift. On the other hand, elements with variable sizes employ a different approach. Here, the physical dimensions of the element are altered to provide varying phase shifts. The underlying principle is that resonant elements of different sizes yield different reflected phases. By adjusting the size of the element, precise control over the phase shift can be achieved. Regarding elements with variable rotation angles, this technique utilizes circular polarization antenna elements. When such an element is rotated around its origin by a certain angle, the radiated phase is modified by twice the amount of the rotation angle. The advancement or delay of the phase depends on the direction of rotation [70]. These techniques offer distinct means of manipulating the phase in reflective metasurfaces, enabling precise control and customization of the reflected waves according to specific design requirements.

When designing a reflective metasurface, it is important to obtain the phase shift characteristic curve of the metasurface element. Simple structures like "V" or "dipole" shaped elements can be studied using classical analytical analysis to calculate their phase characteristics. However, for more complex unit structures, it becomes nearly impossible to obtain the phase shift characteristics analytically. In such cases, electromagnetic numerical simulation software like HFSS or CST can be utilized to calculate the electromagnetic characteristics of the metasurface unit. An effective approach for extracting the reflection phase shift characteristic curves of reflective metasurface elements is to employ the infinite period method. This can be accomplished by using periodic boundary conditions with Floquet port excitation, as illustrated in Fig. 7.18. In a large array with regular spacing, the behavior of all the elements (except those near the edges) is approximately the same. In an infinite array, the behavior of each element is identical due to the absence of edges or variations in the arrangement. Consequently, the fundamental properties of the elements in a very large regular array can be represented by an element in the corresponding infinite array. This infinite array approach has been established as the preferred basis for designing elements in large finite arrays. By employing numerical simulations and the infinite period method, the phase shift characteristics of complex metasurface elements can be accurately analyzed and utilized for the design and optimization of reflective metasurfaces in practical applications.

Figure 7.19 illustrates the characteristics of a typical reflective metasurface element. In general, a high-quality unit should possess a phase shift coverage equal to or greater than 360 degrees, allowing for phase compensations ranging from 0 to 360 degrees. Additionally, the corresponding reflection amplitude should be better than −1 dB to ensure efficient radiation of most of the electromagnetic power. Furthermore, it is crucial to select the element with a smooth and linear phase curve to have better fabrication tolerance and bandwidth range.

In the design of reflective metasurfaces, it is essential to determine the required phase compensation at specific positions on the metasurface. To provide a more visual representation of the design methodology, let's consider the example of designing a reflective metasurface for generating a directional beam, as depicted in Fig. 7.20.

The generalized Snell's law can be given by [11]

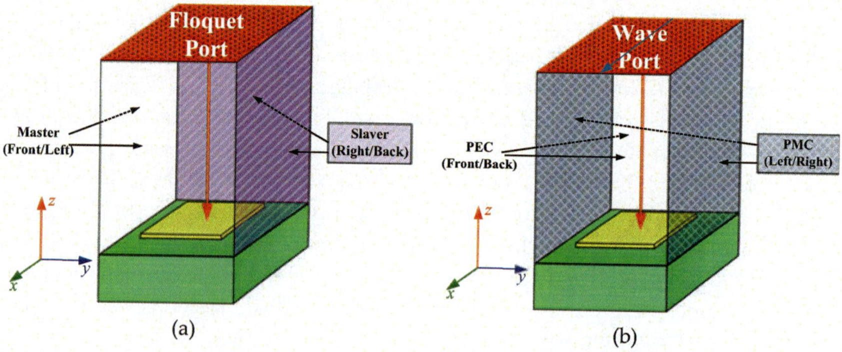

Fig. 7.18 **a** Periodic boundary conditions with Floquet port excitation, **b** PEC-PMC boundary conditions with wave port excitation.

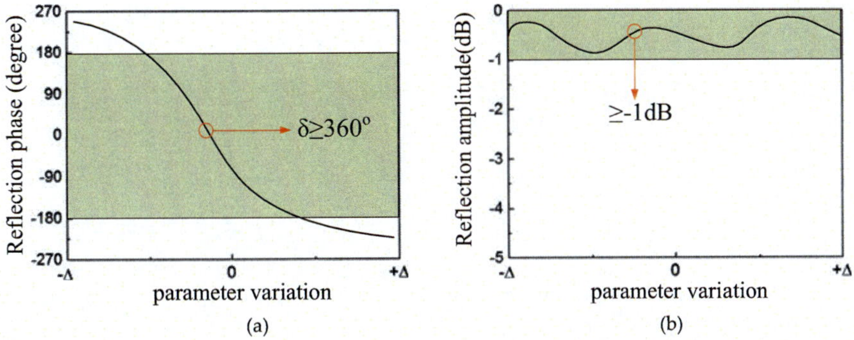

Fig. 7.19 Basic characteristics required for a typical reflective metasurface element, **a** reflection phase, **b** reflection amplitude

Fig. 7.20 Geometric diagram of reflective metasurface for directional beam

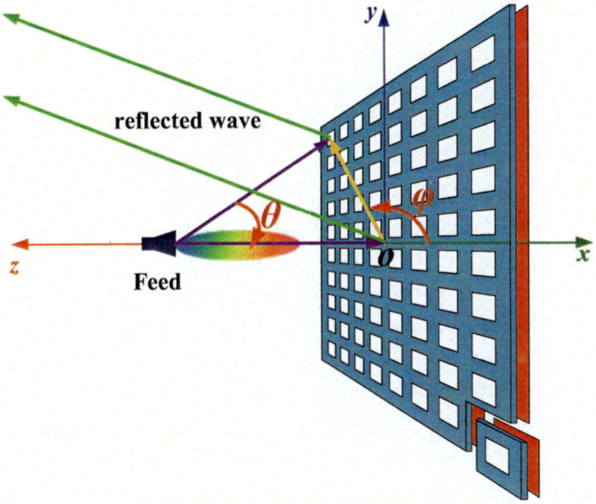

$$\begin{cases} \sin\theta_r\cos\varphi_r - \sin\theta_i\cos\varphi_i = k_0\dfrac{\partial\Phi}{\partial x} \\ \sin\theta_r\sin\varphi_r - \sin\theta_i\sin\varphi_i = k_0\dfrac{\partial\Phi}{\partial y} \end{cases} \tag{7.8}$$

According to the geometric relationship, one has

$$\begin{cases} \sin\theta_i\cos\varphi_i = -\dfrac{\sqrt{x^2+y^2}}{\sqrt{x^2+y^2+F^2}}\dfrac{x}{\sqrt{x^2+y^2}} = -\dfrac{x}{\sqrt{x^2+y^2+F^2}} \\ \sin\theta_i\sin\varphi_i = -\dfrac{\sqrt{x^2+y^2}}{\sqrt{x^2+y^2+F^2}}\dfrac{y}{\sqrt{x^2+y^2}} = -\dfrac{y}{\sqrt{x^2+y^2+F^2}} \end{cases} \tag{7.9}$$

Substituting (7.9) into (7.8), (7.8) can be rewritten as

$$\begin{cases} \dfrac{\partial \Phi}{\partial x} = k_0\left(\sin\theta_r \cos\varphi_r + \dfrac{x}{\sqrt{x^2+y^2+F^2}}\right) \\ \dfrac{\partial \Phi}{\partial y} = k_0\left(\sin\theta_r \sin\varphi_r + \dfrac{y}{\sqrt{x^2+y^2+F^2}}\right) \end{cases} \tag{7.10}$$

Thus, the phase compensation required by any position on the metasurface is obtained as

$$\Phi(x, y) = k_0\left(x\sin\theta_r\cos\varphi_r + y\sin\theta_r\sin\varphi_r + \sqrt{x^2+y^2+F^2}\right) + C \tag{7.11}$$

In this specific design scenario, we have set the operating frequency to 5.8 GHz, and a feed horn is positioned in the axial center direction of the metasurface. The feed-to-metasurface distance is denoted as $F = 0.4$ m. At any given point on the metasurface, the incident wave from the feed arrives with different angles of incidence (θ_i, φ_i). Assuming that the direction of the reflected beam is $(\theta_r, \varphi_r) = (30°, 0°)$, the medium surrounding the metasurface is air, characterized by a refractive index of $n_i = 1$. The parameter k_0 represents the wave number in free space, while the position (x, y) indicates the coordinates of a specific point on the metasurface. In terms of dimensions, the metasurface has a size of 500 mm × 500 mm, and the spacing between individual elements is given by $D = 25$ mm. The phase compensation distribution for each position on the metasurface can be calculated by substituting these parameters into Eq. (7.11), as depicted in Fig. 7.21. To realize the desired metasurface performance, it is necessary to select an appropriate physical unit that corresponds to the phase shift distribution obtained from Fig. 7.21. This one-to-one correspondence between the designed phase shift distribution and the physical unit selection enables the realization of the desired metasurface functionality. By following this design approach and utilizing the information provided in Fig. 7.21, the reflective metasurface can be accurately designed for beam steering applications at the specified frequency of 5.8 GHz.

In the previous discussion, we introduced the phase gradient method as a way to calculate the phase compensation distribution on the metasurface. This method is advantageous when the geometric relationship is clear and straightforward. However, when dealing with highly complex geometric relationships, calculating the phase distribution using the gradient phase method becomes more challenging. To address this, we now introduce another simple and effective method called the direct phase shift method. This approach involves directly calculating the phase shift distribution on the array surface by determining the difference between the target phase and the initial phase. In the same design problem discussed earlier, where the direction of the reflected beam is $(\theta_r, \varphi_r) = (30°, 0°)$, the required phase distribution needed is,

$$\Phi_R = k_0(x\sin\theta_r\cos\varphi_r + y\sin\theta_r\sin\varphi_r) \tag{7.12}$$

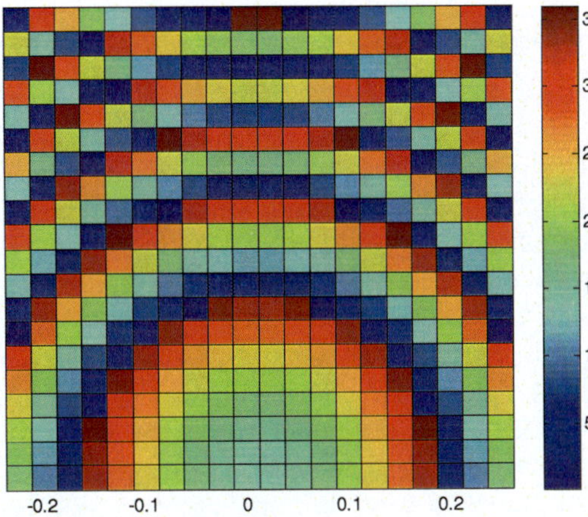

Fig. 7.21 Phase compensation distribution of the designed reflective metasurface for generating directional beam

When the feed horn irradiates the metasurface, the initial phase will be generated as

$$\Phi_I = -k_0\sqrt{x^2 + y^2 + F^2} \tag{7.13}$$

Therefore, the metasurface only needs to provide a phase shift with the following expression

$$\Delta\Phi = k_0(x \sin\theta_r \cos\varphi_r + y \sin\theta_r \sin\varphi_r) + k_0\sqrt{x^2 + y^2 + F^2} \tag{7.14}$$

We find that Eq. (7.14) is consistent with Eq. (7.11), indicating that the two method designs are equivalent. Figure 7.22 shows the calculation process of the direct phase-shift method.

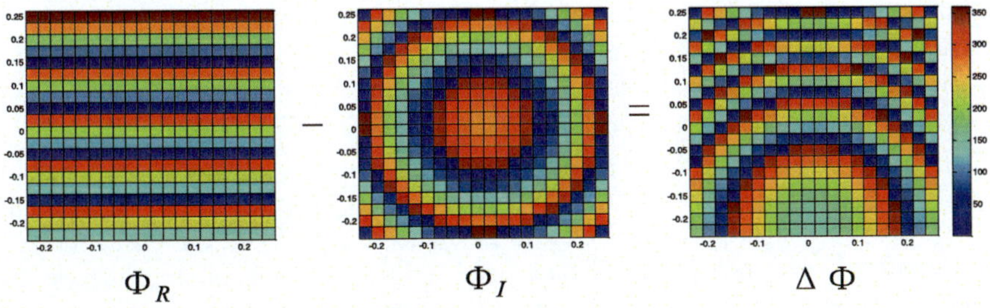

Fig. 7.22 Direct phase method calculation process

7.3.2 Design, Fabrication and Measurement of Reflective Metasurface for Generating Single OAM Vortex Beam

The distinctive properties and potential applications of OAM vortex waves have garnered significant interest in various fields, including optics, atomic and molecular physics [71, 72]. In 2007, the first simulation of radio OAM was conducted, providing a theoretical foundation for concepts related to wireless communication using OAM [73]. Subsequently, a comprehensive system simulation was reported, focusing on the generation of OAM radio beams using a circular antenna array [74]. In 2012, the first experimental implementation of wireless radio transmission using OAM vortex waves demonstrated the potential to increase communication capacity without requiring additional bandwidth [75]. Since then, OAM-based radios and their applications in wireless communication have become a prominent research area. Generating vortex beams carrying OAM in the radio frequency domain is of great significance. Various methods have been explored to achieve twisted radio beams, including the use of SPPs [76], spiral reflectors [77], and antenna arrays [78]. Among these, the SPP has received particular attention due to its wide usage in optics, known for its simple structure and ease of implementation [79, 80]. However, a significant drawback of the SPP is its large beam divergence angle, which hampers long-distance transmission in the low-frequency radio domain. Additionally, dielectric reflection can negatively impact performance, posing further limitations on its application in the radio frequency domain. To address these challenges, a specially designed parabolic antenna can be utilized. This antenna is bent into a spiral curved surface, effectively acting as a reflective phase-revolving plate. Leveraging the beam concentration effect of parabolic antennas, divergent OAM waves can be concentrated. However, generating OAM waves with different mode numbers remains challenging due to the complex structure and operating principles involved in such specialized antennas. Finding an effective method to generate radio frequency vortex beams carrying OAM is crucial for advancing research and applications in this domain. Overcoming the limitations of existing approaches will pave the way for new developments in wireless communication systems utilizing OAM.

In addition to parabolic antennas, array antennas can also be effectively used to generate radio beams carrying OAM. However, it is important to implement a complex feeding system in order to achieve a rotating phase front in the radio beam. Proper phase relationships between the radiating elements must be ensured, along with consistent power delivery, to guarantee the purity of the OAM modes. The number of elements in the antenna array and their mutual coupling also play a role in generating OAM beams. To generate twisted radio beams with more mode numbers, a greater number of antenna elements is required. However, this increases the complexity of the hardware design and debugging process. Metasurfaces have emerged as a promising solution for generating optical vortices based on the generalized laws of reflection and refraction [11]. Subsequently, the generation of optical OAM at visible wavelengths using plasmonic metasurfaces was proposed [81]. Researchers also successfully created a surface plasmon vortex carrying OAM in a gold metasurface under linearly polarized optical excitation [82]. These developments highlight the potential of metasurfaces in generating OAM-carrying beams at optical frequencies. Extending these concepts to the radio frequency domain holds promise for enabling advanced applications in wireless communication systems.

In our proposed study [83], we introduce a reflective metasurface that is specifically designed, fabricated, and experimentally demonstrated to generate OAM vortex waves in the radio frequency domain. We have derived a theoretical formula for the phase-shift distribution, which is then utilized to design the metasurface capable of producing vortex radio waves. The reflective metasurface offers several advantages for OAM generation, including the avoidance of transmission loss and the flexibility of phase control. The metasurface is composed of sub-wavelength elements, and it is not equipped with power division transmission lines. Instead, an illuminating feed antenna is employed to spatially illuminate these elements, causing the incident field to scatter and generating a reflective wave. This reflective wave in the far-field zone exhibits a rotating phase front characterized by $\exp(jl\varphi)$, where l represents the topological charge associated with the desired OAM mode. To illustrate the setup, Fig. 7.23 provides a schematic representation of the OAM generator, showing the metasurface, a metallic ground plate, and the illuminating feed antenna. Through the careful design and implementation of this reflective metasurface, we can achieve the generation of OAM vortex waves in the radio frequency domain without the need for power division transmission lines and with precise control over the phase distribution on the reflective surface.

Consider a reflective metasurface, as shown in Fig. 7.23, which consists of $m \times n$ elements that are illuminated by a feed source located at the position vector $\vec{r_f}$. The reradiated electric field from the metasurface in an arbitrary direction $\hat{u}$, will be

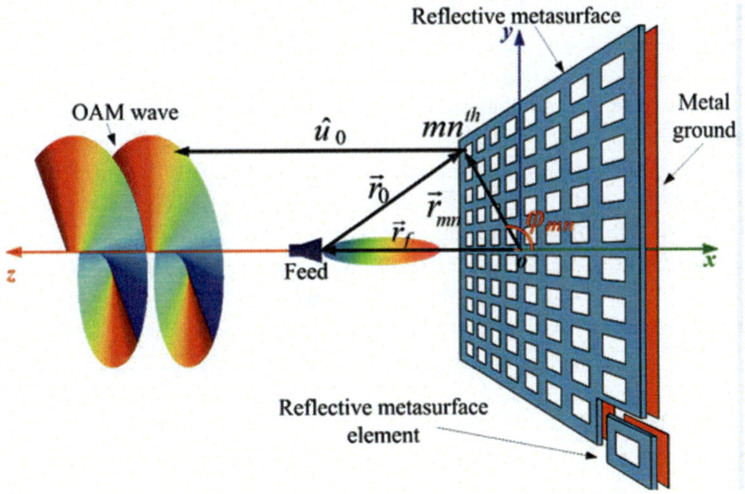

Fig. 7.23 Configuration of OAM-generating reflective metasurface [83]

of the form [76]

$$E(\hat{u}) = \sum_{m=1}^{M}\sum_{n=1}^{N} F\left(\vec{r_{mn}} \cdot \vec{r_f}\right) A\left(\vec{r_{mn}} \cdot \hat{u}_0\right) A\left(\hat{u}_0 \cdot \hat{u}\right)$$
$$\cdot \exp\left\{ jk_0\left[\left|\vec{r_{mn}} - \vec{r_f}\right| + \vec{r_{mn}} \cdot \hat{u}\right] + j\phi^c_{mn}\right\} \tag{7.15}$$

where F is the feed pattern function, A is the reflective metasurface element pattern function, $\vec{r_{mn}}$ is the position vector of the mnth element, and $\hat{u}_0$ is the desired main-beam pointing direction of the reflective metasurface. The phase-shift required at each reflective element for an OAM vortex wave in the desired direction $\hat{u}_0$ can be obtained by

$$\phi^c_{mn} = l\varphi_{mn} - k_0\left[\left|\vec{r_{mn}} - \vec{r_f}\right| + \vec{r_{mn}} \cdot \hat{u}_0\right], \quad l = 0, \ \pm 1, \ \pm 2, \ldots \tag{7.16}$$

where l is the desired OAM mode number, φ_{mn} is the azimuthal angle of the mnth element, k_0 is the propagation constant in vacuum, and ϕ^c_{mn} is the required compensating phase of the mnth reflective element on the metasurface.

The compensating phase distribution described in Eq. (7.16) can be calculated using the electromagnetic field superposition principle, taking into account the mutual coupling among the elements in the metasurface. Accounting for the mutual coupling is crucial in achieving a high-quality reflective OAM beam. By considering different OAM mode numbers, the sub-wavelength metasurface elements can be designed to generate a quasi-continuous spatial phase change.

To validate the prototype system and illustrate its functionality, an OAM-generating design with a specific mode number, $l = 2$, is presented. In this design, the feed point is positioned in the front of the metasurface with $\vec{r_f} = (0, 0, 0.4)$m along the axis. It is important to note that an offset feeding configuration is unnecessary owing to the formation of an amplitude null at the center of the OAM vortex beam. By ensuring a normal incidence configuration, the impact of this block effect on the reflective OAM vortex wave is minimized.

Next, the phase-shift compensated at each element to produce an OAM beam in a given direction should be determined. Considering the coordinate system shown in Fig. 7.23, the progressive phase distribution on the surface to generate a vortex beam in the direction of $\hat{u}_0 = (0, 0, 1)$ can be expressed as $\phi_R = l\varphi_{mn}$. On the other hand, as a result of illumination from a feed horn, the phase correction factor on the plane is $\phi_I = k_0|\vec{r_{mn}} - \vec{r_f}|$. Hence, the phase-shift required by the metasurface is extracted as $\phi^c_{mn} = \phi_R - \phi_I$. The calculation process and phase distribution compensated on the reflective metasurface for OAM with $l = 2$ is illustrated in Fig. 7.24.

The unit cell of the metasurface is designed and shown in Fig. 7.25a [83]. Each element consists of F4B ($\varepsilon_r = 2.65$) substrate with three dipoles printed on the top surface. By controlling the length of the dipoles and the ratio between the center dipole and the two satellite dipoles, the unit cell has a continuous reflection phase

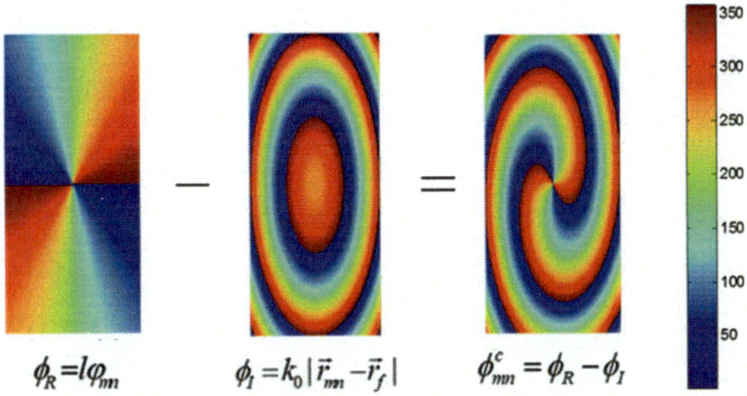

Fig. 7.24 The calculation process and phase distribution compensated on the reflective metasurface [83]

range more than 360° [85]. The simulation model and result confirm the element's good reflection phase response, as shown in Fig. 7.25.

A reflective metasurface was designed to generate an OAM beam at 5.8 GHz. It consisted of a 20 × 20 array of tri-dipole elements arranged in a 50 cm × 50 cm square layout (Fig. 7.26). A horn antenna fed the metasurface at a distance of 0.4 m. Numerical simulations (Fig. 7.27) confirmed the metasurface's ability to effectively generate OAM vortex waves with different mode numbers. This has promising applications in beam manipulation and OAM-based communication systems.

The designed and fabricated prototype of the OAM-generating reflective metasurface with $l = 2$ was successfully measured in an experiment. Figure 7.27b displays the prototype, while the near-field planar scanning technique was employed to measure the OAM vortex wavefront. The experimental setup, shown in Fig. 7.28, operated at a frequency of 5.8 GHz and detected the vertical polarization component of the reflected electric field, E_v, using a standard measuring probe. During the measurement, the near-field sampling plane was set at $z = 3.0$ m, and both the magnitude and phase of E_v, were measured on a sampling grid with a period of 10 mm. Figure 7.29d, e present the measured magnitude and phase distributions of E_v, respectively. Comparing these measurements with the simulated OAM spatial field distributions shown in Fig. 7.29a, b, successful generation and measurement of the OAM vortex wave with $l = 2$ can be observed. The simulated and measured results exhibit good agreement, with only slight shielding in the measured phase distribution caused by the supporting structure of the feeding horn. The key feature of the spatial phase distribution, resembling a perfect doughnut-shaped intensity map, is readily discernible. Additionally, the simulated and measured radiation patterns in the far-field zone are depicted in Fig. 7.29c, f respectively, clearly displaying an amplitude null at the center of the beam, consistent with the characteristics of a vortex optic beam with OAM mode.

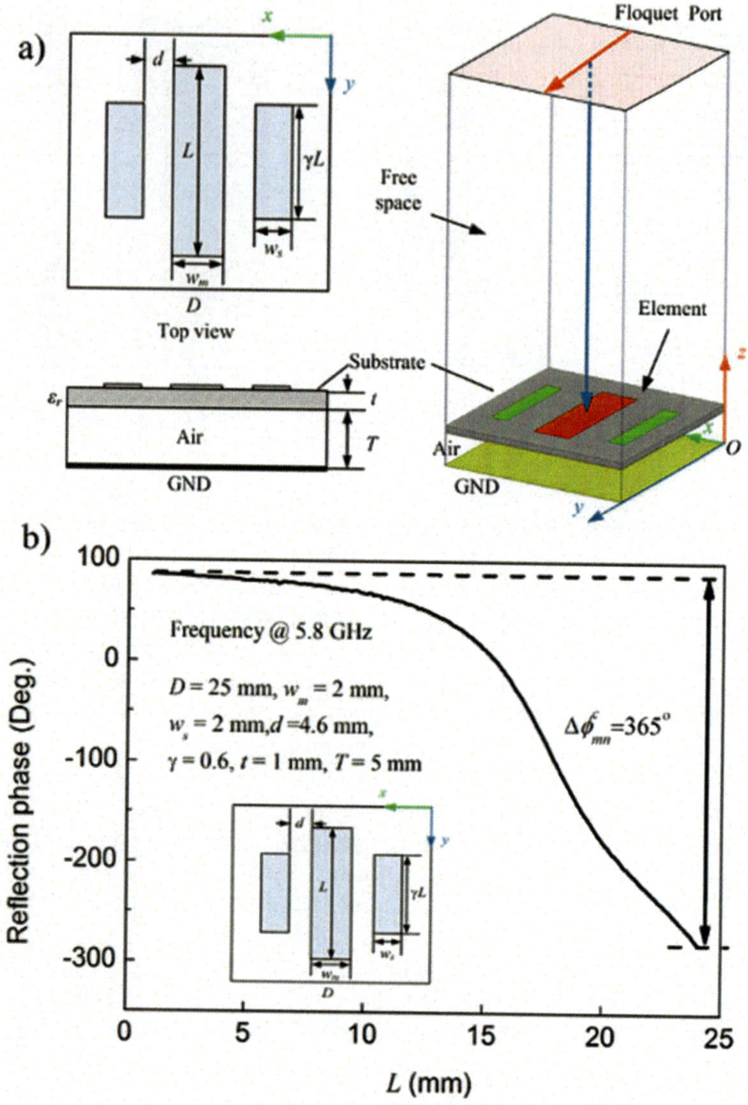

Fig. 7.25 **a** Geometry of tri-dipole element and **b** the reflection coefficient phase versus the length of the center dipole at 5.8 GHz. The thickness of the substrate is 1 mm. Behind the substrate is an air layer of 5 mm, then a metal ground is at the bottom. The dipole width is 2 mm, and the length ratio between the main dipole and two satellite dipoles γ is 0.6 [83]

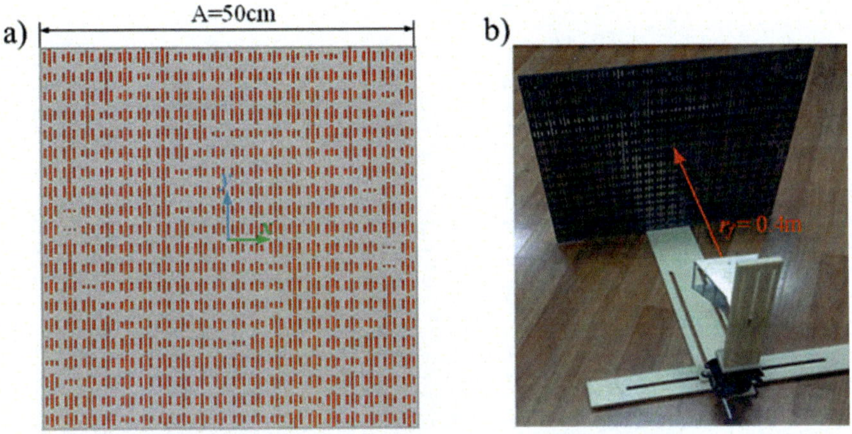

Fig. 7.26 **a** Layout of the designed metasurface for the OAM vortex beam with $l = 2$ and **b** the fabricated prototype of the OAM-generating reflective metasurface with feeding horn antenna under normal incidence [83]

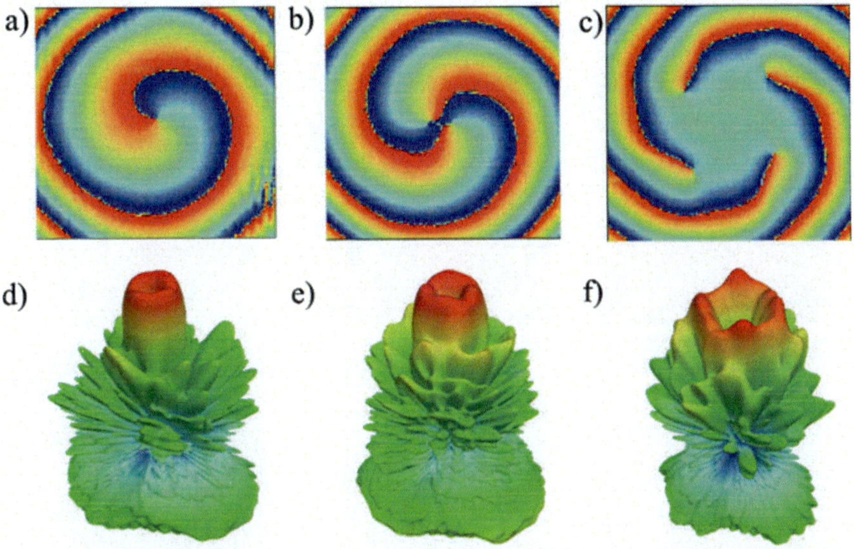

Fig. 7.27 Numerical simulation results of OAM vortex wave with different mode numbers generated by reflective metasurfaces. Wavefront phase characteristics on the observational reference plane at $z = 3.0$ m with **a** $l = 1$, **b** $l = 2$, and **c** $l = 4$ OAM beams and corresponding far-field radiation patterns with **d** $l = 1$, **e** $l = 2$, and **f** $l = 4$ [83]

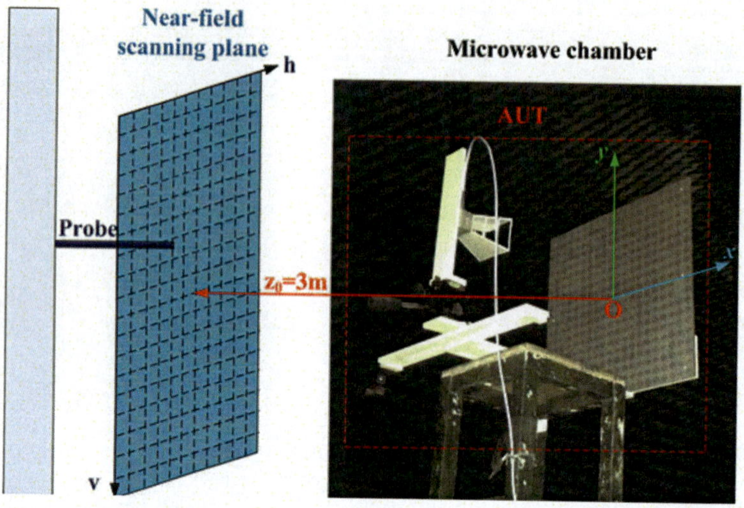

Fig. 7.28 Experimental system configuration for the OAM vortex wave measurement using near-field scanning technique [83]

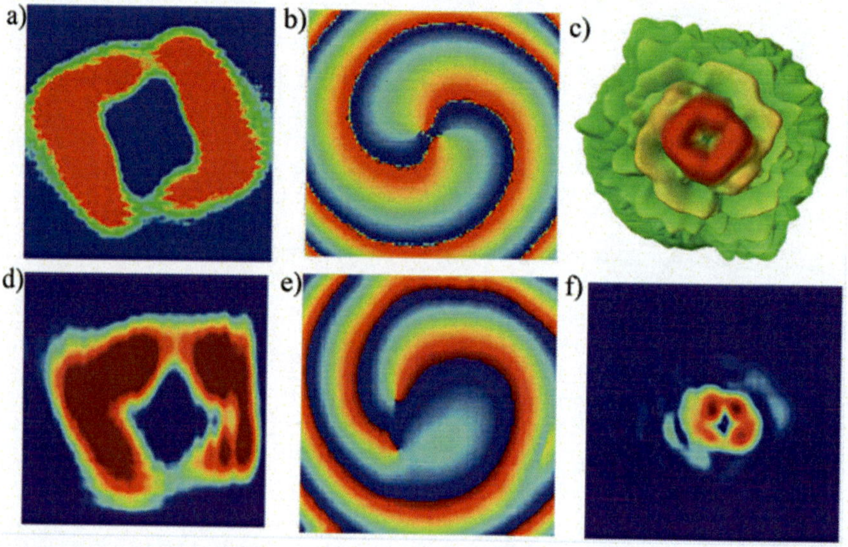

Fig. 7.29 Comparison of simulation and experiment results of the OAM-generating prototype, **a** and **b** simulated electrical field E_v magnitude and phase distributions on the near-field sampling plane at $z = 3.0$ m perpendicular to the beam axis, respectively, **c** top view of the simulated radiation pattern in the far-field zone, **d** and **e** measured electrical field E_v magnitude and phase distributions on the near-field sampling plane at $z = 3.0$ m, and **f** top view of the measured radiation pattern in the far-field zone [83]

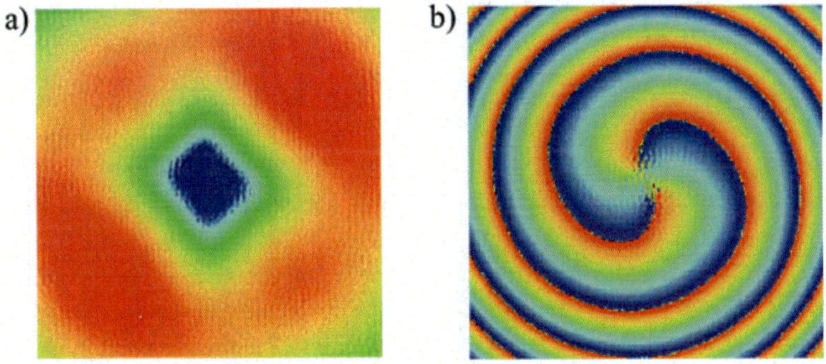

Fig. 7.30 **a** Simulated electrical field E_y magnitude and **b** phase distributions on the near-field sampling plane at $z = 10.0$ m perpendicular to the beam axis [83]

To assess the stability of the propagating vortex wave, simulations were conducted to examine the E-field intensity and phase at a distance of $z = 10.0$ m. Figure 7.30a reveals a larger doughnut-shaped E-field intensity map, indicating the persistence of the vortex wave during propagation. Additionally, Fig. 7.30b demonstrates the retained clarity of the spatial phase distribution. These outcomes suggest that the vortex beams generated by the reflective metasurfaces, with their OAM, exhibit remarkable robustness and reliability, even over extended propagation distances.

Through our experiments, we have successfully demonstrated the effective generation of vortex beams with OAM in the radio-frequency domain using reflective metasurfaces. This achievement is made by the configuration of sub-wavelength metasurface elements, enabling the generation of a continuously distributed spatial phase and resulting in vortex phase wavefronts. As a result, the realization of OAM vortex waves with various mode numbers becomes feasible. It is important to note that while our current design focuses on single linear polarization, the design methodology can be expanded to incorporate dual linear polarization and circular polarization states. This can be achieved by utilizing different reflective elements that correspond to the regulation of spin angular momentum (SAM). By employing this approach, the generation of vortex radio waves with different mode numbers becomes much more accessible. Consequently, this provides a straightforward method for generating OAM vortex waves in radio and microwave wireless communication applications. Overall, our findings open up new possibilities for harnessing the potential of reflective metasurfaces in generating and manipulating vortex beams, offering promising avenues for advancing radio and microwave communication systems.

7.3.3 *Generating Multiple OAM Vortex Beams Using a Metasurface in Radio Frequency Domain by Reflective Metasurface*

Building upon the previous discussion, we now showcase the capability of utilizing a single metasurface to generate multiple radio OAM vortex beams with different mode numbers in various directions. This approach represents a significant advancement in extending the coverage of OAM wireless communication. Drawing from our previous work, we have derived theoretical principles that serve as guidelines for designing metasurfaces capable of producing multiple vortex radio waves, irrespective of whether they possess the same modes or not. By leveraging these theoretical principles, we have laid the groundwork for designing metasurfaces that can generate multiple vortex radio waves simultaneously. This advancement holds immense potential for enhancing wireless communication systems by significantly expanding the range of OAM-based applications. In summary, our work demonstrates the feasibility of generating multiple radio OAM vortex beams using a single metasurface, offering a promising avenue for advancing OAM wireless communication. Our theoretical principles provide a valuable foundation for designing metasurfaces capable of producing multiple vortex radio waves, paving the way for future developments in this field.

Consider a metasurface, as shown in Fig. 7.31, consisting of $M \times N$ elements illuminated by a feeding source at the position vector $\vec{r_{mn}}$. Let the kth desired beam direction be specified by a unit vector $\hat{u}_k$. An effective approach for multi-vorticity metasurface design is by using the superposition of the aperture fields associated with each OAM vorticity on the surface. To generate k vortex beams with a single feeding, the tangential electric field on the metasurface can be written as

$$E = \sum_{m=1}^{M}\sum_{n=1}^{N}\left\{F(\vec{r_{mn}} \cdot \vec{r_f})A(\vec{r_{mn}} \cdot \hat{u}_0)A(\hat{u}_0 \cdot \hat{u})\sum_k \exp\left[j(k_0\vec{r_{mn}} \cdot \hat{u}_k \pm l_k\Phi_k)\right]\right\} \tag{7.17}$$

where F is the feed pattern function, A is the element pattern function, $\vec{r_{mn}}$ is the position vector of the mnth element, l_k is the desired OAM mode number of the kth beam, and Φ_k is azimuth angle in the normal plane of $\hat{u}_k$. As a result, the phase-shift required at each reflective metasurface element can be obtained as

$$\phi_{mn}^c = -k_0\left|\vec{r_{mn}} - \vec{r_f}\right| + \arg\left\{\sum_k \exp\left[j(k_0\vec{r_{mn}} \cdot \hat{u}_k \pm l_k\Phi_k)\right]\right\} \tag{7.18}$$

ϕ_{mn}^c is the required compensating phase of the mn^{th} reflective element on the metasurface.

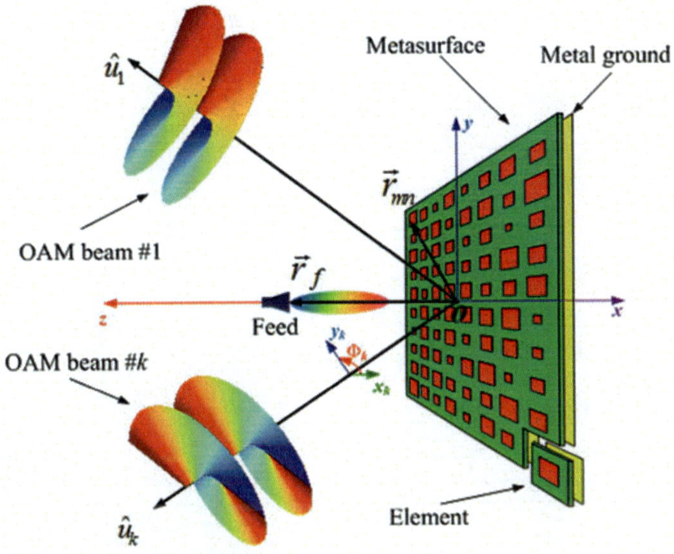

Fig. 7.31 Configuration of multi-beam OAM-generating metasurface [86]

In this section, we have employed the square patch as the fundamental unit cell of the metasurface. Each individual element consists of a substrate made of F4B ($\varepsilon_r = 2.2$) with a thickness of 1 mm, on which square patches are printed on the top surface. There is an air layer with a thickness of 2.7 mm located behind the substrate. By manipulating the side length of the patches, we can achieve continuous reflection phase behavior within the unit cell. To analyze the reflection phase characteristics, we conducted finite-element full-wave analysis using the computational model depicted in Fig. 7.32a. The resulting phase of the reflected field, obtained by applying HFSS at a frequency of 5.8 GHz, is shown in Fig. 7.32b. By obtaining the desired reflection phases, we can then select different patch dimensions that fulfill the necessary phase compensation ϕ_{mn}^c requirements for generating multi-beam OAM vortex waves.

Based on the reflected phase characteristics of the square patch element illustrated in Fig. 7.32b and the design formula (Eq. 7.18) for compensating the phase, a metasurface comprising 20 × 20 elements can be designed to generate two OAM beams operating at 5.8 GHz [86]. The metasurface layout consists of a square array measuring 50 cm × 50 cm. A horn antenna is employed as the feeding source at normal incidence, positioned at a distance of 0.4 m from the metasurface. The numerical simulation results depicted in Fig. 7.33 demonstrate that the proposed metasurface is effective in generating various radio vortex waves with distinct mode numbers and directions.

To further validate the proposed theoretical method and analyses, a prototype of an OAM-generating metasurface was designed, fabricated, and measured. The prototype consists of two beams, each generating a vortex beam with a mode of $l = 2$. One beam is directed towards ($\theta_1 = +30°$, $\varphi_1 = 0°$), and the other beam is

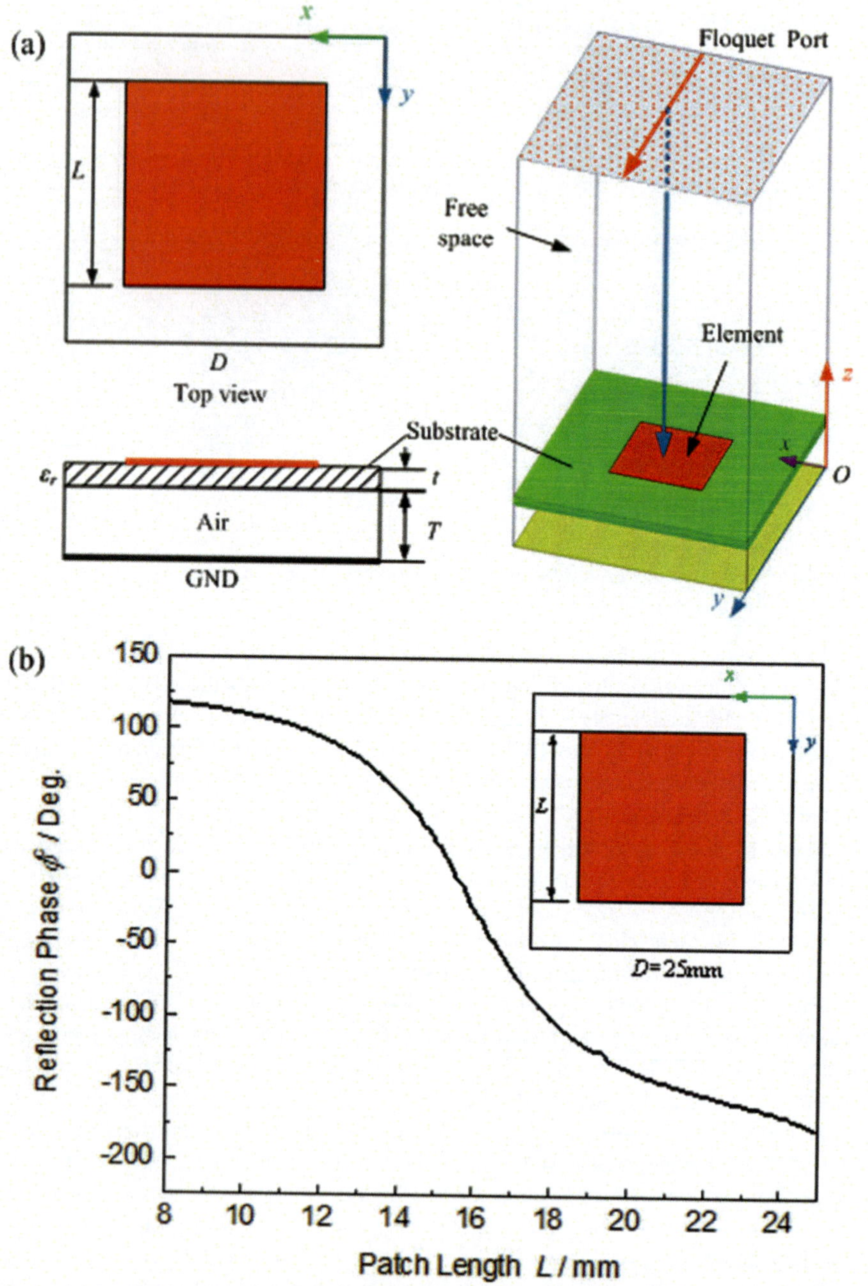

Fig. 7.32 **a** Geometry of square patch element and the infinite periodic model based on the finite element algorithm for calculating the reflection phase, and **b** the phase of the reflection coefficient versus the patch length at 5.8 GHz. The thickness of the substrate is 1 mm. Behind the substrate is an air layer of 2.7 mm, then a metal ground is at the bottom [86]

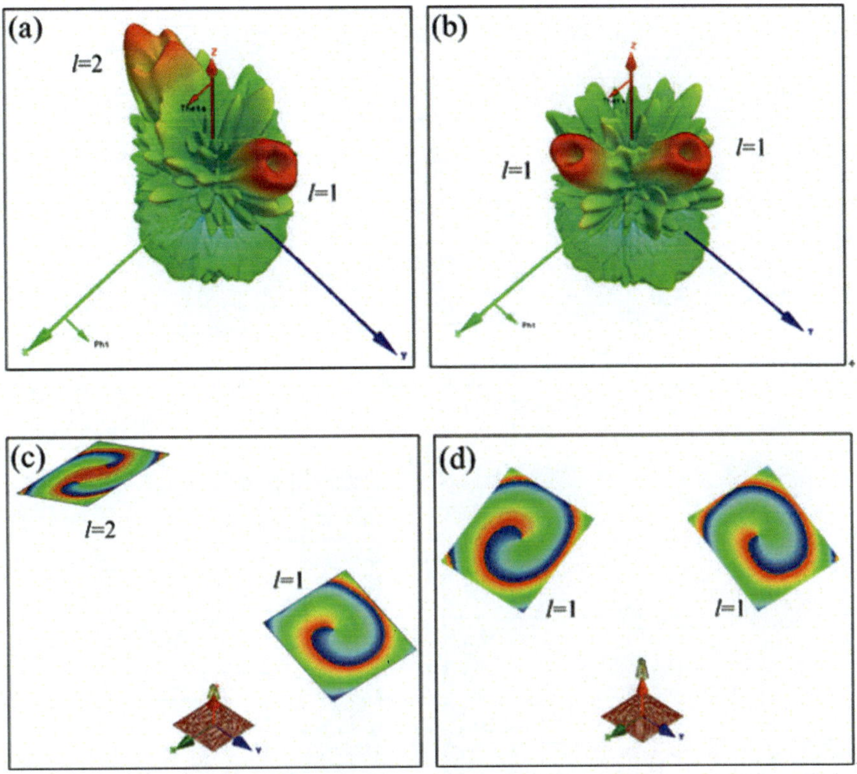

Fig. 7.33 Numerical simulation results of radio OAM vortex waves with different mode numbers in different directions generated by the proposed metasurfaces. **a** far-field 3D radiation patterns of vortex waves with $l_1 = 1$ ($\theta_1 = +30°$, $\varphi_1 = +90°$) and $l_2 = 2$ ($\theta_2 = +30°$, $\varphi_2 = -90°$) modes, **b** far-field 3D radiation patterns of vortex wave with $l_1 = 1$ ($\theta_1 = +30°$, $\varphi_1 = 0°$) and $l_2 = 1$ ($\theta_2 = +30°$, $\varphi_2 = +90°$) modes. **c** and **d** near-field wavefront phase characteristics on the observational reference plane at $z = 3.0$ m corresponding to **a** and **b**, respectively [86]

directed towards ($\theta_2 = -30°$, $\varphi_2 = 0°$). The reflection phase required for simultaneously generating the two OAM vortex waves was calculated using Eq. (7.18) and illustrated in Fig. 7.34a. The layout of the metasurface, shown in Fig. 7.34b, was then designed using a square-patch array with dimensions of 50 cm × 50 cm. Numerical simulation results, depicted in Fig. 7.35, reveal the presence of two doughnut-shaped intensity maps on the ($\theta_1 = +30°$) and ($\theta_2 = -30°$) planes, along with their corresponding spiral phase distributions. These results demonstrate the effective and simultaneous generation of two radio vortex waves using the proposed metasurface.

In addition, a prototype of an OAM-generating metasurface with two beams was designed and fabricated. To measure the OAM vortex wavefront, the near-field planar scanning technique was employed. The experimental setup, as shown in Fig. 7.36, operated at a frequency of 5.8 GHz. A standard measuring probe was used to detect the vertical polarization component of the reflected electric field, E_y. The near-field

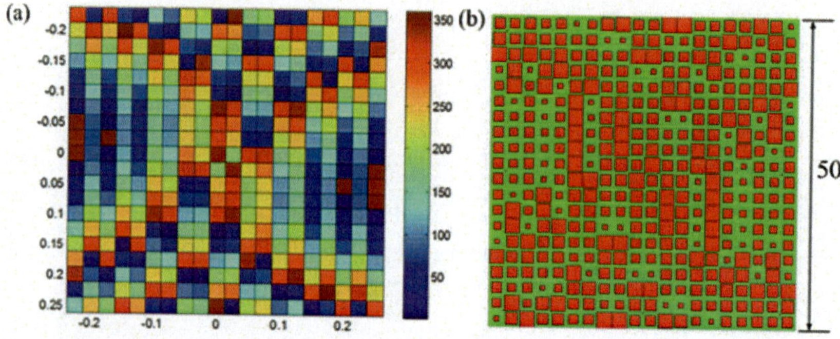

Fig. 7.34 **a** Reflection phase distribution on the metasurface required for generating two OAM vortex waves, and **b** geometry implementation view of square elements array [86]

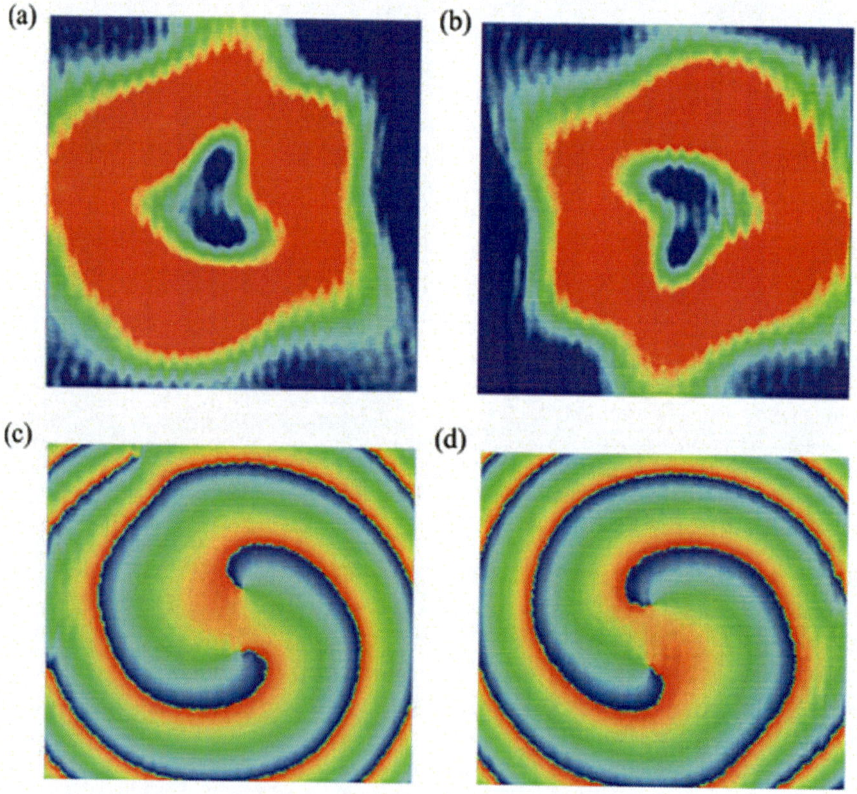

Fig. 7.35 Electrical field E_y characteristics of two OAM vortex beams on the observational planes, **a** magnitude distribution on $\theta = +30°$ plane and **b** magnitude distribution on $\theta = -30°$ plane, **c** phase distribution on $\theta = +30°$ plane, and **d** phase distribution on $\theta = -30°$ plane. The near-field observational plane at $z = 3$ m is perpendicular to the z-axis [86]

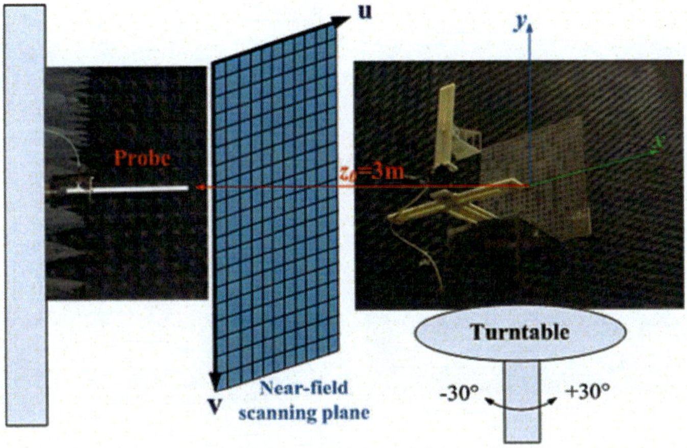

Fig. 7.36 Experimental system configuration for the OAM vortex wave measurement using near-field scanning technique [86]

sampling plane was positioned at $z = 3.0$ m, and the magnitude and phase of E_v were measured on this plane with a sampling grid period of 20 mm. To verify the two OAM beams, measurements were conducted when the turntable rotated at $+30°$ and $-30°$, respectively. The measured magnitude and phase distributions of E_v are illustrated in Figs. 7.37a–d. Comparing these results with the simulated OAM spatial field distributions shown in Figs. 7.35a–d, it can be observed that two OAM vortex waves were successfully generated and measured. The simulated and measured results exhibit good agreement, with only minor phase distribution deformations due to the rotation angle error of the turntable. The primary characteristic of the spatial phase distribution, a perfect doughnut-shaped intensity map, is clearly identifiable in both simulated and measured data.

The simulated 3-D radiation pattern is depicted in Fig. 7.38a, while the comparison between simulated and measured *xoz*-plane radiation patterns in the far-field zone is shown in Fig. 7.38b. It is evident from the comparison that there are two amplitude nulls in the azimuth ($\theta = \pm 30°$), indicating the effective and simultaneous generation of two radio vortex waves. This experimental validation demonstrates that the proposed metasurface can be utilized to generate multiple vortex beams with OAM simultaneously in the radio-frequency domain. With this configuration, it becomes easier to produce multiple vortex radio waves in different directions, each with different OAM modes. This advancement provides an effective solution for expanding the coverage of OAM vortex waves in radio and microwave wireless communication applications.

The operational bandwidth of the proposed reflective metasurface for generating OAM vortex waves in the radio frequency range was analyzed and is illustrated in Fig. 7.39. The phase wavefront variation at different frequencies is shown in Fig. 7.39a, indicating that the operational bandwidth of the metasurface for OAM

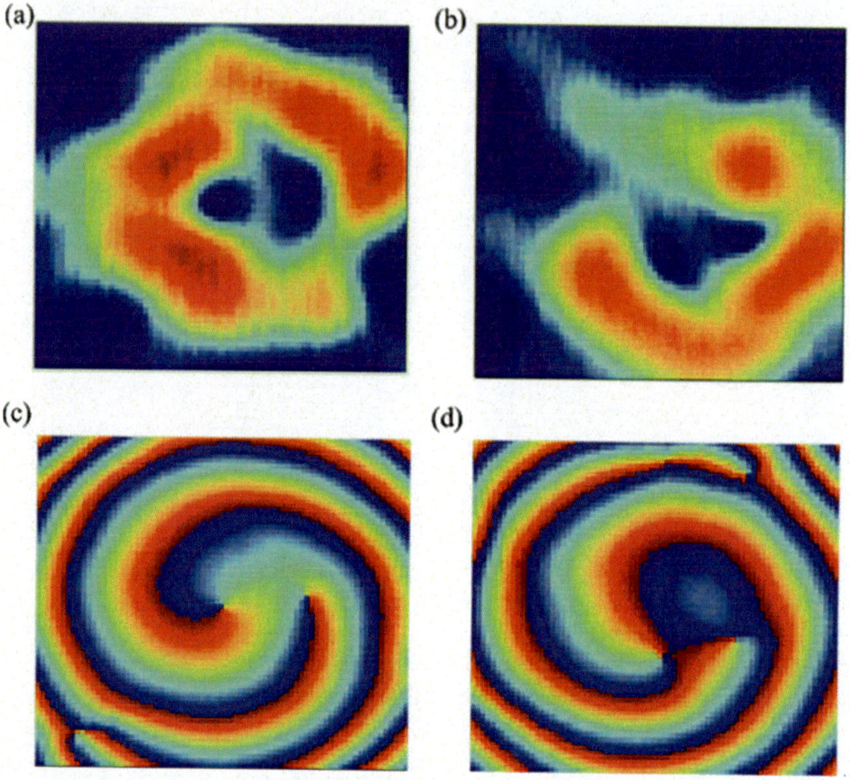

Fig. 7.37 Measured electrical field E_y characteristics of two OAM vortex beams on the observational planes, **a** magnitude distribution on $\theta = +30°$ plane and **b** magnitude distribution on $\theta = -30°$ plane, **c** phase distribution on $\theta = +30°$ plane, **d** phase distribution on $\theta = -30°$ plane. The near-field sampling plane at $z = 3$ m is perpendicular to the z-axis [86]

generation extends approximately from 5.5 to 6.5 GHz. This bandwidth characteristic is also evident when examining the far-field radiation patterns, as depicted in Fig. 7.39b. Thus, the proposed reflective metasurface exhibits a considerable operational bandwidth for generating OAM vortex waves within the specified frequency range.

7.3.4 Generating Dual-Polarized and Dual-Mode OAM Vortex Beam by Reflective Metasurface

In the previous study, the generation of OAM vortex beams with a single linear polarization was demonstrated using a metasurface. In this part, the focus is on the

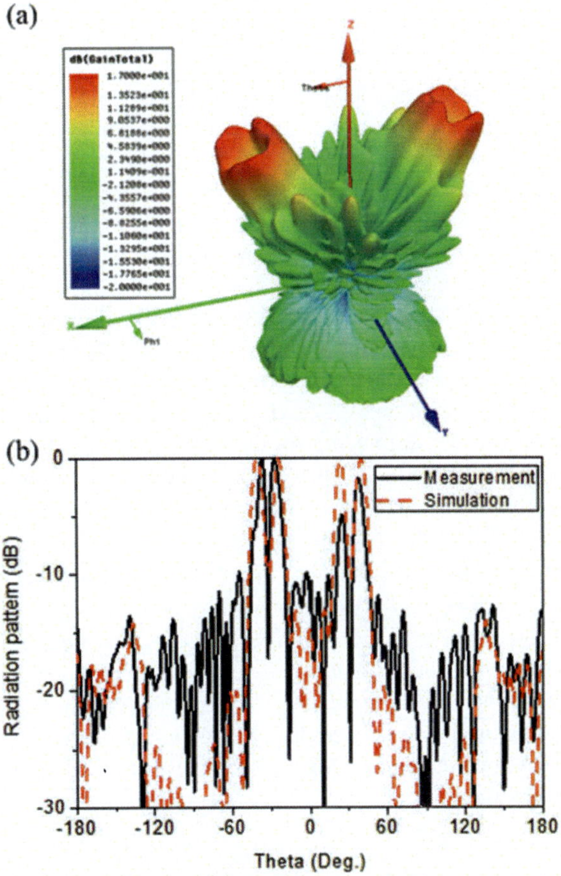

Fig. 7.38 **a** Simulated far-field 3-D radiation pattern, and **b** comparison of simulated and measured radiation patterns on *xoz*-plane in the far-field zone [86]

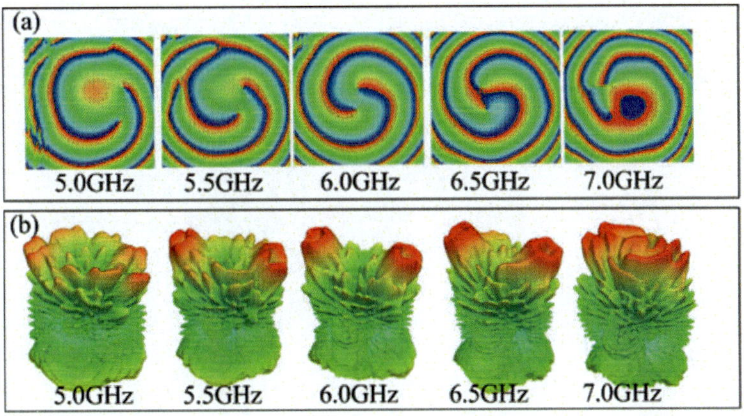

Fig. 7.39 **a** Simulated near-field phase distribution and **b** far-field radiation patterns at different frequencies [86]

generation of diverse OAM vortex radio beams with dual-modes and linear dual-polarization, utilizing a single reflective metasurface. The proposed configuration enables operation in both vertical and horizontal polarizations. It is worth noting that dual-polarized antennas are commonly employed in 802.11n devices to facilitate multi-thread communication. This allows the device to simultaneously transmit and receive data on different polarizations. The advantage of using the proposed metasurface is that it enables multi-thread communication through a single device, as opposed to having separate and orthogonal polarization OAM-generating antennas for the same functionality. To design the metasurface capable of producing the dual-mode and dual-polarization vortex radio waves, crossed-dipoles with different arm lengths are introduced as the metasurface elements. Both numerical simulations and experiments are conducted to validate the accuracy and effectiveness of the theoretical analysis and design. The proposed reflective metasurface holds the potential for application in the polarization multiplexing of OAM communication systems, offering enhanced functionality and versatility in the field of OAM-based communication.

In Fig. 7.40, a metasurface is depicted, comprising of $M \times N$ elements. This metasurface is illuminated by a feeding source located at a specific position vector $\vec{r}_f$. Each element on the metasurface consists of cross-dipole elements, and there are no power division transmission lines present. The feeding source spatially illuminates these metasurface elements. The design of the metasurface aims to scatter the incident wave, which is linearly polarized, and generate a reflection wave with varying rotating phase fronts and different polarized components. The specific configuration of the metasurface elements enables the generation of the reflected wave with the desired properties.

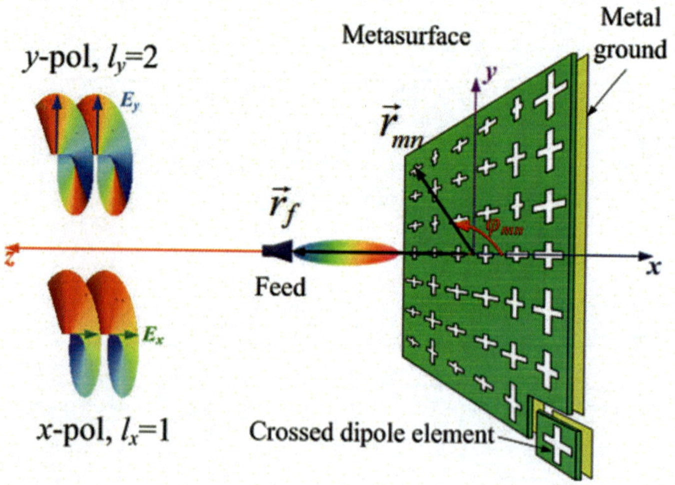

Fig. 7.40 Configuration of the dual-polarized and dual modes OAM-generating metasurface [87]

The reradiated electric fields in two orthogonal polarizations from the metasurface in an arbitrary direction, $\hat{u}$, will be of the form

$$\begin{cases} E_x(\hat{u}) = \sum_{m=1}^{M}\sum_{n=1}^{N} F\left(\vec{r}_{mn}\cdot\vec{r}_f\right)A\left(\vec{r}_{mn}\cdot\hat{u}_0\right)A\left(\hat{u}_0\cdot\hat{u}\right) \\ \quad\cdot\exp\left\{jk_0\left[\left|\vec{r}_{mn}-\vec{r}_f\right|+\vec{r}_{mn}\cdot\hat{u}\right]+j\phi_{mn}^X\right\} \\ E_y(\hat{u}) = \sum_{m=1}^{M}\sum_{n=1}^{N} F\left(\vec{r}_{mn}\cdot\vec{r}_f\right)A\left(\vec{r}_{mn}\cdot\hat{u}_0\right)A\left(\hat{u}_0\cdot\hat{u}\right) \\ \quad\cdot\exp\left\{jk_0\left[\left|\vec{r}_{mn}-\vec{r}_f\right|+\vec{r}_{mn}\cdot\hat{u}\right]+j\phi_{mn}^Y\right\} \end{cases} \tag{7.19}$$

where F is the feeding pattern function, A is the reflective metasurface element pattern function, $\vec{r}_{mn}$ is the position vector of the mnth element, k_0 is the propagation constant in vacuum, and $\hat{u}_0$ is the desired main-beam direction of the reflective metasurface. To produce the reflected vortex waves, the phase-shift required by each element for the OAM vortex beams in x-polarization and y-polarization can be obtained by

$$\begin{cases} \phi_{mn}^X = l_X\varphi_{mn} - \left(k_0\left|\vec{r}_{mn}-\vec{r}_f\right|+\vec{r}_{mn}\cdot\hat{u}_0\right), & l_X = 0,\ \pm1,\ \pm2, \ldots \\ \phi_{mn}^Y = l_Y\varphi_{mn} - \left(k_0\left|\vec{r}_{mn}-\vec{r}_f\right|+\vec{r}_{mn}\cdot\hat{u}_0\right), & l_Y = 0,\ \pm1,\ \pm2, \ldots \end{cases} \tag{7.20}$$

where l_X and l_Y are the desired OAM mode numbers for x- and y-polarizations components, respectively. φ_{mn} is the azimuthal angle of the mn^{th} element, and $\{\phi_{mn}^X,\phi_{mn}^Y\}$ are the required compensating phase on different polarization components of the mn^{th} element on the metasurface, respectively.

The proposed metasurface adopts the use of cross-dipole elements. Each element is comprised of a substrate, F4B ($\varepsilon_r = 2.65$, $\tan\delta = 0.003$), with a thickness of 1 mm. The cross-dipole is printed on the top surface of the substrate. Behind the substrate, there is an air layer with a thickness of 5 mm. To achieve the desired functionality, the horizontal and vertical arm lengths of the cross-dipole elements can be independently controlled. This allows for the unit cell to exhibit a continuous reflection phase behavior, providing a phase range of over 300° for both the x and y polarizations. A simulation based on an infinite periodic model, utilizing the HFSS simulation software with a finite-element algorithm, was performed to analyze the reflection phase characteristics of the cross-dipole elements. The results are presented in Fig. 7.41a. In Fig. 7.41b, c, the phase and magnitude of the reflection E-field are shown as a function of the length of the cross-dipole element in the y-polarization at an operational frequency of 5.8 GHz. It is observed that the length of the cross-dipole element in the x-polarization has a minimal impact on the reflection characteristics. This can be attributed to the orthogonality of the cross-dipole element. Specifically, the reflection phase characteristics in the x-polarization are nearly identical to those in the y-polarization. These findings emphasize the controllable and independent nature

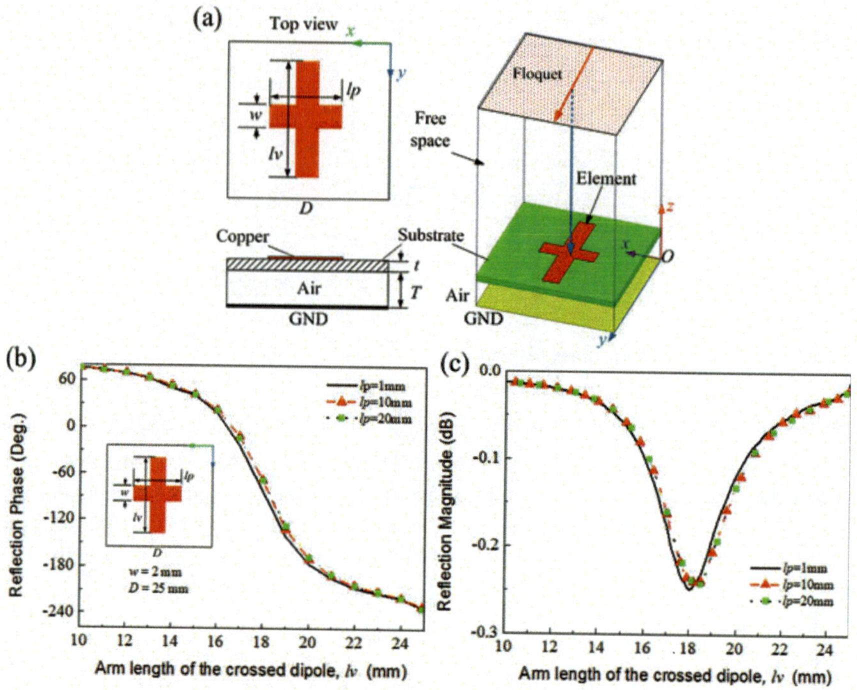

Fig. 7.41 **a** Geometry of the printed cross-dipole metasurface element and simulation model for reflection phase characteristics, **b** the phase and **c** the magnitude of the reflection coefficient versus the length l_v of the cross-dipole at 5.8 GHz. The thickness of the substrate is 1 mm. Behind the substrate is an air layer of 5 mm, and a metal ground is at the bottom [87]

of the cross-dipole elements in achieving the desired reflection phase characteristics for polarization-dependent beam manipulation and generation.

To validate the design, a single metasurface capable of generating a dual-polarized OAM vortex beam is presented. In this case, the x-polarized vortex wave has an OAM-mode number of $l_X = 1$, while the y-polarized vortex wave has an OAM-mode number of $l_Y = 2$. The feeding point is positioned in front of the metasurface with $\vec{r}_f = (0, 0, 0.4)$m along the z-axis. It is important to note that a centered feeding configuration is advantageous as it ensures that the amplitude null of the OAM vortex beam is at the center. This minimizes the blocking effect of the feeding antenna on the reflected OAM vortex wave. Considering normal incidence, the feeding antenna has the least impact on the reflection OAM vortex wave. Once the desired distribution of reflection phases $\{\phi_{mn}^X, \phi_{mn}^Y\}$ is determined based on Eq. (7.20), as illustrated in Figs. 7.42a, b, the length of the crossed-dipole elements can be adjusted to achieve the required phase compensation. The implementation of the x-polarized dipoles and y-polarized dipoles is shown in Figs. 7.42c, d, respectively. To generate a dual-polarized OAM beam using a single metasurface, the incoming waves from the feed should produce both x-polarized and y-polarized electric field (E-field) components.

Therefore, a 45° linearly polarized electromagnetic wave emerges from the feed, as depicted in Fig. 7.42e. Finally, the metasurface, consisting of 20 × 20 elements and utilizing cross-dipole elements, is constructed as a square array with dimensions of 50 cm × 50 cm. The fabricated prototype of the metasurface is shown in Fig. 7.42f. This demonstration highlights the capability of a single metasurface to generate a dual-polarized OAM vortex beam, enabling enhanced functionality and versatility in beam manipulation and communication applications.

The performance of the dual-mode and dual-polarization OAM-generating metasurface was analyzed using full-wave electromagnetic simulations based on the finite-element method. The results of the numerical simulations are presented in Fig. 7.43. These results demonstrate the effective generation of a dual-polarized OAM vortex wave using the proposed single metasurface. In Fig. 7.43, the x-polarized E-field magnitude and phase distributions in the near-field zone at $z = 3.0$ m are shown in Fig. 7.43a, c, respectively. It can be observed that the OAM mode is $l = 1$. Similarly, the y-polarized E-field magnitude and phase distributions in the near-field zone at $z = 3.0$ m, with an OAM mode of $l = 2$, are shown in Fig. 7.43b, d, respectively.

To validate the performance of the dual-polarized OAM-generating metasurface, a prototype is designed, fabricated, and measured. The experimental setup for the measurement is depicted in Fig. 7.44. The near-field planar scanning technique is employed to measure the OAM vortex wavefront. In the experimental setup, a horn antenna is utilized as the feeding source. The horn antenna is positioned at normal incidence with respect to the metasurface but is rotated 45° along the z-axis. This configuration ensures the production of both x-polarized and y-polarized electric field (E-field) components. The distance between the feeding horn and the metasurface is set to 0.4 m.

The measurement of the dual-polarized OAM vortex wavefront was performed using a near-field planar scanning technique in a microwave chamber operating at a frequency of 5.8 GHz. During the experiment, a standard measuring probe antenna was employed to detect the vertical (y-polarization) and horizontal (x-polarization) components of the reflected electric field, denoted as E_v and E_h, respectively. The near-field sampling plane was positioned at $z = 3.0$ m, and the magnitude and phase of E_h and E_v were measured on this sampling plane. The near-field sampling grid period was set at 20 mm. The measured magnitude and phase distributions of E_v and E_h are shown in Figs. 7.45a–d, respectively. Comparing these measured results with the simulated OAM vortex field distributions shown in Figs. 7.43a–d, it can be observed that the dual-polarized and dual-mode OAM vortex wave was successfully generated and measured. The most prominent feature of the spatial phase distribution is clearly discernible. The measured results exhibited good agreement with the simulated results. A characteristic pattern of the field intensity, resembling a doughnut shape, can be observed in both Figs. 7.43 and 7.45. This doughnut-like distribution of electric field magnitudes is a typical signature of an OAM vortex beam. It is worth noting that higher OAM modes correspond to larger divergence angles, consistent with theoretical results in optics. These experimental measurements validate the

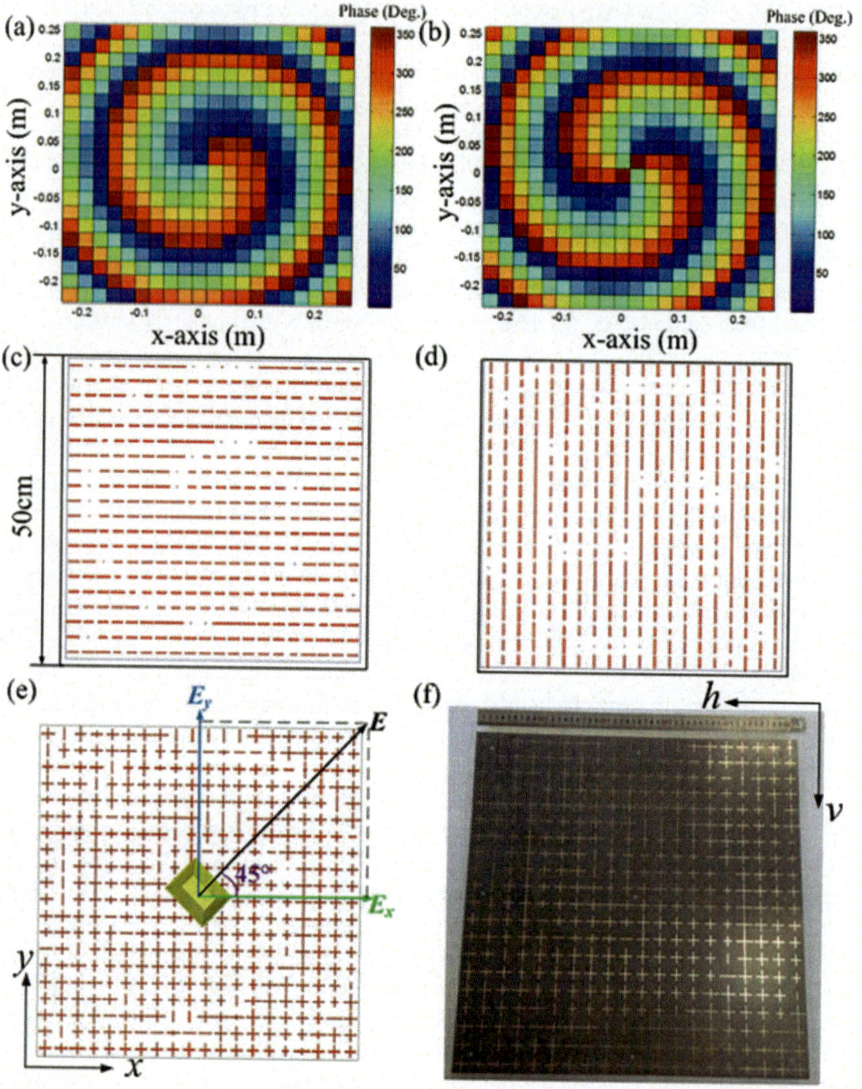

Fig. 7.42 The reflective phase distribution required for **a** x-pol OAM vortex beam with $l_X = 1$ and **b** y-pol OAM beam with $l_Y = 2$. **c** the geometry implementation of x-polarized dipoles and **d** y-polarized dipoles. **e** geometry implementation of the metasurface based on cross-dipole elements for dual-mode and dual-polarization OAM vortex wave, and **f** fabricated prototype of the proposed metasurface [87]

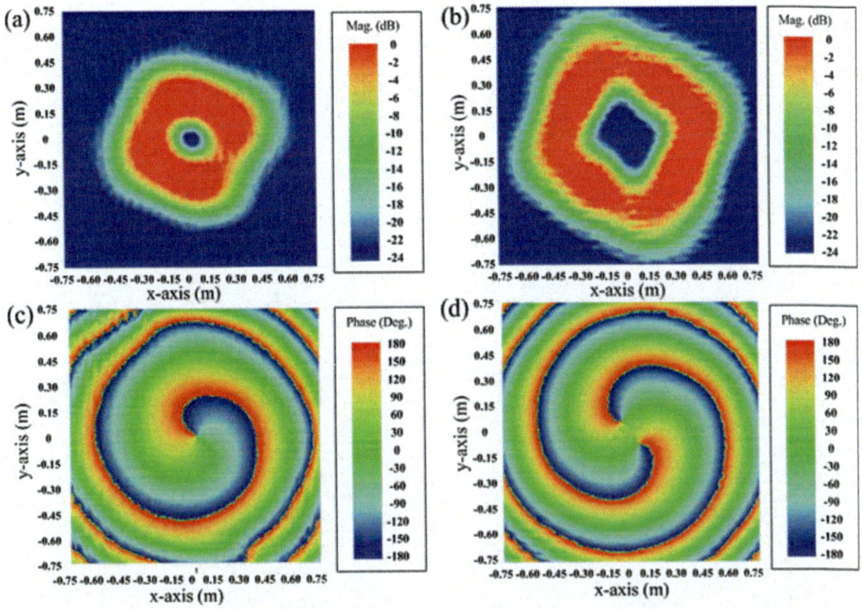

Fig. 7.43 Simulated electrical field distribution in the near-field zone of $z = 3.0$ m, **a** magnitude distribution of x-polarization with $l_X = 1$, **b** magnitude distribution of y-polarization with $l_Y = 2$, **c** phase distribution of x-polarization, and **d** phase distribution of y-polarization [87]

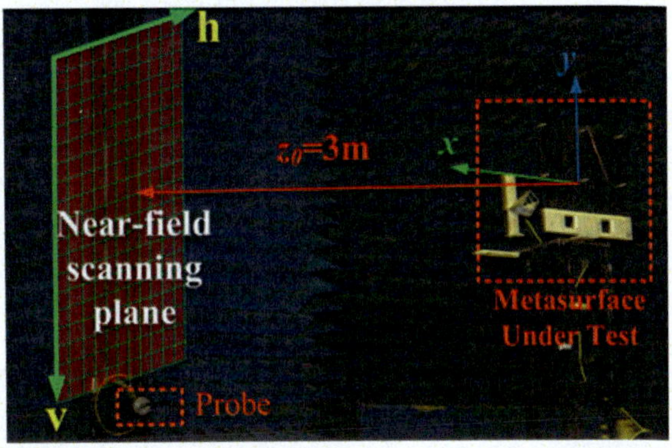

Fig. 7.44 Experimental system configuration for dual-polarization and dual-mode OAM vortex wave measurement using near-field scanning technique [87]

successful generation of a dual-polarized and dual-mode OAM vortex wave using the designed metasurface. The agreement between the simulated and measured results further confirms the accuracy and effectiveness of the metasurface in generating OAM vortex beams in the microwave frequency range.

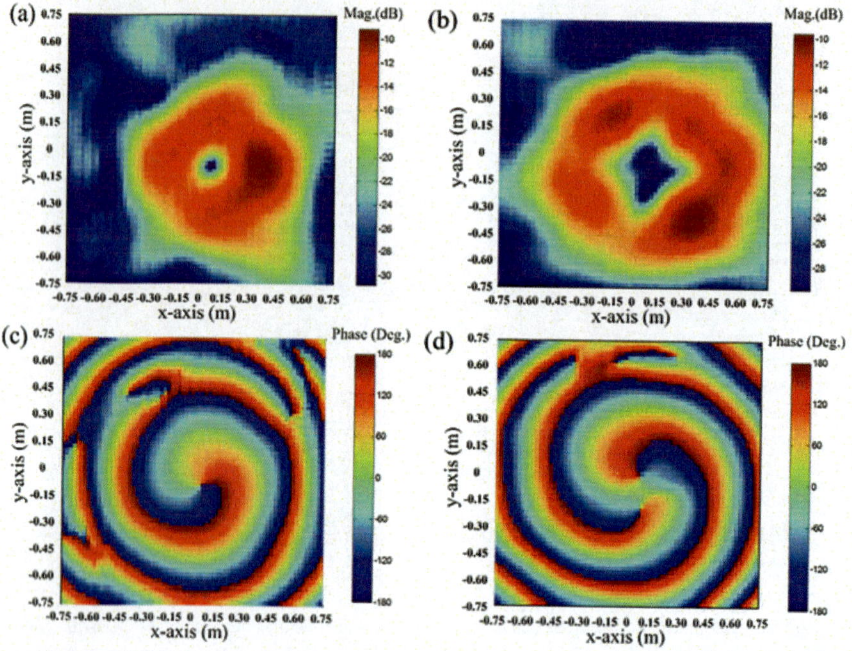

Fig. 7.45 Measured electrical field distribution in the near-field zone of $z = 3.0$ m, **a** magnitude distribution of horizontal polarization (x-polarization), **b** magnitude distribution of the vertical polarization (y-polarization), **c** phase distribution of horizontal polarization, and **d** phase distribution of vertical polarization [87]

Indeed, based on the experimental results, it is evident that the single metasurface can effectively generate dual-polarized vortex waves with OAM in the radio-frequency domain. The configuration of crossed-dipole metasurface elements allows for the generation of vortex phase wavefronts with different polarizations. This capability opens up exciting possibilities in OAM radio wave transmission. By utilizing the metasurface, it becomes feasible to transmit OAM radio waves with two orthogonal linear polarizations using a single device. This is advantageous as it enables the simultaneous transmission of two different signals at the same frequency without any interference. This breakthrough has significant implications for the field of OAM communication. With the ability to employ orthogonal polarizations, it facilitates polarization multiplexing, where multiple signals can be transmitted simultaneously and independently using the same frequency resources. This provides enhanced capacity and spectral efficiency in OAM-based communication systems, paving the way for new applications and advancements in wireless communication technology.

7.4 Design and Applications of Transmission Metasurface: Generating Orbital Angular Momentum Vortex Wave in Radio Frequency

In transmitarray or transmission metasurfaces [88–93], the objective is to achieve phase modulation to launch arbitrary electromagnetic beams, while keeping the transmission amplitude of each element fixed. This is commonly accomplished by utilizing transmission elements with variable phase shifts and low transmission loss. Typically, the phase shift of the transmission element is designed to cover a range of 360°, while ensuring that the transmission coefficient remains within −3 dB. To achieve this, multi-layer frequency selective surfaces with varying sizes are often employed. The use of multiple layers allows for different phase shift ranges to be covered, as illustrated in Fig. 7.46. It can be observed that in order to design a desired transmission metasurface, three or more layers are usually required. This is because the combination of these layers enables the coverage of a phase shift range exceeding 300°, which meets the majority of demands for phase modulation. By employing such multi-layer designs, transmitarrays and transmission metasurfaces can achieve the desired phase modulation to generate a variety of electromagnetic beams.

7.4.1 *Generating OAM Vortex Beam by Phase Modulated Transmission Metasurface*

The usage of a phase-modulated surface (PMS) or transmission metasurface is a common method for generating high-gain beams. By designing the phase shift of the transmission metasurface appropriately, OAM vortex beams can be launched. In Fig. 7.47, a typical design schematic for a transmission metasurface is shown, specifically aimed at generating OAM vortex beams.

Firstly, the phase distribution of the feed horn antenna can be written as

$$\phi_0 = k_0\left[\left|\vec{r_{ij}} - \vec{r_h}\right| + \vec{r_{ij}} \cdot \hat{z}\right] \tag{7.21}$$

And the phase shift of each element on the transmission metasurface for generating the OAM vortex beam can be expressed as:

$$\phi_{i,j} = n\varphi_{i,j} - \phi_0 = n\varphi_{i,j} - k_0\left[\left|\vec{r_{ij}} - \vec{r_h}\right| + \vec{r_{ij}} \cdot \hat{z}\right] \tag{7.22}$$

where $\hat{z}$ is the propagation direction of the vortex beam, $\varphi_{i,j}$ is the azimuth angle of (i, j)th element on the transmission metasurface, and n is the mode number of the OAM vortex beam. Here we use a four-layer double-ring element as the transmission metasurface unit cell, as shown in Fig. 7.47. To generate OAM vortex beams with

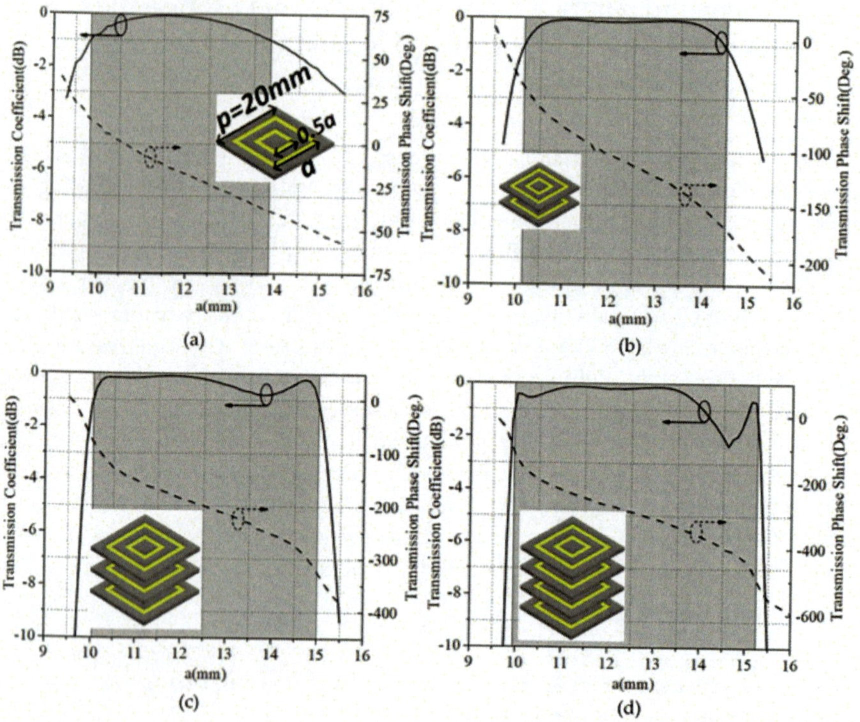

Fig. 7.46 Transmission coefficient of the multi-layer FSS with variable size a, **a** single-layer FSS, **b** double-layer FSS, **c** triple-layer FSS, **d** quad-layer FSS

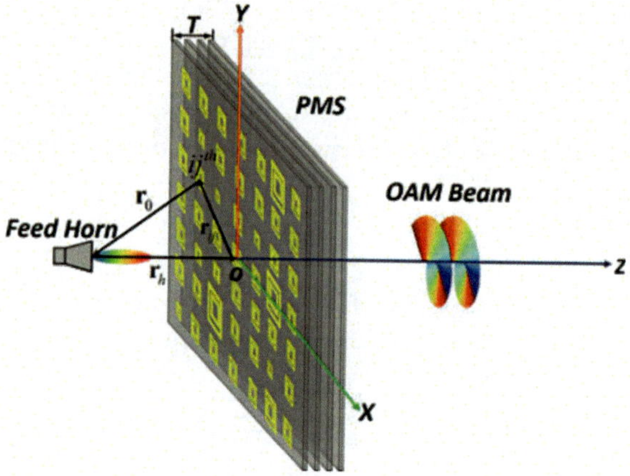

Fig. 7.47 Transmission metasurface design schematic for generating OAM vortex beam

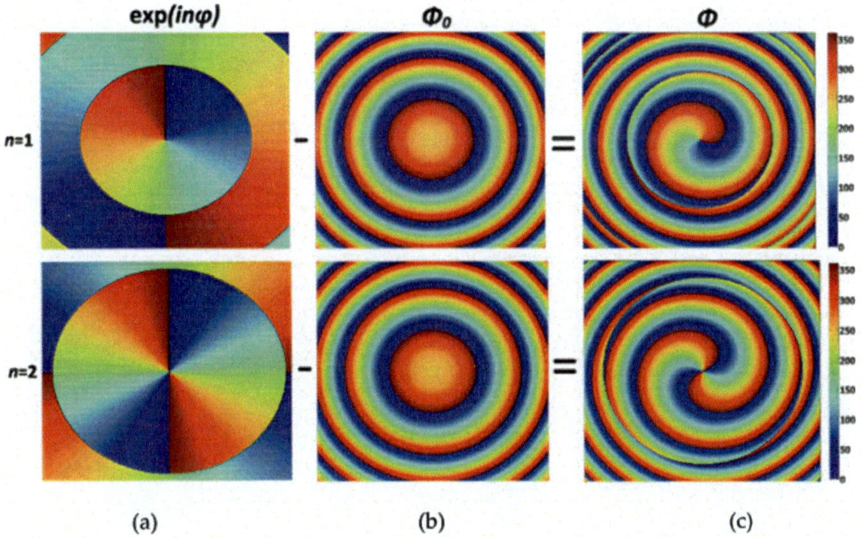

Fig. 7.48 Phase distribution of transmission metasurface for generating OAM vortex beam, **a** phase distribution of OAM vortex beam, **b** phase distribution of horn antenna, **c** phase distribution of transmission metasurface

the mode numbers of $n = 1$ and $n = 2$, the calculated phase shift for the transmission metasurface is shown in Fig. 7.48. After calculating the phase shift for each element on the transmission metasurface, the topology of the practical transmission metasurface could be obtained, as shown in Fig. 7.49.

In the simulations presented in Fig. 7.50a, the E-field distributions of a transmission metasurface generating a conventional OAM vortex beam with a mode number of $n = 1$ are depicted. The observational transmission region, located away from the transmission metasurface, spans from 16.7 wavelengths to 83 wavelengths. To illustrate the transmitted field behavior, the simulated E-field distributions on several observational planes placed at distances of 1 m (33 wavelengths), 2 m (66 wavelengths), and 2.5 m (83 wavelengths) from the metasurface are illustrated in Fig. 7.50b, c. Figures 7.50 and 7.51 demonstrate the effective generation of an OAM vortex beam by the transmission metasurface. Additionally, simulated results for the transmission metasurface generating an OAM mode of $n = 2$ vortex beam are provided in Figs. 7.52 and 7.53. These figures further confirm the ability of the transmission metasurface to generate an OAM vortex beam with a mode number of $n = 2$.

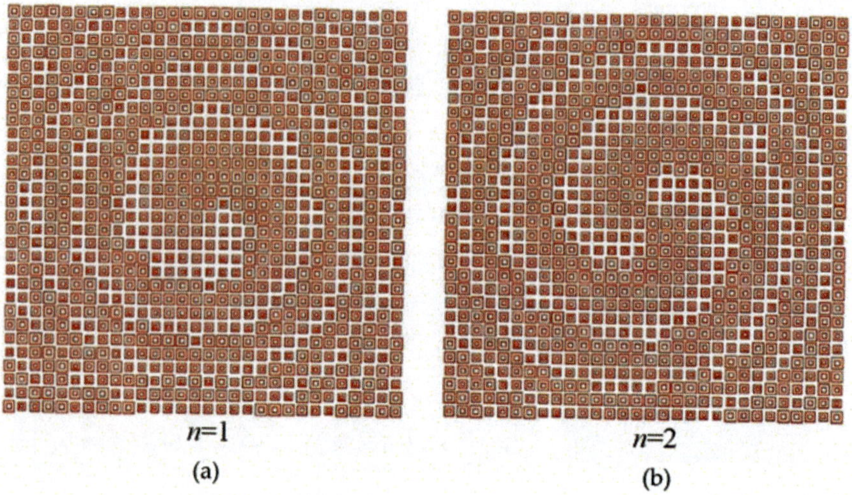

Fig. 7.49 Practical topology of transmission metasurface for generating OAM vortex beam, **a** mode number of $n = 1$, **b** mode number of $n = 2$

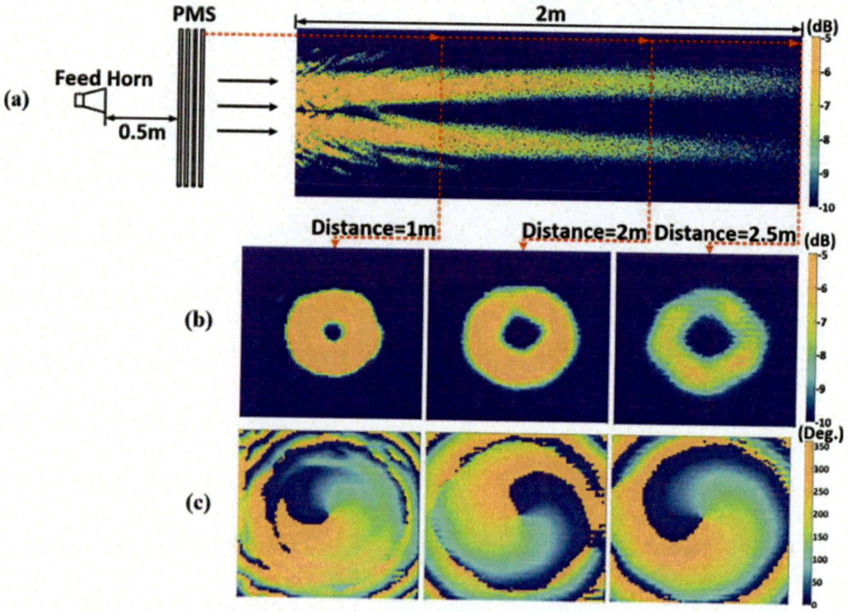

Fig. 7.50 **a** E-field transmission characteristics of the transmission metasurface for OAM mode of $n = 1$ vortex beam, **b** amplitude and **c** phase distributions of E-field on the different observational planes at distances of 1.0 m, 2.0 m and 2.5 m away from the PMS transmission metasurface aperture, respectively

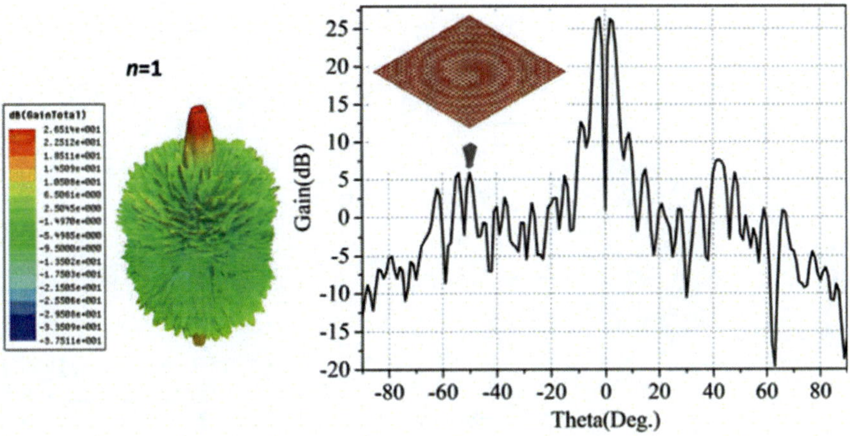

Fig. 7.51 Simulated far-field pattern of the transmitting metasurface for OAM mode of $n = 1$ vortex beam

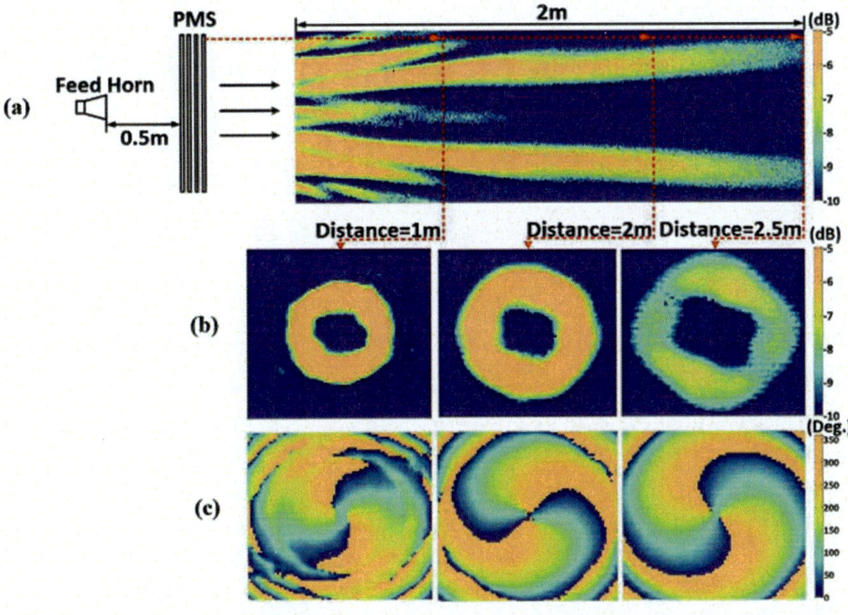

Fig. 7.52 **a** E-field transmission characteristics of the transmission metasurface for OAM mode of $n = 2$ vortex beam, **b** amplitude and **c** phase distributions of E-field on the different observational planes at distances of 1.0 m, 2.0 m and 2.5 m away from the PMS transmit metasurface aperture, respectively

Fig. 7.53 Simulated far-field pattern of transmitting metasurface for OAM mode of $n = 2$ vortex beam

7.4.2 Generation of Bessel Vortex Beam by Amplitude-Phase Modulated Transmission Metasurface

In the free space, the homogeneous wave equation can be expressed as follows:

$$\nabla^2 \vec{E}(\vec{r}, t) - \frac{1}{c^2} \frac{\partial^2}{\partial t^2} \vec{E}(\vec{r}, t) = 0 \tag{7.23}$$

where ∇^2 is the Laplacian operator, and $\vec{r}$ is the position vector, t is the time variable, $\vec{E}$ is the electric field intensity, and c is the propagation speed of light. Consider a Bessel beam where the individual electric field vector $\vec{E}$ expressed in cylindrical coordinates is

$$\vec{E}_n(\vec{r}, t) = \vec{E}_0 J_n(k_\perp \rho) \exp(jn\varphi) \exp\left[j(\omega t - k_z z)\right] \tag{7.24}$$

where J_n is the nth order Bessel function of the first kind, $(k_\perp, k_z)$ are the transverse and longitudinal components of free-space wavenumber, ρ is the radial coordinate of the position, φ is the azimuthal angle coordinate of the position, ω is the angular frequency, and $\vec{E}_0$ is a constant vector. Here z-axis is the direction of propagation. Equation (7.24) shows that the high-order Bessel wave is a weighted superposition of the z component of the OAM eigenmodes. Each exhibits a helical phase front due to the $\exp(jn\varphi)$ phase factor and represents a screw dislocation along the z-axis. In this case, n represents the OAM mode number. For $n = 0$, the phase front is flat, which is also corresponding to the zeroth-order Bessel beam.

It can be seen from Eq. (7.24) that in order to generate a high-order Bessel vortex beam, the amplitude distribution on the transmitting amplitude-phase modulated surface (APMS) must follow $J_n(k_\perp \rho)$ function and the phase compensation of the array complies with $n\varphi$. When $n = 1$, Eq. (7.24) represents first-order Bessel beam or J_1 beam, and so on. When the amplitude distribution on the launcher is uniform, and the phase shifting still follows $n\varphi$, the conventional vortex wave with OAM can be generated but without non-diffraction transmission characteristics. Most of the reported works discussed zeroth-order Bessel beam (J_0) launchers that only need to modulate the amplitude distributions since the phase shifting is $n\varphi = 0$. However, for J_1,J_2 and other high-order Bessel beams, the launcher must modulate amplitude and phase simultaneously.

Consider a multilayer APMS transmission metasurface [94] composed of $I \times J$ elements, as shown in Fig. 7.54. It is illuminated by a standard horn antenna at the position $\vec{r_h}$, and $\vec{r_{ij}}$ is the position vector of ij^{th} element on the surface. The produced Bessel beam propagates along the z-axis by default.

Then the amplitude distribution and phase-shift on each APMS elements for Bessel beam can be written as

$$A_{ij} = J_n(k_\perp |\vec{r_{ij}}|)/D_{horn}, \; n = 0, 1, 2, ..., \tag{7.25}$$

$$\phi_{ij} = n\varphi_{ij} - k_0(|\vec{r_{ij}} - \vec{r_h}|), \; n = 0, 1, 2, ..., \tag{7.26}$$

where n represents the order of standard Bessel function of the first kind, which is also the mode number of OAM vortex wave, φ_{ij} is the azimuthal angle of the ij^{th} element, k_0 is the wavenumber in free space, $k_\perp$ is the transverse wavenumber, and D_{horn} is the E-field amplitude distribution of the horn antenna. It is worth mentioning

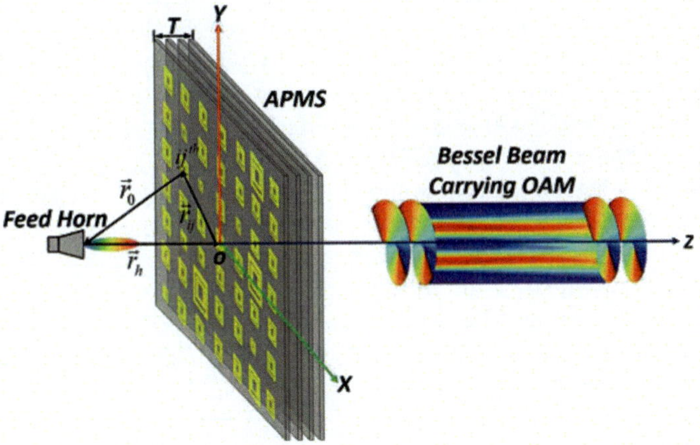

Fig. 7.54 Configuration of multilayer APMS transmission metasurface generating high-order Bessel vortex beam carrying OAM [94]

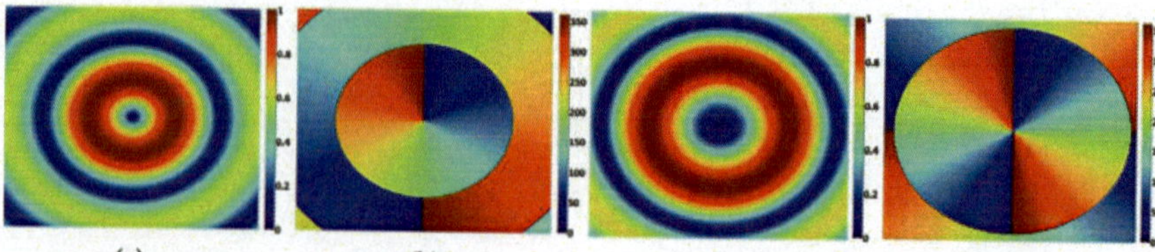

Fig. 7.55 **a** E-field amplitude and **b** phase distributions of the first-order Bessel vortex beam on the plane of $z = T$, **c** E-field amplitude and **d** phase distributions of the second-order Bessel vortex beam on the plane of $z = T$ [94]

that if the transverse wavenumber of the beam is increased, a narrower beam width can be obtained. Inevitably, the non-diffracting transmission distance for the Bessel beam will decrease accordingly. Here we choose $k_{\perp} = 0.1k_0$.

Figure 7.55a–d depict the calculated normalized E-field amplitude and phase distributions on the plane of $z = T$ (T represents the thickness of the APMS) for typical first and second-order Bessel vortex beams. These distributions are obtained using Eqs. (7.25) and (7.26) and correspond to vortex waves carrying OAM modes of $n = 1$ and $n = 2$, respectively. It is evident from the figures that the E-field amplitude distribution of the Bessel vortex beam needs to continuously vary from 0 to 1, while the phase should cover a range of 0° ~ 360°. In practical implementation, the transmission coefficient of the APMS element can be adjusted within a fixed interval, where the interval itself is determined by the dimensional variation increment of the transmission elements. By selecting these sparse amplitude values, it becomes possible to cover a phase-shift range of 360°.

To address the issue of achieving a high phase range, a four-layer conformal square-loop (FCSL) element is proposed. This multi-layer structure involves each layer being printed on 1 mm thick substrates with a relative permittivity of 2.65 and a loss tangent of 0.005, as illustrated in Fig. 7.56. By utilizing four layered surfaces, a phase range of over 300° can be obtained. Typically, when transmission metasurfaces solely possess a phase-shifting capability, four identical layers are required to achieve a phase range exceeding 300 degrees. These four identical layers are separated by an air gap of approximately a quarter wavelength, denoted as $h + t \approx \lambda/4$. Here, h represents the determined height and t is the thickness of the substrate. In this design, however, the aim is to simultaneously achieve a sufficient modulation range of the amplitude and phase. To accomplish this, the four layers of the FCSL element are divided into two groups, with each group consisting of two identical layers. The geometrical parameters of these two groups can be varied independently, providing greater flexibility in achieving the desired modulation characteristics. The FCSL elements are then cascaded with an air spacing of $h = 5.5$ mm, as illustrated in Fig. 7.56b.

Figure 7.56c, d display color maps representing the amplitude and phase distributions of the FCSL element with different combinations of a_1 and a_2 under normal incidence conditions. In Fig. 7.56e, the variation curve of transmission coefficient

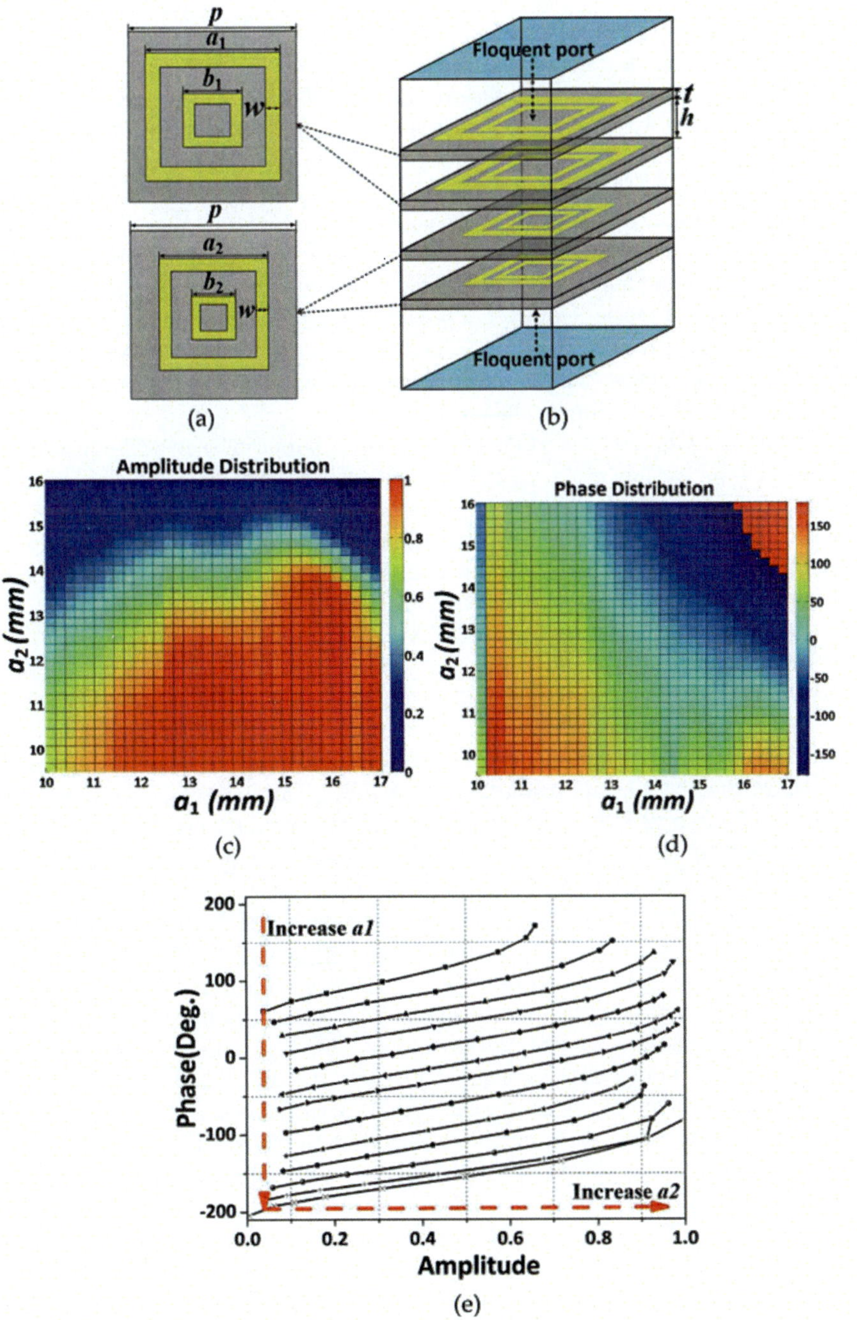

Fig. 7.56 **a** Geometry of the dual conformal square-loop elements, **b** infinite periodic unit model of the FCSL element, and the color maps of **c** transmission coefficient amplitude and **d** phase distributions with changing a_1 and a_2 of the two groups of layers, **e** the variation curve of transmission coefficient amplitude and phase responses with varying a_1 and a_2 [94]

amplitude and phase responses versus a_1 and a_2 at 10 GHz is depicted. In Fig. 7.56a, specific parameter values are chosen, namely $b_1 = 0.5a_1, b_2 = 0.5a_2, w = 2.5$ mm and $P = 20$ mm. The parameter a_1 is varied from 10 to 17 mm in increments of 0.1 mm, while a_2 changes within the same interval as a_1, ranging from 9.5 to 16 mm. From Figs. 7.56c, e, it is apparent that by independently controlling the parameters a_1 and a_2, the amplitude of the transmission coefficient of the FCSL element can be adjusted smoothly from 0 to 1. Additionally, the transmission phase-shift exhibits a range of approximately 300° at different transmission coefficient amplitudes. The distributions of amplitude and phase in the designed APMS transmission metasurface appear to be sparse. However, a phase range of 300° is generally sufficient to meet the phase requirements for most practical designs. These results demonstrate that the FCSL element, with its independent control of a_1 and a_2, enables precise modulation of both amplitude and phase.

Using the FCSL element as the building block, a four-layer APMS transmission metasurface consisting of 28 × 28 FCSL elements has been designed. The objective of this design is to generate a second-order Bessel vortex beam with an OAM mode of $n = 2$ at 10 GHz. The overall dimensions of the designed APMS transmission metasurface are 560 mm × 560 mm × 20.5 mm. To validate the design, a prototype of the APMS transmission metasurface has been fabricated, as depicted in Fig. 7.57. For the experimental setup, a standard X-band horn antenna is employed as the feeding source, positioned at a distance of 0.5 m away from the APMS transmission metasurface.

To verify the characteristics of the non-diffracting Bessel vortex beam generated by the APMS transmission metasurface system, simulations and measurements of E-field distributions are conducted on observational planes located at different distances from the APMS transmission metasurface aperture. The distances chosen for observation are 1.0 m, 2.0 m, and 2.5 m. The measurement is performed using near-field planar scanning techniques at 10 GHz, as illustrated in Fig. 7.58. In the measurement setup, the vertical polarization component of the transmitted electric field (E-field) is detected using a standard waveguide probe. The measured sampling grid length is 10 mm, and the size of the scanning plane is 0.8 m × 0.8 m. Figure 7.59a displays the simulated E-field distributions of the Bessel vortex beam generated by the APMS transmission metasurface. Figure 7.59b, c present the simulated E-field amplitude and phase distributions, respectively, on the observational planes at the three different distances from the APMS transmission metasurface system. These simulated results show that the pseudo-non-diffracting E-field transmission distance can exceed 83 wavelengths (2.5 m). To validate the simulation results, the measured amplitude and phase distributions of the E-field component are depicted in Fig. 7.59d–e. The measured results are found to be in good agreement with the simulated ones, confirming the accuracy of the design. Furthermore, in Figs. 7.59b, d, it can be observed that the E-field intensity exhibits a doughnut-shaped pattern. As the measurement distance increases, the energy distribution remains primarily confined within the scanning plane, indicating the non-diffracting behavior of the generated Bessel vortex beam.

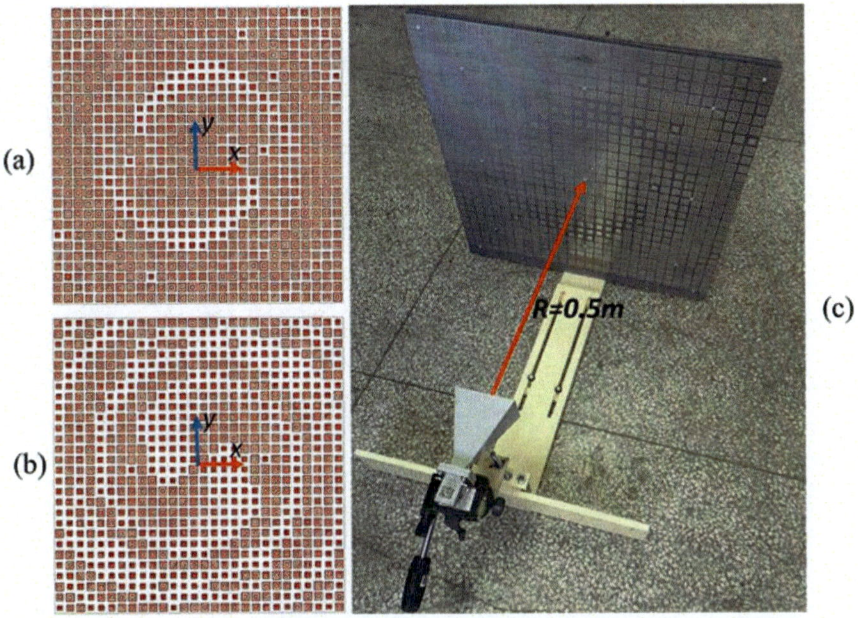

Fig. 7.57 Top view of **a** first group layers and **b** second group layers of APMS transmission metasurface, **c** the fabricated APMS transmission metasurface system generating the second-order Bessel vortex beam with OAM mode of $n = 2$ [94]

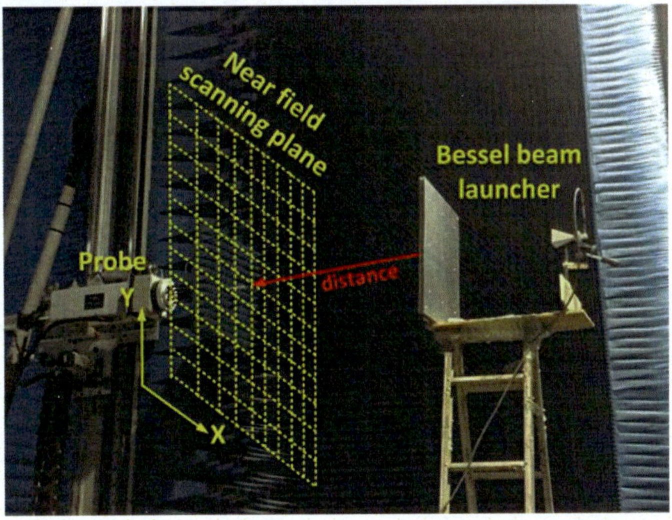

Fig. 7.58 Near field scanning measurement system for the Bessel vortex beam carrying OAM [94]

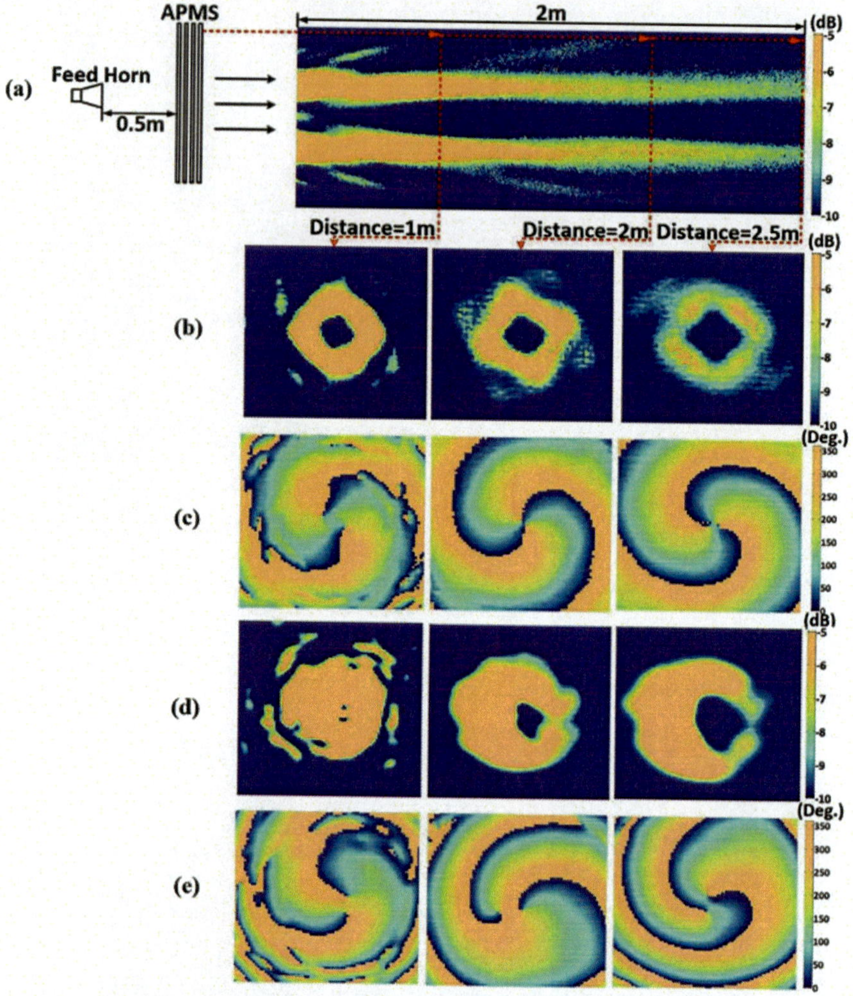

Fig. 7.59 **a** E-field transmission characteristics of the second-order Bessel vortex beam with OAM mode of $n = 2$, **b** simulated E-field amplitude distributions in dB, and **c** phase distributions on the different observational planes at distance of 1 m (33 wavelengths), 2 m (66 wavelengths) and 2.5 m (83 wavelengths), respectively, **d** measured E-field amplitude distributions in dB and **e** phase distributions on the different observational planes at distances of 1 m (33 wavelengths), 2 m (66 wavelengths) and 2.5 m (83 wavelengths) away from the APMS transmission metasurface aperture, respectively [94]

It should be noted that the measurement setup is subject to fabrication tolerances and slight obliqueness in the placement of the APMS transmission metasurface, which can result in asymmetry in the beam spots compared to the simulated results.

Despite these factors, the measurements confirm that the Bessel vortex beam generated by the APMS transmission metasurface maintains its diffraction-less characteristics for distances longer than 83 wavelengths. Furthermore, from Fig. 7.59b, d, it is evident that the Bessel vortex beam with an OAM mode of $n = 2$ is successfully generated. Moreover, as the detecting distance increases, the Bessel vortex beam with OAM becomes more stable, indicating the robustness of the APMS transmission metasurface in producing the desired beam characteristics.

7.5 Conclusion

The chapter provides a comprehensive overview of planar electromagnetic metasurfaces, specifically for reflective and transmissive metasurfaces. These metasurfaces are powerful tools for manipulating electromagnetic waves and offer the ability to adjust electromagnetic wavefronts according to specific application requirements. The primary focus of this chapter is the generation of electromagnetic vortex waves with OAM using metasurfaces. The chapter extensively reviews the properties of OAM vortex waves in optics, as well as their applications in receivers and transmitters. Additionally, the chapter explores the generation of OAM vortex waves in the radio frequency domain utilizing reflection and transmission metasurfaces. Through the use of reflection and transmission metasurfaces, an effective method for OAM-based radio frequency communication is established. These metasurfaces enable precise control over electromagnetic waves, providing a promising avenue for the development and implementation of advanced communication systems that leverage OAM vortex waves.

References

1. S.B. Glybovski, S.A. Tretyakov, P.A. Belov, Y.S. Kivshar, C.R. Simovski, Metasurfaces: from microwaves to visible. Phys. Rep. **634**, 1–72 (2016)
2. X. Ni, Broadband light bending with plasmonic nanoantennas. Science **335**, 427 (2012)
3. D.M. Pozar, S.D. Targonski, H.D. Syrigos, Design of millimeter wave microstrip reflectarrays. IEEE Trans. Antennas Propag. **45**, 287–296 (1997)
4. C.G.M. Ryan, M.R. Chaharmir, J. Shaker, J.R. Bray, A wideband transmitarray using dual-resonant double square rings. IEEE Trans. Antennas Propag. **58**, 1486–1493 (2010)
5. Y. Ye, S. He, 90° polarization rotator using a bilayered chiral metamaterial with giant optical activity. Appl. Phys. Lett. **96**, 203501 (2010)
6. D. Germain, D. Seetharamdoo, S. Nawaz Burokur, A. De Lustrac, Phase-compensated metasurface for a conformal microwave antenna. Appl. Phys. Lett. **103**, 124102 (2013)
7. S. Saadat, H. Mosallaei, E. Afshari, Radiation-efficient 60 GHz on-chip dipole antenna realised by reactive impedance metasurface. IET Microwaves Antennas Propag. **7**, 98–104 (2013)
8. C. Jansen, I.A.I. Al-Naib, N. Born, M. Koch, Terahertz metasurfaces with high Q-factors. Appl. Phys. Lett. **98**, 051109 (2011)
9. M. Farmahini-Farahani, H. Mosallaei, Birefringent reflectarray metasurface for beam engineering in infrared. Opt. Lett. **38**, 462–464 (2013)

10. Z. Wang, S. He, Q. Liu, W. Wang, Visible light metasurfaces based on gallium nitride high contrast gratings. Opt. Comm. **367**, 144–148 (2016)
11. N. Yu, P. Genevet, M.A. Kats, F. Aieta, J.P. Tetienne, F. Capasso, Z. Gaburro, Light propagation with phase discontinuities: generalized laws of reflection and refraction. Science **334**, 333–337 (2011)
12. F. Aieta, P. Genevet, M.A. Kats, N. Yu, R. Blanchard, Z. Gaburro, F. Capasso, Aberration-free ultrathin flat lenses and axicons at telecom wavelengths based on plasmonic metasurfaces. Nano Lett. **12**, 4932–4936 (2012)
13. X. Ni, A.V. Kildishev, V.M. Shalaev, Metasurface holograms for visible light. Nat. Commun. **4**, 657–678 (2013)
14. Z.J. Wong, X. Ni, M. Mrejen, Y. Wang, X. Zhang, An ultrathin invisibility skin cloak for visible light. Science **349**, 1310–1314 (2015)
15. L. Allen, M.W. Beijersbergen, R.J.C. Spreeuw, J.P. Woerdman, Orbital angular momentum of light and the transformation of Lagurre-Gaussian laser modes. Phys. Rev. A **45**(11), 8185–8189 (1992)
16. D.G. Grier, A revolution in optical manipulation. Nature **424**(6950), 810–816 (2003)
17. M.F. Andersen, C. Ryu, P. Cladé, V. Natarajan, A. Vaziri, K. Helmerson, W.D. Phillips, Quantized rotation of atoms from photons with orbital angular momentum. Phys. Rev. Lett. **97**(17), 170406 (2006)
18. G. Molina-Terriza, J.P. Torres, L. Torner, Twisted photons. Nat. Phys. **3**(5), 305–310 (2007)
19. M. Pascucci, G. Tessier, V. Emiliani, M. Guillon, Superresolution imaging of optical vortices in a speckle pattern. Phys. Rev. Lett. **116**(9), 093904 (2016)
20. G. Gibson, J. Courtial, M.J. Padgett, M. Vasnetsov M, V. Pas'ko, S.M. Barnett, S. Franke-Arnold, Free-space information transfer using light beams carrying orbital angular momentum. Opt. Express **12**(22), 5448–5456 (2004)
21. J. Wang, J.Y. Yang, I.M. Fazal, N. Ahmed, Y. Yan, H. Huang, Y. Ren, Y. Yue, S. Dolinar, M. Tur, A.E. Willner, Terabit free-space data transmission employing orbital angular momentum multiplexing. Nat. Photonics **6**(7), 488–496 (2012)
22. E. Willner, H. Huang, Y. Yan, Y. Ren, N. Ahmed, G. Xie, C. Bao, L. Li, Y. Cao, Z. Zhao, J. Wang, M.P.J. Lavery, M. Tur, S. Ramachandran, A.F. Molisch, N. Ashrafi, S. Ashrafi, Optical communications using orbital angular momentum beams. Adv. Opt. Photon. **7**(1), 66–106 (2015)
23. Z. Guo, Z. Wang, M.I. Dedao, K. Guo, The orbital angular momentum encoding system with radial indices of Laguerre-Gaussian beam. IEEE Photon. J. **10**(5), 1–11 (2018)
24. Z. Liu, S. Yan, H. Liu, X. Chen, Superhigh-resolution recognition of optical vortex modes assisted by a deep-learning method. Phys. Rev. Lett. **123**(18), 183902 (2019)
25. M. Okida, T. Omatsu, M. Itoh, T. Yatagai, Direct generation of high power Laguerre-Gaussian output from a diode-pumped Nd:YVO(4) 1.3-mum bounce laser. Opt. Express **15**(12), 7616–7622 (2007)
26. P. Miao, Z. Zhang, J. Sun, W. Walasik, S. Longhi, N.M. Litchinitser, L. Feng, Orbital angular momentum microlaser. Science **353**(6298), 464–467 (2016)
27. Z. Zhang, K. Gui, C. Zhao, H. Zhang, Y. Xing, Direct generation of vortex beam with a dual-polarization microchip laser. IEEE Photon. Tech. Lett. **31**(15), 1221–1224 (2019)
28. M. Uchida, A. Tonomura, Generation of electron beams carrying orbital angular momentum. Nature **464**(7289), 737–739 (2010)
29. J. Wang, A. Cao, M. Zhang, H. Pang, S. Hu, Y. Fu, L. Shi, Q. Deng, Study of characteristics of vortex beam produced by fabricated spiral phase plates. IEEE Photon. J. **8**(2), 6500209 (2016)
30. H. Wei, A.K. Amrithanath, S. Krishnaswamy, 3D Printing of micro-optic spiral phase plates for the generation of optical vortex beams. IEEE Photon. Tech. Lett. **31**(8), 599–602 (2019)
31. J. Arlt, K. Dholakia, L. Allen, M.J. Padgett, The production of multiringed Laguerre-Gaussian modes by computer-generated holograms. J. Mod. Opt. **45**(6), 1231–1237 (1998)
32. V. Carpentier, H. Michinel, J.R. Salgueiro, D. Olivieri, Making optical vortices with computer-generated holograms. Am. J. Phys. **76**(10), 916–921 (2008)

33. S. Li, Z. Wang, Generation of optical vortex based on computer-generated holographic gratings by photolithography. Appl. Phys. Lett. **103**(14), 141110 (2013)
34. S. Tao, W. Lee, X. Yuan, Dynamic optical manipulation with a higher-order fractional Bessel beam generated from a spatial light modulator. Opt. Lett. **28**(20), 1867–1869 (2003)
35. Y. Ohtake, T. Ando, N. Fukuchi, N. Matsumoto, H. Ito, T. Hara, Universal generation of higher-order multiringed Laguerre-Gaussian beams by using a spatial light modulator. Opt. Lett. **32**(11), 1411–1413 (2007)
36. N. Matsumoto, T. Ando, T. Inoue, Y. Ohtake, N. Fukuchi, T. Hara, Generation of high-quality higher-order Laguerre-Gaussian beams using liquid-crystal-on-silicon spatial light modulators. J. Opt. Soc. Am. A **25**(7), 1642–1651 (2008)
37. R.J. Hernández-Hernández, R.A. Terborg, I. Ricardez-Vargas, K. Volke-Sepúlveda, Experimental generation of Mathieu-Gauss beams with a phase-only spatial light modulator. Appl. Opt. **49**(36), 6903–6909 (2010)
38. E. Karimi, B. Piccirillo, E. Nagali, L. Marrucci, E. Santamato, Efficient generation and sorting of orbital angular momentum eigenmodes of light by thermally tuned q-plates. Appl. Phys. Lett. **94**(23), 231124 (2009)
39. E. Karimi, S.A. Schulz, I. DeLeon, H. Qassim, J. Upham, R.W. Boyd, Generating optical orbital angular momentum at visible wavelengths using a plasmonic metasurface. Light: Sci. Appl. **3**(5), e167 (2014)
40. X. Liu, J. Deng, M. Jin, Y. Tang, X. Zhang, K. Li, G. Li, Cassegrain metasurface for generation of orbital angular momentum of light, Appl. Phys. Lett. **115**(22), 221102 (2019)
41. M. Mirhosseini, O.S. Magaña-Loaiza, M.N. O'Sullivan, B. Rodenburg, M. Malik, M.P.J. Lavery, M.J. Padgett, D.J. Gauthier, R.W. Boyd, High-dimensional quantum cryptography with twisted light. New J. Phys. **17**(3), 033033 (2015)
42. J. Leach, M.J. Padgett, S.M. Barnett, S. Franke-Arnold, J. Courtial, Measuring the orbital angular momentum of a single photon. Phys. Rev. Lett. **88**(25), 257901 (2002)
43. H. Wei, X. Xue, J. Leach, M.J. Padgett, S.M. Barnett, S. Franke-Arnold, E. Yao, J. Courtial, Simplified measurement of the orbital angular momentum of single photons. Opt. Comm. **223**(1–3), 117–122 (2003)
44. J. Leach, J. Courtial, K. Skeldon, S.M. Barnett, S. Franke-Arnold, M.J. Padgett, Interferometric methods to measure orbital and spin, or the total angular momentum of a single photon. Phys. Rev. Lett. **92**(1), 013601 (2004)
45. M. Harris, C.A. Hill, P.R. Tapster, J.M. Vaughan, Laser modes with helical wave-fronts. Phys. Rev. A **49**(4), 3119–3122 (1994)
46. V. Basistiy, M.S. Soskin, M.V. Vasnetsov, Optical wavefront dislocations and their properties. Opt. Comm. **119**(5–6), 604–612 (1995)
47. H. Sztul, R. Alfano, Double-slit interference with Laguerre-Gaussian beams. Opt. Lett. **31**(7), 999–1001 (2006)
48. D.P. Ghai, P. Senthilkumaran, R.S. Sirohi, Single-slit diffraction of an optical beam with phase singularity. Opt. Lasers Eng. **47**(1), 123–126 (2009)
49. C. Guo, L. Lu, H. Wang, Characterizing topological charge of optical vortices by using an annular aperture. Opt. Lett. **34**(23), 3686–3688 (2009)
50. J.M. Hickmann, E.J.S. Fonseca, W.C. Soares, S. Chávez-Cerda, Unveiling a truncated optical lattice associated with a triangular aperture using light's orbital angular momentum. Phys. Rev. Lett. **105**(5), 053904 (2010)
51. Y. Taira, S. Zhang, Split in phase singularities of an optical vortex by off-axis diffraction through a simple circular aperture. Opt. Lett. **42**(7), 1373–1376 (2017)
52. G. Berkhout, M. Lavery, J. Courtial, M. Beijersbergen, M. Padgett, Efficient sorting of orbital angular momentum states of light. Phys. Rev. Lett. **105**(15), 153601 (2010)
53. N. Zhang, X. Yuan, R.E. Burge, Extending the detection range of optical vortices by Dammann vortex gratings. Opt. Lett. **35**(20), 3495–3497 (2010)
54. K. Dai, C. Gao, L. Zhong, Q. Na, Q. Wang, Measuring OAM states of light beams with gradually-changing-period gratings. Opt. Lett. **40**(4), 562–565 (2015)

55. D. Hebri, S. Rasouli, M. Yeganeh, Intensity-based measuring of the topological charge alteration by the diffraction of vortex beams from amplitude sinusoidal radial gratings. J. Opt. Soc. Am. B **35**(4), 724–730 (2018)
56. Y. Zhang, P. Li, J. Zhong, S. Qi, X. Guo, D. Wu, S. Liu, J. Zhao, Measuring singularities of cylindrically structured light beams using a radial grating. Appl. Phys. Lett. **113**(22), 221108 (2018)
57. Y. Li, J. Deng, J. Li, Z. Li, Sensitive orbital angular momentum (OAM) monitoring by using gradually changing-period phase grating in OAM-multiplexing optical communication systems. IEEE Photon. J. **8**(2), 7902306 (2016)
58. R. Chen, X. Zhang, Y. Zhou, H. Ming, A. Wang, Q. Zhan, Detecting the topological charge of optical vortex beams using a sectorial screen. Appl. Opt. **56**(16), 4868–4872 (2017)
59. H. Ma, X. Li, Y. Tai, H. Li, J. Wang, M. Tang, Y. Wang, J. Tang, Z. Nie, In situ measurement of the topological charge of a perfect vortex using the phase shift method. Opt. Lett. **42**(1), 135–138 (2017)
60. Z. Liu, S. Gao, W. Xiao, J. Yang, X. Huang, Y. Feng, J. Li, W. Liu, Z. Li, Measuring high-order optical orbital angular momentum with a hyperbolic gradually changing period pure-phase grating. Opt. Lett. **43**(13), 3076–3079 (2018)
61. P. Vaity, J. Banerji, R.P. Singh, Measuring the topological charge of an optical vortex by using a tilted convex lens. Phys. Lett. A **377**(15), 1154–1156 (2013)
62. V.V. Kotlyar, A.A. Kovalev, A.P. Porfirev, Astigmatic transforms of an optical vortex for measurement of its topological charge. Appl. Opt. **56**(14), 4095–4104 (2017)
63. Volyar, M. Bretsko, Y. Akimova, Y. Egorov, Measurement of the vortex and orbital angular momentum spectra with a single cylindrical lens. Appl. Opt. **58**(21), 5748–5755 (2019)
64. Y. Li, Y. Han, Z. Cui, J. Wang, W. Zhao, Simultaneous identification of the azimuthal and radial mode indices of Laguerre-Gaussian beams using a spiral phase grating. J. Phys. D Appl. Phys. **53**(8), 085106 (2020)
65. Trichili, C. Rosales-Guzmán, A. Dudley, B. Ndagano, A.B. Salem, M. Zghal, A. Forbes, Optical communication beyond orbital angular momentum. Sci. Rep. **6**, 27674 (2016)
66. Krenn, R. Fickler, M. Fink, J. Handsteiner, M. Malik, T. Scheidl, R. Ursin, A. Zeilinger, Communication with spatially modulated light through turbulent air across Vienna. New J. Phys. **16**(11), 113028 (2014)
67. Y. Li, Y. Han, Z. Cui, Measuring the topological charge of vortex beams with gradually changing-period spiral spoke grating. IEEE Photon. Technol. Lett. **32**(2), 101–104 (2020)
68. V. Lerner, D. Shwa, Y. Drori, N. Katz, Shaping Laguerre-Gaussian laser modes with binary gratings using a digital micromirror device. Opt. Lett. **37**(23), 4826–4828 (2012)
69. J. Huang, J.A. Encinar, *Reflectarray Antennas*. Wiley (2007)
70. P. Nayeri, Y. Fan, A.Z. Elsherbeni, *Reflectarray Antennas: Theory, Designs, and Applications*. Wiley (2018)
71. G. Gibson, J. Courtial, M. Padgett, M. Vasnetsov, V. Pas'Ko, S. Barnett, S. Franke-Arnold, Free-space information transfer using light beams carrying orbital angular momentum. Opt. Express **12**, 5448–5456 (2004)
72. J. Wang, J.Y. Yang, I.M. Fazal, N. Ahmed, Y. Yan, H. Huang, Y. Ren, Y. Yue, S. Dolinar, M. Tur, A.E. Willner, Terabit free-space data transmission employing orbital angular momentum multiplexing. Nat. Photonics **6**, 488–496 (2012)
73. B. Thidé, H. Then, J. Sjöholm, K. Palmer, J. Bergman, T. D. Carozzi, Y. N. Istomin, N. H. Ibragimov, R. Khamitova, Utilization of photon orbital angular momentum in the low-frequency radio domain. Phys. Rev. Lett. **99**, 087701 (2007)
74. S.M. Mohammadi, L.K.S. Daldorff, J.E.S. Bergman, R.L. Karlsson, B. Thidé, K. Forozesh, T.D. Carozzi, B. Isham, Orbital angular momentum in radio—a system study. IEEE Trans. Antennas Propag. **58**, 565–572 (2010)
75. F. Tamburini, E. Mari, A. Sponselli, F. Romanato, T. Bo, A. Bianchini, L. Palmieri, C.G. Someda, Encoding many channels in the same frequency through radio vorticity: first experimental test. New J. Phys. **14**, 811–815 (2011)

76. R. Niemiec, C. Brousseau, K. Mahdjoubi, O. Emile, A. Menard, Characterization of an OAM Flat-Plate Antenna in the Millimeter Frequency Band. IEEE Antennas Wirel. Propag. Lett. **13**, 1011–1014 (2014)
77. F. Tamburini, E. Mari, T. Bo, C. Barbieri, F. Romanato, Experimental verification of photon angular momentum and vorticity with radio techniques. Appl. Phys. Lett. **99**, 204102–204103 (2011)
78. Q. Bai, A. Tennant, B. Allen, Experimental circular phased array for generating OAM radio beams. Electron. Lett. **50**, 1414–1415 (2014)
79. E. Brasselet, M. Malinauskas, A. Žukauskas, S. Juodkazis, Photopolymerized microscopic vortex beam generators: Precise delivery of optical orbital angular momentum. Appl. Phys. Lett. **97**, 211108–211108-3 (2010)
80. G.A. Turnbull, D.A. Robertson, G.M. Smith, L. Allen, M.J. Padgett, Generation of free-space Laguerre-Gaussian modes at millimetre-wave frequencies by use of a spiral phaseplate. Opt. Comm. **127**, 183–188 (1996)
81. E. Karimi, S. A. Schulz, I. D. Leon, H. Qassim, J. Upham, R.W. Boyd, Generating optical orbital angular momentum at visible wavelengths using a plasmonic metasurface. Light Sci. Appl. **3**, e167 (2014)
82. C.F. Chen, C.T. Ku, Y.H. Tai, P.K. Wei, H.N. Lin, C.B. Huang, Creating optical near-field orbital angular momentum in a gold metasurface. Nano Lett. **15**, 2746–2750 (2015)
83. S. Yu, L. Li, G. Shi, C. Zhu, X. Zhou, Y. Shi, Design, fabrication, and measurement of reflective metasurface for orbital angular momentum vortex wave in radio frequency domain. Appl. Phys. Lett. **108**, 121903 (2016)
84. J. Huang, J. Encinar, *Reflectarray Antennas*. Wiley-IEEE Press (2008)
85. L. Li, Q. Chen, Q. Yuan, K. Sawaya, T. Maruyama, T. Furuno, S. Uebayashi, Novel broadband planar reflectarray with parasitic dipoles for wireless communication applications. IEEE Antennas Wirel. Propag. Lett. **8**, 881–885 (2009)
86. S. Yu, L. Li, G. Shi, C. Zhu, Y. Shi, Generating multiple orbital angular momentum vortex beams using a metasurface in radio frequency domain. Appl. Phys. Lett. **108**, 241901 (2016)
87. S. Yu, L. Li, G. Shi, Dual-polarization and dual-mode orbital angular momentum radio vortex beam generated by using reflective metasurface. Appl. Phys. Express **9**, 082202 (2016)
88. A. Yu, F. Yang, A.Z. Elsherbeni, J. Huang, Transmitarray antennas: an overview. IEEE Antennas Propag. Soc. (APS/URSI) Symp., Jul (2011)
89. A.H. Abdelrahman, F. Yang, A.Z. Elsherbeni, Transmission phase limit of multilayer frequency selective surfaces for transmitarray designs. IEEE Trans. Antennas Propag. **62**(2), 690–697 (2014)
90. A.H. Abdelrahman, A.Z. Elsherbeni, F. Yang, Transmitarray antenna design using cross slot elements with no dielectric substrate. IEEE Antennas Wirel. Propag. Lett. **13**, 177–180 (2014)
91. P. Nayeri, F. Yang, A. Z. Elsherbeni, Design of multifocal transmitarray antennas for beamforming applications. IEEE Antennas Propag. Soc. Int. Symp. (AP-S), 1672–1673 (2013)
92. A. Clemente, L. Dussopt, R. Sauleau, P. Potier, P. Pouliguen, Wideband 400-element electronically reconfigurable transmitarray in X band. IEEE Trans. Antennas Propag. **61**(10), 5017–5027 (2013)
93. C.G.M. Ryan, M.R. Chaharmir, J. Shaker, J.R. Bray, Y.M.M. Antar, A. Ittipiboon, A wideband transmitarray using dual-resonant double square rings. IEEE Trans. Antennas Propag. **58**(5), 1486–1493 (2010)
94. N. Kou, S. Yu, L. Li, Generation of high-order Bessel vortex beam carrying orbital angular momentum using multilayer amplitude-phase-modulated surfaces in radiofrequency domain. Appl. Phys. Express **10**(1), 016701 (2016)

Chapter 8
Invisible Cloak Design and Application of Metasurfaces on Microwave Absorption and RCS Reduction

Yan Shi, Xiangyu Cao, Sijia Li, and Long Li

Abstract Increasing attention to metamaterials has been attracted by researchers due to their unusual electromagnetic properties. One of the intriguing metamaterial devices is the cloak. In the first part of this chapter, we first review two popular methods, including the coordinate transformation method and scattering cancellation approach to design invisibility and illusion cloaking devices. The coordinate transformation-based method as an intuitive and visual approach, offers unprecedented flexibility in the design of the complementary cloaking devices. The scattering cancellation-based cloaking design can be analytically achieved in the Mie series method and characteristic mode (CM) method framework. With some selected examples, we illustrate the invisibility and illusion capabilities of the cloaking devices designed by two methods. On the other hand, metasurface-based microwave absorption and radar cross section (RCS) reduction have attracted much attention in the field of stealth for safety. Especially, the antenna designs with low RCS characteristics become crucial for stealth platform communication. Thus the RCS reduction of antennas without compromising their radiation characteristics has been a topic of immense strategic interest. The second part of this chapter gives some interesting examples of metasurfaces, i.e., perfect metamaterial absorber (PMA) and artificial magnetic conductor (AMC). Experimental verification of their predicted behaviors has been obtained. These antennas with metasurface have led to a wide variety of in-band and broadband RCS reductions.

Keywords Cloak · Coordinate transformation · Scattering cancellation · Complementary · Mie series · Characteristic mode (CM) · Stealth · Microwave absorption · Radar cross section (RCS) reduction · Artificial magnetic conductor (AMC)

Y. Shi · L. Li (✉)
School of Electronic Engineering, Xidian University, Xi'an 710071, Shaanxi, China
e-mail: lilong@mail.xidian.edu.cn

Y. Shi
e-mail: shiyan@mail.xidian.edu.cn

X. Cao · S. Li
Air Force Engineering University, Xi'an 710077, Shaanxi, China

L. Li et al. (eds.), *Electromagnetic Metamaterials and Metasurfaces: From Theory To Applications*, https://doi.org/10.1007/978-981-99-7914-1_8

8.1 Introduction

Metamaterial is a category of artificial materials composed of subwavelength inclusions with rationally designed shapes, sizes, and compositions. The homogenized material properties can be exhibited by engineering the unit cell of metamaterial that is generally much smaller than the operating wavelength. Metamaterials are typically created by designing the inclusions to have resonant properties near a particular wavelength λ of interest through either planar or volumetric loadings of space [1]. They have provided extremely interesting flexibility in the engineering design processes for a variety of electromagnetic, acoustic, elastic, and thermal wave applications. Metasurface, which is the two-dimensional metamaterial, has opened up a number of remarkable new approaches to manipulating EM waves. The metasurfaces' souls are amazing and unique features for controlling electromagnetic wave magnitudes, phases, and polarizations [2]. Due to the powerful and multiform functionalities of the metasurface, several electromagnetic prototypes have been developed for radiating and scattering applications on antennas operating from UHF to optical frequencies. Experimental verification has been achieved due to the excellent performance characteristics, such as profile decrease, bandwidth improvement [3], gain enhancement [4], and radar cross section (RCS) reduction of antenna systems with metasurface [5, 6].

An invisibility cloak is a device that makes the shielding object invisible. The concept of the invisible cloak, often used in fantasy or science fiction, has attracted much attention in the scientific communities since transformation electromagnetics, or transformation optics, was introduced as a powerful and systematic framework to manipulate electromagnetic waves in 2006 [7, 8]. In transformation electromagnetics, a coordinate transformation is employed to map a given geometric space into a desirable, distorted geometric space, and under the coordinate transformation, the transformation of the electromagnetic fields between two spaces is likewise achieved. To realize the coordinate transformation, the relationship of material parameters between two spaces can be set up according to the form invariance of Maxwell's equations. Following the transformation electromagnetics, the experimental realization of a cylindrical cloak at microwave frequency was reported [9]. Afterward, various theoretical verification and experimental realization on invisibility cloaks [10–18] have been developed. Especially in the conventional cloaking design, the coordinate transformation is implemented to guide the electromagnetic wave bent smoothly around the object and restore the waveform outside the cloak. However, the communication between the object and outer space is unfeasible. In 2009, a cloak was proposed to hide an object exterior to the cloaking device [19]. By introducing a double negative (DNG) material called complementary medium and inserting a mirror object related to the target object in the core medium, the external target object can be cloaked according to the coordinate transformation. Various complementary invisibility cloaks with different shapes and sizes have been designed [20–25].

The coordinate transformation provides an intuitive approach to designing the invisibility cloak, but the materials in the designed cloaks are generally strongly anisotropic and inhomogeneous. These complexities of the resulting material parameters limit the practical realization of these cloaks. Alternatively, the scattering cancellation method [26–36] has been proposed to achieve electromagnetic transparency using a shell with appropriately designed permittivity and permeability. Using Mie scattering expansion, the target object with canonical shapes, such as an infinite cylinder and a sphere, can become invisible by covering a homogeneous coating in the long-wavelength limit [26–34]. Recently, a characteristic mode-based method in terms of scattering cancellation has been developed to design the cloaking device for the arbitrarily-shaped target object [36].

RCS reduction attracts much interest in some stealth platforms where low RCS design is urgently required for safety. Microstrip antennas and waveguide slot antennas are usually stealth platform communication devices due to their high radiation efficiency, strong directivity, easy integration, and compact structure. However, the antennas also contribute greatly to the overall RCS, so the stealth performance of the platform will be destroyed, and the stealth system will be worthless. Some different methods have been proposed in the literature to reduce antenna RCS, such as the shaping of the radiation patch [37], adopting radar-absorbing material [38], and passive or active cancellation [39], etc. However, these methods compromise the radiation characteristics of antennas. With the development of metasurface, the metasurface-inspired engineering of antennas and their performance characteristics have provided an alternative approach to addressing the pressing issue of RCS reduction [40–42].

This chapter first elaborates on the cloak design in Sect. 8.2. Two kinds of design methods are given, including the coordinate transformation method and the scattering cancellation method. Some 2D and 3D cloaks with arbitrary shapes are designed following two design methods. Next, the metasurface application on RCS reduction for the antenna is reviewed in Sect. 8.3. The interesting metasurfaces of perfect metamaterial absorber (PMA) and artificial magnetic conductor (AMC) and their application on RCS reduction of the antenna will be covered in more detail. We comprehensively discussed the performance characteristics, such as absorption of PMA, the gain of the antenna, and RCS reduction of the antenna.

8.2 Invisible Cloak Design

8.2.1 Coordinate Transformation-Based Complementary Cloak

According to the coordinate transformation [7, 8], when a space is transformed into another space of different shape and size by using the transformation relation of $\mathbf{x}' = \mathbf{x}'(\mathbf{x})$, the permittivity tensor $\overline{\varepsilon'}$ and permeability tensor $\overline{\mu'}$ in the transformed space or physical space X′ can be expressed in terms of the permittivity tensor $\overline{\varepsilon}$ and permeability tensor $\overline{\mu}$ in the original space X as

$$\overline{\varepsilon'} = \frac{\Lambda\overline{\varepsilon}\Lambda^T}{\det\Lambda}, \quad \overline{\mu'} = \frac{\Lambda\overline{\mu}\Lambda^T}{\det\Lambda}, \tag{8.1}$$

in which the Jacobian transformation matrix is denoted as

$$\Lambda_{pq} = \frac{\partial p}{\partial q}(p = x', y', z'; q = x, y, z), \tag{8.2}$$

and $\det(\Lambda)$ is determinant of Λ.

8.2.1.1 Minimized Complementary Cloak

The contentional invisibility cloak based on the complementary medium [19] consists of three regions, i.e., a restoring region, a complementary medium region, and a region to be hidden, as shown in Fig. 8.1. With the complementary media, a region to be hidden as well as its interior object are optically "canceled" at a certain frequency, and meanwhile, a complementary "image" of the object, which is called as the "anti-object", emerges in the complementary media. With a dielectric core material, the correct optical path in canceled space is restored.

To reduce the size of the complementary cylinder cloak, a part of the cylinder cloak, i.e., a circular sector with a central angle of 2θ centered at the origin O, is kept and two triangle sectors placed at both sides of the circular sector are introduced, as shown in Fig. 8.1. In this scenario, regions 1, 2, and 3 are the core region, the complementary region, and the region to be hidden, respectively. Note that the complementary region is entirely surrounded by the core region, and the region to

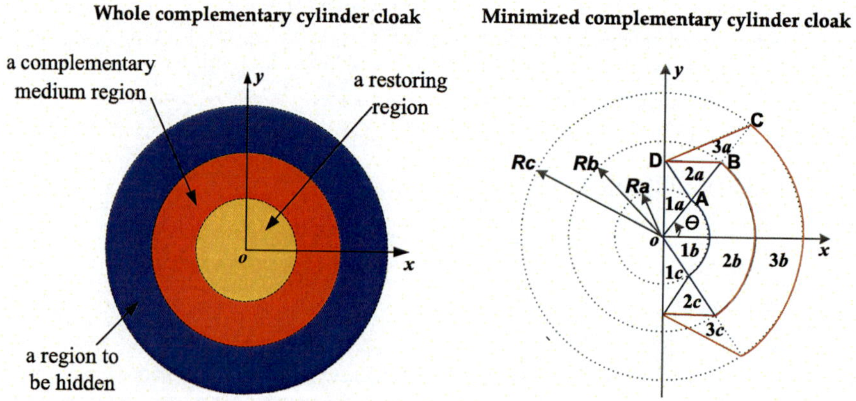

Fig. 8.1 The complementary cloak device [20]

be hidden. In each region, there is a circular part denoted as part *b* and two triangle parts denoted as parts *a* and *c*. In part *b*, a linear coordinate transformation along the radius direction is used from the part 3*b* to the part 2*b*, i.e.,

$$r = kr' + m, \quad \theta = \theta', \quad z = z', \tag{8.3}$$

where $k = (Rc\text{-}Rb)/(Ra\text{-}Rb)$ and $m = Rb\cdot(Ra\text{-}Rc)/(Ra\text{-}Rb)$. Here, *Ra*, *Rb*, and *Rc* are the radius of the core layer, the outer radius of the complementary layer, and the outer radius of the air layer, respectively. According to Eq. (8.1), the material parameters in the part 2*b* are obtained as

$$\begin{cases} \varepsilon_r' = \mu_r' = (kr' + m)\big/ kr' \\ \varepsilon_\theta' = \mu_\theta' = kr'\big/ (kr' + m). \\ \varepsilon_z' = \mu_z' = k(kr' + m)\big/ r' \end{cases} \tag{8.4}$$

A compressing transformation along the radius direction is utilized from the circular sector to the part 1*b* with the following formulas

$$r = r''Rc\big/ Ra, \quad \theta = \theta'', \quad z = z''. \tag{8.5}$$

The resultant material parameters in the part 1*b* are

$$\begin{cases} \varepsilon_r'' = \mu_r'' = 1 \\ \varepsilon_\theta'' = \mu_\theta'' = 1 \\ \varepsilon_z'' = \mu_z'' = (Rc\big/ Ra)^2 \end{cases}. \tag{8.6}$$

In part *a*, the transformations from part 3*a* to part 2*a* and from the triangle sector to part 1*a* can be uniformly expressed as

$$\begin{cases} x' = ax + by + c \\ y' = dx + ey + f \end{cases}, \tag{8.7}$$

where the transformation coefficients *a*, *b*, *c*, *d*, *e*, *f* can be determined according to the coordinate mapping relationship. For example, we map three points C, B, and D to three points A, B, D, respectively, and thus we have $a = 1$, $b = \cot(\theta) \cdot (Ra - Rc)\big/ (Rc - Rb)$, $d = 0$, and $e = (Ra - Rb)\big/ (Rc - Rb)$. Furthermore, the material parameters in part 2*a* are

$$\overline{\varepsilon}' = \overline{\mu}' = \begin{pmatrix} \frac{1+b^2}{e} & b & 0 \\ b & e & 0 \\ 0 & 0 & \frac{1}{e} \end{pmatrix}. \tag{8.8}$$

Similarly, the material parameters in part 1*a* are obtained as

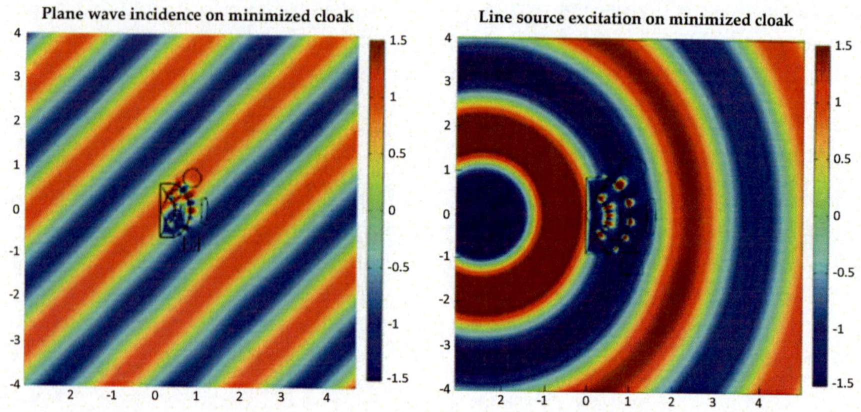

Fig. 8.2 The electric field distributions for invisibility of the minimized complementary cloak [20]

$$\overline{\varepsilon}' = \overline{\mu}' = \begin{pmatrix} a & d & 0 \\ d & \frac{1+d^2}{a} & 0 \\ 0 & 0 & \frac{1}{a} \end{pmatrix}, \tag{8.9}$$

in which $a = Ra/Rc$, $b = 0$, $d = \tan\theta \cdot (Ra - Rc)/Rc$, and $e = 1$. Due to symmetry, the material parameters in part a are the same as those in part c. Note that the material parameters are homogeneous in the whole cloak except the part $2b$. Moreover, the whole cloak except part $1b$ consists of anisotropic materials. Figure 8.2 shows the invisible performance of the minimized complementary cloak. A circular nonmagnetic object with relative permittivity of 2, a non-dielectric rectangle object with a relative permeability of 2, and a ring object with relative permittivity of 2 and relative permeability of 2 are simultaneously placed in the region exterior to the cloak, i.e., the region to be hidden shown in Fig. 8.1. We can observe that all these objects are well hidden. Moreover, these objects can receive the incident electromagnetic wave to realize the communication between the object and outer space. Figure 8.3 demonstrates the illusion effect of the minimized complementary cloak. The cloak generates a near-field distribution similar to that caused by a nonmagnetic circular object with a relative permittivity of 2.

8.2.1.2 Illusion Complementary Cloak

From the above discussions, we can know that the cloak can produce an illusion effect by placing an object in the core region of the complementary cloaking device. Figure 8.4 gives a schematic diagram of the illusion complementary cloak. Consider a complementary cloak with a trapezoid coordinate transformation given by Eq. (8.7), which is called a "true cloak". An object denoted as a "core cloak" is inserted into the core region of the true cloak to generate an "illusion cloak". To make a target

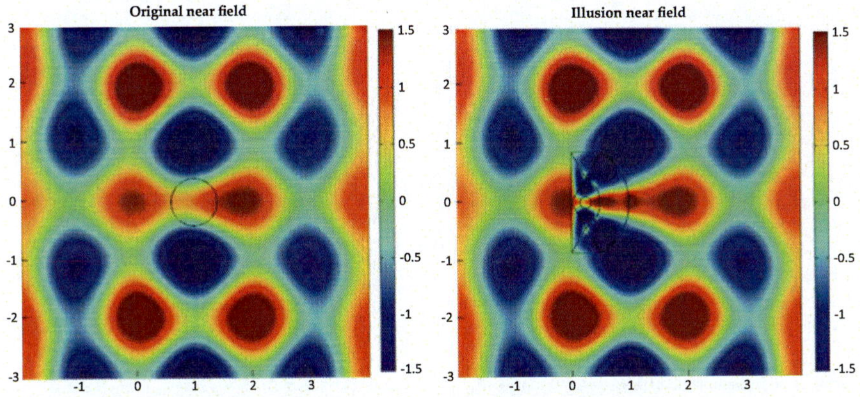

Fig. 8.3 The electric field distribution for illusion of the minimized complementary cloak [20]

object invisible, a fictitious object as an anti-object is placed in the complementary region of the illusion cloak, and thus the resulting true anti-object is inserted into the complementary region of the core cloak. With the illusion complementary cloaking device, the object can be flexibly located outside the invisible region of the true cloak. Moreover, the size of the anti-object in the core cloak can be reduced by using the illusion complementary cloaking device.

Figure 8.5 gives an illusion complementary cloak. A plane wave with a frequency of 600 MHz is incident on the cloaking device. Good invisibility can be obtained with or without the target object. Moreover, the sizes of the required anti-object are greatly reduced along with both the x and y directions. According to Fig. 8.5, we can see that the target object to be hidden can be arbitrarily placed.

Fig. 8.4 Schematic diagram of the illusion complementary cloak [21]

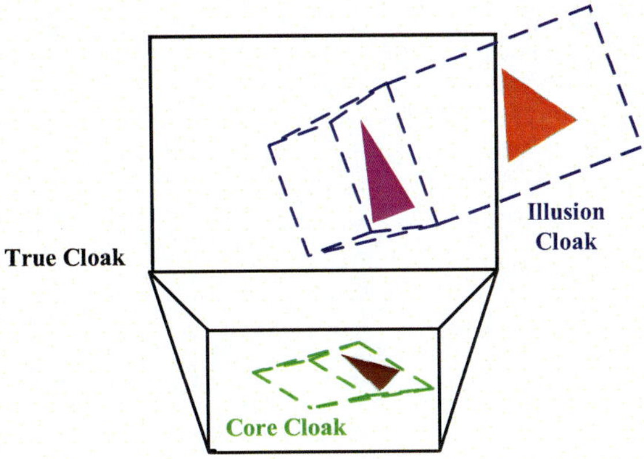

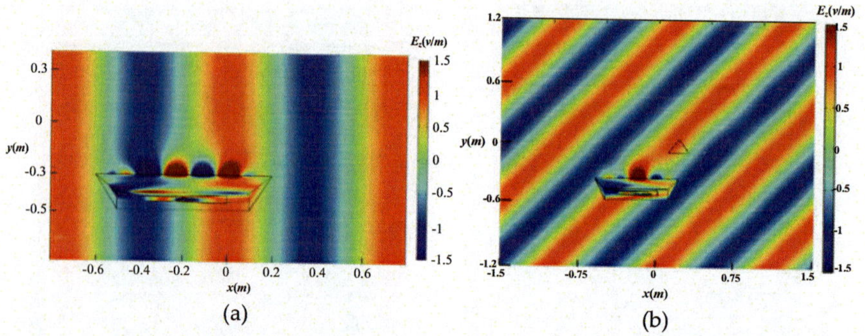

Fig. 8.5 Electric field distribution of the illusion complementary cloak. **a** without object. **b** with object [21]

8.2.1.3 Three-Dimensional Complementary Cloak

The above discussions are about the two-dimensional complementary cloak. We can easily extend them to a three-dimensional case. Consider a three-dimensional arbitrary complementary invisibility cloak in the spherical coordinate system, as shown in Fig. 8.6. Assuming that the boundary of the core region, the inner and outer boundaries of the exterior invisibility shell have the same shapes, we have $R_a(\theta,\varphi) = mR_b(\theta,\varphi)$ $(0 < m < 1)$ and $R_c(\theta,\varphi) = nR_b(\theta,\varphi)$ $(n > 1)$.

With a folding coordinate transformation along the radial direction from the region to be hidden into the complementary region, which is similar to the two-dimensional case given by Eq. (8.3), the material parameters of the complementary region can be obtained as

Fig. 8.6 A three-dimensional arbitrary complementary cloak [22]

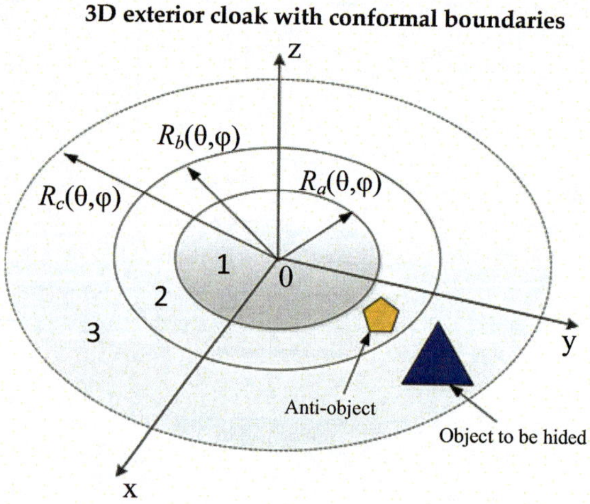

$$\varepsilon'_{rr} = \mu'_{rr} = \frac{(r' - k_2)^2 + \left(\frac{(m-1)n}{1-n} \cdot \frac{\partial R_b}{\partial \theta}\right)^2}{k_1 r'^2} + \frac{\left(\frac{(m-1)n}{\sin\theta(1-n)} \cdot \frac{\partial R_b}{\partial \phi}\right)^2}{k_1 r'^2}, \tag{8.10}$$

$$\varepsilon'_{r\theta} = \varepsilon'_{\theta r} = \mu'_{r\theta} = \mu'_{\theta r} = \frac{(m-1)n}{k_1 r'(1-n)} \cdot \frac{\partial R_b}{\partial \theta}, \tag{8.11}$$

$$\varepsilon'_{r\phi} = \varepsilon'_{\phi r} = \mu'_{r\phi} = \mu'_{\phi r} = \frac{(m-1)n}{k_1 r' \sin\theta(1-n)} \cdot \frac{\partial R_b}{\partial \phi}, \tag{8.12}$$

$$\varepsilon'_{\theta\phi} = \varepsilon'_{\phi\theta} = \mu'_{\theta\phi} = \mu'_{\phi\theta} = 0, \tag{8.13}$$

$$\varepsilon'_{\theta\theta} = \varepsilon'_{\phi\phi} = \mu'_{\theta\theta} = \mu'_{\phi\phi} = \frac{1}{k_1}, \tag{8.14}$$

where

$$\begin{aligned} k_1 &= \frac{R_b - R_a}{R_b - R_c} = \frac{1-m}{1-n}, \\ k_2 &= \frac{R_c - R_a}{R_c - R_b} R_b = \frac{m-n}{1-n} R_b. \end{aligned} \tag{8.15}$$

Similar to Eq. (8.5), the material parameters in the core region can be obtained by compressing the whole region into the inner core region as

$$\varepsilon''_{rr} = \varepsilon''_{\theta\theta} = \varepsilon''_{\phi\phi} = \mu''_{rr} = \mu''_{\theta\theta} = \mu''_{\phi\phi} = n/m, \tag{8.16}$$

$$\begin{aligned} \varepsilon'_{r\theta} &= \varepsilon'_{\theta r} = \mu'_{r\theta} = \mu'_{\theta r} = \varepsilon'_{r\phi} \\ &= \varepsilon'_{\phi r} = \mu'_{r\phi} = \mu'_{\phi r} = \varepsilon'_{\theta\phi} = \varepsilon'_{\phi\theta} = \mu'_{\theta\phi} = \mu'_{\phi\theta} = 0. \end{aligned} \tag{8.17}$$

Note that the complementary region is composed of the materials with anisotropic, negative material parameters due to $k_1 < 0$, while there are the materials with isotropic, positive material parameters in the core region.

As shown in Fig. 8.7, a three-dimensional double-cone complementary cloak is designed. Here $R_b(\theta, \phi) = a/(\sin\theta + |\cos\theta|)$, $R_a = 0.5R_b$, $R_c = 2R_b$, and $a = 0.6$ m. A cubic object with $\varepsilon_r = 3$ and $\mu_r = 1$ is located in the outer vacuum shell, and an anti-object is placed in the complementary shell to cloak the target cube. A line electric current source with the operating frequency of 300 MHz is used, and good invisibility can be observed with the use of the double-cone complementary cloak according to Fig. 8.7.

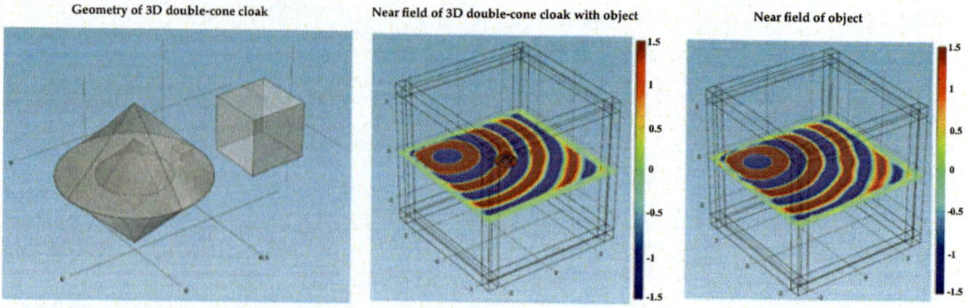

Fig. 8.7 A three-dimensional double-cone complementary cloak for the invisibility of a cubic object [22]

8.2.2 Scattering Cancellation-Based Cloak

Coordinate transformation provides an extremely intuitive visual approach to designing invisible and illusion cloaks. However, the material distributions are generally complex, being inhomogeneous and anisotropic in both ε and μ. By comparison, the scattering cancellation method can achieve electromagnetic transparency using a homogeneous shell with appropriately designed permittivity and permeability. In this section, we start from the cloaking design of the canonically shaped objects based on the Mie series approach, and then we develop a theoretical approach for achieving invisibility and illusion of three-dimensional arbitrary-shaped objects based on characteristic mode theory.

8.2.2.1 Inhomogeneous Cloak for Canonically Shaped Object

Let us consider electromagnetic scattering from an inhomogeneous sphere embedded in free space. Assume that a transverse magnetic (TM) plane wave is normally incident on an inhomogeneous target. Here the time-harmonic $e^{j\omega t}$ convention is used. The inhomogeneous sphere is divided into n thin layers of piecewise homogeneous layers. In order to realize the illusion and invisibility of the inhomogeneous object, the object is covered by a slab of homogeneous material, as shown in Fig. 8.8.

Assume that the nonmagnetic constitutive relation of each layer in spherical coordinates can be expressed as

$$\overline{\varepsilon_i} = \begin{bmatrix} \varepsilon_{ir} & & \\ & \varepsilon_{i\theta} & \\ & & \varepsilon_{i\theta} \end{bmatrix}. \tag{8.18}$$

The incident wave can be written in terms of a linear superposition of spherical waves as

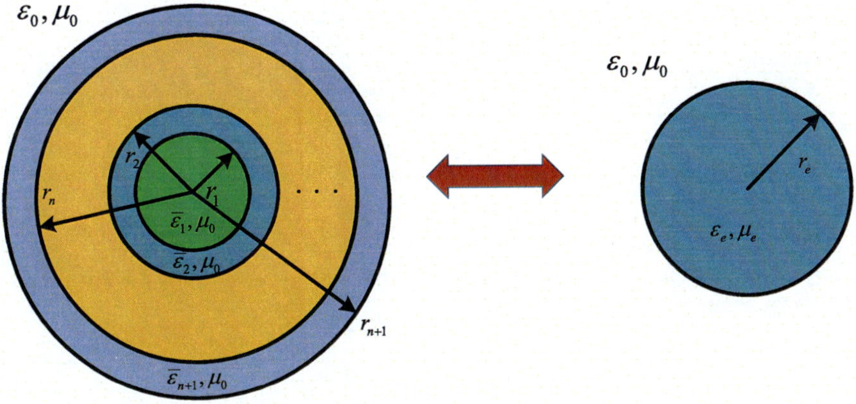

Fig. 8.8 Illusion and invisibility of an n-layer inhomogeneous sphere

$$E_\theta^{inc} = E_0 \frac{\cos\varphi}{k_0 r} \sum_{n=1}^{\infty} j^{-n}(2n+1)\hat{J}_n(k_0 r) P_n(\cos\theta), \quad (8.19)$$

in which $\hat{F}_l(x) = \sqrt{\pi x / 2} F_l(x)$ and $P_n(\cos\theta)$ is nth-order Legendre polynomial. Following the Mie series expansion procedure [43], the scattering field is solved as

$$E_\theta = E_0 \frac{\cos\varphi}{\omega} \sum_{m=1}^{\infty} j^{-m} \frac{2m+1}{m(m+1)} D_m \hat{H}_m^{(1)}(k_0 r) P_m^1(\cos\theta), \quad (8.20)$$

where D_m are the scattering coefficients, which can be determined by implementing the boundary conditions of the continuous tangential components of the electric and magnetic fields. Here $P_n^1(\cos\theta) = dP_n(\cos\theta) / d\theta$, $k_i = \omega\sqrt{\mu_0 \varepsilon_{i\theta}}$, $im = \sqrt{2m(m+1)AR_i + 0.25} - 0.5 (i = 1, 2, 3)$, and $AR_i = \varepsilon_{i\theta} / \varepsilon_{ir}$. The RCS and scattering cross section (SCS) are expressed in terms of the scattering coefficients D_m respectively as

$$\sigma = \lim_{r\to\infty}\left[4\pi r^2 \frac{|E^{sc}|^2}{|E^{inc}|^2}\right], \quad C_{sca} = \frac{2\pi}{k_0^2} \sum_{m=1}^{\infty} (2m+1)|D_m|^2. \quad (8.21)$$

To achieve the illusion, the resultant scattering fields from the coated sphere are similar to those of a homogeneous sphere, as shown in Fig. 8.8. Following the Mie series expansion method, the scattering coefficient S_m of the homogeneous sphere for the TM wave can be obtained as

$$S_m = \frac{\begin{vmatrix} \hat{J}_m(k_e r_e) & \hat{J}_m(k_0 r_e) \\ \frac{k_e}{\omega\varepsilon_e}\hat{J}'_m(k_e r_e) & \frac{k_0}{\omega\varepsilon_0}\hat{J}'_m(k_0 r_e) \end{vmatrix}}{\begin{vmatrix} \hat{J}_m(k_e r_e) & -\hat{H}_m^{(1)}(k_0 r_e) \\ \frac{k_e}{\omega\varepsilon_e}\hat{J}'_m(k_e r_e) & -\frac{k_0}{\omega\varepsilon_0}\hat{H}_m^{(1)\prime}(k_0 r_e) \end{vmatrix}}. \tag{8.22}$$

In the long wavelength limit, the SCS given by Eq. (8.21) is dominated by $m = 1$ term. Therefore, the illusion conditions for the n-layer anisotropic sphere can be obtained by setting $D_1 = S_1$. Specifically, the illusion condition for the n-layer inhomogeneous sphere can be derived as

$$\left(\frac{r_{n+1}}{r_n}\right)^{2t_{n+1}+1} = \frac{\varepsilon_{(n+1)\theta} - \frac{1}{2}(1+t_{n+1})\varepsilon_e^{(n)}}{\varepsilon_{(n+1)\theta} + \frac{1}{2}t_{n+1}\varepsilon_e^{(n)}} \cdot \frac{\begin{array}{l} r_{n+1}^3(2\varepsilon_0+\varepsilon_e)(\varepsilon_{(n+1)\theta} + \frac{1}{2}t_{n+1}\varepsilon_0) \\ +r_e^3(\varepsilon_0-\varepsilon_e)(\varepsilon_{(n+1)\theta} - t_{n+1}\varepsilon_0) \end{array}}{\begin{array}{l} r_{n+1}^3(2\varepsilon_0+\varepsilon_e)[\varepsilon_{(n+1)\theta} - \frac{1}{2}(1+t_{n+1})\varepsilon_0] \\ +r_e^3(\varepsilon_0-\varepsilon_e)[\varepsilon_{(n+1)\theta} + (1+t_{n+1})\varepsilon_0] \end{array}}, \tag{8.23}$$

in which

$$\varepsilon_e^{(n)} = \frac{2\left(\frac{r_n}{r_{n-1}}\right)^{2t_n+1} - 2\frac{\varepsilon_{n\theta} - \frac{1}{2}(1+t_n)\varepsilon_e^{(n-1)}}{\varepsilon_{n\theta} + \frac{1}{2}t_n\varepsilon_e^{(n-1)}}}{(1+t_n)\left(\frac{r_n}{r_{n-1}}\right)^{2t_n+1} + t_n\frac{\varepsilon_{n\theta} - \frac{1}{2}(1+t_n)\varepsilon_e^{(n-1)}}{\varepsilon_{n\theta} + \frac{1}{2}t_n\varepsilon_e^{(n-1)}}}\varepsilon_{n\theta}, \tag{8.24}$$

$$t_i = \sqrt{2AR_i + 0.25} - 0.5\ (i = 1, 2 \cdots n+1). \tag{8.25}$$

It is worthwhile pointing out that invisibility conditions can be easily derived according to the illusion conditions only by replacing ε_e and r_e in Eq. (8.23) by ε_0 and r_{n+1}. Especially, when $\varepsilon_{ir} = \varepsilon_{i\theta}$, the illusion and invisibility conditions given by Eq. (8.23) are reduced to the isotropic case. Similarly, we consider an n-layer inhomogeneous cylinder. In ith layer whose radius is r_i, permittivity and permeability tensors in cylindrical coordinates are expressed respectively as

$$\overline{\varepsilon_i} = \begin{bmatrix} \varepsilon_{ir} & & \\ & \varepsilon_{i\theta} & \\ & & \varepsilon_0 \end{bmatrix}, \quad \overline{\mu_i} = \begin{bmatrix} \mu_0 & & \\ & \mu_0 & \\ & & \mu_{iz} \end{bmatrix}. \tag{8.26}$$

Following the above Mie series procedure, the illusion condition for the inhomogeneous anisotropic cylinder can be derived as

$$\left(\frac{r_{n+1}}{r_1}\right)^2 = \frac{(\mu_{2z}-\mu_{1z})+(\mu_{3z}-\mu_{2z})\left(\frac{r_2}{r_1}\right)^2 + (\mu_{4z}-\mu_{3z})\left(\frac{r_3}{r_1}\right)^2+\cdots+(\mu_{(n+1)z}-\mu_{nz})\left(\frac{r_{n-1}}{r_1}\right)^2 + (\mu_e-\mu_0)\left(\frac{r_e}{r_1}\right)^2}{\mu_{(n+1)z}-\mu_0} \tag{8.27}$$

$$\left(\frac{r_{n+1}}{r_n}\right)^{2t_{n+1}} = \frac{\varepsilon_{(n+1)\theta}-t_{n+1}\varepsilon_e^{(n)}}{\varepsilon_{(n+1)\theta}+t_{n+1}\varepsilon_e^{(n)}}\frac{r_{n+1}^2(\varepsilon_{(n+1)\theta}+\varepsilon_0 t_{n+1})(\varepsilon_e+\varepsilon_0) - r_e^2(\varepsilon_{(n+1)\theta}-\varepsilon_0 t_{n+1})(\varepsilon_e-\varepsilon_0)}{r_{n+1}^2(\varepsilon_{(n+1)\theta}-\varepsilon_0 t_{n+1})(\varepsilon_e+\varepsilon_0) - r_e^2(\varepsilon_{(n+1)\theta}+\varepsilon_0 t_{n+1})(\varepsilon_e-\varepsilon_0)}, \tag{8.28}$$

in which

$$\varepsilon_e^{(n)} = \frac{\frac{\varepsilon_{n\theta}+\varepsilon_e^{(n-1)}t_n}{\varepsilon_{n\theta}-\varepsilon_e^{(n-1)}t_n} - \left(\frac{r_{n-1}}{r_n}\right)^{2t_n}}{\frac{\varepsilon_{n\theta}+\varepsilon_e^{(n-1)}t_n}{\varepsilon_{n\theta}-\varepsilon_e^{(n-1)}t_n} + \left(\frac{r_{n-1}}{r_n}\right)^{2t_n}}\frac{\varepsilon_{n\theta}}{t_n}, \quad \varepsilon_e^{(1)} = \begin{cases} \frac{\varepsilon_{1\theta}}{t_1} & \textit{the core is dielectric} \\ 0 & \textit{the core is conductor} \end{cases}. \tag{8.29}$$

The invisibility conditions of an inhomogeneous anisotropic cylinder can be obtained by substituting the equations of $\varepsilon_e = \varepsilon_0$, $\mu_e = \mu_0$, and $r_{n+1} = r_e$ into Eqs. (8.27) and (8.28). By covering a coating with the material parameters given by Eqs. (8.23), (8.27), and (8.28), the invisibility and illusion phenomena of the inhomogeneous anisotropic cylindrical and spherical objects can be achieved. On the other hand, given the radius of the coating, its different anisotropic parameters generate different invisibility and illusion effects. Hence, we can find the optimal invisibility and illusion performance of the inhomogeneous anisotropic object by optimizing AR_{n+1} of the coating. In order to measure the optimal invisibility and illusion effects, the corresponding evaluation functions are defined as

$$\sigma_1 = \frac{1}{4\pi}\int_0^{\pi}\int_0^{2\pi} |\sigma_c(\theta,\varphi)-\sigma_i(\theta,\varphi)|\sin\theta d\varphi d\theta \quad \text{for illusion,} \tag{8.30}$$

$$\sigma_2 = \frac{1}{4\pi}\int_0^{\pi}\int_0^{2\pi} |\sigma_c(\theta,\varphi)|\sin\theta d\varphi d\theta \quad \text{for invisibility.} \tag{8.31}$$

Here $\sigma_c(\theta,\varphi)$ and $\sigma_i(\theta,\varphi)$ represent the RCSs of the coated cylinder/sphere and the corresponding illusion object, respectively. Therefore, the optimal illusion and invisibility problems become

$$\min_{AR_{n+1}} \sigma_1 \quad \text{for optimal illusion,} \tag{8.32}$$

$$\min_{AR_{n+1}} \sigma_2 \quad \text{for optimal invisibility.} \tag{8.33}$$

Figure 8.9 shows the design of the illusion cloaking device for an inhomogeneous anisotropic sphere composed of three nonmagnetic homogeneous layers. The parameters of the inhomogeneous sphere are as follows: $\varepsilon_{1r} = 40\varepsilon_0/3$, $\varepsilon_{1\theta} = \varepsilon_{1\varphi} = 5\varepsilon_0$, $\varepsilon_{2r} = 2\varepsilon_0/0.28$, $\varepsilon_{2\theta} = \varepsilon_{2\varphi} = 2\varepsilon_0$, $\varepsilon_{3r} = 6\varepsilon_0/0.28$, $\varepsilon_{3\theta} = \varepsilon_{3\varphi} = 6\varepsilon_0$, $r_1 = \lambda_0/29$, $r_2 = 2\lambda_0/29$, and $r_3 = 3\lambda_0/29$. An anisotropic coating with $\varepsilon_{4r} = 9\varepsilon_0/0.055$, $\varepsilon_{4\theta} = \varepsilon_{4\varphi} = 9\varepsilon_0$, and $r_4 = 4\lambda_0/29$ is designed according to Eq. (8.23) so that the resultant RCS of the coated sphere is similar to that of an isotropic sphere with $\varepsilon_e = 4.13\varepsilon_0$ and $r_e = 4.5\lambda_0/29$. Near magnetic field distributions and RCSs at 500 MHz in XOY and XOZ planes between the coated inhomogeneous sphere and the illusion sphere are given to demonstrate the good illusion effect.

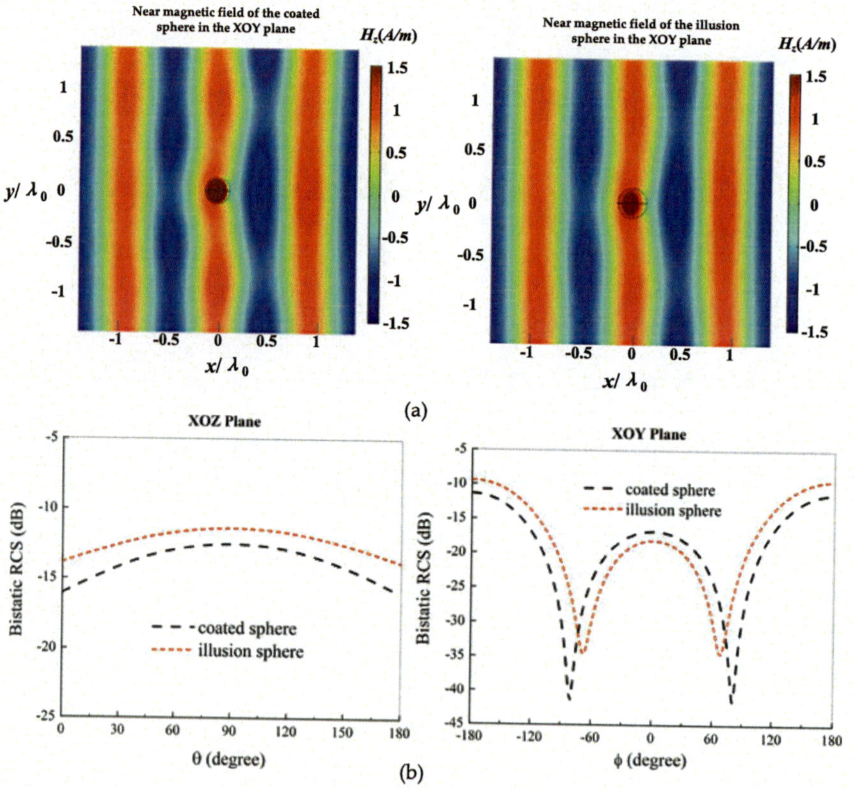

Fig. 8.9 Illusion for a three-layer inhomogeneous sphere. **a** magnetic field distribution at 500 MHz. **b** bistatic RCSs at 500 MHz in XOY and XOZ planes [34]

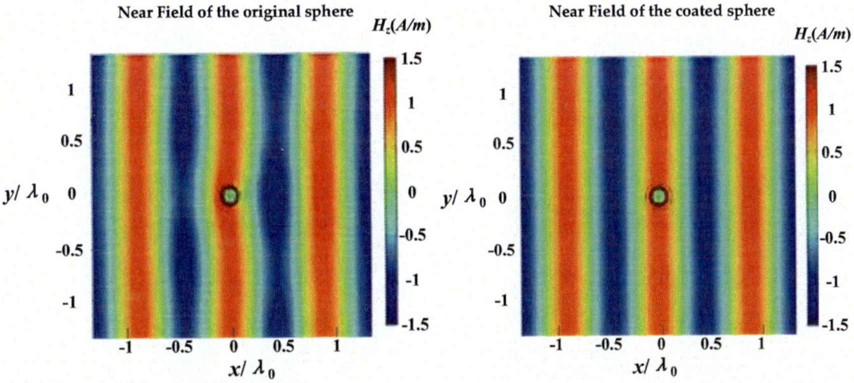

Fig. 8.10 Invisibility for a two-layer inhomogeneous sphere [34]

Next, an invisibility cloak of a two-layer inhomogeneous sphere with a PEC core is designed. The parameters of the inhomogeneous sphere are $\varepsilon_{2r} = 35\varepsilon_0/1.68$, $\varepsilon_{2\theta} = \varepsilon_{2\varphi} = 35\varepsilon_0$, $r_1 = \lambda_0/14$, and $r_2 = 1.5\lambda_0/14$. To reduce the RCS of the inhomogeneous sphere, a shell with $\varepsilon_{3r} = 0.34\varepsilon_0$, $\varepsilon_{3\theta} = \varepsilon_{3\varphi} = 0.24\varepsilon_0$, and $r_3 = 2\lambda_0/14$ is designed according to Eq. (8.23). Figures 8.10 and 8.11 illustrate the near magnetic field distributions at 600 MHz in XOY plane and RCSs at 600 MHz in XOZ and XOY planes for TM and transverse electric (TE) polarized incident waves. It can be seen that with the designed coating, the invisibility of the inhomogeneous sphere is achieved.

In order to further decrease the RCS of the inhomogeneous sphere, the electric anisotropy ratio AR of the coating is optimized according to Eq. (8.33). It can be seen from Fig. 8.12 that the smallest RCS can be obtained when *AR* equals 3.01, which corresponds to the minimal σ_2. Compared with the previous design with $AR = 0.71$, RCS with the optimized *AR* decreases 20 dB at $\theta = 0°$ and 54 dB at $\theta = 180°$.

8.2.2.2 Cloak for Arbitrarily Shaped Object

Analytical expressions for the Mie scattering coefficients are limited to canonical shapes, such as an infinite cylinder and a sphere. In this section, we develop the analytical formulas for illusion and invisibility cloaking devices of arbitrarily shaped objects.

To solve the radiation and scattering problems of an object with arbitrary shape and relative permittivity of ε_r, we start from the discrete dipole approximation (DDA) method [44–49]. The dielectric scatterer is discretized into *N* small cubic elements with the size length of *d*. Each element is approximately regarded as a point dipole

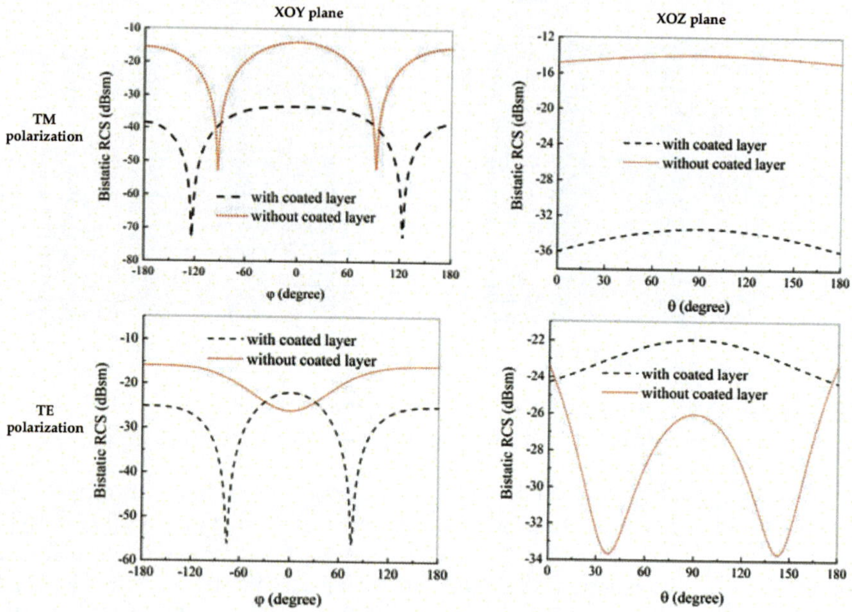

Fig. 8.11 Comparison of bistatic RCS between the coated and uncoated spheres [34]

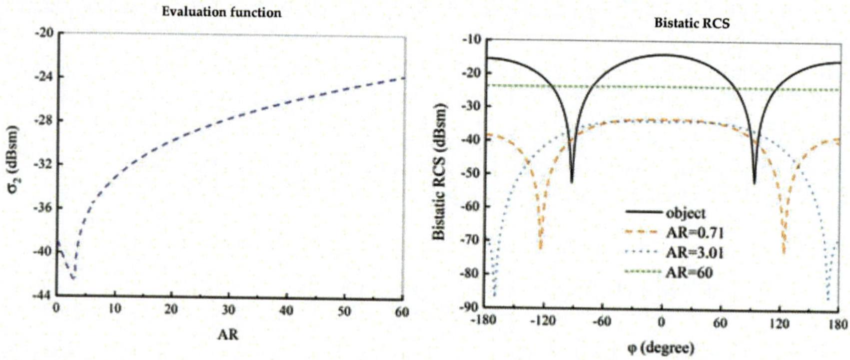

Fig. 8.12 Optimal invisibility effect of an inhomogeneous sphere [34]

with the dipole moment $\vec{p}_k$. At each dipole, there are field contributions from other re-radiating dipoles in addition to the incident field $\vec{E}^{inc}$. Therefore, we have

$$\vec{E}_i^{exc} = \vec{E}_i^{inc} + \sum_{\substack{k=1 \\ k \neq i}}^{N} \overline{G}\left(\vec{r}_i, \vec{r}_k \cdot \vec{p}_k\right) \quad (i = 1, \ldots, N), \tag{8.34}$$

in which

$$\overline{G}\left(\vec{r},\vec{r'}\right)=[k^2\overline{I}+\nabla\nabla]\frac{e^{-jkR}}{R}$$
$$=\frac{e^{-jkR}}{R}\left[k^2(\overline{I}-\frac{\widehat{R}\widehat{R}}{R^2})-\frac{1+jkR}{R^2}\left(\overline{I}-3\frac{\widehat{R}\widehat{R}}{R^2}\right)\right], \quad (8.35)$$

$$\hat{R}=\vec{R}/R=\left(\vec{r}-\vec{r'}\right)/\left(\left|\vec{r}-\vec{r'}\right|\right). \quad (8.36)$$

The field $\vec{E_i}^{exc}$ at ith dipole results in a dipole moment $\vec{p_i}$ with the polarizability α_i, which is given by

$$\vec{p_i}=\alpha_i\vec{E_i}^{exc}. \quad (8.37)$$

Substituting Eq. (8.37) into Eq. (8.34), we can set up a system of 3N complex linear equation

$$\overline{A}\cdot\overline{p}=\overline{E}^{inc}, \quad (8.38)$$

in which $\overline{A}$ is the impedance matrix with the diagonal element of $\alpha_{m'(m)}^{-1}$. Here $m'(m)$ is the index of the cubic element corresponding to the mth row of the matrix system. There are some formulations for the calculation of the dipole polarizability α_i. The well-known dipole polarizability is Clausius–Mossotti (CM) polarizability, i.e.,

$$\alpha_i^{CM}=\frac{3d^3}{4\pi}\frac{\varepsilon_i-1}{\varepsilon_i+2}. \quad (8.39)$$

The CM polarizability is valid in the infinite wavelength limit. The polarizability called lattice dispersion relation (LDR) is regarded as the CM polarizability plus high-order corrections with the following expression

$$\alpha_i^{LDR}=\frac{\alpha_i^{CM}}{1+(\alpha_i^{CM}/d^3)\left[(b_1+m^2b_2+m^2b_3S)(kd)^2+j(2/3)(kd)^3\right]}, \quad (8.40)$$

in which $b_1=-1.8915316$, $b_2=0.1648469$, $b_3=-1.7700004$, and S is a function of the propagation and polarization of the incident wave, which is given as

$$S=(u_x^{inc}k_x^{inc})^2+(u_y^{inc}k_y^{inc})^2+(u_z^{inc}k_z^{inc})^2, \quad (8.41)$$

where k_t^{inc} and u_t^{inc} $(t=x,y,z)$ are components of the incident vector $\vec{k}^{exc}$ and the polarization vector $\vec{u}^{inc}$ of the incident field.

Once the dipole moments are obtained from Eq. (8.38), the RCS for the scattering vector $\vec{k}^{sca} = k\hat{k}^{sca}$ can be calculated as

$$\sigma = \lim_{r\to\infty} 4\pi r^2 \left|\frac{\vec{E}^{sca}}{\vec{E}^{inc}}\right|^2 = 4\pi k^4 \left|\sum_{m=1}^{N} e^{jk\vec{r}_m\cdot\hat{k}^{sca}} \vec{p}_m\right|^2. \tag{8.42}$$

To provide deep insights into the analysis of scattering and radiation problems, the characteristic mode (CM) method [50–52] is used. The matrix $\overline{A}$ in Eq. (8.38) is split into its real and imaginary parts as

$$\overline{A} = \overline{R} + j\overline{X}. \tag{8.43}$$

If $\overline{X}$ is not zero, following Harrington and Mautz's procedure, we introduce the following generalized eigenvalue equation

$$\overline{A}\cdot\overline{q}_n = \nu_n \overline{R}\cdot\overline{q}_n, \tag{8.44}$$

in which ν_n are eigenvalues and $\overline{q}_n$ are eigenfunctions. Combining Eqs. (8.43) and (8.44), we have

$$\overline{X}\cdot\overline{q}_n = \lambda_n \overline{R}\cdot\overline{q}_n, \tag{8.45}$$

with $\nu_n = 1 + j\lambda_n$.

If $\overline{X}$ is equal to zero, the generalized eigenvalue equation is rewritten as

$$\overline{R}\cdot\overline{q}_n = \lambda_n\overline{q}_n. \tag{8.46}$$

Since $\overline{R}$ and $\overline{X}$ are real symmetric matrices, all λ_n and $\overline{q}_n$ are real. Moreover, the eigenfunctions $\overline{q}_n$ are orthogonal with respect to the matrices $\overline{R}$, $\overline{X}$, and $\overline{A}$. When the eigenfunctions $\overline{q}_n$ are normalized with respect to the matrix $\overline{R}$, we have

$$\overline{q}_m^T\cdot\overline{A}\cdot\overline{q}_n = \begin{cases}(1+j\lambda_n)\delta_{mn} & \overline{X}\neq 0\\ \lambda_n\delta_{mn} & \overline{X}=0\end{cases}. \tag{8.47}$$

Therefore, we shall call $\overline{q}_n$ the characteristic dipole moments. The dipole moments due to the impressed field $\vec{E}^{inc}$ can be written as a linear superposition of the characteristic dipole moments as

$$\overline{p} = \sum_l \alpha_l \overline{q}_l, \tag{8.48}$$

where the modal expansion coefficients α_l can be calculated as

$$\alpha_l = \begin{cases} \dfrac{V_l^{inc}}{1 + j\lambda_l} & \overline{X} \neq 0 \\ \dfrac{V_l^{inc}}{\lambda_l} & \overline{X} = 0 \end{cases}. \tag{8.49}$$

Here $V_l^{inc} = \overline{q}_l^T \cdot \overline{E}^{inc}$. Inserting Eq. (8.49) into Eq. (8.48), the RCS can be re-expressed as

$$\sigma = \begin{cases} 4\pi k^4 \left| \sum_l \dfrac{V_l^{inc} V_l^{sca}}{1 + i\lambda_l} \right|^2 & \overline{X} \neq 0 \\ 4\pi k^4 \left| \sum_l \dfrac{V_l^{inc} V_l^{sca}}{\lambda_l} \right|^2 & \overline{X} = 0 \end{cases}. \tag{8.50}$$

According to Eq. (8.50), we can know that the eigenvalue λ_n ranging from $-\infty$ to $+\infty$ is very important for the scattering phenomenon. Eigenfunctions with $\lambda_n < 0$ store electric energy, whereas eigenfunctions with $\lambda_n > 0$ store magnetic energy. Eigenfunctions with $\lambda_n = 0$ scatter most efficiently, while eigenfunctions with $\lambda_n = \infty$ become trivial modes which does not result in scattering problem.

To make dielectric object with an arbitrary shape and relative permittivity ε_1 disguised as another object with relative permittivity ε_e, a dielectric shell with relative permittivity ε_2 covers the object so that the shape of the resultant coated object is same as that of the illusion object, as shown in Fig. 8.13.

According to the DDA method and the CM approach, the scattering fields from the coated object and the illusion object are the same when the eigenvalues λ_n for the coated object are the same as those for the illusion object. However, this rigorous condition is almost impossible to achieve. A reasonable approximate to the rigorous condition is to keep the summation of all eigenvalues the same for the coated and illusion objects, which means that the trace of the system matrices for the coated object is the same as that of the illusion object. If we ignore the non-diagonal elements in the system matrices for the coated and illusion objects, we can obtain

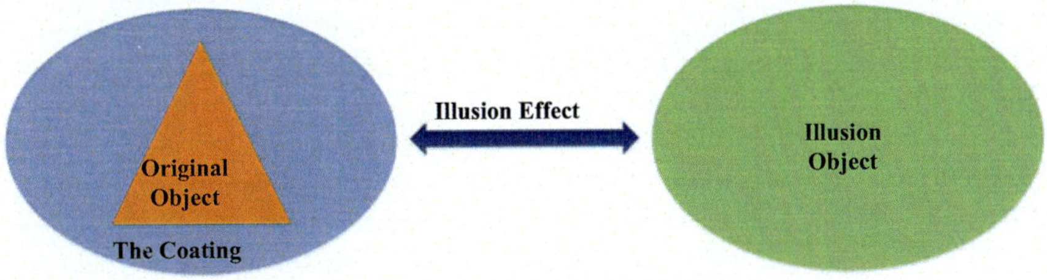

Fig. 8.13 Schematic diagram of the illusion cloak

$$\frac{N_1}{\beta_1} + \frac{N_2}{\beta_2} = \frac{N}{\beta_e}, \tag{8.51}$$

in which

$$\beta_s = \frac{4\pi}{3d^3}\frac{\varepsilon_s + 2}{\varepsilon_s - 1} + \frac{k^2}{d}(b_1 + b_2\varepsilon_s) \quad (s = 1, 2, e). \tag{8.52}$$

Here N is the total number of the dipole moments for both the coated and illusion objects, and N_1 and N_2 are the numbers of the dipole moments for the original object and the coating shell, respectively.

Solving Eq. (8.51), the relative permittivity of the coating shell can be obtained as

$$\varepsilon_2 = \frac{3\beta_2 d^3 - 3k^2(b_1 - b_2)d^2 - 4\pi}{6b_2k^2d^2} - \frac{1}{6b_2k^2d^2}\left\{9d^6\begin{bmatrix}(b_1+b_2)k^2d^2\\ -18(b_1+b_2)k^2\beta_2 d^5\\ +\pi k^2d^2(24b_1 - 120b_2)\\ +9\beta_2^2 - 24\pi\beta_2 d^3 + 16\pi^2\end{bmatrix}\right\}^{\frac{1}{2}}, \tag{8.53}$$

in which

$$\beta_2 = \frac{N_2\beta_1\beta_e}{(N\beta_1 - N_1\beta_e)}. \tag{8.54}$$

The invisibility condition can be obtained when the relative permittivity ε_e of the illusion object in Eq. (8.53) is set as 1.

Consider a goblet-shaped object with relative permittivity of $\varepsilon_1 = 2.5$, as shown in Fig. 8.14. In order to make the object invisible, a coating with the relative permittivity of $\varepsilon_2 = 0.76$ is designed according to Eq. (8.53). As shown in Fig. 8.14a, with the designed coating, the RCS reductions of 5.8 dB at $\phi = 0^\circ$ in the XOY plane and at $\theta = 90^\circ$ in the XOZ plane are obtained. To further reduce the scattering field of the coated object, the optimization procedure given by Eq. (8.33) is implemented to find the relative permittivity of the coating for minimizing the RCS. The optimized relative permittivity of the coating is $\varepsilon_2 = 0.45$. By covering the optimized coating, the RCS reductions of 14 dB at $\phi = 0^\circ$ in the XOY plane and at $\theta = 90^\circ$ in the XOZ plane are achieved. Figure 8.14b illustrates the comparison of the near-field distribution between the sole object and the coated object with the optimized coating. A good invisibility is observed. Figure 8.15 shows the RCS comparisons between the sole object and the coated object with the optimized coating for the obliquely incident plane waves with TE and TM polarizations, respectively. It can be seen that whatever the polarization of the incident wave is, the RCS of the object at $\theta = 0^\circ$ is reduced over 10 dB.

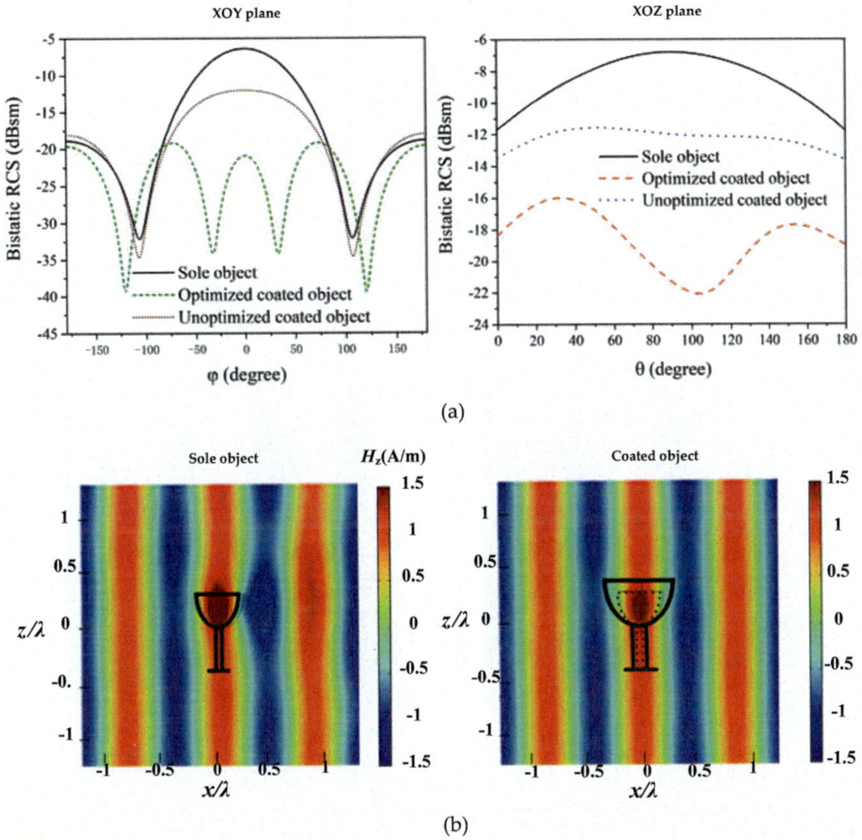

Fig. 8.14 The invisibility of a goblet-shaped object. **a** bistatic RCS comparison between the coated object and the sole object. **b** near-field distribution comparison between the sole object and the coated object [36]

8.3 Metasurface-Based RCS Reduction of Antennas

8.3.1 Microwave Absorber Designs Based on Metasurfaces

A perfect metamaterial absorber with excellent absorption and ultrathin microstructure in Fig. 8.16 was first proposed and validated by Landy et al. in 2008 [53]. The PMA displays great absorption capability. The designed idea of the PMA is to adjust the effective $\varepsilon(\omega)$ and $\mu(\omega)$ independently by varying the dimensions of the electric resonant component and magnetic resonant component in the unit cell to match the effective impedance of PMA to free space and achieve a large resonant dissipation at the meantime. Thus, wave transmission and reflection are minimized simultaneously, and absorption is maximized. Since then, the PMA has become a

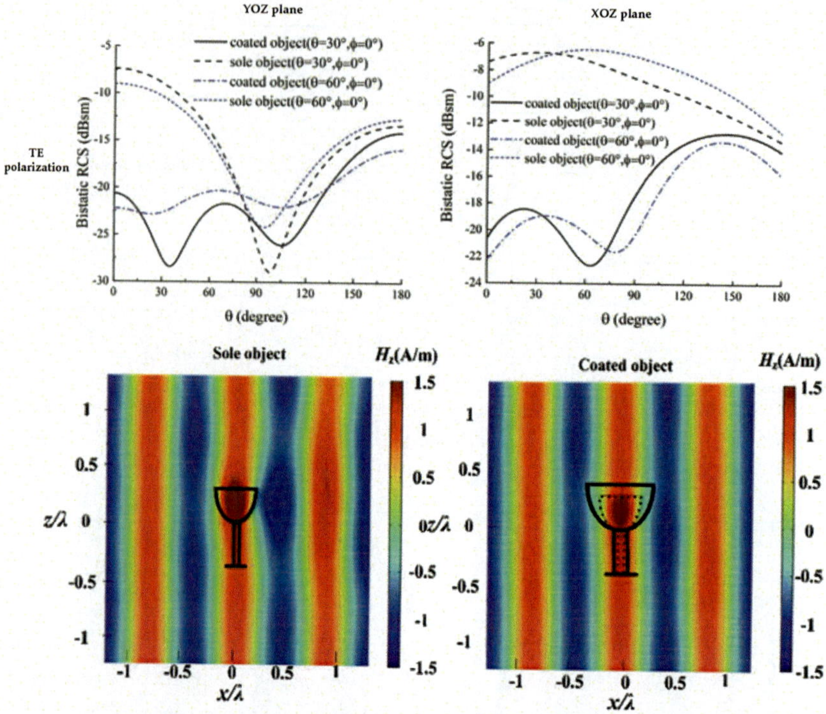

Fig. 8.15 RCS comparison between the coated object and the sole object for obliquely incident wave [36]

potential aspect in the research of metamaterials, and its research was flourishing to achieve wide incident angle absorption [54], polarization-insensitive absorption [55], multi-band absorption [56], broadband absorption [57], and tunable absorption [58]. Several techniques have been devoted to broadening the absorption bandwidth of PMA. For example, the bandwidth of microstructure can be broadened by superimposing the different resonance modes of the metamaterial absorber array with the aperiodic arrangement. Moreover, the fractal and multi-layer microstructures have been demonstrated to increase the bandwidth for PMA. Furthermore, the magnetic medium and hybrid substrate entirely absorbed the electromagnetic wave in a wide bandwidth due to producing an analog of electromagnetically induced transparency. As efficacious techniques, loading lumped elements, plasmonic Brewster funneling, and strong coupling effects combined with the metamaterial have been applied to design the broadband metamaterials.

Some kinds of representative microwave absorber designs based on metasurfaces are introduced here. Through the refined design of the planar topology of the metasurface, the PMAs could realize the same and even better microwave absorption performance and achieve wide-angle and polarization-insensitive characteristics meanwhile. Additionally, the defaults of the conventional absorbers, such as

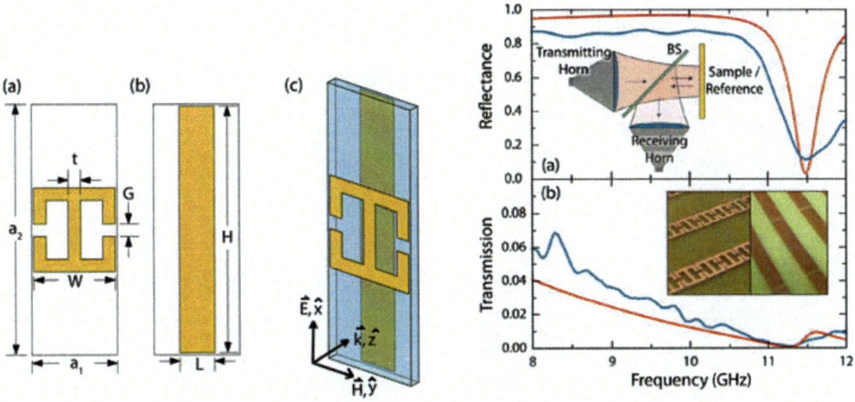

Fig. 8.16 Perfect metamaterial absorber and its performance [53]

the narrowband issue of the Salisbury screen and the high-profile issue of Jaumann absorbers, could be improved by introducing the structural design of the metasurface. The reflectance and transmission of microwave absorber at a certain frequency ω can be defined as $R(\omega) = |S_{11}|^2$, $T(\omega) = |S_{21}|^2$, respectively. The absorptivity could be calculated by

$$A(\omega) = 1 - R(\omega) - T(\omega) \tag{8.55}$$

where the $T(\omega) = |S_{21}|^2$ could be regarded as 0 when the proposed absorber is backed by a metallic sheet.

8.3.1.1 Metamaterial Absorber with a Tetra-Arrow Resonator Structure

Firstly a new metamaterial absorber with a tetra-arrow resonator (TAR) structure that can operate at three different resonant modes is introduced [59]. The compact metamaterial absorber with a simple geometry consists of two metallic layers separated by a lossy dielectric spacer, as shown in Fig. 8.17a. The top layer is set in a periodic pattern of TAR, and the bottom one is solid metal. Interacting with electromagnetic waves, three different resonant modes of the TAR absorber can be excited, respectively. A full-wave electromagnetic simulation was performed using an infinite period model shown in Fig. 8.17b. Therefore, by adjusting the structure parameters of TAR, we can obtain a dual-band, polarization-insensitive, wide incident angle, and ultra-thin ($\lambda/69$ at low-frequency resonance and $\lambda/54$ at high-frequency resonance) absorber whose absorptivity is near perfect at 6.16 GHz and 7.9 GHz. Moreover, a single band ultra-miniature absorber can be achieved, whose cell periodic length is about $\lambda/14$ and thickness only $\lambda/74$. Its absorptivity also comes near perfection at the lowest resonant modes (2.06 GHz). The simulated results of tri-band absorption can be found in Fig. 8.18a. Figures 8.18b and 8.19 show the wide-angle characteristic and the polarization-insensitive characteristic of the TAR absorber by full-wave simulation.

The surface current and equivalent circuit analyses are introduced in detail in [59]. When the surface impedance of the absorber $Z_{in} = \eta_0(1 + S_{11})/(1 - S_{11})$

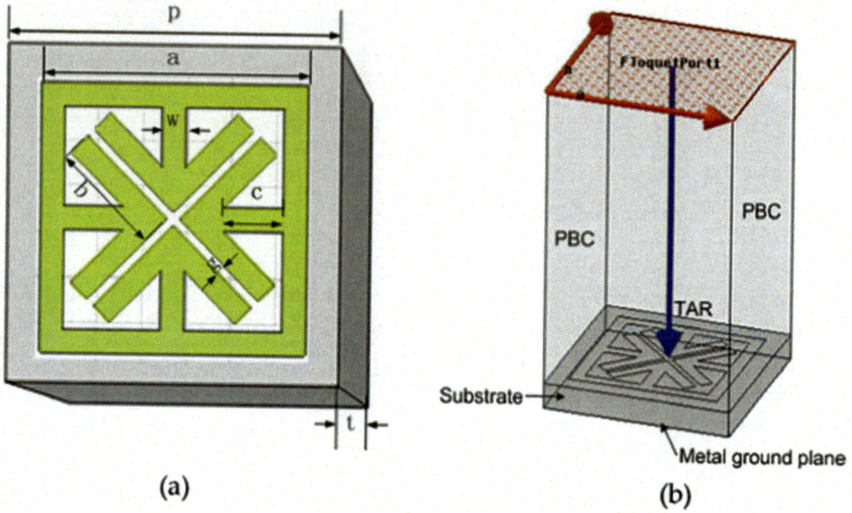

Fig. 8.17 **a** Geometry of the TAR absorber unit cell, **b** infinite periodic model based on finite element algorithm used for calculating the absorptivity, in which periodic boundary conditions (PBC) are placed around TAR cell to model infinite TAR absorber [59]

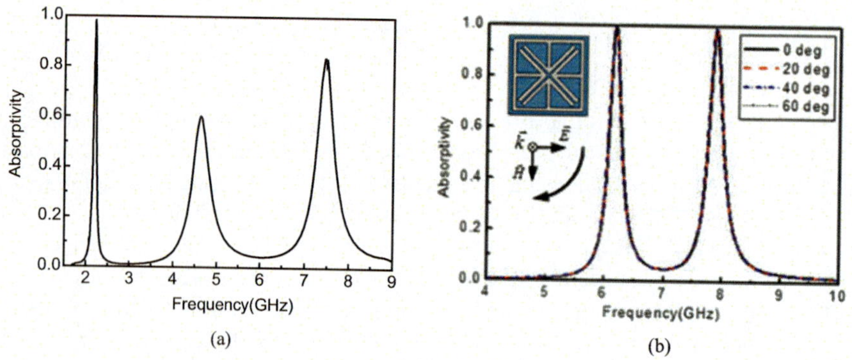

Fig. 8.18 **a** Simulated absorptivity of the TAR absorber with three resonant modes. The structure parameters of the absorber are $p = 10.2$, $a = 10$, $b = 5.5$, $c = 3.34$, $w = 0.5$, $g = 0.28$, and $t = 2$ (Unit: mm); **b** simulated absorptivity of the TAR absorber with dual-band and ultra-thin configuration. The parameters of the absorber are $p = 12$, $a = 10$, $b = 5.5$, $c = 3.34$, $w = 0.5$, $g = 0.28$, and $t = 0.7$ (Unit: mm). Simulated absorption for different polarization angles for the normal incidence, which shows that the TAR structure is polarization insensitive [59]

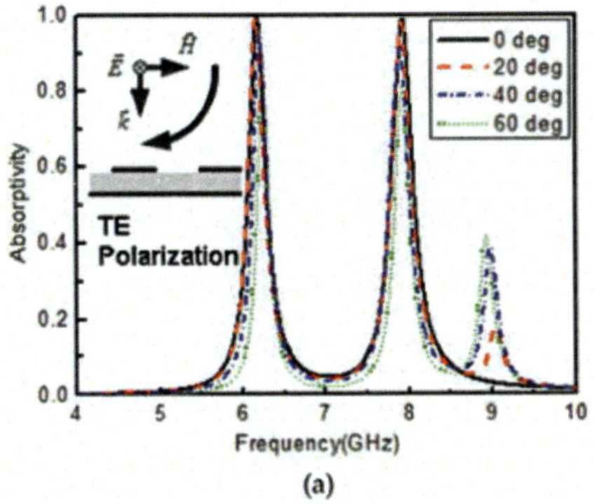

Fig. 8.19 **a** Simulated absorptivity for different incidence angles for the TE polarization. **b** simulated absorptivity for different incidence angles for the TM polarization [59]

is matched to the free-space impedance η_0, as shown in Fig. 8.20a, the incidence wave can get into the absorber without reflection and be dissipated in the lossy dielectric substrate. The contribution of dielectric loss in the dielectric substrate for absorption is more significant than that of the ohmic losses in metal, which is verified in Fig. 8.20b. The prototype and the measurement results of the two cases are depicted in Fig. 8.21, from which we can find that more than 90% absorptivity could be achieved at three resonant modes. The good agreement of the simulation with the experiment results verifies the practicability and credibility of the proposed TAR absorber. The frequency shift of the absorptivity peak is mainly due to the mismatching tolerance.

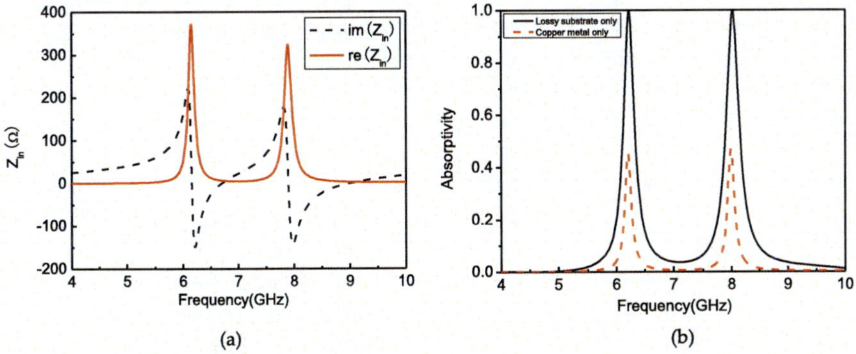

Fig. 8.20 **a** Simulated effective surface impedance of the dual-band absorption; **b** comparison of the contribution of losses for the absorptivity of the TAR absorber in dual-band absorbing cases: one is with a lossy substrate ($\varepsilon_r = 4.4$ and $\tan\delta = 0.02$) and lossless metal (perfect electric conductor), and the other is with lossy metal (Copper with $\sigma = 5.8 \times 10^7\ S/m$) and lossless substrate ($\varepsilon_r = 4.4$ and $\tan\delta = 0$) [59]

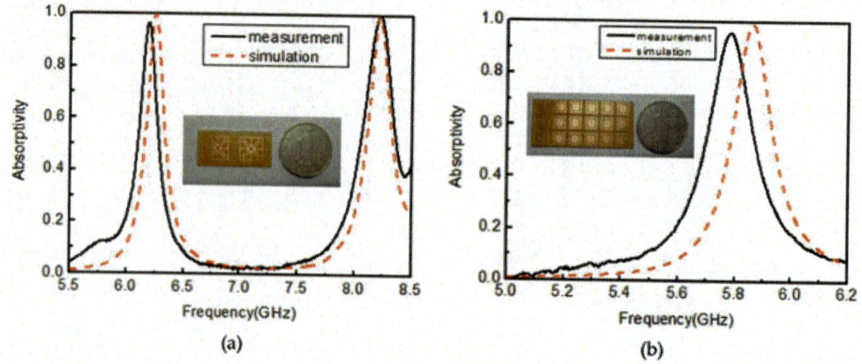

Fig. 8.21 Simulated and measured absorptivity for the TAR absorbers with **a** dual-band operation. The parameters of the TAR unit cell are $p = 15.8$ mm, $a = 11$ mm, $b = 5.6$ mm, $c = 3.5$ mm, $w = 0.55$ mm, $g = 0.57$ mm, $t = 1$ mm; **b** single-band TAR unit cell is ultra-miniature, whose parameters are $p = 7.38$ mm, $a = 5.8$ mm, $b = 2.4$ mm, $c = 1.29$ mm, $w = 0.5$ mm, $g = 0.21$ mm, $t = 1$ mm [59]

8.3.1.2 An Ultra-Wideband, Polarization-Insensitive, and Wide-Angle Thin Absorber

An ultra-wideband, polarization-insensitive, and wide-angle thin absorber is proposed, which consists of a three-layer resistive metasurface with three resonant modes [60]. The total thickness of the designed absorber is 3.8 mm, which is only 0.09λ at the lowest frequency. The bandwidth of the absorption ratio over 90% is from 7.0 to 37.4 GHz. The fractional absorption bandwidth is about 137%. As shown in Fig. 8.22, the proposed metamaterial absorber element is composed of three-layer resistive films, three-layer dielectric substrates, and a metallic ground plane. The metallic ground plate is made of copper with a conductivity of $\sigma = 5.8 \times 10^7$ s/m and a thickness of 0.017 mm. These three-layer dielectric substrates are F4B-2 substrates, which have a relative permittivity of 2.65 and a loss tangent of 0.001. The surface square resistance value of the resistive metasurface is $R_s = 100$ Ω/sq. The optimized parameters of the metamaterial absorber in full-wave simulation with the periodic boundary conditions (PBCs) and Floquet port by High-Frequency Structure Simulator (HFSS) are listed in [60].

The three-layer structure of the metamaterial absorber is shown in Fig. 8.23a. There is a similarity between the behavior of the propagation of uniform plane waves at multilayer interfaces and that of multistage transmission lines. Hence, the equivalent transmission line circuit model is used to explain the absorbing mechanism, as shown in Fig. 8.23b. The subwavelength periodic resistive metasurfaces can be equivalent to serial RLC circuits, and the substrates are equivalent to transmission lines. Three serial resonant modes can be optimized for ultra-wideband absorbing design.

According to the equivalent circuit model shown in Fig. 8.23b, we can obtain the total reflection coefficient as follows:

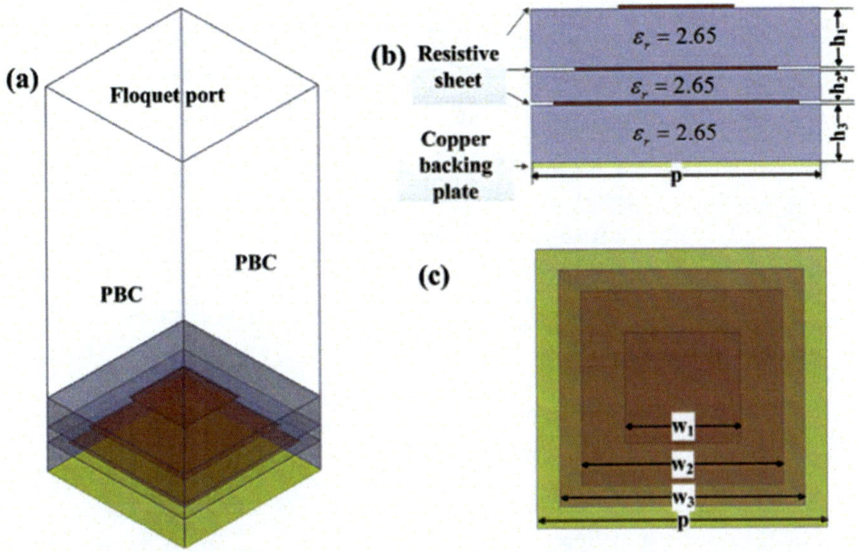

Fig. 8.22 Geometry of the proposed metamaterial absorber unit cell: **a** perspective view, **b** side view, and **c** top view [60]

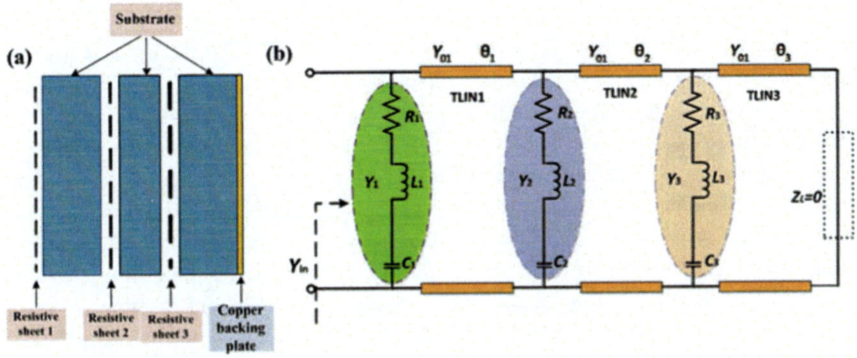

Fig. 8.23 **a** Three-layer absorbing structure and **b** equivalent transmission line circuit of the proposed absorber with three resonant modes [60]

$$\Gamma = \frac{Y_0 - Y_{in}}{Y_0 + Y_{in}} \tag{8.56}$$

where Y_0 is the wave admittance of free space, and Y_{in} is the total equivalent admittance for the three-layer absorber, which can be described by the equivalent network parameters $[A]$ as

$$Y_{in} = \frac{A_{21}Z_L + A_{22}}{A_{11}Z_L + A_{12}} = \frac{A_{22}}{A_{12}} \quad (Z_L = 0) \tag{8.57}$$

By cascading each $[A]$, we can obtain the total network parameter $[A]$ as follows:

$$A = \begin{bmatrix} 1 & 0 \\ Y_1 & 1 \end{bmatrix} \begin{bmatrix} \cos\theta_1 & \frac{j\sin\theta_1}{Y_{01}} \\ jY_{01}\sin\theta_1 & \cos\theta_1 \end{bmatrix} \begin{bmatrix} 1 & 0 \\ Y_2 & 1 \end{bmatrix}$$
$$\begin{bmatrix} \cos\theta_2 & \frac{j\sin\theta_2}{Y_{01}} \\ jY_{01}\sin\theta_2 & \cos\theta_2 \end{bmatrix} \begin{bmatrix} 1 & 0 \\ Y_3 & 1 \end{bmatrix} \begin{bmatrix} \cos\theta_3 & \frac{j\sin\theta_3}{Y_{01}} \\ jY_{01}\sin\theta_3 & \cos\theta_3 \end{bmatrix}$$
$$= \begin{bmatrix} A_{11} & A_{12} \\ A_{21} & A_{22} \end{bmatrix} \tag{8.58}$$

in which

$$Y_i = B_i + jD_i = \frac{1}{R_i + j(\omega L_i - 1/\omega C_i)} \tag{8.59}$$

$$B_i = \frac{R_i\omega^2 C_i^2}{R_i^2\omega^2 C_i^2 + \left(\omega^2 L_i C_i - 1\right)^2} \tag{8.60}$$

$$D_i = -\frac{\omega C_i\left(\omega^2 L_i C_i - 1\right)}{R_i^2\omega^2 C_i^2 + \left(\omega^2 L_i C_i - 1\right)^2} \tag{8.61}$$

and $Y_{01} = Y_0\sqrt{\varepsilon_r}$,$\theta_i = \beta_i h_i$, $\beta_i = 2\pi\sqrt{\varepsilon_r}/\lambda$, $i = 1, 2, 3....$ Using the Eqs. (8.56–8.61), we can minimize the total reflection coefficient (Γ) by optimizing the lumped parameters (R_i, L_i, C_i). The equivalent circuit model can be established and simulated by using Advanced Design System (ADS). The total equivalent impedance is calculated and shown quantitatively in Fig. 8.24a. It can be seen that the equivalent impedance of the multi-layer absorber approaches the wave impedance of free space in the frequency range from 7.0 to 37.4 GHz. By comparing the reflection coefficients calculated by HFSS full-wave simulation and the ADS circuit simulation, as shown in Fig. 8.24b, we can see that the equivalent circuit model of three resonant modes is very effective for the proposed ultra-wideband absorber. The results show that they are in good agreement, and three resonant frequencies of 8.4, 24.1, and 34.3 GHz are also observed in Fig. 8.24b. The simulated absorptivity is also given in Fig. 8.24b. It can be seen that the bandwidth of absorptivity over 90% is from 7.0 to 37.4 GHz, and the fractional absorption bandwidth is about 137%. It can be seen that the three-layer resistive metasurface can broaden the absorption bandwidth and adjust the impedance matching. The multi-mode impedance matching theory can be used to explain the wideband absorbing mechanism.

In order to verify the absorptivity of the design, we fabricated the absorber prototype with 160 mm × 160 mm, as shown in Fig. 8.25. The resistive films with a thickness of 0.03 mm were manufactured on the dielectric substrate by the silk printing technique. The comparison of the measured and simulated reflection coefficients of the absorber is shown in Fig. 8.25. It can be seen that they are in good agreement.

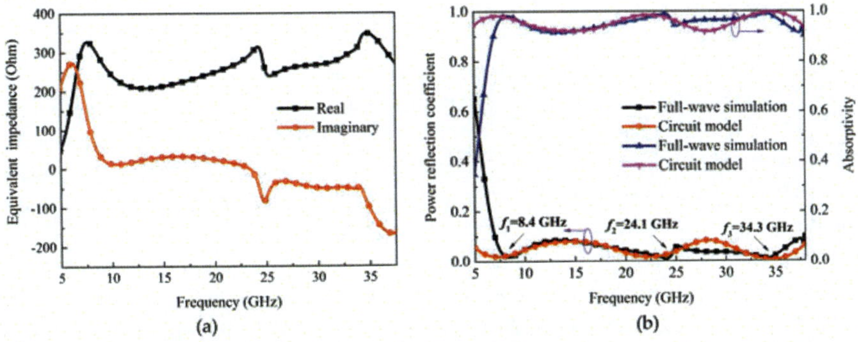

Fig. 8.24 **a** Simulated equivalent impedance of the designed multilayer absorber; **b** comparison of the power reflection coefficient and absorptivity of the equivalent circuit model and full-wave simulation for the designed ultra-wideband absorber [60]

Fig. 8.25 Measured and simulated reflection coefficients of the metamaterial absorber [60]

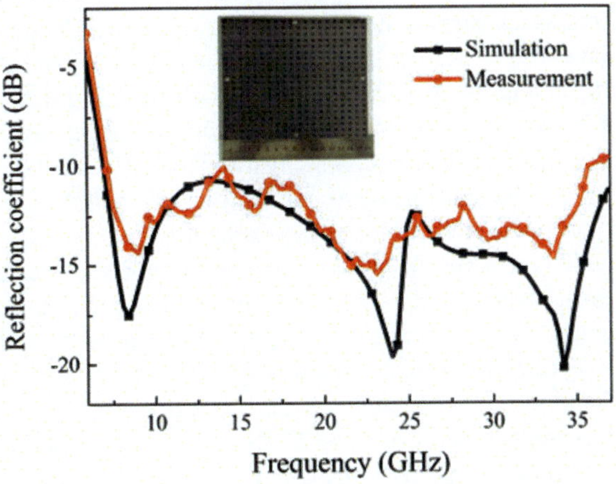

The measured −10 dB reflection coefficient bandwidth, which corresponds to the bandwidth of absorptivity of more than 90%, is in the range of 7.2–35.7 GHz.

8.3.1.3 A Broadband Polarization-Independent and Low-Profile Optically Transparent Metamaterial Absorber

Following the multi-layer metasurface absorber made of conventional metal, e.g., copper, a broadband polarization-independent and low-profile optically transparent metamaterial absorber is designed, consisting of three-layer indium tin oxide (ITO) structure and two-layer soda-lime glass substrate. Taking advantage of the high optical transmittance of the soda-lime glass substrate, the proposed structure has a measured visible light transmittance of 86%, ultraviolet transmittance of 52%, and infrared transmittance of 98%. The total thickness of the designed absorber is

3.8 mm (0.086λ at the lowest operational frequency), which is a very low profile. The bandwidth with absorptivity larger than 85% is from 6.1 to 22.1 GHz.

As shown in Fig. 8.26, the proposed transparent metamaterial absorber element is composed of two-layer dielectric substrates and a three-layer ITO structure, and the three-layer ITO structure is printed on the dielectric substrates. The two-layer dielectric substrates are soda-lime glass designed with a relative permittivity of 5.5, and the thicknesses of the two-layer dielectric substrates are h_1 and h_2, respectively. The three-layer ITO structure consists of three layers of ITO, in which layer 1 possesses the resistance R_{S1}, layer 2 possesses the resistance R_{S2}, and layer 3 possesses the resistance R_{S3}. Layer 1 designed as a square patch with the side length w_1 is printed on the top of the upper substrate. Layer 2 designed as a square patch etched by a cross is printed on the bottom of the upper substrate, in which the square patch is with a side length w_2 and the cross is with a gap g. Layer 3 designed as a large square patch with the side length p is printed on the bottom of the lower substrate. The 1st, 2nd, and 3rd ITO layers are designed as a single-layer impedance surface without thickness and attached to the glass substrates. It should be noted that the 2nd ITO layer can be seen as being sandwiched between the 1st and 2nd glasses, but there are no air gaps between the 1st and 2nd glasses. In the fabrication, the thicknesses of the ITO layers are 10 nm, 23 nm, 185 nm for sheet resistance $R_{S1} = 300\ \Omega/\text{sq}$, $R_{S2} = 80\ \Omega/\text{sq}$, and $R_{S3} = 6\ \Omega/\text{sq}$, respectively. The ITO layers are etched and embedded in the soda-lime glass to maintain a smooth flat interface between the 1st and 2nd glasses during the fabrication process.

To verify the absorptivity of the design, we fabricated the absorber prototype with dimensions of 280 mm × 280 mm, as shown in Fig. 8.27. With the application of the time-domain door, open field test mode was used, as shown in Fig. 8.27. An Anritsu Shockline MS46322A vector analyzer and two pairs of standard gain horn antennas (working in 2–18 GHz, 18–26.5 GHz) are used to cover the operating broadband frequency range from 5 to 25 GHz. The comparison of the measured and simulated S-parameters of the transparent absorber is shown in Fig. 8.28. It can be seen that

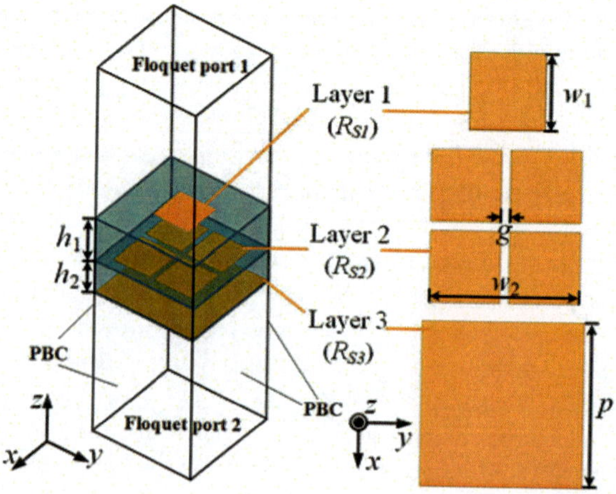

Fig. 8.26 Geometry of the proposed metamaterial absorber unit cell [61]

they are in good agreement. The optical transmittance meter LH1013 we used for the light transmittance measurement can measure the optical transmittance in visible light, ultraviolet, and infrared light. LH1013 adopts the wide spectrum infrared light source, and the measured values can reflect the optical performance of films in the full infrared band. The measurement resolution of LH1013 is 1%, 0.5%, and 0.5% for detecting infrared light, visible light, and ultraviolet, respectively. As shown in Fig. 8.27b, the measuring instrument automatically shows three light transmittance values: visible light, ultraviolet, and infrared. The three kinds of optical transmittance are firstly calibrated to 100% with nothing placed in the measuring instrument. Then the fabricated transparent absorber is placed in the measuring instrument to obtain the measured visible light transmittance of 86%, the ultraviolet transmittance of 52%, and the infrared transmittance of 98%. And thus, the proposed transparent absorber is validated to have high transparency.

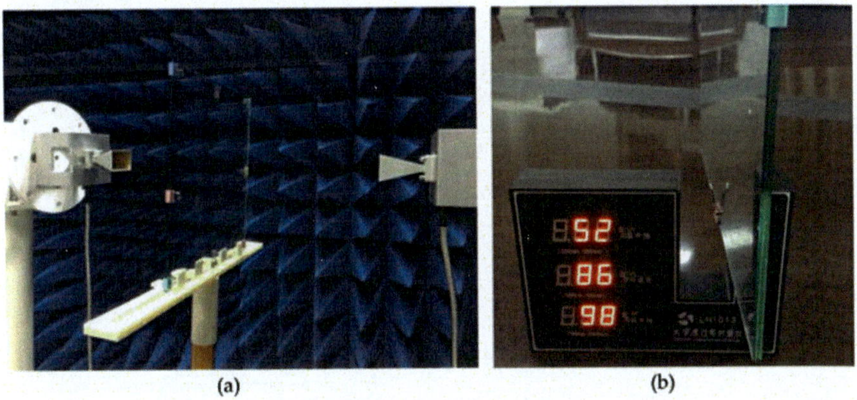

Fig. 8.27 **a** Experimental configuration of space wave measurement method, **b** transparency measurement of the transparent absorber [61]

Fig. 8.28 Measured and simulated S-parameters and absorptivity of the transparent absorber [61]

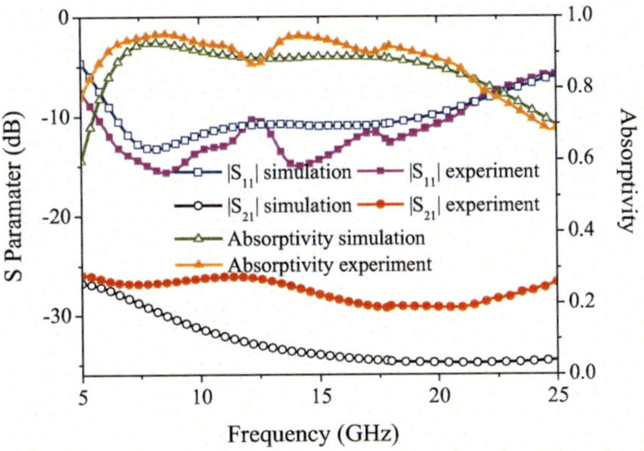

8.3.2 PMA and RCS Reduction of Antennas

The application of perfect metamaterial absorbers is another important aspect of their research. The perfect metamaterial absorbers have been loaded on the waveguide slot antenna or microstrip antenna to reduce the RCS. Several RCS reduction effects can be achieved due to the different loading methods for a perfect metamaterial absorber. Compared to the conventional radar absorbing materials, the metamaterial-inspired antennas exhibit low RCS and maintain or even improve the radiation performance. Consequently, they have been a topic of immense strategic interest for researchers due to their significant advantage. The RCS of a target is the equivalent projected area of a metallic sphere that scatters the same power in the same direction as the target does. Antenna RCS (σ) can be divided into structural mode RCS (σ^{st}) and antenna mode RCS (σ^{an}). Their relationship has been given by [62]

$$\sigma = |\sqrt{\sigma^{st}} + \sqrt{\sigma^{an}} e^{j\varphi}|^2 \tag{8.62}$$

where φ is the phase difference between the two modes. The structural mode RCS depends on the structural characteristic of the target antenna, such as the metal surfaces, corners, edges, and so on. While the antenna mode RCS is related to the radiation characteristics of the target antenna. Power received into the antenna can be reflected by the source impedance connected to the antenna input port. Then, the reflected power reradiates as a source of backscattering. The relationship between the antenna mode scattering and the radiation property of the antenna is given as

$$\sigma_M^{an} = G^2 \Gamma^2 \frac{\lambda^2}{4\pi} \tag{8.63}$$

where σ_M^{an} is the monostatic RCS related to the antenna mode scattering, Γ is the reflection coefficient due to the mismatch between the source impedance and the antenna, G is the antenna gain pattern, and λ is the wavelength. According to Eq. (8.63), we can see that σ_M^{an} is proportional to twice of G. Given that the reflection and gain performance are well guaranteed in the operation band, the total in-band RCS of the antenna can be reduced by restraining the structural mode scattering. The reflectance can also be expressed by

$$|S_{11}|^2 = \frac{|E^r|^2}{|E^i|^2} \tag{8.64}$$

where E^i is the incident field, E^r is the reflected field. According to the definition, RCS is given by

$$\sigma = \lim_{R\to\infty} 4\pi R^2 \frac{|E^s|^2}{|E^i|^2} \tag{8.65}$$

where E^s is the scattered field, and R is the detecting distance. For the monostatic RCS of perfect metamaterial absorber with $T(\omega) = 0$ under normal incidence, $E^s = E^r$, and so RCS can be rewritten by

$$\sigma = \lim_{R\to\infty} 4\pi R^2(1 - A) \tag{8.66}$$

For the perfectly electric conduct (PEC) or copper, A_{pec} is equal to zero. Thus the RCS reduction of a perfect metamaterial absorber compared with a PEC plate with comparable dimensions can be obtained by

$$\Delta\sigma = -10log(1 - A) \text{ dB} \tag{8.67}$$

From Eq. (8.67), it can be deduced that RCS reduction quickly goes up with the increase of absorbance. If A is 50%, the RCS reduction is only 3 dB, while if A is 90%, the RCS reduction reaches to 10 dB. It is worth noting that the foregoing results are all obtained under ideal conditions. In this section, the part of the metal ground planes of a waveguide slot antenna and a ridged waveguide slot antenna array are covered by a perfect metamaterial absorber, the σ^{st} can be reduced due to the high absorption of the perfect metamaterial absorber.

8.3.2.1 Ultra-Thin PMA and Application on RCS Reduction of Waveguide Slot Antenna

The ultra-thin perfect metamaterial absorber is composed of two metallic layers separated by a lossy dielectric spacer [63]. The top layer consists of an etched oblique 45°cross-gap patch set in a periodic pattern, and the bottom one is solid metal. The unit cell geometry is shown in Fig. 8.29. The lossy dielectric spacer is an FR4 substrate with relative permittivity $\varepsilon_r = 4.4$ and loss tangent tan$\delta = 0.02$. The optimized geometries parameters are $p = 9$ mm, $w_1 = 8$ mm, $w_2 = 1.6$ mm, $l = 7.5$ mm, $t =$ 0.5 mm. The thickness of the PMA is about 0.01λ at 5.75 GHz. The metal portions of the PMA are modeled as lossy copper with a conductivity $\sigma = 5.8 \times 10^7$ S/m.

The fabricated device of a PMA, as shown in Fig. 8.29, was implemented using the commonly printed circuit board fabrication method. The simulated and measured absorptivities of the ultra-thin perfect metamaterial absorber are given in Fig. 8.30. It can be seen that the measured maximum absorptivity is 98.8% at 5.75 GHz with a full width at half maximum of 220 MHz (5.64–5.86 GHz), and the simulated maximum absorptivity is still 99.8%. The good agreement of the simulation with the experiment verifies the practicability and credibility of the ultra-thin PMA. It is necessary to note that the contribution of dielectric loss in dielectric spacer for absorption is more significant than that of the losses in metal. According to the theory of equivalent circuit, Fig. 8.31 shows the extracted real and imaginary part of effective impedance z under normal incidence. The ultra-thin PMA achieves near perfect impedance matched to the free space where the real part of effective impedance z is near unity,

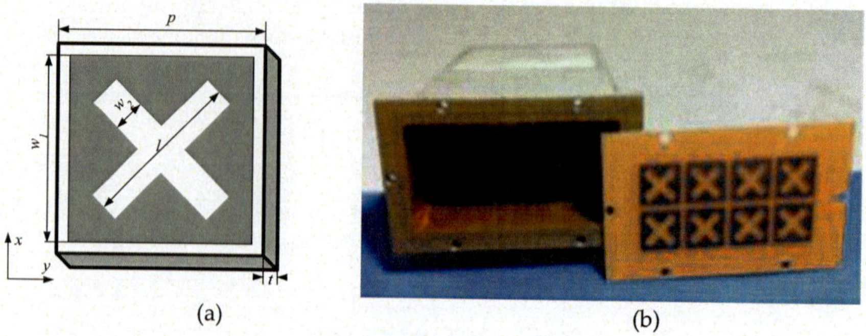

Fig. 8.29 **a** Configuration of the ultra-thin PMA and **b** its fabricated device [63]

Re(z)≈1, and the imaginary part is minimized, Im(z)≈0, at the absorptive peak, which displays nearly perfect absorption. Thus the impedance matched to the free space ensures the reflection of the incident wave at the interface between the free space and the PMA to be small. The absorber also shows better polarization-insensitive and wide-angle absorption for TM mode than for TE mode for all angles of incidence [63].

The ultra-thin perfect metamaterial absorber has been loaded on a waveguide slot antenna to reduce RCS. The parameters of the antenna are: slot length $L = 25.6$ mm, width $W = 2$ mm. The antenna is fed from the rear of one C-band standard waveguide with a tee junction whose broad wall is 40.4 mm and the narrow wall is 20.2 mm. The size of the total antenna aperture is 135 mm × 135 mm. To reduce the RCS of the slot antenna, the PEC ground plane of the antenna is covered with the ultra-thin PMA, as shown in Fig. 8.32. To lead the electromagnetic wave to radiate into the outer space as much as possible and reduce the coupling between the PMA and the slot simultaneously, some distance between them must be left. The measured reflection coefficient is given in Fig. 8.33. Similarly, the PMA has a very slight effect

Fig. 8.30 Comparison of absorptivity for simulation and measurement [63]

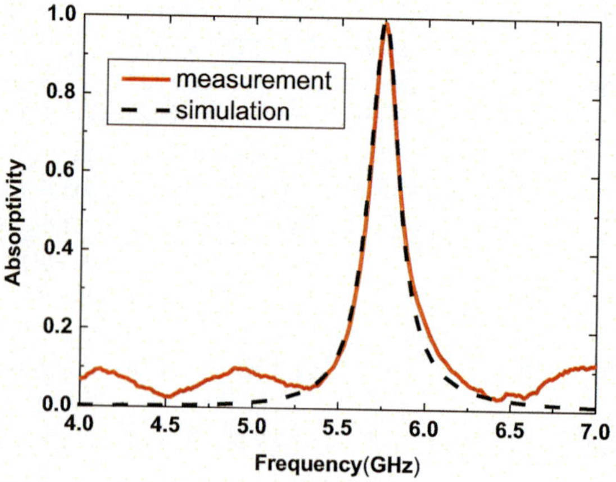

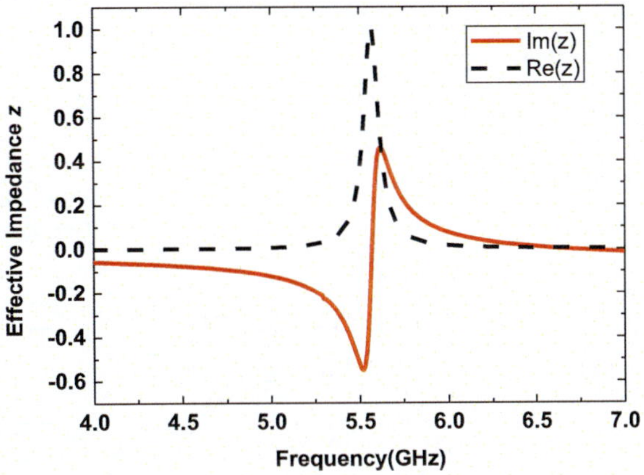

Fig. 8.31 The real part and imaginary part of the retrieved effective impedance, *z*, from the simulation data under normal incidence [63]

on the radiation of the antenna. The center resonance frequency shifts from 5.55 to 5.75 GHz. The small error between simulated and measured results is due to the following: on one hand, the simulated model is finite in size, and the measurement is in the free space, and on the other hand, there exist some fabricated errors of the antenna.

The comparison of the measured patterns for the waveguide slot antennas with the different ground planes of the same size was measured in Fig. 8.34. The measurement was taken at 5.75 GHz, which is in the absorption band of the practical perfect metamaterial absorber. The forward gain of the antenna with the PMA is 0.6 dB lower than that of the antenna with the PEC ground plane. This shows that the performance of the waveguide slot antenna covered with the PMA is slightly degraded due to the interaction of the metamaterial structure with the slot. The measured results demonstrate that the radiation characteristics of slot antenna with PMA are preserved basically.

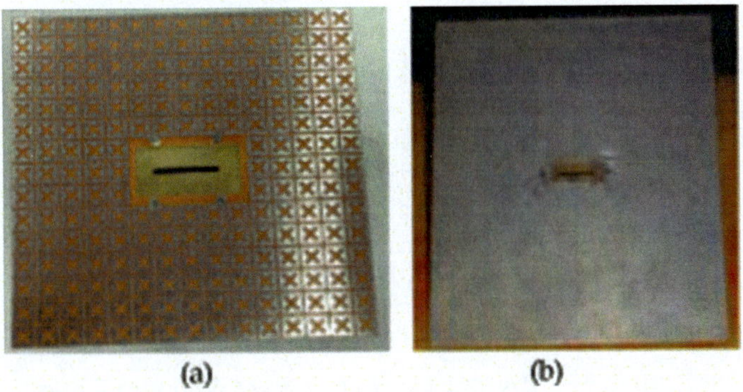

Fig. 8.32 Photograph of **a** waveguide slot antenna with PMA and **b** the common waveguide slot antenna [63]

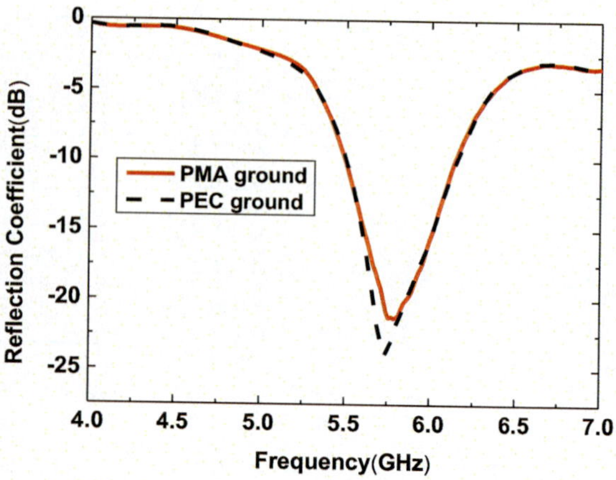

Fig. 8.33 Comparison of the measured reflection coefficient of waveguide slot antenna with PEC and perfect metamaterial absorber ground [63]

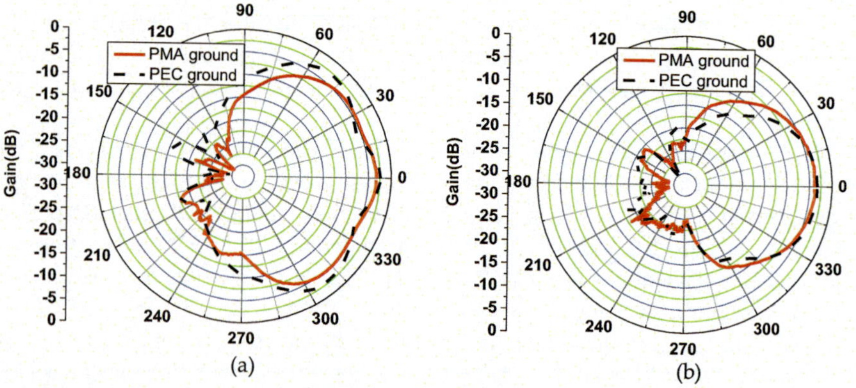

Fig. 8.34 Comparison of the measured radiation pattern of waveguide slot antenna with PEC and PMA grounds **a** E-plane, **b** H-plane [63]

Figure 8.35 gives the monostatic RCS of two antennas with the PMA and PEC grounds in the same size. The incident wave is perpendicular to the ground plane of the antenna. It is evident RCS reduction above 7 dB from 5.6 to 5.87 GHz, and the peak RCS in front direction has 14 dBsm reductions at 5.75 GHz, corresponding to the region of high absorptivity observed in Fig. 8.30. The measured results demonstrate that the PMA can absorb the incident wave effectively, and the waveguide slot antenna with the PMA has a low RCS characteristic.

8.3.2.2 Application of PMA on RCS Reduction of Array Antenna

Similarly, a PMA with a maximum absorptivity of 99.5% at 3.20 GHz has been designed to load on a waveguide slot array antenna. The practical antenna arrays

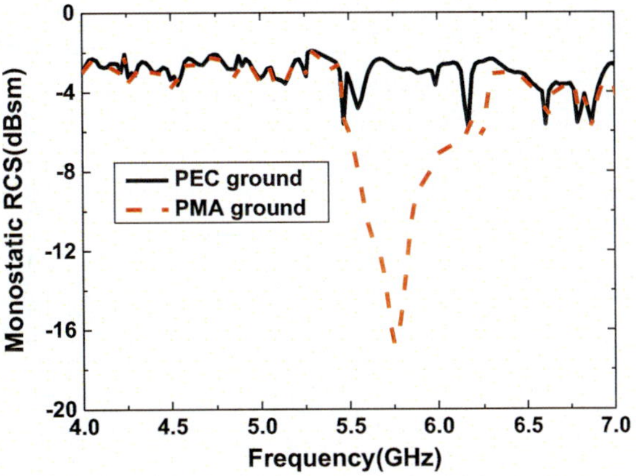

Fig. 8.35 Comparison of the measured monostatic RCS of waveguide slot antenna with PEC and PMA grounds [63]

with and without PMA are depicted in Fig. 8.36. Rectangular slots were cut in the middle of the broadside of the waveguide. The ridged waveguide slot array antenna with 8 × 10 slot elements was fed from the rear with a tee junction. The total size of the PEC plane of the slot array is 606 mm × 402 mm. To reduce the slot array's RCS, the array antenna's metal ground is covered with a PMA. To avoid destroying the aperture field of the slot, the PMA is used only between slots in the E-plane direction. Furthermore, to lead electromagnetic waves to radiate into outer space as much as possible and reduce the coupling between the PMA and the slot at the same time, some space between them must be left.

The measured and simulated reflection coefficients are given in Fig. 8.37, which present excellent impedance matching around 3.195 GHz. The simulated bandwidth for $|S_{11}| \leq 10$ dB is from 3.165 to 3.23 GHz, while the measured is 3.16 to 3.245 GHz. The measured reflection coefficients of the waveguide slot antenna arrays with and without metamaterial absorber keep very well, which agree well with the simulated ones. It indicates that loading the PMA does not destroy the performance of the array antenna because of the appropriate distance between the absorber and slot.

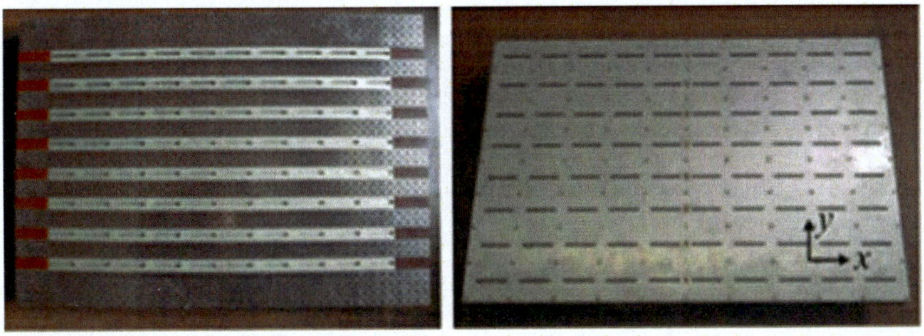

Fig. 8.36 Photographs of waveguide slot antenna array with and without PMA

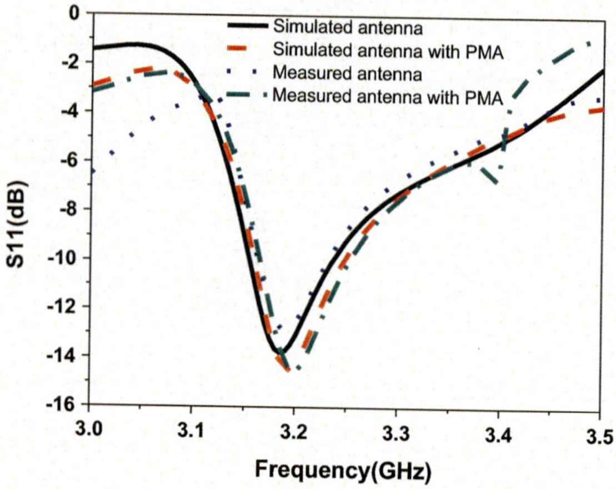

Fig. 8.37 Comparison of the measured reflection coefficient of waveguide slot antenna arrays with and without PMA

The comparison of the patterns of the two antenna arrays is depicted in Fig. 8.38. Similar radiation patterns in both E- and H-planes between the two antenna arrays are obtained, and the simulated forward gain of the array with the PMA is 0.15 dB lower than that of the array with the PEC ground plane. Compared with the simulated results, the measured forward gain of the array with the PMA is 0.77 dB higher than that of the array with the PEC ground plane. The result is that the PMA suppresses surface waves. The measured results agree well with the simulated ones, demonstrating that the loaded PMA has little influence on the antenna array radiation performance.

The scattering properties of the PMA-loaded antenna array are compared with those of the reference structure with the PEC ground plane. Both two arrays are terminated with a matched load. The results are obtained by reflection measurements on the manufactured prototypes in an anechoic chamber. Measurement circumstance and setup are shown in Fig. 8.39. Two horn antennas are utilized as transmitter

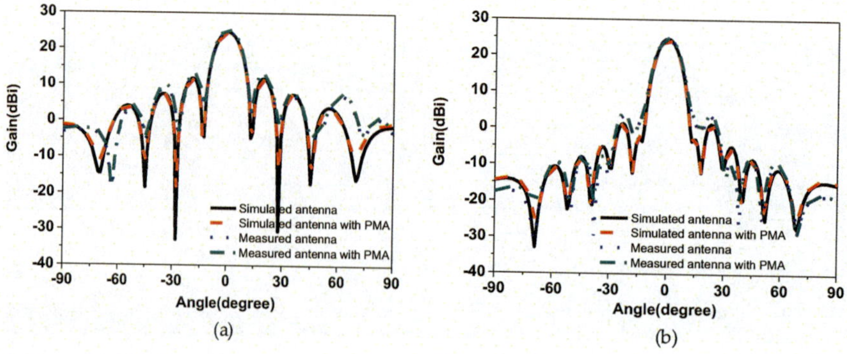

Fig. 8.38 Comparison of radiation patterns of waveguide slot antenna arrays with and without PMA ground **a** E-plane. **b** H-plane

and receiver, respectively. The monostatic behavior of the antenna array has been evaluated by means of the horn antenna reflection coefficient characterization. The incoming EM wave illuminates the antenna array placed on the rotation equipment, and the scattering wave for each of the rotating angles can be received by the receiving horn. To avoid additional scattering, the rotation equipment is stacked with circular foam. The precisely known RCS calibration metal sphere is used to eliminate the frequency response errors in the test system before measuring the antenna array RCS. Note that time-domain gating has been applied to filter out any undesired reflection; only the reflection coming from the array is considered.

The variation of the RCS as a function of frequency under normal incidence for the two configurations has been reported in Fig. 8.40a, b. For the incident wave with horizontal polarization, it is the scattering of the PEC plane of the antenna array that mainly contributes to the whole RCS of the array because the incoming EM wave cannot induce the slot voltage, so the frequency response curve of RCS for the array is similar to that for PEC plane, as shown in Fig. 8.40a. It can be seen that there is a strong in-band RCS reduction above 6 dB from 3.16 to 3.255 GHz, and the peak RCS in the front direction has 9.734 dB reduction at 3.21 GHz due to strong absorbance, as shown in Fig. 8.40a. As to the vertical polarization case, the frequency response curve of RCS is fluctuant for the incoming EM wave that motivates the slot voltage, as depicted in Fig. 8.40b. At the same time, we can see that there is obvious in-band RCS reduction above 6 dB from 3.175 to 3.25 GHz, and the peak RCS in the front direction has 9.462 dB reduction at 3.195 GHz.

According to Eq. (8.65), it is concluded that the antenna mode scattering is not reduced by loading the PMA because the radiation performance of the antenna with PMA is guaranteed. Consequently, the antenna array RCS reduction is also considered as expected in the band since the structural mode RCS is restrained by loading a PMA at the vertical polarization case. In addition, we also can see that the RCS reduction value of the antenna array does not strictly correspond to the value calculated by Eq. (8.67) at the maximum absorptivity. It is worth noting that the calculated value is based on the ideal absorptivity obtained by simulating an infinite period PMA unit. The RCS reduction value of the antenna array with PMA is affected by discontinuously covering period PMA. It is of obvious merit that the design of low RCS

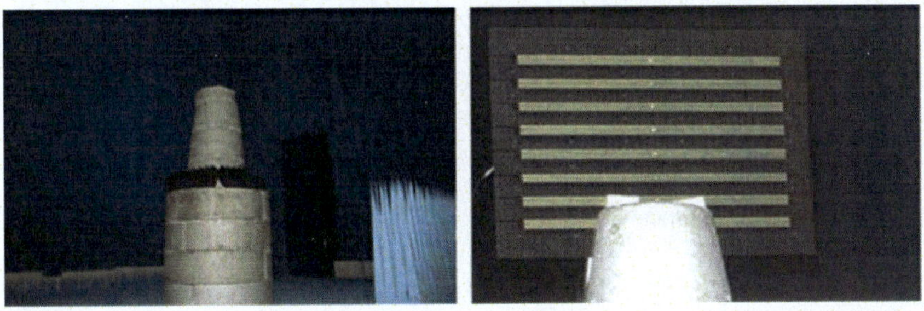

Fig. 8.39 Measurement circumstance and setup

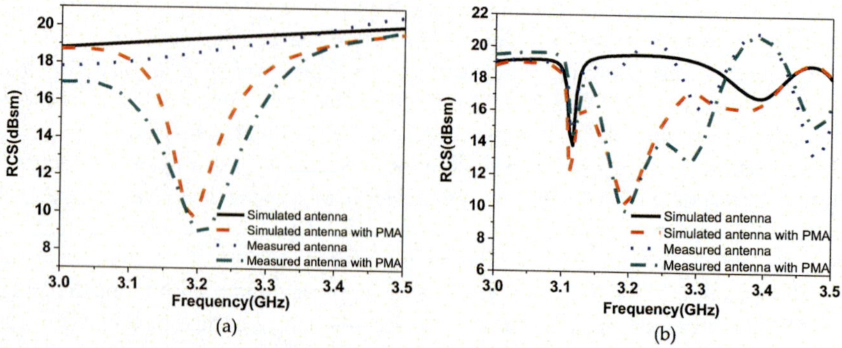

Fig. 8.40 Comparison of monostatic RCS of waveguide slot antenna arrays with and without PMA ground. **a** horizontal polarization. **b** vertical polarization

antenna array can be easily achieved by loading a PMA in a common antenna array without resorting to a complete redesign. In addition, this method can be regarded as a complementarity for frequency selective surface radome application in stealth technology.

8.3.2.3 Fractal Tree PMA and Application on RCS Reduction of Microstrip Antenna

To broaden the absorption bandwidth of PMA, a fractal perfect metamaterial absorber has been proposed based on a tree-shaped microstructure [64]. As shown in Fig. 8.41, the fractal tree perfect metamaterial absorber (FT-PMA) is composed of two-layer substrates, a three-dimensional fractal metal tree, four lumped resistances, and metallic ground without a pattern. FR4 is used as the substrate with a thickness of 2.0 mm. The three-dimensional fractal tree microstructure is shown in Fig. 8.41b, c. The metal is copper with a conductivity of 5.8×10^7 S/m, and its thickness is 0.036 mm. The width of the fractal tree in the top layer is 0.6 mm ($w_1 = 0.6$ mm), and its length is 9.0 mm ($l_1 = 9.0$ mm). The four lumped resistances have been used in the fractal tree microstructure. In Fig. 8.41c, the width of the copper is 0.6 mm ($w_2 = 0.6$ mm), and the length is 8.5 mm ($l_2 = 8.5$ mm). The radius of the cavity for the fractal tree is 0.3 mm in the top and bottom layers. The four lumped resistances are all selected as R = 200 Ω. The present FT-PMA device was fabricated and measured using the free-space test method in a microwave anechoic chamber. The devices were easily implemented using the commonly printed circuit board fabrication method on two substrates with a thickness of 2 mm, as shown in Fig. 8.42.

As shown in Fig. 8.43, the experimental broadband absorption of 86.9% could be achieved from 4.82 to 12.23 GHz with an absorptivity larger than 0.9 for normal incidence. Measurements were in good agreement with the results obtained from simulations. Moreover, it was obvious that the proposed broadband metamaterial absorber performed polarization-insensitive absorption.

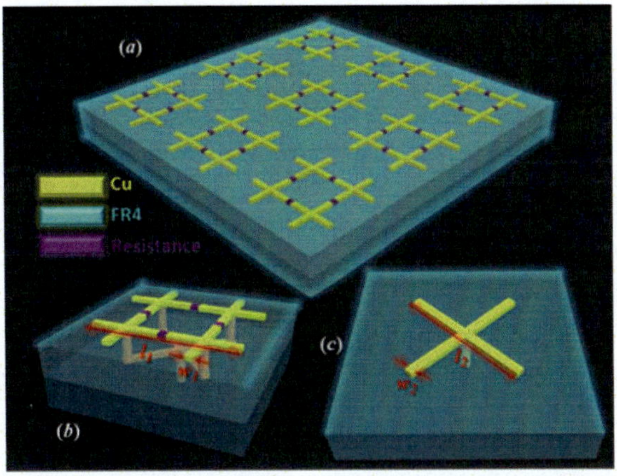

Fig. 8.41 **a** Geometry of the FT-PMA. **b** unit cell of FT-PMA based on the three-dimensional fractal metal tree microstructure. **c** bottom layer of the FT-PMA [64]

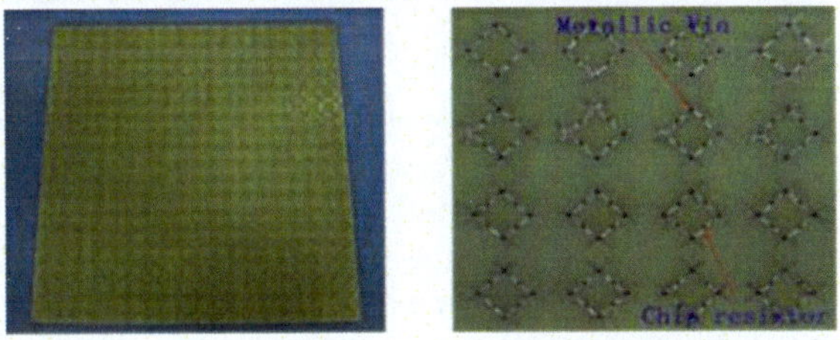

Fig. 8.42 Photographs of FT-PMA device [64]

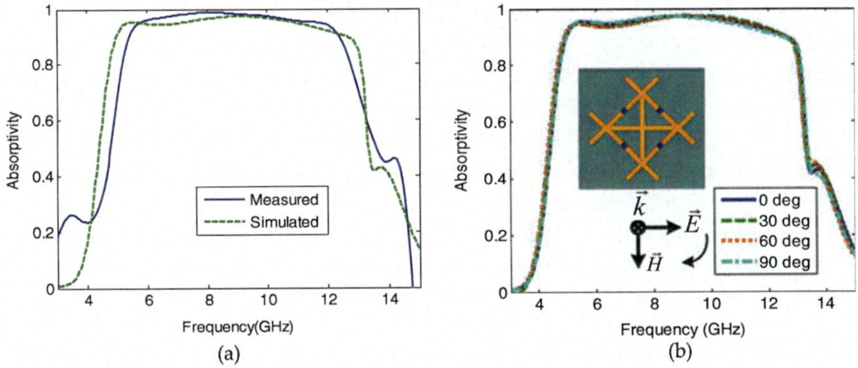

Fig. 8.43 **a** Comparison of measured and simulated absorption for FT-PMA. **b** simulated absorption with different polarized angles

The fractal tree perfect metamaterial absorber provides an important way to reduce the RCS of antennas in a wide bandwidth. Broadband RCS reduction and gain enhancement microstrip antenna has been illustrated using a shared aperture artificial composite metamaterial (SA-ACM) based on the FT-PMA [65]. The SA-ACM is composed of FT-PMA and partial reflection surface, where the FT-PMA and

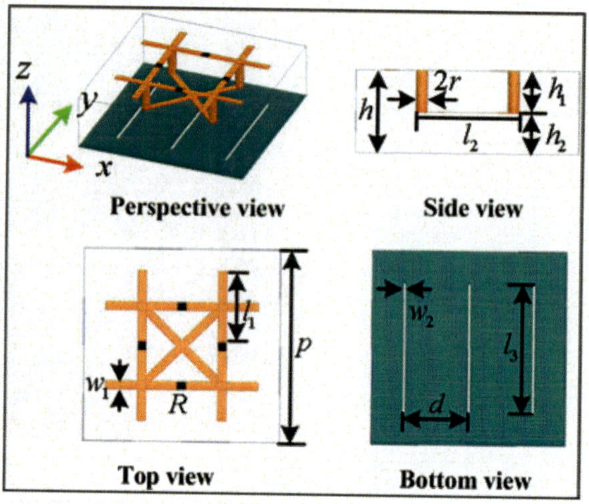

Fig. 8.44 Geometry of the shared aperture artificial composite metamaterial. The optimized parameters are: $h = 4, h_1 = 2, h_2 = 2, p = 12, l_1 = 4.4, l_2 = 5.55, l_3 = 8, w_1 = 0.5, w_2 = 0.15, d = 4, r = 0.3$ (units: mm) [65]

partial reflection surface share the same aperture in the vertical dimensionality. The schematic diagram and geometric parameters of the SA-ACM unit cell are depicted in Fig. 8.44. The partial reflection surface consists of etched parallel slots in a metallic plane. The SA-ACM, as the role of superstrate, is applied to the microstrip antenna. The Fabry-Perot resonator cavity constructed by the partial reflection surface and the metallic ground of the microstrip antenna can achieve high gain, while the FT-PMA can obtain the low RCS characteristic by absorbing the incident wave.

For practical applications, efficient absorbers are required to absorb as much energy as possible and be insensitive to incident directions. Therefore, the performance of the current SA-ACM is evaluated at various incident angles. The simulated absorptivity as a function of frequency and incident angle for x and y polarizations is shown in Fig. 8.45. It can be seen that the SA-ACM has a high absorptivity over a wide incident angle in the wide band range. The reflection (S_{22}) characteristics of the SA-ACM under x polarization illumination are shown in Fig. 8.46. It can be seen that the reflection amplitude is over 0.91 from 9.7 to 10.2 GHz, while the corresponding reflection phase increases from 168.7° to 177.4°. The high reflection coefficient and positive phase gradient indicate that the partial reflection surface combined with the metallic ground plane could be applied to construct a Fabry-Perot cavity. In addition, the resonance at 10 GHz is caused by the coupling between structures at the upper and bottom layers owing to the thin dielectric slab (less than the quarter wavelength). To further understand the coupling, the electric field distributions in the xoz and yoz planes are depicted in Fig. 8.47. It is obvious that the strong electric field resonance is produced under the x polarization, making some electromagnetic waves transmit through the thin dielectric slab, whereas the coupling is not induced under the y polarization. It also indicates that SA-ACM is suitable for a linearly polarized antenna.

A traditional microstrip antenna resonating at 10.0 GHz is designed. The substrate for the antenna is FR-4 with a thickness of $d = 2$ mm. The SA-ACM is adopted as the

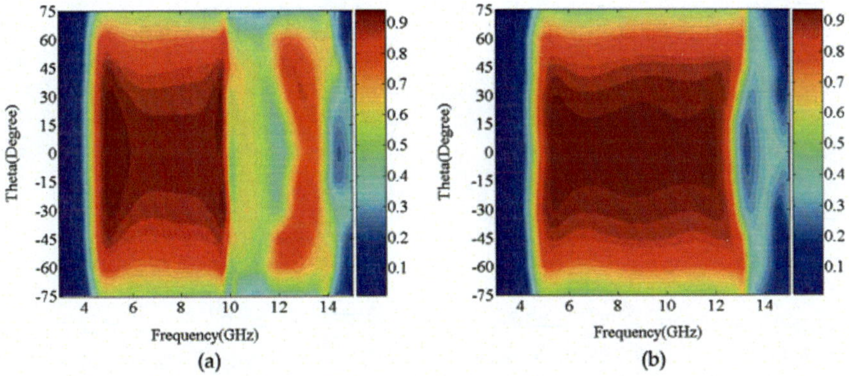

Fig. 8.45 The simulated absorption spectra of SA-ACM. **a** *x* polarization, **b** *y* polarization [65]

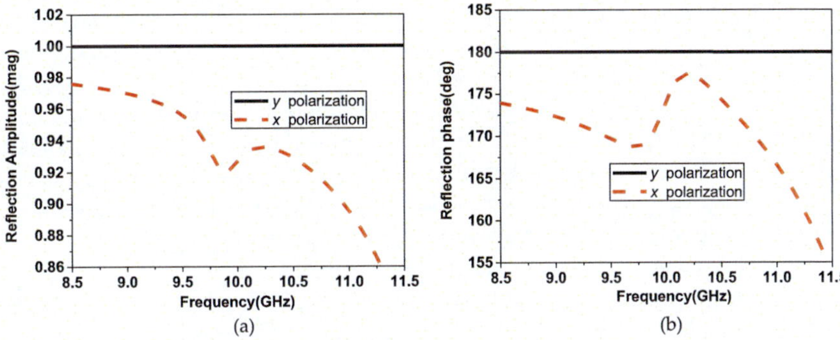

Fig. 8.46 The reflection (S_{22}) characteristics of SA-ACM. **a** reflection amplitude. **b** reflection phase [65]

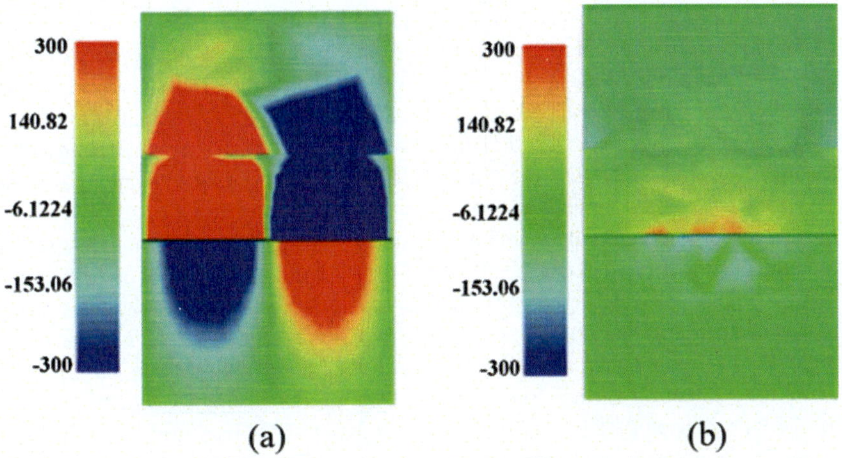

Fig. 8.47 The electric field distribution on the cross section of SA-ACM. (a) *xoz* plane. (b) *yoz* plane [65]

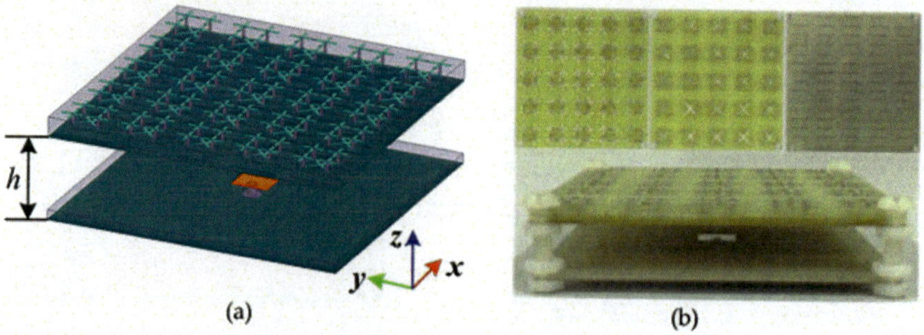

Fig. 8.48 The schematic view of the proposed SA-ACM-based antenna. **a** schematic view of the SA-ACM antenna. **b** photographs of the SA-ACM antenna [65]

superstrate above the antenna, and the geometrical model of the proposed antenna is shown in Fig. 8.48a. The whole SA-ACM is composed of 5 × 5 cells, and the lateral dimension of the antenna is 60 mm × 60 mm. Based on the simulation results of the SA-ACM, the reflection phase of the partial reflection surface is about $\varphi_1 = 174°$ at 10 GHz, while $\varphi_2 = 180°$ is for the metallic ground plane. The calculated value of the antenna cavity distance L is optimized to be 14.1 mm. The proposed antenna is fabricated, and its photography is shown in Fig. 8.48b. Four Nylon spacers are utilized to support the SA-ACM above the microstrip antenna.

The measured and simulated reflection coefficients are shown in Fig. 8.49a. The simulated bandwidth of the original antenna is from 9.5 to 10.9 GHz, while the antenna with SA-ACM is 9.6 to 10.7 GHz. Compared with the original antenna, the simulated center resonance frequency with SA-ACM shifted from 10 to 10.2 GHz. The fractional bandwidth of the antenna is 14%, but the fractional bandwidth of the antenna with SA-ACM is 10.8%. The smaller bandwidth of the antenna with SA-ACM corresponding to the original antenna is caused by the high Q factor of the resonator. The curve of antenna gain is shown in Fig. 8.49b. The gain of the original antenna is only about 4 dB from 9.5 to 10.9 GHz. The gain is obviously improved when the SA-ACM is employed above the microstrip antenna. The maximum gain enhancement reaches about 6.6 dB around 10.2 GHz, and the 3 dB gain bandwidth is from 9.7 to 10.7 GHz. The measured reflection coefficients of the antennas with and without SA-ACM agree well with the simulated ones.

The measurement of radiation patterns of the antennas with and without SA-ACM were taken in the principal cuts, i.e., E-and H-planes, which correspond to the *xoz* and *yoz* planes, respectively. Figure 8.50 depicts the comparison of co-polarization and cross-polarization radiations of the antennas with and without SA-ACM. The main lobe of the traditional antenna is very broad in both E and H planes. The introduction of a partial reflection surface makes the antenna produce a highly directive beam. Consistent radiation patterns are achieved at 9.7 and 10.2 GHz. It can be observed that the cross-polarization of the antenna with SA-ACM in the E-plane is less than −40 dB while it is less than −25 dB in H-plane, which is similar to the original

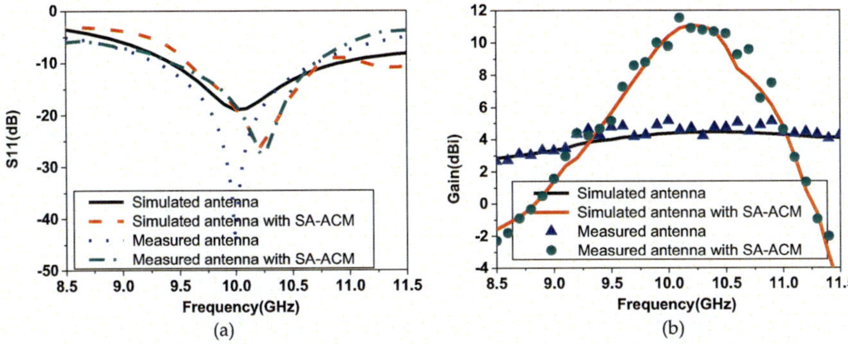

Fig. 8.49 Comparison of reflection coefficients and gain of the antennas with and without SA-ACM. **a** the reflection coefficient. **b** the gain [65]

antenna. It indicates that the cross-polarization of the antenna radiation pattern with the employment of SA-ACM does not deteriorate.

The scattering properties of the loaded SA-ACM antenna are compared with those of the original microstrip antenna. The results are obtained by reflection measurements on the manufactured prototypes in an anechoic chamber. The experimental variation of the RCS is reported in Fig. 8.51a, b under normal incidence for the two configurations. For the incident wave with horizontal polarization, the RCS of the traditional patch antenna is increased from −14 to −4 dBsm when the incident frequency varies from 3 to 15 GHz. However, the loaded SA-ACM can reduce the RCS in broadband covering from 3 to 15 GHz. It can be seen that there is a strong RCS reduction of over 10 dB from 3.8 to 5 GHz, from 6.5 to 9.8 GHz, and from 12.3 to 12.9 GHz. In addition, the RCS reduction is not obvious from 9.9 to 12.2 GHz because the antenna mode scattering is enhanced due to the increased gain, and the restraint of structural mode scattering is impaired owing to the drop of absorptivity. As to the vertical polarization case, the RCS reduction is also considered as expected due to strong absorbance, and especially the RCS reduction over 10 dB is from 5.2 to 11.9 GHz, which is similar to the frequency band of absorptivity above 90%. The variation of the RCS is also investigated as a function of angle in (−90°, +90°) at 8 GHz, as shown in Fig. 8.52. With the employment of the SA-ACM, the RCS of the antenna is significantly reduced in the angular region $-60^\circ \leq \theta \leq 60^\circ$ for both horizontal and vertical polarizations at two orthogonal planes. The peak RCS in the front direction has 10.8 dB reduction. It means that most of the incident wave is absorbed by the FT-PMA rather than scattered in other directions.

8.3.3 AMC and RCS Reduction of Antenna

PEC and AMC are combined for destructive interference at boresight, so backscatter energy is dispersed rather than absorbed. However, the band of RCS reduction is

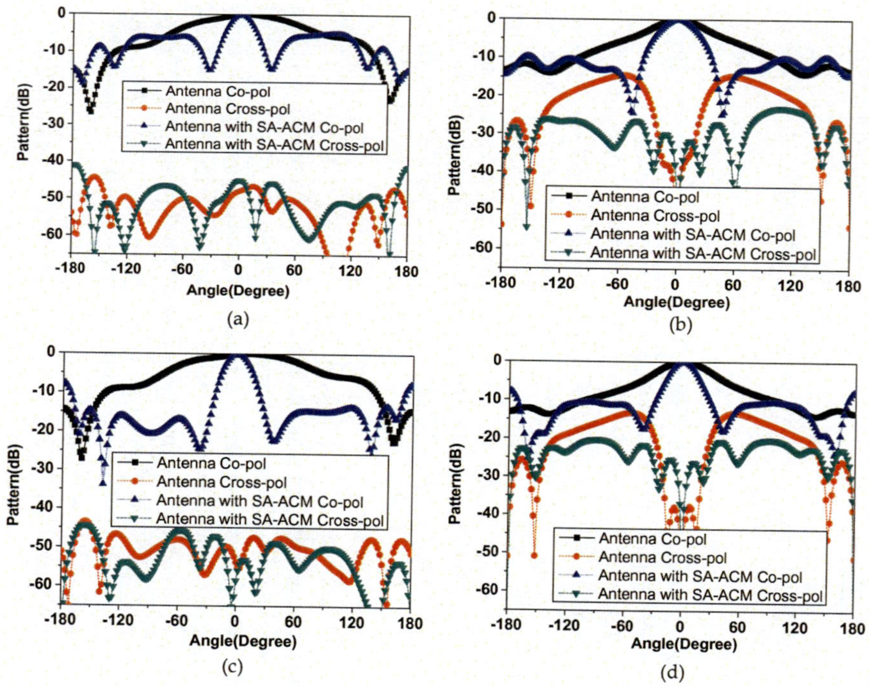

Fig. 8.50 Comparison of radiation patterns of antennas with and without SA-ACM **a** E-plane at 9.7 GHz. **b** H-plane at 9.7 GHz. **c** E-plane at 10.2 GHz. **d** H-plane at 10.2 GHz [65]

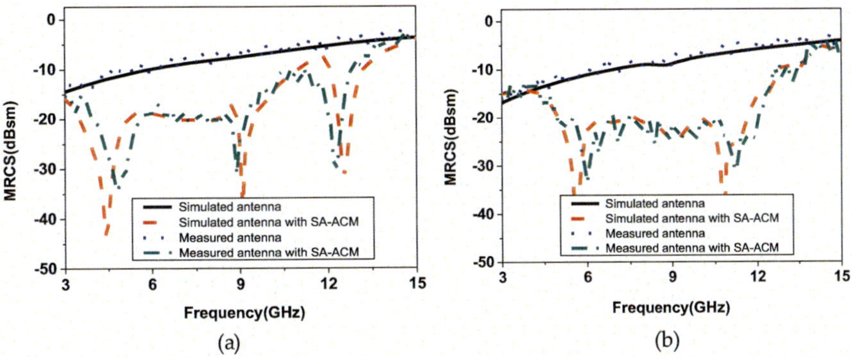

Fig. 8.51 Comparison of monostatic RCS of the antennas with and without SA-ACM. **a** horizontal polarization (*x* polarization in the simulation of the unit) **b** vertical polarization (*y* polarization in the simulation of the unit) [65]

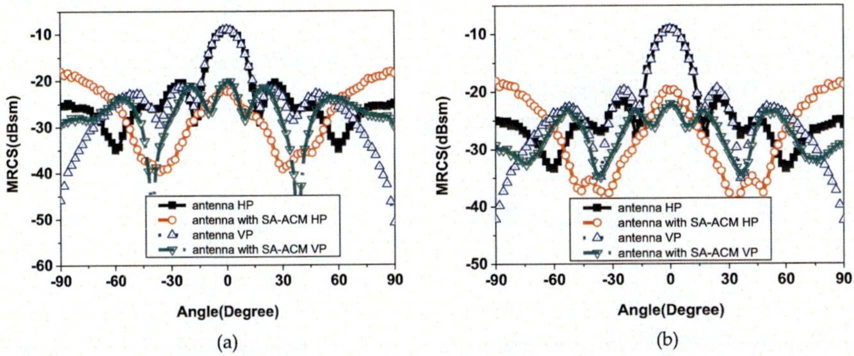

Fig. 8.52 Comparison of measured monostatic RCS of the antennas with and without SA-ACM as a function of angle under normal incidence. **a** *xoz*- plane **b** *yoz*-plane [65]

still restricted by the in-phase reflection of AMC, which only appears when the structure resonates. Solutions that combine AMCs of different sizes or different geometries have been proposed to overcome this disadvantage [66]. The reflection phases of each AMC are elaborately designed and cooperate to satisfy the phase cancellation criterion. As the phase difference no longer depends on resonance, effective cancellation can be achieved over broadband. So far, the phase difference is generally produced by at least two different AMC structures.

The principle of broadband RCS reduction lies in the backscatter cancellation, which depends on the phase difference of AMCs. For example, the polarization-dependent AMCs are orthogonally arranged in a fence-like macroscopic layout. Figure 8.53 shows the schematic diagram of the proposed metasurface. The metasurface is composed of several stripes. The adjacent two stripes contain identical polarization-dependent AMC cells but 90° rotation. When a plane wave normally impinges on the metasurface, the total reflected energy is a summation of the reflection from all the AMC stripes. Assuming that both the AMCs of orthogonal directions show equal reflection pattern, according to standard array theory, for an antenna array with $M \times N$ elements, the total reflection can be represented by [67]

$$E_r = \mathrm{EP} \cdot \mathrm{AF}_x \cdot \mathrm{AF}_y \tag{8.68}$$

$$\mathrm{AF}_x = \sum_{\mathrm{m}=1}^{M} e^{j[(m-1)(kd \sin\theta \cos\varphi + \delta_x)]} \tag{8.69}$$

$$\mathrm{AF}_y = \sum_{n=1}^{N} e^{j[((n-1)(kd \sin\theta \sin\varphi + \delta_y)]} \tag{8.70}$$

where EP is the element pattern that can be approximated to $\cos\theta$. The AF_x and AF_y are the array factors of the x- and y-directions, respectively. k is the wavenumber. d is the distance between adjacent elements. The progressive phase shifts between the elements along x- and y-directions are denoted as δ_x and δ_y. The maximum reflection direction (θ_m, φ_m) can be deduced from Eqs. (8.68–8.70)

$$\theta_m = \pm \arcsin\left(\frac{\sqrt{\delta_x^2 + \delta_y^2}}{kd}\right), \quad \varphi_m = \arctan\left(\frac{\delta_y}{\delta_x}\right) \tag{8.71}$$

Especially for the fence-like structure, the elements along the x-direction are the same, so that $\delta_x = 0$. Thus, the maximum reflection direction can be predicted by

$$\theta_{\mathrm{m}} = \pm \arcsin\left(\frac{\delta_y}{kd}\right), \quad \varphi_m = 90^\circ \tag{8.72}$$

To evaluate the reduction of reflection in a normal direction, we simply consider two adjacent elements along y direction ($\theta = 0°$, $\varphi = 90°$ and $N = 2$). The Eq. (8.64)

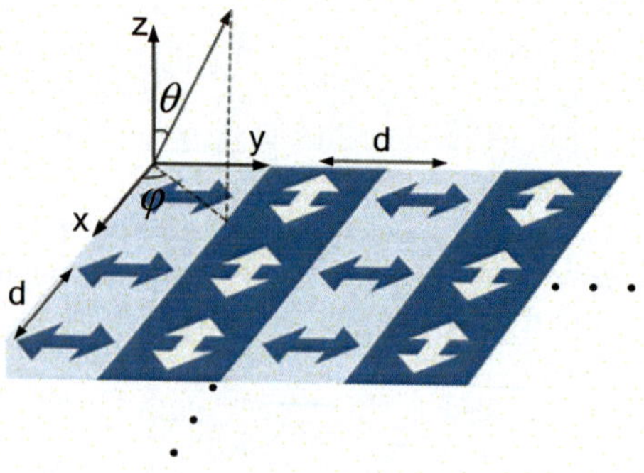

Fig. 8.53 Schematic diagram of the fence-like metasurface. The arrows represent the arranging orientation of AMC [67]

can be simply given as

$$E_r = \text{EP} \cdot (1 + e^{j\delta_y}) \tag{8.73}$$

One can see that when $\delta_y = \pm 180°$, the reflection is totally canceled out. However, as the reflection phase varies with frequency, the 180° phase difference is unstable over a wide band. Usually, a 10 dB RCS reduction is set as a criterion compared to a same-size PEC surface, that is

$$10 \log\left(|E_r|^2 / |\text{E}_{pec}|^2\right) \leq -10 \text{ dB} \tag{8.74}$$

Hence, the effective reflection phase difference is deduced as

$$143° \leq |\delta_y| \leq 217° \tag{8.75}$$

Here, we set 180° ± 30° as the criterion for the following analysis. Noted that the reduction is dependent on a dynamic variation of phase difference instead of precise value, and the working band is expected to expand by a large margin.

8.3.3.1 Polarization-Dependent AMC and Application on RCS Reduction of Antenna

Conventional AMC structures usually consist of symmetrical geometry, so they have identical phase responses for a normal incident plane wave of arbitrary polarization. For the phase cancellation case, at least two unequal phases (φ_1 and φ_2) should be introduced according to Eq. (8.73). Therefore, two AMC structures should be carefully designed. Here, the simplest rectangular patch cell is designed to validate the design of polarization-dependent AMC [67]. Figure 8.54a shows the cell geometry. The top layer is a rectangular metallic patch where *px* and *py* represent the length of the edge along the *x*- and *y*-axis, respectively. The dielectric substrate is F4B-2 with a constant of 2.65 and loss tangent of 0.001, which has a dimension of 10 mm × 10 mm × 3 mm. The bottom is covered by a full metallic layer so that plane waves cannot penetrate. Figure 8.54b shows the reflection magnitudes and phases under *x*- and *y*-polarization incidences. The magnitudes maintain 0.995 over 4–8 GHz for both polarizations, implying that the energy is almost reflected without absorption. This result meets the requirement of the elements in the phase cancellation principle. For the rectangular patch structure, the inductive and capacitive components are mainly relative to the length of the patch along the polarization direction and the gap between adjacent patches, respectively. Due to the asymmetric geometry, the reflection phase shows obvious dependence on the polarization of the incident wave. For *x*- polarization incidence, the zero value appears at 4.56 GHz, while for *y*-polarization case, the in-phase reflection shifts upward to 6.92 GHz.

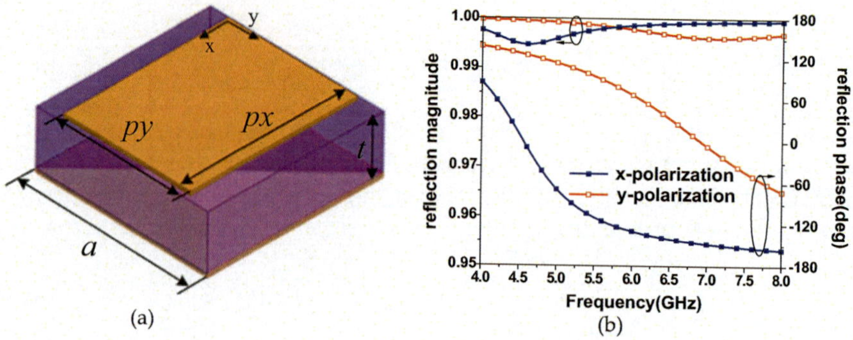

Fig. 8.54 **a** Geometry of rectangular patch cell and **b** its reflection characteristics for *x* and *y*-polarized plane wave [67]

The fence-like metasurface is shown in Fig. 8.55a. It consists of six stripes. Each stripe contains 24 × 4 optimized polarization-dependent AMC cells. The cells in the adjacent two stripes are orthogonally arranged to yield phase differences. The total size of the metasurface is 240 mm × 240 mm × 3 mm. Figure 8.55b shows an RCS comparison between the metasurface and a same-size metallic surface with plane wave normally impinging. Obvious reduction is obtained from 4 to 8 GHz for both polarizations, but the curves are different due to the asymmetrical macroscopic layout. For the *x*-polarized case, the RCS gets a nearly 10 dB reduction from 4.72 to 7.02 GHz. The maximum reduction of 39 dB is obtained at 5.10 GHz. The deterioration around 6.10 GHz is attributed to the strong coupling between adjacent stripes. For the *y*-polarized case, the 10 dB reduction band is 5.08–6.47 GHz, and the maximum reduction of 24 dB occurs at 5.68 GHz.

Figure 8.56 shows the structure and device of a metasurface-slot antenna (MS-slot antenna). The antenna aperture is a 120 mm × 120 mm fence-like metasurface with three stripes. The cells in the middle stripe are orthogonal to those in the other

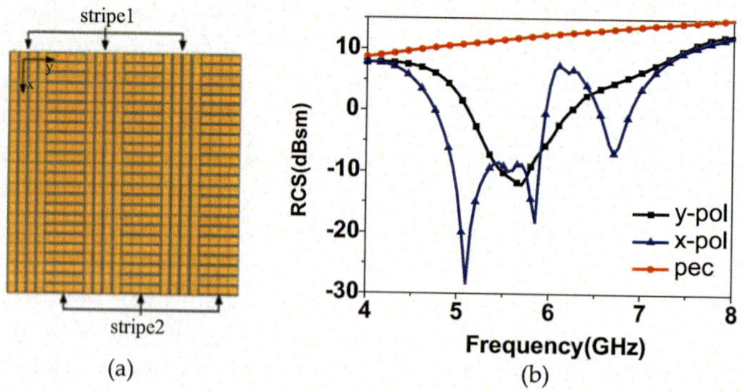

Fig. 8.55 **a** Layout of fence-like metasurface and **b** simulated RCS results for normal incidence [67]

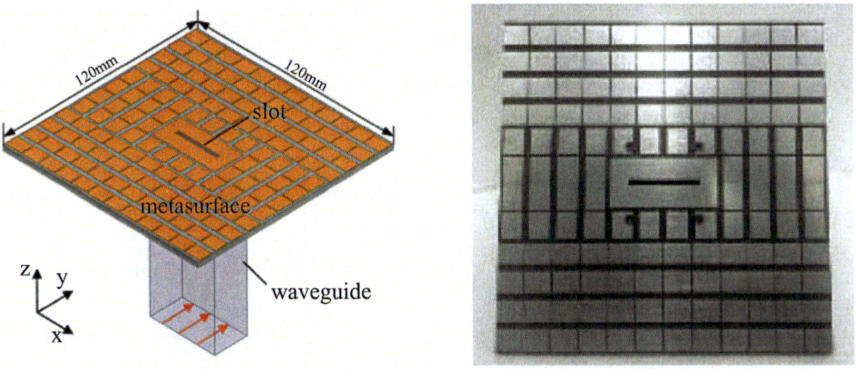

Fig. 8.56 Layout of the metasurface-slot antenna and photograph of the metasurface-slot antenna device [67]

two. The radiating source of the antenna is a slot of 25.6 mm × 2 mm, which is positioned at the center and fed from the rear of one C-band standard waveguide. To compare and analyze the influence of the metasurface, an antenna with the same-size full metallic ground is set as a reference.

The S_{11} parameter obtained by vector network analyzer Agilent N5230C is shown in Fig. 8.57a. The −10 dB band ranges from 5.42 to 5.84 GHz and is obtained for the MS-slot antenna due to the coupling between the metasurface and slot. The band of the MS-slot antenna is 70 MHz broader than that of the reference antenna. Figure 8.57b shows the boresight gain versus frequency. The MS-slot antenna has a stable gain over the operating band and keeps about 3 dB higher than the value of the reference antenna. Figure 8.58 shows the comparison of the measured patterns at 5.58 GHz. It can be observed that the MS-slot antenna has a more concentrated beam at boresight so that the gain gets enhanced. Figure 8.59 shows the RCS reduction of the MS-slot antenna compared to the reference antenna with the metallic ground. The 6 dB reflection reduction band is from 4.35 to 7.80 GHz (56.8% bandwidth). Measured results verify the successful application of metasurface for antenna gain enhancement and RCS reduction.

8.3.3.2 PID-AMC and Application on RCS Reduction of Antenna Array

An approach devoted to achieving ultra-wideband RCS reduction of a waveguide slot antenna array is proposed while maintaining its radiation performance. Three kinds of artificial magnetic conductors tiles consisting of three kinds types of basic units resonant at different frequencies are designed and arranged in a novel quadruple-triangle-type configuration to perform a composite planar metasurface. The metasurface is characterized by a low radar signature feature over an ultra-wideband based on the principle of phase cancellation. Measured results demonstrate that after applying the

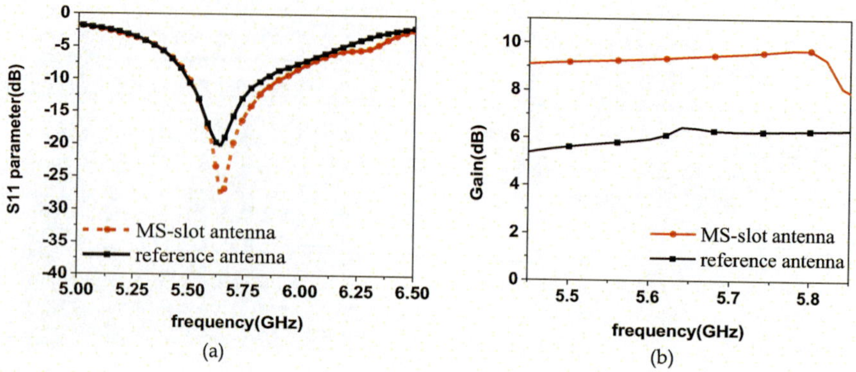

Fig. 8.57 Comparison of measured performance between the metasurface antenna and reference antenna. **a** S parameter **b** gain [67]

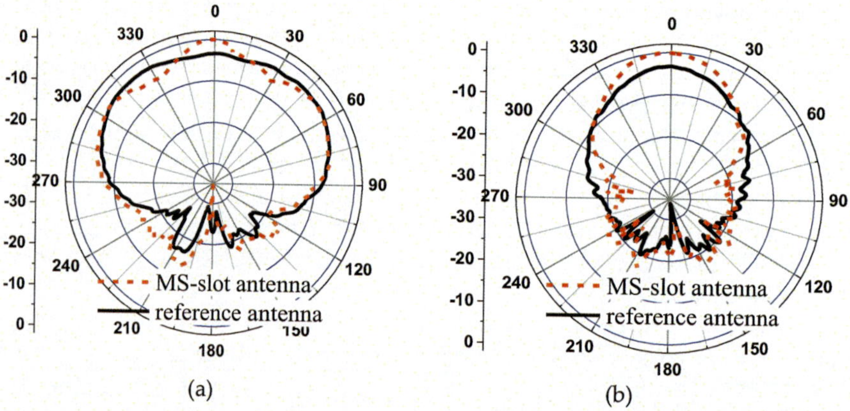

Fig. 8.58 Comparison of measured radiation pattern between the metasurface antenna and reference antenna. **a** E-plane **b** H-plane [67]

composite metasurface to cover part of the antenna array, an ultra-wideband RCS reduction involving in-band and out-of-band is achieved for co- and cross-polarized incident waves energy cancellation, while the radiation performance is well retained.

Three different AMC tiles that resonate at different frequencies are illustrated to achieve ultra-wideband RCS reduction [68]. Each single sub-unit of the three different AMC tiles with detailed dimensions is depicted in Fig. 8.60a, denoted as AMC1, AMC2, and AMC3, respectively. Each of the three sub-units is composed of two metallic layers separated by an exactly same-sized substrate with a dielectric constant of 2.65 and a loss tangent of 0.002. Moreover, despite of the top metallic layers with different shapes, all of the three sub-units are backed by a full metallic ground to ensure that no plane wave can penetrate. The quadruple-triangle arrangement is given in Fig. 8.60b.

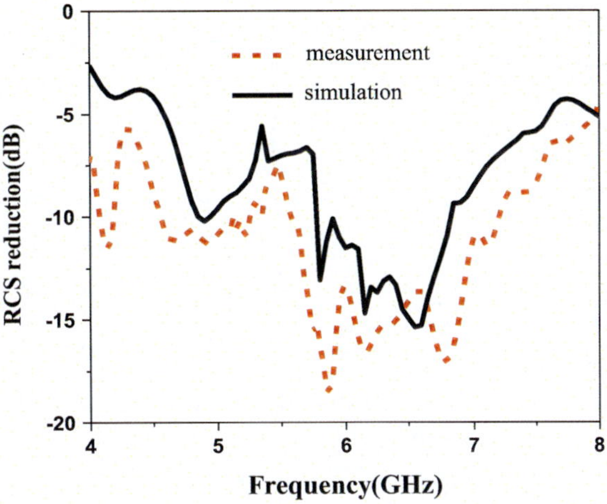

Fig. 8.59 RCS reduction compared with reference antenna [67]

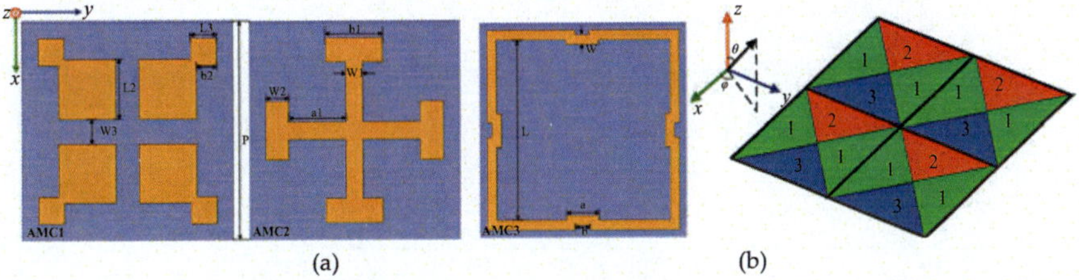

Fig. 8.60 **a** Geometry and dimensions of AMC1, AMC2 and AMC3. **b** schematic view of quadruple-triangle arrangement. $P = 9$ mm, $a = 1.4$ mm, $a1 = 2.45$ mm, $b = 0.6$ mm, $b1 = 2.4$ mm, $b2 = 0.86$ mm, $W = 0.4$ mm, $W1 = 0.7$ mm, $W2 = 1$ mm, $W3 = 1$ mm, $L1 = 7.6$ mm, $L2 = 2.44$ mm and $L3 = 1.1$ mm [68]

Figure 8.61 shows the reflection magnitudes and phases under the *x*- and *y*-polarized incidences for each sub-unit. The magnitudes maintain 0.98 from 2 to 21.5 GHz for both polarizations, indicating that the energy is almost reflected without absorption. These results meet the requirement for the phase cancellation principle. One can observe that AMC1 exhibits only one 0° reflection phase point at 10.9 GHz, while AMC2 and AMC3 demonstrate dual 0° reflection phase points at 7.15 GHz and 18 GHz, 3.98 GHz, and 19.18 GHz, respectively. The different resonant statuses yield the destructive phase difference ranging over an ultra-wideband, as depicted in Fig. 8.62. For normal incidence, the phase difference covering a range of 180° ± 30° between every two of the three sub-units nearly ranges from 3.98 to 18.84 GHz except from 4.2 to 6.9 GHz.

Considering their applications in the antenna array, the three different AMC sub-units are arranged in a quadruple-triangle-type chessboard configuration to adapt themselves to the limited room between array elements. To balance the RCS reduction performance in the low- and high-frequency bands and maximize the RCS reduction

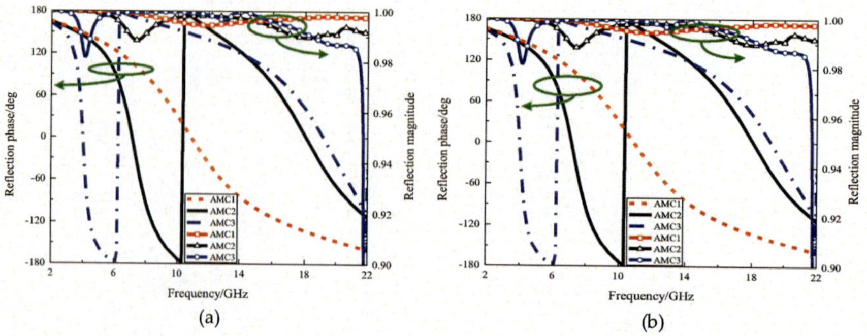

Fig. 8.61 Reflection characteristics of AMC1, AMC2 and AMC3 under **a** x-polarized incidence **b** y-polarized incidence [68]

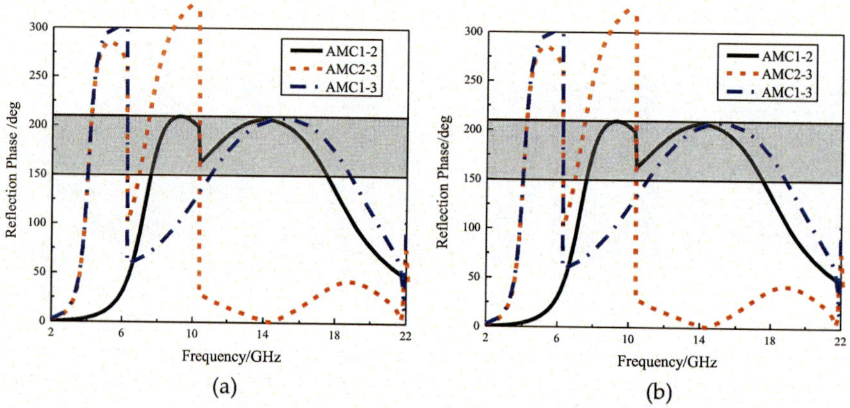

Fig. 8.62 Reflection phase differences between every two of the three AMC units under **a** x-polarized incidence; **b** y-polarized incidence [68]

performance, based on the simulated reflection phase differences in Fig. 8.62, the AMC1 is chosen as a mediator, which accounts for two-quarters of the quadruple-triangle configuration. In this way, the AMC1 tiles can always be surrounded by AMC2 and AMC3 tiles to yield the required phase difference in a wide frequency band as wide as possible, as depicted in Fig. 8.60b. Moreover, the proposed arrangement can significantly diminish EM echo in the specular direction and redistribute backscattered energy more evenly uniformly in the rear hemisphere space to benefit the bi-static detection case. It is worth pointing out that the three AMC tiles can be arranged in other alternative configurations.

Considering the limited space to accommodate the metasurface, each AMC tile contains four sub-units when applied to the waveguide slot antenna array (WGSAA). The fabricated prototype is shown in Fig. 8.63, with seven identical metasurface bars fabricated and mounted on the bare antenna. Each bar consists of 4×68 sub-units with a size of 36×615 mm^2. With respect to the antenna with metasurface, the measured S_{11} of the metasurface-based antenna array is 3.165 ~ 3.297 GHz, 24 MHz broader than that of the bare antenna, as depicted in Fig. 8.64. The radiation patterns of the two antennas at 3.2 GHz in Fig. 8.65 show that the loading of the metasurface makes the antenna gain increase by 0.77 dBi in the E-plane.

To experimentally evaluate the scattering performance, the results of RCS reduction are given in Fig. 8.66. A continuous 6-dB RCS reduction is achieved from 10.12 GHz to 18 GHz, 7.2 GHz to 18 GHz for cross- (E-field along the *y*-axis) and co-polarized (E-field along the *x*-axis) incidences, respectively. Meanwhile, in-band RCS reduction is achieved for co-polarized incidence with maximum RCS reduction reaching up to −5.8 dB. This is believed to be due to the co-polarized incidence inducing the slot voltage and then leading to RCS reduction based on phase cancellation.

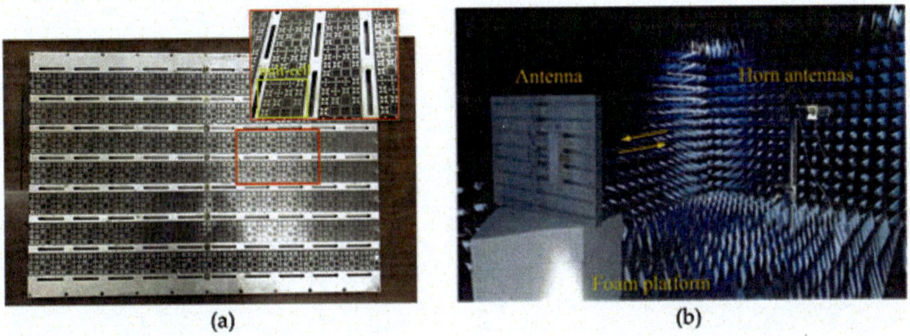

Fig. 8.63 Photograph of **a** fabricated WGSAA loaded with MS and **b** measurement setup [68]

Fig. 8.64 Comparisons between measured and simulated S_{11} parameters [68]

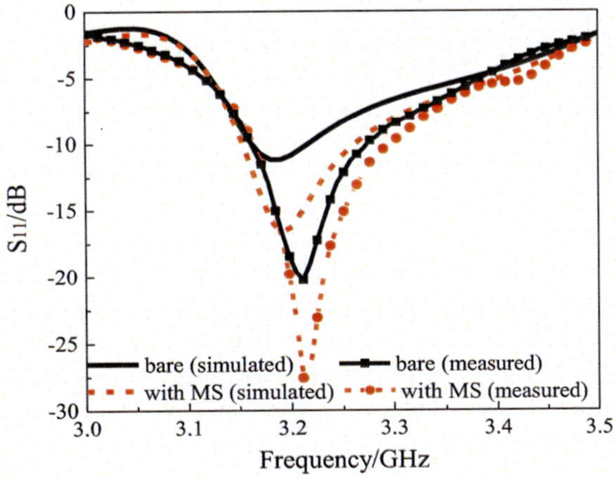

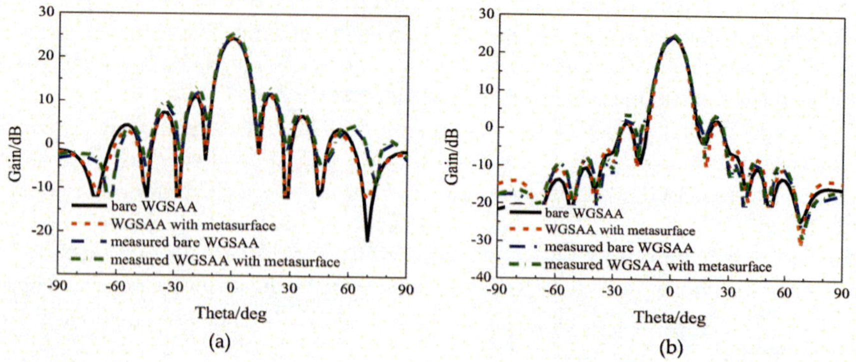

Fig. 8.65 Comparisons between measured and simulated radiation patterns in **a** E-plane and **b** H-plane [68]

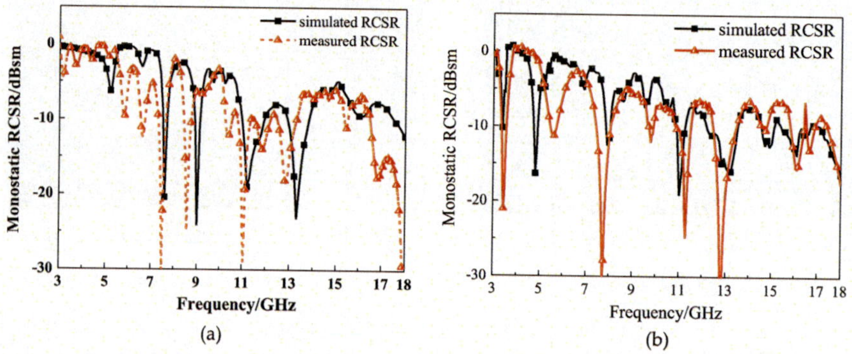

Fig. 8.66 Measured and simulated RCS reduction compared to the bare antenna for **a** cross-polarization and **b** co-polarization [68]

8.3.3.3 Low Scattering Microstrip Antenna Array Using Coding AMC Ground

To guarantee radiation properties, both AMC and antenna should work in different frequencies. The goal is to design a low-profile microstrip antenna array at 3 GHz that minimizes the increase of RCS within the X-band (8.12 GHz) when added to a platform already structured [69]. The coding AMC ground, which consists of two different structures, is designed utilizing the phase cancellation principle. A low scattering antenna array with coding AMC ground, the antenna out-of-band RCS is reduced owing to the broadband diffusion property of AMC ground. Adopting this method will not influence antenna radiation property, and the aperture is not increased.

Two AMC cells shown in Fig. 8.67a are both three-layer structures and designed as follows. Jerusalem crosses metal patch constitutes the AMC1 cell. The square ring with gaps and patches constitutes AMC2 cell. They are printed on dielectric slab PTFE with a thickness of $h = 1.5$ mm. The underside of the substrate is copper without a pattern. Optimization parameters are as follows: $a = 2.4$ mm, $b = 2.4$ mm, $c = 3.5$ mm, $d = 3.2$ mm, $L = 9$ mm. Figure 8.68 is the reflection phase of two AMC structures with the upper substrate. The antenna substrate is PTFE with a dielectric constant 2.65 and loss tangent 0.001, which is the same as the antenna array. The thickness of the substrate is 3 mm. The results of the reflection phase are given in Fig. 8.68. It can be seen that the two curves both shift to a lower frequency, which is because the upper substrate becomes a part of AMC and the whole thickness is increasing. Two AMC structures yield an effective phase difference from 6 to 14 GHz, especially in the X band. According to the concept of coding metasurface, the coding AMC ground should reduce the RCS of the antenna array by optimizing the layout of two AMC structures.

Two AMC structures are respectively nominated as '0' and '1' digital elements. To satisfy the periodic boundary in the element simulation, a lattice that contains 5 × 5 identical unit cells is generated, and the whole AMC ground consists of 4 × 4 lattices.

$$F(\theta, \varphi) = \sum_{m,n} A_{m,n} e^{j\varphi_{\theta,\varphi}} e^{j\varphi_{mn}\pi} \tag{8.76}$$

$$\begin{aligned}\varphi_{\theta,\varphi} = {} & \frac{2\pi}{\lambda} x_m (\sin\theta \cos\varphi - \sin\theta_{inc} \cos\varphi_{inc}) \\ & + \frac{2\pi}{\lambda} y_n (\sin\theta \sin\varphi - \sin\theta_{inc} \sin\varphi_{inc})\end{aligned} \tag{8.77}$$

where $A_{mn} = 1$, φ_{mn} is "0" or "1" element in the matrix, θ_{inc} and φ_{inc} are elevation and azimuth angles of the incident wave. In addition, $x_m = [m\text{-}0.5(M + 1)]d$, $y_n = [n\text{-}0.5(N + 1)]d$, and $d = 0.3\lambda_{10\text{GHz}}$, respectively. To redirect the scattering energy in all directions, the peak value of the scattering field is an objective function. Particle

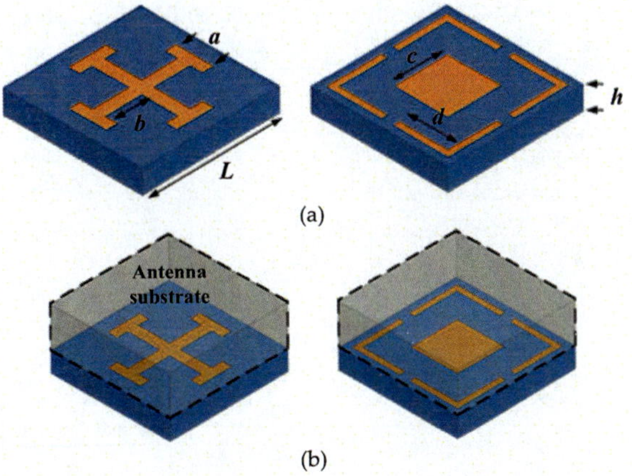

Fig. 8.67 **a** Structures of two AMC cells and **b** AMC structures with upper antenna substrate [69]

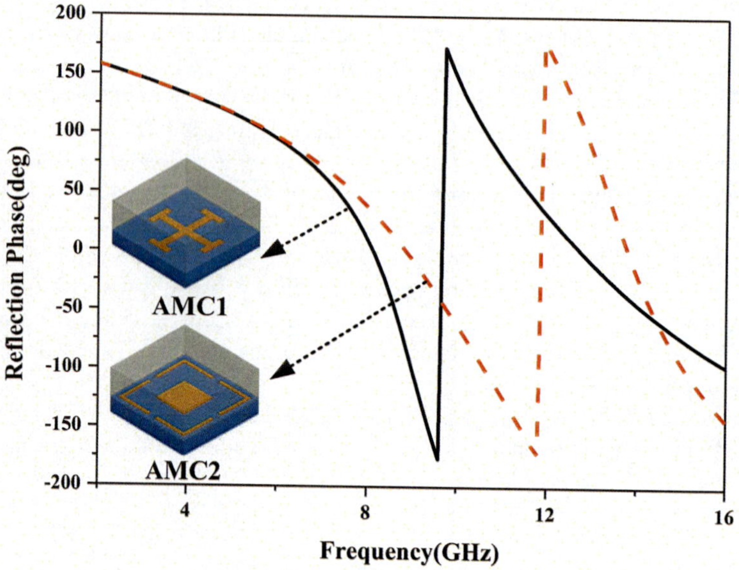

Fig. 8.68 Reflection phases of two AMCs with upper antenna substrate [69]

swarm optimization (PSO) algorithm is used to design the optimal layout of AMC ground, shown in Fig. 8.69. The AMC ground is loaded under 2 × 2 microstrip antenna array, where *e* and *La* are widths of square metal patches and antenna element. The radiation patches are fed by coaxial probes from the bottom of the substrate through SMA connectors, as shown in Fig. 8.70.

The fabricated antenna array is shown in Fig. 8.71. The whole array consists of two parts, the up layer is a 2 × 2 antenna array radiation patch, and the sub-layer is coding AMC ground. Two parts are connected by plastic screws. Figure 8.72 gives the simulated and measured reflection coefficients of the proposed low-scattering antenna array. The impedance bandwidth of the antenna array loaded in the coding AMC ground is from 2.99 to 3.16 GHz. Radiation patterns at 3.05 GHz are also

Fig. 8.69 Geometry of optimal AMC ground, **a** distribution of two elements, **b** detailed configuration [69]

0	1	0	1
0	0	1	0
1	1	0	1
1	0	1	0

(a)

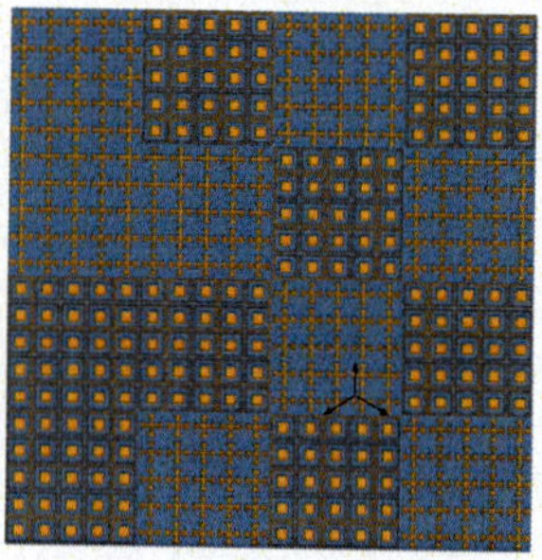

(b)

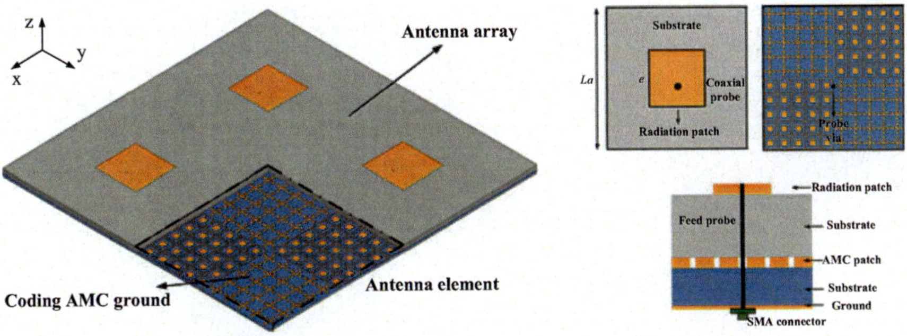

Fig. 8.70 Geometry of low scattering antenna array, $La = 90$ mm, $e = 27$ mm [69]

measured in the anechoic chamber. A 1–4 power divider is exploited to distribute the input power equally. The results are shown in Fig. 8.73a, b, and both are in good agreement with simulated ones. The maximum gains both maintain at 14.13 dBi around. Meanwhile, two arrays have nearly identical radiations in *xoz* and *yoz* planes.

Fig. 8.71 Prototypes of fabricated low scattering antenna array [69]

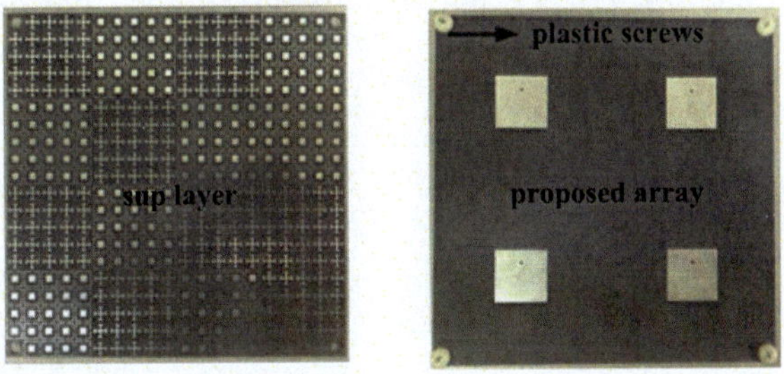

Fig. 8.72 Comparison between simulated and measured reflection coefficients [69]

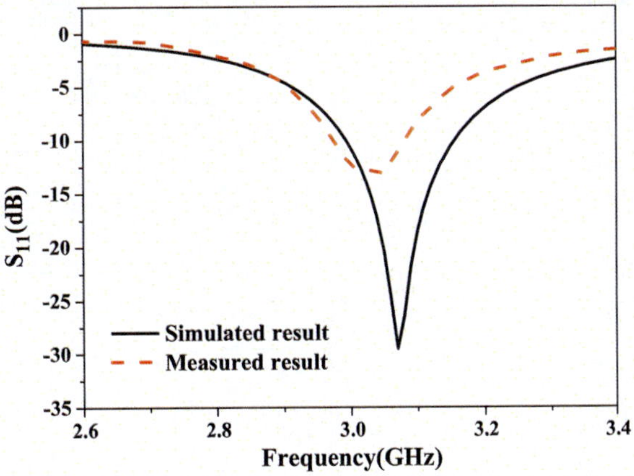

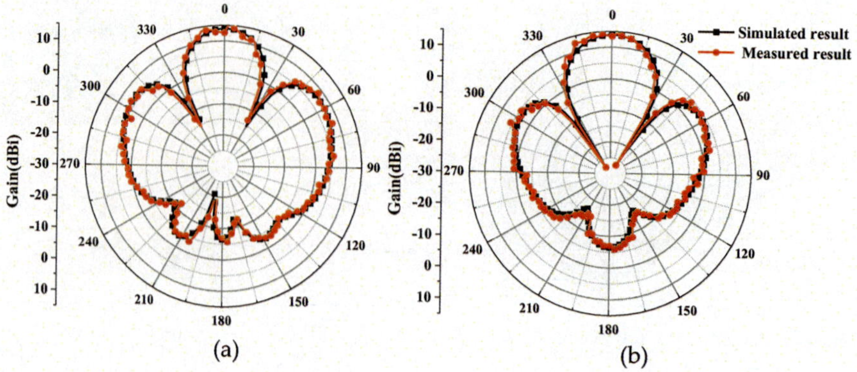

Fig. 8.73 Simulated and measured radiation patterns at 3.05 GHz for **a** *xoz* plane and **b** for *yoz* plane [69]

Restricted by the experimental condition, the monostatic RCS reduction of the antenna array loaded the coding AMC ground has been measured in Fig. 8.74. For the *x*-polarized wave, the measured and simulated results show a good agreement with each other. For the *y*-polarized wave, there is a discrepancy from 12 to 14 GHz. The cause of the error is that the simulations employ ideal incident plane waves and the tolerances of the fabrication and measurement.

Fig. 8.74 Measured monostatic RCS reduction for both polarizations under normal direction [69]

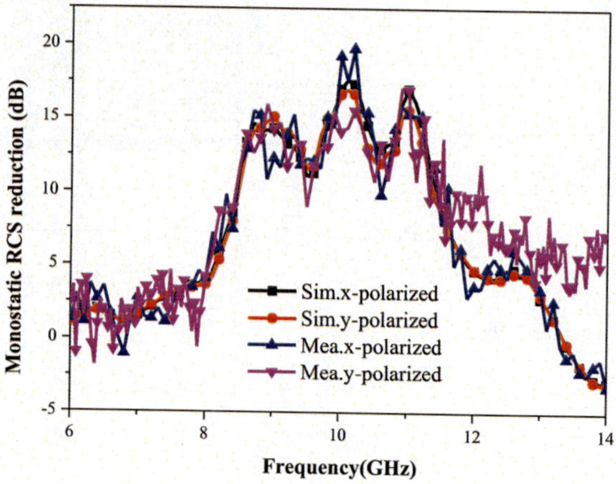

8.4 Conclusion

In this chapter, we have discussed two parts, i.e., cloaking devices for invisibility and metasurface-based devices for RCS reduction. With the coordinate transformation approach, three kinds of complementary cloaks, including minimized cloak, illusion cloak, and three-dimensional cloak, are designed to achieve perfect transparency and illusion. The main challenge in the complementary cloaks is the complexity of the material parameters. Alternatively, the scattering cancellation method provides a simple way to achieve the cloaking devices by wrapping a shell with the designed homogeneous material. The Mie series expansion method for the canonical shapes and the characteristic mode method for an arbitrarily-shaped object, respectively, are used to derive analytical formulas for the invisibility and illusion cloaks. Meanwhile, we highlight that optimal invisibility and illusion performance can be approximately achieved by optimizing the material parameter of the covering. On the other hand, the perfect metamaterial absorber and artificial magnetic conductor have been loaded on the microstrip antenna array and waveguide slot antenna array to reduce their radar cross section. Some excellent loading methods improve the scattering performance and enhance the radiation.

References

1. S. Li, X.-Y. Cao, J. Gao, T. Liu, Y.-J. Zheng, Z. Zhang, Analysis and design of three layers perfect metamaterial-inspired absorber based on double split-serration-rings structure. IEEE Trans. Antennas Propag. **63**(11), 5155–5160 (2015)
2. S. Li, J. Gao, X.-Y. Cao, Z. Zhang, T. Liu, Y.-J. Zheng, C. Zhang, G. Zheng, Hybrid metamaterial device with wideband absorption and multiband transmission based on spoof surface plasmon polaritons and perfect absorber. Appl. Phys. Lett. **106**(18), 181103 (2015)
3. T. Liu, X. Cao, J. Gao, Q. Yang, W. Li, Design of miniaturized broadband and high gain metamaterial patch antenna. Microw. Opt. Tech. Lett. **53**(12), 2858–2861 (2011)
4. Y. Zheng, J. Gao, X. Cao, S. Li, Wideband RCS reduction and gain enhancement microstrip antenna using chessboard configuration superstrate. Microw. Opt. Technol. Lett. **57**(7), 1738–1741 (2015)
5. Y. Zheng, J. Gao, Y. Zhou, X. Cao, H. Yang, S. Li, T. Li, Wideband gain enhancement and RCS reduction of Fabry-Perot resonator antenna with chessboard arranged metamaterial superstrate. IEEE Trans. Antennas Propag. **66**(2), 590–598 (2018)
6. X. Liu, J. Gao, X. Cao, Y. Zhao, S. Li, A coding diffuse metasurface for RCS reduction. IEEE Antennas Wirel. Propag. Lett. **16**, 724–727 (2017)
7. J.B. Pendry, D. Schurig, D.R. Smith, Controlling electromagnetic fields. Science **312**, 1780–1782 (2006)
8. U. Leonhardt, Optical conformal mapping. Science **312**, 1777–1780 (2006)
9. D. Schurig, J.J. Mock, B.J. Justice, S.A. Cummer, J.B. Pendry, A.F. Starr, D.R. Smith, Metamaterial electromagnetic cloak at microwave frequencies. Science **314**(5801), 977–980 (2006)
10. S.A. Cummer, B.I. Popa, D. Schurig, D.R. Smith, J.B. Pendry, Full wave simulations of electromagnetic cloaking structures. Phys. Rev. E **74**, 036621 (2006)
11. D. Schurig, J.B. Pendry, D.R. Smith, Calculation of material properties and ray tracing in transformation media. Opt. Express **14**, 9794–9804 (2006)

12. W.X. Jiang, T.J. Cui, X.M. Yang, Q. Cheng, R. Liu, D.R. Smith, Invisibility cloak without singularity. Appl. Phys. Lett. **93**, 194102 (2008)
13. R. Liu, C. Ji, J.J. Mock, J.Y. Chin, T.J. Cui, D.R. Smith, Broadband ground-plane cloak. Science **323**, 366–369 (2009)
14. J. Li, J.B. Pendry, Hiding under the carpet: a new strategy for cloaking. Phys. Rev. Lett. **101**, 203901 (2008)
15. H. Chen, X. Luo, H. Ma, C.T. Chen, The anti-cloak. Opt. Express **16**, 14603–14608 (2008)
16. G. Castaldi, I. Gallina, V. Galdi, A. Alù, N. Engheta, Cloak/anticloak interactions. Opt. Express **17**, 3101–3114 (2009)
17. T.J. Cui, D.R. Smith, R.P. Liu, *Metamaterials Theory, Design, and Applications* (Springer, New York, 2010)
18. D.H. Werner, D. Kwon, *Transformation Electromangetics and Metamaterials* (Springer, London, 2014)
19. Y. Lai, H.Y. Chen, Z.Q. Zhang, C.T. Chan, Complementary media invisibility cloak that cloaks objects at a distance outside the cloaking shell. Phys. Rev. Lett. **102**, 093901 (2009)
20. Y. Shi, W. Tang, C.H. Liang, A minimized invisibility complementary cloak with a composite shape. IEEE Antennas Wirel. Propag. Lett. **13**, 1800–1803 (2014)
21. Y. Shi, L. Zhang, W. Tang, L. Li, C. H. Liang, Design of a minimized complementary illusion cloak with arbitrary position. Int. J. Antennas Propag., Article ID 932495 (2015)
22. Y. Shi, W. Tang, L. Li, C.H. Liang, Three-dimensional complementary invisibility cloak with arbitrary shapes. IEEE Antennas Wirel. Propag. Lett. **14**, 1550–1553 (2015)
23. F.F. Huo, L. Li, T. Li, Y.M. Zhang, C.H. Liang, External invisibility cloak for multi-objects with arbitrary geometries. IEEE Antennas Wirel. Propag. Lett. **13**, 273–276 (2014)
24. C.F. Yang, J.J. Yang, M. Huang, Z. Xiao, J.H. Peng, An external cloak with arbitrary cross section based on complementary medium and coordinate transformation. Opt. Express **19**, 1147–1151 (2011)
25. Y. Lai, J. Ng, H.Y. Chen, D.Z. Han, J.J. Xiao, Z.Q. Zhang, C.T. Chan, Illusion optics: the optical transformation of an object into another object. Phys. Rev. Lett. **102**, 253902 (2009)
26. Y.R. Padooru, A.B. Yakovlev, P.Y. Chen, A. Alú, Analytical modeling of conformal mantle cloaks for cylindrical objects using sub-wavelength printed and slotted arrays. J. Appl. Phys. **112**, 034907 (2012)
27. A. Alú, N. Engheta, Achieving transparency with plasmonic and metamaterial coatings. Phys. Rev. E **72**, 016623 (2005)
28. A. Alú, N. Engheta, Polarizabilities and effective parameters for collections of spherical nanoparticles formed by pairs of concentric double-negative, single-negative, and/or double-positive metamaterial layers. J. Appl. Phys. **97**, 094310 (2005)
29. Y.X. Ni, L. Gao, C.W. Qiu, Achieving invisibility of homogeneous cylindrically anisotropic cylinders. Plasmonics **5**, 251–258 (2010)
30. Y. Wu, J. Li, Z.Q. Zhang, C.T. Chan, Effective medium theory for magnetodielectric composites: Beyond the long-wavelength limit. Phys. Rev. B **74**, 085111 (2006)
31. Z.H. Jiang, D.H. Werner, Quasi-three-dimensional angle-tolerant electromagnetic illusion using ultrathin metasurface coatings. Adv. Funct. Mater. **24**, 7728–7736 (2014)
32. N. Xiang, Q. Cheng, H.B. Chen, J. Zhao, W.X. Jiang, H.F. Ma, T.J. Cui, Bifunctional metasurface for electromagnetic cloaking and illusion. Appl. Phys. Express **8**, 092601 (2015)
33. L. Zhang, Y. Shi, C.H. Liang, Achieving illusion and invisibility of inhomogeneous cylinders and spheres. J. Opt. **18**, 085101 (2016)
34. L. Zhang, Y. Shi, C.H. Liang, Optimal illusion and invisibility of multilayered anisotropic cylinders and spheres. Opt. Express **24**, 23333–23352 (2016)
35. S. Tricrico, F. Bilotti, A. Alú, L. Vegni, Plasmonic cloaking for irregular objects with anisotropic scattering properties. Phys. Rev. E **81**, 026602 (2010)
36. Y. Shi, L. Zhang, Cloaking design for arbitrarily shape objects based on characteristic mode method. Opt. Express **25**, 32263–32279 (2017)
37. H.-K. Jang, W.-J. Lee, C.-G. Kim, Design and fabrication of a microstrip patch antenna with a low radar cross section in the X-band. Smart Mater. Struct. **20**, 015007 (2011)

38. S. Genovesi, F. Costa, A. Monorchio, Low profile array with reduced radar cross section by using frequency selective surfaces. IEEE Trans. Antennas Propag. **60**(5), 2327–2335 (2012)
39. F. Costa, S. Genovesi, A. Monorchio, A frequency selective absorbing ground plane for low-RCS microstrip antenna arrays. Prog. Electromag. Res. **126**, 317–332 (2012)
40. H. Yang, F. Yang, S. Xu, Y. Mao, M. Li, X. Cao, J. Gao, A 1-Bit 10×10 reconfigurable reflectarray antenna: design, optimization, and experiment. IEEE Trans. Antennas Propag. **64**(6), 2246–2254 (2016)
41. S. Li, G. Jun, X. Cao, Z. Zhao, Z. Di, Broadband and high-isolation dual-polarized microstrip antenna with low radar cross section. IEEE Antennas Wirel. Propag. Lett. **13**(1), 1413–1416 (2014)
42. S. Li, X. Cao, J. Gao, Q. Zheng, H. Yang, Broadband and miniaturization of tilted beam antenna using CSRR splits and flower-spiral structure. Microw. Opt. Technol. Lett. **56**(1), 27–31 (2014)
43. J.M. Jin, *Theory and Computation of Electromagnetic Field* (John Wiley, New Jersey, 2010)
44. E.M. Purcell, C.R. Pennypacker, Scattering and adsorption of light by nonspherical dielectric grains. Astrophys. J. **186**, 705–714 (1973)
45. B.T. Draine, The discrete-dipole approximation and its application to interstellar graphite grains. Astrophys. J. **333**, 848–872 (1988)
46. B.T. Draine, J.J. Goodman, Beyond clausius-mossotti—wave-propagation on a polarizable point lattice and the discrete dipole approximation. Astrophys. J. **405**, 685–697 (1993)
47. B.T. Draine, P.J. Flatau, Discrete-dipole approximation for scattering calculations. J. Opt. Soc. Am. A **11**, 1491–1499 (1994)
48. F.M. Kahnert, Numerical methods in electromagnetic scattering theory. J. Quant. Spectrosc. Radiat. Transf. **79**, 775–824 (2003)
49. M.A. Yukin, A.G. Hoekstra, The discrete dipole approximation: an overview and recent developments. J. Quant. Spectrosc. Radiat. Transf. **106**, 558–589 (2007)
50. R.J. Garbacz, R.H. Turpin, A generalized expansion for radiated and scattered fields. IEEE Trans. Antennas Propag. **19**, 348–358 (1971)
51. R.F. Harrington, J.R. Mautz, Theory of characteristic modes for conducting bodies. IEEE Trans. Antennas Propag. **19**, 622–628 (1971)
52. R.F. Harrington, J.R. Mautz, J.Y. Chang, The theory of characteristic modes for dielectric and magnetic bodies. IEEE Trans. Antennas Propag. **20**, 194–198 (1972)
53. N.I. Landy, S. Sajuyigbe, J.J. Mock, D.R. Smith, W.J. Padilla, Perfect metamaterial absorber. Phys. Rev. Lett. **100**(20), 207402 (2008)
54. S. Li, J. Gao, X.-Y. Cao, G. Zheng, Polarization-insensitive and thin stereometamaterial with broadband angular absorption for the oblique incidence. Appl. Phys. A **119**(1), 371–378 (2015)
55. S. Li, G. Jun, X.-Y. Cao, Z. Zhang, Loading MPA method for RCS reduction based on the surface current distribution of guidewave slot array antennas. IET Microw. Antennas Propag. **9**(2), 175–181 (2015)
56. S.-J. Li, P.-X. Wu, H.-X. Xu, Y.-L. Zhou, X.-Y. Cao, J.-F. Han, C. Zhang, H.-H. Yang, Z. Zhang, Ultra-wideband and polarization-insensitive perfect absorber using multilayer metamaterials, lumped resistors, and strong coupling effects. Nanoscale Res. Lett. **13**, 386 (2018)
57. S. Li, X. Cao, J. Gao, Z. Zhang, Y. Zheng, D. Zhang, Multiband and broadband polarization-insensitive perfect absorber devices based on a tunable and thin double split-ring metamaterial. Opt. Express **23**(3), 3523–3533 (2015)
58. S. Li, J. Gao, X. Cao, W. Li, Z. Zhang, D. Zhang, Wideband, thin and polarization-insensitive perfect absorber based the double octagonal rings metamaterials and lumped resistances. J. Appl. Phys. **116**(4), 043710 (2014)
59. L. Li, Y. Yang, C. Liang, A wide-angle polarization-insensitive ultra-thin metamaterial absorber with three resonant modes. J. Appl. Phys. **110**(6), 207402 (2011)
60. L. Li, Z. Lv, Ultra-wideband polarization-insensitive and wide-angle thin absorber based on resistive metasurfaces with three resonant modes. J. Appl. Phys. **122**(5), 055104 (2017)
61. L. Li, R. Xi, H. Liu, Z. Lv, Broadband polarization-independent and low-profile optically transparent metamaterial absorber. Appl. Phys. Express **11**, 052001 (2018)

62. Y. Zheng, J. Gao, L. Xu, X. Cao, T. Liu, Ultra-wideband and polarization independent radar cross section reduction with composite artificial magnetic conductor surface. IEEE Antennas Wirel. Propag. Lett. **16**, 1651–1654 (2017)
63. T. Liu, X. Cao, J. Gao, Q. Zheng, W. Li, H. Yang, RCS reduction of waveguide slot antenna with metamaterial absorber. IEEE Trans. Antenna Propag. **61**(3), 1479–1484 (2013)
64. T. Chen, S.-J. Li, X.-Y. Cao, J. Gao, Z.-X. Guo, Ultra-wideband and polarization-insensitive fractal perfect metamaterial absorber based on a 3-dimensional fractal tree microstructure with multi-modes. Appl. Phys. A (Mat. Sci. & Process.) **125**(4), 232–239 (2019)
65. W. Li, X. Cao, J. Gao, Z. Zhang, L. Cong, Broadband RCS reduction and gain enhancement microstrip antenna using shared aperture artificial composite material based on quasi-fractal tree. IET Microw. Antennas Propag. **10**(4), 370–377 (2016)
66. A.Y. Modi, C.A. Balanis, C.R. Birtcher, H.N. Shaman, Novel design of ultrabroadband radar cross section reduction surfaces using artificial magnetic conductors. IEEE Trans. Antenna Propag. **65**(10), 5406–5417 (2017)
67. Y. Zhao, X. Cao, J. Gao, X. Yao, T. Liu, W. Li, S. Li, Broadband low-RCS metasurface and its application on antenna. IEEE Trans. Antenna Propag. **64**(7), 2954–2962 (2016)
68. L. Cong, X. Cao, T. Song, J. Gao, Ultra-wideband low-RCS metasurface and its application on waveguide slot antenna array. Chin. Phys. B **27**(11), 114101 (2018)
69. C. Zhang, J. Gao, X. Cao, L. Xu, J. Han, Low scattering microstrip antenna array using coding artificial magnetic conductor ground. IEEE Antennas Wirel. Propag. Lett. **17**, 869–872 (2018)

Chapter 9
Metasurface-Based Wireless Power Transfer System

Shixing Yu, Pei Zhang, Hao Xue, and Long Li

Abstract Wireless power transfer (WPT) technology, originally proposed by Nikola Tesla, has regained prominence and garnered significant attention in academia and industry in recent years. WPT can generally be categorized into magnetic induction and magnetic coupling resonances (MCRs) for short-distance applications (typically less than 1 m), and microwave or laser radiation for long-distance applications (typically greater than 1 m). However, there remains a crucial challenge in improving the transmission distance and efficiency of WPT systems. Fortunately, metasurfaces (MSs) offer a promising solution to address these issues. This chapter introduces two metasurface-based WPT systems, namely a magnetic coupling resonance WPT system and a microwave radiation WPT system. In the first system, highly sub-wavelength magnetic negative (MNG) metasurfaces and double negative (DNG) metasurfaces are designed and integrated into the WPT system to enhance its efficiency. The tunneling effect of equivalent epsilon-near-zero (ENZ) metamaterials in the MCR-WPT system is uncovered, and a theoretical analysis using the effective medium model is proposed to study the WPT behavior. In the second system, a general synthesis procedure is outlined to design a reflective metasurface that enables high-efficiency WPT through near-field focusing, while accommodating desired multi-feed and multi-focus characteristics. Leveraging metasurface element design and phase synthesis techniques, the planar reflective metasurface controls the transmission of electromagnetic waves from specific sources, enabling the formation of near-field focusing beams towards desired destinations in the near-field region. This achieves wireless power multi-feed synthesis and multi-focus allocation with high efficiency. Two element structures are introduced to form arrays: the tri-dipole structure with single-polarization characteristics and the cross-dipole structure with

S. Yu
College of Big Data and Information Engineering, Guizhou University, Guiyang 550025, China
e-mail: sxyu1@gzu.edu.cn

P. Zhang
The 28th Research Institute of China Electronics Technology Group Corporation, Nanjing 310100, China

H. Xue · L. Li (✉)
School of Electronic Engineering, Xidian University, Xi'an 710071, Shaanxi, China
e-mail: lilong@mail.xidian.edu.cn

L. Li et al. (eds.), *Electromagnetic Metamaterials and Metasurfaces: From Theory To Applications*, https://doi.org/10.1007/978-981-99-7914-1_9

dual-polarization characteristics. Several metasurface prototypes are designed, fabricated, and experimentally measured for various scenarios. The stability and feasibility of the near-field focusing reflective metasurface for practical WPT applications are demonstrated through analysis and comparison of measured results with simulation data.

Keywords Wireless power transfer (WPT) · Magnetic resonant coupling · Metasurfaces (MSs) · Magnetic negative (MNG) · Double negative (DNG) · Epsilon-near-zero (ENZ) · Tunneling effect · Microwave radiation · Reflective metasurface · Near-field focusing · Multi-feed · Multi-focus

9.1 Introduction

Wireless power transfer (WPT) technology was originally proposed by Nikola Tesla [1–3] during the early twentieth century. Currently, there are three primary methods employed for WPT: inductive coupling, resonant coupling, and radiant transfer. Inductive coupling is an efficient technique that has been widely utilized for short-distance WPT [4, 5]. Resonant coupling, based on the principles of electromagnetic resonance, enables non-radiant WPT through near-field resonant coupling. Typically, it allows for transfer distances several times greater than the size of the transmitting device [6, 7]. Radiant WPT relies on the transmission of electromagnetic waves or lasers, and is particularly suitable for long-distance applications such as space-based transfers and satellite solar power stations. In order to facilitate advancements in various fields, including micro-robotics, medical treatment, mining, and portable electronic devices, the development of an efficient and compact WPT system is crucial. However, each of the aforementioned methods has its limitations when it comes to practical usage, whether it be the system's efficiency or size.

Figure 9.1 [8, 9] illustrates the schematic circuit for magnetic resonant coupling WPT. The source coil is characterized by parameters L_S and C_S, while the device coils possess parameters L_D and C_D. R_S and R_D represent the resistances of the source and device coils, respectively. Typically, the resonant frequencies of both the source and device coils are aligned to be identical. Furthermore, the operating frequency is adjusted to match the resonant frequencies $f = 1/2\pi\sqrt{L_S C_S} = 1/2\pi\sqrt{L_D C_D}$ of the coils. It is important to note that this mechanism involves a compromise between the transfer distance and transfer efficiency.

In recent years, there have been advancements in enhancing the efficiency of WPT through the use of metamaterials with μ-response [10, 11]. In 2011, a power relay system based on a near-field metamaterial super lens was proposed by Smith et al., who also conducted a comprehensive analysis of this system. Their findings indicated that even with a realistic magnetic loss tangent of approximately 0.1, the power transfer efficiency could be significantly improved by using the metamaterial super lens, particularly when the load exceeds a specific threshold value [12]. However, the current unit size of metamaterials is around 1/10 resonant wavelength, making

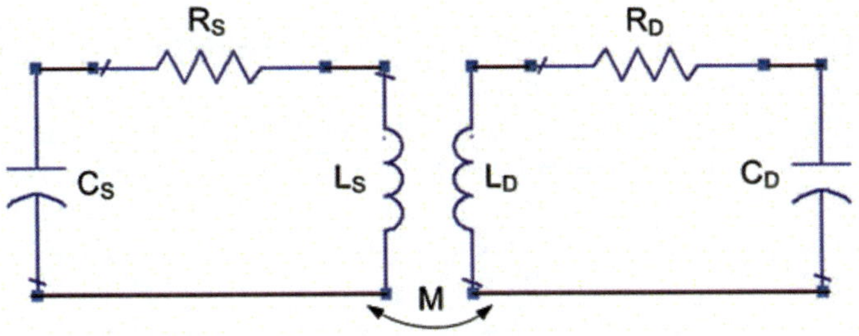

Fig. 9.1 Equivalent schematic circuit of the magnetic resonant coupling WPT [9]

them impractical for use at low frequencies. Therefore, the main challenges in WPT systems using metasurfaces (or metamaterials) revolve around achieving small-sized metasurfaces that operate effectively at radio frequencies with minimal loss. Wherein it is crucial to consider the influence of metamaterial properties on transfer efficiency.

In Sect. 9.2, we first introduce a highly sub-wavelength negative refractive index (NRI) metasurface that operates at radio frequencies, specifically in the HF band [13]. This NRI metamaterial employs a dual-layer design comprising planar spirals and meandering lines covering metallic strips, thereby satisfying the Lorentz-Drude model [14–16]. Such metamaterials have the potential to enhance the transfer efficiency of the WPT system. Without the covering strips, the corresponding metamaterial exhibits the property of negative permeability (MNG) rather than negative refractive index, which can also contribute to improving WPT efficiency. Therefore, we compare the WPT system utilizing the NRI metamaterial versus the system employing the MNG metamaterial. Additionally, we propose another type of metamaterial based on the concept presented in reference [17], which possesses negative permeability and will enhance the WPT system in this section. Experimental results demonstrate that the property of negative permeability and magnetic resonant coupling lead to efficiency enhancement. If the system is well matched, the efficiency can be approximated by the square of $|S_{21}|$. Hence, we utilize full-wave, finite-element-based simulation software HFSS and simulator ADS [18, 19] to compare S_{21} among different systems. The measured results validate that the miniaturized WPT system incorporating these proposed metamaterials can significantly improve transfer efficiency and operating distance [9, 20].

The emergence of metamaterials has opened up new possibilities for designing materials with controllable permittivity and permeability, allowing precise control over the propagation of electromagnetic waves [16, 21]. Theoretically, through the artificial design of appropriate structures, it is possible to achieve effective permittivity or permeability values equal to zero at specific frequencies, as substantiated by numerous research studies [22]. One of the most captivating phenomena associated with zero-index metamaterials (ZIM) is the tunneling effect [23]. The work of Silveirinha and Engheta demonstrated that epsilon-near-zero (ENZ) metamaterials can be employed as packing materials in narrow channels between waveguides to achieve super-coupling and squeezing of wave energy [24, 25]. This microwave experiment verified the occurrence of super-coupling [26–28]. In 2007, Soljačić et al. proposed a novel technique that utilized strongly coupled magnetic resonant helix coils to enable power delivery over a distance of 2 m with an efficiency of nearly 40%, successfully lighting a 60 W bulb [6]. The coupled magnetic resonant wireless power transfer system consists of multiple objects with the same intrinsic frequency, and the classical circuit analysis theory, employing lumped circuit elements (L, C, and R) models, was used to describe this system [29].

In Sect. 9.2.4, a novel perspective is introduced for analyzing resonant WPT systems, revealing their equivalence to ENZ metamaterial structures. Firstly, a multi-port network theory is employed to analyze the WPT system, resulting in a generalized resonance equation that determines the system's resonant frequency. Secondly, by examining the S-parameters of a practical WPT system's equivalent two-port network, unique behaviors of S_{11} and S_{21} are observed during resonant operation. Thirdly, under quasi-static assumptions, effective media parameters are retrieved from the S-parameters [30, 31], indicating that the effective permittivity (ε_r) of the WPT system approaches zero at the resonance frequency. Essentially, this suggests that the magnetic resonant WPT system can be regarded as an ENZ metamaterial according to the effective medium model. The high-efficiency power transmission in the WPT system is attributed to the tunneling effect of ENZ materials. Overall, this research offers a fresh perspective on the magnetic resonant WPT theory that can serve as a foundation for designing novel structures for efficient WPT systems over medium distances in future applications [32].

Section 9.3 specifically focuses on designs utilizing reflective metasurfaces to enable microwave radiant WPT. By controlling the wavefront phase of the metasurfaces, an electromagnetic near-field focusing (NNF) effect is achieved, facilitating high-efficiency and long-distance WPT. The NNF property is associated with the Fresnel and near regions of antennas, allowing the convergence of electromagnetic waves from the transmitting source to a specific point within a near-field region defined by the outer boundary of $2D^2/\lambda$ [33, 34]. Over the years, NNF has been implemented through various antenna structures, including parabolic reflectors [35], dielectric lens antennas [36], microstrip phased arrays [37–40], planar Fresnel zone plate (FZP) lenses [41, 42], among others. However, challenges such as the processing difficulties of parabolic reflectors, complexity of microstrip arrays, and low-efficiency of planar FZP lenses have hindered the widespread adoption of NNF for WPT applications. In recent times, a simpler alternative solution to phased arrays has emerged in the form of microstrip leaky-wave antennas [43, 44]. Nevertheless, further exploration is required to develop a transmitter that can balance efficiency and cost considerations in WPT systems.

On the other hand, the practical requirements of WPT demand greater diversification and flexibility. Consequently, researchers have devoted significant attention to enhancing control over focused beams. In 2013, a notable contribution involved utilizing Gaussian arrays to control amplitude excitation, reducing unwanted secondary lobes. By employing higher-order Bessel beam methods, this approach successfully achieved multi-focused beams in the near-field using a circular antenna array operating at 20 GHz [45]. In 2016, a Ka-band NFF array employing substrate integrated waveguide (SIW) technology was proposed. Here, metallic circular holes were designed as phase shifters and radiating elements, resulting in a side-lobe level below −18 dB through careful dimension arrangements of these holes [46]. Subsequently, in 2017, researchers introduced a reconfigurable holographic metasurface aperture to enable dynamic NFF capabilities [47]. Additionally, in 2018, a microstrip array was presented, capable of achieving a steerable focal distance ranging from 78 to 249 mm as the frequency varied between 9.25 and 10.5 GHz [48]. These advancements contribute to better control and manipulation of focused beams in WPT, addressing the need for increased versatility and adaptability in practical applications.

Since the beginning of this century, research on metasurfaces has introduced a groundbreaking method for regulating electromagnetic waves in an organized manner [49–53]. By arranging sub-wavelength electromagnetic structures periodically or aperiodically, the amplitude, phase, and polarization properties can be effectively manipulated. Each component of the metasurface is phase-compensated based on its optical path difference in relation to an in-phase point in space. The metasurface enables control over the beam direction, thereby focusing the power within a designated aperture in the near-field region. One notable application was the proposal of an NFF reflective array for a 2.4 GHz radio frequency identification (RFID) system, achieving a focus range of 2 m [54]. The application of NFF metasurfaces has also been explored by scholars in the fields of underwater ultrasonic waves [55] and mid-infrared waves [56]. In Sect. 9.3.2, the research and design of reflective metasurfaces enable the realization of highly efficient WPT by regulating the transmission of electromagnetic waves from specific feeds. This regulation creates near-field focusing beams directed towards specific destinations. A design procedure for a near-field focusing reflective metasurface with multi-feed and multi-focus characteristics is presented. The key feature of this design is the incorporation of elements with phase adjustment capability, allowing for precise near-field focusing through phase synthesis techniques. Four different types of focusing scenarios are explored: single-feed and single-focus (SFSF) case, single-feed and dual-focus (SFDF) case, dual-feed and single-focus (DFSF) case, and dual-feed and dual-focus (DFDF) case. In Sect. 9.3.3, the implementation of near-field focusing metasurfaces with 20 × 20 elements at ISM-5.8 GHz is discussed, specifically focusing on single-focus and two-focus scenarios, namely point-to-point (P2P) and point-to-multipoint (P2M) WPT. The planar near-field measurement technique is employed to scan the focus plane and validate the proposed approach and prototype. The measured results of the two metasurfaces are reported and compared with simulated performance, demonstrating excellent near-field focusing achieved by the proposed tri-dipole metasurfaces. Section 9.3.4 introduces a dual-polarization NFF reflective metasurface featuring a cross-dipole structure, enabling independent polarization control. In conjunction with the multi-beam reflection method, this design addresses the requirements of WPT for multi-focus power allocation and multi-source power synthesis. Two NFF reflective metasurfaces operating at 10 GHz with dimensions of 390 mm × 390 mm are designed for single-feed single-focus and single-feed dual-focus scenarios in different polarizations. A WPT system is established to verify the effectiveness of

NFF. Finally, the focusing performance for long-distance WPT is further analyzed through full-wave simulation, shedding light on various characteristics of NFF.

9.2 Highly Sub-Wavelength Metamaterials for MCR-WPT System

Metamaterials are artificial materials that possess unique properties not found in conventional natural materials. One notable type of metamaterial is double negative (DNG) materials, which exhibit simultaneous negative permeability and negative permittivity, resulting in a NRI. Veselago systematically introduced DNG metamaterials back in 1968 [57], and they are commonly known as left-handed materials (LHM) or NRI materials that have garnered considerable attention in the scientific and engineering communities. These metamaterials showcase exceptional electromagnetic properties, including the reversal of Snell's law, abnormal Doppler effects, and the reversal of the Vavilov-Cherenkov effect [58, 59]. Utilizing NRI metamaterials in electronic devices can effectively enhance their performance.

However, there remains a noticeable lack of demonstrated metamaterials operating at low frequencies, particularly around 100 MHz or lower. This frequency range encompasses widely-used devices such as televisions, radios, and wireless power transfer systems [5]. The unit cell size of metamaterials is typically around $\lambda/10$, which becomes impractical for low-frequency applications, where λ represents the operating wavelength. To address this challenge, Ziolkowski et al. proposed lumped element-based negative-index metamaterials operating at ultra-high frequencies (UHF) [5]. By employing lumped elements, they successfully achieved highly sub-wavelength unit cells approximately 75 times smaller than the operating wavelength at 400 MHz. Additionally, Chen et al. proposed extremely sub-wavelength planar magnetic metamaterials with negative permeability properties [17]. Hence, it is evident that designing DNG or NRI metamaterials for lower frequencies remains a persistent challenge.

This section introduces highly sub-wavelength double negative metamaterials specifically designed for high-frequency (HF) operation [13]. The unit cell is composed of a two-layered structure. One side features a planar spiral that realizes the MNG portion based on the Lorentz model, while the other side utilizes a meander line to achieve the electric negative (ENG) properties. It is crucial to ensure that two narrow metallic strips make contact with the meander line to simulate the Drude model. The interaction between the MNG and ENG components of the unit cell yields a DNG material, which significantly contributes to the miniaturization of the proposed metamaterials.

Furthermore, a simple probe method is employed to conveniently measure the transmission and reflection characteristics of the metamaterials, facilitating the retrieval of their parameters from experimental results. This measurement technique offers convenience in obtaining valuable information about the metamaterials.

Notably, the simulation results obtained from the transverse electromagnetic (TEM) waveguide exhibit good agreement with those obtained using the probe method.

In summary, the development of highly sub-wavelength double negative metamaterials operating at HF frequencies presents a promising avenue for addressing the challenges associated with designing DNG or NRI metamaterials for lower frequencies.

9.2.1 Design of Highly Sub-Wavelength DNG Metamaterials

To achieve a material that exhibits both negative electric permittivity and negative magnetic permeability simultaneously, Pendry [58] initially proposed a composite structure comprising long wire arrays and split ring resonator (SRR) arrays. This composite structure combines the negative permittivity derived from wires based on the Drude model and the negative permeability derived from SRRs based on the Lorentz model. As a result, it can provide a narrow frequency band with negative permeability when subjected to specific polarization of incident electromagnetic (EM) waves. The first experimental demonstration of such a LHM was carried out by Smith et al. [15, 16].

To further advance the applications of DNG metamaterials in radio frequencies, we have designed a highly sub-wavelength DNG metamaterial that possesses miniaturization advantages and operates at HF ranges. This metamaterial consists of a dual-layer structure created by etching both sides of a substrate, referred to as MS. On the front side of the MS, an etched square spiral connected to a circular patch is present, while on the back side, a meander line is etched. These front and back patterns are depicted in Fig. 9.2a, b, respectively.

When an incident EM wave with a magnetic field H is perpendicular to the plane of the spiral, circular currents are induced on the spiral, and charges accumulate across the gaps in the spiral arms. Consequently, each individual spiral acts as a serial RLC resonant circuit, following the frequency-dispersive Lorentz model, thereby generating an effective negative permeability. On the other hand, when the meander line is excited by an electric field oriented perpendicular to the long arms of the meander line, it exhibits inductance behavior. The characteristics of the meander line comply with the Drude model if each electrically small copper strip, serving as a transmission line, terminates in a short circuit. To fulfill this requirement, two narrow metallic strips are coated on the edges of the MS, and the meander line is extended to make contact with these strips. Thus, this type of metamaterial still adheres to the Lorentz-Drude model. Additionally, the strong coupling between the topological structures on both sides contributes to the miniaturization of the proposed metamaterials [13].

The DNG metamaterials possess the unique property of exhibiting effective negative permeability around the magnetic resonant frequency, as well as effective negative permittivity from DC to the plasma frequency. Consequently, these metamaterials can be adjusted to achieve a NRI over a frequency range where both permeability and

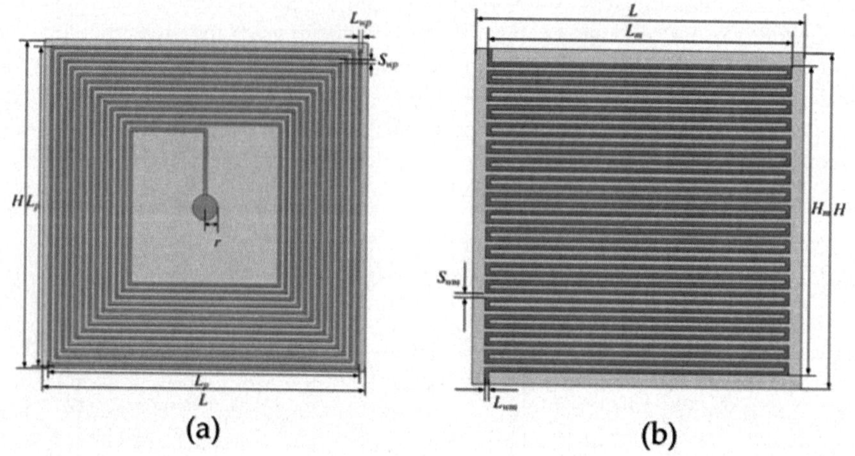

Fig. 9.2 Designed DNG metamaterials, **a** front view of square spirals pattern, **b** back view of meander lines pattern [13]

permittivity exhibit negative values simultaneously. The parameters for the proposed structure are provided in Table 9.1. The board has a thickness of 1 mm, while the copper foil measures 0.017 mm in thickness. The surface of the copper is plated with silver. The width of the metallic strips for coating is several times larger than the substrate thickness. Theoretically, the length of these metallic strips can be as long as the substrate itself, but for experimental convenience, we have made it slightly greater than the length of the substrate. The substrate is composed of a dielectric material with a relative permittivity of 2.6 and a loss tangent of 0.015.

Simulating the unit cell of such metamaterials can be conveniently achieved using the TEM waveguide method [18, 30]. However, measuring these metamaterials poses challenges due to the requirement for a large experimental setup, especially at low operation frequencies. In this case, a simple probe method was employed to measure the reflection and transmission coefficients. The probe is connected at one end to a metallic strip, and at the other end, it is connected to the other metallic strip. This configuration ensures that the electric field is forced to be perpendicular to the metallic strips, allowing for accurate measurements. For reference, the front view and back view photographs of the measurement system used for the fabricated DNG metamaterials are presented in Fig. 9.3a, b, respectively.

Table 9.1 Parameters of DNG metamaterials (unit: mm)

H	L	r	L_p	L_{wp}
78	78	3	75	0.63
S_{wp}	L_m	H_m	L_{wm}	S_{wm}
0.96	76.8	76.8	0.9	0.6

(a) (b)

Fig. 9.3 Fabricated the unit cell of DNG metamaterials and probes experiment, **a** front view and **b** back view [13]

9.2.2 *Characteristics of Highly Sub-Wavelength DNG Metamaterials*

To validate and assess the reliability of this approach, we conducted a comparison between the estimated effective medium parameters obtained from the TEM waveguide method and the probes method, as depicted in Fig. 9.4. It is evident from the results that the data acquired through the probes method closely aligns with those obtained using the TEM waveguide method. An important observation is that the TEM waveguide method operates with a characteristic impedance of 377 Ω at the ports, whereas the probes method operates at 50 Ω. Consequently, the effective permittivity and permeability estimated by each method need to be normalized to their respective characteristic impedances. For the measurement of DNG metamaterials at lower frequencies, the monopole probes method proves to be effective.

Figure 9.5 illustrates the distribution of electric and magnetic fields' amplitude on the mid-plane of the metamaterial substrate at the magnetic resonant frequency of 13.4 MHz. It is evident that a significant resonant coupling behavior exists between the topological structures on both sides. The unit cell size of DNG metamaterials is approximately 280 times smaller than the operational wavelength at 13.4 MHz. This strong coupling between the topological structures on both sides plays a pivotal role in achieving the miniaturization of the proposed metamaterials.

9.2.3 *WPT System Integrating with Metamaterials*

In this section, we present an enhanced WPT system that incorporates coupling rings with high sub-wavelength MSs [9, 20]. Placing one or more MSs between

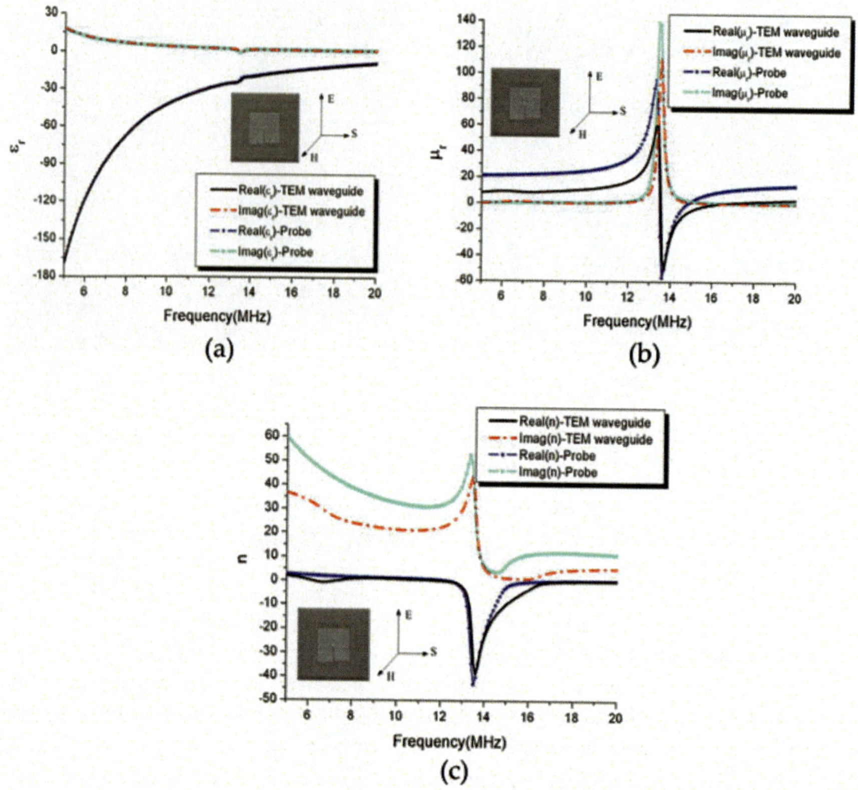

Fig. 9.4 Comparison of the retrieved results of the DNG metamaterials obtained by the TEM waveguide method and the monopole probes method, **a** effective permittivity, **b** effective permeability, and **c** effective refractive index 13]

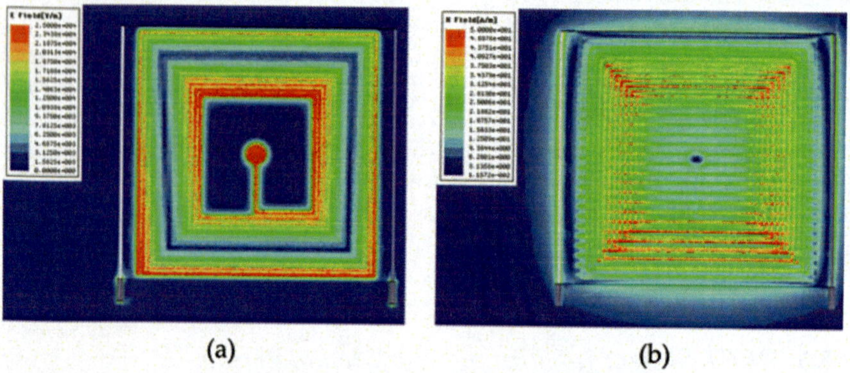

Fig. 9.5 Simulated amplitude distributions of electric and magnetic fields on the mid-plane of the DNG metamaterial substrate at the magnetic resonant frequency of 13.4 MHz, **a** E-field and **b** H-field [13]

the two coupling rings results in a significant increase in power transfer efficiency. This improvement can be attributed to a change in the coupling mechanism, wherein the introduction of one or more NRI/MNG MSs between the rings enables resonant coupling. Resonant coupling exhibits high efficiency within the mid-range frequency. Consequently, incorporating one or more MSs between the two rings leads to a noteworthy enhancement in both transfer distance and efficiency.

9.2.3.1 WPT System with One MS

A. WPT System with One NRI MS

Figure 9.6 illustrates the WPT system model using one NRI MS. The configuration consists of a single cell positioned between the two coupling rings. Additionally, Fig. 9.6 displays a photograph of the measurement system. In this setup, R represents the outer radius of the copper ring, r_{in} corresponds to the diameter of the copper ring, and D signifies the distance between the two rings. Likewise, d_1 and d_2 indicate the distances between the MS and each of the two rings, respectively. For the specific scenario presented, R is measured as 35 mm, r_{in} as 2 mm, while d_1, d_2, and D remain variable. The fabrication of the NRI MS was carried out.

Figure 9.7a shows the measured results of the variation of S parameters with frequency at the distance of 80 mm when the NRI MS is put just in the middle of the two copper rings, i.e., $D = 80$ mm, $d_1 = 40$ mm, and $d_2 = 40$ mm. Figure 9.7b shows the comparison of measured S_{21} of the proposed WPT system with and without NRI MS when the distance varies at the frequency of 14.6 MHz. In the measurement, the copper rings were also symmetrically located and symmetrically moved. Figure 9.8a shows the magnetic field of the new WPT system with one NRI MS at the NRI frequency of the MS, and Fig. 9.8b shows the magnetic field at the frequency which is not the resonant frequency of the MS. So it can be seen from Figs. 9.7 and 9.8 that the new WPT system works in the mechanism of resonant coupling. And we can

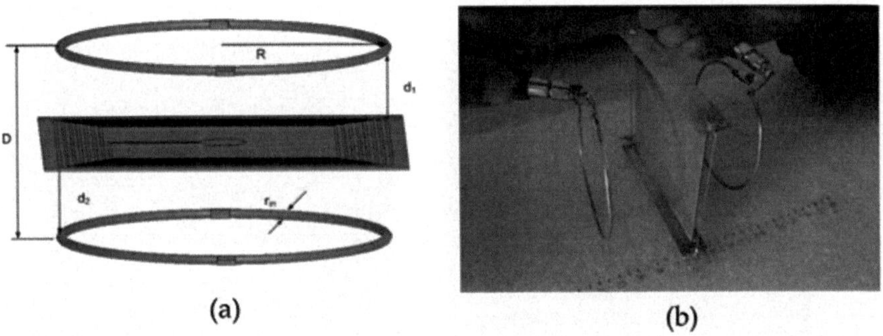

Fig. 9.6 **a** WPT system with one NRI MS, **b** photograph of the measuring system with one NRI WPT system [9]

also find from Fig. 9.7 that the efficiency is considerably improved compared with the WPT system without MSs.

B. **WPT System with One MNG MS**

In the previous scenario, if the two coated metallic strips are removed from the NRI MS, the MS would retain the property of negative permeability alone. We refer to these modified metamaterial structures as SMMNG (Spirals-Meander MNG) metamaterials, denoted as such because one side of the MS features a spiral pattern while the other side exhibits a meandering line pattern. Placing the SMMNG MS between the two coupling rings yields the similar WPT model as depicted in Fig. 9.6a. Figure 9.9 presents a comparison of the measured S_{21} values for the WPT system with an NRI MS and the WPT system with the SMMNG MS at a frequency of 14.6 MHz while varying the distance between the rings. It is evident from Fig. 9.9 that the

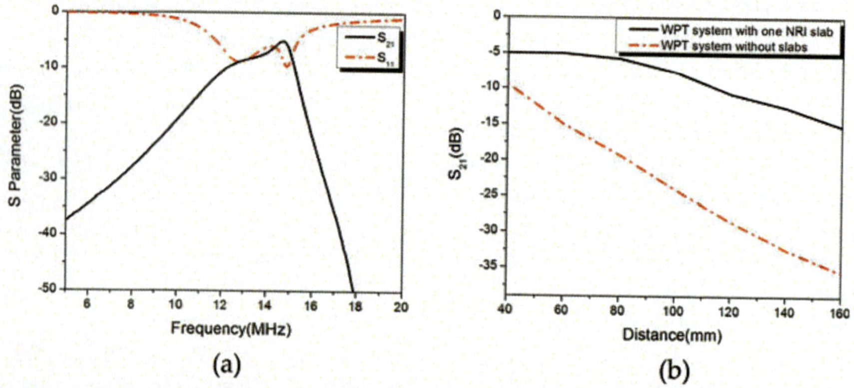

Fig. 9.7 **a** Characteristics of S parameters with the frequency of the WPT system with one NRI MS when the distance of the two copper rings is 80 mm, **b** comparison of measured S_{21} of the WPT system with and without NRI MS when the distance varies [9]

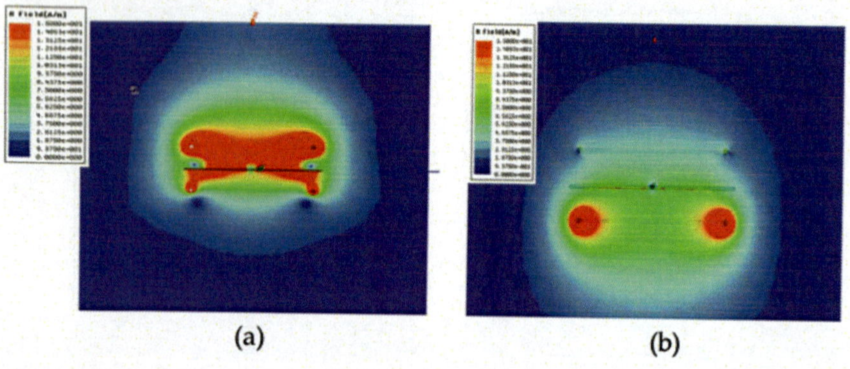

Fig. 9.8 Magnetic field of the WTP system with one NRI MS, **a** magnetic field at the NRI frequency of the MS, **b** magnetic field at the non-resonant frequency of the NRI MS [9]

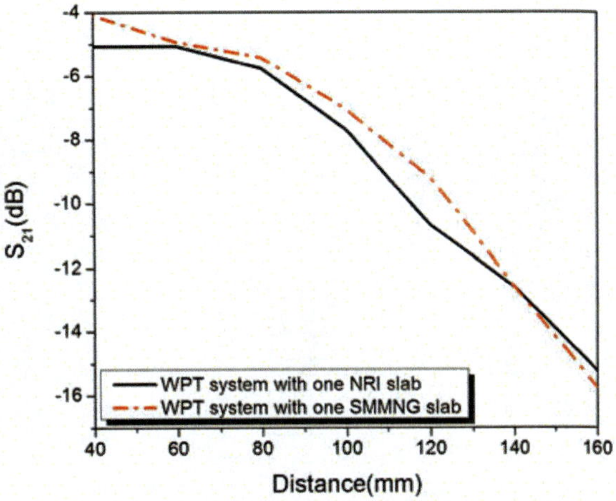

Fig. 9.9 Comparison of measured S_{21} of the WPT system with NRI MS and the SMMNG MS when the distance varies [9]

enhancement in WPT efficiency in both cases is nearly identical. This observation highlights that the significant contribution to the high transfer efficiency stems from the negative permeability characteristic, enabling the WPT system to operate based on magnetic resonance.

In order to validate the improvement achieved by utilizing MNG metamaterials in the WPT system, we employed a different type of MNG metamaterial structure initially proposed by Chen et al. [16], as illustrated in Fig. 9.10a. These MNG metamaterials feature a dual-layer design, with square spirals etched on both sides. Considering the operating frequency of the WPT system, we designed and fabricated the SSMNG (Spiral-Spiral MNG) MS, showcased in Fig. 9.10b. The two spirals on opposing sides are intentionally arranged in an anti-symmetrical configuration.

The effective parameters for relative permittivity and relative permeability, denoted as ε_r and μ_r, respectively, were obtained using retrieval techniques [19, 30]. The results are presented in Fig. 9.11a, b accordingly. During the calculation of the effective material properties, the proposed MSs were found to possess a periodicity of 78 mm. It is observable that the SSMNG MSs exhibit an effective negative permeability. The specifications of the designed structure are itemized in Table 9.2. The substrate board has a thickness of 1 mm, while the copper foil is 0.017 mm thick. The copper surface is plated with silver. The substrate itself is composed of a dielectric material with a relative permittivity of 2.6 and a loss tangent of 0.015.

In our setup, we integrate a single SSMNG MS into the WPT system, as depicted in Fig. 9.12a. Only one cell is positioned between the two coupling rings, with the parameters of the coupling rings remaining consistent with Case A. Figure 9.12b showcases the measured results of the WPT system with and without the SSMNG MS, while symmetrically placing the two copper rings at a frequency of 14.6 MHz. It is apparent that the inclusion of one SSMNG MS in the WPT system significantly enhances the power transfer efficiency. This finding substantiates the fact that

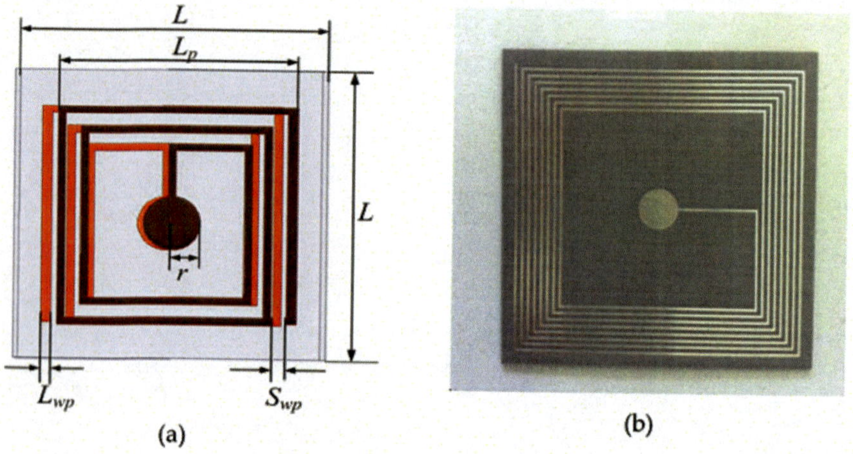

Fig. 9.10 **a** Solid view of the SSMNG metasurface geometry, **b** top view of the fabricated SSMNG MS [9]

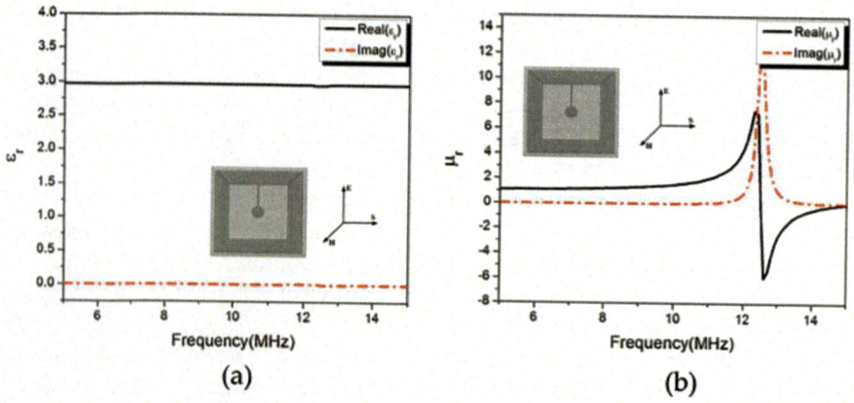

Fig. 9.11 **a** Effective permittivity and **b** effective permeability of the designed SSMNG MSs [9]

Table 9.2 Parameters of the SSMNG metamaterials (UNIT: mm)

L	L_p	L_{wp}	S_{wp}	r
78	70.4	0.6	1	5

the WPT system operates through magnetic resonance coupling by incorporating negative permeability metamaterials between the two coupling rings.

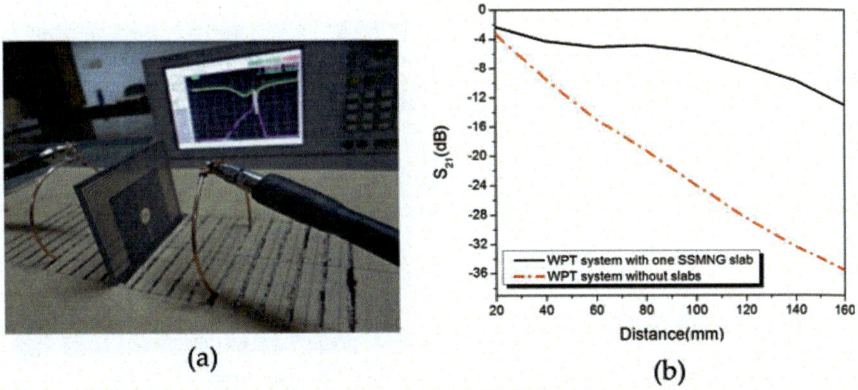

Fig. 9.12 **a** Photograph of the measured WPT system with one SSMNG MS, **b** comparison of the measured S_{21} of the WPT system with and without one SSMNG MS when the distance varies [9]

9.2.3.2 WPT System with Multiple MSs

A. **WPT System with Two MSs**

In this approach, we consider the placement of two MSs between the two rings. For illustrative purposes, we exclusively analyze the WPT system utilizing two SMMNG MSs. This strategy holds practical value as it allows for easy integration of the MSs with the source and load devices. The parameters of the MS, source ring, and load ring remain consistent with those discussed previously. The model of the WPT system with two SMMNG MSs is depicted in Fig. 9.13a, and a photograph of the experimental setup is presented in Fig. 9.13b. In Fig. 9.14, the measured S_{21} results of the proposed WPT system, at a frequency of 14.6 MHz, are shown when the distance from the MS to the copper ring remains fixed, denoted as $d_1 = d_2$.

The analysis depicted in Fig. 9.14 illustrates the substantial enhancement in transmission efficiency achieved by utilizing a WPT system with two SMMNG MSs. Notably, it is observed that efficiency is diminished when the values of d_1 and d_2 are either excessively small or large. The reduced efficiency for small d_1 and d_2 values arises from the compromised matched condition, whereas the decreased efficiency for large d_1 and d_2 values stems from the diminished coupling between the MS and the copper ring. Thus, it is crucial to ensure efficient performance even for small d_1 and d_2 values. To achieve this objective, the matched network can be optimized to maximize transmission efficiency.

B. **WPT System with Three MSs**

In this WPT system configuration (SM-SS-SM), two SMMNG MSs and one SSMNG MS are employed. The system setup can be visualized in Fig. 9.15a. Comparing the measured results of S_{21} at 14.6 MHz between the system with three MSs and those with two MSs or without any MSs, as depicted in Fig. 9.15b, it becomes evident that the system with three MSs exhibits significantly higher efficiency. Specifically,

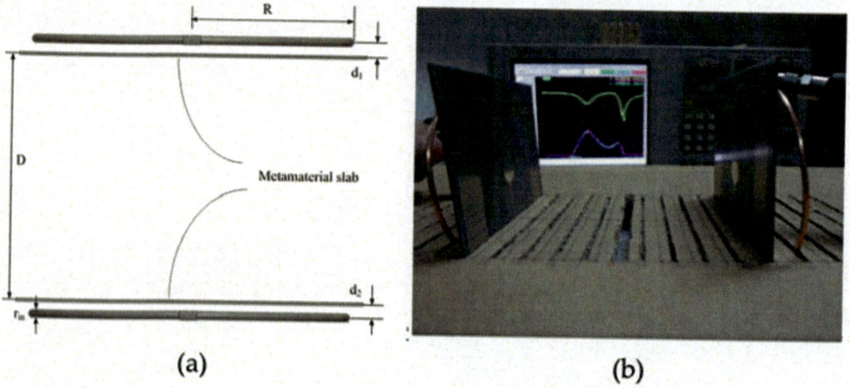

Fig. 9.13 **a** WPT system with two MSs, **b** photograph of the WPT system with two SMMNG MSs [9]

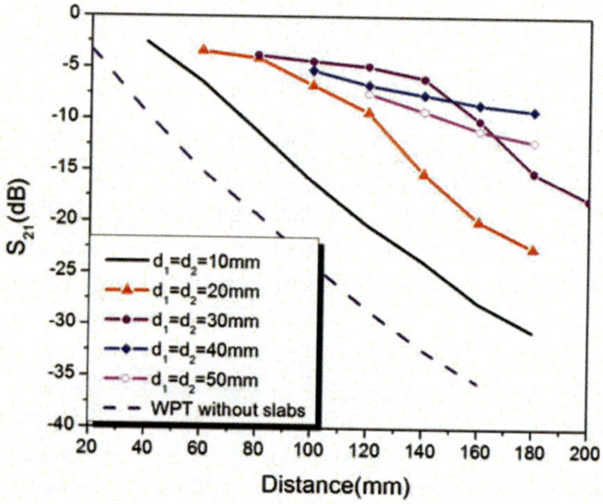

Fig. 9.14 Measured S_{21} of the WPT system with two SMMNG MSs and that of the WPT system without MSs, and the distance between the copper ring and the MS varies from 10 to 50 mm [9]

when the distance between the source and load rings is 140 mm, the transmission coefficient S_{21} of the WPT system with three MNG MSs (SM-SS-SM) experiences a notable gain of 27.3 dB, improving from −32.4 dB without any MSs to −5.1 dB with the three MNG MSs incorporated.

9.2.4 Tunneling Effect of Equivalent ENZ Metasurfaces in Magnetic Resonant WPT System

In this study, we present a thorough investigation into the utilization of effective ENZ metamaterials in magnetic resonant WPT systems. The WPT system is analyzed

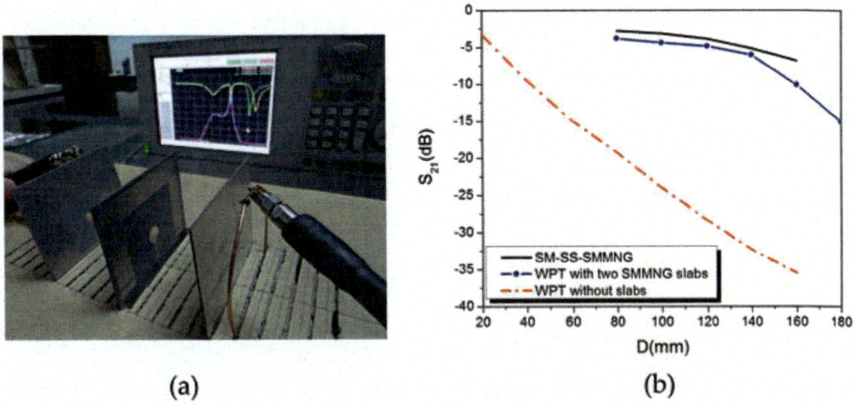

Fig. 9.15 **a** Photograph of the WPT system with three MSs, one SSMNG MS and two SMMNG MSs, **b** comparison of measured S_{21} of the WPT system with three MSs and the system with two MSs and without MSs [9]

using an equivalent network and an effective medium model, revealing the significant impact of ENZ metamaterials on power transfer efficiency. Theoretical analysis demonstrates that the enhanced efficiency can be attributed to the tunneling effect facilitated by the equivalent ENZ metamaterials. To validate the proposed model and investigate the phenomenon of super-coupling, simulations and experiments are conducted on a practical WPT system employing planar spiral coils. This research offers a novel perspective on magnetic resonant WPT theory, paving the way for the design of new structures aimed at achieving efficient WPT over intermediate distances.

9.2.4.1 Generalized Resonance and Equivalent Medium Analysis of WPT System

A complex resonant open system can be described by using a generalized multiport network, as shown in Fig. 9.16, which $[a] = [a_1, a_2, \cdots, a_n]^T$ is a normalized incident wave vector and $[b] = [b_1, b_2, \cdots, b_n]^T$ is a normalized reflection wave vector [60].

For a multiport network system, it obeys the law

$$-\oiint_{\Omega} \frac{1}{2}\left(\vec{E} \times \vec{H}^*\right) \cdot \hat{n} ds = -\sum_{i=1}^{n} \int_{T_i} \frac{1}{2}(\vec{E_t} \times \vec{H_t}^*) \cdot \hat{n} ds$$
$$= P_L + j2\omega(W_m - W_e) \quad (9.1)$$

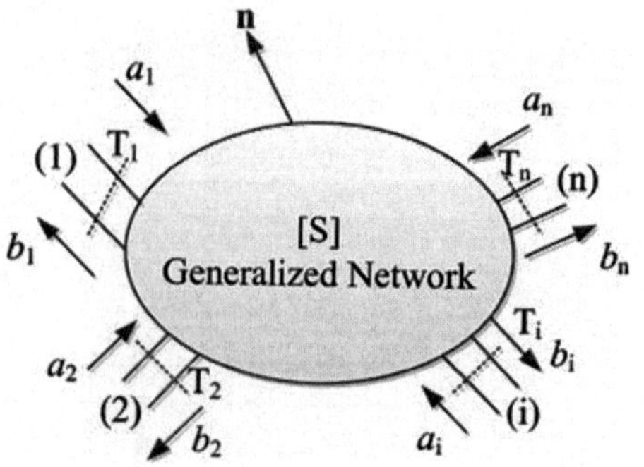

Fig. 9.16 Complex multiport generalized [S] network ($T_1, T_2, \ldots, T_n$ are reference planes)

where $P_L = \frac{1}{2}\iiint\limits_v \sigma\left|\vec{E}\right|^2 dV$, $W_m = \frac{1}{4}\iiint\limits_v \mu\left|\vec{H}\right|^2 dV$, and $W_e = \frac{1}{4}\iiint\limits_v \varepsilon\left|\vec{E}\right|^2 dV$ stand for the power dissipated within the system, the magnetic field energy and the electric field energy stored in the system, respectively. Using the matrix form, we can easily derive

$$\frac{1}{2}[I]^+[V] = P_L + j2\omega(W_m - W_e) \tag{9.2}$$

where $[I]$ and $[V]$ are the equivalent current and voltage vectors of ports. The symbol of $[]^+$ represents the Hermitian transpose operation. Further, the relationship between $[a]$, $[b]$, and $[I]$, $[V]$ can be expressed by the following equations

$$\begin{cases} [V] = [a] + [b] \\ [I]^+ = [a]^+ - [b]^+ \end{cases} \tag{9.3}$$

Consequently, Eq. (9.2) can be denoted as

$$\left([a]^+[a] - [b]^+[b]\right)/2 + j\mathrm{Im}\left([a]^+[b]\right) = P_L + j2\omega(W_m - W_e). \tag{9.4}$$

In Eq. (9.4), the former is the real part, and the latter is the imaginary part. The following equations can be obtained

$$\begin{cases} [a]^+\left([U] - [S]^+[S]\right)[a] = 2P_L \\ \mathrm{Im}\left([a]^+[S][a]\right) = 2\omega(W_m - W_e) \end{cases} \tag{9.5}$$

Notice that the matrix equation of the scattering parameter $[b] = [S][a]$ has been considered in the above equations and $[U]$ is a unitary matrix. For a complex resonant WPT system, we can define the resonance condition by the energy balance between the electric field energy and the magnetic field energy, namely,

$$W_m = W_e \tag{9.6}$$

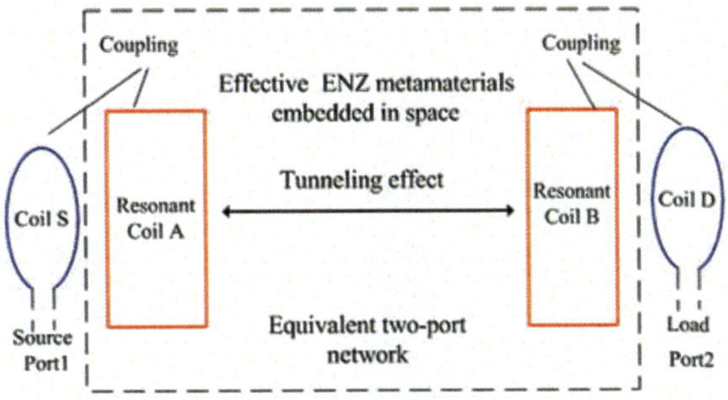

Fig. 9.17 Equivalent two-port network of a resonant WPT system

As a result, we obtained the generalized resonance equation as

$$\mathrm{Im}\big([a]^{+}[S][a]\big) = 0 \tag{9.7}$$

The conventional magnetic resonant WPT system can be equivalent to a two-port network, as shown in Fig. 9.17.

In the magnetic resonant WPT system, the resonance frequency varies with the coupling coefficients between the transmitter and receiver coils. The highest power transmission efficiency is achieved at the system's resonance frequencies, where the S-parameters of the equivalent network must satisfy the generalized resonance equation. Additionally, under the quasi-static assumption, the effective permittivity and permeability can be determined using effective medium theory [61, 62]. Figure 9.17 illustrates the resonant coils A and B as metamaterial elements embedded in free space. The Nicholson-Ross-Weir (NRW) approach, proposed by Nicholson, Ross, and Weir [31, 63, 64], is an effective method to retrieve the effective constitutive media parameters for such symmetrical resonant structures. Significantly, the effective permittivity (ε_r) of the WPT system approximates zero under the generalized resonance condition. Hence, the magnetic resonant WPT system can essentially be regarded as an ENZ metamaterial structure.

9.2.4.2 Tunneling Effect of WPT System

In this section, we construct a WPT system using a pair of planar spiral coils printed on an FR4 substrate with a thickness of 1 mm. The substrate dimensions are set to 300 mm in length (a) and 350 mm in width (b). At the center of the coil, a printed square loop with dimensions $c = 140$ mm is placed for excitation or reception, connected to the port through metallic vias and parallel transmission lines on the backside. The resonant coil consists of 11.75 turns, with an initial side length of 153 mm and an end side length of 268 mm. The simulation model of the WPT system is illustrated in Fig. 9.18. Both the transmitter and receiver coils are of equal size and separated by a distance of 0.5 m.

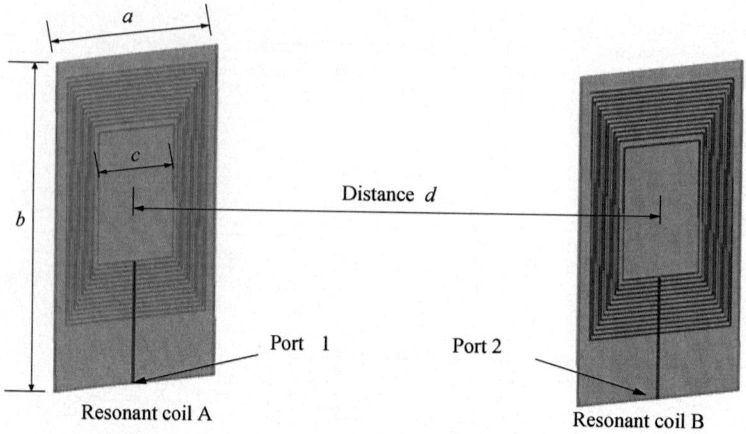

Fig. 9.18 WPT system using a pair of printed planar spiral coils

The NRW method can be employed to extract the effective media parameters for this symmetrical structure. Initially, the composite terms are introduced by

$$V_1 = S_{21} + S_{11}, \quad V_2 = S_{21} - S_{11} \tag{9.8}$$

where S_{11} and S_{21} are the equivalent network S-parameters. Derive the following quantities,

$$Y = \frac{1 - V_1 V_2}{V_1 - V_2} = \frac{1 + \Gamma^2}{2\Gamma} \tag{9.9}$$

$$\Gamma = Y \pm \sqrt{Y^2 - 1} \tag{9.10}$$

when the electrical thickness of the media is not too large, i.e., $kd \leq 1$ and $|\Gamma| < 1$ where $k = \omega\sqrt{\varepsilon_r \mu_r}/c = k_0\sqrt{\varepsilon_r \mu_r}$, and d is the thickness of the media. In the WPT system, d represents the distance between two spiral coils that are very small compared to the operational wavelength. Therefore, the effective medium parameters can be expressed as follows,

$$k = \frac{1}{jd}\left(\frac{(1 - V_1)(1 + \Gamma)}{1 - \Gamma V_1}\right) \tag{9.11}$$

$$\mu_r = \frac{2}{jk_0 d}\frac{1 - V_2}{1 + V_2} \tag{9.12}$$

$$\varepsilon_r = \left(\frac{k}{k_0}\right)^2 \frac{1}{\mu_r} \tag{9.13}$$

where $k_0 = 2\pi / \lambda_0$, and λ_0 represents the free space wavelength. To validate the ENZ metamaterial model of the magnetic resonant WPT system, both full-wave simulations and experiments are conducted. In the experiment, we utilize a vector network analyzer (Agilent FieldFox 9918A) to measure the S-parameters of the planar spiral coils WPT system. The transmitter coil is connected to port 1 via a coaxial cable, while the receiver coil is connected to port 2 via another coaxial cable. The network analyzer is accurately calibrated. Figure 9.19a presents a perspective view of the experimental measurement setup. The simulated and measured S-parameters are depicted in Fig. 9.19b, demonstrating efficient power transfer at about 8.20 MHz. The full-wave simulation results closely match the measurement results. Figure 9.20 illustrates the comparison of the vector magnetic field distributions of the WPT system at resonant and non-resonant states. It is evident that a strong magnetic coupling exists at the operating frequency. The measured real and imaginary parts of S_{11} and S_{21} in the practical WPT system are displayed in Fig. 9.21a, b, respectively. Figure 9.21a shows that the imaginary part of S_{11} converges to zero at 8.12 MHz, which corresponds to the generalized resonant condition mentioned in Eq. (9.7). Similarly, the NRW method is utilized to extract the effective medium parameters, as shown in Fig. 9.21c, d. Notably, the effective permittivity approximates zero at 8.12 MHz, which closely aligns with the point of maximum power transmission. In this scenario, the system can be considered as an equivalent ENZ metamaterial structure. The effective permeability, as depicted in Fig. 9.22d, exhibits magnetic resonant behavior.

To investigate the tunneling effect of the planar spiral coils WPT system, a numerical experiment model is established, as illustrated in Fig. 9.22. The simulation setup involves placing resonant coils A and B, which possess similar dimensions to the WPT experiment depicted in Fig. 9.19, within a narrow channel with a separation distance of 0.9 m. The polarization of the incident TEM wave is constrained by the parallel-plate waveguide configuration. In the numerical simulation, the front and

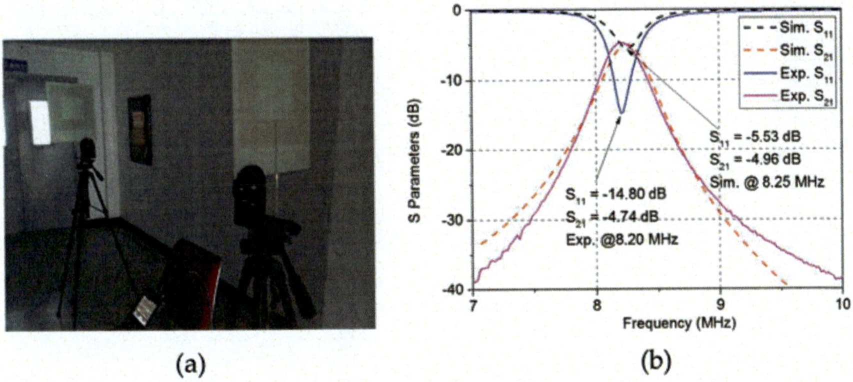

Fig. 9.19 **a** Perspective view of the experimental system, **b** measured results of the magnitude of S-parameters

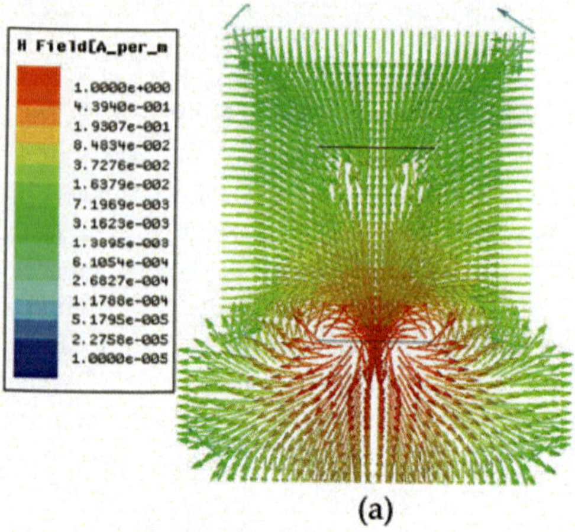

Fig. 9.20 **a** Magnetic field of the WPT system at non-resonance frequency, **b** magnetic field of the WPT system at the resonance frequency

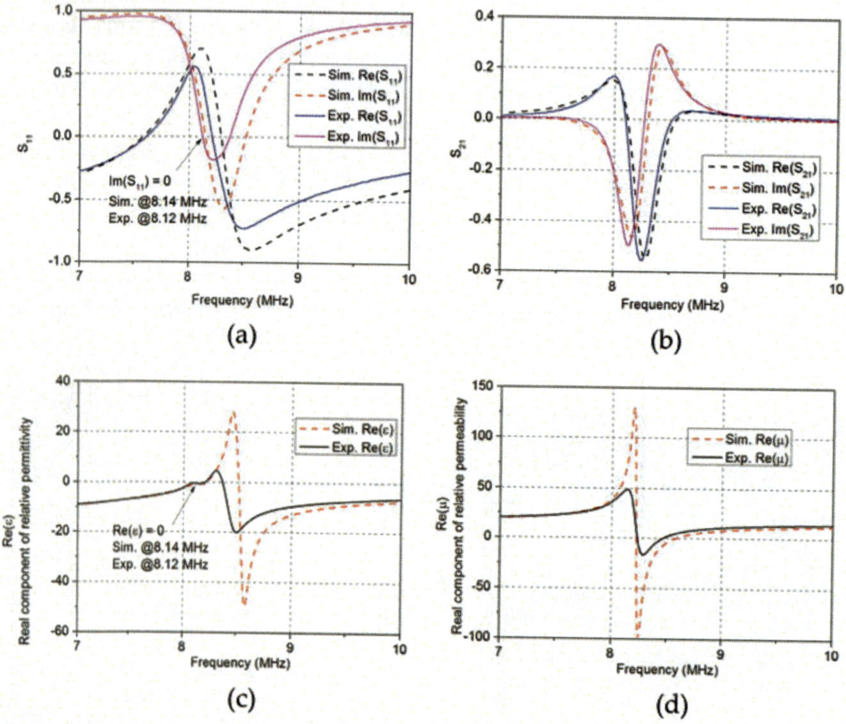

Fig. 9.21 Simulated and measured results of the WPT system, **a** real part and imaginary part of S_{11}, **b** real part and imaginary part of S_{21}, **c** effective relative permittivity, and **d** effective relative permeability

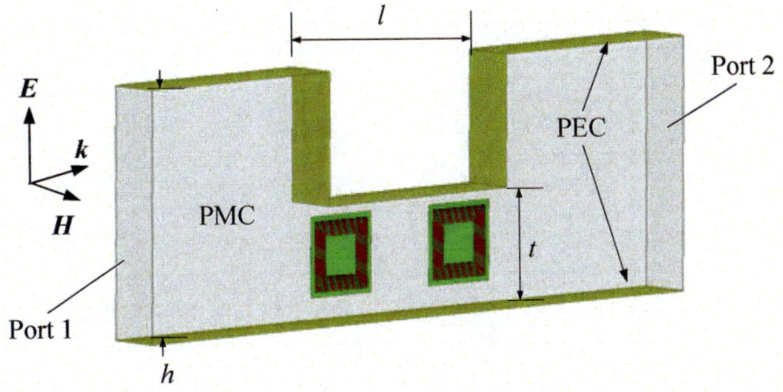

Fig. 9.22 Simulation model of the tunneling effect of the planar spiral coils WPT system, in which $t = 0.45$ m, $l = 0.9$ m, $h = 0.95$ m

back boundaries are designated as perfectly magnetic conductors (PMC), while the top and bottom boundaries are assigned as perfectly electric conductors (PEC). This experimental model aims to analyze the behavior of the WPT system with respect to the tunneling phenomenon.

The simulated Poynting vector is shown in Fig. 9.23a, which reveals the squeezing of the electromagnetic waves through the narrow channel. Because of omitting the one-turn square loop and the parallel-line, the system has a little resonant frequency deviation. It can be seen from Fig. 9.23b that the tunneling effect and super-coupling phenomenon occur at 8.58 MHz with $S_{21} = -5.4$ dB, while S_{21} is -16.5 dB without coils at 8.58 MHz, and when the system is with coils but out of resonant frequency, S_{21} is only about -15 dB between 7.0 and 10.0 MHz.

Based on the aforementioned theoretical analysis, computer simulations, and experiments, a novel theoretical framework is introduced for analyzing resonant WPT systems. This framework incorporates the concept of an equivalent network and employs the effective medium theory. The results reveal that the high transfer efficiency observed in the WPT system can be attributed to the presence of an effective ENZ medium and its tunneling effect. Furthermore, this work provides both theoretical and experimental evidence of electromagnetic tunneling and the super-coupling phenomenon occurring through an ENZ metamaterial at HF band. This research sheds light on the underlying mechanisms responsible for efficient power transfer in WPT systems and showcases the potential applications of ENZ metamaterials in high-frequency electromagnetic interactions.

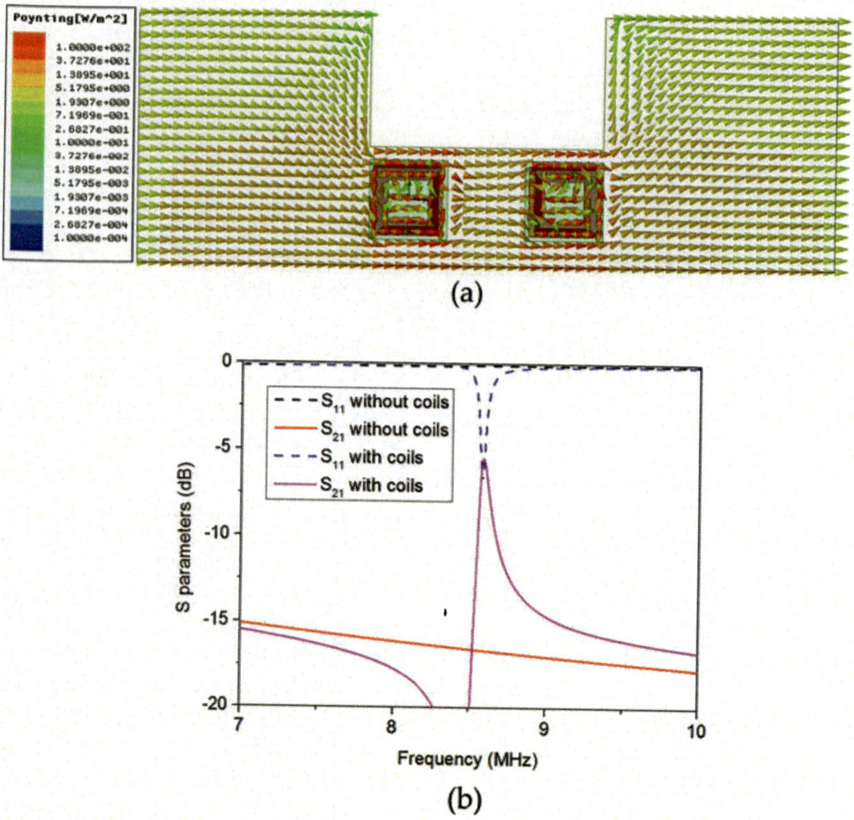

Fig. 9.23 Simulation result of tunneling phenomenon, **a** front view of Poynting vector, **b** comparison of the magnitude of S-parameters with and without coils

9.3 NFF Reflective Metasurfaces for Microwave WPT System

9.3.1 Multi-beam Phase Synthesis Theory of Reflective Metasurface

WPT techniques are designed to efficiently and simultaneously transfer electromagnetic energy to one or multiple devices. In achieving this, metasurfaces play a crucial role by independently and flexibly controlling the phase shift. This capability enables the creation of multiple foci with spatial power combinations using multiple feeds, as explained in this section. This concept can be illustrated through Fig. 9.24. In this setup, the metasurface is illuminated by multiple feeds, resulting in the generation of multiple reflected wave foci at various locations within the near-field region.

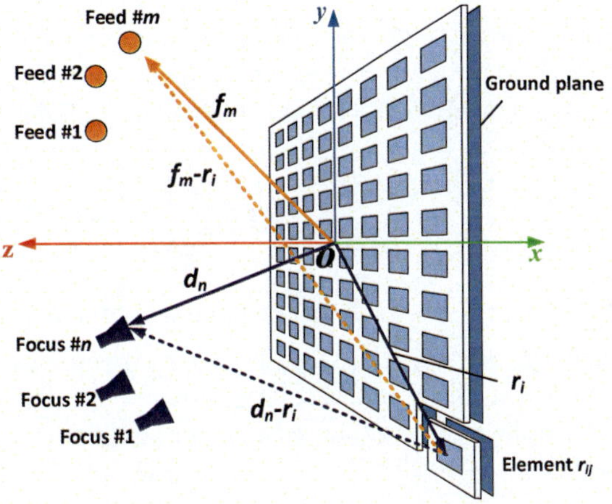

Fig. 9.24 Geometry of reflective metasurface for multi-feed and multi-focus WPT system

By precisely controlling the positions of these foci, wireless power can be effectively transferred to specific locations. The desired field distribution on the metasurface, necessary for generating the multi-focus effect in the near-field region, can be achieved by superimposing the aperture E-field associated with each focus. To create multiple foci at chosen locations, denoted by $\vec{d_n} = (x_n, y_n, z_n)$, where n represents an integer from 1 to the total number of foci (N), the E-field on the reflection metasurface can be expressed as follows:

$$\begin{aligned} E_R(x_i, y_i) &= A_R(x_i, y_i) \cdot \exp(j\varphi_R(x_i, y_i)) \\ &= \sum_{n=1}^{N} A_n(x_i, y_i) \cdot \exp(j\varphi_n(x_i, y_i)) \end{aligned} \tag{9.14}$$

In the expression (9.14), (x_i, y_i) represents the coordinates of the center of the i-th element on the metasurface. $A_n(x_i, y_i)$ and $\varphi_n(x_i, y_i)$ correspond to the required amplitude and phase at the location (x_i, y_i) for the n-th focus. The amplitude, $A_n(x_i, y_i)$, can be uniform or tapered, depending on the specific requirements. On the other hand, the phase distribution, $\varphi_n(x_i, y_i)$, adopts a progressive phase configuration. Consequently, the overall compensated phase distribution across the metasurface is given by

$$\varphi_R(x_i, y_i) = \arg\left\{\sum_{n=1}^{N}\left[D_n \exp\left(-jk_0\left|\vec{d_n} - \vec{r_i}\right|\right)\right]\right\} \tag{9.15}$$

where D_n represents the amplitude of the electric field at the intended focal spot, while k_0 denotes the wave number in free space. Additionally, multiple feeds are employed to illuminate the metasurface, enabling simultaneous spatial power combination at the designated location. To achieve this, the phase correction factor provided in Eq. (9.16) must be chosen so that the cumulative phase from all feeds remains constant across all patches, resulting in a planar phase front. This ensures that the desired spatial power combination is achieved at the intended focal spot.

$$\varphi_F(x_i, y_i) = \arg\left\{\sum_{m=1}^{M}\left[E_m \exp\left(jk_0\left|\vec{f_m} - \vec{r_i}\right|\right)\right]\right\} \tag{9.16}$$

where $E_m(x_i, y_i)$ is the E-field amplitude of the ith element from the feed m at the location $\vec{f_m} = (x_m, y_m, z_m)$. $\varphi_F(x_i, y_i)$ represents the total phase-shift factor of the ith element from all M feeding sources. After performing the summation of the complex field distributions, the required total phase shift on the ith element $\Delta\varphi(x_i, y_i)$ can be obtained as follows:

$$\Delta\varphi(x_i, y_i) = \varphi_R(x_i, y_i) - \varphi_F(x_i, y_i) \tag{9.17}$$

In the case of the multi-focus and multi-feed WPT system, the phase distribution required for the desired focusing effect can be achieved by adjusting the phase shift at each individual element of the metasurface. Conversely, variations in the amplitude distribution across the metasurface have minimal impact on the positions of the foci. The amplitudes of the incident fields on the metasurface elements are primarily determined by the radiation patterns of the feeds. With the exception of a slight reduction caused by element loss, the reflection coefficient amplitude for each element is practically identical. The critical design parameter lies in determining the reflection phases or phase-shifts of each metasurface element. Once the compensating phases for all elements are determined, the scattered electric field from the metasurface at the designated focal spot $\vec{R}$ can be expressed as follows:

$$E\left(\vec{R}\right) = \sum_{m=1}^{M}\sum_{i=1}^{I}\frac{\left|F_m\left(\vec{r_i} - \vec{f_m}\right)\right|}{\left|\vec{r_i} - \vec{f_m}\right|}e^{-jk_0\left|\vec{r_i}-\vec{f_m}\right|}G_i\left(\vec{R_i}\right)\frac{e^{-jk_0\left|\vec{R_i}\right|}}{\left|\vec{R_i}\right|} \tag{9.18}$$

where F_m is the radiation pattern function of the feed m, $G_i\left(\vec{R_i}\right)$ is the pattern function of the ith reflected element, and $\vec{R_i} = \vec{R} - \vec{r_i}$.

9.3.2 Four Types NFF Issues

SFSF Case

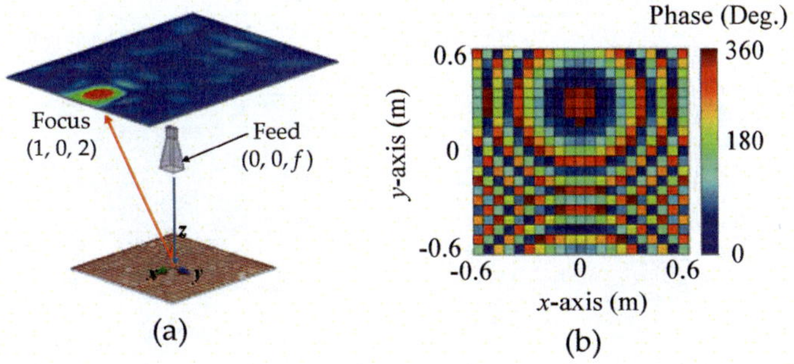

Fig. 9.25 Diagrams of SFSF case ($f = 1$ m), **a** simulation model and E-field intensity on the focal plane, and **b** the required phase-shift distribution on the metasurface to generate a single focus with a single feed obtained by Eq. (9.19) [65]

Let's first focus on generating a single focus with a single feed. In this scenario, we assume that the feeding horn is positioned on the axis, specifically at $\vec{f_m} = (0, 0, f)$ m. We consider normal incidence for this case. Assuming a central working frequency of 2.45 GHz, we can set the focus point at any desired position within the near-field region, such as $\vec{d_n} = (1, 0, 2)$ m. To achieve this, the required phase-shift for the ith element on the metasurface is determined. This phase-shift can be calculated using the following expression. The resulting values are illustrated in Fig. 9.25b, as shown below:

$$\begin{aligned}\Delta\varphi(x_i, y_i) &= k_0\left(\left|\vec{f_m} - \vec{r_i}\right| + \left|\vec{d_n} - \vec{r_i}\right|\right) \\ &= k_0(|(-x_i, -y_i, f)| + |(1 - x_i, -y_i, 2)|)\end{aligned} \tag{9.19}$$

In the design of the metasurface, a periodic element with dimensions of 60 mm $\times$ 60 mm has been selected. This size corresponds to approximately half of the wavelength in free space. Once the distribution of phase-shifts and feeding positions are determined, a metasurface comprising 21 $\times$ 21 elements can be designed and simulated. Each element in the metasurface consists of three metallic dipoles, which are printed on a substrate made of F4B material with a thickness of 1 mm. The F4B substrate has a relative permittivity (ε_r) of 2.65. The dipoles and the ground plane are separated by a 10 mm air layer.

The subsequent section will provide a comprehensive analysis of the element model, offering detailed performance.

SFDF Case

In the case of dual-focus requirements, where two foci need to be generated simultaneously for multi-beam applications, the metasurface can be designed to produce multiple beams in specific directions using a single feed. In this example, we aim to create two symmetrical foci positioned on the YZ plane at coordinates (0, −1, 1) m

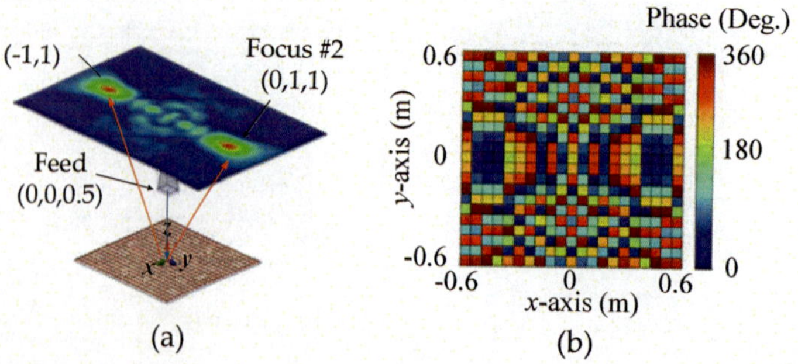

Fig. 9.26 **a** Simulation model of dual-focus and their E-field intensity on the focal plane, and **b** the required phase-shift distribution on the metasurface to generate two foci with a single feed obtained by Eq. (9.20) [65]

and (0, 1, 1) m in the Cartesian coordinate system. The feeding horn is placed along the z-axis at coordinate (0, 0, 0.5) m, as depicted in Fig. 9.26a. To achieve this dual-focus configuration, the phase-shift distribution on the metasurface is determined using Eq. (9.20). The resulting phase-shift distribution, illustrated in Fig. 9.26b, is calculated with a central working frequency set at 2.45 GHz.

$$\Delta\varphi_i = k_0\left|\vec{f}_m - \vec{r}_i\right| - \arg\left\{\sum_{n=1}^{2}\left[D_n \exp\left(-jk_0\left|\vec{d}_n - \vec{r}_i\right|\right)\right]\right\} \tag{9.20}$$

DFSF Case

An additional example of a metasurface is presented, showcasing its capability to function as a spatial power combiner. In this scenario, two feeds are employed to illuminate the metasurface, resulting in a single focus with high power intensity at the designated location. The required phase-shift distribution on the metasurface surface follows a similar pattern as described in Eq. (9.20) owing to the reciprocity theorem

$$\Delta\varphi_i = \arg\left\{\sum_{m=1}^{2}\left[F_m \exp\left(jk_0\left|\vec{f}_m - \vec{r}_i\right|\right)\right]\right\} + k_0\left|\vec{d}_m - \vec{r}_i\right| \tag{9.21}$$

In this particular example, two feeds are symmetrically positioned on the YZ plane. The coordinates of the feeds are (0, sin(π/4), cos(π/4)) m and (0, −sin(π/4), cos(π/4)) m. Additionally, the designed focal spot is located at (0, 0, 1) m. The necessary phase-shift distribution on the metasurface for this configuration is depicted in Fig. 9.27b. Through simulation, it has been observed that the intensity of the electromagnetic field reaches its peak precisely at the position of (0, 0, 1) m when the frequency of 2.45 GHz is applied, thus validating the theoretical design.

DFDF Case

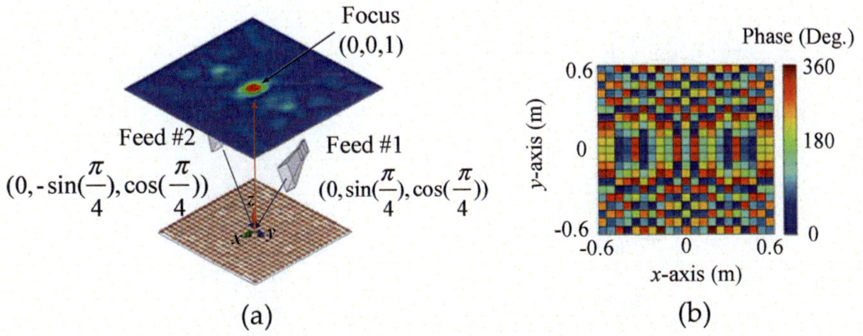

Fig. 9.27 **a** Simulation model and E-field intensity on the focal plane, and **b** required phase-shift distribution on the metasurface to generate a single focus with two feeds [65]

This particular case illustrates a dual-feed and dual-focus metasurface design implemented to focus the near-field at different positions at a frequency of 2.45 GHz, as depicted in Fig. 9.28(a). In this example, two feeds are positioned on the YZ plane with coordinates of (0, sin(π/4), cos(π/4)) m and (0, −sin(π/4), cos(π/4)) m. The two designed focal spots are symmetrically placed on the XZ plane at coordinates of (sin(π/4), 0, cos(π/4)) m and (−sin(π/4), 0, cos(π/4)) m. The necessary phase shift required on the metasurface is determined by Eq. (9.22), and the corresponding calculated results are displayed in Fig. 9.28b. Simulation results shown in Fig. 9.28c reveal that two peaks of field intensity manifest on the XZ plane, precisely at the intended focal spots, thereby confirming the consistency with the predetermined scenario. Moreover, Fig. 9.28d demonstrates the absence of any focus appearing on another plane, providing further verification of the effectiveness and accuracy of the design approach.

$$
\begin{aligned}
\Delta\varphi_i = \arg\left\{\sum_{m=1}^{2}\left[F_m \exp\left(jk_0\left|\vec{f_m} - \vec{r_i}\right|\right)\right]\right\} \\
- \arg\left\{\sum_{n=1}^{2}\left[D_n \exp\left(-jk_0\left|\vec{d_n} - \vec{r_i}\right|\right)\right]\right\}
\end{aligned}
\tag{9.22}
$$

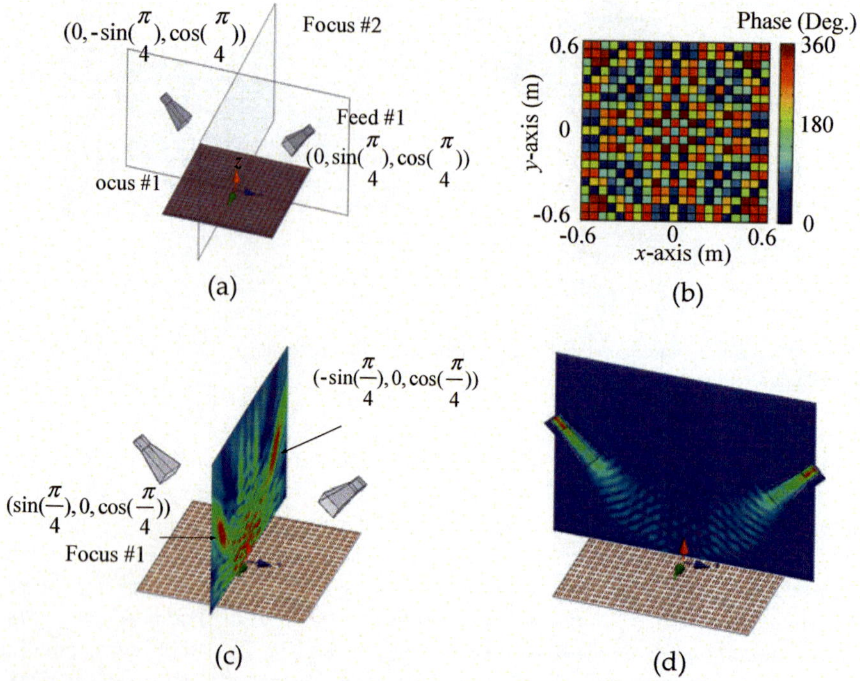

Fig. 9.28 **a** Simulation model of DFDF case, **b** required phase-shift distribution on the metasurface to generate two focus with two feeds, **c** E-field intensity distribution on XZ plane, and **d** E-field intensity distribution on YZ plane [65]

9.3.3 NFF Reflective Metasurface at 5.8 GHz with Tri-Dipole Element

9.3.3.1 Metasurface Element Design

To achieve effective near-field focusing performance in metasurface design, the inclusion of a reflective element is crucial. It is desirable to minimize the slope of the reflection phase curve to ensure that the phase variation remains insensitive to changes in the element's dimensions. Steep phase curves may pose challenges with operational bandwidth and element fabrication tolerance. Additionally, it is essential to ensure that the element exhibits sufficient variation in dimensions to cover a phase range of 360°. A smooth and linear phase variation is also preferred. Considering these factors comprehensively, a printed tri-dipole structure, as illustrated in Fig. 9.30, is selected as the reflective unit element. This structure offers multi-resonant characteristics, enabling a broader phase coverage. The length of the dipoles and the ratio between the center main dipole and the two satellite dipoles are controlled to achieve a continuous reflection phase range exceeding 360°. Each element is fabricated using F4B substrates ($\varepsilon_r = 2.65$), with three dipoles printed on the top surface. The substrate has a thickness of 1.0 mm and is accompanied by a 5.0 mm air layer and a metallic ground at the bottom. The width of the three dipoles is set at 2.0 mm, and their lengths are adjusted to compensate for the desired phase shift. The length ratio between the main dipole and the two satellite dipoles is 0.6 (Fig. 9.29).

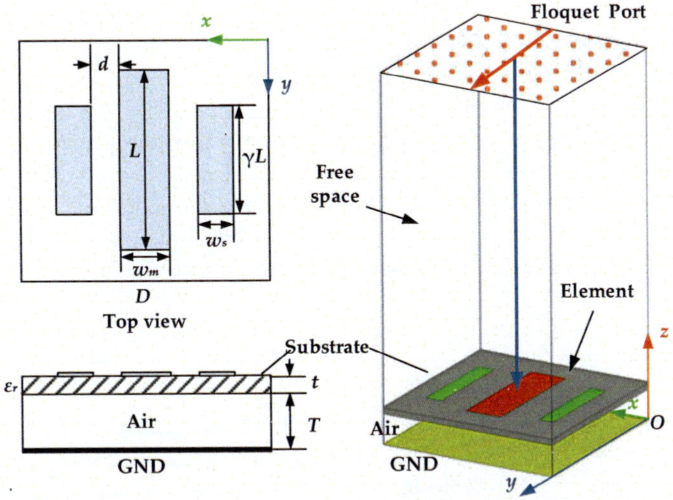

Fig. 9.29 Geometry of tri-dipole element and an infinite periodic model based on the finite element algorithm to calculate the reflection phase of the unit cell in normal incidence, in which periodic boundary conditions are in place around the element [65]

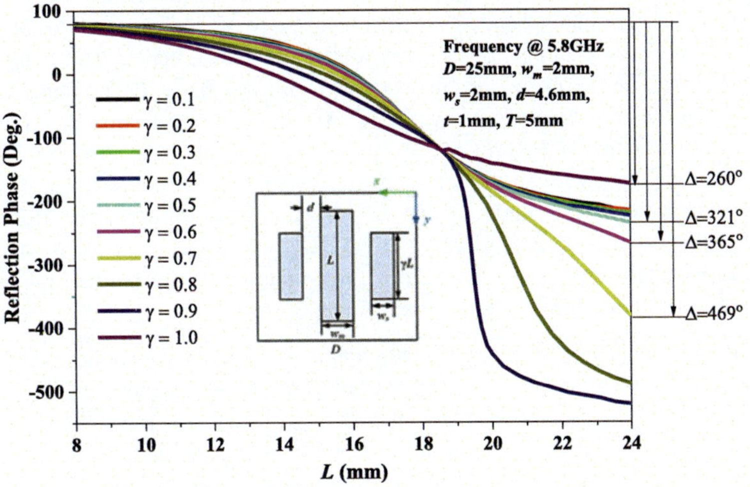

Fig. 9.30 Reflection phase characteristics of tri-dipole element versus the length of the main dipole [65]

Accurately predicting the phase-shift characteristics of scattering elements under various incident wave conditions is crucial for analyzing and designing a P2P/P2M metasurface for WPT. To achieve this, an infinite periodic model utilizing Ansys HFSS is constructed to analyze the reflection phase of the tri-dipole element, as depicted in Fig. 9.30. The model assumes periodic boundary conditions (PBCs) around the unit cell and employs symmetric satellite dipoles and a single linear polarization structure to simplify the analysis. The periodicity is set to 25 mm along both directions, although an asymmetric design or dual-polarization approach can be adopted for specific requirements. Figure 9.30 illustrates the reflection phase characteristic of the unit cell in relation to the length of the main dipole when a plane wave is incident normally on the reflective surface. The operational frequency is set at 5.8 GHz. Simulation results confirm a desirable reflection phase response that covers a range of 360°, which is attributed to the multi-resonance response and the mutual coupling between the main dipole and satellite dipoles.

9.3.3.2 P2P Case

The objective of this section is to design a P2P metasurface for WPT operating at a center frequency of 5.8 GHz. The designed metasurface enables efficient power transfer from the feeding source to the focal point. To avoid blockage caused by the feed horn and its support structures, an offset-fed configuration is employed. The prototype is designed with the feed-horn phase center located at (0, −0.3sin(π/8), 0.3cos(π/8)) m, while the reflective focus is placed at (0, sin(π/8), cos(π/8)) m in the YZ-plane at the center frequency of 5.8 GHz. As previously mentioned, the P2P metasurface consists of a 20 × 20 array of tri-dipole elements. The required phase-shift distribution on the metasurface is illustrated in Fig. 9.31a, while the corresponding length dimensions of each tri-dipole element are calculated and displayed in Fig. 9.31b. The layout of the metasurface, with the 20 × 20 elements etched on the substrate, can be seen in Fig. 9.31c. An observation plane is positioned at $z = 92$ cm (cos(π/8)≈0.92). Figure 9.31d shows the focused electric field intensity at the designated location, verifying the successful transfer of power to the intended focal point and validating the theoretical design. Based on the aforementioned analysis and design, a prototype of the P2P metasurface with the 20 × 20 tri-dipole elements is fabricated and measured. The image of the metasurface prototype, corresponding to Fig. 9.31c, is displayed in Fig. 9.32a.

We conducted near-field scanning measurements in a microwave anechoic chamber to validate the performance of the metasurface. The measurements were performed at 5.8 GHz using a broadband horn antenna (2–18 GHz) as the feeding source. The feed-horn phase center was positioned at (0, −0.3sin(π/8), 0.3cos(π/8)) m in the Cartesian coordinate system, corresponding to an offset angle of 22.5 degrees. The experimental setup for the measurement is depicted in Fig. 9.32b, with both the transmitter (TX) and receiver (RX) utilizing vertical polarization. The vertical component of the reflected electric field was detected using a standard probe operating

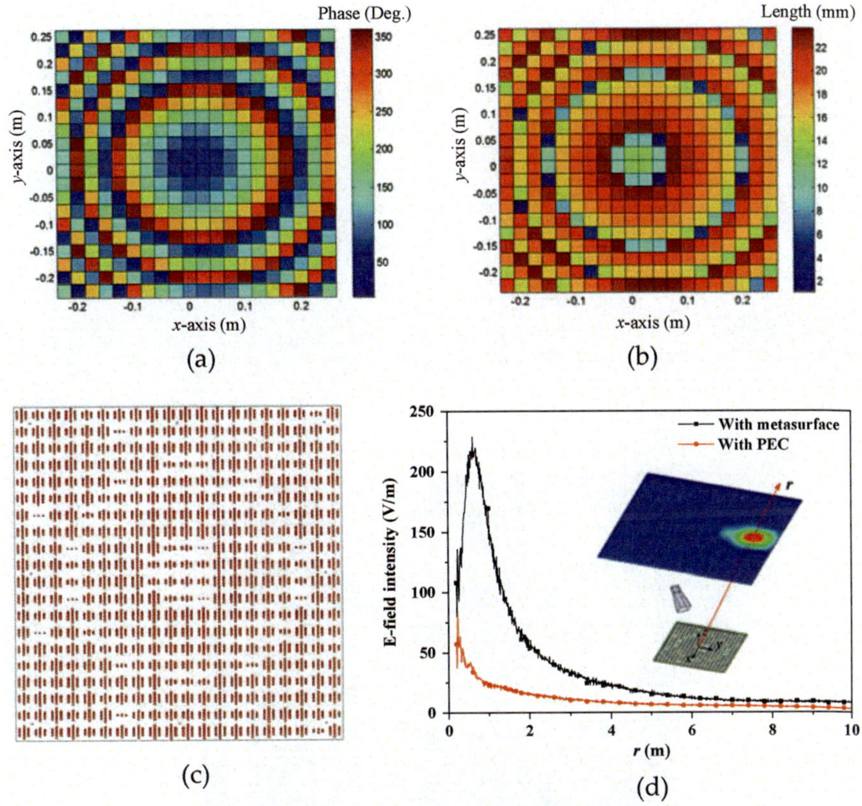

Fig. 9.31 NFF metasurface of 500 mm × 500 mm with 20 × 20 elements for P2P case, **a** required phase-shift distribution on the reflective metasurface, **b** length dimensions of each tri-dipole elements of the metasurface (unit: mm), **c** geometry topology of the designed metasurface with 20 × 20 tri-dipole elements, **d** simulation model and E-field intensity on the focal plane at 5.8 GHz [65]

at 5.8 GHz. The E-field distribution on a 1.1 m × 1.1 m near-field scanning plane was measured at fixed sampling grid points with a lattice period of 10 mm, which was located at a distance of 0.92 m from the metasurface. The approximate position of the focus center was found to be (−8, 33) cm. The spatial map of the measured E-field intensity is displayed in Fig. 9.32c, exhibiting excellent agreement with the simulation results depicted in Fig. 9.31d. This validates the effectiveness of the metasurface design in achieving the desired near-field focusing capability.

9.3.3.3 P2M Case

The subsequent experiment aims to validate the functionality of the P2M metasurface configuration operating at 5.8 GHz. In this case, the metasurface is designed to

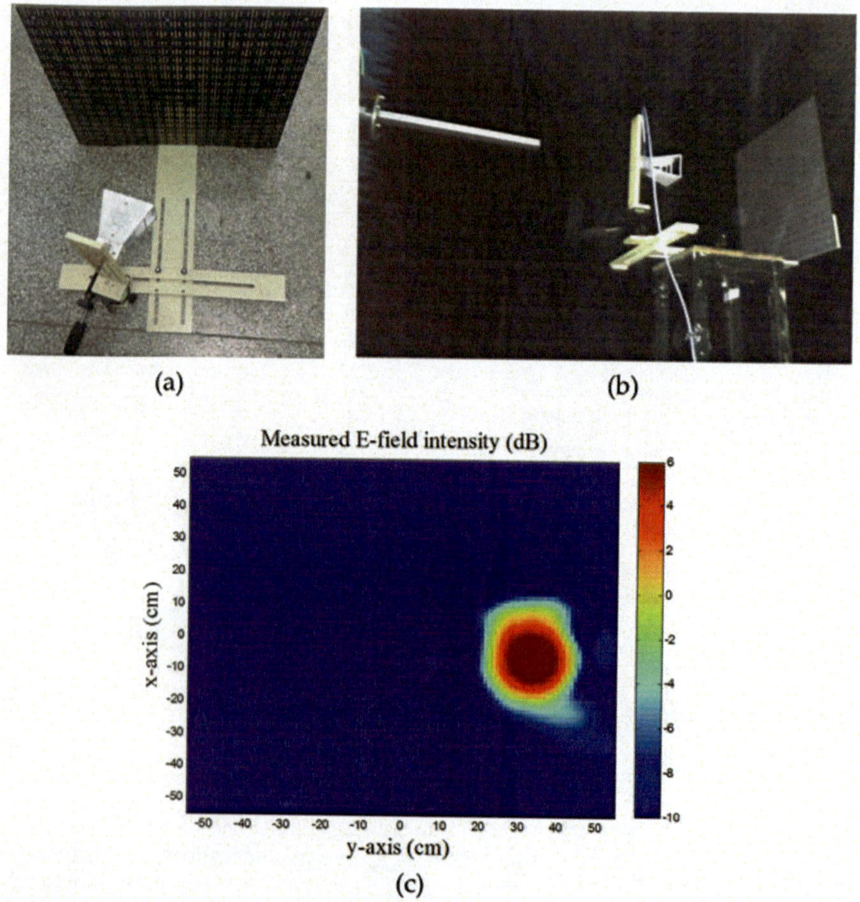

Fig. 9.32 **a** Fabricated P2P metasurface of 20 × 20 elements with an offset feeding horn, **b** measurement system in microwave anechoic chamber, and **c** measured E-field intensity distribution on the 1.1 m × 1.1 m scanning plane with sampling grid points of 10 mm lattice period, and the scanning plane is 0.92 m away from the metasurface [65]

efficiently transfer power to two designated positions. The feeding horn used in this setup is the same as the one described in the P2P case mentioned earlier. However, the difference lies in the new metasurface design, which now has two specific points for power transfer. These two points are symmetrically located at (−0.3, sin(π/8), cos(π/8)) m and (0.3, sin(π/8), cos(π/8)) m, respectively. To achieve this configuration, the required phase-shift distribution on the metasurface can be calculated using Eq. (9.20), and the results are displayed in Fig. 9.33a. Additionally, Fig. 9.33b presents the corresponding length distribution of each tri-dipole element on the metasurface. The layout of the metasurface, consisting of a 20 × 20 array of elements etched onto the substrate, is illustrated in Fig. 9.33(c). Simulation results reveal the transverse electric field distribution on the focal plane, as depicted in Fig. 9.33d, demonstrating that the electric field reaches its maximum intensity precisely at the intended focal spots.

The metasurface prototype, designed to accommodate two specific power transfer points, was fabricated and presented in Fig. 9.34a. To assess its performance, the E-field magnitude distribution was measured using the near-field planar scanning technique, as shown in Fig. 9.34b. One focus center was located at (33, 34) cm, while the other focus center was positioned at (35, −32) cm. When comparing Fig. 9.34b with Fig. 9.34d, a favorable agreement can be observed between the simulated and measured results for the two foci. It is worth noting that due to some blockage introduced by the supporting structure for the feeding horn, the measured intensity of the electric field may lack perfect symmetry. Nonetheless, Fig. 9.34b indicates that the maximum electric field intensities precisely correspond to the designated spots, thus aligning with the simulation results depicted in Fig. 9.33d.

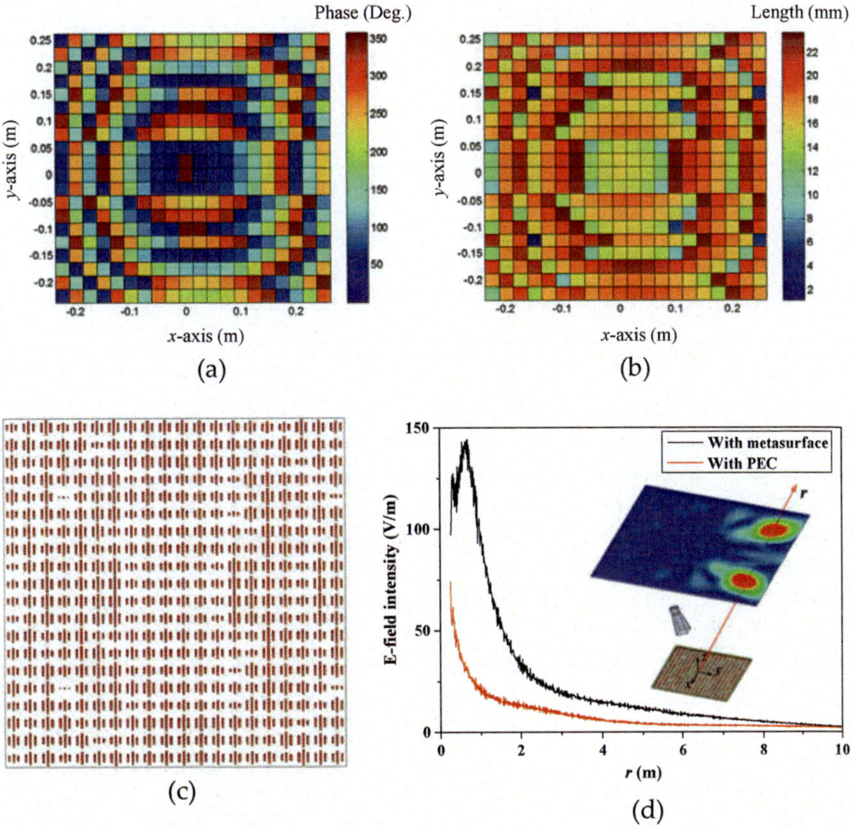

Fig. 9.33 NFF metasurface of 500 mm × 500 mm with 20 × 20 elements for P2M case, **a** required phase-shift distribution on the metasurface, **b** length dimensions of each tri-dipole elements of the metasurface (unit: mm), **c** geometry topology of the designed metasurface with 20 × 20 tri-dipole elements, **d** simulation model and E-field intensity on the focal plane at 5.8 GHz [65]

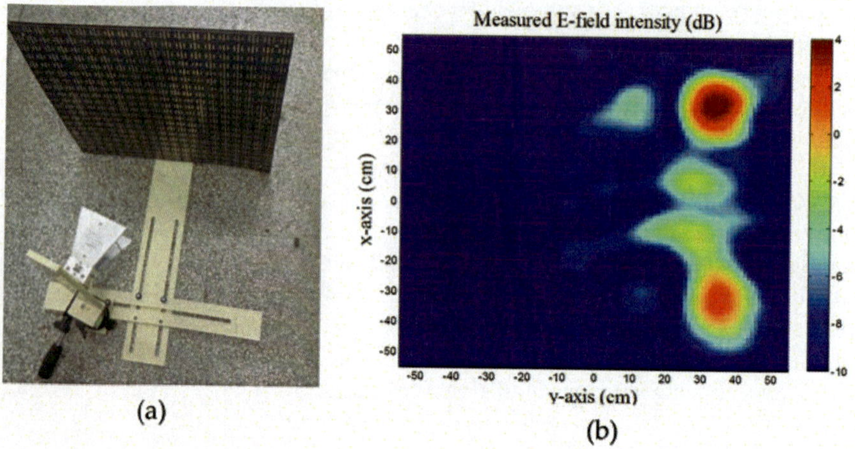

Fig. 9.34 **a** Fabricated dual-focus metasurface of 20 × 20 elements with an offset feeding horn, and **b** measured E-field intensity on the 1.1 m × 1.1 m scanning plane with square grid points of 10 mm lattice period, 0.92 m away from the metasurface [65]

9.3.3.4 Wireless Power Transfer Efficiency Analysis

A good design of a P2P/P2M WPT metasurface will achieve a high WPT efficiency. The WPT efficiency η can be defined as the ratio of the totally focused power (P_2) to the total power radiated by the feeding source (P_0):

$$\eta = \eta_1 \eta_2 = \frac{P_2}{P_0} \tag{9.23}$$

where η_1 represents the power transfer efficiency from the feeding source to the metasurface, i.e., $\eta_1 = P_1/P_0$, where P_1 is the power captured by the metasurface. η_2 represents the near-field transfer efficiency from metasurface to foci, i.e., $\eta_2 = P_2/P_1$, where P_2 is the total power captured by the focusing spots. In this section, the calculation of P_1 and P_2 is performed based on numerical integration using the Poynting theorem. P_1 integrates the Poynting vector over the entire aperture (Γ-plane in Fig. 9.35a) of the metasurface, while P_2 is the sum of the power contained at N foci. The power at each focus is obtained by integrating the Poynting vector over the small slice (Ω-planes in Fig. 9.35a) covering the corresponding focal spot. For example, in the metasurface design illustrated in Fig. 9.31, which operates as an offset-feed and single-focus configuration at a center frequency of 5.8 GHz, the WPT efficiency is calculated and depicted in Fig. 9.35b (labeled as (1-1)). The curve labeled (1-2) corresponds to the two-focus metasurface shown in Fig. 9.33, where the total WPT efficiency is the sum of the efficiencies at the two foci. The curve labeled PEC in Fig. 9.35b represents the WPT efficiency when replacing the metasurface with a PEC plate of the same dimensions. In this case, no power is focused on the designated spots. From Fig. 9.35, it can be observed that the proposed near-field focusing planar metasurfaces achieve a high WPT efficiency of 70% due to the focusing transmission. To further enhance the WPT efficiency, efforts should be made to reduce the spillover power. In the case of (1-1), the bandwidth with 50% WPT efficiency ranges from 5.5 to 6.5 GHz, with a fractional bandwidth of approximately 16.7%. Similarly, in the case of (1-2), the bandwidth with 50% WPT efficiency spans from 5.6 to 6.6 GHz, with a fractional bandwidth of approximately 16.4%.

To assess the near-field focusing characteristics of P2P/P2M WPT metasurfaces, it is important to examine the WPT efficiencies across different observational planes. Taking the example of the designed metasurface shown in Fig. 9.31, where the focus is positioned at $z = 0.92$ m, Fig. 9.35c, d provide insights into the WPT efficiencies. These figures reveal that the WPT efficiency is at its maximum when the observational plane aligns with the designed focusing plane. Furthermore, the efficiency remains relatively high when the observational planes are in close proximity to the focusing plane. However, as the observational planes move farther away from the focusing plane, the efficiency degrades significantly. This observation validates the expected near-field focus characteristics of the metasurface design.

9.3.4 Dual-Polarization Metasurface at 10 GHz with Cross-Dipole Element

9.3.4.1 Dual-Polarization Metasurface Element Design

In this section, the focus is on achieving dual-polarization independent regulation while considering the fabrication cost. To fulfill these requirements, the chosen metasurface element is a single-layer cross-dipole structure, as depicted in Fig. 9.36. This structure enables a phase shift range of approximately 330°, ensuring independent regulation of both polarizations. The cross-shaped metal branch is etched onto the upper layer of the substrate, which is made of F4BM-2 material (with a relative permittivity of 2.2). By varying the period of the element and the substrate thickness, different reflection characteristics can be obtained. Reducing the element period enhances the linearity of the phase-shift curve, but it decreases the overall phase-shift range. On the other hand, increasing the thickness of the dielectric substrate improves the linearity of the phase-shift curve but reduces the phase-shift range. After conducting full-wave simulation optimization, the structural parameters are determined as follows: $D = 15$ mm, $H = 3$ mm, and $W = 1$ mm. The simulation model can be seen in Fig. 9.36.

Another crucial aspect of the chosen element is its capability to independently regulate dual-polarization incident waves. To analyze its performance, we consider the scenario where a wave with y-direction polarization is incident. We examine three different lengths of the element in the fixed x-direction, namely, $L_x = 3$, 8, and 13 mm. The resulting reflection phase characteristics, represented by three-phase shift curves, are displayed in Fig. 9.37a. From the figure, it is evident that changing

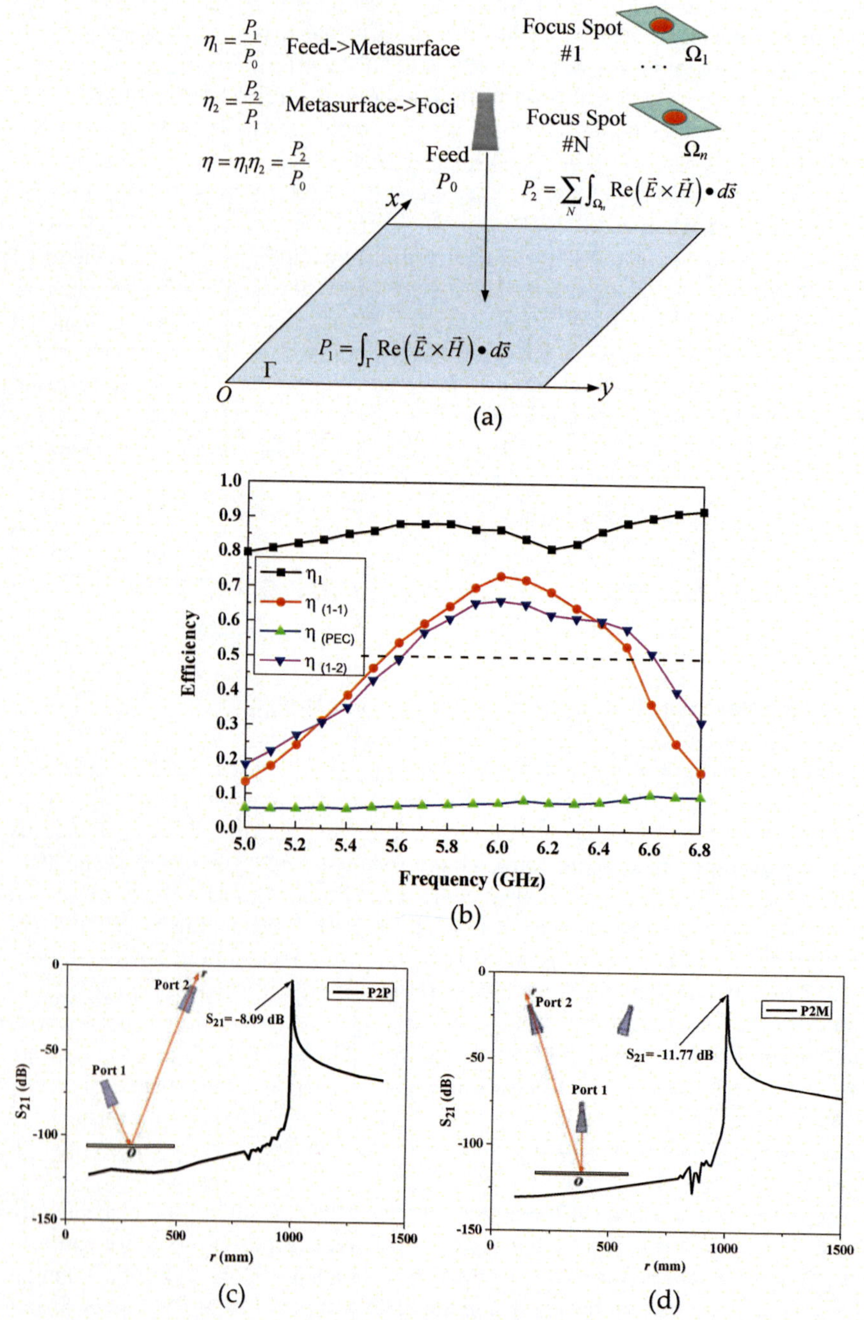

Fig. 9.35 **a** Sketch illustrating the calculation of the WPT efficiencies, **b** WPT efficiency curves corresponding to the different metasurface prototypes shown in Figs. 9.31 and 9.33, respectively, in which the efficiency of a PEC plate with the same dimension is also given for comparison. WPT efficiency curves corresponding to **c** P2P and **d** P2M [65]

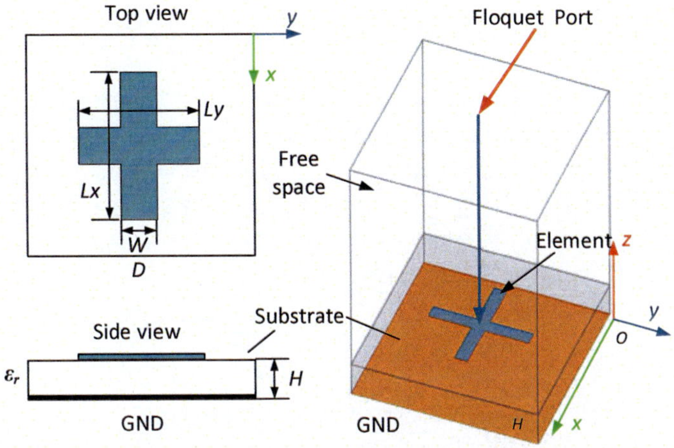

Fig. 9.36 Geometry of single-layer cross-dipole element structure [66]

the length of the element in the x-direction has minimal impact on the reflection phase in the y-direction. Likewise, a similar trend can be observed for x-direction polarization, as depicted in Fig. 9.37b. This observation effectively verifies the dual-polarization independent regulation characteristics of the cross-dipole element. It enables the configuration of different focus positions and focusing functions based on the polarization variation of incident waves. Ultimately, this capability expands the range of applications for reflective metasurfaces.

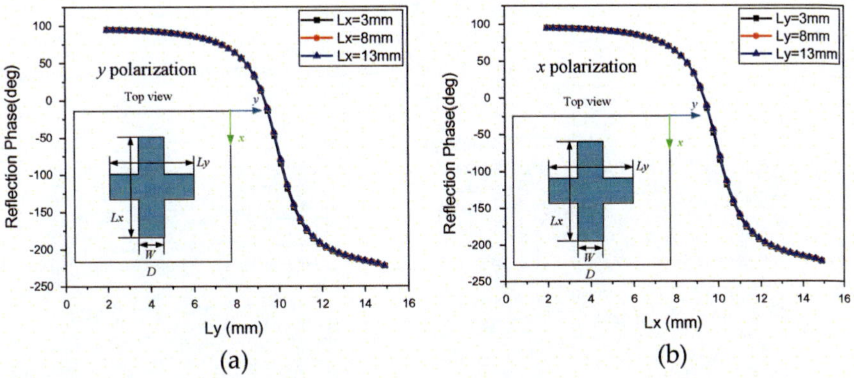

Fig. 9.37 Reflection phase characteristics of polarization-independent regulation: **a** phase shift curve of x-direction element with y-direction polarization excitation; **b** phase shift curve of y-direction element with x-direction polarization excitation [66]

9.3.4.2 Dual-Polarization Reflective Metasurface Case 1: SFSF

To design a reflective metasurface that achieves a SFSF configuration for dual-polarization, we position the feed at $\vec{r_f} = (0, 0, 0.2)$ m. It normally illuminates the reflective metasurface at a frequency of 10 GHz. The focus for y-direction polarization is set at $\vec{r_{dy}} = (0.3, 0, 1)$ m, while the focus for x-direction polarization is set at $\vec{r_{dx}} = (-0.3, 0, 1)$ m. The corresponding phase distribution on the reflective metasurface, as shown in Fig. 9.38, is as follows:

$$\Delta\varphi_{ij}(x_i, y_j) = k_0\left(\left|\vec{r_f} - \vec{r_{ij}}\right| + \left|\vec{r_d} - \vec{r_{ij}}\right|\right) \tag{9.24}$$

The size of the 'cross-dipole' etched on each element depends on the required reflection phase of the metasurface element, which is depicted in Fig. 9.38. The phase-size relationship extracted by element simulation is shown in Fig. 9.37. The corresponding size topology of the designed metasurface can be obtained and observed in Fig. 9.39. With different colors, Fig. 9.39 shows the branch lengths distribution in the y and x directions of the cross-dipoles of the reflective metasurface. The designed reflective metasurface consists of 26 × 26 elements, whose top view of the simulation model is illustrated in Fig. 9.40. Each cross-dipole element of the metasurface is with the same width of 1 mm. The overall size of the designed metasurface is 390 mm × 390 mm, and the thickness is 3 mm. The material of the substrate is F4BM-2 ($\varepsilon_r = 2.2$).

Thanks to the dual-polarization independent regulation capability, the reflective metasurface allows for flexible focus positions for different incident wave polarizations. Figure 9.41 illustrates that as the polarization direction of the incident wave changes, the position of the focus also shifts accordingly, demonstrating the metasurface's ability to independently regulate the focus for dual-polarization. This feature

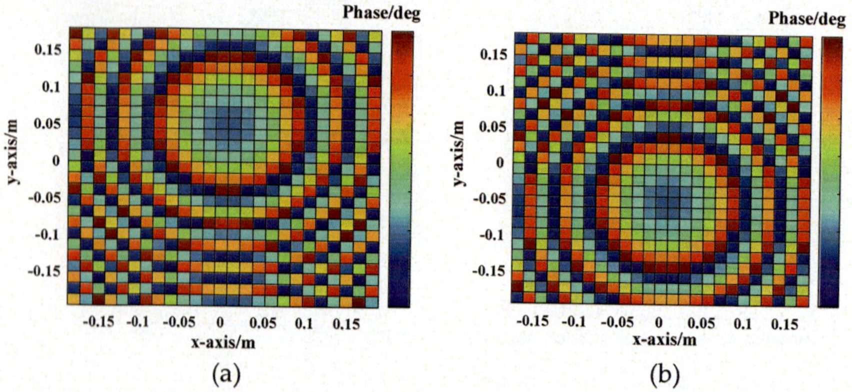

Fig. 9.38 Phase distribution of SFSF reflective metasurface: **a**) y-direction polarization excitation; **b** x-direction polarization excitation [66]

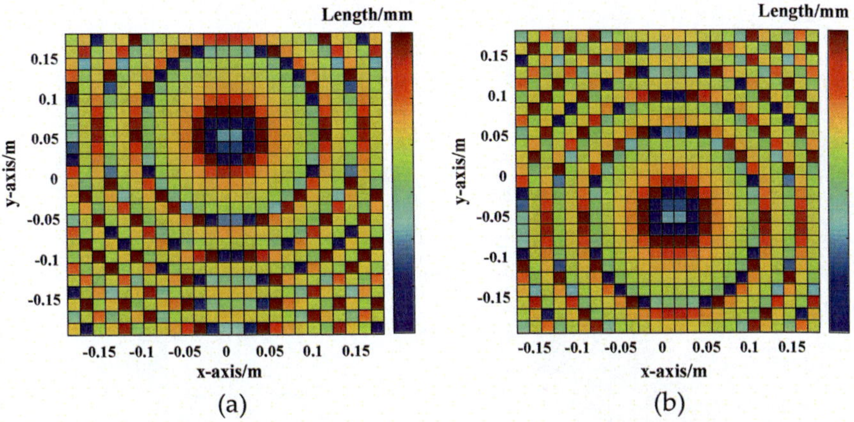

Fig. 9.39 Size topology of SFSF reflective metasurface: **a** y-direction polarization excitation; **b** x-direction polarization excitation [66]

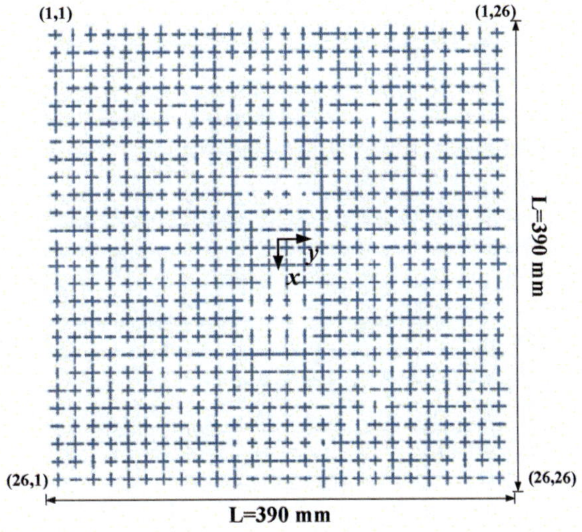

Fig. 9.40 Geometry of SFSF reflective metasurface in dual-polarization [66]

enables versatile control and manipulation of the focal point based on the polarization of the incident waves.

9.3.4.3 Dual-Polarization Reflective Metasurface Case 2: SFDF with Different Power Transfer Ratio

In this case, we aim to design a novel reflective metasurface that achieves a SFDF configuration with dual-polarization. This configuration is intended for various power

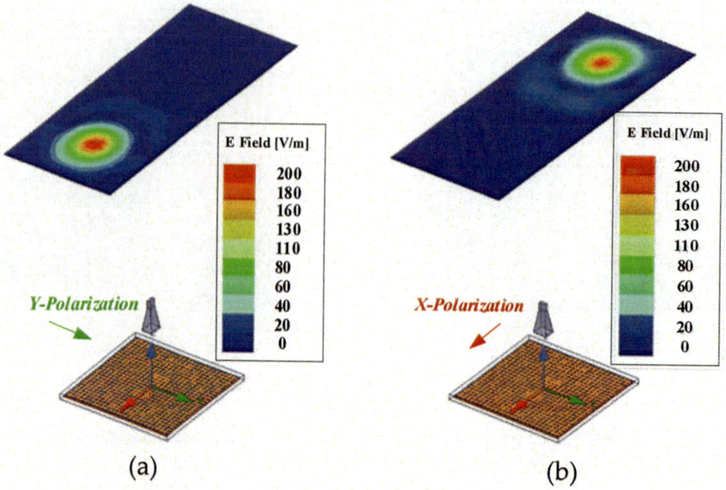

Fig. 9.41 Full-wave simulation results of SFSF metasurface: **a** y-direction polarization excitation; **b** x-direction polarization excitation [66]

transfer systems. The feed is positioned at $\vec{r_f} = (0, 0, 0.2)$ m, and it normally illuminates the metasurface at a frequency of 10 GHz. For the situation involving y-direction polarization, we set two foci $\vec{r_{d1}} = (0.3, 0, 1)$ m and $\vec{r_{d2}} = (-0.3, 0, 1)$ m. The ratio of the E-field amplitude between these two foci is $D_1{:}D_2 = 1{:}1$. The required phase distribution for this configuration is depicted in Fig. 9.42a. Additionally, Fig. 9.42b displays the corresponding size topology in the y-direction polarization of the desired metasurface. Simultaneously, for x-direction polarization, we modify the amplitude ratio of $D_1{:}D_2$ to be 1:1.2. In this case, the two foci are positioned at $\vec{r_{d3}} = (0, 0.3, 1)$ m and $\vec{r_{d4}} = (0, -0.3, 1)$ m. The final design of the reflective metasurface, incorporating both polarizations, is presented in Fig. 9.43.

$$\Delta\varphi_{ij}(x_i, y_j) = k_0\left|\vec{r_f} - \vec{r_{ij}}\right| - \arg\left\{\sum_{m=1}^{2}\left[D_m \exp\left(-jk_0\left|\vec{r_{d_m}} - \vec{r_{ij}}\right|\right)\right]\right\} \quad (9.25)$$

The simulation results with y-direction polarization excitation are presented in Fig. 9.44a. It is evident that the two foci are precisely positioned, and the distribution of E-field intensity is nearly uniform. This indicates successful achievement of equal power distribution in the designed metasurface. To investigate the impact of polarization, we rotate the feed horn by 90° around the z-axis, which corresponds to a rotation on the azimuth plane, without altering the incident angle. Consequently, the incident wave polarization becomes x-direction polarization. The simulation results of this configuration are displayed in Fig. 9.44b. Notably, the two predetermined focus positions remain unchanged, while the distribution of E-field intensity exhibits

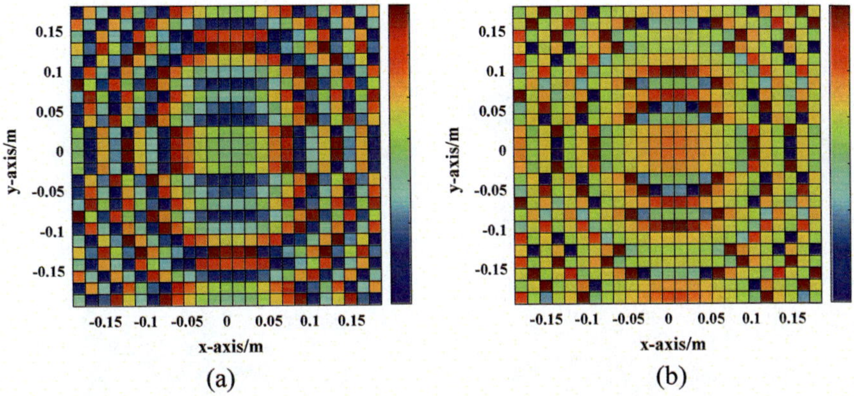

Fig. 9.42 SFDF metasurface in *y*-direction polarization excitation: **a** phase distribution; **b** size topology [66]

Fig. 9.43 Geometry of SFDF reflective metasurface with dual-polarization for different power transfer ratios [66]

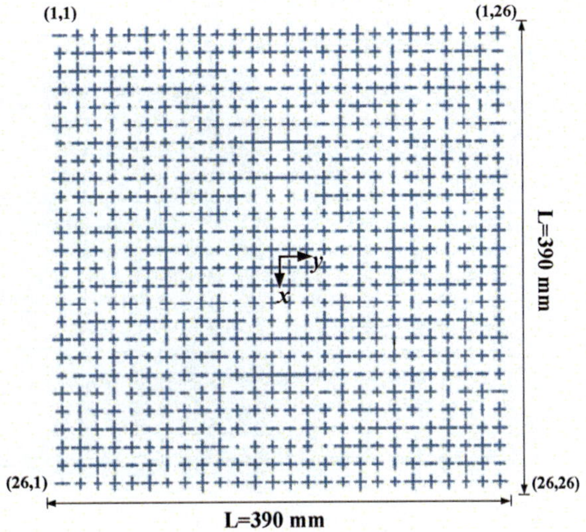

noticeable variations. This demonstrates the realization of unequal power distribution in the dual-focus configuration. In the subsequent step, we rotate the feed horn by 45° around the *z*-axis, allowing for simultaneous excitation of dual-polarization incident waves. The simulation results presented in Fig. 9.45 confirm the attainment of four distinct foci through the implementation of a multi-focus design and the independent regulation of dual-polarization. However, it is important to note that due to the non-ideal nature of the horn feed as a point source, achieving perfect equality between the two polarization components is challenging when rotating the horn by 45°, resulting in some disparities in the intensity distribution of the focal E-fields.

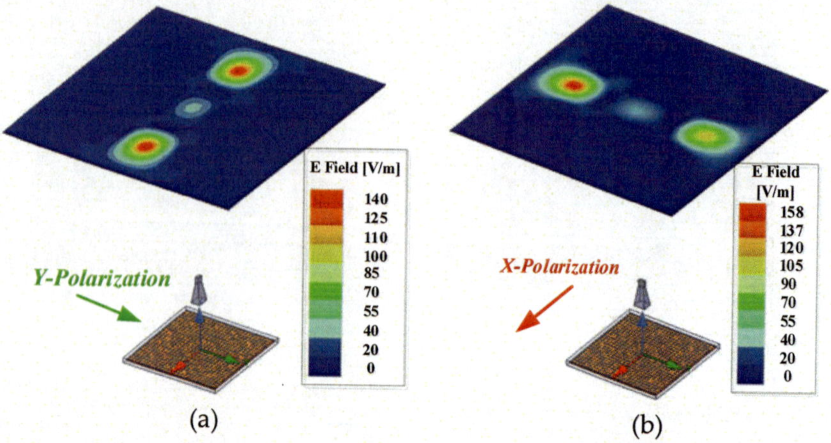

Fig. 9.44 Full-wave simulation results of SFDF metasurface: **a** *y*-direction polarization excitation; **b** *x*-direction polarization excitation [66]

Fig. 9.45 Full-wave simulation results of E-field intensity distribution of SFDF metasurface in dual-polarization [66]

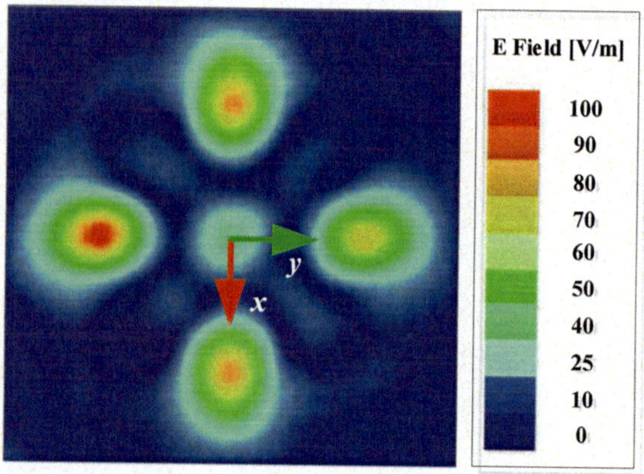

9.3.4.4 Measurements of Dual-Polarization Metasurfaces

Following the full-wave simulation described earlier, the reflective metasurface with consistent dimensions and parameters is fabricated for experimental validation. A broadband horn operating in the 2–18 GHz range is utilized as the feeding source, positioned $\vec{r_f} = (0, 0, 0.2)$ m for normal illumination. The experimental setup is depicted in Fig. 9.46a. The measurements are conducted within a microwave anechoic chamber using planar near-field scanning to assess the designed NFF reflective metasurface. The scanning range spans 1 m × 0.4 m, and the distance between the 10 GHz standard probe and the metasurface is 1 m. Figure 9.46b illustrates the measurement setup. For the SFSF scenario, the reflective metasurface described in Fig. 9.40 is fabricated and subjected to measurement. By rotating the horn around the

z-axis to induce different polarization excitations, we obtain measurement results. Figure 9.47a, b respectively depict the cases of y-direction and x-direction polarizations. Upon comparing these results with Fig. 9.41, it is evident that the measured data align closely with the simulated results, providing further validation for the effectiveness of the NFF design.

The dual-focus focusing system can also be achieved by adjusting the angle around the z-axis of the feed horn. When the feed horn is rotated by 40° on the azimuth plane while keeping the incident angle unchanged, it generates two waves with different intensities and orthogonal polarization directions incident on the metasurface. The measurement results, as shown in Fig. 9.48, confirm the presence of two focal points at the designated positions of $\vec{r_{dx}} = (-0.3, 0, 1)$ m for x-polarization and $\vec{r_{dy}} = (0.3, 0, 1)$ m for y-polarization, respectively. Although the actual measurement is affected by the non-ideal point source nature of the horn and the blockage caused by the support structure, the maximum E-field intensity position aligns with the intended design, thus confirming the successful realization of NFF using the reflective metasurface.

For the SFDF scenario, a reflective metasurface with identical parameters as depicted in Fig. 9.43 was manufactured and measured. The measured E-field distribution for equal-power transfer in the y-direction polarization excitation is illustrated in Fig. 9.49a, while Fig. 9.49b showcases the measured E-field distribution for unequal-power transfer in the x-direction polarization excitation. Remarkably, a close alignment between the measured and simulated results can be noted when compared to Fig. 9.44, affirming the reliability and accuracy of the experimental results.

Fig. 9.46 Measurement system of the designed reflective metasurface, **a** prototype of SFSF dual-polarization metasurface, **b** near-field scanning measurement system [66]

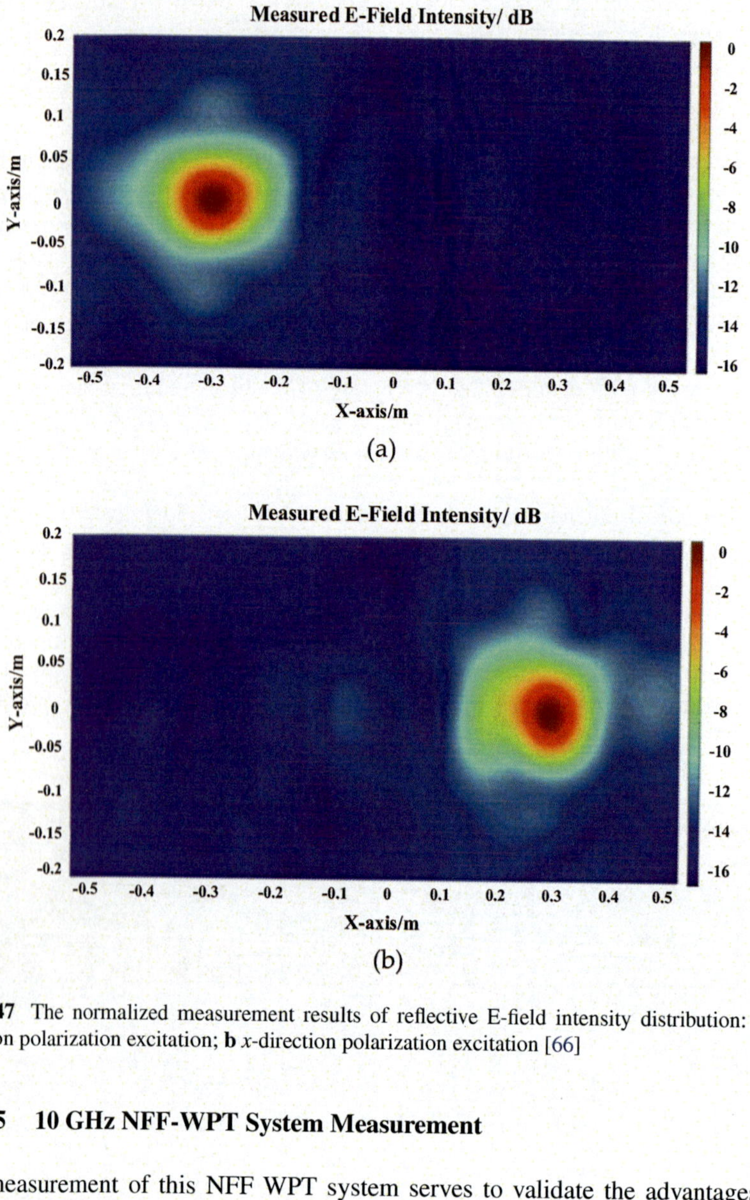

Fig. 9.47 The normalized measurement results of reflective E-field intensity distribution: **a** *y*-direction polarization excitation; **b** *x*-direction polarization excitation [66]

9.3.4.5 10 GHz NFF-WPT System Measurement

The measurement of this NFF WPT system serves to validate the advantages of NFF. The configuration of the transmission system can be observed in Fig. 9.46, and all equipment and parameters remain unchanged from the near-field scanning measurements conducted in case 1, which entails the SFSF WPT system.

To serve as the receiver in the system, a slot-coupling metasurface antenna has been meticulously designed and fabricated, as depicted in Fig. 9.50. The antenna comprises a three-layer structure: the top layer consists of a 4 × 4 mushroom metasurface configuration, the second layer is the slot-coupling ground, and the bottom layer

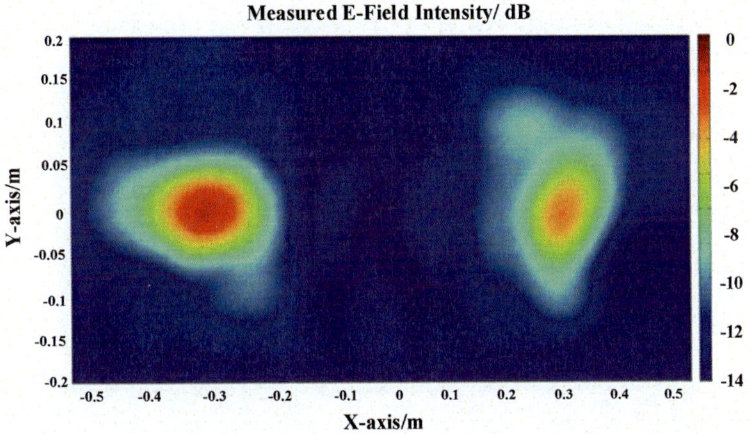

Fig. 9.48 The normalized measurement results of reflective E-field intensity for SFSF reflective metasurface in 40° polarization excitation [66]

constitutes a metal microstrip feedline. The upper layer substrate has a thickness (h_1) of 1.5 mm, while the lower layer substrate has a thickness (h_2) of 0.8 mm. Both substrates are constructed with F4B material, featuring a permittivity of 2.65. The dimensions of the proposed metasurface antenna are 18 mm × 18 mm × 2.3 mm. Figure 9.50 illustrates the geometry of the designed antenna, while Table 9.3 outlines the optimized dimensions. A comprehensive insight into the design process can be found in references [67] and [68]. Additionally, Fig. 9.51 displays the simulated and measured $|S_{11}|$, along with a prototype of the designed receiving antenna.

To measure the received power, the receiving antenna is connected to a power sensor (RS-NRP18S), serving as the load. The receiving antenna is positioned precisely at the predetermined focus position, specifically at $\vec{r_d} = (0.3, 0, 1)$ m. By utilizing the power sensor, the received power by the metasurface antenna can be accurately detected. Notably, the compact size of the receiver, which is smaller than 2 cm, enables it to be entirely situated within the core region of the focusing aperture, where the power density is at its peak. This aspect is crucial in verifying the practical performance of NFF transfer.

To establish a benchmark for assessing the performance of NFF transfer, we define the non-NFF transfer as follows: while keeping the transmitting power and working frequency consistent with the NFF transfer, we position the same feed horn directly at the receiving antenna's location, aligning it with the preset focus employed in the NFF transfer case. By maintaining identical transmission distance and using the same receiving antenna, we can evaluate the focusing transfer effectiveness of NFF by comparing the wireless power received. The measured results, presented in Table 9.4, indicate that the receiving power achieved through NFF is 15 dB higher than that attained through non-NFF transmission at the same distance. To ensure the reliability of this verification, multiple tests were conducted, both with and without the

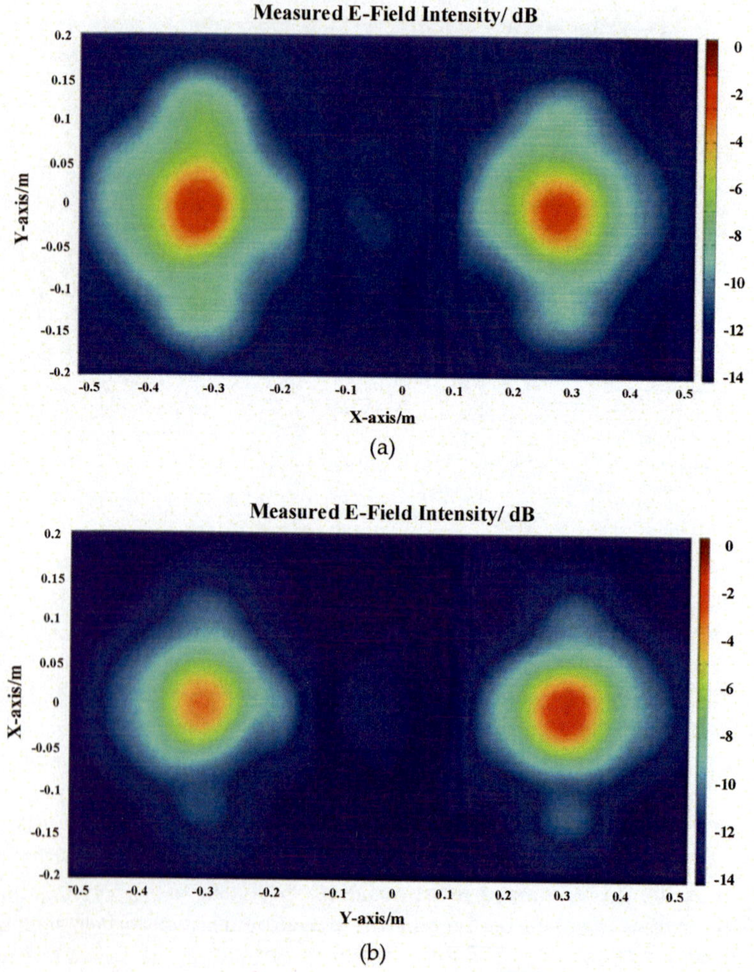

Fig. 9.49 The normalized measurement results of reflective E-field intensity for SFDF metasurface: **a** equal power distribution in y-direction polarization excitation; **b** unequal power distribution in x-direction polarization excitation [66]

power amplifier (+25 dB). Consistently, stable outcomes were obtained. Combining these results with the previous near-field scanning measurements, we can confidently affirm the superiority and high efficiency of the NFF transfer system, enabled by the utilization of reflective metasurfaces in WPT applications.

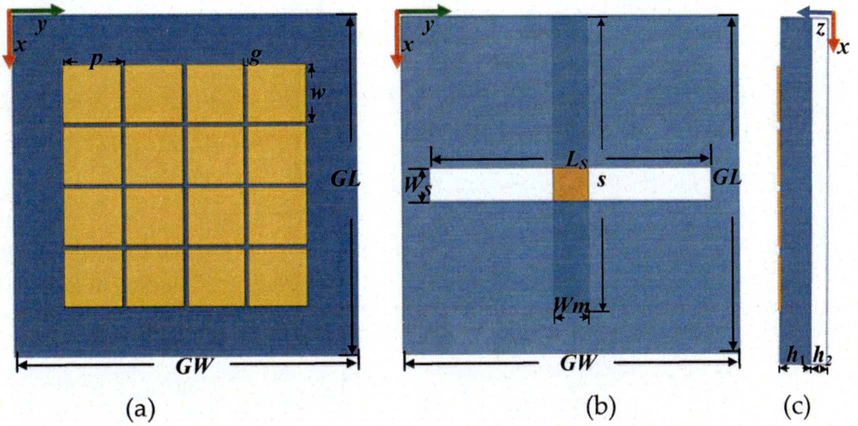

Fig. 9.50 Geometry of the proposed slot-coupling metasurface antenna working at 10 GHz as the receiver of the WPT system measurement: **a** top view of the 4 × 4 mushroom metasurface; **b** top view of the microstrip-line aperture structure; **c** side view [66]

Table 9.3 Dimensions of the designed metasurface antenna (unit: mm)

p	g	w	GL	GW	h_1
3.2	0.2	3	18	18	1.5
Ls	Ws	Wm	S	h_2	
15	1.82	1.85	15.8	0.8	

Fig. 9.51 Measured and simulated reflection coefficient, the prototype of the proposed metasurface antenna, and the proposed antenna connected with an RS-NRP18S power sensor [66]

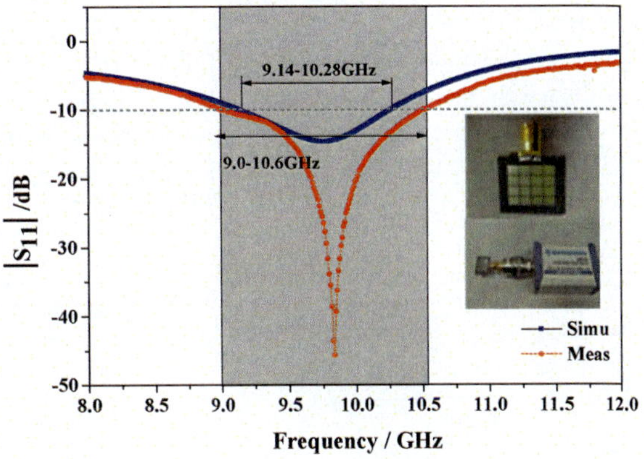

Table 9.4 The receiving power by the metasurface antenna of NFF transfer and non-NFF transfer

	NFF (dBM)	Non-NFF (dBM)
$P_{in} = -2$ dBm	−18 to −20	−35
$P_{in} = -2$ dBm + PA (25 dB)	5	−10

9.3.4.6 Analysis of Dual-Polarization Metasurface

A. **Comparison Between NFF Reflective Metasurface and Traditional Directional-Beam Reflectarray**

The reflective metasurface utilized in the NFF technique, proposed here, distinguishes itself from traditional directional-beam reflectarrays. While the NFF aims at achieving focal points within the near-field region, including the Fresnel region, the traditional directional-beam reflectarray is designed for radiating energy in the far-field region. To further elucidate the characteristic differences between these two approaches, we conducted simulations of their respective far-field radiation patterns. As depicted in Fig. 9.52, the NFF reflective metasurface exhibits a broader main beam in the far-field, with the beam aligning towards the same direction as the focus position. Conversely, the directional-beam reflectarray showcases a narrower main beam in the far-field, accompanied by a more concentrated distribution of E-field intensity. These findings highlight the distinctive behavior and performance characteristics of the NFF reflective metasurface when compared to traditional directional-beam reflectarrays.

Subsequently, a detailed analysis and comparison of the beam-concentrating characteristics between the two cases were conducted in the near-field region using HFSS, as presented in Fig. 9.53. For the NFF metasurface, the focus was designated at $\vec{r_d} = (0, 0, 1)$ m, while the main beam direction was set along the vector $(\theta_r, \varphi_r) = (0°, 0°)$. Consequently, we positioned the observational plane at a distance of 1 m from them, which lies within the near-field region. Through comprehensive full-wave simulations, the E-field intensity distribution on the reference plane was obtained and illustrated in Fig. 9.53. Evidently, the NFF reflective metasurface exhibits a significantly more pronounced E-field intensity distribution, indicative of superior power convergence. Notably, the size of the focusing aperture of the NFF reflective metasurface is smaller than that of the traditional directional beam

Fig. 9.52 The far-field radiation characteristics: **a** NFF reflective metasurface; **b** traditional directional-beam reflectarray [65]

reflectarray. This observation further emphasizes the distinguishing characteristics of NFF transfer, whereby the beam is concentrated primarily within the near-field region before gradually diverging into the far-field region.

B. Analysis of NFF Transfer Efficiency of Reflective Metasurface

This section primarily focuses on discussing the transfer efficiency of the NFF reflective metasurface. The transfer efficiency, denoted as η_{rm}, is determined by the ratio of P_d, which represents the power captured on the reference plane by the focusing aperture, and P_g, the power captured from the feed by the metasurface aperture. It is important to note that the NFF transfer efficiency of the reflective metasurface (η_{rm}) is equivalent to the previously mentioned η_2, both of which reflect the reflective metasurface's ability to converge electromagnetic waves. As illustrated in Fig. 9.54, S_d represents the area of the focusing aperture on the observational plane within the near-field region, while S_g corresponds to the physical aperture of the NFF reflective metasurface. The values of P_d and P_g can be calculated using numerical integration based on the Poynting theorem. Initially, wireless power is transmitted from the feed and illuminates the reflective metasurface. By performing Poynting vector integration over the metasurface aperture, the power captured by the metasurface aperture, P_g, can be determined using the formula presented in Fig. 9.54. Subsequently, as a result of regulation and reflection by the metasurface, the wireless power takes the form of a focused beam, which then illuminates the focusing aperture. P_d can be obtained by performing Poynting vector integration on the focusing aperture in a similar manner.

By means of full-wave simulation, the NFF transfer efficiency curve of the reflective metasurface is obtained and shown in Fig. 9.55. The four curves shown in Fig. 9.55 are respectively: (1) MTS indicates the power capture efficiency of the metasurface, which is defined as the ratio of the power of the NFF metasurface aperture to the power from the horn; (2) 1T1R1P indicates the NFF transfer efficiency in the case of SFSF for single-polarization excitation; (3) 1T2R1P indicates the NFF

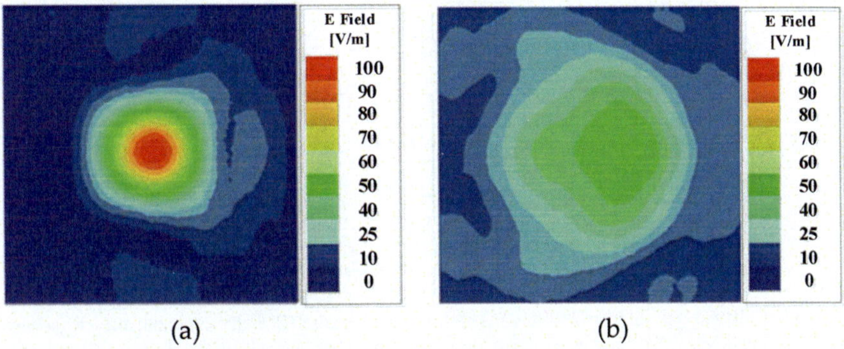

Fig. 9.53 E-field intensity distribution of the reference plane in the near-field region: **a** NFF reflective metasurface; **b** traditional directional-beam reflectarray [66]

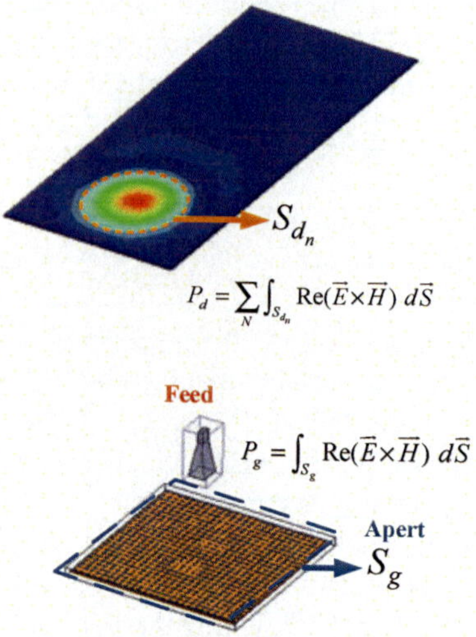

Fig. 9.54 Schematic diagram of NFF transfer efficiency of reflective metasurface calculation [66]

transfer efficiency in the case of SFDF for single-polarization excitation; (4) 1T1R2P indicates the NFF transfer efficiency in the case of SFSF for dual-polarization excitation. Wherein for (3) and (4), the entire NFF transfer efficiency of the metasurface is the sum of the efficiency of each focus.

Observing Fig. 9.55, it can be noted that the 1T1R1P configuration achieves a maximum efficiency η_{rm} of 71.6% at an operating frequency of 10 GHz. The NFF transfer efficiency demonstrates a fractional bandwidth of 50%, spanning from 9.2

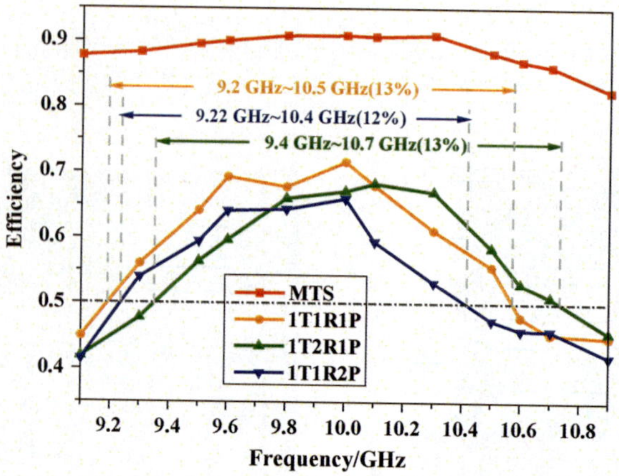

Fig. 9.55 Full-wave simulation results of NFF transfer efficiency of reflective metasurface varying with frequency: MTS: the power capture efficiency of the metasurface; 1T1R1P: the NFF transfer efficiency of SFSF in single-polarization excitation; 1T2R1P: the NFF transfer efficiency of SFDF in single-polarization excitation; 1T1R2P: the NFF transfer efficiency of SFSF in dual-polarization excitation [66]

to 10.5 GHz. Similarly, the 1T2R1P and 1T1R2P curves exhibit a similar trend. Compared to the single focus case, these two configurations experience a minor decline in their maximum efficiencies. Due to the decomposition of the polarization component and the design of multiple beams, some beam dispersion is introduced, resulting in a decrease in the total power on the focusing aperture. For the 1T2R1P case, the fractional bandwidth spans 13% from 9.4 to 10.7 GHz, with a maximum efficiency η_{rm} of 68.3%. On the other hand, the case of 1T1R2P demonstrates a fractional bandwidth of approximately 12%, ranging from 9.2 to 10.4 GHz, with a maximum efficiency η_{rm} of 65.9%. It is noteworthy that all the aforementioned cases guarantee the fractional bandwidth with 50% NFF transfer efficiency, proving to be suitable for WPT systems.

3 Analysis of NFF Transfer Performance

In theory, highly efficient NFF transfer is achievable within the near-field region of the transmitting antenna, within a certain range. However, in practical scenarios, microwaves present challenges in maintaining ideal light wave propagation characteristics. Regardless of the type of antennas used, such as microstrip phased arrays or reflective metasurfaces, it is difficult to fully converge electromagnetic waves to specific ideal points in space. This gives rise to notable phenomena that require attention:

(a) When the transmitting antenna aperture remains constant and the NFF transfer distance increases, even though the focus is still within the near-field region, the beam's focusing ability diminishes. In other words, NFF cannot ensure efficient power transfer throughout the entire near-field region for a transmitting aperture of the same size.
(b) It is worth noting that the E-field intensity at the center of the focusing aperture is generally not the point with the highest E-field intensity in the radial direction of the transmission path. This phenomenon has also been demonstrated in previous studies, specifically in references [33] and [54].

Set the feed horn at $\vec{r_f} = (0, 0, 0.2)$ m and consider the y-direction polarization excitation, and set the focus with different focusing distances. The preset focus is respectively at $\vec{r_{d1}} = (0.3, 0, 0.5)$ m, $\vec{r_{d2}} = (0.3, 0, 1)$ m, $\vec{r_{d3}} = (0.3, 0, 2)$ m, and $\vec{r_{d4}} = (0.3, 0, 5)$ m, which means that the focusing distance in the radial direction is respectively 0.5 m, 1 m, 2 m, and 5 m. A SFSF NFF reflective metasurface with the size of 390 mm × 390 mm (13λ × 13λ) working at 10 GHz is designed.

Figure 9.56 presents the full-wave simulation results of the E-field intensity distribution for the aforementioned four cases, along with the corresponding curves depicting the variation of E-field intensity with distance in the radial direction. Notably, when the focus is set at a radial distance of 0.5 m, the maximum E-field intensity occurs at 0.4 m. Similarly, for focus distances of 1 m, 2 m, and 5 m, the maximum E-field intensity occurs at distances of 0.7 m, 0.8 m, and 1.1 m, respectively. However, it is worth mentioning that in the case where the preset focusing distance is set to 5 m (167λ), the focusing beam may start to diverge before reaching

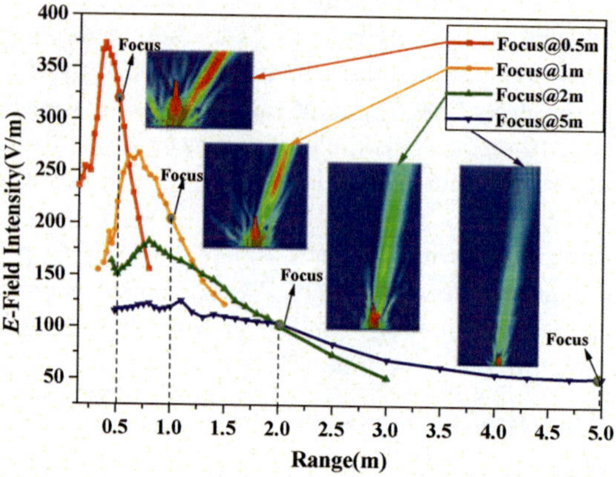

Fig. 9.56 Full-wave simulation results of the E-field intensity distribution of different preset focusing distances and the curves of the E-field intensity varying with distance in the radial direction [66]

the intended focusing aperture. This highlights a constraint between the NFF performance and the electrical size of the transmitting aperture. As the preset focusing distance increases, the difference between the E-field intensity at the center of the focusing aperture and the maximum E-field intensity in the radial direction becomes larger, thus indicating a weaker convergence ability of the beam. Consequently, when the preset focusing distance becomes excessively long, even if the focus is technically still within the near-field region, the beam will exhibit noticeable divergence, rendering the realization of ideal NFF transfer more challenging.

Figure 9.57 illustrates the E-field intensity distribution in the focusing plane for different preset focus distances of the reflective metasurface, specifically at $\vec{r_{d1}} = (0.3, 0, 0.5)$ m, $\vec{r_{d2}} = (0.3, 0, 1)$ m, and $\vec{r_{d3}} = (0.3, 0, 2)$ m. Upon observation, it becomes evident that as the focus gets closer to the reflective metasurface, the focusing beam becomes narrower, resulting in a smaller size for the focusing aperture. Furthermore, the E-field intensity at the center of the focusing aperture also increases proportionally with the proximity of the focus to the reflective metasurface.

However, it is crucial to highlight that the concentrated distribution of E-field intensity on the focusing plane does not necessarily equate to a high NFF transfer efficiency for the main beam on the same plane. By integrating the Poynting vector over the focusing aperture and calculating the NFF transfer efficiency for the aforementioned cases, the results obtained are as follows: 0.5 m (16.7λ)—67.1%, 1 m (33.3λ)—71.6%, and 2 m—63.3%. It is noteworthy that the case with a focus distance of 1 m (33.3λ) achieves the highest NFF transfer efficiency while maintaining a relatively concentrated beam. Conversely, setting the focus at a distance of 0.5 m in close proximity to the reflective metasurface leads to a strong coupling field above the metasurface aperture and results in significant side lobes in other directions. As a consequence, the NFF transfer efficiency decreases. Therefore, it is not accurate to assume that the closer the preset focus distance is to the reflective metasurface, the higher the NFF transfer efficiency of the focusing beam will be.

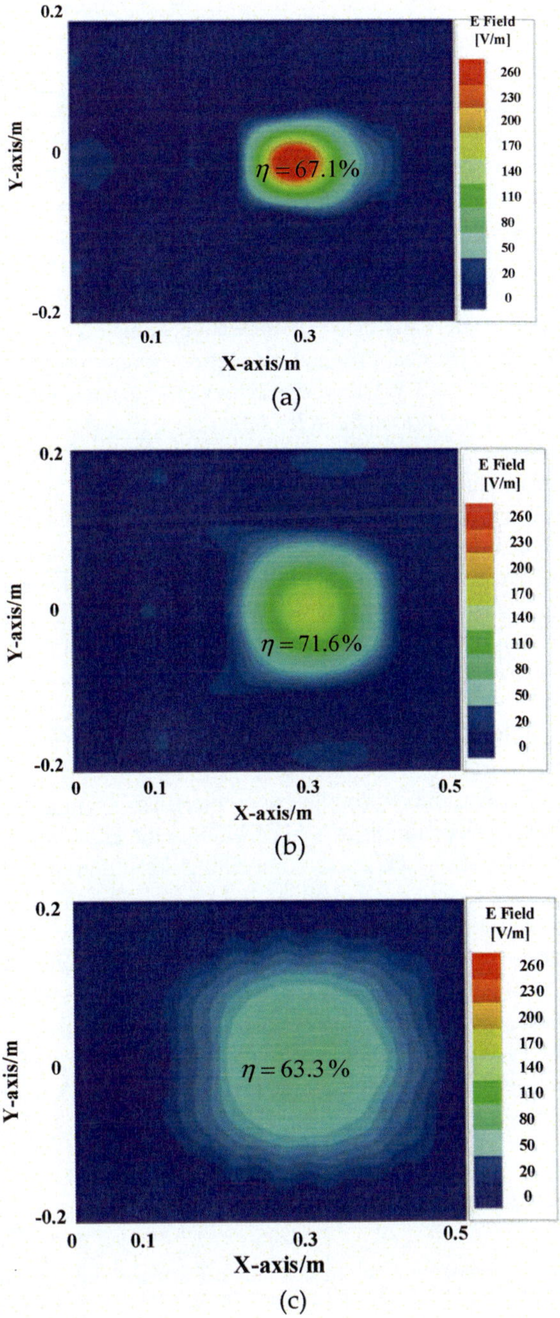

Fig. 9.57 Full-wave simulation results of E-field intensity distribution of the focusing plane for cases that the focus is set at **a** $\vec{r_{d1}} = (0.3, 0, 0.5)$ m, **b** $\vec{r_{d2}} = (0.3, 0, 1)$ m, and **c** $r\ \vec{r_{d3}} = (0.3, 0, 2)$ m [66]

9.4 Conclusion

Wireless power transfer technology has garnered significant attention in both academia and industry, with two key directions emerging: magnetic coupling technology for short-distance applications and microwave radiation technology for long-distance applications. In this chapter, our focus is on breaking the efficiency barriers of WPT systems in these two contexts by introducing electromagnetic metasurfaces.

Initially, we developed highly sub-wavelength NIR and MNG metamaterials to optimize wireless power transfer. Integrating these metamaterials into the WPT system alters the power transfer mechanism to resonant coupling, leveraging the negative permeability and magnetic resonant properties of the metamaterials. As a result, the new WPT system achieves resonance at the frequencies corresponding to the negative refractive index or negative permeability of the metamaterials. Through theoretical analysis, computational simulations, and experimental verification, we present a novel theoretical analysis model for magnetic resonant WPT systems, employing effective network and effective medium theories. Our findings unveil the high power transfer efficiency attributed to the tunneling effect of the equivalent ENZ metamaterials. Additionally, we validate the electromagnetic tunneling phenomenon through planar spiral coils, constructing an ENZ metamaterial at the HF band (shortwave), using both numerical and experimental investigations. This research offers fresh insights into magnetic resonant WPT theory and proposes new structures for efficient mid-distance WPT systems. The derived generalized resonant equation can be utilized for the design of innovative multi-input and multi-output (MIMO) wireless power transfer systems.

Furthermore, we outline a comprehensive synthesis procedure to design reflective metasurfaces aiming to achieve highly efficient wireless power transfer through near-field focusing, while accommodating desired multi-feed and multi-focus characteristics. By employing element designs based on the 'tri-dipole' structure with single-polarization characteristics and the 'cross-dipole' structure with dual-polarization characteristics, we design and fabricate NFF reflective metasurfaces operating at 5.8 and 10 GHz. Through detailed analysis and comparison with measured results, we demonstrate the stability and feasibility of utilizing NFF reflective metasurfaces for practical WPT applications. This proposal provides an effective solution for high-power MIMO wireless power diversity and synthesis.

References

1. N. Tesla, Apparatus for transmitting electrical energy, U.S. patent 1, 119, 732 (1914)
2. N. Tesla, Nikola Tesla: Lectures, Patents, Articles, Nikola Tesla Museum, Beograd Yugoslavia, p.A109 (1956)
3. N. Tesla, A. Marincic, V. Popovic, T. Teofilovic, Colorado springs notes. Nolit Beograd, Yugoslavia **3**, 1899–1900 (1978)
4. A. Kurs, A. Karalis, R. Moffatt, J.D. Joannopoulos, P. Fisher, M. Soljačić, Wireless power transfer via strongly coupled magnetic resonance. Science **317**, 83–86 (2007)

5. D. Huang, Y. Urzhumov, D.R. Smith, K.H. Teo, J. Zhang, Magnetic superlens-enhanced inductive coupling for wireless power transfer. J. Appl. Phys. **111**, 064902 (2012)
6. T.A. Vanderelli, J.G. Shearer, J.R. Shearer, Method and apparatus for a wireless power supply, U.S. patent 7,027,311 (2006)
7. A. Erentok, R.W. Ziolkowski, J.A. Nielsen, R.B. Greegor, C.G. Parazzoli, M.H. Tanielian, S.A. Cummer, B.I. Popa, T. Hand, D.C. Vier, Lumped element-based, highly sub-wavelength, negative index metamaterials at UHF frequencies. J. Appl. Phys. **104**, 034901 (2008)
8. B.L. Cannon, J.F. Hoburg, D.D. Stancil, S.C. Goldestein, Magnetic resonant coupling as a potential means for wireless power transfer to multiple small receivers. IEEE Trans. Power Electron. **24**(7), 1819–1825 (2009)
9. Y.C. Fan, L. Li, S.X. Yu, C. Zhu, C.H. Liang, Experimental study of efficient wireless power transfer system integrating with highly sub-wavelength metamaterials. Prog. Electromag. Res. **141**, 769–784 (2013)
10. J. Choi, C. Seo, High-efficiency wireless energy transmission using magnetic resonance based on metamaterial with relative permeability equal to -1. Prog. Electromag. Res. **106**, 33–47 (2010)
11. B. Wang, K.H. Teo, T. Nishino, W. Yerazunis, J. Barnwell, J. Zhang, Experiments on wireless power transfer with metamaterials. Appl. Phys. Lett. **98**, 254101 (2011)
12. Y. Urzhumov, D.R. Smith, Metamaterial-enhanced coupling between magnetic dipoles for efficient wireless power transfer. Phys. Rev. B **83**, 205114 (2011)
13. L. Li, Y.C. Fan, S.X. Yu, C. Zhu, C.H. Liang, Design, fabrication and measurement of highly sub-wavelength double negative metamaterials at HF frequencies. J. Appl. Phys. **113**, 213712 (2013)
14. J.B. Pendry, Extremely low frequency plasmons in metallic microstructures. Phys. Rev. Lett. **76**, 4773 (1996)
15. R.A. Shelby, D.R. Smith, S. Schultz, Experimental verification of a negative index of refraction. Science **292**, 77–79 (2001)
16. D.R. Smith, W.J. Padilla, D.C. Vier, Composite medium with simultaneously negative permeability and permittivity. Phys. Rev. Lett. **84**(18), 4184–4187 (2000)
17. W.C. Chen, C.M. Bingham, K.M. Mak, N.W. Caira, W.J. Padilla, Extremly sub-wavelength planar magnetic metamaterials. Phys. Rev. B **85**, 201104 (2012)
18. High Frequency Structure Simulator, Ansys Corporation
19. Advanced Design System, Agilent Corporation
20. Y.C. Fan, L. Li, Efficient wireless power transfer system by using highly sub-wavelength negative-index metamaterials. In: IEEE International Wireless Symposium, Beijing, Apr. 14–18 (2013)
21. J.B. Pendry, A.J. Holden, D.J. Robbins, W.J. Stewart, Magnetism from conductors and enhanced nonlinear phenomena. IEEE Trans. Microw. Theory Tech. **47**, 2075–2084 (1999)
22. M.G. Silveirinha, N. Engheta, Design of matched zero-index metamaterials using nonmagnetic inclusions in epsilon-near-zero media. Phys. Rev. B **75**, 075119 (2007)
23. M.G. Silveirinha, A. Alù, B. Edwards, N. Engheta, Overview of theory and applications of epsilon-near-zero materials. In: URSI General Assembly, Chicago, IL (2008)
24. M.G. Silveirinha, N. Engheta, Tunneling of electromagnetic energy through subwavelength channels and bends using ε-near-zero materials. Phys. Rev. Lett. **97**, 157403 (2006)
25. M.G. Silveirinha, N. Engheta, Theory of supercoupling, squeezing wave energy, and field confinement in narrow channels and tight bends using ε near-zero metamaterials. Phys. Rev. B **76**, 245109 (2007)
26. Q. Cheng, R. Liu, D. Huang, T.J. Cui, Circuit verification of tunneling effect in zero permittivity medium. Appl. Phys. Lett. **91**, 234105 (2007)
27. B. Edwards, A. Alù, M.E. Young, M.G. Silveirinha, Experimental verification of epsilon-near-zero metamaterial coupling and energy squeezing using a microwave waveguide. Phys. Rev. Lett. **100**, 033903 (2008)
28. R. Liu, Q. Cheng, T. Hand, J.J. Mock, T.J. Cui, S.A. Cummer, D.R. Smith, Experimental demonstration of electromagnetic tunneling through an epsilon-near-zero metamaterial at microwave frequencies. Phys. Rev. Lett. **100**, 023903 (2008)

29. A.P. Sample, D.A. Meyer, J.R. Smith, Analysis, experimental results, and range adaptation of magnetically coupled resonators for wireless power transfer. IEEE Trans. Industr. Electron. **58**, 544–554 (2011)
30. D.R. Smith, D.C. Vier, T. Koschny, C.M. Soukoulis, Electromagnetic parameter retrieval from inhomogeneous metamaterials. Phys. Rev. E **71**, 036617 (2005)
31. R.W. Ziolkowski, Design, fabrication, and testing of double negative metamaterials. IEEE Trans. Antennas Propag. **51**, 1516–1529 (2003)
32. S.X. Yu, L. Li, Effective epsilon-near-zero metamaterials in magnetic resonant wireless power transfer system. In: 2015 Cross-Strait Quad-Regional Radio Science and Wireless Technology Conference (CSQRWC), August 12–15, Xi'an, China (2015)
33. J.S. Lii, Properties of focused apertures in the fresnel region. IRE Trans. Antennas Propag. **10**(4), 399–408 (1962)
34. R.C. Hansen, Focal region characteristics of focused array antennas. IEEE Trans. Antennas Propag. **33**(12), 1328–1337 (1985)
35. L. Shafai, A.A. Kishk, A. Sebak, Near field focusing of apertures and reflector antennas. In: WESCANEX 97: Communications, Power and Computing. Conference Proceedings, pp. 246–251 (2002)
36. J. Bor, S. Clauzier, O. Lafond, M. Himdi, 60 GHz foam-based antenna for near-field focusing. Electron. Lett. **50**(8), 571–572 (2015)
37. A. Buffi, P. Nepa, G. Manara, Design criteria for near-field-focused planar arrays. IEEE Antennas Propag. Mag. **54**(1), 40–50 (2012)
38. F. Tofigh, J. Nourinia, M.N. Azarmanesh, K.M. Khazaei, Near-field focused array microstrip planar antenna for medical applications. IEEE Antennas Wirel. Propag. Lett. **13**(13), 951–954 (2014)
39. R. Siragusa, P. Lemaitre-Auger, S. Tedjini, Tunable near-field focused circular phase-array antenna for 5.8-GHz RFID applications. IEEE Antennas Wirel. Propag. Lett. **10**(1), 33–36 (2011)
40. K.D. Stephan, J.B. Mead, D.M. Pozar, L. Wang, J.A. Pearce, A near field focused microstrip array for a radiometric temperature sensor. IEEE Trans. Antennas Propag. **55**(4), 1199–1203 (2007)
41. S. Karimkashi, A.A. Kishk, Focusing properties of Fresnel zone plate lens antennas in the near-field region. IEEE Trans. Antennas Propag. **59**(5), 1481–1487 (2011)
42. I.V. Minin, O.V. Minin, *Basic Principles of Fresnel Antenna Arrays*. Lecture Notes in Electrical Engineering, vol. 19 (2008)
43. A.J. Martínez-Ros, J.L. Gómez-Tornero, J. Monzó-Cabrera, Microwave near-field focusing properties of width-tapered microstrip leaky-wave antenna. IEEE Trans. Antennas Propag. **61**(6), 2981–2990 (2013)
44. J.L. Gómez-Tornero, D. Blanco, E. Rajo-Iglesias, N. Llombart, Holographic surface leaky-wave lenses with circularly-polarized focused near-fields—part I: concept, design and analysis theory. IEEE Trans. Antennas Propag. **61**(7), 3475–3485 (2013)
45. P. Lemaitre-Auger, R. Siragusa, C. Caloz, D. Kaddour, Circular antenna arrays for near-field focused or multi-focused beams. In: URSI International Symposium on Electromagnetic Theory, pp. 425–428 (2013)
46. Y.J. Cheng, F. Xue, Ka-band near-field-focused array antenna with variable focal point. IEEE Trans. Antennas Propag. **64**(5), 1725–1732 (2016)
47. O. Yurduseven, D.L. Marks, J.N. Gollub, D.R. Smith, Design and analysis of a reconfigurable holographic metasurface aperture for dynamic focusing in the Fresnel zone. IEEE Access **5**, 15055–15065 (2017)
48. P.F. Li, S.W. Qu, S. Yang, Y.A. Liu, Q. Xue, Near-field focused array antenna with frequency-tunable focal distance. IEEE Trans. Antennas Propag. **66**(7), 3401–3410 (2018)
49. N. Yu, P. Genevet, M.A. Kats, F. Aieta, J.P. Tetienne, F. Capasso, Z. Gaburro, Light propagation with phase discontinuities: generalized laws of reflection and refraction. Science **334**(6054), 333–337 (2011)

50. A.V. Kildishev, A.M. Shaltout, V.M. Shalaev, Evolution of photonic metasurfaces: from static to dynamic. J. Opt. Soc. of Am. B **33**(3), 501 (2016)
51. Y. Yuan, K. Zhang, X. Ding, B. Ratni, S.N. Burokur, Q. Wu, Complementary transmissive ultra-thin meta-deflectors for broadband polarization-independent refractions in the microwave region. Phot. Res. **7**(1), 80–88 (2019)
52. Z. Wang, X. Ding, K. Zhang, B. Ratni, S.N. Burokur, X. Gu, Q. Wu, Huygens metasurface holograms with the modulation of focal energy distribution. Adv. Optical Mat. **6**(12) (2018)
53. X. Ding, F. Monticone, K. Zhang, L. Zhang, D. Gao, S. N. Burokur, A. de Lustrac, Q. Wu, C.-W. Qiu, A. Alu, Ultrathin Pancharatnam-Berry metasurface with maximal cross-polarization efficiency. Adv. Mat. **27**(7), 1195–1200 (2015)
54. H.T. Chou, T.M. Hung, N.N. Wang, H.H. Chou, C. Tung, P. Nepa, Design of a near-field focused reflectarray antenna for 2.4 GHz RFID reader applications. IEEE Trans. Antennas Propag. **59**(3), 1013–1018 (2011)
55. X. Wu, X. Xia, J. Tian, Z. Liu, W. Wen, Broadband reflective metasurface for focusing underwater ultrasonic waves with linearly tunable focal length. Appl. Phys. Lett. **108**(16), 163502(1–5) (2016)
56. S. Zhang, M.H. Kim, F. Aieta, A. She, T. Mansuripur, I. Gabay, M. Khorasaninejad, D. Rousso, X. Wang, M. Troccoli, High efficiency near diffraction-limited mid-infrared flat lenses based on metasurface reflectarrays. Lasers Electro-Opt., 18024–18034 (2016)
57. V.G. Veselago, The electrodynamics of substances with simultaneously negative values of permittivity and permeability. Sov. Phys. Usp. **10**, 5–13 (1968)
58. J.B. Pendry, Negative refraction makes a perfect lens. Phys. Rev. Lett. **85**, 3966–3969 (2000)
59. D.R. Smith, How to build a superlens. Science **308**, 502 (2005)
60. R.E. Collin, *Foundations for Microwave Engineering* (McGraw Hill Book Company, Inc., New York, 1966)
61. D.R. Smith, S. Schultz, Determination of effective permittivity and permeability of metamaterials from reflection and transmission coefficients. Phys. Rev. B **65**, 195104 (2002)
62. J. Choma, W.K. Chen, *Feedback Networks: Theory and Circuit Application* (World Scientific Publishing Company, Singapore, 2007)
63. A.M. Nicolson, G.F. Ross, Measurement of the intrinsic properties of materials by time domain techniques. IEEE Trans. Instrument. Measure. **IM-19**, 377–382 (1970)
64. W.B. Weir, Automatic measurement of complex dielectric constant and permeability at microwave frequencies. Proc. IEEE **62**, 33–36 (1974)
65. S. Yu, H. Liu, L. Li, Design of near-field focused metasurface for high-efficient wireless power transfer with multifocus characteristics. IEEE Trans. Industr. Electron. **66**, 3993–4002 (2019)
66. P. Zhang, L. Li, X.M. Zhang, H.X. Liu, Y. Shi, Design, measurement, and analysis of near-field focusing reflective metasurface for dual-polarization and multi-focus wireless power transfer. IEEE Access **7**, 110387–110399 (2019)
67. W. Liu, Z.N. Chen, X. Qing, Metamaterial-based low-profile broadband mushroom antenna. IEEE Trans. Antennas Propag. **62**(3), 1165–1172 (2014)
68. Z. Wu, L. Li, X. Chen, K. Li, Dual-band antenna integrating with rectangular mushroom-like superstrate for WLAN applications. IEEE Antennas Wirel. Propag. Lett. **15**, 1269–1272 (2016)

Chapter 10
Rectifying Metasurfaces for Wireless Energy Harvesting System

Xuanming Zhang, Long Li, and Pei Zhang

Abstract In this chapter, an overview of wireless power transfer (WPT) and wireless energy harvesting (WEH) technologies, followed by a detailed examination of the utilization of metamaterials and metasurfaces in WPT and WEH. Firstly, we discusses metasurfaces for ambient energy harvesting (AEH), encompassing their recent progress, discussion, and potential future opportunities. Then, AEH systems based on metasurfaces and antennas are compared. In addition, to enhancing the effectiveness of energy collection systems, metasurfaces also broaden their applicability to various types of devices. The next step was to present two cases in which metasurfaces were designed for AEH. One of the metasurfaces features a closed-ring butterfly array and is wide-angle and polarizated-angle-independent, while the other metasurface is dual-band, rectifying integrated, and insensitive to the polarization angle. To simplify the ambient energy harvester's structure, a novel surface-mount component incorporating method is developed. The metasurface's high-impedance characteristics and its multi-mode resonance make it possible to directly eliminate the matching network between it and the non-linear rectifier. The above designs have been tested, and the excellent performances proved the huge application prospect of metasurface in the field of WEH.

Keywords Wireless power transfer (WPT) · Wireless energy harvesting (WEH) · Ambient energy harvesting (AEH) · Wide-angle · Polarization-angle-independent · Rectifying integrated

---

X. Zhang
School of Electronic Engineering, Xi'an University of Posts and Telecommunications, Xi'an 710121, China

L. Li (✉)
School of Electronic Engineering, Xidian University, Xi'an 710071, China
e-mail: lilong@mail.xidian.edu.cn

P. Zhang
The 28Th Research Institute of China Electronics Technology Group Corporation, Nanjing 310100, China

L. Li et al. (eds.), *Electromagnetic Metamaterials and Metasurfaces: From Theory To Applications*, https://doi.org/10.1007/978-981-99-7914-1_10

10.1 Introduction

Energy is a constant need for humanity. Wireless power transfer (WPT) technology enables the transmission of power without the need for traditional physical connections [1–4]. Over the past few decades, the advancement of WPT technology has transitioned from theoretical validation to becoming commercially available by means of iterative enhancements in wireless communication and wireless sensing [5–7].

In today's world, the Internet of Things (IoT) is being positioned as a future trend with the help of artificial intelligence (AI) and fifth-generation communication (5G) technology. Mobile terminals and wireless sensor networks (WSNs) that are intelligent, miniaturized, and low-power are extensively distributed on a global scale. The adaptability and sustainability of a wireless device's performance in diverse and intricate settings is vital. By using ambient energy harvesting technology, low-power devices can be powered continuously without relying on batteries, which have a limited life cycle [8, 9]. Various types of energy can be gathered from the surroundings, encompassing solar, heat, movement, and wireless RF/microwave energy. Several factors, such as weather conditions, time of day, and site location, impact the availability of these types of energy in the actual world. In spite of these constraints, electromagnetic (EM) radiation has managed to overcome them due to the remarkable progress in wireless technology [10–12]. Various forms of electromagnetic (EM) energy are present in the surroundings based on frequency spectrum assignments, encompassing RF, microwave [8–12], infrared thermal radiation [13–15], and solar energy [12, 16]. In the industrial and commercial sectors, solar energy has proven to be a reliable source of energy. Nevertheless, it encounters several notable obstacles, such as poor effectiveness (e.g., as low as 23%) and restricted accessibility in daylight and favorable conditions. Infrared thermal radiation can gather heat transfer from the earth to the frigid outer space, which is roughly equivalent to the solar radiation received by the earth. Despite its potential as a sustainable energy option, the practical application of this renewable source remains limited as it has solely been examined in theory [17]. The concept of RF ambient energy harvesting (AEH) was proposed during the early 1990s. Until now, the distribution of signals in the surroundings primarily consisted of wireless communication bands like television, 5G and Wi-Fi, exhibiting a typical power density ranging from 2 $\mu W/m^2$ to 10 mW/m^2 [8, 9, 11]. Considering factors such as time, weather, and location, it is discovered that the RF offers the widest range of practicality when compared to solar and infrared thermal radiation.

The RF energy present in the surroundings is gradually expanding its reach to encompass all weather conditions, throughout the day, and across a wide range of frequencies. This ubiquitous presence of RF energy allows low-power devices to consistently gather energy from radio waves in their vicinity. In other words, the collection of ambient RF power is an unavoidable direction in the low-power advancement of WPT and wireless energy harvesting (WEH) systems [1, 9]. Its primary uses currently include powering IoT sensor nodes, consumer electronics, implant devices, and similar applications. Hence, the primary emphasis of this chapter lies

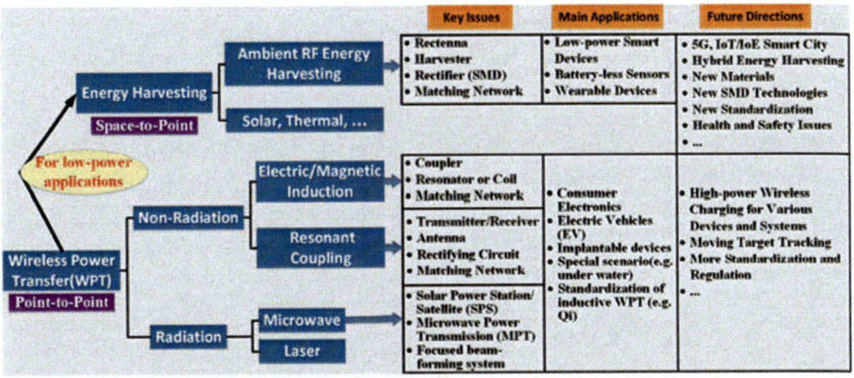

Fig. 10.1 The classification, attributes, significant concerns, primary uses, and prospective paths of WPT and AEH [71]

in the field of harvesting technologies for ambient RF energy. The classification, attributes, significant concerns, primary uses, and prospective paths of WPT and AEH is presented in Fig. 10.1.

Different approaches were commonly used for the application of WPT and AEH. The primary purpose of WPT is to transmit power in a specific direction, typically with a power range exceeding the Watt-level. Transmitter/receiver and transmission medium need to be considered systematically. Typically, it is characterized by a narrow bandwidth and possesses significant power. Nevertheless, AEH generally exhibits wide bandwidth and possesses low energy consumption. The main emphasis is on the collection of energy from space to a single point, particularly in the receiving component, covering a power spectrum ranging from milliwatts to microwatts or lower. One of the most crucial components for the research on WPT and AEH is the rectenna [18, 19]. Currently, the rectenna is extensively utilized in numerous applications of WPT or AEH [20–28].

Conversely, metamaterials (MMs) have garnered significant interest across various domains [29]. Established in 2011, the metasurface (MS) is a crucial plane or quasi-plane metamaterial structure [30]. Due to its exceptional performance and uncomplicated design, the MS have been widely used in applications ranging from microwave to optical wavelengths. Previously, MS/MM has been utilized in the context of near-field WPT systems, including resonant coupling systems, medical implantable devices, and near-field focused transmitters [31–36]. Consequently, MS enhances the close-range properties of transmission to boost the efficiency and range of the WPT system [37, 38].

10.2 Metasurfaces for AEH: The State of the Art, Challenges, and Future Prospects

According to recent studies, MS has been reported to improve the performance of AEH. There are two approaches to enhance the antenna performance: either by incorporating the concept of MS into the rectenna or by utilizing a MS directly as an power harvester instead of the antenna [39–41]. The second idea is derived from the absorber [42]. The objective of the power receiver is to optimize the obtained RF power and direct it towards the rectifier. As a result, RF energy from space is converted into circuit power instead of wasting energy within the structure.

In 2012, the utilization of metamaterial particles harvester was developed [43]. The harvester is a flat 9 × 9 array with split-ring resonator (SRR) cells. At a frequency of 5.8 GHz, the SRR stores RF energy by using a resistive load placed at each unit cell's gap. In this case, the MM particles acts as the energy harvester as an alternative to the traditional antenna. It is noted that the independent MM particles harvester and the MS array harvester exhibit fundamental distinctions. The latter can fully manipulate the EM wave to enhance the performance of AEH, and thus become the hot research area. The flat SRR design increases the attenuation of signal transmission in the absence of a ground plane, resulting in an efficiency of less than 80% for energy harvesting, which is dependent on the angle of incidence. Additionally, previous research has been expanded to include the optical spectrum [44, 45]. In 2015, a more advanced energy harvester called the electric-inductive-capacitive (ELC) structure was introduced [46]. The ELC structure utilized an extra via to connect the top metal surface structure and the bottom metal grounding, effectively directing surface power into the load resistance. The ELC harvester, which achieves a harvesting efficiency of up to 97%, exhibits a performance similar to that of the ideal EM absorber. As a result, the subsequent energy harvester has primarily been enhanced based on this fundamental framework. Afterwards, a 11 × 11 ground-backed complementary split-ring (G-CSRR) array harvester was introduced as an enhancement to the single-element G-CSRR harvester. This improved harvester exhibits superior efficiency and broader frequency range when compared to the patch antenna array operating at the same operating frequency and occupying the same physical space [47, 48].

Then, a G-CSRR structure with a broader frequency range (each array cell having four ports) was developed to enhance the bandwidth of the MS receiver [49]. The MS characterized by a circular configuration, was designed as being capable of efficiently collecting energy in a wide range of frequencies, spanning from 6.2 to 21.4 GHz [50]. Broadband or multi-band harvesters for ambient RF energy harvesting have the potential to gather and store a greater amount of energy across a wider range of frequencies compared to narrow-band harvesters.

To enhance the capture capacity of electromagnetic waves from unknown directions and positions, the receiver also faces notable challenges regarding polarization insensitivity and incident angle insensitivity. Currently, the tri-band SRR harvester with wide-angle and polarization-angle-insensitive characteristics was proposed [51]. The multi-polarization-angle energy harvesting was also achieved by another

"Ω" pattern MS design [52]. Both of these structures had four receiving ports, which greatly enhances the property of the harvster but it also increases the complexity of the collector. Afterwards, simplified single-band collectors were suggested to attain efficient AEH and decrease the receiving port with a straightforward MS design [53, 54]. In reality, a recently proposed triple-band butterfly-shaped closed-ring MS with sub-wavelength multi-mode has shown for achieving efficient AEH [55]. The suggested MS was simple, compact, and realized polarization and incident angle insensitive, yet it encountered a challenge with high impedance characteristics at the receiving ports.

The MS collector design has also incorporated several fascinating characteristics [56–59]. An all-metal SRR with three-dimensional characteristics was presented to decrease the loss of dielectric and enhance both the efficiency and bandwidth. The MS design with the coded closed rings could completely automate the design of the collector, resulting in enhanced efficiency and improved impedance matching of the receiving ports. Utilizing flexible materials in a MS has the potential to improve the adaptability of the surrounding harvesters.

The MS harvester mentioned earlier is without any rectification. In fact, the rectified 5 × 1 array SRR structure at 900 MHz was introduced in 2013, achieving a maximum RF-DC efficiency of 36% at 24 dBm [60]. Subsequently, a ELC MS with the rectifying circuit was designed in a three-layer configuration. In this configuration, all unit cells are linked to a rectifying circuit through a microstrip line network (consisting of 8 × 8 array and one receiving port). The harvester rectifying efficiency at 10 dBm is up to 67% at 2.45 GHz [61]. The RF-DC efficiency at 12 dBm of the sandwich MS was 40% using the ring resonator [62]. By modifying the configuration of single-polarized cells, a comparable rectifying MS collector was successfully developed, attaining an efficiency of 70% at 9 dBm [63]. A new three-layer design without a RF combining network has recently been suggested, resulting in an RF-DC efficiency at 5 mW/cm^2 of 66.9% at 2.45 GHz [64].

The integration of the rectifying and matching network into the MS can lead to increased losses and manufacturing errors. Consequently, the existing rectified MS structure becomes more complex. In this case, integrating the MS and diode in a coplanar design proves to be an efficient design. The I-shaped MS were introduced in 2014, involving the coplanar MS embedded with a diode for the solar power satellite (SPS) [65]. At 0 dBm, the efficiency of RF-DC was merely 28%. Subsequently, a coplanar rectifying frequency selective surface (FSS) design with diodes was introduced, which remains unaffected by polarization [66]. However, as FSS dimensions increase, RF-DC efficiency decreases, posing a challenge for matching impedances. An MS concept with a single layer of rectifying has been introduced recently [67]. The composition of the structure includes a cut-wire MS-integrated diode, and a wire with high inductance was utilized to link all units without the power-combining network. At 0 dBm, the simulation result shows a rectifying efficiency of only 50%. A recent development in the field of rectennas demonstrated that the integrated diodes' antenna harvesting surfaces could achieve high efficiency at medium/high power levels [68–70]. The combination of the rectifier and MS in the integrated AEH harvester design offers a valuable concept.

During recent research and progress, the AEH MS harvesters have undergone significant changes. Compared to single-band designs, they have become multi-band or wideband, and also become multipolarized. Additionally, they have been of wide-angle and polarization-angle-insensitive characteristics. These advancements enable them to effectively capture power from the surrounding RF power sources. Figure 10.2. summarizes the notable advancements and progress of the AEH MS harvester. Every iteration of the MS collector is equipped with different structural configurations aimed at enhancing performance and incorporating additional functionalities. In spite of this, MS collectors mostly rely on the EM resonance. Below, we discuss the future outlook for MS receiver/collector in ambient energy harvesting applications.

In comparison to the traditional antenna, the MS receiver has the ability to optimize the AEH efficiency per unit physical area of the MS. However, the majority of current MS collectors and energy harvesting receivers require an input power of over 0 dBm. Therefore, the advancement of effective energy harvesting technology continues to be a difficult task for low power density situations in real-life settings [1, 61, 64]. Furthermore, by enhancing the research and development of the MS harvester, it is anticipated that not only will the semiconductor components for low power be upgraded, but also the performance will be enhanced in terms of compactness, effectiveness, and adaptability. As a result, efficient AEH can be achieved in diverse environments as needed. Advances in MS, including design ideas, manufacturing techniques, and numerous applications, will continuously be used to develop MS harvesters.

MS has achieved remarkable advancements in numerous areas. The latest research progress suggests that MSs are transitioning towards becoming smaller, adaptable, programmable, and digitalized [71–84]. The advancement of intelligent MSs has improved the MS's capacity to control EM waves at a more advanced level. This not only enhances the scope of research in MSs, including tracking and imaging, but also holds importance for other applications of MSs. Furthermore, these innovative ideas of MS offer valuable insights and opportunities for the advancement of AEH applications.

Significant impacts have been discovered in the MS harvester due to the latest advancements in rectenna and rectifier circuits. The performance of the MS receiver/collector can also be improved by incorporating additional distinctive characteristics like flexible substrates, reconfigurable or adjustable designs, and quantum driving algorithms [85–87]. By combining the MS with antennas and various harvesters, it can achieve multiple functions at the same time [88, 89]. Additionally, this could potentially create immense prospects for the diversified growth of MSs and the implementation of energy in real-world scenarios.

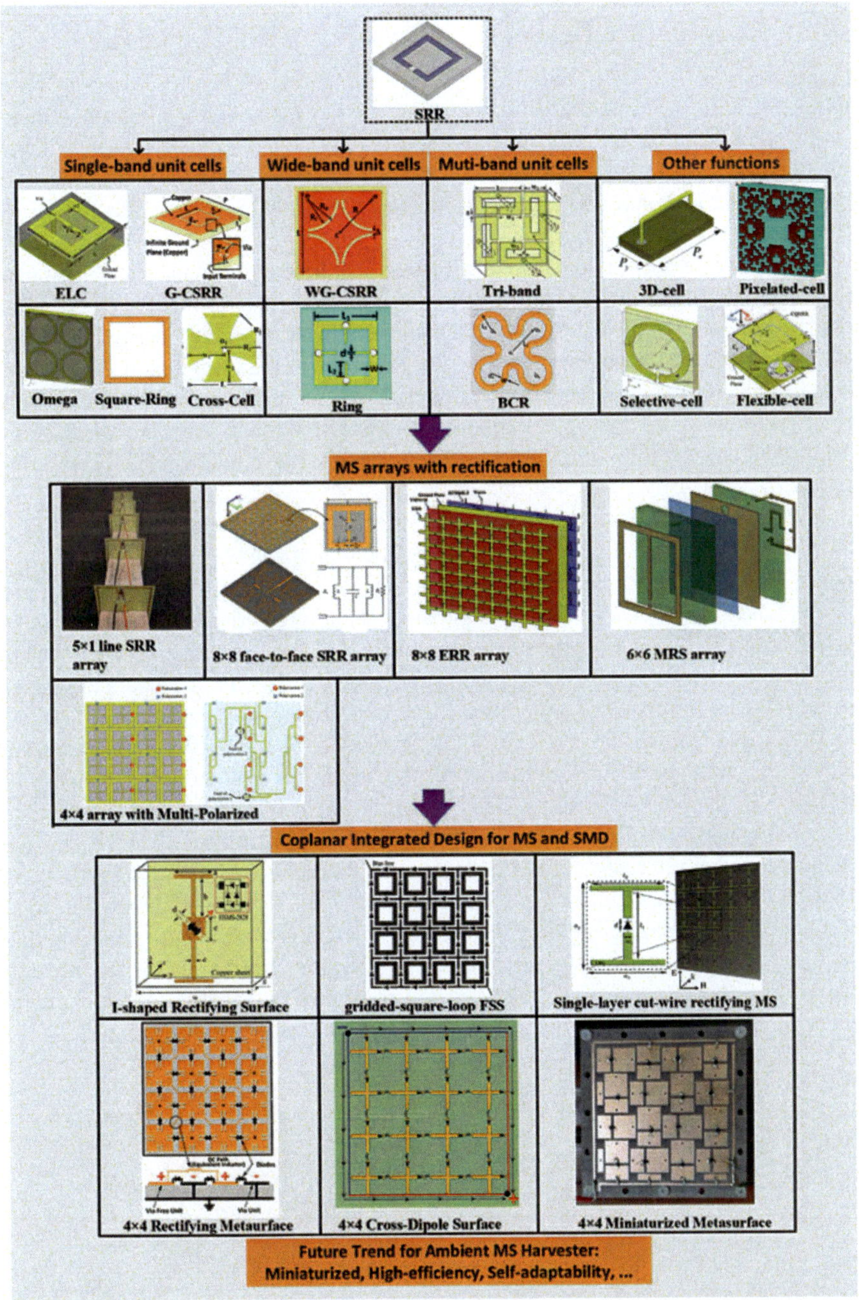

Fig. 10.2 The MS harvester for AEH has made considerable advancements, undergone evolution, and shown promising future trends

10.3 Metasurface Design for Wireless Energy Harvesting

In the preceding section, the advancement, difficulties, and outlook regarding metasurfaces for the collection of ambient radio frequency energy are discussed in detail. This section presents an illustration of the design and validation of metasurfaces for WEH using a tri-band, compact, incident angle, and polarization angle insensitive metasurface as an example [55].

The design concept is presented in Fig. 10.3, where an original square resonant ring is transformed into a symmetric butterfly closed-ring resonantor (BCR). The fundamental square closed-loop can accomplish absorption and energy harvesting with wide-angle and arbitrary-polarized capabilities in the single band. The square loop has transformed into a Minkowski fractal configuration, enabling the attainment of absorption across multiple frequency bands and allowing for miniaturization. Additional optimization of the fractal structure can achieve the butterfly closed-ring resonator.

It is note that we have chosen to adopt the closed ring of the MS instead of the split ring form. Upon the surface of the MS being exposed to an incident wave, a pair of identical circular currents are induced in the same orientation. These currents subsequently flow towards the load situated at the receiving port. Regardless of polarization and orientation, the surface current is able to travel along the loop towards the collection point. The BCR's geometry is depicted in Fig. 10.4. The upper surface of the BCR is comprised of a modified Minkowski fractal configuration.

The closed ring wire has a width of 1.84 mm, while the BCR cell has a period of $P = 26.9$ mm corresponding to about $\lambda_0/12$ at the lowest operating frequency of 900 MHz (λ_0 is free-space wavelength). A metallic ground is located on the opposite side of the F4B dielectric substrate with the thickness of 4.0 mm. The metalized via positioned at the corner of the BCR is linked to the receiving port to gather and transmit the incident electromagnetic energy. A resistor with a resistance of 2776 Ω is connected to the receiving port between the via and ground. Please be aware that there are two metalized vias in the unit cell that have central symmetry. To transmit and collect energy, one via is linked to the receiving port, while the other metal via is connected directly to the ground to enhance the equivalent inductance of the

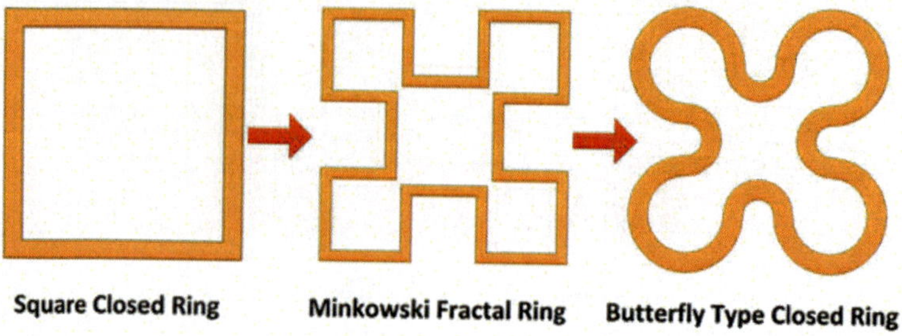

Fig. 10.3 Metasurface element with multimode resonanor designed for energy harvesting [55]

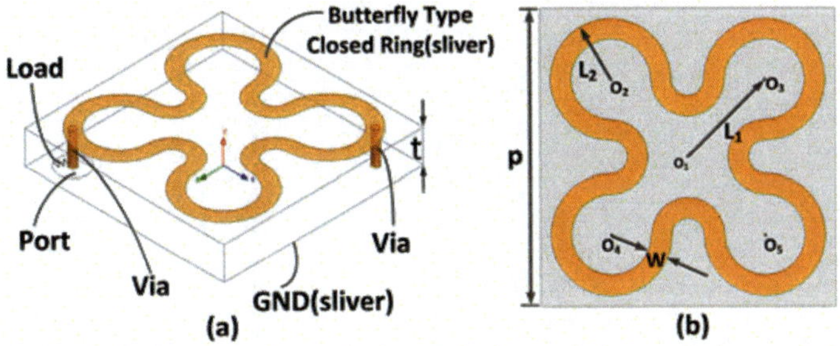

Fig. 10.4 The BCR's geometry, **a** side view and **b** top view [55]

BCR. This elongates the path for surface current transmission and allows for the miniaturization of the entire structure.

In order to demonstrate the effectiveness of the proposed design, Floquet ports and periodic boundary conditions were used to simulate the various polarization and incident angle conditions. It is important to note that the BCR array's working band and performance are determined by the parameters of its structure. The final BCR model was achieved by simultaneously optimizing the load resistance, the geometric topology, and the dielectric material. One can calculate the harvesting efficiency for the metasurface harvester using the following method.

$$\eta = \frac{P_{load}}{P_{received}} \times 100\% \tag{10.1}$$

in which P_{load} represents the power obtained by the receiving port, while $P_{received}$ denotes the overall power harvested at the aperture of the entire metasurface. The calculation of $P_{received}$ can be determined by integrating the Poynting vector over the aperture area of the metasurface. The BCR metasurface's energy harvesting efficiency is displayed in Fig. 10.5 when a plane wave is incident at a perpendicular angle. The BCR metasurface exhibits three resonance points for energy harvesting achieving efficiencies of 70% at 900 MHz, 80% at 2.6 GHz, and 82% at 5.7 GHz, correspondingly.

The simulated S_{11} of the single BCR unit and the BCR array are compared in Fig. 10.6. The proposed BCR unit is a resonant element capable of operating in three different frequency ranges. Nevertheless, the single BCR unit resonates at higher frequencies. The BCR array design achieves the power collection by means of the interaction among the units enabling the attainment of performance that is insensitive to polarization, wide-angle, miniaturized, and multi-band.

At three resonant frequencies, the surface current distributions of the BCR are displayed in Fig. 10.7. It should be noted that various resonant modes on the BCR stimulate varying levels of current intensity and path distributions. Three resonant modes produce surface currents that flow into a single harvesting port at their maximum capacity. The BCR experiences the excitation of the 1st anti-circulating

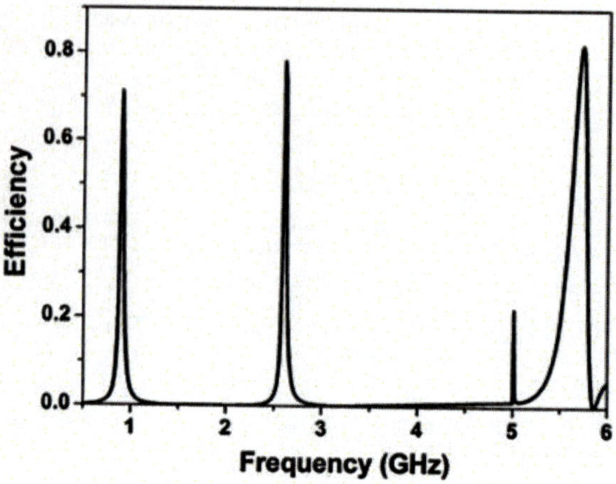

Fig. 10.5 The efficiency of the BCR metasurface under normal incidence [55]

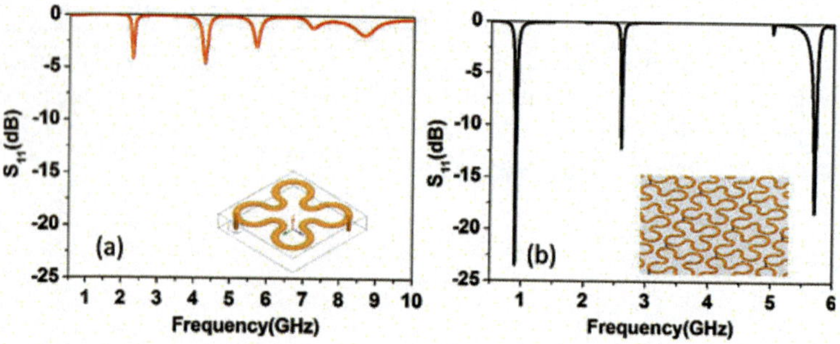

Fig. 10.6 The S_{11} of **a** the BCR single unit and **b** the BCR array [55]

current at the resonance mode of 900 MHz. This current follows its longest path, and ultimately flows through the metalized via to the receiving port. The 2nd surface current with a half-arc path generates the resonance mode at 2.6 GHz. The 3rd surface current creates a resonance mode at a high frequency of 5.7 GHz, characterized by quarter arc paths. By adjusting the wire's length, width, and period, it is possible to flexibly control the three resonant frequencies. The efficiency for various polarization angles under normal incidence are illustrated in Fig. 10.8. The findings indicate that the BCR exhibits favorable polarization angle stability.

For both transverse electric (TE) and transverse magnetic (TM) polarizations, it is important to take into account the stability of the polarization and the incident angle for the BCR. The efficiencies for EM waves with TE and TM polarizations in the three bands are depicted in Fig. 10.9a–f. The BCR maintains a high efficiency for both TE and TM polarizations. As the angle of incidence increases, there is a slight change in frequency within the high frequency range caused by the alteration of the current path within the structure. An adaptive rectification circuit can be utilized to make the corresponding adjustments (Fig. 10.10).

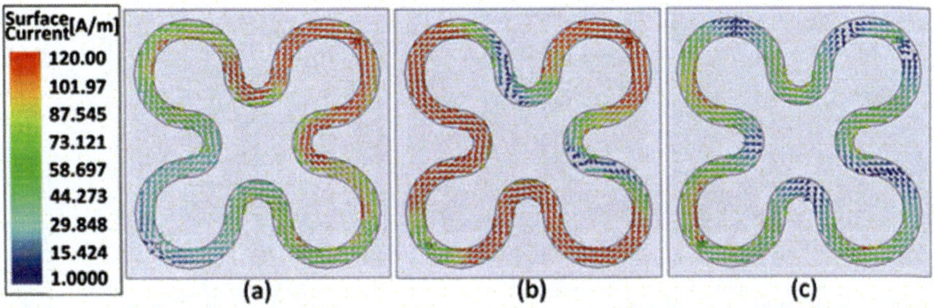

Fig. 10.7 Simulated surface current on the BCR at **a** 1st resonance mode (0.9 GHz), **b** 2nd resonance mode (2.6 GHz), and **c** 3rd resonance mode (5.7 GHz) [55]

Fig. 10.8 Efficiency for different polarization angles under normal incidence [55]

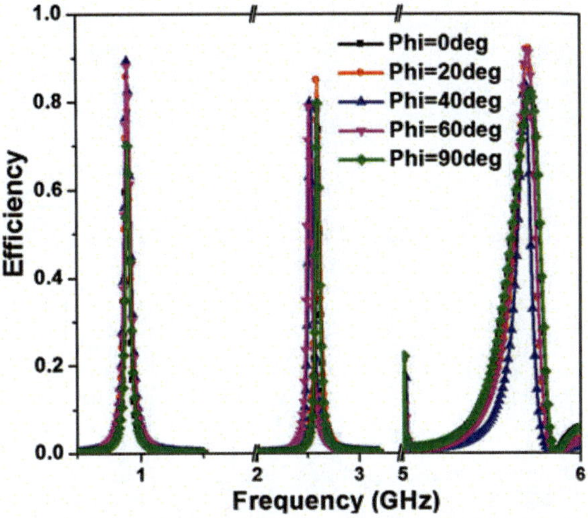

In order to confirm the effectiveness of the BCR design, we created a 7 × 7 BCR array with identical parameters to those used in the simulation, as depicted in Fig. 10.11. Since every unit cell in the 7 × 7 grid possesses a collection point to gather energy simultaneously, the responses of the central cell represent those of other unit cells in a larger array without edge effects. Hence, the measurement of the total power extraction can be estimated. Here, we measured the efficiency of the central unit by utilizing a spectrum analyzer. In addition, we measured the efficiency separately at each individual band due to the wide range of the entire frequency spectrum.

The testing efficiency of the fabricated BCR is displayed in Fig. 10.12. It gives the efficiencies at various polarizations and incident angles. The results of the experiment demonstrate that the BCR metasurface, which was specifically designed, effectively captures ambient electromagnetic waves with different polarizations and incident angles across all three frequency bands concurrently. A frequency shift occurs in the

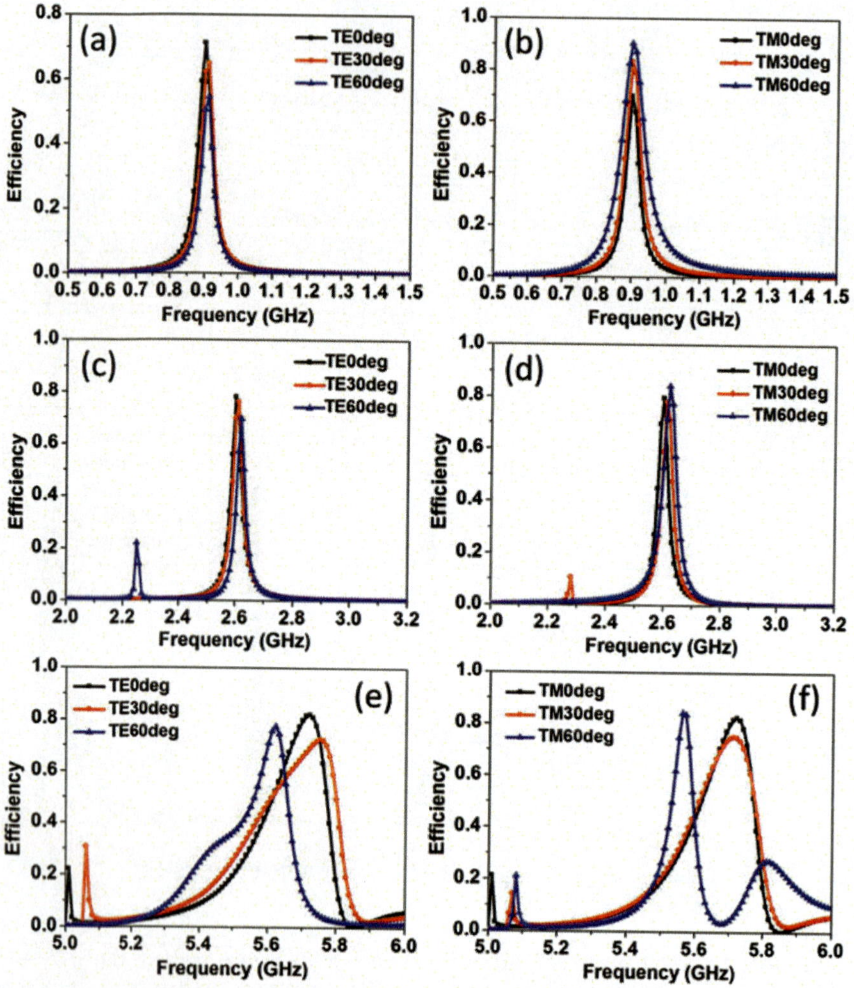

Fig. 10.9 Efficiencies for the incident waves with different polarizations and incident angles, **a** TE-polarized at 0.9 GHz; **b** TM-polarized at 0.9 GHz; **c** TE-polarized at 2.6 GHz; **d** TM-polarized at 2.6 GHz; **e** TE-polarized at 5.7 GHz; and **f** TM-polarized at 5.7 GHz [55]

high-frequency range when the angle of incidence exceeds 60°. Furthermore, the measurement results align well with the simulation ones depicted in Fig. 10.9. The practicality and validity of the proposed BCR array are confirmed.

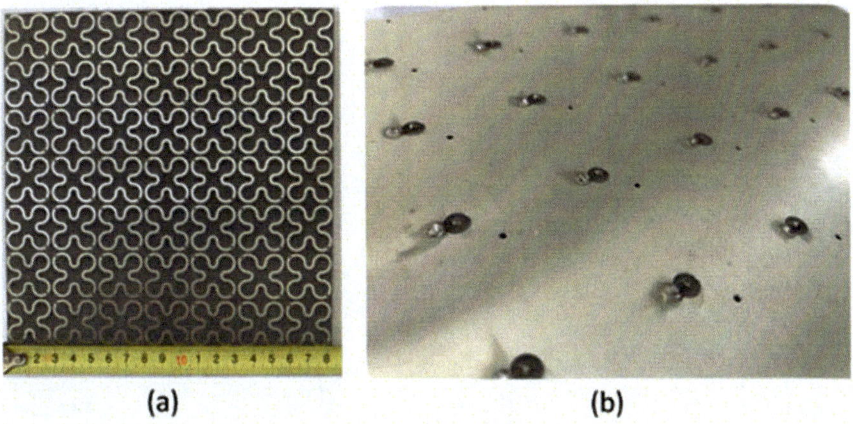

Fig. 10.10 Manufactured BCR array, **a** top view and **b** bottom view [55]

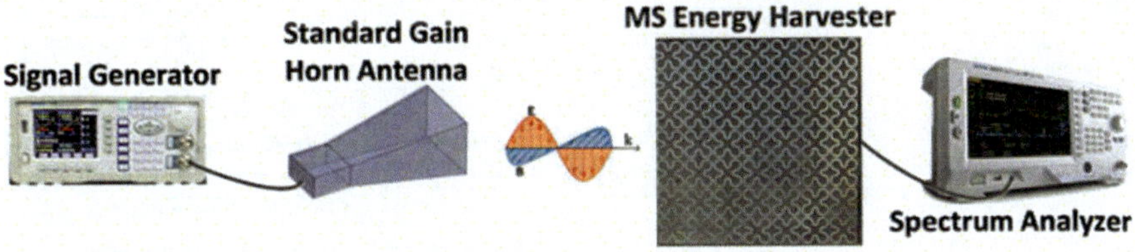

Fig. 10.11 Testing setup for the fabricated BCR array [55]

10.4 Rectifying Metasurface Design for Wireless Energy Harvesting

The preceding section introduces the metasurface design for the AEH. It is worth mentioning that the above design can only achieve RF power acquisition. To achieve full wireless energy harvesting, specifically the collection of DC energy, a subsequent design of rectifying network is required. The 'antenna (array) + rectification (array)' terminals which can realize RF harvesting and DC conversion, is called rectenna (Array). In Fig. 10.13, two configurations are commonly introduced in a typical rectenna array. The first one is that the RF power received by rectenna array is first combined and then rectified. The second one is initially rectified and ultimately combined. In other words, the rectifier is connected to the antenna arrays prior to being DC combined.

However, all of the above configurations necessitate the use of supplementary RF or DC power combining networks and impedance matching networks in the current rectenna array designs. Moreover, when the size of the antenna array expands, it becomes considerably difficult to design a matching network that can adjust to various impedance ranges. Additionally, the power combining network becomes

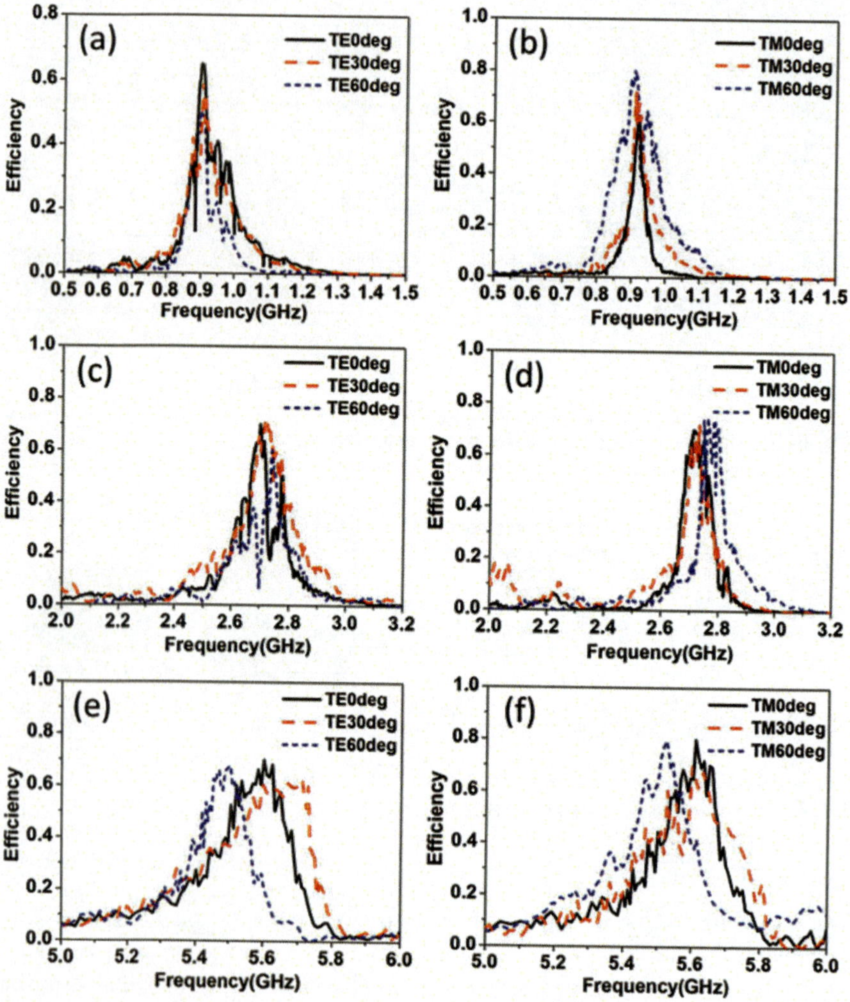

Fig. 10.12 Measured efficiencies of the fabricated BCR array with various polarizations and incident angles, **a** TE-polarized at 0.9 GHz; **b** TM-polarized at 0.9 GHz; **c** TE-polarized at 2.6 GHz; **d** TM-polarized at 2.6 GHz; **e** TE-polarized at 5.7 GHz; and **f** TM-polarized at 5.7 GHz [55]

both extensive and intricate, resulting in potential drawbacks such as increased loss, elevated cost, and mismatch issues.

In recent years, metasurface integrated rectifier structures have received a great deal of attention [90–94]. In this section, a unique and compact dual-band design of rectifying metasurface (RMS) array is presented, which is wide-angle and independent of polarization [90]. The structure of the RMS system is shown in Fig. 10.14. The configuration consists of only three parts: a MS with built-in diodes, a DC filter and a load. Due to the strong multiple resonances at the working operating bands,

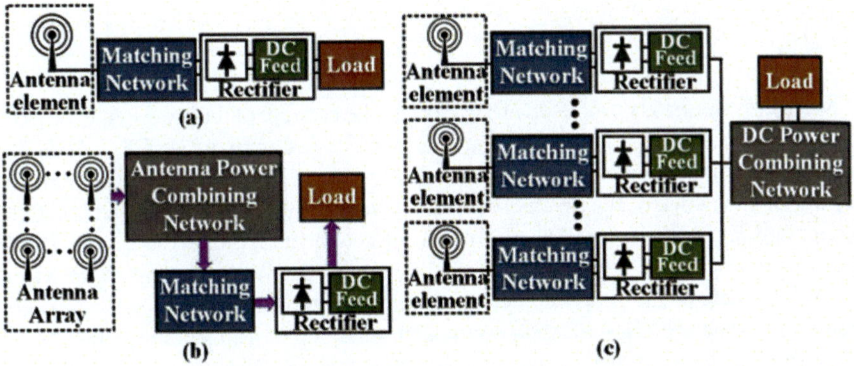

Fig. 10.13 The setup of (**a**) a typical single rectenna system, (**b**) a conventional rectenna array system that combines the first and last rectification, and (**c**) a rectenna array system that combines the first rectification and last combination in a traditional manner [90]

the MS offers a greater flexibility in achieving adjustable high-impedance properties compared to the antenna. By appropriately combining the MS and built-in diodes, the RMS can be fully integrated without requiring RF or power combining networks and impedance matching networks, thus allowing for a straightforward and compact design. Moreover, the proposed RMS has the capability to achieve a broad range of incident angles and is not affected by polarization, all while maintaining a high level of harvesting efficiency. In comparison to traditional rectennas, the proposed RMS demonstrates exceptional efficiency due to its compact and simple design, as well as its cost-effectiveness in production. Additionally, it has the ability to adjust to different input power sources and even diodes, which makes it highly appropriate for fulfilling the self-adjusting power supply needs of compact IoT sensors.

Fig. 10.14 The structure of the proposed rectifying metasurface setup [90]

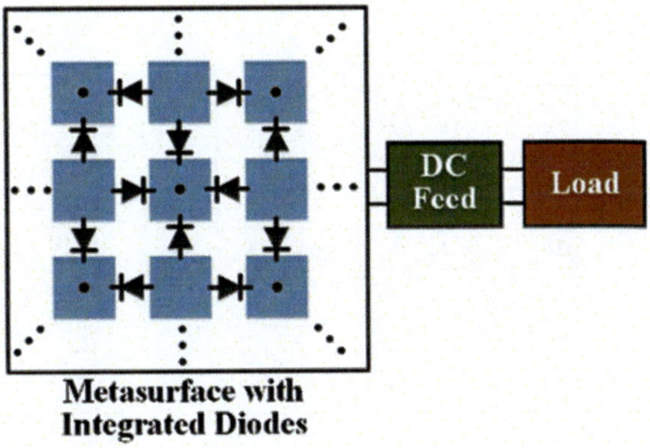

10.4.1 RMS Unit Cell Design

The RMS structure is introduced in Fig. 10.15. The concept of the proposed geometric form stems from the photonic-bandgap metasurface design, so called PBG. The PBG unit consists of a metal patch in the shape of a square and metal wires that connect the center of each edge of the patch. Therefore, since the PBG array structure functions as an equipotential surface, every unit is interconnected. The PBG configuration possesses a wide range of stopbands. On the other hand, the objective of the MS is to collect energy. The proposed design, inspired by the PBG, consists of a periodic array on a square plane with each component made up of a square metal pad.

Instead of being directly connected in PBG structure, the receiving ports of the proposed RMS are connected in series via metal lines. The receiving ports have the ability to perfectly match the input impedance of the diode build-in the MS units. It is important to highlight that the receiving ports are positioned between each unit instead of being inside each unit, because different positions of the port have a significant distinction in their equivalent circuit. To ensure that the diodes are integrated and a continuous DC path is created within the overall MS, the receiving ports are aligned with the MS. The space in RMS unit amplifies the capacitance impact, while the connecting lines introduce an extra inductance influence. Table 10.1 provides a design example.

The MS's behavior is simulated using the commercial software HFSS, and the characteristics of all ports can be calculated by the simulation model of Floquet port. To connect the neighboring RMS units, the lumped impedance ports are utilized. The function of these receiving ports is to examine the performance of the power received by the proposed design, including collection effectiveness and impedance matching.

Figure 10.16 illustrates the input impedance for the receiving port. The working frequencies of the RMS can be acquired to be approximately 2.4 and 5.8 GHz within

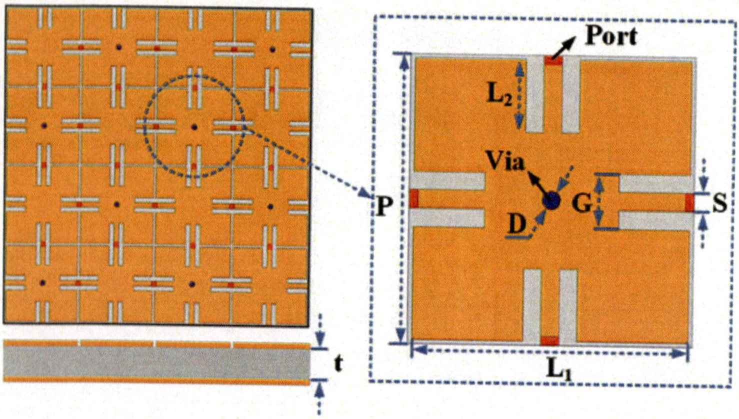

Fig. 10.15 The geometry of the proposed rectifying metasurface [90]

Table 10.1 Summary of metasurface geometry

Parameter	Value (mm)
D	1
G	3
L_1	15.5
L_2	4
P	16
S	1
t	1.27

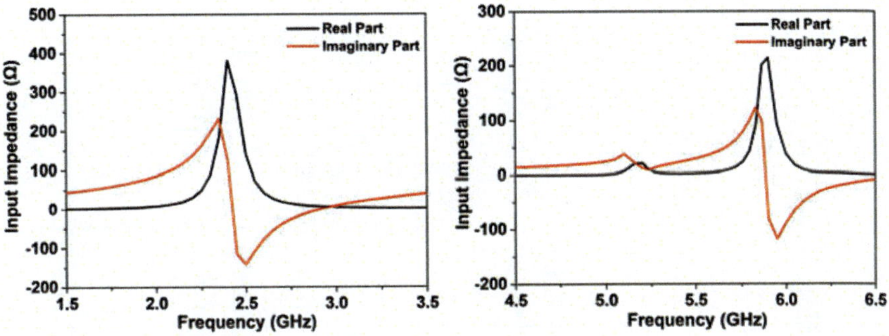

Fig. 10.16 The calculated input impedance for the receiving port in 2.4 and 5.8 GHz working bands [90]

the desired band. The input impedance is 400 + 0 * jΩ (2.4 GHz) and 200 + 0 * jΩ (5.8 GHz), exhibiting a high impedance state. The input impedance of the MS receiving port changes from 0 to 400 Ω for the real part near the two resonant frequencies, while the imaginary part ranges from −180 to 250 Ω.

Figure 10.17 illustrates the analysis of the surface current of the RMS at the two working bands. At 2.4 GHz, The surface current of the RMS flows in the same direction, creating a 1st resonant mode by following the maximum electric current path of the RMS. Currently, the RMS unit cell is relatively tiny in size, having a period of $\lambda_0/8$ (where λ_0 represents the 2.4 GHz wavelength). In the meantime, the opposing surface current of the RMS neighboring components occurs at 5.8 GHz, while the MS unit has a period of $\lambda_1/3$ (where λ_1 represents the 5.8 GHz wavelength). In this case, the RMS is on longer electrically small. The RMS functions in a differential manner. Consequently, the RMS exhibits two frequencies that resonate with significant impedance, functioning in distinct resonant modes.

The total efficiency of the RMS is calculated as follows:

$$\eta_{Total} = \eta_{MS}\eta_{RF-DC} \tag{10.2}$$

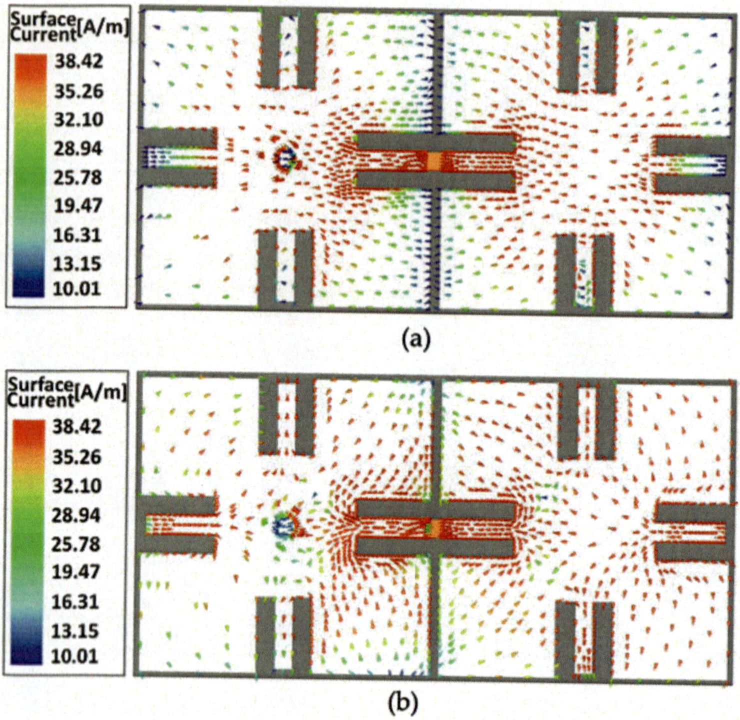

Fig. 10.17 Simulated surface current on the RMS at **a** first working band (2.4 GHz) and **b** second working band (5.8 GHz) [90]

$$\eta_{MS}=\frac{p_{MS-load}}{p_{received}}\times 100\% \tag{10.3}$$

$$\eta_{RF-DC}=\frac{p_{DC-Load}}{p_{MS-load}}\times 100\% \tag{10.4}$$

η_{MS} represents the RF-AC efficiency of the RMS when rectification is not considered. The power harvested by the RMS receiving ports is referred to as $P_{MS\text{-}load}$, while the received power on the overall RMS is $P_{received}$. The RF-DC efficiency of the rectifier is denoted as $\eta_{RF\text{-}DC}$, while the dc power of the output load is represented by $P_{DC\text{-}load}$.

Subsequently, the model can be employed to compute the efficiency performance according to (10.3) for various polarizations and incident angles. Two types of oblique incidences can generally be considered, similar to the examination of the microwave absorber, i.e., TE-polarized incidence and TM polarized incidence.

The efficiency of energy collection by RMS is displayed in Fig. 10.18 for various incident angles, considering both TE and TM polarizations. When the TE polarization is considered, the efficiency at the 2.4 GHz gradually decreases from 92% to 80% as the incidence angle varies from 0° to 60°. Similarly, at 5.8 GHz, the collection efficiency decreases from 88 to 83%. However, the resonant frequency remains almost unchanged for both frequency bands. Consequently, the effective electrical resonance gradually weakens, leading to a tiny decrease in both bandwidth and efficiency. As the incident angle increases, the resonance frequency of the TM polarization slightly shifts due to the enhancement of the equivalent magnetic resonance, while the bandwidth remains nearly constant.

Finally, we determine the energy collection effectiveness with and without the bulid-in via in RMS, as shown in Fig. 10.19. The presence of a small frequency deviation can be observed from the responses of via and non-via RMS units, because the inclusion of the build-in via enhances the impact of equivalent inductance in the RMS. The overall efficiency of the RMS with the via remains above 86%. This performance will be utilized in the subsequent section for the arrangement of the RMS and circuit integration. Hence, the proposed MS unit cell exhibits attributes like elevated input impedance, superior effectiveness, and consistent polarization and incidence stabilities in 2.4 and 5.8 GHz of operational modes.

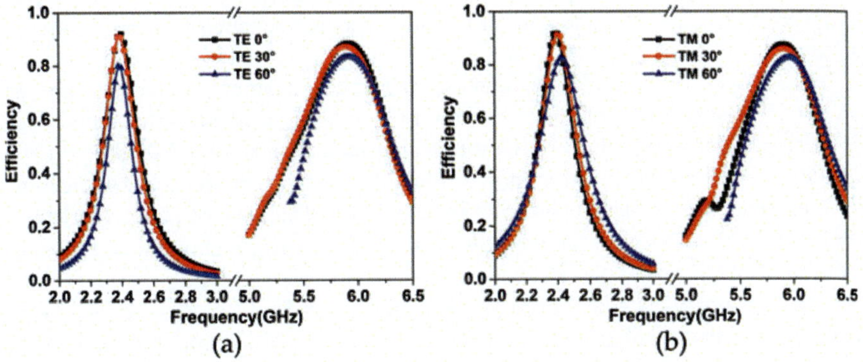

Fig. 10.18 Calculated efficiencies of the RMS with **a** TE and **b** TM polarizations for different incident angles [90]

Fig. 10.19 Calculated efficiencies with and without build-in via [90]

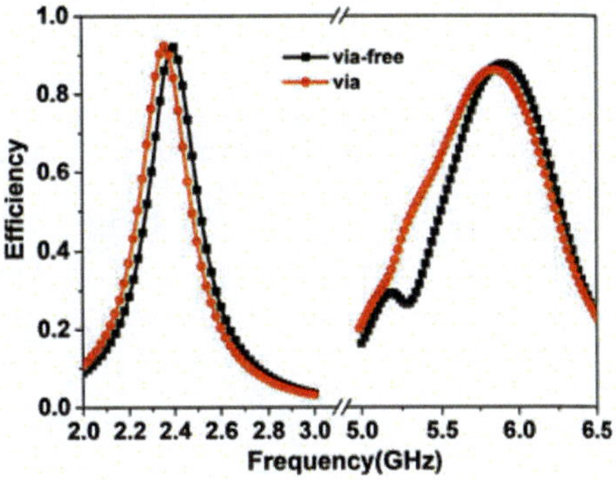

10.4.2 Rectifier Design

One can notice that the RF power is converted to the DC power at the RMS output terminal which is linked in parallel with the build-in diode. Typically, the build-in diodes can be chosen based on the frequency and range of the input power. The function of the RF choke when connected in series is to create a pathway for direct current and prevent the transmission of radio frequency power. On the other hand, the capacitor when connected in parallel serves to stabilize the waveform and store DC energy. The proposed design incorporates a inductor and a capacitor with the value of 47 and 100 nF. The inductor in series and capacitor in shunt arrangement are equivalent to a DC filter. A resistance of 0.7 KΩ is employed for DC power load.

The rectifier circuit branches in the rectenna array can be connected in series, parallel, and cascaded ways. Recent reports generally consider overall parallel and partial series connection methods as optimal choices for enhancing total efficiency and reducing ohmic losses. Figure 10.20b shows the RMS design with the parallel connection of the rectifier and unit cells, aiming to combine the structure and enhance efficiency.

The ADS software utilizes the harmonic balance simulator to model the proposed rectifier. The simulated efficiencies of multi-channel and single-channel rectifiers at 2.4 GHz are depicted in Fig. 10.21, illustrating their performance in relation to the incident power. As an illustration, we select the SMS-7630 diode, while keeping all other parameters unchanged. The increase in the number of parallel channels leads to a slight improvement in rectification efficiency, highlighting the benefits of a multi-channel parallel connection.

The input impedances at 0 dBm (diode model: SMS-7630), 5 dBm (diode model: HSMS-2850) and 10 dBm (diode model: HSMS-2860) of the proposed rectifier for the various diodes are depicted in Fig. 10.22. By utilizing various diodes, the input impedance of proposed rectifier can be conjugately matched to the impedance of the RMS receiving port while keeping the topology and dimensions unchanged.

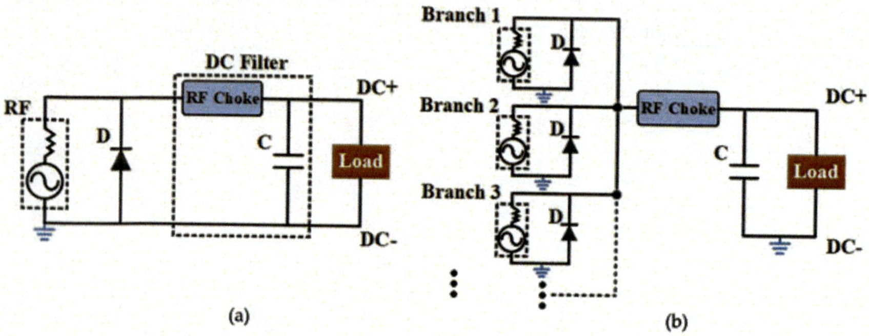

Fig. 10.20 The diagram illustrates **a** the class-F rectifier and **b** the rectifier with multiple parallel channels [90]

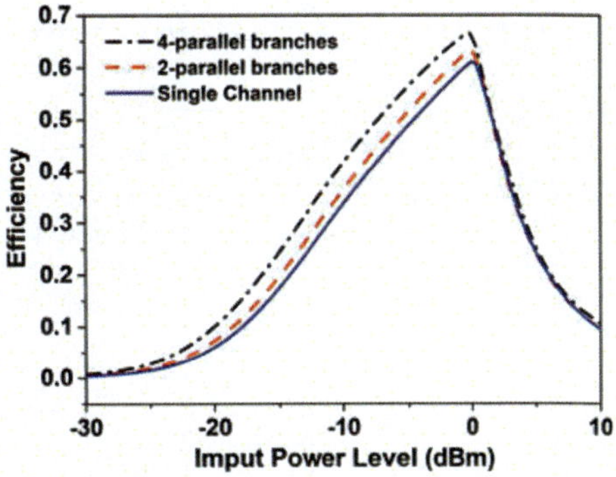

Fig. 10.21 The efficiency of a parallel multi-channel and single-channel rectifier [90]

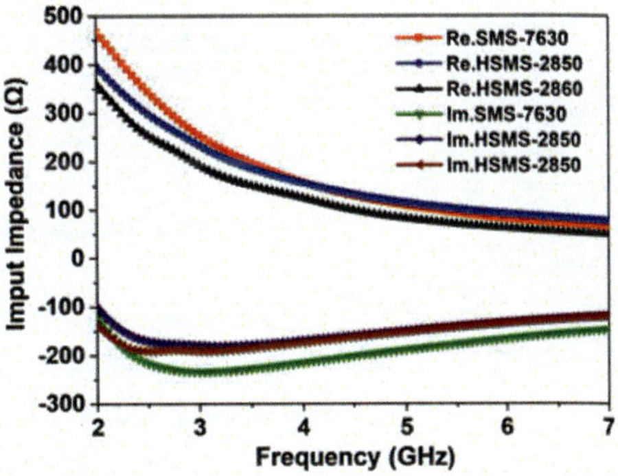

Fig. 10.22 The input impedance at 0 dBm (diode model: SMS-7630), 5 dBm (diode model: HSMS-2850) and 10 dBm (diode model: HSMS-2860) of the Class-F rectifier for the various diode models [90]

10.4.3 Metasurface Array Design

In this case, we constructed the RMS using an innovative approach that integrates diodes into the RMS, as depicted in Fig. 10.23. The RMS is composed of via and non-via units in the alternating arrangement, and their characteristics have been discussed earlier. Each via unit has four connecting lines that are attached to the positive (+) terminal of the build-in diode, whereas the 4 connecting lines of each non-via unit are connected to the negative (−) terminal of the build-in diode. All via RMS units are linked to the ground (DC–). The units that do not have via connections are connected by thin metal wires that act as high impedance line to create DC paths while preventing alternating current, resulting in the establishment of the positive DC voltage (DC+). Based on the parallel rectifier topology discussed earlier, the diodes' positive and negative terminals are linked to their respective equipotential surfaces, resulting in a novel RMS arrangement mode.

The equivalent circuit of the RMS is shown in Fig. 10.24. A RF source is created at the receiving ports located between every two neighboring cells. Each port in the array is equipped with an equivalent RF voltage source and a diode, forming the final circuit of the rectifier in parallel form. Next, the opposing ends of each corresponding rectification branch on the DC– are interconnected, while the corresponding positive

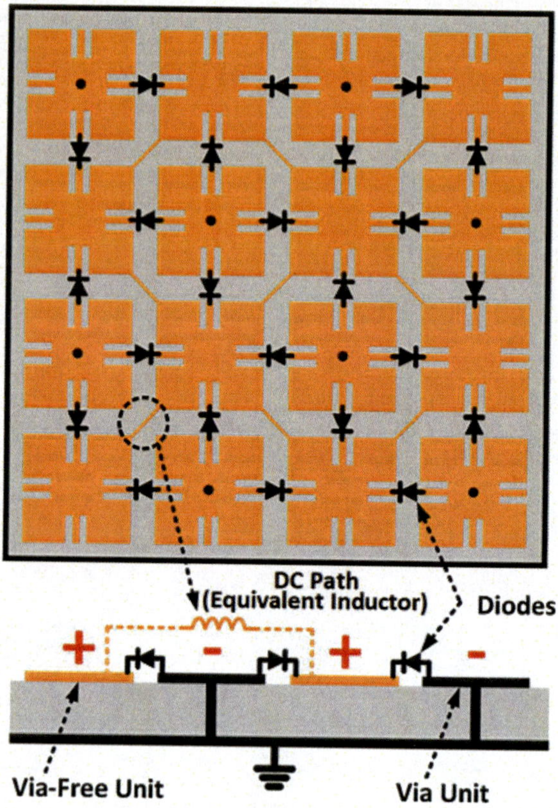

Fig. 10.23 Diagram illustrating the suggested RMS with an innovative approach to integrate diodes into the texture [90]

ends of all rectification lines are also joined together on the DC+. Consequently, the DC+ connects all the corresponding rectification lines in parallel, which are subsequently directed into the DC-filter uniformly for the smooth output of DC waveform and the isolation of RF power from the load.

Co-simulation is performed using HFSS and ADS to investigate the interaction impacts of the EM field and the RF circuit. The ADS utilizes the power source at the frequency domain as the RF source between the RMS units. Next, the touchstone S1P file imports the input impedance of the RF port. This file is then used to calculate the variation of the input impedance of the RMS in the HFSS.

The efficiency and S_{11} of the RMS are determined through co-simulation at various input power levels, as depicted in Fig. 10.25a and b. It should be noted that the efficiency of converting RF to DC is denoted as η_{Total}. The MS efficiency without rectification is demonstrated in Fig. 10.18. The SMS-7630 diode reaches its saturation voltage at approximately 0 dBm, making it the optimal input power level. The two resonant modes yield a total RF-DC efficiency of 66% (2.4 GHz) and 55% (5.8 GHz) at 0 dBm. Additionally, the performance is extended and compared by adding the result obtained at ±3 dBm. The RMS exhibits consistent high rectifying efficiency

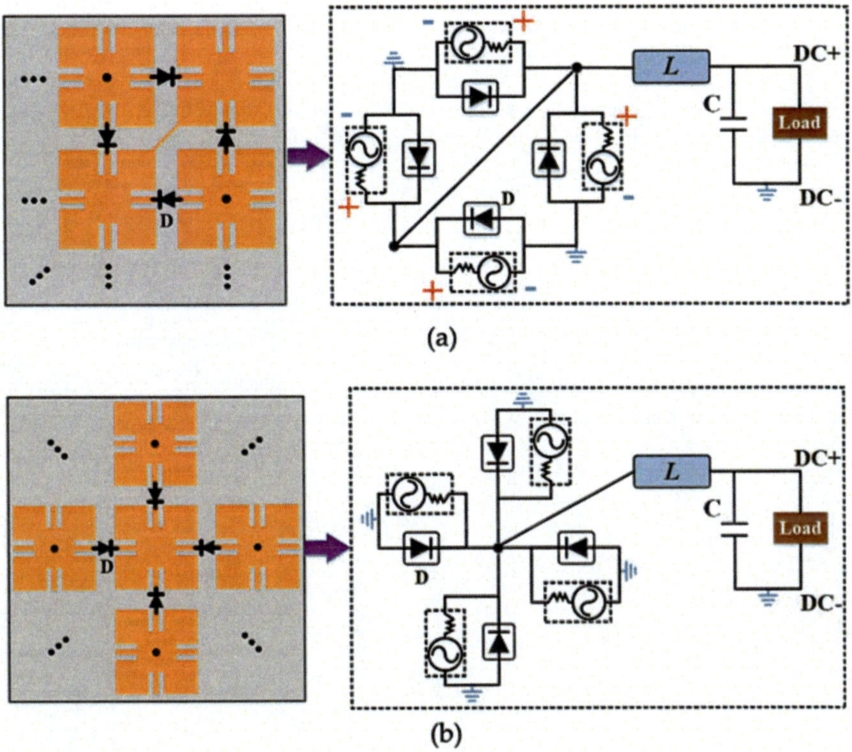

Fig. 10.24. **a** Topology of the equivalent circuit of the RMS **b** another topology of the equivalent circuit of the RMS [90]

(> 50%) for both resonant modes, even when the input power level is increased or decreased two times.

The MS integrated with various diodes is depicted in Fig. 10.26, illustrating the conversion efficiency. The diode functions at its respective optimal power level, while keeping the rest parameters constant. By increasing the input power, the efficiencies in two working bands are enhanced, allowing for optimization of the load to achieve the best efficiency at their respective power level. Therefore, the RMS is a power-adjustable design for varying RF power levels while keeping the parameters of the RMS unchanged and eliminating the need for an extra matching network, which is seldom utilized in the most recent rectenna arrays report.

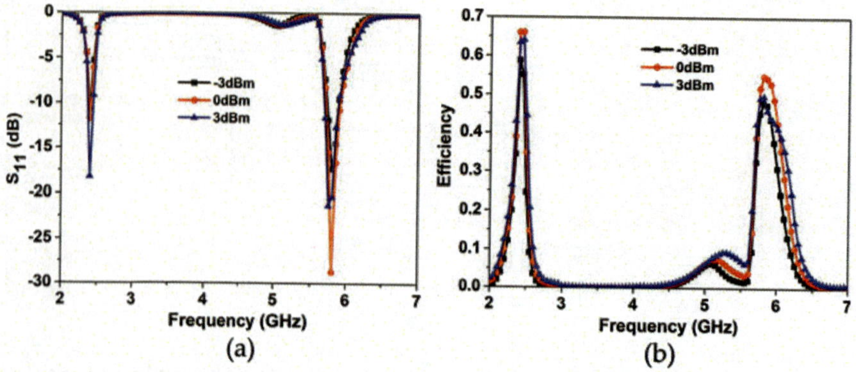

Fig. 10.25 **a** Simulated S_{11} and **b** RF-DC efficiency of the RMS. The diode model is SMS-7630 [90]

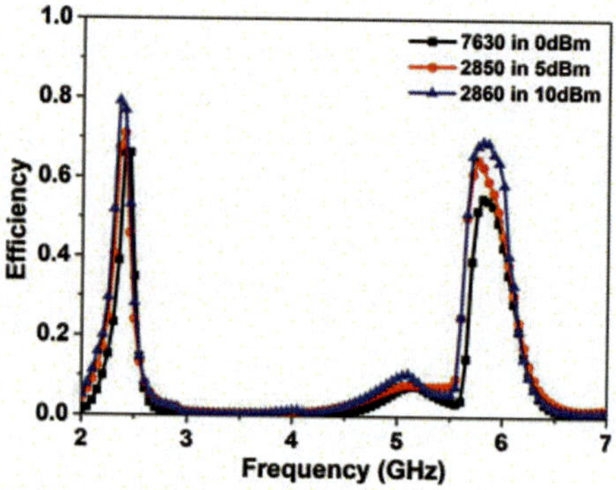

Fig. 10.26 The RF-DC efficiency of the RMS with various diodes at their respective optimal power levels [90]

10.4.4 Measurements and Validations

Figure 10.27 illustrates the arrangement design of the 4 × 4 fabricated RMS. The DC feed circuit consisting of the capacitor and inductor is located in one side of the top substrate. It is connected to the DC+ according to the equivalent RMS circuit.

Furthermore, Fig. 10.28 illustrates the presentation of the measurement setup. The entire test was conducted within a microwave anechoic. The RF power is excited by standard horn antenna. Then, a DC voltmeter is employed for measuring the voltage in RMS DC output. The calculation of measurement efficiency is determined in the following:

$$\eta_{measurement} = \frac{P_{load}}{P_r} \times 100\% \tag{10.5}$$

where P_{load} represents the output DC power:

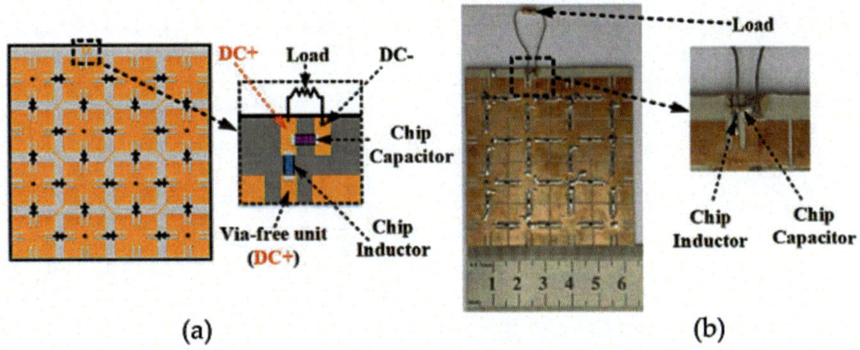

Fig. 10.27 **a** Layout design, **b** fabricated design of proposed RMS [90]

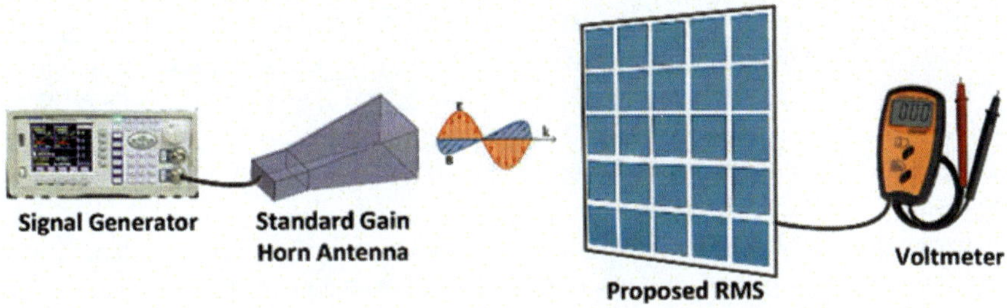

Fig. 10.28 The procedure used to measure the efficiency of the RMS [90]

$$P_{load} = \frac{V_{out}^2}{R_{load}} \tag{10.6}$$

where V_{out} is the tested voltage in RMS DC output and R_{load} is the RMS output resistance. P_r is the RF power received by the RMS surface:

$$P_r = \frac{G P_{input}}{4\pi R^2} \cdot A_e \tag{10.7}$$

where G is the antenna gain, and P_{input} is the input power. R is the distance between the antenna and the RMS. A_e is the effective receiving aperture of the RMS.

The efficiency of RMS under normal incidence is demonstrated in Fig. 10.29 for various incident powers. The maximum test efficiencies in 0 dBm of RMS are 58 and 51%, respectively, working at 2.4 and 5.8 GHz. At 2.4 GHz, the maximum efficiencies for 3 and -3 dBm are 53 and 48% respectively, while at 5.8 GHz, they are 45 and 40%. Typically, the RMS experiences a decrease in efficiency of less than 20% in dual band, when the input power varies within a certain range.

The RMS efficiencies for TE, TM-polarizations and different incident angles are given in Fig. 10.30. It can be concluded that the RMS experiences a minor frequency

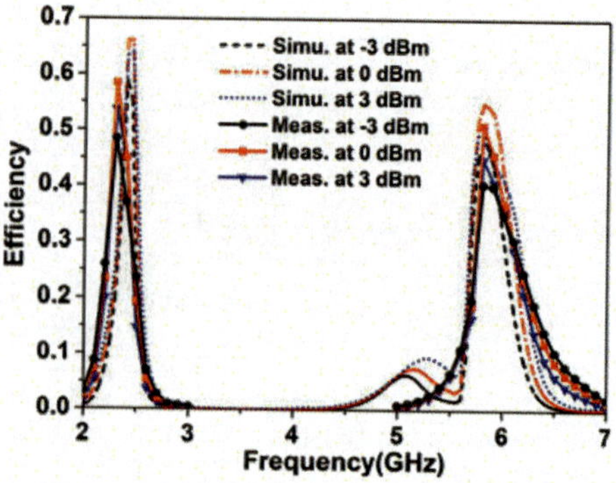

Fig. 10.29 Comparison of the energy collection effectiveness of the produced RMS under varying levels of incoming power. The SMS-7630 is the type of diode [90]

deviation when TE and TM polarizations are at oblique incidences, but this does not significantly impact the harvesting efficiency as the incidence angle increases in both resonant modes. Under various polarization angles φ, the proposed RMS maintained a harvesting efficiency of more than 56% at 2.4 GHz and above 50% at 5.8 GHz. Note that the efficiency mentioned refers to the overall efficiency of RF-DC. The results of the MS without rectification shown in Fig. 10.6 are comparatively different from the results of the MS with rectification.

To verify the RMS for the integration of various diodes, three versions of the RMS with the same dimensions and different diodes are investigated. Figure 10.31 shows the efficiencies with the three different diode models with their optimal input power. By introducing the matching diodes, it is possible to working in two distinct frequency ranges without altering any other parameters. Next, the simulation and measurement of conversion efficiency are conducted while varying the load resistance, as depicted in Fig. 10.32. In this case, the RMS is equipped with an SMS-7630 build-in diode model. With the output impedance of 0.7 kΩ, the RMS at two frequency bands can simultaneously achieve a high conversion efficiency (>50%) while also possessing a specific impedance bandwidth. The consistency of the RMS is confirmed by the simulation and experimental findings.

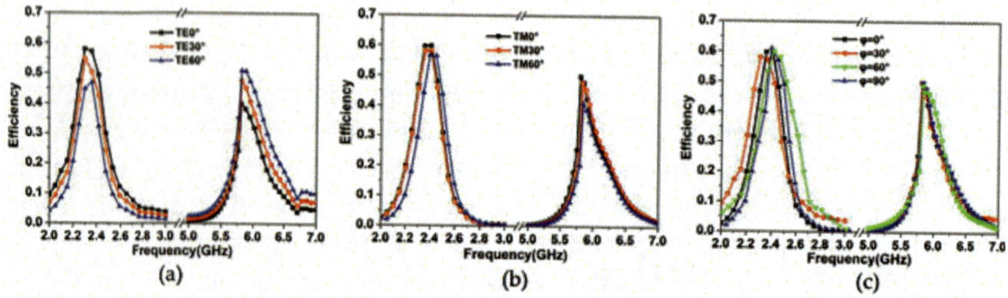

Fig. 10.30 The efficiencies of the RMS for **a** TE, **b** TM polarizations along different incident angles, and **c** various polarization angles under normal incidence [90]

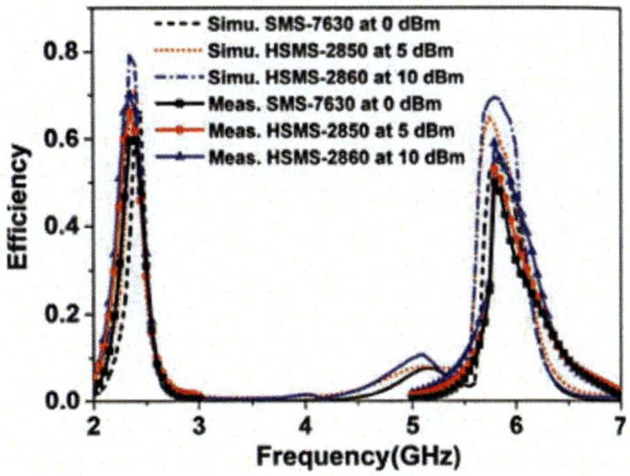

Fig. 10.31 The efficiency of the RMS that was fabricated, tested and compared using various diodes at their optimal input power levels [90]

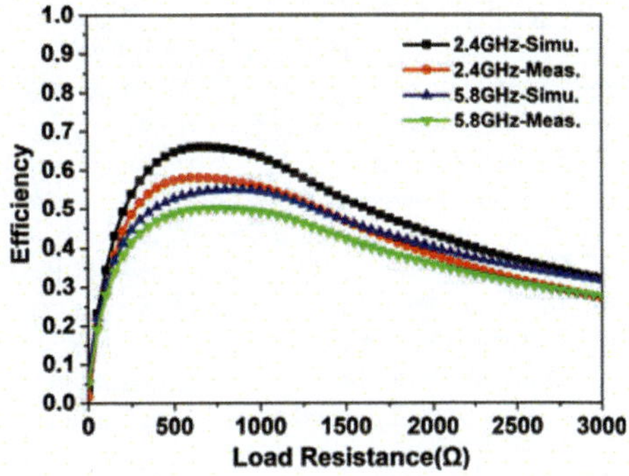

Fig. 10.32 The simulated and measured efficiencies at 0 dBm versus output resistance at two working bands. The diode model is SMS-7630 [90]

10.5 Conclusion

To summarize, metasurfaces have the ability to not just facilitate effective collection of ambient EM/RF energy but also expand the range of potential capabilities for different types of AEH application devices. A versatile method is employed to address various important concerns of WPT and AEH, encompassing effectiveness, range, dimensions, and matching of impedance. In the coming years, we anticipate the establishment of commercial standards for AEH applications, like powering sensor networks, similar to the existing standardization and regulation of WPT. This development will hold immense importance. We are optimistic that the utilization of metasurfaces for the collection of ambient energy will lead to a promising future for everyone.

References

1. C. Song et al., Advances in wirelessly powered backscatter communications: From antenna/RF circuitry design to printed flexible electronics. Proc. IEEE **110**(1), 171–192 (2022)

2. J. Zhou, P. Zhang, J. Han, L. Li, Y. Huang, Metamaterials and metasurfaces for wireless power transfer and energy harvesting. Proc. IEEE **110**(1), 31–55 (2022)
3. S. Tian, X. Zhang, X. Wang, J. Han, L. Li, Recent advances in metamaterials for simultaneous wireless information and power transmission. Nanophotonics **11**(9), 1697–1723 (2022)
4. N. Shinohara, Power without wires. IEEE Microw. Mag. **12**(7), S64–S73 (2011)
5. K. Wu, D. Choudhury, H. Matsumoto, Wireless power transmission, technology, and applications. Proc. IEEE **101**(6), 1271–1275 (2013)
6. B. Strassner, K. Chang, Microwave power transmission: historical milestones and system components. Proc. IEEE **101**(6), 1379–1396 (2013)
7. M. Song, P. Belov, P. Kapitanova, Wireless power transfer inspired by the modern trends in electromagnetics. Appl. Phys. Rev. **4**, 021102 (2017)
8. M. Piñuela, P.D. Mitcheson, S. Lucyszyn, Ambient RF energy harvesting in urban and semi-urban environments. IEEE Trans. Microw. Theory Tech. **61**, 2715–2726 (2013)
9. S. Hemour, K. Wu, Radio-frequency rectifier for electromagnetic energy harvesting: development path and future outlook. Proc. IEEE **102**, 1667–1691 (2014)
10. S. Hemour, Y. Zhao, C.H.P. Lorenz, D. Houssameddine, Y. Gui, C. Hu, K. Wu, Towards low-power high-efficiency RF and microwave energy harvesting. IEEE Trans. Microw. Theory Tech. **62**, 965–976 (2014)
11. S. Kim, R. Vyas, J. Bito, K. Niotaki, A. Collado, A. Georgiadis, M.M. Tentzeris, Ambient RF energy-harvesting technologies for self-sustainable standalone wireless sensor platforms. Proc. IEEE **102**, 1649–1666 (2014)
12. K. Niotaki, A. Collado, A. Georgiadis, S. Kim, M.M. Tentzeris, Solar/electromagnetic energy harvesting and wireless power transmission. Proc. IEEE **102**, 1712–1722 (2014)
13. S. Byrnes, R. Blanchard, F. Capasso, Harvesting renewable energy from earth's mid-infrared emissions. Proc. Natl. Acad. Sci. U.S.A. **111**, 3927–3932 (2014)
14. S. Buddhiraju, P. Santhanam, S. Fan, Thermodynamic limits of energy harvesting from outgoing thermal radiation. Proc. Natl. Acad. Sci. U.S.A. **115**, E3609–E3615 (2018)
15. W. Li, S. Fan, Nanophotonic control of thermal radiation for energy applications. Opt. Express **26**, 15995–16021 (2018)
16. A. Polman, H. Atwater, Photonic design principles for ultrahigh-efficiency photovoltaics. Nat. Mater. **11**(3), 174 (2012)
17. A. Sharma, V. Singh, T. Bougher, B. Cola, A carbon nanotube optical rectenna. Nat. Nanotechnol. **10**, 1027–1032 (2015)
18. Y.J. Ren, K. Chang, 5.8-GHz circularly polarized dual-diode rectenna and rectenna array for microwave power transmission. IEEE Trans. Microw. Theory Tech. **54**, 1495–1502 (2006)
19. A. Massa, G. Oliveri, F. Viani, P. Rocca, Array designs for long-distance wireless power transmission: state-of-the-art and innovative solutions. Proc. IEEE **101**, 1464–1481 (2013)
20. V. Kuhn, C. Lahuec, F. Seguin, C. Person, A multi-band stacked RF energy harvester with RF-to-DC efficiency up to 84%. IEEE Trans. Microw. Theory Tech. **63**, 1768–1778 (2015)
21. C. Song, Y. Huang, P. Carter, J. Zhou, S. Yuan, Q. Xu, K. Muayad, A Novel six-band dual CP rectenna using improved impedance matching technique for ambient RF energy harvesting. IEEE Trans. Antennas Propag. **64**, 3160–3171 (2016)
22. C. Song, Y. Huang, J. Zhou, J. Zhang, S. Yuan, P. Carter, A high-efficiency broadband rectenna for ambient wireless energy harvesting. IEEE Trans. Antennas Propag. **63**, 3486–3495 (2015)
23. C. Song, Y. Huang, J. Zhou, P. Carter, S. Yuan, Q. Xu, Z. Fei, Matching network elimination in broadband rectennas for high efficiency wireless power transfer and energy harvesting. IEEE Trans. Industr. Electron. **64**, 3950–3961 (2017)
24. C. Song, Y. Huang, P. Carter, J. Zhou, S. Joseph, G. Li, Novel compact and broadband frequency-selectable rectennas for a wide input-power and load impedance range. IEEE Trans. Antennas Propag. **66**, 3306–3316 (2018)
25. C. Song, A. Lopez-Yela, Y. Huang, Novel quartz clock with integrated wireless energy harvesting and sensing functions. IEEE Trans. Industr. Electron. **66**, 4042–4053 (2019)
26. Y. Huang, N. Shinohara, T. Mitani, A constant efficiency of rectifying circuit in an extremely wide load range. IEEE Trans. Microw. Theory Tech. **62**, 986–993 (2014)

27. Q.W. Lin, X.Y. Zhang, Differential rectifier using resistance compression network for improving efficiency over extended input power range. IEEE Trans. Microw. Theory Tech. **64**, 2943–2954 (2016)
28. Z. Liu, Z. Zhong, Y.X. Guo, Enhanced dual-band ambient RF energy harvesting with ultra-wide power range. IEEE Microw. Wirel. Compon. Lett. **25**, 630–632 (2015)
29. R.A. Shelby, D.R. Smith, S.C. Nemat-Nasser, S. Schultz, Microwave transmission through a two dimensional, isotropic, left-handed metamaterial. Appl. Phys. Lett. **78**, 489 (2001)
30. C.L. Holloway, E.F. Kuester, J.A. Gordon, J. O'Hara, J. Booth, D.R. Smith, An overview of the theory and applications of metasurfaces: the two-dimensional equivalents of metamaterials. IEEE Antennas Propag. Mag. **54**, 10–35 (2012)
31. B. Wang, K.H. Teo, T. Nishino, W. Yerazunis, J. Barnwell, J. Zhang, Experiments on wireless power transfer with metamaterials. Appl. Phys. Lett. **98**, 254101 (2011)
32. A.L.A.K. Ranaweera, T.P. Duong, J.W. Lee, Experimental investigation of compact metamaterial for high efficiency mid-range wireless power transfer applications. J. Appl. Phys. **116**, 043914 (2014)
33. K. Agarwal, R. Jegadeesan, Y.X. Guo, N.V. Thakor, Wireless power transfer strategies for implantable bioelectronics: methodological review. IEEE Rev. Biomed. Eng. **10**, 136–161 (2017)
34. L. Li, H. Liu, H. Zhang, W. Xue, Efficient wireless power transfer system integrating with metasurface for biological applications. IEEE Trans. Industr. Electron. **65**, 3230–3239 (2018)
35. S. Yu, H. Liu, L. Li, Design of near-field focused metasurface for high-efficient wireless power transfer with multifocus characteristics. IEEE Trans. Industr. Electron. **66**, 3993–4002 (2019)
36. P. Zhang, L. Li, X.M. Zhang, H.X. Liu, Y. Shi, Design, measurement, and analysis of near-field focusing reflective metasurface for dual-polarization and multi-focus wireless power transfer. IEEE Access **7**, 110387–110399 (2019)
37. L. Li, Y.C. Fan, S.X. Yu, C. Zhu, C.H. Liang, Design, fabrication and measurement of highly sub-wavelength double negative metamaterials at HF frequencies. J. Appl. Phys. **113**, 213712 (2013)
38. Y.C. Fan, L. Li, S.X. Yu, C. Zhu, C.H. Liang, Experimental study of efficient wireless power transfer system integrating with highly sub-wavelength metamaterials. Prog. Electromagn. Res. **141**, 769–784 (2013)
39. N. Zhu, R.W. Ziolkowski, H. Xin, A metamaterial-inspired, electrically small rectenna for high-efficiency, low power harvesting and scavenging at the global positioning system L1 frequency. Appl. Phys. Lett. **99**, 114101 (2011)
40. D. Ferreira, L. Sismeiro, A. Ferreira, R.F. Caldeirinha, T.R. Fernandes, I. Cuiñas, Hybrid FSS and rectenna design for wireless power harvesting. IEEE Trans. Antennas Propag. **64**, 2038–2042 (2016).
41. Z. Chen, B. Guo, Y. Yang, C. Cheng, Metamaterials-based enhanced energy harvesting: a review. Physica B: Condens. Matter **438**, 1–8 (2014)
42. N.I. Landy, S. Sajuyigbe, J.J. Mock, D.R. Smith, W.J. Padilla, Perfect metamaterial absorber. Phys. Rev. Lett. **100**, 207402 (2008)
43. O.M. Ramahi, T.S. Almoneef, M. Alshareef, M.S. Boybay, Metamaterial particles for electromagnetic energy harvesting. Appl. Phys. Lett. **101**, 173903 (2012)
44. M.R. Alshareef, O.M. Ramahi, Electrically small resonators for energy harvesting in the infrared regime. J. Appl. Phys. **114**, 223101 (2013)
45. M.R. Alshareef, O.M. Ramahi, Electrically small particles combining even- and odd-mode currents for microwave energy harvesting. Appl. Phys. Lett. **104**, 253906 (2014)
46. T.S. Almoneef, O.M. Ramahi, Metamaterial electromagnetic energy harvester with near unity efficiency. Appl. Phys. Lett. **106**, 153902 (2015)
47. B. Alavikia, T.S. Almoneef, O.M. Ramahi, Electromagnetic energy harvesting using complementary split-ring resonators. Appl. Phys. Lett. **104**, 163903 (2014)
48. B. Alavikia, T.S. Almoneef, O.M. Ramahi, Complementary split ring resonator arrays for electromagnetic energy harvesting. Appl. Phys. Lett. **107**, 033902 (2015)

49. B. Alavikia, T.S. Almoneef, O.M. Ramahi, Wideband resonator arrays for electromagnetic energy harvesting and wireless power transfer. Appl. Phys. Lett. **107**, 243902 (2015)
50. H. Zhong, X. Yang, X. Song, Z. Guo, F. Xu, Wideband metamaterial array with polarization independent and wide incident angle for harvesting ambient electromagnetic energy and wireless power transfer. Appl. Phys. Lett. **111**, 213902 (2017)
51. H. Zhong, X. Yang, C. Tan, K. Yu, Triple-band polarization-insensitive and wide-angle metamaterial array for electromagnetic energy harvesting. Appl. Phys. Lett. **109**, 253904 (2016)
52. S. Shang, S. Yang, M. Shan, J. Liu, H. Cao, High performance metamaterial device with enhanced electromagnetic energy harvesting efficiency. AIP Adv. **7**, 105204 (2017)
53. F. Yu, X. Yang, H. Zhong, C. Chu, S. Gao, Polarization-insensitive wide-angle-reception metasurface with simplified structure for harvesting electromagnetic energy. Appl. Phys. Lett. **113**, 123903 (2018)
54. X. Zhang, H. Liu, L. Li, Electromagnetic power harvester using wide-angle and polarization-insensitive metasurface. Appl. Sci. Basel **8**, 497 (2018)
55. X. Zhang, H. Liu, L. Li, Tri-band miniaturized wide-angle and polarization-insensitive metasurface for ambient energy harvesting. Appl. Phys. Lett. **111**, 071902 (2017)
56. S. Wang, P. Xu, W. Geyi, Z. Ma, Split-loop resonator array for microwave energy harvesting. Appl. Phys. Lett. **109**, 203903 (2016)
57. S. Shang, S. Yang, J. Liu, M. Shan, H. Cao, Metamaterial electromagnetic energy harvester with high selective harvesting for left- and right-handed circularly polarized waves. J. Appl. Phys. **120**, 045106 (2016)
58. B. Ghaderi, V. Nayyeri, M. Soleimani, O.M. Ramahi, Pixelated mMetasurface for dual-band and multi-polarization electromagnetic energy harvesting. Sci. Rep. **8**, 13227 (2018)
59. A. Ghaneizadeh, K. Mafinezhad, M. Joodaki, Design and fabrication of a 2D-isotropic flexible ultra-thin metasurface for ambient electromagnetic energy harvesting. AIP Adv. **9**, 025304 (2019)
60. A.M. Hawkes, A.R. Katko, S.A. Cummer, A microwave metamaterial with integrated power harvesting functionality. Appl. Phys. Lett. **103**, 163901 (2013)
61. P. Xu, S.Y. Wang, W. Geyi, Design of an effective energy receiving adapter for microwave wireless power transmission application. AIP Adv. **6**, 105010 (2016)
62. M.E. Badawe, T.S. Almoneef, O.M. Ramahi, A metasurface for conversion of electromagnetic radiation to DC. AIP Adv. **7**, 035112 (2017)
63. T.S. Almoneef, F. Erkmen, O.M. Ramahi, Harvesting the energy of multi-polarized electromagnetic waves. Sci. Rep. **7**, 14656 (2017)
64. X. Duan, X. Chen, L. Zhou, A metamaterial electromagnetic energy rectifying surface with high harvesting efficiency. AIP Adv. **6**, 125020 (2016)
65. R. Wang, D. Ye, S. Dong, Z. Peng, Y. Salamin, F. Shen, J. Huangfu, C. Li, L. Ran, Optimal matched rectifying surface for space solar power satellite applications. IEEE Trans. Microw. Theory Tech. **62**, 1080–1089 (2014)
66. S. Keyrouz, G. Perotto, H.J. Visser, Frequency selective surface for radio frequency energy harvesting applications. IET Microw. Antennas Propag. **8**, 523–531 (2014)
67. G.T.O. Tékam, V. Ginis, J. Danckaert, P. Tassin, Designing an efficient rectifying cut-wire metasurface for electromagnetic, energy harvesting. Appl. Phys. Lett. **110**, 083901 (2017)
68. T.S. Almoneef, F. Erkmen, M.A. Alotaibi, O.M. Ramahi, A new approach to microwave rectennas using tightly coupled antennas. IEEE Trans. Antennas Propag. **66**, 1714–1724 (2018)
69. F. Erkmen, T.S. Almoneef, O.M. Ramahi, Electromagnetic energy harvesting using frequency-selective surfaces. IEEE Trans. Microw. Theory Tech. **66**, 2433–2441 (2018)
70. A.Z. Ashoor, O.M. Ramahi, Polarization-independent cross-dipole energy harvesting surface. IEEE Trans. Microw. Theory Tech. **67**, 1130–1137 (2019)
71. L. Li, X. Zhang, C. Song, Y. Huang, Progress, challenges, and perspective on metasurfaces for ambient radio frequency energy harvesting. Appl. Phys. Lett. **116**, 060501 (2020)
72. T.J. Cui, M.Q. Qi, X. Wan, J. Zhao, Q. Cheng, Coding metamaterials, digital metamaterials and programmable metamaterials. Light Sci. Appl. **3**, e218 (2014)

73. L.H. Gao, Q. Cheng, J. Yang, S.J. Ma, J. Zhao, S. Liu, H.B. Chen, Q. He, W.X. Jiang, H.F. Ma, Q.-Y. Wen, L.-J. Liang, B.-B. Jin, W.-W. Liu, L. Zhou, J.-Q. Yao, R.-H. Wu, and T.-J. Cui, Broadband diffusion of terahertz waves by multi-bit coding metasurfaces. Light Sci. Appl. **4**, p. e324 (2015)
74. T.J. Cui, S. Liu, L. Li, Information entropy of coding metasurface. Light Sci. Appl. **5**, p. e16172 (2016)
75. S. Liu, T.J. Cui, L. Zhang, Q. Xu, Q. Wang, X. Wan, J.Q. Gu, W.X. Tang, M.Q. Qi, J.G. Han, W.L. Zhang, X.Y. Zhou, Q. Cheng, Convolution operations on coding metasurface to reach flexible and continuous controls of terahertz beams. Adv. Sci. **3**, 1600156 (2016)
76. L. Zhang, X. Wan, S. Liu, J.Y. Yin, Q. Zhang, H.T. Wu, T.J. Cui, Realization of low scattering for a high-gain Fabry-Perot antenna using coding metasurface. IEEE Trans. Antennas Propag. **65**, 3374–3383 (2017)
77. L. Li, T.J. Cui, W. Ji, S. Liu, J. Ding, X. Wan, Y.B. Li, M. Jiang, C. Qiu, S. Zhang, Electromagnetic reprogrammable coding-metasurface holograms. Nat. Commun. **8**, 197 (2017)
78. L. Zhang, X.Q. Chen, S. Liu, Q. Zhang, J. Zhao, J.Y. Dai, G.D. Bai, X. Wan, Q. Chen, G. Castaldi, V. Galdi, T.J. Cui, Space-time-coding digital metasurfaces. Nat. Commun. **9**, 4334 (2018)
79. J.Y. Dai, J. Zhao, Q. Cheng, T.J. Cui, Independent control of harmonic amplitudes and phases via a time-domain digital coding metasurface. Light Sci. Appl. **7**, 90 (2018)
80. L. Li, Y. Shuang, Q. Ma, H. Li, H. Zhao, M. Wei, C. Liu, C. Hao, C. W. Qiu, T.J. Cui, Intelligent metasurface imager and recognizer. Light Sci. Appl. **8**, p. 97 (2019)
81. Q. Ma, G. Bai, H.B. Jing, C. Yang, L. Li, T.J. Cui, Smart metasurface with self-adaptively reprogrammable functions. Light Sci. Appl. **8**, p. 98 (2019)
82. J. Han, L. Li, G. Liu, Z. Wu, Y. Shi, A wideband 1-bit 12×12 reconfigurable beam-scanning reflectarray: design, fabrication, and measurement. IEEE Antennas Wirel. Propag. Lett. **18**, 1268–1272 (2019)
83. G. Liu, H. Liu, J. Han, Y. Mu, L. Li, Reconfigurable metasurface with polarization-independent manipulation for reflection and transmission wave fronts. J. Phys. D. Appl. Phys. **53**, 045107 (2020)
84. L. Li, Y. Li, Z. Wu, F. Huo, Y. Zhang, C. Zhao, Novel polarization-reconfigurable converter based on multil-ayer frequency-selective surfaces. Proc. IEEE **103**(7), 1057–1070 (2015)
85. X. Zhang, J. Grajal, J.L. Vazquez-Roy, U. Radhakrishna, X. Wang, W. Chern, L. Zhou, Y. Lin, P.C. Shen, X. Ji, X. Ling, A. Zubair, Y. Zhang, H. Wang, M. Dubey, J. Kong, M. Dresselhaus, T. Palacios, Two-dimensional MoS2-enabled flexible rectenna for Wi-Fi-band wireless energy harvesting. Nature **566**, 368–372 (2019)
86. T. Ngo, A.D. Huang, Y.X. Guo, Analysis and design of a reconfigurable rectifier circuit for wireless power transfer. IEEE Trans. Industr. Electron. **66**, 7089–7098 (2019)
87. K. Paul, A.K. Sarma, Fast and efficient wireless power transfer via transition less quantum driving. Sci. Rep. **8**, 4134 (2018)
88. F.H. Lin, Z.N. Chen, Low-profile wideband metasurface antennas using characteristic mode analysis. IEEE Trans. Antennas Propag. **65**, 1706–1713 (2017)
89. A. Collado, A. Georgiadis, Conformal hybrid solar and electromagnetic (EM) energy harvesting rectenna. IEEE Trans. Circ. Syst. I Regul. Pap. **60**, 2225–2234 (2013)
90. L. Li, X. Zhang, C. Song, W. Zhang, T. Jia, Y. Huang, Compact dual-band, wide-angle, polarization-angle-independent rectifying metasurface for ambient energy harvesting and wireless power transfer. IEEE Trans. Microw. Theory Tech. **69**(3), 1518–1528 (2021)
91. K. Lee, S.K. Hong, Rectifying metasurface with high efficiency at low power for 2.45 GHz band. IEEE Antennas Wirel. Propag. Lett. **19**(12), 2216–2220 (2020)
92. C. Wang, J. Zhang, S. Bai, X. Zhu, Z. Zheng, A harmonic suppression energy collection metasurface insensitive to load and input power for microwave power transmission. IEEE Trans. Microw. Theory Tech. **70**(8), 4036–4044 (2022)

93. Y. Wei et al., A multiband, polarization-controlled metasurface absorber for electromagnetic energy harvesting and wireless power transfer. IEEE Trans. Microw. Theory Tech. **70**(5), 2861–2871 (2022)
94. F. Yu, X.-X. Yang, Z. Yi, S. Gao, Miniaturized polarization-insensitive rectifying metasurface for powering internet of things devices. IEEE Antennas Wirel. Propag. Lett. **22**(7), 1637–1641 (2023)

Chapter 11
Information Metamaterials and Metasurfaces

Zhang Jie Luo and Tie Jun Cui

Abstract Over a long time since the proposal of metamaterials, their exotic EM properties have been conventionally characterized by continuous effective medium parameters. That is to say, the theories and technologies for tailoring EM waves still stay on the physical level. In 2014, this limitation was broken by the emergence of digital coding metamaterial and the development of metamaterial is thus extended from the physical world to the digital world, bringing forth a brand-new concept—information metamaterial. Characterized by binary codes and equipped with digital hardware, the information metamaterial can do far more than just the manipulation of EM waves; through direct interactions and operations with digital signals within EM fields, information processing, transmission, or even recognition can be possible. The significance to the EM community is manifested by the emergence of new types of components and devices and the proposal of information theories, operational theorems, and the transformation of system architectures for wireless communication. Today, the information metamaterial has grown up gradually to become an independent subject of metamaterials. This chapter presents basic concepts, theories, principles, techniques, and typical applications of the information metamaterials and metasurfaces in detail, from which readers can understand the topic. Finally, the chapter is closed by summarizing the development track of the information metamaterial.

Keywords Digital coding metamaterial · Information metamaterials and metasurfaces · Information processing · Transmission · Recognition · Information theories · Operational theorems

Z. J. Luo · T. J. Cui (✉)
State Key Laboratory of Millimeter Waves, Southeast University, Nanjing 210096, China
e-mail: tjcui@seu.edu.cn

L. Li et al. (eds.), *Electromagnetic Metamaterials and Metasurfaces: From Theory To Applications*, https://doi.org/10.1007/978-981-99-7914-1_11

11.1 Introduction

Due to stability, preciseness, and security advantages, digital technologies have become dominant in information processing, transmission, and storage. Traditionally, information operations on the digital level and electromagnetic (EM) wave manipulations on the physical level are realized separately through different hardware modules and designed by different professionals. The emergence of digital coding metamaterials recently provides a new perspective on this long-standing issue. Different from the conventional metamaterials that are described by continuous effective medium parameters (permittivity or permeability), discrete digital codes like "1" and "0" are adopted to represent the distinct EM responses of meta-atoms that comprise the digital coding metamaterial. In this way, the digital world and EM physical world are linked together by the single hardware, and novel inspirations are thus offered for the designs and applications of the metamaterials [1, 2]. For example, the convolution operation and addition theorem are proposed for designing a metamaterial with multiple arbitrary scattering beams; the information entropy is proposed to estimate quantitatively the information carried by a coding metamaterial, which helps realize new imaging systems and wireless communication systems. By programming the meta-atoms of the digital metamaterial, many intriguing breakthroughs in wireless communication, harmonic manipulation, imaging, nonreciprocity, nonlinear control, et al., are achieved. Integrated with sensors and detectors, the metamaterial is further endowed with the ability to sense the environment or their postures and can make self-decision to adjust their features without human instructions. A much more exciting development is the intelligent metamaterial that is integrated with deep learning technology, which has been put forward for image reconstruction and gesture recognition.

The coding metamaterial, digital metamaterial, programmable metamaterial, and smart intelligent metamaterial can all be categorized as information metamaterial [3]. To cover the main picture of the information metamaterials, this chapter will start in Sect. 11.2 with the concept and design of the coding metamaterials, which is followed by the applications in beam manipulations, such as diffuse scattering, deflection, vortex beam generation, et al. In Sect. 11.3, we focus on the development of programmable metamaterials, including the designing principles and applications in the fields of beam manipulation, harmonic control, nonreciprocity, nonlinear control, holography imaging, and wireless communication with new systematic architectures. Afterward, smart metamaterials are presented in Sect. 11.4 with self-decided EM wave manipulation functions and remote intelligent imaging and recognition. Operational theorems and information theories of the information metamaterial are discussed in Sects. 11.5 and 11.6, respectively. Finally, the current development of information metamaterial is summarized, and its further directions are briefly expected.

11.2 Coding Metamaterials and Metasurfaces

11.2.1 Basic Concept and Design

Since the realization of the metamaterials at the beginning of this century [4], its unnatural EM properties are usually characterized by the continuous effective medium parameters, i.e., permittivity, permeability, or refraction index, and so they can be classified as the analogy type. Inspired by the binary idea of the digital system and digital signal processing, the concept of digital metamaterials was raised by Cui [5] and Engheta [6] independently. Although sharing the same terminology, quite distinct ideas were adopted by the two groups. The concept of "metamaterial bit" proposed by Engheta is still described by the permittivity. With two metamaterial bits, an EM metamaterial with the desired permittivity can be synthesized to achieve certain functions, such as a digital convex lens, digital graded-index flat lens, and digital hyperlens. In sharp contrast, the description using the effective medium parameters was abandoned by Cui. Any discrete properties of the meta-atom that comprises a metamaterial can be represented using finite binary digital codes. A typical 1-bit coding strategy is to conduct with two meta-atoms with opposite phases, which are encoded as "0" or "1", respectively, which is illustrated in Fig. 11.1a.

A simple subwavelength square patch structure is implemented as the typical coding meta-atom. The metallic patch with the variable size is etched on the top side of a dielectric substrate (with a thickness of 1.964 mm, a side length of 5 mm, a dielectric constant of 2.65, and a loss tangent of 0.001). Due to the metallic ground on the backside of the structure, its reflection magnitude is larger than 0.85. When the side lengths of the top patches are set to be 4.8 and 3.75 mm for two meta-atoms, respectively, their reflection phase difference ranges from 135° to 200° in the band of 8.1~12.7 GHz, with exactly 180° at 8.7 and 11.5 GHz. Therefore, these two meta-atoms are coded as "0" and "1", respectively, disregarding their absolute phases, as plotted in Fig. 11.1b. Because the structure is ultrathin compared with the operational wavelength, the metamaterials are also named metasurfaces. In the remainder of the chapter, we'll use the terminology "metasurfaces" regarding ultrathin metamaterials.

There are at least three advantages of this concept compared with the conventional analogy of metamaterials. First, the meta-atoms do not have to operate near the resonant condition, which is what most of the analog metamaterials do, thus resulting in possibilities of less loss and broader bandwidth. Second, only one parameter, e.g., S_{11} phase, is considered during the designing process of the meta-atom, which is a more flexible and simpler task for designers; For the analogy metamaterial, people need to study a set of parameters to extract the target one, i.e., permittivity, permeability, or refraction index [7], which is very complicated. Third, it is natural to connect the concept of digital meta-atom code with digital science, so a lot of mature digital ideas and technologies can be borrowed to trigger more advanced functions and applications of the metasurface. In particular, by purposely arranging the binary coding meta-atoms according to certain sequences, the metamaterial can manipulate the EM wave and bring many intriguing phenomena in a much more efficient and simpler

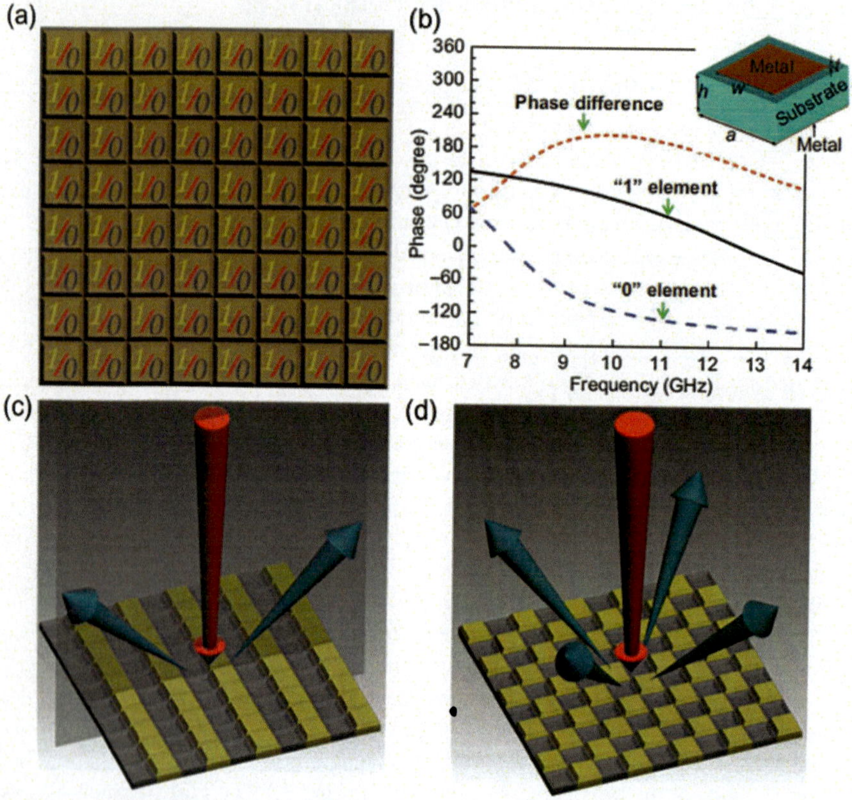

Fig. 11.1 **a** The concept of the 1-bit coding metamaterial, whose meta-atoms are characterized by "0" or "1" code. **b** The reflection phases of the "0" and "1" meta-atoms. Their phase difference is around 180° over a wide frequency region. **c** and **d** The scattering effects of two coding metamaterials with the periodic coding sequences of 010101…/010101… and 010101…/101010…/010101…/101010…, respectively [5]

way. Furthermore, through the coding representation of the meta-atoms, a series of operational theorems, such as the convolution operation [8], addition theorem [9], information entropy [10], and theories [11], are also developed. In this way, not only the wave manipulation but also the potential applications in the information community are synthesized as a micro-system, which paves a route towards a series of new architectures for communication, imaging, and other intriguing applications.

Although proposed aiming at the microwave spectrum, the digital coding metasurface has been extended very soon to terahertz and even the acoustic community also. This chapter will focus on the works on the microwave and terahertz spectrums.

11.2.2 Spatial Coding Metamaterials

One of the exciting applications of the digital coding metamaterial is EM wave manipulation by spatially distributing the binary elements into a 2D lattice with particular sequences. When the encoded meta-atoms are distributed on the metasurface with the periodic coding sequence of 010101.../010101..., the plane wave normally illustrated on the surface is deflected to two main symmetrical directions, as shown in Fig. 11.1c [5]. If the sequence switches to periodic 010101.../101010.../010101.../101010... along the two directions, or the checkerboard arrangement, the beam will be redirected to four symmetric orientations, as illustrated in Fig. 11.1d. Assuming that the reflection magnitudes of the meta-atoms are unity, the far-field scattering pattern of the coding metasurface (with $N \times N$ meta-atoms) under normal incidence can be quantitatively calculated by:

$$f(\theta,\varphi) = f_e(\theta,\varphi)\sum_{m=1}^{N}\sum_{n=1}^{N}\exp\{-j\{\varphi(m,n) + kD\sin\theta[(m-1/2)\cos\varphi + (n-1/2)\sin\varphi]\}\} \tag{11.1}$$

Here $\varphi(m, n)$ is the reflection phase of a certain meta-atom at the position (m, n). Considering only the relative phase difference for simplicity, the phase of the coding element "0" can be set as 0°, and that of the element "1" can be set as 180°. $f_e(\theta, \varphi)$ is the field function of a meta-atom. D is the period of the meta-atoms, $k = 2\pi/\lambda$ is wave number in free space, θ and φ are the elevation and azimuthal angles, respectively. The directivity function of the metasurface is calculated as:

$$Dir(\theta,\varphi) = \frac{4\pi|f(\theta,\varphi)|^2}{\int_0^{2\pi}\int_0^{\pi/2}|f(\theta,\varphi)|^2\sin\theta d\theta d\varphi} \tag{11.2}$$

Clearly, the scattering behavior of the metasurface can be easily controlled by delicately designing the coding sequence. As for the aforementioned periodic coding distributions, the deflected beam directions can be predicted by simplifying Eq. (11.1) as [12],

$$\varphi_1 = \pm\arctan\frac{\Gamma_x}{\Gamma_y},\ \varphi_2 = \pi \pm \arctan\frac{\Gamma_x}{\Gamma_y} \tag{11.3}$$

$$\theta = \arcsin(\lambda\sqrt{\frac{1}{\Gamma_x^2} + \frac{1}{\Gamma_y^2}}) \tag{11.4}$$

where Γ_x and Γ_y are physical periods of the coding elements along the x- and y-axes, respectively. In addition to these simple symmetric deflections caused by the periodic coding patterns, more advanced scattering beams can also be achieved by using the convolution operation and addition theorem, which will be discussed in Sect. 11.5.

11.2.2.1 Diffusing Effect

From Eqs. (11.1) and (11.2), it is apparent that the digital coding metasurface offers an approach to reallocating the EM energy in space purposely through the interaction between the EM field and the metamaterial. Therefore, it is reasonable to predict that, by particular designs of the coding sequences, it is possible to redirect the incident wave into numerous directions and hence cause the diffusion effect. Based on the energy conservation principle, the energy in a certain direction is thus decreased, which can be exploited for the radar-cross-section (RCS) reduction of a surface. The conventional method for the RCS reduction is to bend the EM wave around the target [13, 14], or suppress the magnitude of the incident wave using an absorber [15, 16]. In addition to them, the digital coding metasurface provides a new mechanism for the target.

To prove this idea, a digital coding metasurface consisting of 8 × 8 lattices, each of which includes 7 × 7 "0" or "1" coded elements, is designed, as shown in Fig. 11.2a and b [5]. The elements are exactly the aforementioned ordinary square patch meta-atoms. From the results plotted in Fig. 11.2b, one can easily observe the wideband monostatic RCS reductions obtained from full-wave simulations and measurements. In fact, the performance of the RCS reduction is relevant to the number of the lattices N and the coding sequence. The optimized codes for different N are listed in the literature [5], and it is clear that a better performance can be achieved with a larger N and the corresponding lattice sequence. Besides, the property of the coding metasurface is tolerant of the size of the lattice D. When D/λ varies from 0.6 to 3.0, the RCS reduction remains almost unchanged, which indicates that the metasurface can work in a wide bandwidth. Although the theoretical phase difference between the "0" and "1" coded meta-atoms should be 180°, it shows that at least 10-dB RCS reduction can still be realized when the phase difference ranges from 145° to 215°. This also guarantees the wideband property of the coding metasurfaces.

Although it has been demonstrated that the RCS reduction can be realized by using the digital coding metasurface, it is quite a challenge to choose a proper distribution of the coded lattice of meta-atoms or the coding sequence. So far, many brute-force numerical optimization tools have been utilized for this purpose, such as hybrid algorithms, meta-atom-swarm algorithms, simulated annealing algorithms, and genetic algorithms. However, as the electrical size of the metasurface enlarges, the optimization process becomes increasingly complicated, and the time consumption and requirements of the computational resources would turn out to be unacceptable. Under this context, Moccia et al. studied the theoretical scaling law of the RCS reduction and derived the bound to assess an optimized coding sequence for the digital coding metasurface [16]. Furthermore, a simple and deterministic designing strategy was proposed for the coding sequence to yield suboptimal RCS reduction results that were comparable to the aforementioned bound. As suggested by the researchers, the suboptimal results were also comparable with the brute-force optimized ones while causing an almost negligible burden on the calculation for the metasurface with even a large dimension. A coding metasurface working at X-band was designed, simulated, fabricated, and measured to validate the theoretical derivations. The aforementioned

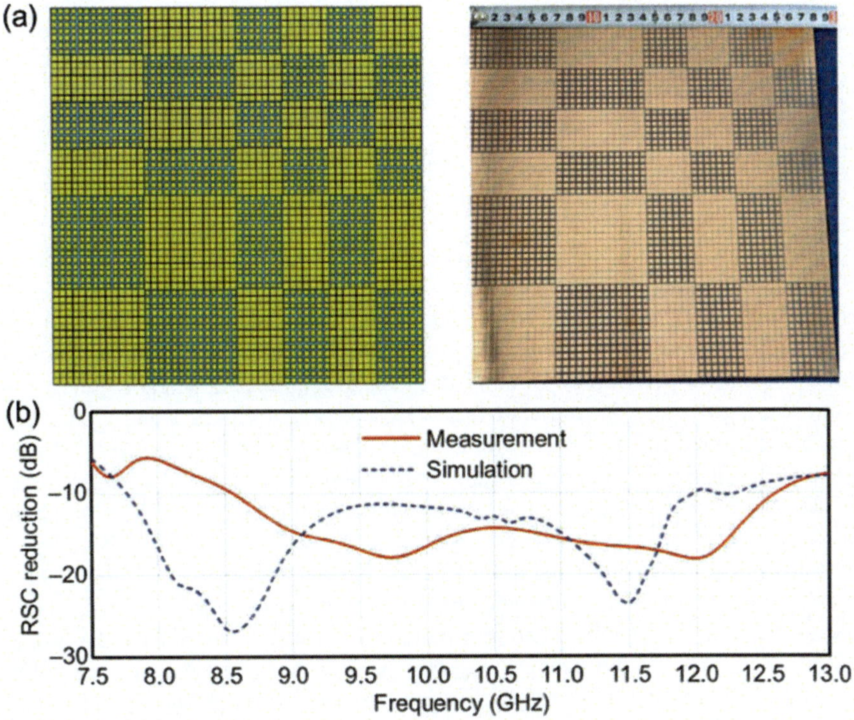

Fig. 11.2 **a** The simulation model and a photograph of the digital coding metasurface with 8 × 8 lattices. Each lattice includes 7 × 7 "0" or "1" coded elements. **b** Simulated and measured RCS reduction of the metasurface over the wide bandwidth [5]

ordinary square-patch meta-atoms were adopted, and the measured results were in fair agreement with the simulated ones, showing the effectiveness of the theoretical predictions.

11.2.2.2 Anisotropic Beam Deflection

The meta-atom for the digital metamaterial under current consideration is isotropic, which means that its EM responses to orthogonally polarized incident waves are identical. Here we would like to introduce the concept of the anisotropic coding metamaterial. It is composed of anisotropic meta-atoms, and its features can be changed by altering the polarization of the wave that impinges on the surface. Liu et al. have conducted a representative study on this topic with deep and comprehensive analysis [17]. One should notice that this work aims at the terahertz spectrum, and the designing philosophy is the same as in the microwave band.

To begin with, a 1-bit anisotropic meta-atom is designed with the configuration shown in Fig. 11.3a. It is a subwavelength structure with a dumbbell-shaped metallic patch on the top and a reflecting ground on the other side, making sure that the reflection magnitude is close to 1 and the phase is determined by the shape of the dumbbell shape. Here the reflection coefficient is expressed by a tensor

$$\overline{R}_{mn} = \begin{bmatrix} \hat{x} R_{mn}^x & 0 \\ 0 & \hat{y} R_{mn}^y \end{bmatrix} \tag{11.5}$$

where R_{mn}^x and R_{mn}^y are the reflection coefficients of a certain meta-atom under the x and y polarizations, and (m, n) indicates its position on the coding metasurface. It is easy to understand that the meta-atom degenerates back to isotropic when $R_{mn}^x = R_{mn}^y$. Due to the anisotropy, the reflection coefficients under the x and y polarizations are independently adjusted by tuning the four structural parameters of the meta-atoms h_1, h_2, w_1, w_2. The simulated reflection phase curves for x and y polarized waves are plotted in Fig. 11.3b with $h_1 = 45\ \mu\text{m}$, $h_2 = 20\ \mu\text{m}$, $w_1 = 37.5\ \mu\text{m}$, $w_2 = 18.5\ \mu\text{m}$. The dielectric constant is set as $\varepsilon = 3.0$, and the loss tangent is set as $\tan\delta = 0.03$. It is observed that, in the vicinity of 1 THz, the reflection phase difference experienced by the x and y polarized waves is close to 180°. The digital state for the x-polarization is defined as "1", and the digital state for the y-polarization is defined as "0"; in other words, the anisotropic meta-atom is referred to as "1/0" coding element. The numbers before and after the slash symbol indicate the coding states under x and y polarizations, respectively. By rotating the meta-atom by 90°, the features under the orthogonal polarizations are switched, and thus this meta-atom is defined as "0/1" coding element. Two isotropic meta-atoms are also designed for the coding metasurface by simply replacing the dumbbell shape of the top patch with the square ones. When the dimension of the patch $a = 45\ \mu\text{m}$, the reflection phases of the x- and y-polarized waves are similar to the red curve shown in Fig. 11.3b, and thereby the coding state of the meta-atom is named "0/0"; when $a = 30\ \mu\text{m}$, the phases of the orthogonally polarized waves are similar to the green curve, and hence the coding state is "1/1". By purposely arranging these anisotropic meta-atoms, the scattering behaviors of the metasurface under the two polarizations are decoupled and hence can be designed independently.

The first example of an anisotropic digital coding metasurface is built by repeating a 2D coding matrix:

$$M_1^{1\text{-bit}} = \begin{pmatrix} 0/0 & 0/1 \\ 1/0 & 1/1 \end{pmatrix} \tag{11.6}$$

where the row indicates the x-direction, and the column indicates the y-direction. That is to say, the coding sequence under the x polarization is "010101…" in the y-direction, and it remains unchanged in the x-direction; the sequence under the y-polarization is "010101…" in the x-direction, and it keeps unchanged in the other direction. An often-used trick should be emphasized during the coding metasurface design. In order to reduce the unwanted coupling effect between the neighboring but different coding elements, the same $N \times N$ meta-atoms are employed to comprise a

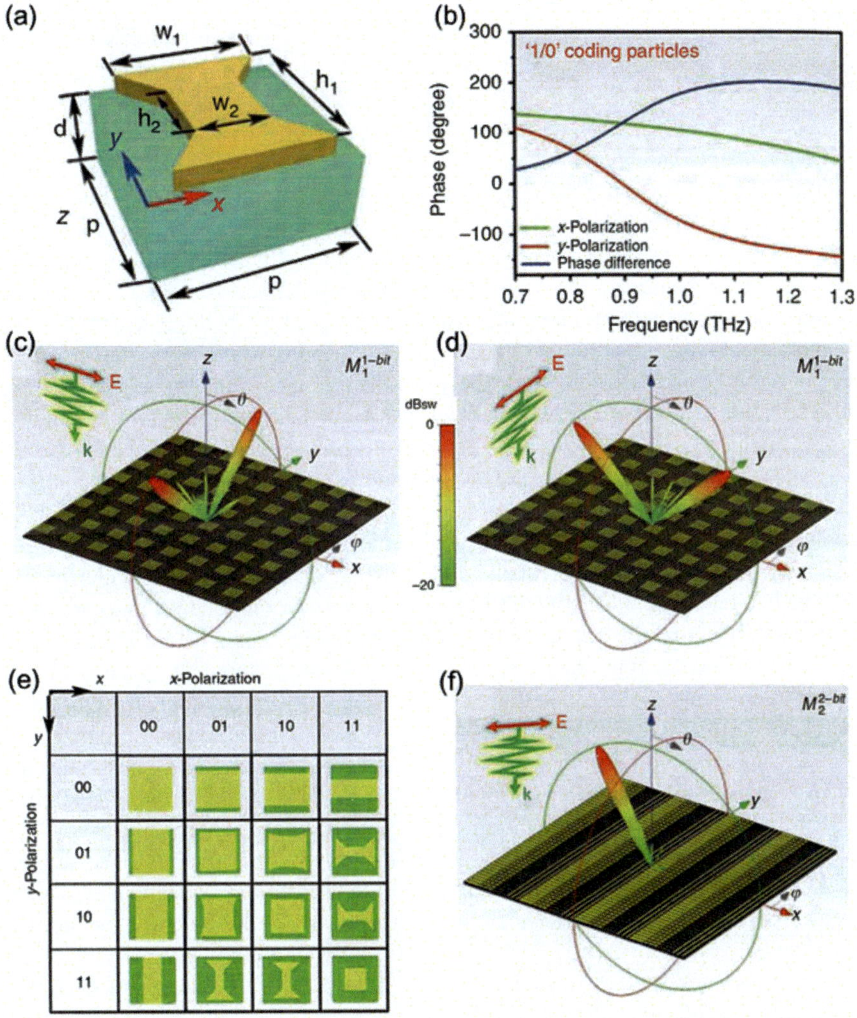

Fig. 11.3 **a** The dumbbell-shape configuration of the meta-atom of the 1-bit anisotropic metasurface. **b** The simulated reflection phase curves of the meta-atom under the x- and y-polarizations. **c** The simulated far-field scattering pattern of the first metasurface under x-polarization, showing two split deflected beams on the yoz-plane. **d** The simulated far-field scattering pattern of the first metasurface under the y-polarization, showing two split deflected beams on the xoz-plane. **e** The meta-atoms used in the 2-bit anisotropic metasurface and their coding states. **f** Simulated 3D far-field scattering pattern for the metasurface encoded with coding matrix $M_2^{2\text{-bit}}$ when the incident terahertz wave is linearly polarized by 45° with respect to the x-axis [17]

lattice, which is called "super cell", and the metasurface is composed of such coding super cells. The dimension of the super cell is not set randomly; instead, it should be created according to specific beam-manipulation requirements, which will be discussed later. N is chosen to be 4 in this example, and the metasurface consists of 16×16 super cells.

The far-field scattering pattern of the first anisotropic metasurface under x and y-polarizations are obtained through the full-wave simulations using CST, as shown in Fig. 11.3c and d. It can be seen that the x-polarized incident wave is split into two symmetric beams on the yoz-plane, and the y-polarized wave is split into two symmetric ones on the xoz-plane. These phenomena can be explained in the following mathematical way. As predicted by Eq. (11.3), Γ_x is infinite for the x-polarization, so $\varphi_1 = \varphi_2 = \pm\pi/2$, indicating the two split beams on the yoz-plane. For the y-polarization, Γ_y is infinite, so $\varphi_1 = 0$, $\varphi_2 = \pi$, indicating the two symmetric beams on the xoz-plane. By inserting the periods of the coding sequence Γ_x and Γ_y into Eq. (11.4), one can get the elevation angle of the beams, which is 48° in this case. Simulated near-field results are also provided in the literature for those who are interested, revealing the same physical behaviors as the far-field ones illustrate. When the polarization angle is rotated by 45° with respect to the x-axis, which means that the incidence includes both the x and y polarization components, one can easily predict that the metasurface can yield four symmetric beams. More interestingly, by changing the rotating angle, the energy proportion of the orthogonal polarization is tuned, resulting in the deflected beams' alterable magnitudes. In addition to the beam-deflecting effect, the diffusing effect under y-polarization is realized by adopting another coding sequence, while the aforementioned split beams under the other polarization are maintained. The two examples above clearly demonstrate the independent manipulations of the orthogonal polarization.

The powerful and flexible beam-controlling capability is further demonstrated by the 2-bit anisotropic coding metasurfaces. A set of 2-bit anisotropic meta-atoms are designed, which can provide four reflection phases independently for the two polarized waves, i.e., 0°, 90°, 180°, and 270°, corresponding to "00", "01", "10", and "11" coding states, as listed in Fig. 11.3e. With certain coding sequences, the independent scattering effect of the metasurface on the x and y polarized waves obeys the generalized Snell's law [18], just like its 1-bit counterpart, but with more flexibility. Besides the beam deflections discussed before, by designing the 90° phase difference under the x and y polarizations for each meta-atom, it is possible to create a free-background reflection-type quarter-wave plate. When the incident polarization is 45° with respect to the x-axis, the metasurface can produce a circularly polarized beam. Moreover, if the metasurface is encoded with the matrix

$$M_2^{2\text{-bit}} = \begin{pmatrix} 00/01 & 01/10 & 10/11 & 11/00 \\ 00/01 & 01/10 & 10/11 & 11/00 \\ 00/01 & 01/10 & 10/11 & 11/00 \\ 00/01 & 01/10 & 10/11 & 11/00 \end{pmatrix} \tag{11.7}$$

which means that the phase gradient is created for both polarizations along the x-direction, and the circularly polarized beam is further deflected along the x-axis, as shown in Fig. 11.3f.

11.2.2.3 Beam-Editing Based on Polarization Bit and Orbital-Angular-Momentum-Mode Bit

From the above examples of digital coding metasurfaces, one may notice that the digital information is embedded into the metasurfaces by spatially arranging the coding elements according to specific digital sequences. Conventionally, the meta-atoms are isotropic and coded using the phase states, and the information is manifested by the direction of the main beam provided by the metasurface [5]. Theoretically, the information can be recovered by recognizing the direction, which is the origin of a new wireless communication system that will be presented in Sect. 11.3.5. However, the same beam direction could be generated by different coding sequences, as illustrated in Fig. 11.4a, and if the number of receivers is too small to tell the difference between the different sequences, the information of the sequences may be lost. The issue can be mitigated by the anisotropic design mentioned above since the orthogonal polarizations are dealt with independently, but the transmission capability is still underexplored.

Under this context, Ma et al. propose a vector method to write information into the metasurface by employing the polarization and orbit angular momentum (OAM) bits at the same time, which can be totally received by the receiver without loss [19]. The concept is exhibited in Fig. 11.4b, with a reflective circular coding metasurface. For one thing, a polarization converter transforms a linearly polarized incidence into co- and cross-polarizations. In addition, the two polarizations are endowed with different OAM modes due to the specific arrangement of the phase-coding elements. One should notice that both the reflected polarizations and the OAM modes can be specifically predesigned as desired, thus bringing about a huge orthogonal space that is immune to information loss or cross-talk. Besides the greater channel capacity, the method is believed to be more secure for communications.

Two aspects are emphasized in this work, i.e., the designs of the meta-atoms and their arrangement, which are absolutely different from the conventional digital coding metasurfaces. The structure of the meta-atom is plotted in Fig. 11.4d. Of great concern are two parameters that determine the properties of the meta-atom, i.e., the rotation angle of the central metallic θ and the open angle of the symmetric split ring α. When θ is $\pm$ 45°, the polarization of the incident wave is supposed to be converted to the corresponding cross-polarization by the meta-atom; when θ is 0° or 90°, the original polarization of the incident wave is maintained as reflected by the meta-atom. The open angle of the symmetric split ring α, on the other hand, controls the reflection phase provided by the meta-atom. While α is varies from 30° to 150°, it is found that the reflection phase ranging from 0° to 360° is provided for the cross-polarized reflected wave when $\theta = \pm 45°$ and the co-polarized reflected wave when θ is 0° or 90°, respectively. As examples, eight values of α and four

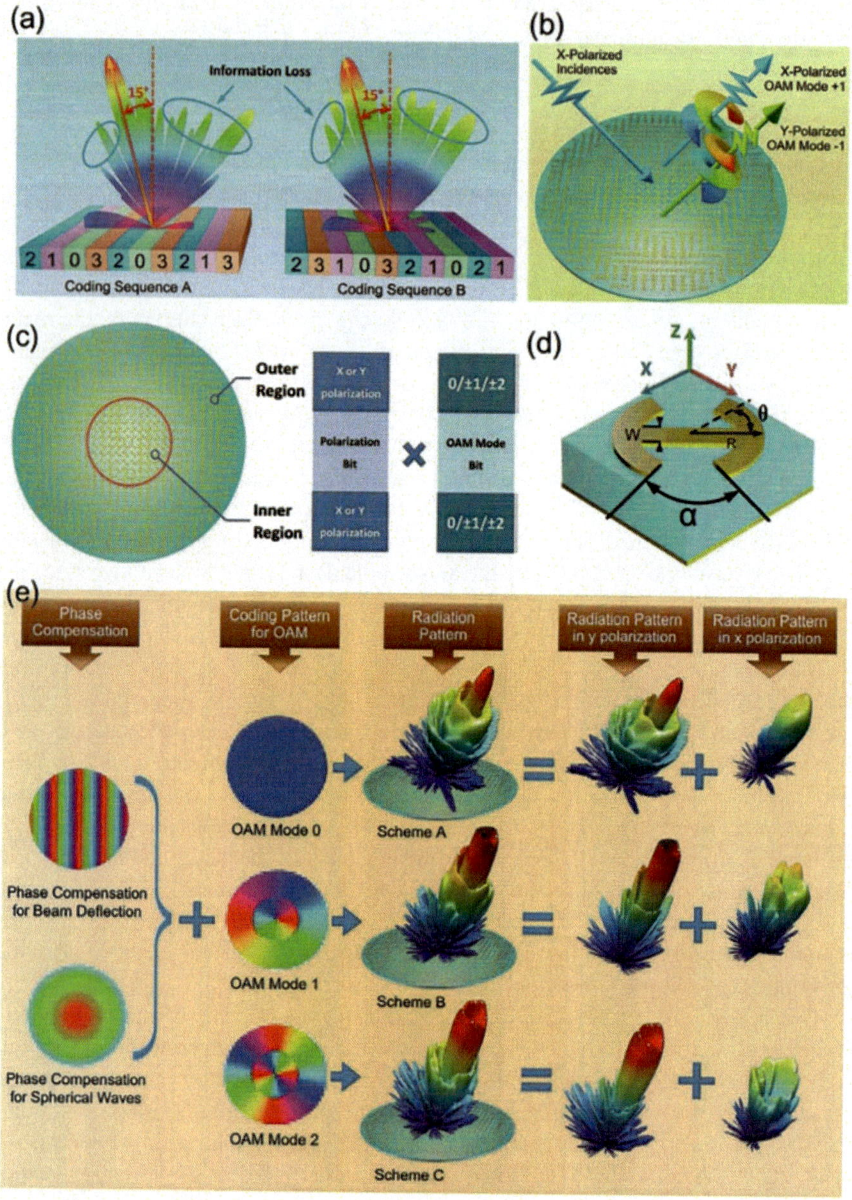

Fig. 11.4 **a** The illustration of information loss by encoding phase sequences on the metasurface. **b** Conceptual illustration of the vector beam modulator based on the coding metasurface. Information can be encoded on the metasurface in the orthogonal polarizations and OAM modes. **c** The layout of the circular metasurface that orthogonally encode information in the polarization bit and OAM-mode bit in the inner and outer regions, respectively. **d** Configuration of the meta-atom. **e** The operating mechanism of the metasurface with the three schemes and the simulated far-field scattering beams [19]

values of θ are selected, and thus a total of 32 phase statuses are ready for the coding metasurface design.

The other point in this work is the distribution of the meta-atoms on the surface, which determines the polarization bit and OAM bit encoded into the metasurface. Figure 11.4c shows the layout of the circular metasurface that includes two regions, i.e., the inner round region and the outer annular region. Each region can be designed to convert the polarization of the incident wave or not, offering two polarization bits. For each polarization, five OAM modes (0, $\pm$ 1, and $\pm$ 2 modes) are ready to be chosen for the OAM bit. This means that ten orthogonal states in total are obtained for each region, and the $(C_{10}^{1} + C_{10}^{2})$ metasurfaces can generate different situations.

Three typical schemes of the meta-atom distributions are selected to demonstrate the idea. The inner region is encoded for the cross-polarization bit, and the outer region is encoded for the co-polarization bit. The OAM modes (or bit) for the three schemes are 0, $\pm$ 1, and $\pm$ 2 modes, respectively. Figure 11.4e gives the generations of the three schemes and the corresponding simulated far-field scattering beams. Due to the tilted conical corrugated horn as the feed, the phase compensations for beam deflection and spherical waves are also considered for the meta-atom designs and their arrangement, which is not the point here. Observing the results, it is clear that OAM modes 0 under both the *x*- and *y*-polarizations are generated by Scheme A, OAM modes +1 and −1 carried in the *x* and *y*-polarizations are provided by Scheme B, and OAM modes +2 and -2 carried in the *x* and *y*-polarizations are provided by Scheme C, which is recognized by the amplitude nulls in the patterns. Simulated phase distributions of the three schemes and measured results of Scheme B are also presented in the literature, which is not shown here.

11.2.2.4 Multifunctional Coding Metasurfaces

As modern systems are becoming more and more compact and integrated, electronic devices are required to be embedded with increasing numbers of functions. There are two main types of multifunctional EM devices in terms of the function-triggering factor. The first type is triggered by the inherent properties of incident EM waves, such as direction [20], polarization [12, 21], magnitude [22, 23], waveform [24], frequency [25, 26], et al. When one or several properties change, the function of the device is switched. The other type is activated by external variables such as light intensity [27], the posture of the device [28], digital signal [29], and so on. The latter ones typically rely on active components; instead, the former ones can be realized using passive configurations, hence with lower cost and easier fabrication. The aforementioned anisotropic coding metasurface can be regarded as a bi-functional device depending on the polarization state of the incident wave. Here a passive digital coding metasurface with three independent functionalities is proposed by Zhang et al., which can be triggered by switching the direction or polarization of the incident wave [30]. Specifically, *y*-polarized waves propagating along + *z* direction are reflected by the metasurface with a deflected angle, and the same polarized wave with the opposite direction is reflected and diffused when impinging on the metasurface. On the other hand, x-polarized waves transmit through the metasurface and generate a vortex beam.

The three functions are achieved by taking advantage of a meta-atom with the following characteristics. When EM waves impinge an array of such meta-atoms, y-polarized waves from opposite sides are reflected with the reflection phases that cover 360°; x-polarized waves can propagate through with the transmission phase that covers 360°; more importantly, the reflection phases for the two directions and the transmission phase should be tuned independently by changing the structural parameters. The configuration of the meta-atom shown in Fig. 11.5a is proposed. It is an anisotropic structure with five metallic layers. Attention should be paid to the middle metallic layer with a grating along the y-direction that plays a crucial role in reflecting the y-polarization and transmitting the x-polarization. The transmission phase can be tuned by adjusting the structural parameter T_x, and the reflection phases for the waves coming from the two sides are dependent on R_{y1} and R_{y2}, respectively. Because the coverages of the phases are close to 360°, many different functions can be realized through specific arrangements of the meta-atoms.

A coding metasurface with 30×30 is designed by the researchers, which contains three coding patterns for the three independent functions mentioned above. The first function, i.e., beam deflection for y-polarized waves propagating along $+z$ direction, is realized by creating a periodic gradient coding pattern F_1, as shown in Fig. 11.5b. The second function, i.e., diffusing effect for the y-polarization along $-z$ direction, is realized by paving the opposite-phase meta-atom with an optimized pattern F_2, as shown in Fig. 11.5b. To obtain the third function F_3, the transmitted OAM beam for the x-polarization, the metasurface is divided into 16 sectors, as plotted in Fig. 11.5b. The transmission phase of the meta-atoms in each sector is 45°, and the OAM with mode 2 is generated by the successively rotated phase gradient. It should be kept in mind that these phase requirements are met by the single layout of the metasurface; in other words, for each meta-atom, the three different phase responses are integrated by delicately tuning the critical structural parameters. The simulated results of the functions are given in Fig. 11.5c–f. It is observed that the y-polarized waves propagating along $+z$ and $-z$ directions are anomalously deflected and diffused, respectively, by the metasurface. When the coding metasurface is illuminated by the x-polarized wave, the vortex beam carrying the OAM mode 2 is demonstrated by the ring-shaped magnitude profile of the far-field pattern and the spiral-like phase distribution of the near-field result, as shown in Fig. 11.5e and f. More details about the simulations and measurements are offered in the paper [30].

11.2.2.5 Frequency-Dependent Dual-Functional Coding Metasurfaces

Besides the above polarization- and direction-dependent coding metasurfaces, a frequency-dependent bi-functional coding metasurface at the terahertz spectrum is proposed by Liu et al. [25]. The concept is illustrated in Fig. 11.6a and b, showing that the scattering performance of the metasurface changes according to the frequency of the incident wave. At the high frequency, the coding sequence of the meta-atoms

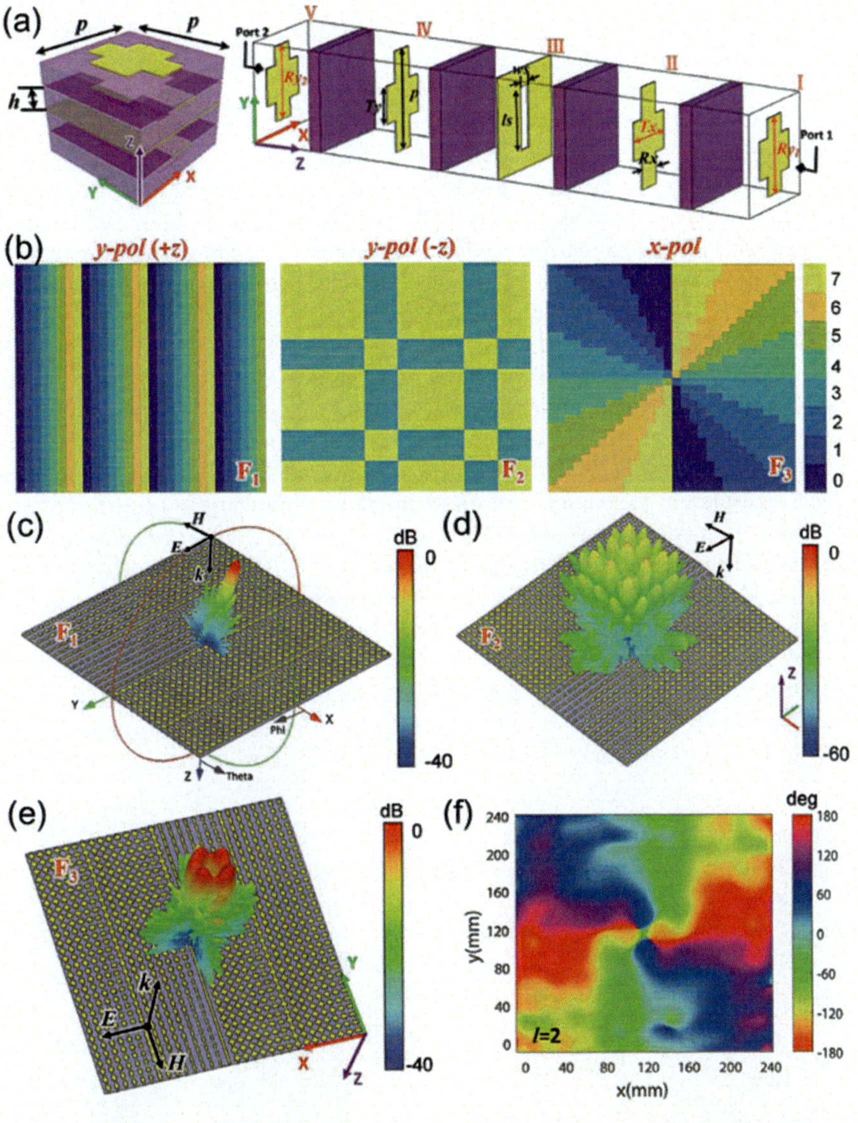

Fig. 11.5 **a** Configuration of the meta-atom of the multi-functional metasurface. **b** The coding patterns of the metasurface under incidences with different directions and polarizations. **c**–**e** The simulated 3D far-field patterns of the deflection under the y-polarization along $+z$ direction, the diffusing effect under the y-polarization along $-z$ direction, the vortex beam carrying the OAM mode 2 under the x-polarization along $-z$ direction, respectively. **f** The simulated phase distribution of the E_x component on the xoy plane cutting at the $z = -200$ mm, with an area of 240 mm × 240 mmm [30]

on the surface is alternative "0" and "1", resulting in the split symmetric scattering beams; at the low frequency, the coding pattern changes to the checkerboard, which scatters the incident beam into four symmetric beams.

The meta-atom in this work contains two electric liquid crystal resonators and a metallic ground layer, which are separated by two polyimide spacers, as plotted in Fig. 11.6c. Among the structural parameters, the heights of the resonators, respectively h_1 and h_2, are the keys that impact the reflection phase of the meta-atom over the spectrum. The 1-bit dual-frequency coding metasurface adopts four structures to realize the "0" and "1" coding states at the low and high frequencies, respectively, i.e. "0/0", "0/1", "1/0", "1/1" elements. The bits before and after the slash symbol indicate the digital states at the lower and higher frequencies, respectively. Specifically, the heights h_1/h_2 for the top and bottom resonators are chosen to be 31 μm /21.5 μm, 20 μm /30 μm, 38.5 μm /30 μm, and 15 μm/35 μm, respectively, for the four coding meta-atoms. The phase responses of the four coding elements at the two frequency points, 0.79 THz and 1.19 THz, are plotted in Fig. 11.6d. Without the need to consider the absolute phase values, the phase difference between the "0" and "1" elements at 0.78 THz is about 177°, which is denoted by the red triangles. The phase difference between the "0" and "1" elements at 1.19 THz is about 165°, which is denoted by the blue spots. In this way, the coding bits can be freely chosen, and various performances can be realized independently at the two frequencies.

Three different coding sequences are chosen to evaluate the performance of the dual-band coding metasurface. Each sequence can be regarded as a combination of two subcoding sequences at the lower and high frequencies, which are named as S_L and S_H. For the first coding sequences S_1, S_{1L} and S_{1H} are designed as "0101..."

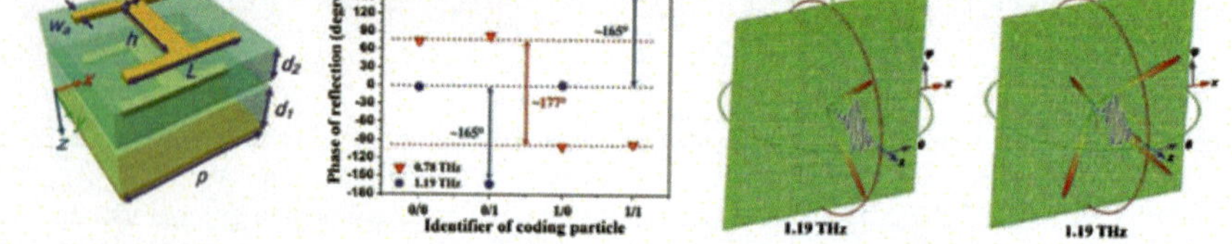

Fig. 11.6 (**a**, **b**) Conceptual illustrations of the dual-band coding metasurface. The coding distributions at higher and lower frequencies are different, resulting in the dual and four deflected beams, respectively. **c** 3D view of the meta-atom. **d** The absolute values of the reflection phases of the four coding particles at 0.78 (red triangle) and 1.19 THz (blue round dot). **e** and **g** Simulated 3D far-field scattering patterns of the metasurfaces with the coding sequence S_1 at 0.78 and 1.19 THz, respectively. (f) and (h) Simulated 3D far-field scattering patterns of the metasurfaces with the coding sequence S_2 at 0.78 and 1.19 THz, respectively [25]

along the x and y directions, respectively. Here the numbers of the meta-atoms for the super cell in S_{1L} and S_{1H} are set as 4 and 3, respectively. The simulated 3D far-field scattering patterns under y-polarization at 0.78 THz and 1.19 THz are shown in Fig. 11.6e and g. It clear that the incident wave at the lower frequency is split into two symmetric beams on the xoz-plane. At the higher frequency of 1.19 THz, in contrast, the split beams are converted to the yoz-plane because the variation direction of the sequence is switched to the y-direction. The elevation angles of the split beams are in accordance with the theory predicted by Eq. (11.3) together with the structural parameters. The second coding sequence S_2 is the combination of the checkerboard distribution S_{2H} [1, 0; 0, 1] at the high frequency and S_{2L}, which is the same as S_{1L}. As expected, four symmetric oblique beams are obtained at 1.19 THz, as shown in Fig. 11.6f and h, with the elevation angle calculated using Eq. (11.4) and azimuthal angle yielded from Eq. (2.3). Comparing the results of sequences S_1 and S_2, it is obvious that the same scattering patterns at the lower frequency are maintained, not affected by the sequence change at the higher frequency, implying the excellent isolation between the digital states at the two frequencies. To further demonstrate the conclusion, the third coding sequence S_3 is employed with the same "0101..." pattern along the x-direction for S_{3L} and S_{3H}, but with different numbers of meta-atom in the super cell as 5 and 2, respectively. As predicted by Eqs. (11.3) and (11.4), two symmetric split beams are generated at the two frequencies, but different elevation angles are obtained due to the different periods of the sequences. The results of coding sequences S_1 and S_3 are verified by the experimental measurement conducted in the terahertz region.

11.2.2.6 Frequency Coding Metasurfaces

Based on the work on the frequency-dependent coding metasurface, Wu et al. created a frequency coding theory for the topic [26]. The core of the theory is to digitalize the phase response sensitivity of meta-atom over frequency. Together with the aforementioned spatial coding theory, the frequency-spatial coding metamaterial is proposed, which can manipulate the EM wave with multiple functions using a single digital coding metasurface without changing the spatial coding pattern. The meta-atoms used in this type of metasurface share almost the same phase response at an initial frequency, yet with different frequency-dependency. In other words, their phases separate as the frequency moves, hence giving birth to a new coding pattern at a new frequency that controls the EM wave differently, as illustrated in Fig. 11.7a.

In order to digitally encode a meta-atom, a coding strategy is created in the frequency domain. Taking a 1-bit case as an example, a meta-atom could be encoded as "0-0," "0-1," "1-0," or "1-1." The former digit is the spatial bit at the initial frequency, where "0" and "1" imply the two cases with opposite phase responses. The latter bit is the frequency coding that represents the phase sensitivity level of the meta-atom, i.e., "0" and "1" denote low and high phase sensitivities, respectively. Mathematically, the phase response of a meta-atom over the frequency can be expressed by Taylor's series

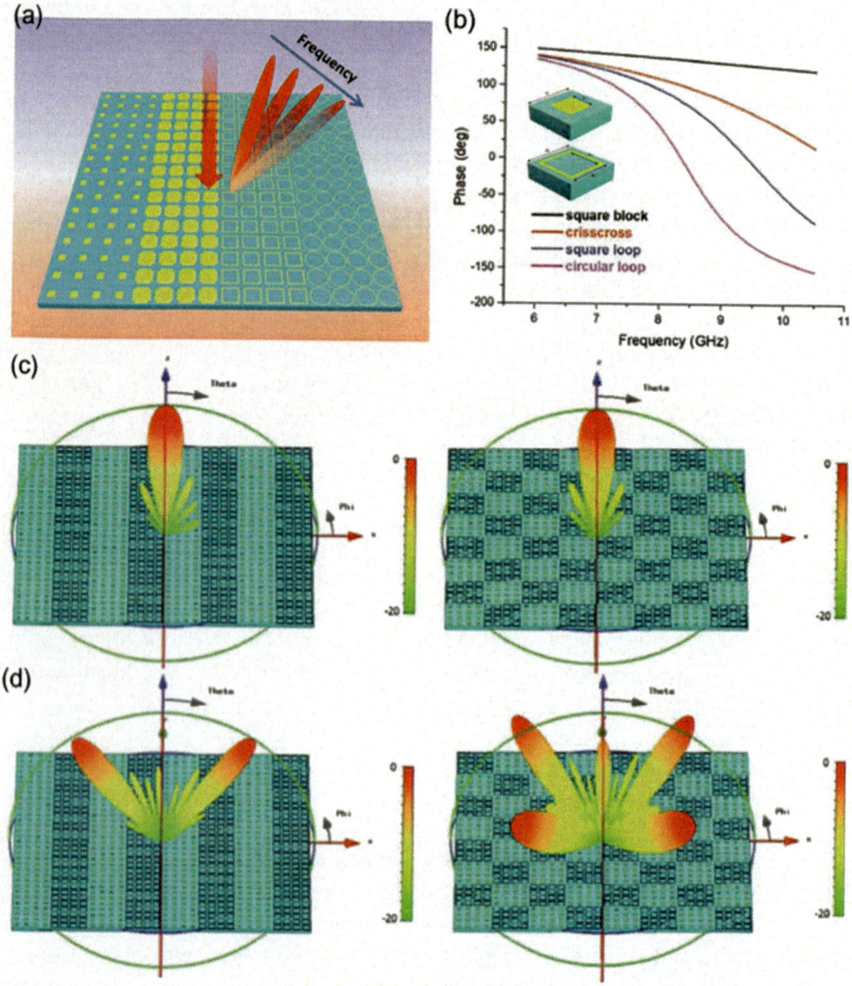

Fig. 11.7 **a** Conceptual illustration of the frequency coding metasurface. As the frequency changes, different wave-deflection angles are obtained. **b** The reflection phase curves versus frequency of two reflective meta-atoms with a square block or a square loop top patterns. **c** The normal reflections of two metasurfaces at the initial frequency of 6.0 GHz. **d** At a higher frequency of 10.5 GHz, two and four split beams are generated by the two metasurfaces [25]

$$\varphi(f) = \alpha_0 + \alpha_1(f - f_0) + \alpha_2(f - f_0)^2 + \cdots + \alpha_n(f - f_0)^n + \alpha_{n+1}(f')(f_0 \le f' \le f) \tag{11.8}$$

where α_0 is the 0th order phase at the initial frequency f_0, and α_n is the nth order of the phase response over the frequency. For the conventional coding metasurface, only the 0th order phase is utilized. In this work, the first order is also used, i.e. $\varphi(f) \approx \alpha_0 + \alpha_1(f - f_0)$, which means that the phase variation is regarded to be linear as a function of the frequency if the higher orders are ignored. Therefore, both the phase response at the initial frequency α_0 and the phase sensitivity α_1 have to be considered during the meta-atom designing process, which is more complicated than before. In the 1-bit example, two reflective meta-atoms with square block and square loop top patterns are adopted with the phase curves versus frequency shown in Fig. 11.7b (black and blue curves, respectively). It is apparent that the two meta-atoms exhibit almost the same phase at the initial frequency $f_0 = 6.0$ GHz, so they are encoded as "0" for the spatial bit. The linear phase sensitivities can be calculated as

$$\begin{aligned}\alpha_1^{block} &= \left[\varphi^{block}(f_1) - \varphi^{block}(f_0)\right]/(f_1 - f_0) \approx -0/4.5 \text{ (rad/GHz)} \\ \alpha_1^{loop} &= \left[\varphi^{loop}(f_1) - \varphi^{loop}(f_0)\right]/(f_1 - f_0) \approx -\pi/4.5 \text{ (rad/GHz)}\end{aligned} \tag{11.9}$$

and thus α_1^{block} and α_1^{loop} are encoded as "0" and "1" to characterize the frequency responses. Therefore, the square block and square loop meta-atoms are represented as "0-0" and "0-1" elements, respectively.

Here the 1-bit case is employed to illustrate the application of the frequency-spatial coding metasurface. The first metasurface is designed with the coding sequence "0-0," "0-1," "0-0," "0-1" along the *x*-axis only, and the second one is encoded with the sequence "0-0," "0-1," "0-0," "0-1" along both the *x*- and *y*-axis (checkerboard). The number of meta-atoms in a super cell is 4 × 4. At the initial frequency 6.0 GHz, the phase difference between the two meta-atoms is almost zero, so the two metasurfaces can be regarded with homogeneous phase distribution that generates almost the same normal reflections, as shown in Fig. 11.7c. At the frequency f_1 10.5 GHz, the phase difference becomes almost 180°, and the incident wave is deflected anomalously according to the generalized Snell's law. The two metasurfaces yield two and four split beams, respectively, as can be found in Fig. 11.7d. It should be noticed that while the frequency moves from the initial one to the higher one, the scattering patterns change gradually with shrinking normal reflections and growing deflected beams.

The same group of researchers continue the investigation and brought about the deeper concept of space-frequency-domain gradient metasurface [31]. In addition to the phase response of the metasurface element expressed by Eq. (11.8), the phase gradient of a group of elements with the shared phase difference can be given as

$$\Psi(f) = \gamma_0 + \gamma_1(f - f_0) + \gamma_2(f - f_0)^2 + \cdots + \gamma_n(f - f_0)^n + \gamma_{n+1}(f')^{n+1} \tag{11.10}$$

which denotes the *n*th-order phase difference. If $\gamma_1, \gamma_2, \ldots \gamma_n$ are zero, the phase difference among the group of elements remains the same, which is expressed as

$$\Psi_s = \varphi(f_0) = \gamma_0 \tag{11.11}$$

It is clear that Ψ_s is the phase gradient of the elements at the initial frequency, which is defined as the space-domain gradient. On the other hand, the frequency-domain gradient is defined to be the phase gradient over the frequency, which is

$$\Psi_f = \frac{\partial \Psi(f)}{\partial f} = \gamma_1 + 2 \times \gamma_2(f - f_0) + \cdots + n \times \gamma_n(f - f_0)^{n-1} \tag{11.12}$$

So, the phase gradient over the operational band is determined by the space gradient and the frequency gradient simultaneously, as expressed as

$$\Psi = \Psi_s + \int_{f_0}^{f} \Psi_f \mathrm{d}f \tag{11.13}$$

Equation (11.13) provides a more comprehensive perspective to understand the phase pattern on a coding metasurface in space and frequency domain. Specifically, at the initial frequency f_0, the phase gradient on the surface is constant, which is decided by the space gradient only. Owing to the elements that behave diversely over the whole frequency band, frequency gradient occurs, and its accumulation has a remarkable effect on the phase pattern on the surface, leading to different manipulating effects on EM waves. In addition, one may notice that the space gradient and frequency gradient are orthogonal vectors in space; as a result, they can be considered as independent components during the designing process of the meta-atom along the orthogonal directions.

According to the generalized Snell's law, the orientation of the beam deflected by a metamaterial can be predicted according to [18]

$$\theta = \arcsin\left(\frac{\lambda}{2\pi} \times \frac{\mathrm{d}\varphi}{\mathrm{d}r}\right) \tag{11.14}$$

which is the origin of Eq. (11.4). Here $\mathrm{d}\varphi/\mathrm{d}r$ is the phase gradient on the metasurface, which is considered as constant for a metasurface with only the space gradients. In general, θ is a function of the frequency of an ordinary metasurface, but not much attention is paid to it. For the space-frequency-domain gradient metasurface here, delicate efforts are put into the frequency gradient so that the beam deflection is under control over the whole frequency band. By inserting Eq. (11.13) into (11.14), we get

$$\theta = \arcsin\left(\frac{\lambda}{2\pi} \times \frac{\mathrm{d}\varphi}{\mathrm{d}r}\right) = \arcsin\left[\frac{c}{2\pi f d} \times \left(\Psi_s + \int_{f_0}^{f} \Psi_f \mathrm{d}f\right)\right] \tag{11.15}$$

where d is the length of the super cell, and c is the velocity of light in space. It is thus apparent that by purposely designing the frequency gradient, the scanning trajectory of the EM wave can be achieved.

Besides the beam scanning, a more intriguing demonstration of the space-frequency-domain gradient metasurface is the continuous vortex-mode transformation. It is known that vortex beams carry OAM with azimuthal phase modulation of

$m\phi$, where ϕ is the azimuthal angle, and m is the topological charge (TC). Mathematically, a vortex beam with TC $= m$ can be generated by modulating the azimuthal phase distribution

$$\Phi(x, y) = m \times \varphi = m \times \arctan(y/x) \tag{11.16}$$

At the initial frequency f_0, the vortex beam with TC $= m_0$ is obtained by aligning the spiral-like space gradient

$$\Psi_s(\varphi) = 2\pi \cdot m_0/n \tag{11.17}$$

where n is the number of equally divided sectors of the metasurface along the azimuthal direction. Like the aforementioned beam scanning case, the introduction of the frequency gradient, $\Psi_f(\varphi)$, accounts for the transformation of the TC of the vortex beam in the frequency domain; in other words, the OAM mode changes along with the frequency. Assuming that the frequency gradient is also equally divided into n sectors, the TC can be written as

$$m(f) = m_0 + \frac{n}{2\pi} \times \int_{f_0}^{f} \Psi_f \mathrm{d}f \tag{11.18}$$

As a consequence of the spatial and frequency gradients, the vortex beams can be continuously manipulated throughout the frequency band. An example is raised by the researchers with $\Psi_s(\varphi) = \pi/4$ and $\Psi_f(\varphi) = \pi/20$ from 8 to 13 GHz. As a result, the TC of $m(f) = 1 + (f - 8)/5$ is obtained according to Eq. (11.18), which means that it increases from 1 to 2 while the frequency moves from 8 to 13 GHz, as plotted in Fig. 11.8a.

The vortex beam manipulation is taken as a demonstration example through numerical simulations and measurements. The configuration of the meta-atom is shown in Fig. 11.8b, which consists of four metallic layers spaced by F4B substrates. The three square loops and a reflective ground layer are from the top to the bottom. Eight sets of structural parameters are properly selected, and their phase responses plotted in Fig. 11.8c are achieved, implying the space gradient $\Psi_s(\varphi) \approx \pi/4$ and frequency gradient $\Psi_f(\varphi) \approx \pi/20$ in the frequency band. The metasurface includes 40 × 40 meta-atoms, and eight meta-atom groups are uniformly distributed in the eight sectors, as shown in Fig. 11.8d. Results from numerical simulations and measurements are given in Fig. 11.8e, showing the transformation of the vortex beams that varies from $m = 1$ to 2 gradually with the changing frequency.

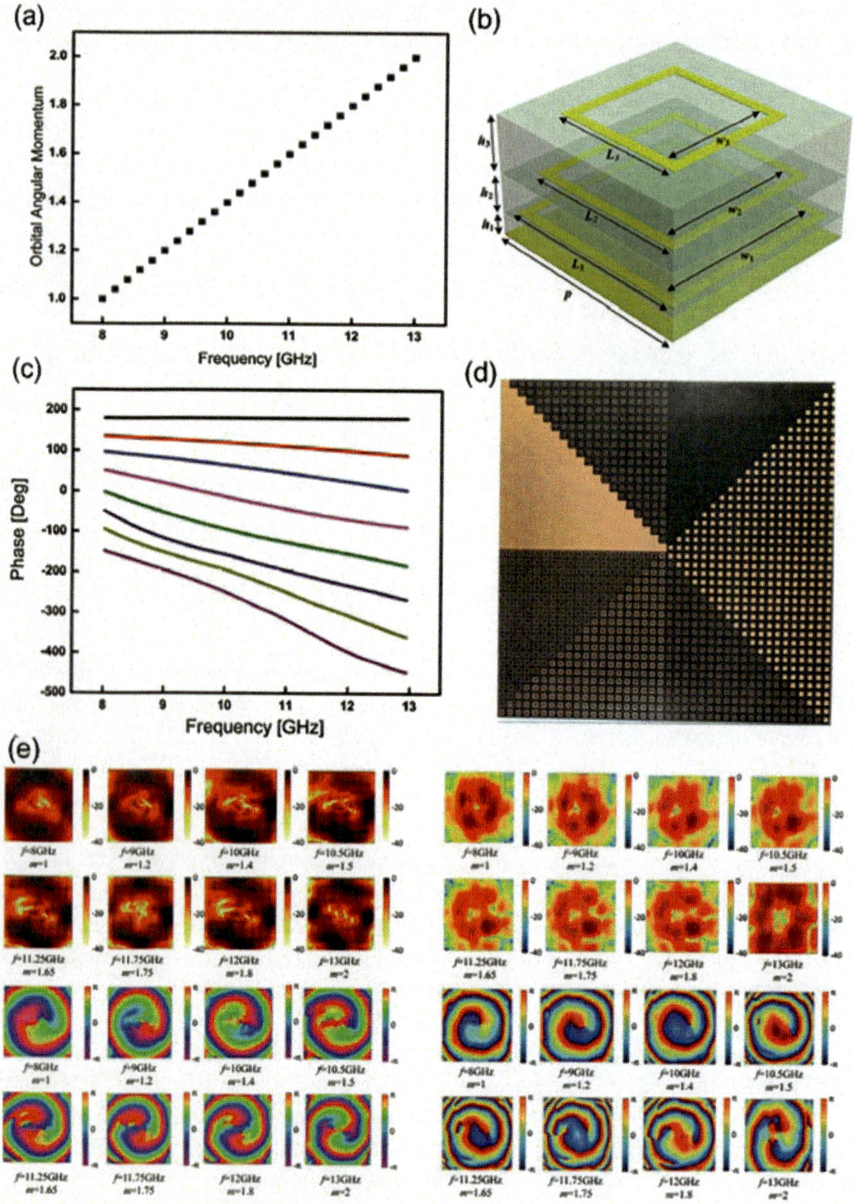

Fig. 11.8 **a** Calculated OAM of the vortex beam as a function of frequency. **b** 3D view of the meta-atom. **c** Simulated phase curves of the selected meta-atoms. **d** A photograph of the fabricated metasurface. **e** Simulated and measured near-fields of the space-frequency-domain gradient metasurface, suggesting the OAM mode transformations from 8 to 13 GHz [31]

11.3 Programmable Metamaterials and Metasurfaces

11.3.1 Basic Principle and Structure

While the meta-atoms are encoded and arranged into an array according to a certain sequence, the information of the sequence is embedded, which is manifested by the scattering properties of the coding metamaterial, such as beam directions or OAM. For the aforementioned passive coding metamaterials, however, their functions are set in stone once the design is finished, which means that the integrated digital information cannot be altered. They help demonstrate the concept, but limitations in real-time information representation, processing, and transmission are also apparent. This issue can be solved by adopting reconfigurable meta-atoms with tunable EM responses. If a metamaterial is composed of such meta-atoms whose coding states are switchable independently and dynamically, information can be easily integrated without having the structural configuration changed. This type of metamaterial is categorized as a programmable metamaterial.

Technically, there are many approaches toward the designs of tunable meta-atoms, including the integration of semiconductor components [5, 22–24], liquid crystals [32], microelectromechanical systems (MEMs) [33], vanadium dioxide [34], reconfigurable cantilevers [35], etc. Considering the response time and cost, semiconductor components such as PIN diodes and varactor diodes are preferred at the microwave frequency spectrum. Figure 11.9a gives a typical electrically tunable meta-atom embedded with a PIN diode [5]. It is a PCB structure with two metallic layers. Two symmetric metallic patterns located on the top with a PIN diode across them. Two pieces of reflecting plates are placed on the bottom. Two metallic vias connect the top and bottom plates. By applying DC voltage across the two plates on the bottom, the working state of the diode can be controlled. When the voltage is 3.3 V, the diode is ON; when the voltage is 0, the diode is OFF. The reflection properties of the meta-atom under the two conditions are simulated using CST, where the diode is modeled using the equivalent circuits shown in Fig. 11.9b. The reflection phases of the meta-atom with the diode ON and OFF are presented in Fig. 11.9c, showing a phase difference of approximately 180° from 8.3 to 8.9 GHz. Therefore, it is reasonable to encode the meta-atom as the "1" element with DC voltage 3.3 V and "0" element with DC voltage 0 V.

The control of the electrically tunable meta-atoms could be implemented using a normal DC voltage source. An alternative method is to use an FPGA module, which is more of a source for programmable controls. A metasurface consisting of 30×30 tunable meta-atoms is designed. To simplify the controlling network, every five columns of the meta-atoms share the same controlling signal, which means that six digits are needed to control the coding states of the meta-atoms. As shown in Fig. 11.9d, four coding sequences, i.e., 000000,111111, 010101, and 001011, are restored in the FPGA hardware in advance and can be triggered by toggling different triggers. The metasurface is connected with the FPGA and thus equipped with four different functions. The scattering far-field patterns of the metasurface under the

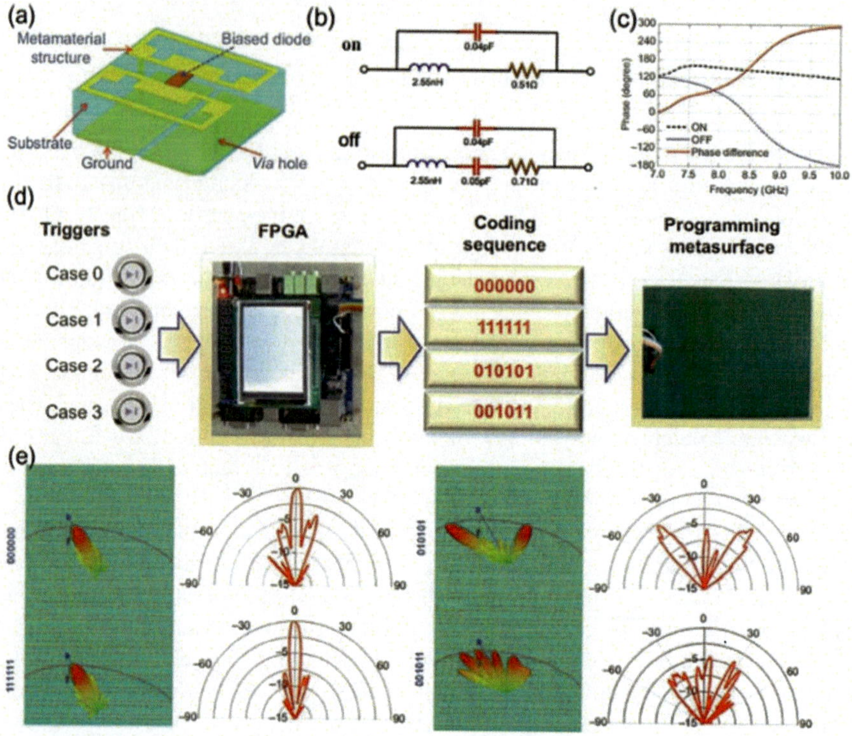

Fig. 11.9 **a** The typical electrically tunable meta-atom embedded with a PIN diode. **b** The equivalent circuit models of the PIN diode used in the tunable meta-atom at the ON or OFF states. **c** Simulated reflection phases of the meta-atom when the diode is ON and OFF, respectively. **d** A flow diagram for realizing a programmable metasurface controlled by the FPGA hardware. **e** Simulated and measured far-field scattering patterns of the 1-bit digital metasurface with various coding sequences [5]

control of the FPGA are measured, which are compared with the simulations in Fig. 11.9e. Very good agreement between the experiment and the simulation serves as strong proof of the feasibility of the programmable metasurface. It should be emphasized that the functions are switched manually in this case, which can also be accomplished automatically with a much higher speed (as quickly as several nanoseconds, depending on the clock speed of the FPGA chip). Thanks to the advantage of the high-speed switching effect, a plethora of intriguing physical phenomena and novel applications are expected by using the programmable metasurface, such as the precise control of harmonics, nonreciprocity, wireless communication, and beyond, which will be covered later.

Attention should be paid to several issues during the design of the programmable metasurface. Firstly, compared with the passive coding digital metasurface, inevitable EM energy loss would be induced by the utilization of tuning components with parasitic resistance, especially at the resonant frequency. This could be mitigated by delicately optimizing the structure and using decent components with a smaller resistance. Secondly, because the performance of a real diode declines as the frequency increases, a proper type should be carefully chosen considering the frequency spectrum of interest. Thirdly, since all the meta-atoms are supposed to be controlled independently, the biasing network would become dramatically complex as the array scales up. One possible solution is to distribute the biasing wires on extra layers, yet increasing the complexity and cost of the fabrication. The utilization of wireless controlling methods, such as light signals, could also be the alternative solution [27].

In the following sections, several typical programmable metamaterials are presented, including time-domain metamaterial, space–time metamaterial, and nonlinear metamaterial, accounting for applications such as harmonic controls, nonreciprocity, and nonlinearity controls. Besides, several new wireless communication systems based on programmable metamaterials have been introduced. Furthermore, a digitally reconfigurable holography imaging technique is discussed.

11.3.2 Time-Domain Metamaterials

The generation of harmonics has been attracting wide attention for years. In optics, it relies on the interactions between the high-intensity laser and nonlinear materials. In the microwave region, expensive and complex components such as amplifiers and phase shifters are required for accurate controls of amplitudes and phases of harmonics. The programmable metamaterial offers a novel route towards the goal by taking advantage of its reconfigurability in the time domain by periodically switching the phase responses of the meta-atoms. In this case, the metamaterial is also called the time-domain programmable metamaterial.

11.3.2.1 Independent Control of Harmonic Amplitudes and Phases

Before the introduction of intriguing physical phenomena, it is necessary to discuss the mechanics behind harmonic generation in a mathematical way [36]. Considering an example in which EM waves illuminate a reflective time-domain programmable metasurface. The temporal expressions of the incident wave, reflected wave, and reflection coefficient are $E_i(t)$, $E_r(t)$, and $\Gamma(t)$, respectively, and their relationship is $E_r(t) = \Gamma(t) * E_i(t)$. Assuming $E_i(t) = e^{-j\omega_c t}$, where ω_c is the angular frequency, the reflected wave in the frequency domain is

$$E_r(\omega) = \Gamma(\omega) * [\delta(\omega - \omega_c)] = \Gamma(\omega - \omega_c) \tag{11.19}$$

Here $\delta(\omega - \omega_c)$ is the Dirac delta function with angular frequency ω_c. If the reflection coefficient $\Gamma(t)$ is time-invariant, the reflection only contains the frequency of the incident wave ω_c; while if $\Gamma(t)$ is a periodic signal with a period T, it can be decomposed into the superposition of numerous harmonically related complex exponentials

$$\Gamma(t) = \sum_{k=-\infty}^{+\infty} a_k e^{jk\omega_0 t} \tag{11.20}$$

where $\omega_0 = 2\pi/T$. In the frequency domain, the reflection can be expressed as

$$E_r(\omega) = 2\pi \sum_{k=-\infty}^{+\infty} a_k E_i(\omega - k\omega_0) \tag{11.21}$$

where a_k is the coefficient of the kth harmonic component. Equation (11.21) implies that a series of new spectrums are generated because of the periodic change of the reflection coefficient. The frequencies of the harmonics are $k\omega_0$, and k is a non-zero integer. Here the amplitude of the coefficient A is constant, and the two states of the reflection phase are ϕ_1 and ϕ_2. The phase response is described by the periodic square wave

$$\Gamma(t) = Ae^{j\left\{\phi_1 + (\phi_2 - \phi_1)\sum_{n=-\infty}^{+\infty}[\varepsilon(t-nT) - \varepsilon(t-\frac{T}{2}-nT)]\right\}} \tag{11.22}$$

Here $\varepsilon(t - nT)$ is the unit step function shifted by nT. The harmonic coefficient a_k can be calculated from Eqs. (11.20)–(11.22)

$$a_k = \begin{cases} A\cos\dfrac{\phi_2 - \phi_1}{2} e^{j\frac{\phi_2+\phi_1}{2}}, k = 0 \\ \dfrac{2A}{k\pi}\sin\dfrac{\phi_2 - \phi_1}{2} e^{j\frac{\phi_2+\phi_1}{2}}, k = \pm 1, \pm 3, \pm 5... \\ 0, k = \pm 2, \pm 4 \pm 6... \end{cases} \tag{11.23}$$

The Fourier transform of Eq. (11.23) yields

$$\begin{aligned} E_r(\omega) &= 2\pi A\cos\frac{\phi_2 - \phi_1}{2} e^{j\frac{\phi_2+\phi_1}{2}} E_i(\omega) \\ &+ \sum_{m=-\infty}^{+\infty} \frac{4A}{2m-1} \sin\frac{\phi_2 - \phi_1}{2} e^{j\frac{\phi_2+\phi_1}{2}} E_i[\omega - (2m-1)\omega_0] \end{aligned} \tag{11.24}$$

Equations (11.23) and (11.24) indicate that only synchronous component ($k = 0$) and odd harmonics ($k = \pm 1, \pm 3, \pm 5...$) exist in the reflected waves, and the amplitudes and phases of these components are related to ϕ_1 and ϕ_2. As examples, Fig. 11.10a gives the measured reflection spectra under the conditions where ϕ_1/ϕ_2 = 0°/180°, 90°/250°, and 180°/270° (obtained by implying different reverse biasing voltages on the varactors of meta-atoms, which will be described later). It can be seen that while the phase difference $\Delta\phi = \phi_1 - \phi_2$ approaches 180°, the intensity

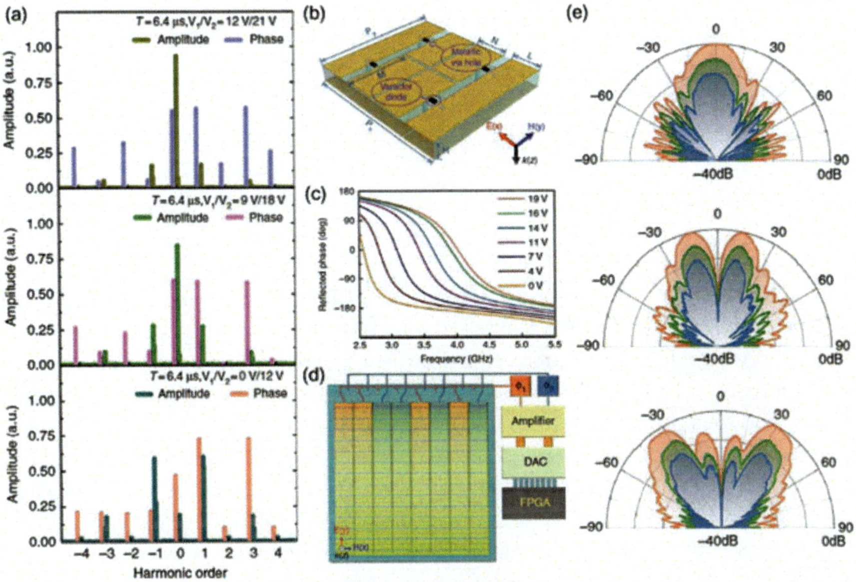

Fig. 11.10 **a** Measured harmonic intensities/phases distributions of the time-domain programmable metasurface at 3.7 GHz, with the modulation period $T = 6.4\,\mu$s and different biasing voltages V_1/V_2. **b** 3D view of the meta-atom. **c** Simulated reflection phase curves of the meta-atom as a function of the biasing voltage. **d** System of the time-domain programmable metasurface. **e** Measured E-plane scattering patterns of the +1st order harmonic under different coding sequences of the metasurface, '00000000', '00001111' and '00110011', respectively. The red, green, and blue lines, respectively demonstrate the variance of the scattering magnitude due to the different voltage pairs A1, A2, and A3 [36]

of the synchronous component is reduced, and those of the harmonics are enhanced. When the phase difference is 180°, the intensity of the synchronous component is almost zero, and the harmonics of ±1st order reach 0.6366, accounting for 81.05% of the EM energy.

On the other hand, it is found in Eqs. (11.23), (11.24), and Fig. 11.10a that the phases of the harmonics are also dependent on ϕ_1 and ϕ_2, which means that their amplitudes and phases are still coupled. To decouple the amplitude and phase, an extra time delay t_0 in the time-varying reflection coefficient is proposed, through which the harmonic phase can be tuned without impacting the amplitude. t_0 in the time domain brings about a phase shift $e^{-jk\omega_0 t_0}$ to the kth order harmonic of $\Gamma(t - t_0)$ in the frequency domain. Equation (11.24) can be rewritten as

$$E_r(\omega) = 2\pi A \cos\frac{\phi_2 - \phi_1}{2} e^{j\frac{\phi_2+\phi_1}{2}} E_i(\omega) + \sum_{m=-\infty}^{+\infty} \frac{4A}{2m-1} \sin\frac{\phi_2 - \phi_1}{2} e^{j[\frac{\phi_2+\phi_1}{2} - (2m-1)\omega_0 t_0]} E_i[\omega - (2m-1)\omega_0] \tag{11.25}$$

The amplitude and phase of the kth order harmonic are rewritten as

$$E_r(k\omega_0+\omega_c)=|A|\angle\Phi$$
$$=\begin{cases}\left|A\cos\frac{\phi_2-\phi_1}{2}\right|e^{j\left\{\frac{\phi_2+\phi_1}{2}+\pi\left[\varepsilon\left(\frac{\phi_2-\phi_1-\pi}{2}\right)+\varepsilon\left(\frac{\phi_1-\phi_2-\pi}{2}\right)\right]\right\}}, & k=0\\ \left|\frac{2A}{k\pi}\sin\frac{\phi_2-\phi_1}{2}\right|e^{j\left\{\frac{\phi_2+\phi_1}{2}-\pi[\varepsilon(\phi_2-\phi_1)+\varepsilon(k)]-k\omega_0 t_0\right\}}, & k=\pm1,\pm3,\pm5\ldots\\ 0, & k=\pm2,\pm4,\pm6\ldots\end{cases} \tag{11.26}$$

In practice, after a proper choice of ϕ_1 and ϕ_2 for the harmonic amplitude, one can further adjust the phase by picking a time delay t_0.

The above mathematical derivations are implemented by a real time-domain programmable metasurface. The meta-atom configuration shown in Fig. 11.10b is utilized. Besides the metallic patches on the top and bottom layers, the key of the structure lies in the four varactor diodes that locate across the side rectangular patches and central rectangular patches on the top layer. The patches on the top and bottom are connected by metallic via holes. By applying DC biasing voltage between the patches on the bottom layer, the capacitance of the varactors is controlled, and thus the reflection phase of the meta-atom is tuned. Different from the meta-atoms loaded with PIN diodes, which have only two operating states, the phase response of this meta-atom is continuous as a function of the biasing voltage. The varactor SMV 2019 is chosen in this work, which is modeled as a serial RLC circuit in the full-wave simulations using CST. The simulated reflection phase curves of the meta-atom as a function of the biasing voltage is given in Fig. 11.10c. While the voltage varies from 19 to 0 V, the resonance of the meta-atom.moves downward from about 4 GHz to 2.6 GHz. At the interested frequency of 3.7 GHz, the phase changes linearly and covers a range of about 270°. In the following experiments, a series of biasing voltages are chosen, i.e., 0, 3, 6, 9, 12, 15, 18, and 21 V, which, in principle, correspond to the phase values 0°, 10°, 30°, 90°, 180°, 210°, 250°, and 270°, respectively. The sketch of the metasurface is presented in Fig. 11.10d, which consists of 7 × 8 meta-atoms. Each column is controlled by the same biasing signal. An FPGA, a digital-analog conversion module, and an analog amplifier module are utilized to provide the DC biasing signals for the varactors. Square wave controlling signals are generated with different voltages, periods, and time delays. The operation frequency is $f_0 = 3.7$ GHz.

The first experiment is conducted to demonstrate the nonlinear generation capability of the time-domain programmable metasurface. Three pairs of biasing voltages, i.e., $V_1/V_2 = 0$ V/12 V (A_1), 9 V/18 V (A_2), and 12 V/21 V (A_3), are selected for the phase change in the time domain, theoretically corresponding to phases $\phi_1/\phi_2 = 0°$/180°, 90°/250°, 180°/270°, respectively. The measured amplitudes and phases of the synchronous and harmonic components with the modulation period $T = 6.4$ μs are plotted in Fig. 11.10a. Although not quite consistent with the theoretical predictions, the generation of the harmonics is obvious. While the phase difference is approaching 180°, the suppression of the synchronous component and the enhancement of the harmonics are clearly in good agreement with the theory. Additionally, as the period of the square wave decreases from 6.4 to 1.6 μs, the frequency gap between the adjacent harmonics is increased from 156.25 to 625 kHz, with the intensities unchanged during the process.

Basically, there are two ways to demonstrate the control of the harmonic phase using the time delay. The direct method is to examine the phase through the fast Fourier transformation of the echo signal from the metasurface, as shown in Fig. 11.10a. The other one is to check the scattering phenomenon of the metasurface. If the phase gradient for the harmonic is achieved on the surface, anomalous beam manipulations can be observed, or the ordinary reflection would be obtained. As an example in the experiment, the scattering property of the +1st order harmonic phase is measured with voltage pair of the switching function A_1 (0 V/12 V, corresponding to $\Delta\phi = 180°$) and modulation period $T = 6.4$ μs (corresponding to 3.70015625 GHz for +1st order harmonic). By choosing the time shift t_0 = T/2 = 3.2 μs, a phase difference of 180° can be imposed on the harmonic compared with the case with $t_0 = 0$. Therefore, the digital element "0" is defined with $t_0 = 0$, and element "1" is defined with $t_0 = T/2 = 3.2$ μs for the harmonic. With the coding sequences "00000000", "00001111", and "00110011", the measured scattering far-field patterns are plotted in Fig. 11.10e, from which the beam splitting phenomena are clearly obtained from the latter two sequences. The elevation angles of the beams are in accordance with the theoretical predictions, thus demonstrating the phase gradients provided by the time shifts.

On the other hand, the control of the harmonic amplitude can be checked by varying the voltage combination, which is also illustrated in Fig. 11.10e as the voltages change from A_1 to A_3. It can be seen that the decay rate of the beams increases from 0 to 10 dB, while the directions of the beams remain almost unaffected. It should be mentioned that, not just the +1st order, but all the harmonics are dependent simultaneously on the voltage pair and time delay. Furthermore, by adopting a denser division of the time delay within a period, multiple-bit coding meta-atoms can be realized for more flexible controls of the harmonics. Taking +1st order harmonic, for instance, the time delay t_0 of 0 (0 μs), $T/8$ (0.8 μs), $T/4$ (1.6 μs), 3 $T/8$ (2.4 μs), $T/2$ (3.2 μs), 5 $T/8$ (4 μs), 3 $T/4$ (4.8 μs), 7 $T/8$ (5.6 μs) lead to the phase of 0, π/4, π/2, 3π/4, π, 5π/4, 3π/2, 7π/4, respectively, which can be encoded as elements "0", "1", "2", "3", "4", "5", "6", and "7" in the 3-bit coding metasurface. In this way, more precise manipulation of the harmonic can be realized.

11.3.2.2 Phase Modulation

A more general mathematical derivation for the time-domain programmable metamaterial that controls the harmonics is presented by Zhao et al. [37]. Again, starting from the wave $E_i(t)$ that impinges on a metasurface with the reflectivity $\Gamma(t)$, the reflected wave $E_r(t) = \Gamma(t) * E_i(t)$, whose Fourier transform can be written as

$$E_r(f) = \frac{1}{2\pi} E_i(f) * \Gamma(f) \tag{11.27}$$

The reflectivity is defined to be a periodic function of time that combines scaled and shifted pulses over a period

$$\Gamma(t) = \sum_{m=0}^{L-1} \Gamma_m g(t - m\tau), \qquad (0 < |t| < T) \tag{11.28}$$

where $g(t)$ is a period pulse function with period T, L is a positive integer larger than zero, $\tau = T/L$ is the pulse width, and Γ_m is the reflectivity over the interval $(m-1)\tau < t < m\tau$. In each period,

$$g(t) = \begin{cases} 1, & 0 < t < \tau \\ 0, & \text{otherwise} \end{cases} \tag{11.29}$$

It can be unfolded into Fourier series as

$$g(t) = \sum_{k=-\infty}^{+\infty} c_k e^{jk2\pi t/T} = \sum_{k=-\infty}^{+\infty} c_k e^{jk2\pi f_0 t} \tag{11.30}$$

This means that the pulse function can be decomposed into infinite numbers of harmonics, in the frequency domain, with the harmonic frequencies kf_0, where k is an integer indicating the harmonic order. Therefore,

$$\begin{aligned} \Gamma(t) &= \sum_{m=0}^{L-1} \Gamma_m g(t - m\tau) = \sum_{m=0}^{L-1} \Gamma_m \sum_{k=-\infty}^{+\infty} c_k e^{-jk2\pi f_0 t} \\ &= \sum_{k=-\infty}^{+\infty} c_k \cdot \left(\sum_{m=0}^{L-1} |\Gamma_m| e^{-j\frac{2km\pi}{L}}\right) e^{-jk2\pi f_0 t} \end{aligned} \tag{11.31}$$

The Fourier series coefficients c_k are calculated by

$$\begin{aligned} c_k &= \frac{1}{T}\int_0^T g(t) e^{-j2\pi kt/T} dt \\ &= \sum_{m=0}^{L-1} \frac{1}{T} \int_{m\tau}^{(m+1)\tau} e^{-j2\pi kt/T} dt \\ &= \sum_{m=0}^{L-1} \int_{m\tau/T}^{(m+1)\tau/T} e^{-j2\pi kt/T} d\frac{t}{T} \\ &= \sum_{m=0}^{L-1} \int_{m/L}^{(m+1)/L} e^{-j2\pi kt} dt \end{aligned}$$

$$= \sum_{m=0}^{L-1} \frac{1}{-j2\pi k} \cdot (-2j \sin \frac{k\pi}{L}) \cdot e^{-jk\pi \frac{2m+1}{L}}$$

$$= \sum_{m=0}^{L-1} \frac{1}{L} \sin c(\frac{k\pi}{L}) \cdot e^{-jk\pi \frac{2m+1}{L}} \tag{11.32}$$

Therefore, the Fourier transform coefficient of the reflection a_k in Eq. (11.20) is

$$a_k = \sum_{m=0}^{L-1} \frac{\Gamma_m}{L} \sin c(\frac{k\pi}{L}) \cdot e^{-jk\pi \frac{2m+1}{L}}$$

$$= \sum_{m=0}^{L-1} \Gamma_m e^{\frac{-j2k\pi m}{L}} \cdot \frac{1}{L} \sin c(\frac{k\pi}{L}) e^{\frac{-jk\pi}{L}} = TF \cdot UF \tag{11.33}$$

where

$$TF = \sum_{m=0}^{L-1} \Gamma_m e^{\frac{-j2k\pi m}{L}} \tag{11.34}$$

$$UF = \frac{1}{L} \sin c(\frac{k\pi}{L}) e^{\frac{-jk\pi}{L}} \tag{11.35}$$

Here a_k indicates the normalized amplitude and phase of the kth harmonic $f_c \pm kf_0$. Equation (11.33) implies that a_k can be regarded as the product of two parts, i.e., the time factor *TF* and the unit factor *UF*. The *TF* is related to the time-coding strategy with the modulated reflectivity within a period, and the *UF* is the Fourier series coefficient of the basic pulse with pulse width τ. Therefore, the amplitude and phase of the harmonics can be controlled by the modulation of the reflectivity in the time domain, including amplitude modulation (AM) and phase modulation (PM). Compared with PM, AM is limited by the following three shortcomings. Firstly, the variation of the reflection magnitude means that the metasurface dissipates EM energy during a certain interval in a period, thus lowering the efficiency. Secondly, the 0th harmonic, i.e., synchronous component, cannot be suppressed totally because it cannot be canceled by the reflectivity with a constant phase. Thirdly, harmonics generated by AM are symmetrically distributed in the frequency spectrum with respect to the central frequency; in other words, + kth and −kth harmonics have equal amplitude. This may lead to a waste of energy if one of them is unwanted in a communication application.

Because PM is implemented by varying the phase state of the reflectivity in the time domain, with unity amplitude, the above limitations of AM can be avoided. As examples, several time-coding sequences are adopted to demonstrate the concept and the resulting harmonic amplitude distributions are presented in Fig. 11.11. In the 1-bit coding sequences, digital 0 means that the reflection phase during a time interval is 0°, and digital 1 represents the phase 180°. Similarly, in the 2-bit coding sequences, digital 00, 01, 10, and 11 indicate the reflection phases of 0°, 90°, 180°, and 270°, respectively, during a time interval. The first sequence is "010101..." with the period $T = 1$ μs and interval number $L = 2$ (Fig. 11.11a). The corresponding harmonic intensity distribution is shown in Fig. 11.11b. Because of the opposite phases of the reflectivity during a period, the synchronous incident component, as well as the even-order harmonics, is totally canceled, and a large portion of the energy is transferred to ± 1th harmonics. When the coding sequence is 2-bit "00-01-00-01..." with $T = 1$ μs and $L = 2$ (Fig. 11.11c), the harmonic intensities are obtained as shown in Fig. 11.11d. Different from the first example, the synchronous component remains because the phases are not exactly opposite to cancel it. Moreover, one may also observe that the harmonics caused by the 1-bit sequences are symmetrically distributed in the frequency spectrum. But if the time-coding sequences "00-01-10-11" (Fig. 11.11e) and "11-10-01-00" (Fig. 11.11g) are employed, asymmetric harmonics shown in Fig. 11.11f and h can be obtained. It is clear that most of the energy is converted to +1th and −1th harmonics, respectively, and the synchronous component is totally suppressed. In fact, if a 3-bit sequence with a smoother phase gradient in the time domain is adopted, the harmonic conversion efficiency will be as high as 95% [38].

11.3.3 Space–Time Metamaterials and Metasurfaces

11.3.3.1 Harmonic Control

In the last work, all the meta-atoms on the metasurface behave identically, controlled by the same time-domain coding sequence. Although the EM energy is redistributed as harmonics in the frequency spectrum, they are not distinguished in the space domain, i.e., all the harmonics are reflected normally by the metasurface. Under this context, Zhang et al. propose a space-time-coding digital metasurface [39], every single meta-atom of which possesses a reflection coefficient varying independently with its unique time-coding sequence. Amplitude and phase arrangements on the metasurface at harmonic frequencies are highly reconfigurable, thus enabling flexible manipulations of the EM wave in both frequency and space domains.

Assuming that the number and dimension of meta-atoms along the x-direction are N and d_x, respectively, and those along the y-direction are M and d_y, respectively, the time-domain far-field pattern scattered by the space-time-coding digital metasurface can be written as follows based on Eq. (11.1)

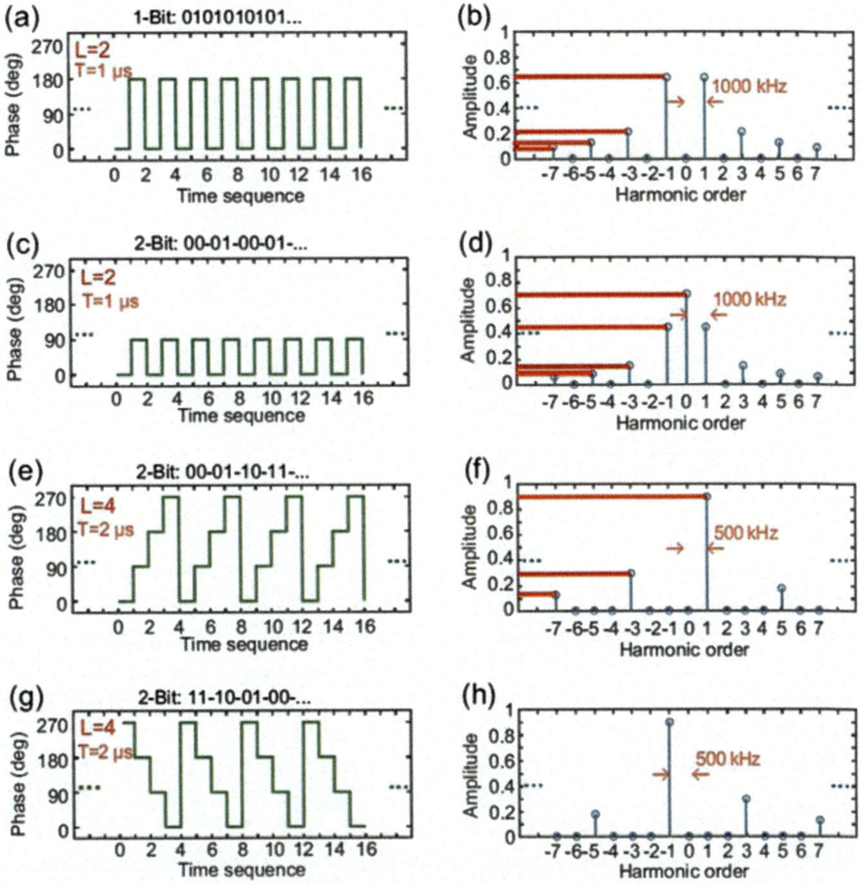

Fig. 11.11 **a** and **b** 1-bit PM coding 01010101... with $L = 2$ and $T = 1$ μs. **c** and **d** 2-bit PM coding 00-01-00-01-... with $L = 2$ and $T = 1$ μs. **e** and **f** 2-bit PM coding 00-01-10-11-... with $L = 4$ and $T = 2$ μs. **g** and **h** 2-bit PM coding 11-10-01-00-... with $L = 4$ and $T = 2$ μs [37]

$$f(\theta, \varphi, t) = \sum_{q=1}^{N} \sum_{p=1}^{M} E_{pq}(\theta, \varphi) \Gamma_{pq}(t)$$
$$\exp\{-jk_c[d_x(p-1)\sin\theta\cos\varphi + d_y(q-1)\sin\theta\sin\varphi]\} \quad (11.36)$$

where $E_{pq}(\theta, \varphi)$ is the far-field pattern function, at the central frequency f_c, of the coding meta-atom located at the position (p, q). k_c=$2\pi/\lambda_c$ is the wave number at f_c. $\Gamma_{pq}(t)$ is the time-modulated reflection coefficient of the (p, q)th meta-atom, which is a periodic function of time

$$\Gamma_{pq}(t) = \sum_{m=0}^{L-1} \Gamma_{pq}^{m} g(t - m\tau), \quad (0 < |t| < T) \tag{11.37}$$

Similar to Eq. (11.33), the Fourier series coefficient of $\Gamma_{pq}(t)$ can be expressed as

$$a_{pq}^{k} = \sum_{m=0}^{L-1} \frac{\Gamma_{pq}^{m}}{L} \sin c\left(\frac{k\pi}{L}\right) \cdot e^{-jk\pi \frac{2m+1}{L}} \tag{11.38}$$

Here k is the order of the harmonics, L is the number of intervals during a period. Therefore, the far-field scattering pattern of the kth harmonic is written as

$$F_k(\theta, \varphi) = \sum_{q=1}^{N} \sum_{p=1}^{M} E_{pq}(\theta, \varphi)$$
$$\exp\{-jk_c[d_x(p-1)\sin\theta\cos\varphi + d_y(q-1)\sin\theta\sin\varphi]\} a_{pq}^{k} \tag{11.39}$$

Here PM is adopted with unity reflection amplitude. A 3D space-time-coding matrix with dimensions (8, 8, 8) is considered, which means that the metasurface consists of 8×8 meta-atoms, and there are 8-time intervals during a period for each meta-atom. The reflection phase of each meta-atom is encoded by a 1-bit coding strategy, i.e., $0°$ and $180°$ are respectively encoded as digits "0" and "1".

If the time-coding sequence for each meta-atom is precisely designed, the scattering performances of the harmonics can be realized. Take the sequence shown in Fig. 11.12a and b, for instance. Under this sequence, each column of elements in the x-direction has the same digital code, and the phases of each column of elements in the y-direction are delayed by a time interval successively. In this way, phase gradients are formed at each harmonic frequency on the surface, as plotted in Fig. 11.12c. In fact, this strategy shares the same idea with the harmonic phase control using a time delay [36]. For a time delay of t_q here, the phase change for kth harmonic is $2\pi k f_0 t_q$. Observing the phase gradient pattern of the harmonics, it is found that the phase variation among the eight elements in the y-direction becomes increasingly dramatic as the order of harmonics increases. Based on Eq. (11.4), it is reasonable to see that the deflected angle increases as the harmonic changes from 0th through − 3rd (+3rd), as displayed in Fig. 11.12d. However, due to the unbalance of the two digits within a period, a large portion of incident energy still remains synchronous, as illustrated in Fig. 11.12c. To solve this problem, a binary meta-atom swarm optimization is utilized for the time-coding sequences, and the power levels of the harmonics are uniform.

Besides the harmonic beam-deflection phenomena, the space–time metasurface can also be utilized for building a multi-bit programmable strategy without a complex biasing controlling system. From Fig. 11.12c, it is observed that the phase of 0th harmonic, i.e., the synchronous signal, remains constant, which accounts for the scattering pattern pointing to the broadside shown in Fig. 11.12d. Here assuming k in Eq. (11.38) is 0, the argument of a_{pq}^{0} is

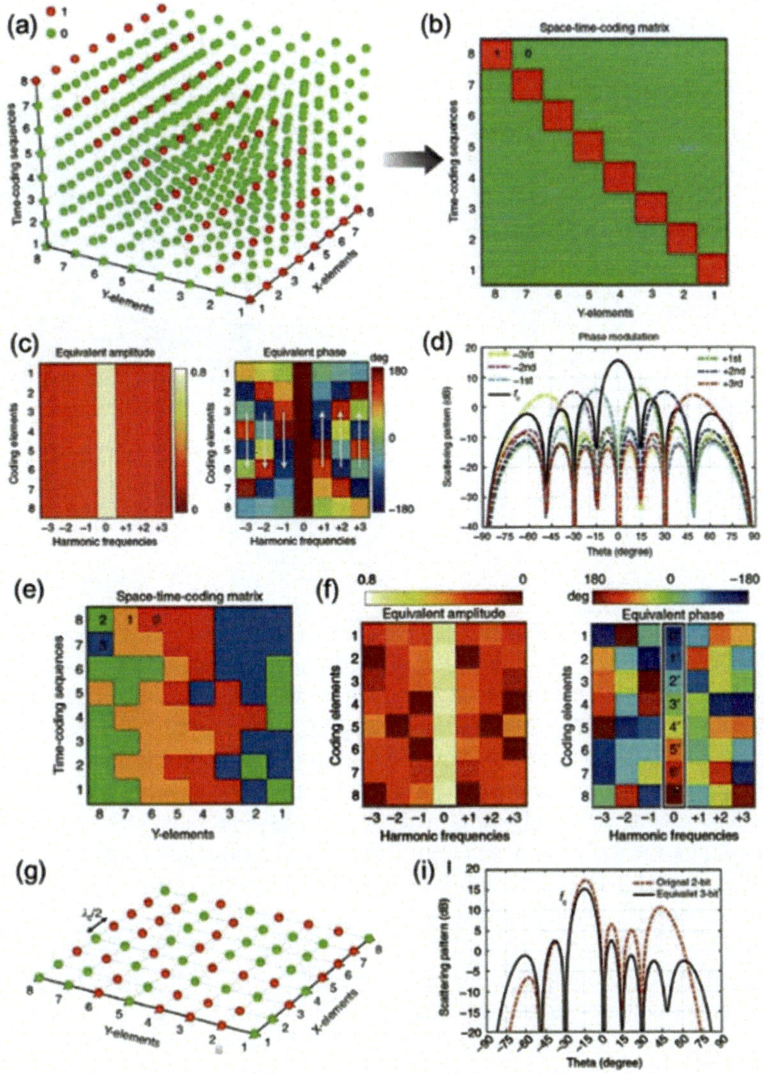

Fig. 11.12 **a** and **b** The 1-bit time-coding matrix used for the space-time-coding metasurface. **c** Equivalent magnitudes and phase gradients formed at each harmonic frequency on the metasurface with the 1-bit matrix. **d** The deflected angle of each harmonic. **e** The 2-bit time-coding matrix is used for generating 3-bit coding distribution. **f** Equivalent magnitudes and phase gradients formed at each harmonic frequency on the metasurface with the 2-bit matrix. **g** The BPSO-optimized space-coding sequence for the low-RCS property. **h** The BPSO-optimized space-coding and random time-coding sequence for the low-RCS property. **i** The comparison between the deflected patterns generated by directly a 2-bit space-coding sequence and an equivalent 3-bit equivalent space coding sequence. **j** The 2D far-field low-RCS pattern of +1st harmonic [39]

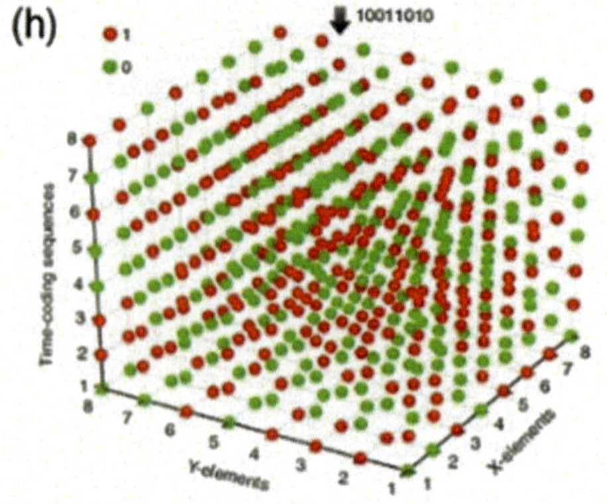

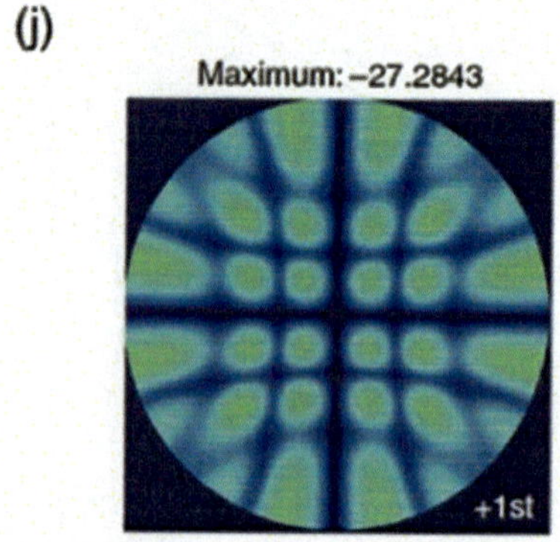

Fig. 11.12 (continued)

$$\varphi_{pq}^{0} = \arg\left(\sum_{m=0}^{L-1} \frac{\Gamma_{pq}^{m}}{L}\right) \tag{11.40}$$

It is apparent that, for the phase of synchronous component, only 0°and 180° are provided by a 1-bit time-coding strategy; in contrast, a much larger range covering 360° is obtained by a 2-bit time-coding sequence, which is promising to build a 3-bit equivalent space coding distribution on the surface at the central frequency. This principle provides a new approach towards multi-bit programmable metasurface designs with more precise manipulations of EM waves yet relatively simpler biasing circuitry. In an illustrative example, the array with 8×8 elements is employed. The eight columns along the *y*-direction are controlled by 2-bit space-time-coding sequences, which are created by using an in-house MATLAB code based on Eq. (11.38). The sequences are illustrated in Fig. 11.12e, where the red, yellow, green, and blue dots represent 0, 1, 2, 3 digits (0°, 90°, 180°, 270°), respectively. Observing the eight sets of the equivalent amplitudes and phases shown in Fig. 11.12f, it is clear that most of the EM energy is conserved at the central frequency, and their phase gradient from −180° to 180° results in 3-bit coding "0" (−135°), "1" (−90°), "2" (−45°), "3" (0°), "4" (45°), "5" (90°), "6" (135°), and "7" (180°). According to the generalized Snell's law, the phase gradient on the surface offers a deflected angle of 14.5°, as shown in Fig. 11.12i. Although the same deflected beam can also be achieved by a 2-bit space-domain sequence "00112233", it is clear that the equivalent 3-bit case exhibits a much lower sidelobe at the cost of a little gain reduction, which is attributed to the loss due to harmonics.

The space–time metasurface is also useful in the RCS reduction by redistributing the EM energy in both space and frequency domains, which is more advanced than the conventional low-RCS coding metasurface that scatters the wave only in the space domain. The idea starts from a normal space-coding metasurface with a binary particle swarm optimization (BPSO)-optimized sequence shown in Fig. 11.12g, which, as expected, redirects the incident wave into numerous orientations and yields a scattering pattern with much lower backscattered power compared with a normal metallic plate. Then, a random time-coding sequence "10011010" is applied to this space-coding sequence, i.e., the time digit "1" means that the space-coding sequence during the time step keeps unchanged, and the time digit "0" means that every single digit of the space-coding sequence during the time step becomes the opposite state, as illustrated in Fig. 11.12h. This operation makes the EM energy converted to all the harmonic bands, and at each harmonic frequency, the energy is distributed uniformly in all directions. For example, Fig. 11.12j displays the 2D far-field scattering pattern of +1st harmonic.

11.3.3.2 Nonreciprocity

In addition to the harmonic manipulation, the space–time metasurface also offers a practical solution to nonreciprocity. According to the time-reversal symmetry, the properties of a normal system should remain the same if the time variable is flipped. For an EM device, this means that the responses experienced by EM waves are identical when the source and detector are exchanged. This is equivalent to "reciprocity." In some scenarios, the reciprocity is required to be broken so that the input–output properties are asymmetric. Conventional nonreciprocal devices, e.g., isolators and circulators, rely on gyrotropic materials combined with a static magnetic field, which are bulky, heavy, and lossy, so it is difficult to fit them in the metasurface. This problem urges investigations on magnetless technology for nonreciprocity. Currently, the methodologies can be divided into three main categories, i.e., linear time-invariant method, linear time-variant method, and nonlinear method [40]. The time-variant nonreciprocity based on time-modulated devices has been demonstrated in the designs of circulators and antennas [41, 42]. Here, the space–time metasurface breaks the reciprocity in both space and frequency domains, which includes only a programmable metasurface and an FPGA [43].

Figure 11.13a shows that a wave with an elevation angle θ_1 and frequency f_1 is anomalously deflected by the metasurface with an elevation angle θ_2 and frequency f_2. In a time-reversal scenario where the wave at frequency f_2 reversely impinges on the metasurface, the deflected wave is supposed to be at frequency f_1 and elevation angle θ_1 if the metasurface is reciprocal. However, owing to the nonreciprocal property of the metasurface controlled by the space–time-coding signal, the deflected angle in the time-reversal case is $\theta_3 \neq \theta_1$ at the frequency $f_3 \neq f_1$. In this illustrative case, a reflective space–time metasurface with N columns of meta-atoms is considered. The dimension of each meta-atom is d. The reflection coefficients of the meta-atoms in each column are modulated uniformly by a periodic function with time step L. The reflection coefficient of the pth elements $\Gamma_p(t) = \sum_{m=1,2,\ldots,L} \Gamma_p^m g_p^n(t)$, where $g_p^n(t)$ is the impulse function with period T_0. Assuming a transverse-magnetic polarized plane wave that impinges on the metasurface with an elevation angle θ_i and azimuthal angle $\varphi = 0°$, the time-domain far-field scattering pattern can be described from Eq. (11.36) as

$$f(\theta,t)=\sum_{p=1}^{N}E_p(\theta)\Gamma_p(t)\exp\left\{-ik_c[d(p-1)(\sin\theta+\theta_i)]\right\} \tag{11.41}$$

where k_c=$2\pi/\lambda_c$ is the wave number at f_c, $E_p(\theta)=\cos\theta$ is the scattering pattern of the pth coding element at f_c. Therefore, the far-field scattering pattern of the kth harmonic is written as

$$F_k(\theta)=\sum_{p=1}^{N}E_p(\theta)\exp\left[-j2\pi d(p-1)(\frac{\sin\theta}{\lambda_r}+\frac{\sin\theta_i}{\lambda_c})\right]a_p^k \tag{11.42}$$

where $\lambda_r=c/(f_c+mf_0)$ is the wavelength of kth harmonic, and the Fourier series coefficient of $\Gamma_p(t)$

$$a_p^k=\sum_{m=0}^{L-1}\frac{\Gamma_p^m}{L}\sin\mathrm{c}(\frac{k\pi}{L})\cdot e^{-jk\pi\frac{2m+1}{L}} \tag{11.43}$$

Here $N=16$, $L=4$, and the adopted 2-bit space-time-coding matrix is sketched in Fig. 11.13b. Along the time axis, the reflection phase of each meta-atom progressively shifted by 90° in the four-time intervals within a period. Along the space axis, the phase shift between the neighboring meta-atoms is also 90°. The spatiotemporal phase gradient distribution is clearly formed that is able to convert a large portion of EM energy into the +1st harmonic frequency f_c+f_0, as shown in Fig. 11.13c. At the kth harmonic frequency f_c+kf_0, corresponding to the wave number $k_c+k\Delta k_0=2\pi(f_c+kf_0)/c$, the phase difference of neighboring meta-atoms is calculated as

$$\Delta\psi_k=-2\pi kf_0\cdot\frac{T_0}{4}=-\frac{k\pi}{2} \tag{11.44}$$

Therefore, the phase gradient of +1st harmonic is

$$\frac{\partial\psi}{\partial x}=\frac{\Delta\psi_1}{d}=-\frac{\pi}{2d} \tag{11.45}$$

This can be observed in Fig. 11.13d. Assuming the frequency of the incident wave is f_c, and the incident angle is θ_1, the relation between the incident angle and the +1st reflected angle is θ_2 [44]

$$(k+\Delta k)\sin\theta_2=k\sin\theta_1+\frac{\partial\psi}{\partial x} \tag{11.46}$$

If the metasurface is illuminated by a wave from the other side of the normal vector with the elevation angle θ_2 and frequency f_c+f_0, the dominant reflection occurs at the 1st harmonic with the frequency f_c+2f_0, and the relation between the reflection angle θ_3 and θ_2 is

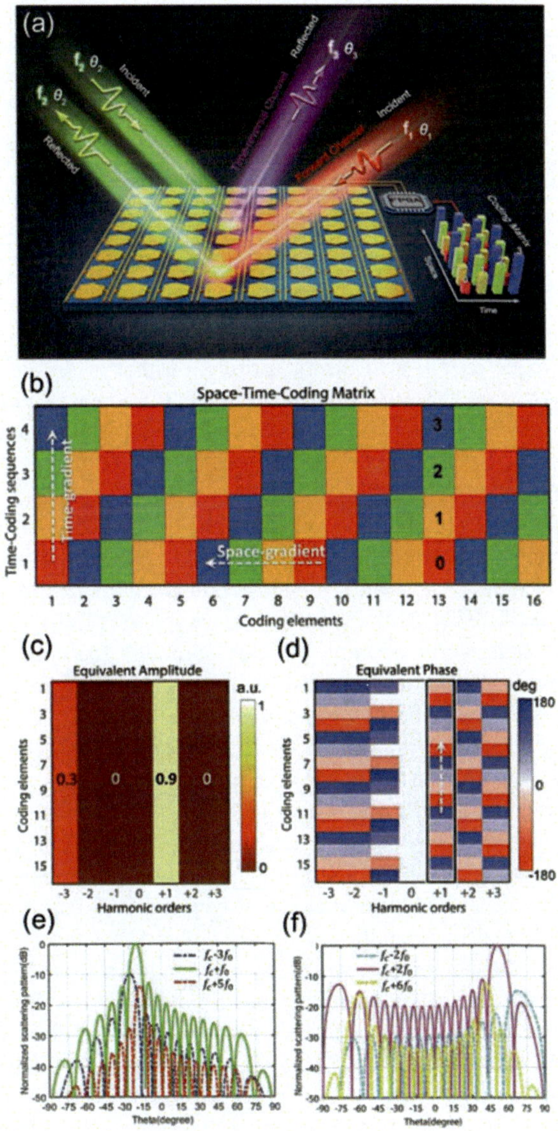

Fig. 11.13 **a** Concept illustration of the nonreciprocity based on the space-time-coding digital metasurface. **b** The 2-bit space-time-coding matrix. **c** Equivalent amplitudes of the harmonics of the coding elements. **d** Equivalent phases of the harmonics of the coding elements. **e** Calculated scattering patterns of each harmonic by the metasurface with $f_c = 5$ GHz, $f_0 = 250$ MHz, $d = \lambda_c/2$, $\theta_1 = 60°$. **f** The scattering patterns of each harmonic in the time-reversal scenario [43]

$$(k + 2\Delta k)\sin\theta_3 = (k + \Delta k)\sin\theta_2 - \frac{\partial\psi}{\partial x} \tag{11.47}$$

From Eqs. (11.46) and (11.47), we have

$$\sin\theta_2 = \frac{k\sin\theta_1 + \frac{\partial\psi}{\partial x}}{k + \Delta k} = \frac{\sin\theta_1 - \frac{\lambda_c}{4d}}{1 + \frac{f_0}{f_c}} \tag{11.48}$$

$$\sin\theta_3 = \frac{k}{k + 2\Delta k}\sin\theta_1 = \frac{\sin\theta_1}{1 + \frac{2f_0}{f_c}} \tag{11.49}$$

A deviation factor that describes the angular separation between the time-reversed reflected and incident waves is defined as

$$\delta = |\sin\theta_3 - \sin\theta_1| = \frac{\sin\theta_1}{1 + \frac{f_c}{2f_0}} \tag{11.50}$$

It means that the reflected wave in this time-reversal scenario occurs at the frequency $f_c + 2f_0$ with a direction that is different from the original incident direction if $\theta_1 \neq 0$. It is apparent that θ_3 increases with the ratio f_0/f_c and θ_1.

Figure 11.13e displays calculated patterns of each harmonic created by the metasurface when $f_c = 5$ GHz, $f_0 = 250$ MHz, $d = \lambda_c/2$, $\theta_1 = 60°$, from which it is observed that the dominant +1st harmonic $f_c + f_0$ is stronger than the others by almost 10 dB, and the reflection angle is 20.3°. In the time-reversal scenario shown in Fig. 11.13f, the +1st harmonic $f_c + 2f_0$ is also dominant, and the reflected angle becomes 51.2°, which are in accordance with the theoretical angles $\theta_2 = 20.40°$ and $\theta_3 = 51.93°$ calculated by Eqs. (11.48) and (11.49).

It is worth pointing out that, although theoretically demonstrated, the nonreciprocal phenomenon is difficult to be verified in practice using the PIN-diode-embedded metasurface hardware. Observing Eq. (11.50), the angular separation δ can be enlarged by increasing the incident angle θ_1 or decreasing the ratio f_c/f_0. Because the modulation frequency f_0 is limited by the switching speed of the PIN diode and the frequency of the controlling module, which is much smaller than f_c, the first option is not effective. On the other hand, lowering f_c brings about the increase of working wavelength, which requires larger metasurface sizes and larger antenna-metasurface distances for planar incident wavefronts and causes a poorer absorbing effect of the anechoic chamber. Therefore, only nonreciprocal phenomena in the frequency domain can be observed easily through the current space-time programmable metasurface integrated with the commercial PIN diodes. For example, when $f_c = 9.5$ GHz, $f_0 = 1.25$ MHz, $d = 14$ mm, and $\theta_1 = 34°$, the dominant reflected beam with $\theta_2 = 0.27°$ at $f_c + f_0 = 9.50125$ GHz is obtained, and the time-reversal reflected beam occurs at $f_c + 2f_0 = 9.5025$ GHz with $\theta_3 = 33.7°$. Apparently, the frequency difference can be easily detected by a high-precision spectrum analyzer, but the angular separation θ_3 -θ_1 is hardly measured.

11.3.4 *Programmable Nonlinear Metamaterials and Metasurfaces*

11.3.4.1 Reconfigurable Nonlinearity

Emerging at the end of the last century, the development of nonlinear metamaterials at the microwave region was inspired by the highly resonant structure, such as the split ring resonator (SRR) [45]. Although a plethora of nonlinear microwave metamaterials has been extensively studied, such as resonance shifting [46, 47], absorption [48–50], nonreciprocity [40, 51, 52], wave mixing [53], harmonics generation [54], self-focusing [55], etc., a long-standing challenge remains on the reconfigurability of these devices, thus hindering the extension of their applications. Under this context, a novel digitally reconfigurable nonlinear metasurface is proposed based on an active, programmable nonlinear mechanism [56]. For one thing, the nonlinearity does not rely on the resonant condition, and the nonlinear threshold can be flexibly adjusted; more importantly, the operating states of the meta-atom, i.e., linear or nonlinear, can be digitally switched, and thus the spatial distribution of the linear and nonlinear meta-atoms is dynamically reconfigured in a programmable manner. In this way, the concept of the programmable metasurface is extended to the nonlinear community. An example is illustrated in Fig. 11.14a, showing the intensity-dependent beam deflections pointing to digitally-defined orientations.

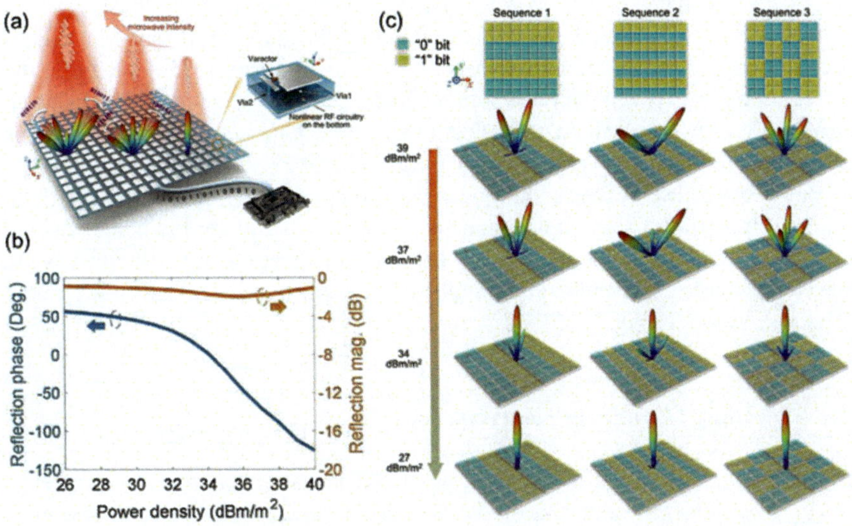

Fig. 11.14 **a** Conceptual illustration of the digitally reconfigurable nonlinear metasurface. **b** The nonlinear property of the meta-atom as a function of the power density. **c** With the power density decreasing from 39 to 27 dBm/m^2, the normalized theoretical far-field scattering patterns on the *uv*-plane at 5.20 GHz, showing the nonlinear beam deflection phenomena [56]

The proposed metasurface consists of the active nonlinear meta-atom sketched in Fig. 11.14a. It is a three-layer PCB structure that consists of the top metallic patches (including the varactor), the reflecting ground on the middle layer, and the circuitry on the bottom layer. Two metallic vias connect the top patches and the bottom circuit. In the nonlinear mode, the circuit senses the incident EM intensity and generates a direct-current (DC) signal back to the varactor on the top. Since the performance of the meta-atom depends on the capacitance of the varactor, it can thus be altered by the variation of the incident intensity. The nonlinear phase response of the meta-atom is displayed in Fig. 11.14b, implying that the reflection phase at 5.20 GHz decreases from 56° to − 125° when the power density on the surface is enhanced. Thanks to the active operating mechanism, the nonlinearity of the meta-atom can be dynamically controlled by applying digital signals to the ENANBLE pin of the circuit. With the digit "1", the meta-atom works in the nonlinear mode with the power-dependent phase feature; with the digit "0", the nonlinearity is deactivated, and the reflection phase remains constant. In this way, the spatial distribution of the nonlinear/linear meta-atoms on the metasurface can be reconfigured for a variety of wave-manipulation effects.

A metasurface with 16 × 16 meta-atoms is built, and 2 × 2 particles are grouped as a super particle. Three commonly used digital sequences Sequence 1 ("00110011" along the *y*-axis), Sequence 2 ("01010101" along the *y*-axis), and Sequence 3 ("00110011" along the *x*- and *y*-axes) are chosen, as shown in Fig. 11.14c. Based on the nonlinear phase data (Fig. 11.14b), the power-dependent far-field scattering patterns are calculated based on Eq. (11.1), as plotted in Fig. 11.14c. Observing each column with a typical sequence, the anomalous deflected beams gradually reduce, and the normal reflection increases when the intensity diminishes. This is because the phase difference between the linear and nonlinear particles decreases. Finally, when the power density is the lowest, the reflections become dominant, and similar patterns for the three sequences are obtained (see the last row of Fig. 11.14c). This is because the phase difference between the linear and nonlinear meta-atoms is negligible, leading to the almost same homogeneous phase distribution on the panel, despite their different nonlinear/linear arrangements. By changing the digital sequence, the deflected patterns under the strong illumination are flexibly altered, as shown in the different columns of Fig. 11.14c.

11.3.4.2 Digital Nonlinear Nonreciprocity

If not particularly designed, a normal metamaterial obeys the time-reversal symmetry, which means that its input and output are reciprocal. In Sect. 11.3.3.2, we have discussed the nonreciprocity achieved by the space-time programmable metasurface, which can be categorized as the time-variant type. Besides the time-variant and time-invariant methods, a nonreciprocal metamaterial can also be realized by nonlinearity [29]. Traditionally, nonlinear nonreciprocal metasurfaces rely on the Fano resonance. Owing to the asymmetric configuration of the metasurface under the resonant condition, forward and backward incident waves would experience different transmission rates at a fixed frequency. This kind of nonreciprocity is characterized by several restrictions, such as narrow bandwidth, fixed threshold, and hysteresis. Here, based on the concept of programmable metamaterial, a digital nonreciprocal solution is proposed, which serves as a promising candidate to overcome the above limitations. Moreover, its transmission property is digitally defined such that it can be customized flexibly for diverse requirements.

The proposed nonlinearity is based on an "analog-digital-analog" mechanism, which is realized by a tunable transmissive surface integrated with two EM detectors and a digital controlling module. The digital module contains an FPGA and two analog–digital–analog (ADA) modules, each of which includes an analog-to-digital (AD) submodule and a digital-to-analog (DA) submodule. The meta-atom adopted by the transmissive surface is plotted in Fig. 11.15g, which contains four metallic layers separated by three substrates and a varactor on the top layer along the x-axis. The transmission rate of the meta-atom is adjustable by the reverse biasing voltage across the varactor, as demonstrated in Fig. 11.15h. The EM detector is composed of a receiving antenna and a detecting circuit, as shown in Fig. 11.15i. To capture the incoming waves from the forward and backward directions, two detectors are placed facing opposite directions, respectively. The digital controlling module is shown in Fig. 11.15j.

The nonlinear process is unfolded into several successive active and digital steps, as displayed in Fig. 11.15f. Forward and backward EM waves are captured by the two detectors, respectively, and their intensities are converted into analog DC signals. The digital signals that carry the intensity and direction information of the EM waves are generated by the AD submodules. Preloaded with a code that describes the required nonreciprocal function, the FPGA reads the input digital signals and generates a controlling digital signal, which is later converted into a controlling DC by a DA submodule. Under the control of the DC signal, the nonreciprocal property is thus realized by the transmissive surface.

Figure 11.15 illustrates five nonreciprocal transmission functions that are exhibited by a single metasurface and dynamically switchable., i.e., EM diode functions with reversible directions (Fig. 11.15a and b), unidirectional limiting function (Fig. 11.15c), and tunable thresholds (Fig. 11.15d and e). In analogy with the electronic diode used in a circuit, the EM diode is a nonreciprocal device that allows the unidirectional propagation of EM waves stronger than a certain threshold. The EM limiter refers to the device that permits EM waves lower than a threshold. These features cannot be integrated into a conventional device, but they can be easily integrated into the proposed metasurface. Moreover, not only the threshold but also the propagation direction can be conveniently set, offering a much more flexible solution for wave manipulation.

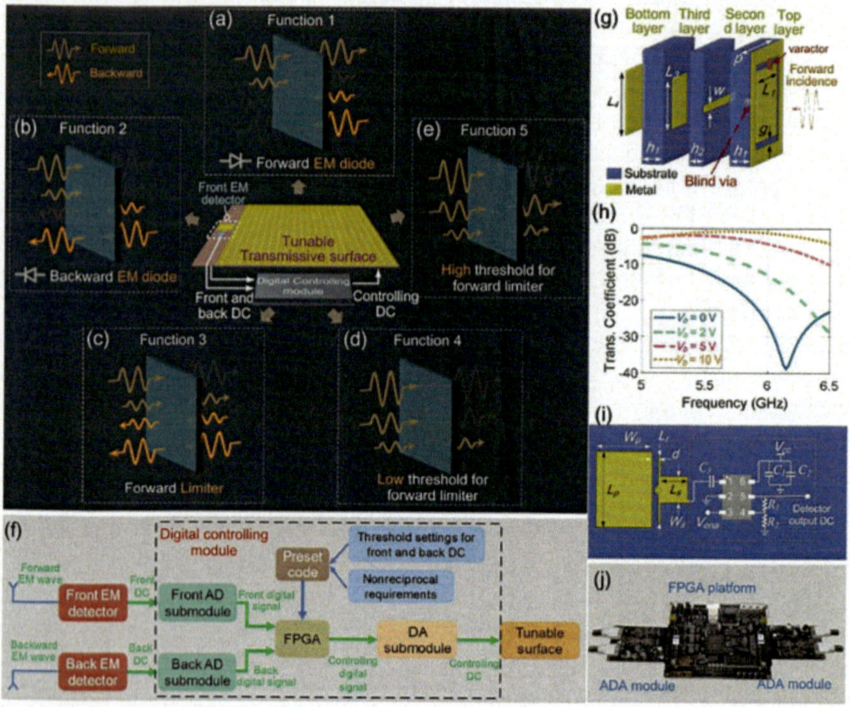

Fig. 11.15 **a–e** The five customized nonreciprocal functions of the nonlinear metasurface. **f** The digital nonreciprocal process of the proposed metasurface. **g** 3D view of the meta-atom of the tunable transmissive surface. **h** Measured transmission phase of the meta-atom as a function of the reverse biasing voltage across the varactor. **i** The configuration of the EM detector. **j** A photograph of the digital controlling module [29]

11.3.5 New Wireless Communication System

In a modern wireless communication system, the transmission rate and bit error rate are determined by digital modulations. The well-known digital modulation technologies include amplitude-shift keying (ASK), frequency-shift keying (FSK), and phase-shift keying (PSK), which modulate digital baseband signals to the amplitude, frequency, and phase of carrier waves, respectively, as sketched in Fig. 11.16b. Because the frequency of the digital signals is too low for radiation, they are supposed to be converted to analog signals through a digital-to-analog converter (DAC), which is then modulated to high-frequency carrier waves by a mixer. Before being radiated by an antenna, the high-frequency signals are amplified by a power amplifier. The process is briefly sketched in Fig. 11.16a, where some devices, such as digital up converter (DUC) is omitted. Here, several novel wireless communication systems are introduced based on the programmable metasurface, i.e., directly digital modulation (DDM), binary frequency-shift keying (BFSK) system, and quadrature phase-shift keying (QPSK). Unlike the traditional systems that require the aforementioned devices, these new systems only contain FPGA and programmable metamaterial. Therefore, the hardware cost is dramatically reduced, and the loss or distortion of information could be mitigated.

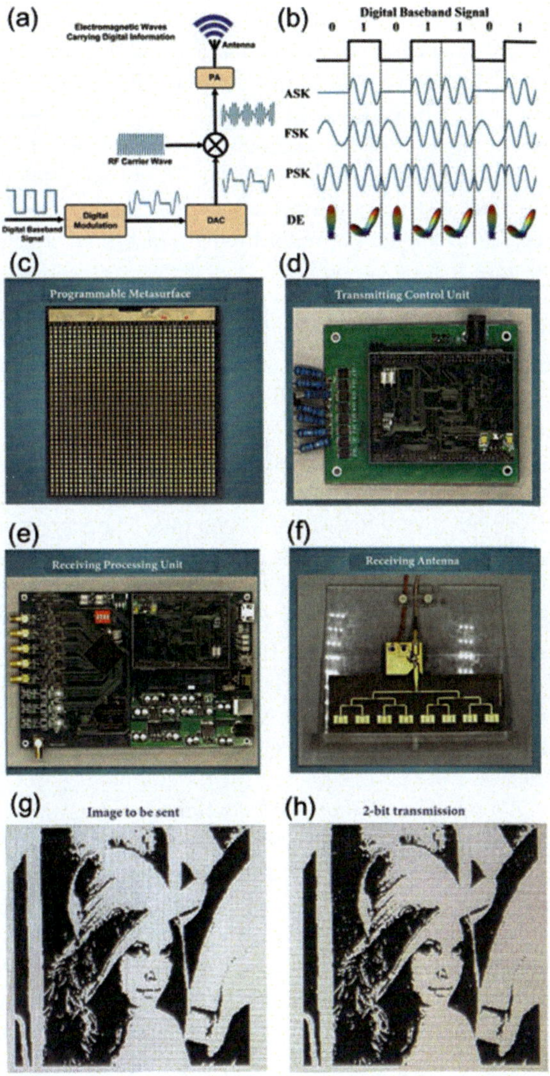

Fig. 11.16 **a** Schematic of the traditional wireless communication system. **b** ASK, FSK, and PSK that are mostly used in traditional wireless systems, and the DDM based on radiation pattern modulation. **c–f** Photographs of the DDM wireless communication system, including the fabricated reprogrammable metasurface, transmitting control unit, receiving processing unit, and receiving antenna. **d** The original image to be transmitted. **b** The received image under the 2-bit-symbol transmission mode [57]

11.3.5.1 Directly Digital Modulation

Different from ASK, FSK, or PSK, Cui et al. proposed to modulate the digital information directly by the far-field radiation pattern generated by the metasurface [57]. As shown in the last row of Fig. 11.16a, the vertical single beam represents the digital code "0", and the dual split beams represent the digital code "1". If receivers in the far-field recognize the two patterns, the transmitted digital signal can be recovered according to the mapping relation between the pattern and digital coding sequences of the metamaterial. In order to obtain the correct information at the receiving terminal, multiple receivers are supposed to be distributed so that the shape of the far-field pattern can be recognized. Signal loss of one or more key receivers would cause the failure of information recovery. For this reason, the DDM architecture is a secret communication system by nature. In contrast, for conventional wireless communication systems, the transmitted information can be recovered by any receivers located in the area covered by the transmitter, so specific technologies are needed for information encryption.

Two concepts should be defined here, i.e., information code and hardware code. The binary code to be sent out is called "information code". The digital sequence used to control the metamaterial for the generation of a particular far-field pattern is called "hardware code". For example, the digital baseband signal 0 or 1 is the information code, and the coding sequences "00000" and "01010" that generate the single and dual beams, respectively, are the hardware codes. Although the coding strategy using the single or dual beams is straightforward, it is a 1-bit information code with a very low transmission rate. The increasing number of far-field patterns with different shapes leads to a larger bit number for information code and thus a higher data transmission rate. On the other end, a denser distribution of receivers offers a better read for the far-field pattern, yet at the cost of higher expenses and more complex algorithms. Besides, the number of transmission bits is also impacted by the background noise. Therefore, to have a higher transmission rate and robustness with limited system complexity, a channel estimation algorithm, and an optimization algorithm are proposed to effectively increase the number of available digital states [57].

A prototype of the DDM communication system is constructed to validate the concept. The system includes the transmitting part and receiving parts. The transmitting part comprises a programmable metasurface and a control unit, and the receiving part has the processing unit and the antenna, as shown in Fig. 11.16c and f. All these devices are controlled by a microcontroller unit (MCU). The programmable metasurface contains 35×35 meta-atoms, each of which is embedded with a PIN diode. By tuning the biasing voltage, the reflection phase range of about 180° is realized at 10.15 GHz, with a magnitude larger than 0.98. The meta-atoms are divided into 7 controllable columns (5 columns of meta-atoms form a controllable column), so total $2^6 = 64$ different scattering patterns can be generated by the metasurface. The scattering EM energy is received by the two receiving antennas, sampled by the processing unit, and converted into digital data, from which the original digital information is recovered by the MCU that is integrated with the channel estimation and optimization algorithms. Through this system, a picture is successfully transmitted and recovered, as shown in Fig. 11.16g and h, with an average transmission speed of 124 Byte/s. It should be noticed that the experiment is conducted only using the 1D programmable metasurface, i.e., the meta-atoms are biased in columns; if a 2D programmable metasurface is utilized with every single meta-atom being controllable, the transmission speed can be further enhanced.

11.3.5.2 Binary Frequency-Shift Keying

From Sect. 11.3.2.2, we know that the central frequency of the incident wave can be efficiently converted to the +1st and −1st order harmonics by a programmable metasurface with opposite coding sequences (00-01-10-11 and 11-10-01-00). These harmonics can be employed as two distinct frequencies for designing a BFSK communication system [37]. The system is sketched in Fig. 11.17a, including the FPGA and the metasurface. First, a bitstream, e.g., 01011101, is generated by the FPGA according to the information ready to be sent. This bitstream is called the information code in DDM. Second, every single bit is mapped to a corresponding frequency in BFSK through the coding sequences applied to the programmable metasurface. The coding sequence is the hardware code in DDM. Finally, the EM wave containing the information is sent out by the metasurface.

To validate the idea, the BFSK transmission experiment is conducted in a chamber, as shown in Fig. 11.17b. A soft-defined radio (SDR) platform USRP-2943R is employed as the BFSK receiver and demodulator. The carrier wave at 3.6 GHz is provided by a signal generator and radiated by a wideband ridged horn. Frequencies of the ± 1st harmonics are $f_c \pm 312.5$ kHz. The signals are received by a dipole antenna, which is connected with USRP-2943R. As an example, a color picture is converted into a bitstream, which is successfully transmitted and recovered by the SDR receiver. Figures 11.17(c) and (d) present the recovered pictures with receiving angles $\alpha = 0°$ and 30°, respectively. The stability of the system is proved by the results shown in Fig. 11.17e and f with the presence of an interference signal at f_c + 550 kHz. Measured results show that a transmission rate of 312.5 kbps is reached by the BFSK system.

11.3.5.3 Quadrature Phase-Shift Keying

In order to enhance the communication rate, a QPSK solution is proposed based on the programmable metasurface [58]. The idea is straightforward because phase manipulation is a direct effect of the coding metasurface on the incident wave. Different from the BFSK, every single bit of the transmitted stream is represented directly by the phase state of the metasurface in the phase modulation rather than requiring periodic coding sequences. For example, a coding sequence "1110010011100100" is needed to present the stream bit "0" in BFSK, as shown in Fig. 11.11g; while it contains 16 information bits in the phase modulation. In other words, the hardware

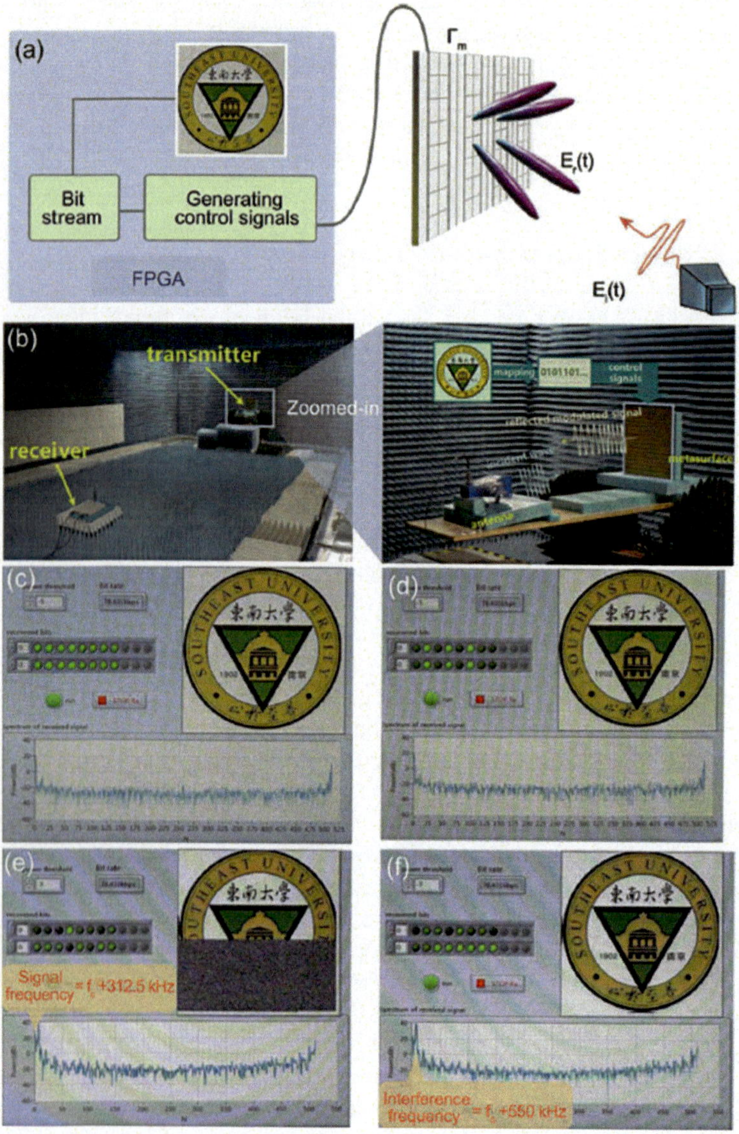

Fig. 11.17 **a** Illustration of the BFSK wireless transmitting terminal based on the programmable metasurface. **b** Photographs of the BFSK wireless communication system. **c** and **d** The received pictures of the BFSK system with receiving angles $\alpha = 0°$ and 30°, respectively. **d** The receiving process of the BFSK system at the presence of the interference signal at $f_c + 550$ kHz, demonstrating the robustness of the system [37]

code and the information code are exactly the same, which leads to a much higher modulation efficiency and thus a higher transmission rate.

A reprogrammable metasurface prototype is fabricated to validate the concept, as shown in the inset of Fig. 11.18a. It consists of 8 × 16 meta-atoms embedded with varactor diodes. By altering the controlling voltage across the varactors, the phase state of the metasurface is adjusted. Measured results in Fig. 11.18a indicate the phase range at 4.0 GHz covering 255° when the voltage changes from 0 to 21 V, with amplitudes larger than 0.56, which is enough for the QPSK system. Considering the loss and requirements, four sets of reflection amplitudes and phases are carefully selected for the QPSK modulation, as shown in Table 11.1. Figure 11.18b presents the constellation diagram with the 2-bit binary digits employed in the QPSK system. The colored scatters in the Figure stand for the constellation points that represent the binary digits 00, 01, 10, 11. With the growth of the voltage, the in-phase and quadrature components along the blue trajectory indicate the real and imaginary parts of the reflection coefficients of the selected points, respectively. That is to say, the reflection magnitude of a point is described by its distance from the origin, and its reflection phase is implied by the angle. It can be seen that the magnitudes are not exactly equal, but the symbols are still useful for information recovery using some mature algorithms.

The QPSK transmission system is constructed as shown in Fig. 11.18c, and the receiving terminal is built based on USRP-2943. To begin with, random binary bit streams at various bit rates are transmitted using the system, which is accurately recovered by the receiving part. In addition, the system concept is also demonstrated by real-time video transmission. Observing the picture of the system shown in Fig. 11.18d, it can be seen that the transmitting and receiving terminals are located at the left and right sides of the table, respectively, with a distance of 2.5 m. A 489p (640 × 480) resolution video is successfully modulated by the proposed metasurface and recovered in real-time by the receiving terminal. The transmission data rate measures 1638.4 kbps, and can be improved further by the optimizations of the waveform of the biasing voltage for the metasurface and phase curves of the metasurface meta-atoms.

11.3.6 Programmable Holography Imaging

As an important imaging technique, holography records the amplitude and phase information of light and thus is helpful for a better reconstruction of the image of objects [59]. Compared with conventional holograms, the ones based on metasurface exhibit higher spatial resolutions, lower noise, higher precision, and higher efficiency [60, 61]. Like other passive metamaterials, however, performances of the conventional holographic metasurfaces are fixed, and very limited images can be generated. The concept of a programmable metasurface provides a reconfigurable solution for microwave holograms [62]. This means that a large number of holographic images can be realized by a single metasurface and dynamically switched in real-time.

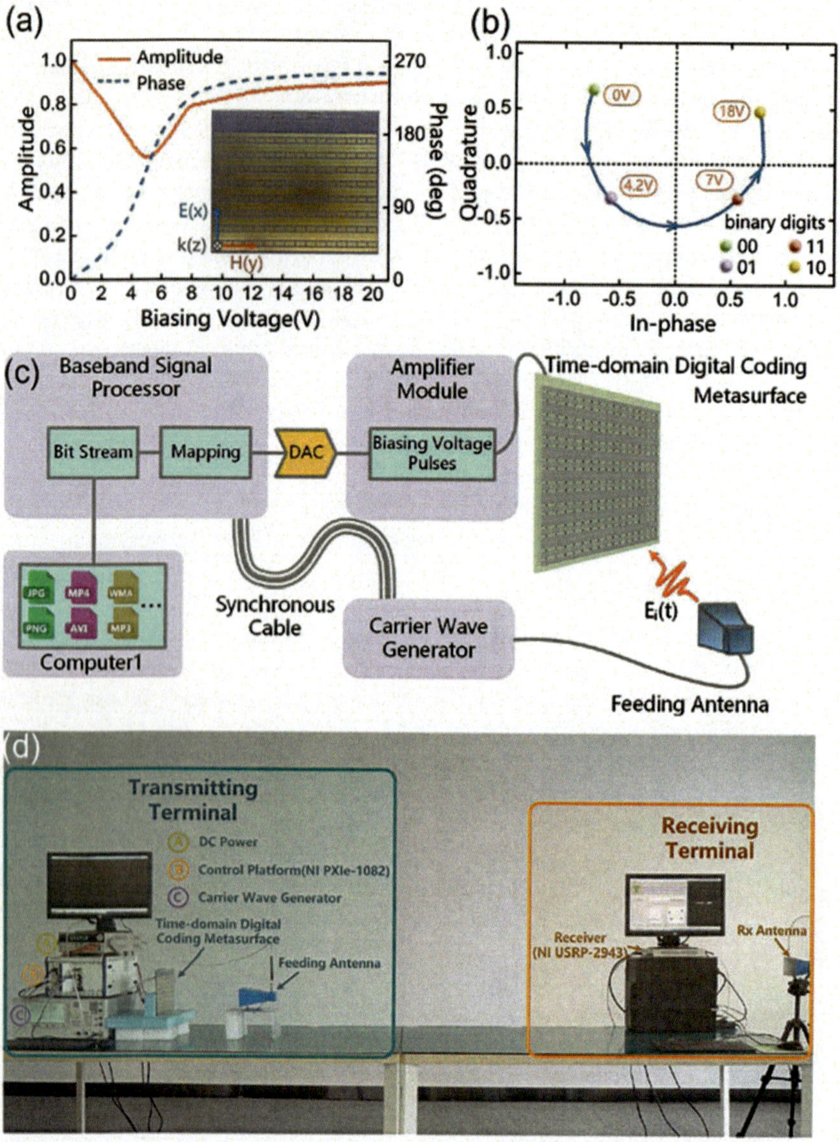

Fig. 11.18 **a** A photograph of the metasurface used in the QPSK and its measured reflection properties. **b** The constellation diagram with the 2-bit binary digits in the QPSK. **c** A plot of the QPSK transmission system based on the programmable metasurface. **d** A photograph of the wireless communication system based on the metasurface [58]

Table 11.1 The mapping relationship among biasing voltages, QPSK symbols, and binary digits

Biasing voltage	0 V	4.2 V	7 V	18 V
QPSK symbol	$1e^j$ (−221.4°)	$0.66e^j$ (−151.2°)	$0.64e^j$ (−28.8°)	$0.89e^j$ (−32.4°)
Binary digits	00	01	11	10

The meta-atom plotted in Fig. 11.19a is employed for the 1-bit programmable hologram metasurface. Under the high and low biasing voltages, the embedded PIN diode works at the "ON" and "OFF" states, respectively. The reflection phase difference of the meta-atom between the two states is almost 180° at 7.8 GHz, so the meta-atom is referred to as the "1" or "0" elements in the two conditions. When the array of PIN diodes is connected to an FPGA, the phase profile on the metasurface is controlled by the digital signal, which accounts for the holographic image provided by the metasurface. To achieve a certain image of an object, the phase profile on the surface is optimized using a modified Gerchberg-Saxton algorithm.

The dynamically programmable holograms are demonstrated experimentally through a metasurface prototype with 20 × 20 meta-atoms. The prototype relates to an FPGA, which is preloaded with the code. Linear-polarized quasi-plane waves illuminate the metasurface, and a standard waveguide probe is utilized to scan the image area in the near-field region with a resolution of 5 × 5 mm^2. Figure 11.19b gives the measured holographic images of three sentences "LOVE PKU! SEU! NUS!", which agree quite well with simulations. The distance for the best results is 400 mm away from the metasurface. Thanks to the digitally-defined reconfigurability, the dynamic hologram can have a relatively wide observation from 400 to 700 mm by adaptively reprogramming the holograms, and the operating frequency bandwidth can be improved further. Figure 11.19c and d present the image of the letter "S" at different observation distances before and after the hologram is adaptively reprogrammed.

11.4 Smart Metamaterials and Metasurfaces

11.4.1 Self-adaptive Smart Metasurfaces

For the aforementioned digital and programmable metamaterials, their functionalities are all based on human instructions and pre-designs. This limitation is broken by the concept of smart metamaterials with self-decision ability [63]. Smart metamaterials can initiatively sense the surrounding information and react to the specific functions according to the pre-designed algorithm, no longer requiring human instructions with the presented architecture. In doing this, a sensing-feedback-reaction loop demands decision-making and acting, which is typically realized by a sensor, a micro-controlling unit (MCU), and an FPGA.

Assuming a scenario with an airborne metamaterial communicating with a satellite, as shown in Fig. 11.20a. The metamaterial can self-adaptively adjust its radiation

Fig. 11.19 **a** Conceptual illustration of the programmable holographic metasurface, including the configuration of the meta-atom, the operating mechanism, and holographic images. **b** Measured holographic images of three sentences "LOVE PKU! SEU! NUS!". **c** and **d** Images of a letter "S" at different observing distances before and after the adaptive reprogramming process, respectively [62]

beam to aim at the satellite when the attitude and flying direction of the airplane change. To accomplish such intelligence, as illustrated in Fig. 11.20b, the gyroscope monitors the moving states of the metasurface and returns the angle data on a 3-axis coordinate to the MCU. Under the control of the MCU, FPGA automatically calculates a coding pattern for desirable focusing directions. The concept is proved by a 2-bit phase-programmable metasurface, whose meta-atom is exhibited in Fig. 11.20c. Two PIN diodes (SMP1320 from SKYWORKS) are integrated on each meta-atom to obtain four distinct phase responses, as plotted in Fig. 11.20c, which are encoded as "00", "01", "10", and "11". Moreover, a fast feedback algorithm is developed to acquire the coding patterns for arbitrary beam-deflection fields according to the data from the sensors. FPGA further executes the bias configuration on the PIN diode array.

Diverse beam functionalities are designed and demonstrated. As illustrated in Fig. 11.20d, automatic single-beam steering to the north-pole direction (i.e., pointing to the satellite) is achieved, no matter how the metasurface rotates along the elevation angle (φ) or along the azimuth angle (θ). Besides, dynamic dual-beam steering is also realized, as depicted in Fig. 11.20e, which can be modulated independently using the fast algorithm. When the metasurface rotates from 0° to 60°, one beam is always directed towards the north pole (beam staring), while the other beam rotates with the metasurface to realize beam scanning, where the included angle between two beams changes from 27° to 87°.

In addition, the presented smart metasurface is promised to extend other sensors on this platform, as depicted in Fig. 11.20f, where the metasurface is supplemented with sensors for light, humidity, height, and heat. Therefore, more dimensions of the sensing functions can be developed as well. Take the light sensor, for example (shown in Fig. 11.20g), when the environment luminance alters, the metasurface switches the scattering pattern from dual-beam to low-scattering on the basis of the pre-defined algorithm.

11.4.2 Intelligent Smart Metasurfaces

The Internet of Things, smart cities and homes are primarily promoted by the development of intelligent devices. The advancement of machine-learning technology also drives the evolution of metamaterial from programmable to intelligent. The first demonstration is proven in the field of imaging and recognizing [64]. As a non-touch approach, radio-frequency technology has become a popular method to locate and track people, recognize gestures, monitor breath, etc., through imaging. However, reconstructing a full-scene image is conventionally a nonlinear EM inverse problem that is inherently time-consuming. Therefore, it is challenging to do it in real-time, and multiple transmitters and/or receivers are needed, making the task expensive and inconvenient. Moreover, the functions of most current devices are set in stone [65–69], which means that they can hardly perform instantly switchable tasks, e.g., adaptively from a full scene of people of interest to his/her subtle body gestures.

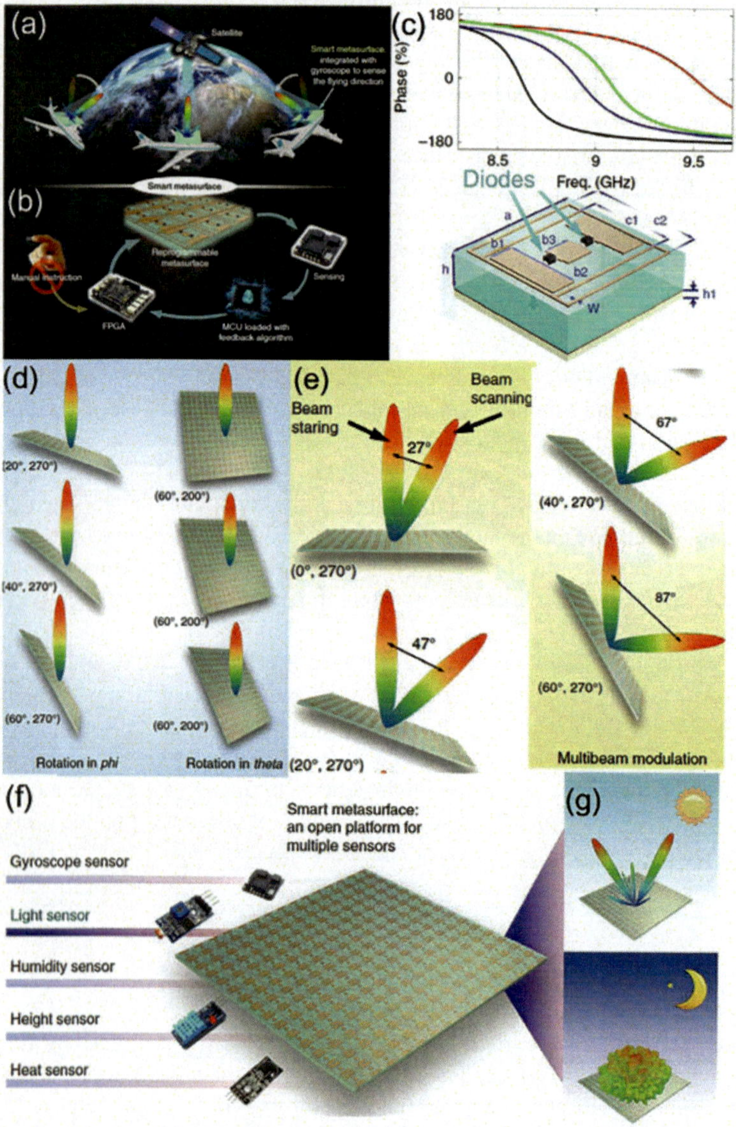

Fig. 11.20 **a** The example situation for self-adaptive smart metasurface. **b** The illustration of a closed-loop system for smart metasurface. **c** The meta-atom used in the 2-bit smart metasurface and its phase responses. **d** The automatic single-beam steering to the north-pole direction with various rotating angles of the metasurface. **e** The dynamic dual-beam steering with one staring to the north pole and the other rotating with the metasurface. **f** Smart metasurfaces integrated with different sensors. **g** The dual-beam and low RCS reactions under different environment luminance [63]

An intelligent metasurface imager and recognizer is proposed by Li et al., which is driven by artificial neural networks (ANN) [64]. Within a meta-system mainly composed of a programmable metasurface, an FPGA, and the ANN platform, only two antennas as transmitter and receiver, respectively (shown in Fig. 11.21b), three different tasks, i.e., high-resolution imaging of a full human body, focusing on an arbitrary local body part, and recognizing body signs and vital signs, are integrated, which can be adaptively and instantly switched smartly and instantly. More interestingly, the proposal's working frequency is around 2.4 GHz; in other words, the commodity Wi-Fi signal can be utilized for the above functions. In fact, the metasurface can work in the active mode or passive mode. As illustrated in Fig. 11.21b, in the active mode, RF signals are sent into the investigated area by Antenna 1 on purpose, bounced back by the subject, and received by Antenna 2. In the passive mode, on the other hand, both antennas are used as receivers to collect the stray Wi-Fi signals in the environment bounced by the subject. The programmable metasurface is a reflective type with 32 × 24 meta-atoms. Every single meta-atom is electrically controlled by a PIN diode (Skyworks SMP1345-079LF). Figure 11.21c gives the simulated and measured reflection properties of the meta-atom when the PIN diode is ON or OFF, from which the phase difference of about 180° is clearly obtained for the programmable task. When controlled by FPGA, the programmable metasurface can dynamically reconfigure its beam patterns as demand.

A cluster of ANNS with three convolutional neural networks (CNNs) is proposed for real-time data processing, and desired images and recognition results are thus mapped with microwave data collected in the environment. Trained with a large number of labeled training samples, the ANNs are able to produce the results within a very short period. The scheme of the three building blocks of the data flow and control flows is shown in Fig. 11.21d. The first CNN, namely IM-CNN-1, processes the microwave data collected by the intelligent metasurface and reconstructs the image of the whole human body. In doing this, the programmable metasurface, controlled by the FPGA, works as a spatial microwave modulator to register the information about the specimen in a compressive-sensing manner. 8×10^4 pairs of labeled training samples are used to train IM-CNN-1, and a high-resolution image of a human body can be instantly reconstructed in less than 0.01 s. A well-known Fast R-CNN [70], is adopted to find the region of interest within the whole image. In order to enable the programmable metasurface to read the desired spots, its radiation wave should be focused on the targets, which is realized by a modified Gerchberg-Saxton (G-S) algorithm. In this way, unwanted interference is eliminated, and the signal-noise ratio of echoes from the interested spots is enhanced by a factor of 20 dB. The second CNN, called IM-CNN-2, is utilized to process microwave data for recognizing hand signs and human breath. 80,000 samples of hand signs are taken to train IM-CNN-2, and the average recognition accuracy of 95% can be achieved. It is found that the hand-sign recognition performance keeps almost stable for multiple test persons even if they are hidden behind a 5-cm-thick wooden wall.

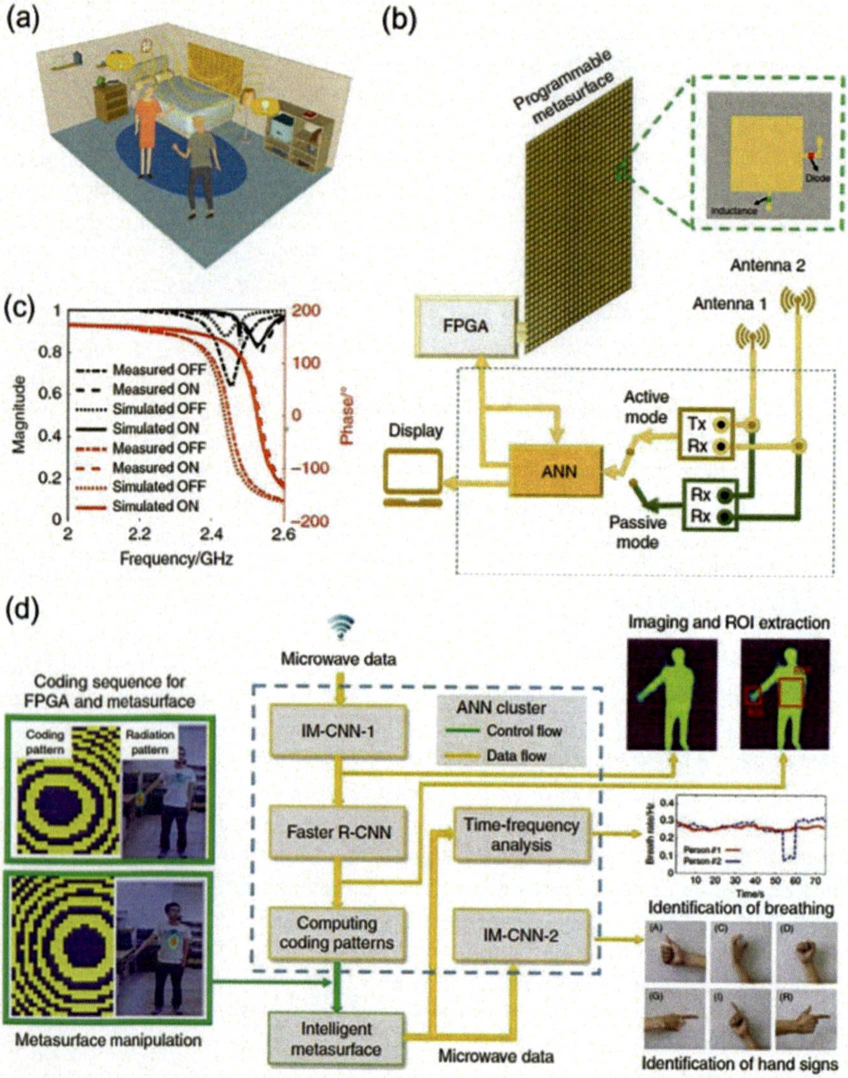

Fig. 11.21 **a** A picture to illustrate the application of the intelligent metasurface, which, decorated as a part of a wall, is utilized to monitor people in an indoor environment by manipulating ambient Wi-Fi signals. **b** The configuration of the intelligent metasurface system and the meta-atom of the programmable metasurface. **c** Simulated and measured reflection magnitudes and phase responses of the meta-atom. **d** Microwave data processing flow by using deep learning CNN cluster, in which the roles of IM-CNN-1, Faster R-CNN, and IM-CNN-2 are described [64]

11.5 Operational Theorems

11.5.1 Convolution Operations on Coding Metasurfaces

It is well known that, through the Fourier transforms, a signal in the time domain can be decomposed into the superposition of frequency-domain signals. In other words, the Fourier transform links the time domain and frequency domain. Under the context of the coding metasurface, the operation links the coding-pattern and scattering-pattern domains [8]. It means that the scattering property of a digital coding metasurface can be predicted by

$$\vec{E}(\theta, \varphi) = jk(\hat{\theta}\cos\varphi - \hat{\varphi}\sin\varphi\cos\theta)P(u, v) \tag{11.51}$$

provided the coding pattern. Here $P(u, v)$ is the 2D Fourier transform of tangential electric field $E(x, y)$ on the coding metasurface

$$P(u, v) = \int_{-\frac{Np}{2}}^{\frac{Np}{2}} \int_{-\frac{Np}{2}}^{\frac{Np}{2}} E(x, y)e^{jk_0(ux+vy)}dxdy \tag{11.52}$$

Comparing the two cases, it is believed that the properties of the Fourier transform can be applied in the design of the coding metasurface, and the convolution operation is proposed aiming to redirect a predesigned scattering beam into a new orientation. In the field of signal processing, if a time-domain signal $f(t)$ is multiplied by a time-shift item $e^{j\omega_0 t}$, the convolution of a frequency-domain signal $f(\omega)$ (corresponding to $f(t)$) and a Dirac-delta function $\delta(\omega - \omega_0)$ can be obtained as

$$f(t) \cdot e^{j\omega_0 t} \Leftrightarrow f(\omega) * \delta(\omega - \omega_0) = f(\omega - \omega_0) \tag{11.53}$$

This implies that, through the convolution in the frequency domain, the frequency components corresponding to the time-domain signal are shifted by ω_0 without distortion. As for the coding metasurface, replacing the t and ω with x_λ and $\sin\theta$, respectively, results in

$$E(x_\lambda) \cdot e^{jx_\lambda \sin\theta_0} \Leftrightarrow E(\sin\theta) * \delta(\sin\theta - \sin\theta_0) = E(\sin\theta - \sin\theta_0) \tag{11.54}$$

where $x_\lambda = x/\lambda$ is the electric length, and θ is the angle with respect to the normal direction. $E(\sin\theta)$ is the scattering pattern of a coding pattern $E(x_\lambda)$. $e^{jx_\lambda \sin\theta_0}$ describes a coding pattern with unity amplitude and gradient digits along a certain direction, whose scattering pattern is a pencil beam with a deflected angle θ_0. Equation (11.54) can be interpreted that the multiplication of a coding pattern $E(x_\lambda)$ by a gradient coding sequence $e^{jx_\lambda \sin\theta_0}$ brings about a deviation of $E(\sin\theta)$ away from its original orientation by $\sin\theta_0$ in the angular coordinate.

In practice, the multiplication of two coding sequences is realized by adding their binary digits at the corresponding position. For example, the multiplication of two 2-bit sequences S_1 (001223300112233…) and S_2 (333222111000333222111000…) results in S_3 (3303000010111112122222323333030000010111…). Another two 2-bit coding metasurfaces and their multiplication are taken to prove the above convolution concept. Their coding patterns are shown in Fig. 11.22a and b. The four different bright levels represent the four digits “0”, “1”, “2”, “3”, from dark to bright, which describe the reflection phases 0°, 90°, 180°, 270°, respectively, of the super cells that comprise the metasurface. The first coding pattern is a cross with coding element “2” at the center surrounded by coding element “0” on the surface. The second coding pattern is a gradient coding sequence “01230123…”. Their scattering patterns are plotted in Fig. 11.22d and e, respectively. Their multiplication results in the coding pattern and the corresponding scattering pattern shown in Fig. 11.22c and f, from which one can easily observe the initial beam of the first cross-coding pattern being tilted from the normal axis to the deflected direction caused by the second gradient pattern. Figure 11.22g through i illustrate the analogical convolution process in the frequency spectra domain. It is clearly seen that the initial spectrum is located around the zero frequency, which is in analogy to the initial scattering pattern (Fig. 11.22d). The ideal Dirac function $\delta(\omega-\omega_0)$ shifts the spectrum to the higher frequency, which mimics the deflecting effect of the tilted pencil beam (Fig. 11.22e) in the scattering pattern.

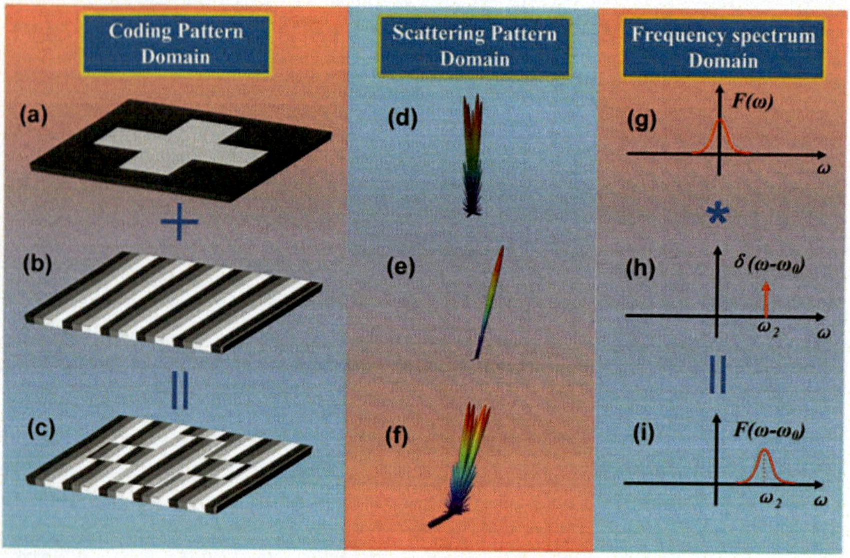

Fig. 11.22 The principle of the scattering-pattern shift in analogy to the Fourier Transform. **a–c** Coding patterns of a cross, gradient distribution, and their modulus, respectively. **d–f** Calculated scattering far-field patterns using FFT of the coding patterns shown in **a–c**, respectively. **g–h** The analogical frequency spectra of the coding patterns in **a–c**, respectively [8]

11.5.2 Addition Theorem on Coding Metasurfaces

One may find that a simple scattering phenomenon, such as the deflected pencil beam, can be easily obtained through a directly designed coding pattern. However, for a complicated scattering pattern like the one shown in Fig. 11.22f, it is challenging to design the coding pattern in demand. A practical alternative to reduce the difficulty is to break the goal down into two or more steps. For example, in the first step, we can have the scattering pattern pointing to the normal direction (Fig. 11.22d), and deviate it then to the predesigned direction by applying the convolution operation of its coding pattern with the gradient coding pattern. As another example, we know that splitting two scattering beams symmetrically on the coordinate planes is easy, but manipulating them individually could be a tough job. To address this issue, a straightforward way is to design the two beams independently and add them together. Inspired by the idea, the addition theorem for the coding metasurface is proposed by Wu et al. [9].

Before the addition theorem, we should introduce the definition of complex digital codes. The conventional coding metamaterial is usually encoded based on the phase responses of the meta-atoms, i.e., 0° and 180° are encoded as 0 and 1 in the 1-bit case, and 0°, 90°, 180°, 270° are encoded as 00, 01, 10, 11 in the 2-bit case. It should be emphasized that the phase here is the absolute phase. In contrast, for the addition theorem, the whole phase term $e^{j\varphi}$ induced by a meta-atom is employed to define the digital states. Therefore, the complex plane is proposed to show the property of the complex digital code. The absolute value of the complex code is always 1, so it locates on a unit circle called the coding circle. Arbitrary-bit digital states can be denoted by unit vectors on the coding circle, and the argument φ is the absolute phase of the code. The 1-bit, 2-bit, and 3-bit coding circles, the complex codes, and their corresponding arguments are plotted in Fig. 11.23a.

Based on the above definitions, the addition operation of two complex digital codes is defined as

$$e^{j\varphi_1} + e^{j\varphi_2} = e^{j\varphi_0} \tag{11.55}$$

Here the magnitudes of the codes are ignored. Equation (11.55) means that the addition of two codes (with arguments φ_1 and φ_2, respectively) leads to a complex code with argument φ_0. The coding circle is of great help in showing the addition operation by the vector superposition principle. The complex codes here are distinguished by the dots above the digits, i.e., the 1-bit complex 0 and 1 are denoted as $\dot{0}_1$ and $\dot{1}_1$, and 2-bit complex codes are denoted as $\dot{0}_2$, $\dot{1}_2$, $\dot{2}_2$, and $\dot{3}_2$ (abbreviations of 00, 01, 10, 11), where the subscripts indicate the order of bit. Two examples are illustrated in Fig. 11.23b, showing that two 2-bit codes $\dot{0}_2 + \dot{1}_2$ results in a 3-bit code $\dot{1}_3$, and $\dot{0}_2 + \dot{3}_2$ goes to $\dot{7}_3$. It is obvious that any higher-bit codes can be realized from lower-bit complex digital codes through addition operations. The physical meanings of the addition operation can be interpreted from microscopic and macroscopic views. On

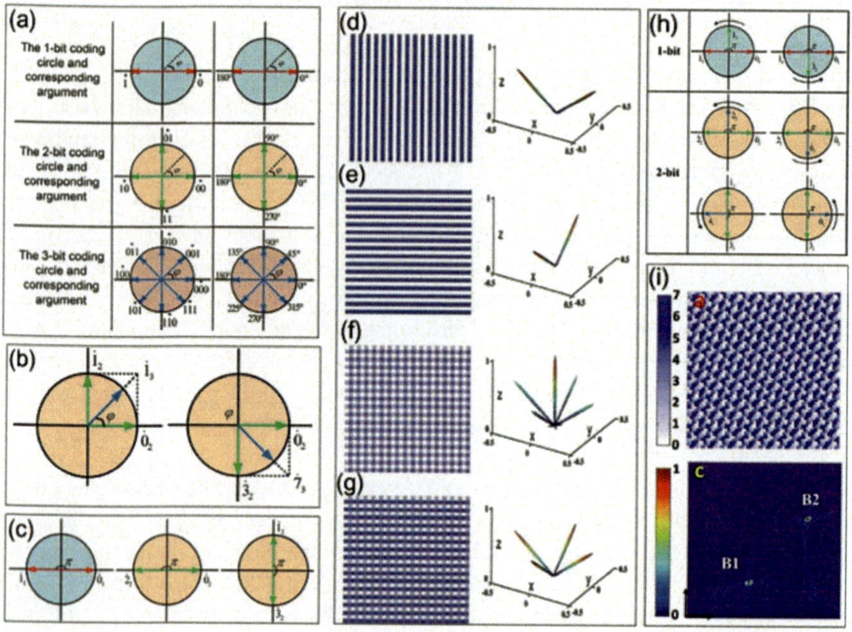

Fig. 11.23 **a** The 1-bit, 2-bit, and 3-bit complex codes and their corresponding arguments in the coding circle. **b** Two typical addition processes of 2-bit complex codes, $\dot{0}_2 + \dot{1}_2 = \dot{1}_3$ and $\dot{0}_2 + \dot{3}_2 = \dot{7}_3$. **c** Three situations of "indefinite coding addition" in the 1-bit and 2-bit addition operations. **d–f** The indefinite coding addition that causes the information loss of the digital codes, leading to the unexpected reflection beam at $\theta = 0°$. **g** The correct result under the regulation by correcting the indefinite coding elements. **h** Solutions for indefinite addition operations in 1-bit and 2-bit complex digital codes. **i** The dual scattering beams, which are difficult to be realized using conventional approaches, can be designed by combining the addition and convolution operations [9]

the meta-atom level, it means the information addition of two complex digital codes; on the metasurface level, it means the superposition of two coding patterns and thus their functions.

We would like to discuss a special case that may happen during the addition operation. If two complex digital codes have reverse directions on the coding circle with a phase difference of 180°, as shown in Fig. 11.23c, the result would be zero with any values of phase based on the parallelogram rule. This situation is called "indefinite coding addition," and the related coding elements are named as "indefinite elements." This indefinite coding addition will cause the information loss of the digital codes, which is illustrated by the example shown in Fig. 11.23d through f. Two complex coding sequences 00110011… along the *x*- and *y*-directions are chosen, which are named P_x and P_y, respectively (Fig. 11.23d and e). The scattering patterns of P_x and P_y are the split two pencil beam on the *xoz*- and *yoz*-planes, respectively, and the addition of the two sequences should result in the superposition of these beams. However, due to the indefinite coding addition $\dot{1}_1 + \dot{0}_1 = \dot{0}_1 + \dot{1}_1 = \dot{1}_2$, an unexpected reflection beam is observed at $\theta = 0°$ (Fig. 11.23f). This is because the case of

$\dot{1}_1 + \dot{0}_1 = \dot{3}_2$ fails in the addition operation. Although the occurrence probability of the indefinite elements decreases as the coding bit increases, this issue should be addressed by setting a regulation since the most commonly used cases are 1-bit, 2-bit, and 3-bit. When two indefinite elements with an angle difference of 180° are added, their sequential order should be considered. Figure 11.23h presents the regulations on 1-bit and 2-bit complex code additions, i.e., $\dot{0}_1 + \dot{1}_1 = \dot{1}_2$, $\dot{1}_1 + \dot{0}_1 = \dot{3}_2$, $\dot{0}_2 + \dot{2}_2 = \dot{2}_3$, $\dot{2}_2 + \dot{0}_2 = \dot{6}_3$, et al. Under this regulation, the expected result shown in Fig. 11.23g is obtained. It should be remarked that the two beams can be manipulated flexibly and independently because they are originated from two different coding sequences.

Furthermore, the combination of the addition and convolution operations can give birth to more advanced manipulations of scattering beams. Take the two scattering beams as examples, one of which is resulted from the convolution operation of the sequences S1 (012301230123...) along the y-direction and S2 (001122330011 2233...) along the x-direction, and the other is the convolution result of the sequences S3 (333222111000333222111000...) along the y-direction and S4 (312031203120...) along the x-direction. The resulting coding metasurface and scattering pattern in Fig. 11.23i can be obtained by adding the sequences after the convolutions. From these digital operations on the digital patterns, the advantages of the coding metamaterials are sufficiently demonstrated in the EM wave manipulations.

11.6 Information Theories

11.6.1 The Entropy of the Information Metamaterials

Since the far-field radiation pattern provided by a coding metamaterial is determined by its coding pattern, it is reasonable to believe that the far-field pattern carries the digital sequences. If someone at a distance senses the shape of the far-field pattern, the digital information can be recovered. Yes, this is the idea of wireless communication, and, as will be seen in Sect. 11.3.5, the metamaterial has been employed as a transmitting device in a new wireless communication system. So, from the information respective, one may wonder how much information a coding metamaterial can carry.

In an information system, messages are generated by transmitters, modulated by channels, and sent to receivers. The information in such a system can be estimated by Shannon entropy [71]. For an information system that employs the coding metamaterial to modulate the channel and enhance the information carrier, a similar concept can be used to quantitatively measure the information carried by the metamaterial with various coding patterns [10], whose idea is illustrated in Fig. 11.24a. In doing this, geometrical entropy and physical entropy are proposed to describe the information of the coding pattern and the far-field pattern of the metamaterial, respectively. The normalized Shannon entropy is used to describe the geometrical information, i.e., coding pattern information, which is defined as

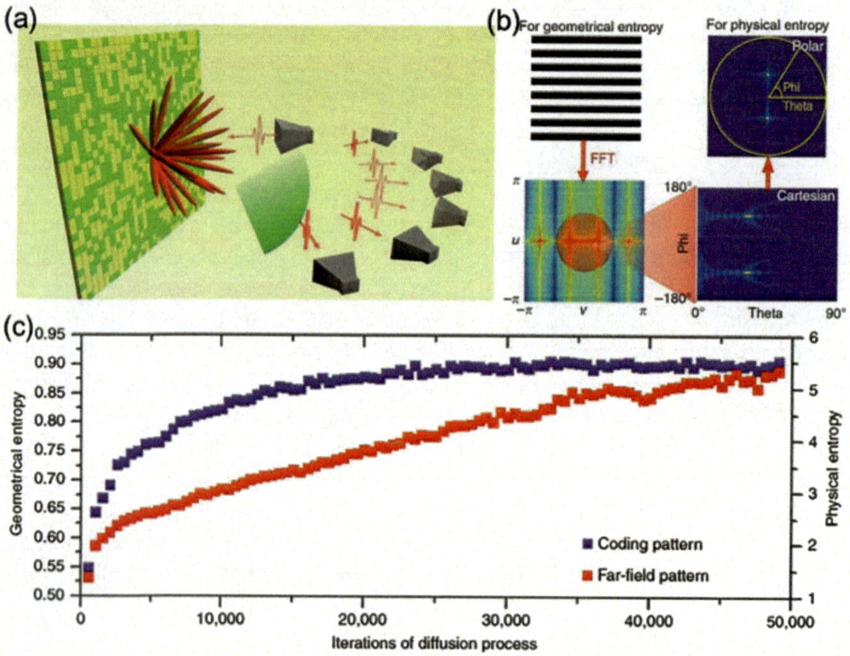

Fig. 11.24 **a** An information system including a reflecting coding metamaterial. **b** The process to obtain the physical entropy from the coding pattern of a metamaterial through FFT and coordinate transformation. **c** The geometrical and physical entropies at all iterations from 1 to 99 (or the step of interchanging operation increases from 0 to 49,500) [10]

$$H_1 = -N^{-1}\sum_x P(x)\log_2 P(x) \tag{11.56}$$

where N is the number of coding units, $P(x)$ is the corresponding probability, and $x \in \{0,\ 1\}^N$. It can be seen that the Shannon entropy defined by Eq. (11.56) implies only the numbers of different coding elements, instead of their spatial arrangement on a surface. Therefore, to estimate the coding pattern with a pair of units, a 2D information entropy is defined as

$$H_2 = -\sum_{i=1}^{2}\sum_{j=1}^{2} P_{ij}\log_2 P_{ij} \tag{11.57}$$

in which i and j indicate the coding unit, P_{ij} is the joint probability of a group G(i, j) that represents two adjacent coding elements. Considering a 1-bit coding metamaterial, there are only four different cases of groups: G (0, 0), G (0, 1), G (1, 0), and G (1, 1). According to Eq. (11.57), the information on the coding pattern is determined by the combination of the four groups distributed on the coding surface. As we know that the far-field pattern can be calculated using the fast Fourier transform

(FFT) of the coding pattern, and the physical entropy, i.e., far-field pattern entropy, can be obtained by following the processes shown in Fig. 11.24b. After the FFT, the far-field pattern of the coding pattern in the 2D polar coordinate system is obtained by a coordinate transformation, and the physical entropy is finally calculated by $H_2 = -\frac{1}{2}\sum_{i=1}^{256}\sum_{j=1}^{256} P_{ij}\log_2 P_{ij}$, where P_{ij} represents the joint probability of a group G (i, j): the gray level i of the current pixel and the gray level j of its adjacent pixel. That is to say, the physical entropy indicates the average amount of information of each pixel in its far-field pattern image. From Fig. 11.24b, it is interestingly found that the physical entropy and geometrical entropy of a coding metamaterial have an approximately monotonic relation while the coding patterns are becoming more diffuse. In other words, the physical entropy increases as its geometrical entropy becomes larger, and they approach close to each other in the end as the iteration of diffusion process continues further, as plotted in Fig. 11.24c.

11.7 Information Theory of Metasurfaces

A general theory of metasurface from the information perspective is proposed to have a deeper understanding of the relation between the information on the metasurface and its radiation pattern in the far-field region [11]. In illustrating this idea, the aperture function $\varphi_{ij}(\boldsymbol{r})$ is adopted to represent the response of a rectangular-shaped meta-atom with the area of $s = a \times b$, which can be expressed as

$$\varphi_{ij}(\boldsymbol{r}) = A_{ij}e^{j\theta_{ij}} \cdot \Pi[\frac{x - a(i-1)}{a}] \cdot \Pi[\frac{y - b(j-1)}{b}] \tag{11.58}$$

where $\Pi(t)$ is the rectangular function, A_{ij} and θ_{ij} are the amplitude and phase responses of the ij^{th} meta-atom, respectively. Thus, the normalized electric-field distribution of the metasurface can be expressed as:

$$\varphi_A(\boldsymbol{r}) = \sum_{i=1}^{N_x}\sum_{j=1}^{N_y}\varphi_{ij}(\boldsymbol{r}) \Big/ (a \cdot b \cdot \sum_{i=1}^{N_x}\sum_{j=1}^{N_y} A_{ij}^2)^{\frac{1}{2}} \tag{11.59}$$

where N_x and N_y are the numbers of meta-atoms along the x and y directions. The denominator term is introduced to normalize the aperture function of the metasurface, such that the square of the normalized function $\varphi_A^2(\mathbf{r}) = \varphi_A^*(\mathbf{r}) \times \varphi_A(\mathbf{r}) = |\varphi_A(\mathbf{r})|^2$ can be considered as a probability density function (PDF) that describes the EM energy distribution on the metasurface plane.

In order to characterize the information of the metasurface and the generated radiation pattern, differential entropy is adopted, such that the entropy of EM energy distribution on the metasurface plane can be expressed as:

$$H(\boldsymbol{r})=H(P_1(\boldsymbol{r}))=H(\varphi_A^2(\boldsymbol{r}))=-\int\int\varphi_A^2(\boldsymbol{r})\ln\varphi_A^2(\boldsymbol{r})dr^2 \tag{11.60}$$

Similar to the definition of observation information in information optics [72], the information of metasurface can be defined as the reduction of entropy from the maximum, which can be derived from Eqs. (11.59) and (11.60) as:

$$I_1=I(\boldsymbol{r})=-\Delta H_1(\boldsymbol{r})=H(\boldsymbol{r})_{\max}-H_1(\boldsymbol{r})=\ln N_xN_y+\sum_{i=1}^{N_x}\sum_{j=1}^{N_y}c_{ij}^2\ln c_{ij}^2 \tag{11.61}$$

where $c_{ij}=\left(A_{ij}^2/\Sigma_{i=1}^{N_x}\Sigma_{j=1}^{N_y}A_{ij}^2\right)^{\frac{1}{2}}$ is the normalized amplitude response of the meta-atom.

Subsequently, the general entropy uncertainty relation between a Fourier-transform pair (such as α and β) is introduced [73], for which the restricted relation can be expressed as:

$$\Delta\boldsymbol{\alpha}+\Delta\boldsymbol{\beta}\geq n(1+\ln\pi) \tag{11.62}$$

where $\Delta\mathbf{T}=-\int P(\tau)\ln P(\tau)d\tau^n$ and the term $P(\tau)$ is the PDF of random variable **T**, and delta **T** is the differential entropy of **T** in n-dimensional space.

The electric far-field $E(k)$ in the wave-vector space (k-space) is the Fourier transform of electric field distribution on the metasurface $\varphi_A(\boldsymbol{r})$. Similarly, the differential entropy of the far-field energy density function in the k-space can be expressed as:

$$H(\boldsymbol{k})=H(P_2(\boldsymbol{k}))=H(f(\boldsymbol{k}))=-\int\int f(\boldsymbol{k})\ln f(\boldsymbol{k})dk^2 \tag{11.63}$$

where $f(k)=\alpha E^2(k)$ is the normalized far-field energy density distribution function, and α is a constant coefficient to normalize the function. Additionally, based on inequality (11.62), the term $H(k)$ would be constrained by the uncertainty relation as:

$$H(\boldsymbol{k})=H(f(\boldsymbol{k}))\geq\ln(\pi\mathrm{e})^2-H(\mathbf{r})=\ln\frac{\pi^2e^2}{ab}+\sum_{i=1}^{N_x}\sum_{j=1}^{N_y}c_{ij}^2\ln c_{ij}^2 \tag{11.64}$$

And the information of radiation pattern (I_2) in the k-space can be defined by the reduction of the wave-vector entropy from the maximum as $I_2=H(\mathbf{k})_{\max}-H(\mathbf{k})$, which satisfies the relation:

$$I_2 = I(\boldsymbol{k}) = H(\boldsymbol{k})_{\max} - H(\boldsymbol{k}) \le \ln\frac{abk^2}{\pi e^2} - \sum_{i=1}^{N_x}\sum_{j=1}^{N_y} c_{ij}^2 \ln c_{ij}^2 \quad (11.65)$$

Thus, the information sum of the metasurface and its radiation pattern can be derived from Eq. (11.61) and inequality (11.65) as:

$$I_1+I_2=I(\mathbf{r}) + I(\boldsymbol{k}) \le \ln(\frac{4\pi \cdot S}{e^2\lambda^2}) \quad (11.66)$$

where $S = N_x \times N_y \times a \times b$ is the size of the metasurface, and λ is the wavelength of the electromagnetic wave. This inequality implies an important fact that the information sum of a metasurface and its radiation pattern has an upper bound, as is schematically shown in Fig. 11.25a.

For verifications, multiple metasurface samples containing 40 × 40 subwavelength particles are analyzed, in which each meta-atom occupies an area of λ/8 × λ/8. The phase distributions of the three metasurface samples are plotted in Fig. 11.25b, e, and h, respectively, and the amplitude distributions are plotted in Fig. 11.25c, f, and i, respectively, in which fifteen cases are considered. The generated far-field patterns with respect to given amplitude and phase modulations are presented in Fig. 11.25d, g, and j, and the calculated results of the I_1 and I_2 are shown in Fig. 11.25k. It can be observed that the calculated results of I_1 and I_2 satisfy the requirement determined by inequality (11.66), showing that the information sum of the above cases is less than the theoretical upper limit of $\ln(4\pi S/e^2\lambda^2)$. In addition, it can be noted from Fig. 11.25k that the information of radiation pattern (I_2) tends to decline as the information of metasurface (I_1) increases, as long as the size of the metasurface is fixed.

11.8 Conclusion

In this chapter, we focus on introducing the basic concept, principle, technology, and typical applications of the information metamaterials and metasurfaces. Looking back on the evolvement of the information metamaterials, the hardware has experienced three main stages. At the very beginning, the proposal of the digital coding metamaterial provides a digital perspective for the scattering manipulation, which also paves a new path to lowering the design complexity of a metamaterial. During this period, various coding metamaterials are invented, such as anisotropic, multi-functional, frequency-dependent, etc. The second stage witnesses the programmability of the metamaterial enabled by the integration of tunable semiconductors and digital controllers, making the real-time control of the status of meta-atoms come true. Benefiting from this, harmonics are generated using a much simpler system,

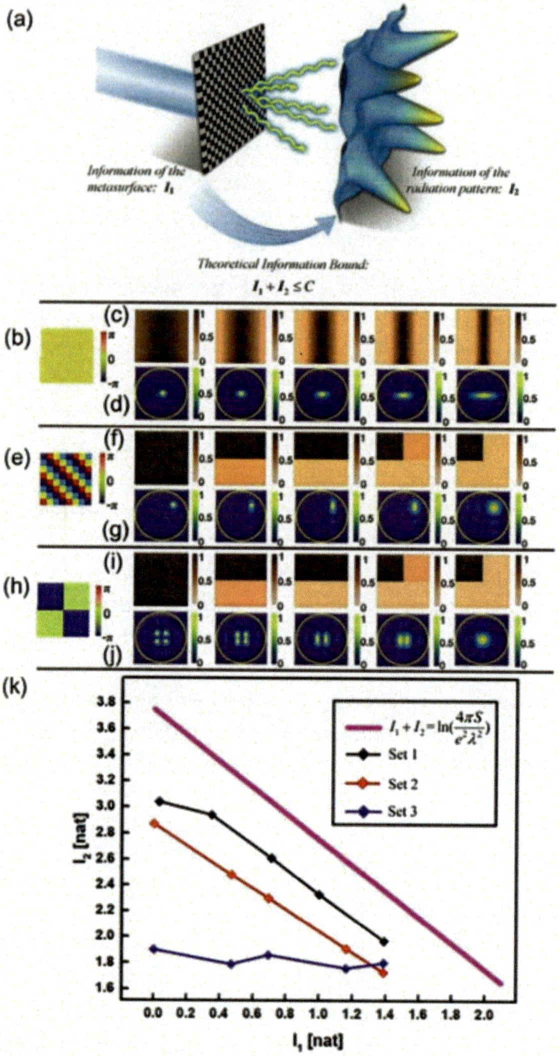

Fig. 11.25 **a** Schematic of information relation between the metasurface and its radiation pattern, showing that their information sum has an upper bound. **b**, **e**, **h** The phase distributions of the three metasurface samples. **c**, **f**, **i** The amplitude distributions of the three metasurface samples. **d**, **g**, **j** The generated far-field patterns of the metasurface samples. **k** The calculated information sum of the three sets of metasurface samples, which are all below the theoretical upper bound [11]

and the modulation in both space and time domains realize independent controls of their amplitude and phase. In addition, the nonreciprocity and nonlinearity are digitally reconfigurable or customizable at microwave frequencies for the first time. In this period, people are also excited to see the proposals of several new wireless communication systems based on programmable metamaterials. The third stage is characterized by the self-adaptively smart metamaterial and intelligent metamaterial. Combined with sensors and feedback algorithms, the smart metamaterial can sense the variables of the environment and make decisions without human intervention. By embedding the deep learning technology, furthermore, a low-cost and real-time intelligent metasurface system for remotely monitoring human movements and recognizing subtle gestures is devised. Apparently, information metamaterial is transforming our conventional acknowledges of metamaterials and EM devices. Judging from the current trend, it would not be surprising to expect many new possibilities if more fancy technologies, e.g., big data, cloud processing, large memory, etc., are adopted. It should be remarked that this concept has already extended to higher frequency spectrums and even the acoustic community, which cannot be covered all here.

Besides hardware development, operational theorems and information theories are also proposed for the information metamaterial. This chapter discusses the convolution operations and the addition theorem for achieving more advanced scattering behaviors. A deeper understanding of the coding metamaterials from the information perspective is provided by the discussions on the information entropy and the information processing capabilities, laying a solid foundation for further research on the field of information metamaterials and its appealing applications in the near future.

References

1. T.J. Cui, Microwave metamaterials. Natl. Sci. Rev. **5**(2), 134–136 (2018)
2. T.J. Cui, L. Li, S. Liu, Q. Ma, L. Zhang, X. Wan, W.X. Jiang, Q. Cheng, Information metamaterial systems. iScience. **23**, 101403 (2020)
3. T.J. Cui, S. Liu, L. Zhang, Information metamaterials and metasurfaces. J. Mater. Chem. C **5**, 3644–3668 (2017)
4. R.A. Shelby, D.R. Smith, S. Schultz, Experimental verification of a negative index of refraction. Science **292**, 77–79 (2001)
5. T.J. Cui , M.Q. Qi , X. Wan , J. Zhao, Q. Cheng, Coding metamaterials, digital metamaterials and programmable metamaterials. Light Sci. Appl. **3**, e218 (2014)
6. C.D. Giovampaola, N. Engheta, Digital metamaterials. Nat. Mater. **13**, 1115–1121 (2014)
7. Z. Szabó, G.-H. Park, R. Hedge, E.-P. Li, A unique extraction of metamaterial parameters based on Kramers-Kronig relationship. IEEE Trans. Microw. Theory Tech. **58**(10), 2646–2653 (2010)
8. S. Liu, T.J. Cui, L. Zhang, Q. Xu, Q. Wang, X. Wan, J.Q. Gu, W.X. Tang, M.Q. Qi, J.G. Han, W.L. Zhang, X.Y. Zhou, Q. Cheng, Convolution operations on coding metasurface to reach flexible and continuous controls of terahertz beams. Adv. Sci. **3**(10), 1600156 (2016)
9. R.Y. Wu, C.B. Shi, S. Liu, W. Wu, T.J. Cui, Addition theorem for digital coding metamaterials. Adv. Opt. Mater. **6**(5), 1701236 (2018)
10. T.J. Cui, S. Liu, L. Li, Information entropy of coding metasurface. Light Sci. Appl. **5**, e16172 (2016)

11. H. Wu, G.D. Bai, S. Liu, L. Li, X. Wan, Q. Cheng, T.J. Cui, Information theory of metasurfaces. Natl. Sci. Rev. **7**, 561–571 (2020)
12. S. Liu, T.J. Cui, Q. Xu, D. Bao, L. Du, X. Wan, W.X. Tang, C. Ouyang, X.Y. Zhou, H. Yuan, H.F. Ma, W.X. Jiang, J. Han, W. Zhang, Q. Cheng, Anisotropic coding metamaterials and their powerful manipulation of differently polarized terahertz waves. Light Sci. Appl. **5**, e16076 (2016)
13. J.B. Pendry, D. Schurig, D.R. Smith, Controlling electromagnetic fields. Science **312**, 1780–1782 (2006)
14. U. Leonhardt, Optical conformal mapping. Science **312**, 1777–1780 (2006)
15. N.I. Landy, S. Sajuyigbe, J.J. Mock, D.R. Smith, W.J. Padilla, Perfect metamaterial absorber. Phys. Rev. Lett. **100**, 207402 (2008)
16. H.T. Chen, J. Zhou, J.F. O'Hara, F. Chen, A.K. Azad, A.J. Taylor, Antireflection coating using metamaterials and identification of its mechanism. Phys. Rev. Lett. **105**, 073901 (2010)
17. M. Moccia, S. Liu, R.Y. Wu, G. Castaldi, A. Andreone, T.J. Cui, V. Galdi, Coding metasurfaces for diffuse scattering: scaling laws, bounds, and suboptimal design. Adv. Opt. Mater. **5**, 1700455 (2017)
18. N.F. Yu, P. Genevet, M.A. Kats, F. Aieta, J.P. Tetienne, F. Capasso, Z. Gaburro, Light propagation with phase discontinuities: generalized laws of reflection and refraction. Science **334**, 333–337 (2011)
19. Q. Ma, C.B. Shi, G.D. Bai, T.Y. Chen, A. Noor, T.J. Cui, Beam-editing coding metasurfaces based on polarization bit and orbital-angular-momentum-mode bit. Adv. Opt. Mater. **5**(23), 1700548 (2017)
20. K. Chen, G. Ding, G. Hu, Z. Jin, C. Qiu, Directional Janus metasurface. Adv. Mater. **32**(2), 201906352 (2020)
21. H.F. Ma, G.Z. Wang, W.X. Jiang, T.J. Cui, Independent control of differently-polarized waves using anisotropic gradient-index metamaterials. Sci. Rep. **4**, 6337 (2014)
22. Z. Luo, X. Chen, J. Long, R. Quarfoth, D. Sievenpiper, Nonlinear power-dependent impedance surface. IEEE Trans. Antennas Propag. **63**(4), 1736–1745 (2015)
23. A. Li, S. Singh, D. Sievenpiper, Metasurfaces and their applications. Nanophotonics **7**(6), 989–1011 (2018)
24. H. Wakatsuchi, J. Long, D.F. Sievenpiper, Waveform selective surfaces. Adv. Funct. Mater. **29**, 1806386 (2019)
25. S. Liu, L. Zhang, Q.L. Yang, Q. Xu, Y. Yang, A. Noor, Q. Zhang, S. Iqbal, X. Wan, Z. Tian, W.X. Tang, Q. Cheng, J.G. Han, W.L. Zhang, T.J. Cui, Frequency-dependent dual-functional coding metasurfaces at terahertz frequencies. Adv. Opt. Mater. **4**(12), 1965–1973 (2016)
26. H. Wu, S. Liu, X. Wan, L. Zhang, D. Wang, L. Li, T.J. Cui, Controlling energy radiations of electromagnetic waves via frequency coding metamaterials. Adv. Sci. **4**, 1700098 (2017)
27. X.G. Zhang, W.X. Tang, W.X. Jiang, G.D. Bai, J. Tang, L. Bai, C.-W. Qiu, T.J. Cui, Light-controllable digital coding metasurfaces. Adv. Sci. **5**, 1801028 (2018)
28. Q. Ma, G.D. Bai, H.B. Jing, C. Yang, L. Li, T.J. Cui, Smart metasurface with self-adaptively reprogrammable functions. Light Sci. Appl. **8**(98), (2019)
29. Z. Luo, M.Z. Chen, Z.X. Wang, L. Zhou, Y.B. Li, Q. Cheng, H.F. Ma, T.J. Cui, Digital nonlinear metasurface with customizable nonreciprocity. Adv. Funct. Mater. **29**, 1906635 (2019)
30. L. Zhang, R.Y. Wu, G.D. Bai, H.T. Wu, Q. Ma, X.Q. Chen, T.J. Cui, Transmission-reflection-integrated multifunctional coding metasurface for full-space controls of electromagnetic waves. Adv. Funct. Mater. **28**, 1802205 (2018)
31. H.T. Wu, D. Wang, X. Fu, S. Liu, L. Zhang, G.D. Bai, C. Zhang, W.X. Jiang, H.L. Jiang, R.Y. Wu, J.W. Wu, T.J. Cui, Space-frequency-domain gradient metamaterials. Adv. Opt. Mater. **6**(23), 1801086 (2018)
32. Q. Wang, X.G. Zhang, H.W. Tian, W.X. Jiang, D. Bao, H.L. Jiang, Z.J. Luo, L.T. Wu, T.J. Cui, Millimeter-wave digital coding metasurfaces based on nematic liquid crystals. Adv. Theory Simul. **2**(12), 1900141 (2019)
33. X. Zhao, G. Duan, A. Li, C. Chen, X. Zhang, Integrating microsystems with metamaterials towards metadevices. Microsyst. Nanoeng. **5**, 5 (2019)

34. Y.-G. Jeong, S. Han, J. Rhie, J.S. Kyoung, J.W. Choi, N. Park, S. Hong, B.J. Kim, H.T. Kim, D.S. Kim, A vanadium dioxide metamaterial disengaged from insulator-to-metal transition. Nano Lett. **15**(10), 6318–6323 (2015)
35. P. Pitchappa, C.P. Ho, L. Cong, R. Singh, N. Singh, C. Lee, Reconfigurable digital metamaterial for dynamic switching of terahertz anisotropy. Adv. Opt. Mater. **4**, 391–398 (2016)
36. J.Y. Dai, J. Zhao, Q. Cheng, T.J. Cui, Independent control of harmonic amplitudes and phases via a time-domain digital coding metasurface. Light Sci. Appl. **7**, 90 (2018)
37. J. Zhao, X. Yang, J.Y. Dai, Q. Cheng, X. Li, N.H. Qi, J.C. Ke, G.D. Bai, S. Liu, S. Jin, A. Alu, T.J. Cui, Programmable time-domain digital-coding metasurface for non-linear harmonic manipulation and new wireless communication systems. Natl. Sci. Rev. **6**, 231–238 (2019)
38. J.Y. Dai, L.X. Yang, J.C. Ke, M.Z. Chen, W.K. Tang, X. Li, M. Chen, Z.H. Wu, Q. Cheng, S. Jin, T.J. Cui, High-efficiency synthesizer for spatial waves based on space-time-coding digital metasurface. Laser Photon. Rev. **14**(6), 2070033 (2020)
39. L. Zhang, X.Q. Chen, S. Liu, Q. Zhang, J. Zhao, J.Y. Dai, G.D. Bai, X. Wan, Q. Cheng, G. Castaldi, V. Galdi, T.J. Cui, Space-time-coding digital metasurfaces. Nat. Commun. **9**, 4334 (2018)
40. C. Caloz, A. Alù, S. Tretyakov, D. Sounas, K. Achouri, Z.-L. Deck-Léger, Electromagnetic nonreciprocity. Phys. Rev. Appl. **10**, 047001 (2018)
41. N.A. Estep, D.L. Sounas, J. Soric, A. Alù, Magnetic-free non-reciprocity and isolation based on parametrically modulated coupled-resonator loops. Nat. Phys. **10**, 923–927 (2014)
42. Y. Hadad, J.C. Soric, A. Alù, Breaking temporal symmetries for emission and absorption. Proc. Nat. Acad. Sci. USA **113**, 3471–3475 (2016)
43. L. Zhang, X.Q. Chen, R.W. Shao, J.Y. Dai, Q. Cheng, G. Castaldi, V. Galdi, T.J. Cui, Breaking reciprocity with space-time-coding digital metasurfaces. Adv. Mater. **31**(41), 1904069 (2019)
44. A. Shaltout, A. Kildishev, V. Shalaev, Time-varying metasurfaces and Lorentz non-reciprocity. Opt. Mater. Express **5**, 2459–2467 (2015)
45. J.B. Pendry, A.J. Holden, D.J. Robbins, W.J. Stewart, Magnetism from conductors and enhanced nonlinear phenomena. IEEE Trans. Microw. Theory Tech. **47**(11), 2075–2084 (1999)
46. B.N. Wang, J.F. Zhou, T. Koschny, C.M. Soukoulis, Nonlinear properties of split-ring resonators. Opt. Express **16**(20), 16058–16063 (2008)
47. A.P. Slobozhanyuk, M. Lapine, D.A. Powell, I.V. Shadrivov, Y.S. Kivshar, R.C. McPhedran, P.A. Belov, Flexible helices for nonlinear metamaterials. Adv. Mater. **25**(25), 3409–3412 (2013)
48. D. Sievenpiper, Nonlinear grounded metasurfaces for suppression of high-power pulsed RF currents. IEEE Antenna Wirel. Propag. Lett. **10**, 1516–1519 (2011)
49. Z. Luo, J. Long, X. Chen, D. Sievenpiper, Electrically tunable metasurface absorber based on dissipating behavior of embedded varactors. Appl. Phys. Lett. **109**, 071107 (2016)
50. A. Li, Z. Luo, H. Wakatsuchi, S. Kim, D. Sievenpiper, Nonlinear, active, and tunable metasurfaces for advanced electromagnetics applications. IEEE Access **5**, 27439–27452 (2017)
51. D.E. Fernandes, M.G. Silveirinha, Asymmetric transmission and isolation in nonlinear devices: why they are different. IEEE Antenna Wirel. Propag. Lett. **17**, 1953–1957 (2018)
52. Y. Shi, Z. Yu, S. Fan, Limitations of nonlinear optical isolators due to dynamic reciprocity. Nat. Photon. **9**, 388–392 (2015)
53. A. Rose, D.A. Powell, I.V. Shadrivov, D.R. Smith, Y.S. Kivshar, Circular dichroism of four-wave mixing in nonlinear metamaterials. Phys. Rev. B **88**, 195148 (2013)
54. D. Filonov, Y. Kramer, V. Kozlov, B.A. Malomed, P. Ginzburg, Resonant meta-atoms with nonlinearities on demand. Appl. Phys. Lett. **109**, 111904 (2016)
55. Z. Luo, X. Chen, J. Long, R. Quarfoth, D. Sievenpiper, Self-focusing of electromagnetic surface waves on a nonlinear impedance surface. Appl. Phys. Lett. **106**, 211106 (2015)
56. Z. Luo, Q. Wang, X.G. Zhang, J.W. Wu, J.Y. Dai, L. Zhang, H.T. Wu, H.C. Zhang, H.F. Ma, Q. Cheng, T.J. Cui, Intensity-dependent metasurface with digitally reconfigurable distribution of nonlinearity. Adv. Opt. Mater. **7**(19), 1900792 (2019)
57. T.J. Cui, S. Liu, G.D. Bai, Q. Ma, Direct transmission of digital message via programmable coding metasurface. Research, 2584509 (2019)

58. J.Y. Dai, W.K. Tang, J. Zhao et al., Wireless communications through a simplified architecture based on time-domain digital coding metasurface. Adv. Mater. Technol. **4**(7), 1900044 (2019)
59. D. Gabor, A new microscopic principle. Nature **161**, 777–778 (1948)
60. X. Ni, A.V. Kildishev, V.M. Shalaev, Metasurface holograms for visible light. Nat. Commun. **4**, 3807 (2013)
61. G. Zheng, H. Mühlenbernd, M. Kenney, G. Li, T. Zentgraf, S. Zhang, Metasurface holograms reaching 80% efficiency. Nat. Nanotechnol. **10**, 308–312 (2015)
62. L. L. Li, T. J. Cui, W. Ji, S. Liu, J. Ding, X. Wan, Y. B. Li, M. Jiang, C.-W. Qiu, S. Zhang, Reprogrammable coding metasurface holograms. Nat. Commun. **8**(197), (2017)
63. Q. Ma, G.D. Bai, H.B. Jing, C. Yang, L. Li, T.J. Cui, Smart metasurface with self-adaptively reprogrammable functions. Light Sci. Appl. **8**, 98 (2019)
64. L. Li, Y. Shuang, Q. Ma, H. Li, H. Zhao, M. Wei, C. Liu, C. Hao, C.-W. Qiu, T. J. Cui, Intelligent metasurface imager and recognizer. Light Sci. Appl. **8**, 97 (2019)
65. J. Xiao, Z. Zhou, Y. Yi, L.M. Ni, A survey on wireless indoor localization from the device perspective. ACM Comput. Surv. **49**, 25 (2016)
66. P. del Hougne, M.F. Imani, M. Fink, D.R. Smith, G. Lerosey, Precise localization of multiple noncooperative objects in a disordered cavity by wave front shaping. Phys. Rev. Lett. **121**, 063901 (2018)
67. M. Mercuri, L.R. Lorato, Y.-H. Liu, F. Wieringa, C.V. Hoof, T. Torfs, Vital-sign monitoring and spatial tracking of multiple people using a contactless radar-based sensor. Nat. Electron. **2**, 252–262 (2019)
68. X. Dai, Z. Zhou, J.J. Zhang, B. Davidson, "Ultra-wideband radar-based accurate motion measuring: human body landmark detection and tracking with biomechanical constraints. IET Radar Sonar Navig. **9**, 154–163 (2015)
69. P.M. Holl, F. Reinhard, Holography of Wi-Fi radiation. Phys. Rev. Lett. **118**, 18390 (2017)
70. S.Q. Ren, K. He, R. Girshick, J. Sun, Faster R-CNN: towards real-time object detection with region proposal networks. IEEE Trans. Pattern Anal. Mach. Intell. **39**, 1137–1149 (2017)
71. C.E. Shannon, A mathematical theory of communication. ACM SIGMOBILE Mobile Comput. Commun. Rev. **5**(1), 3–55 (2001)
72. F.T.S. Yu, *Entropy and information optics*, 2nd edn. (CRC Press, Boca Raton, 2017)
73. I. Białynicki-Birula, J. Mycielski, Uncertainty relations for information entropy in wave mechanics. Commun. Math. Phys. **44**, 129–132 (1975)

Chapter 12
Summary

Long Li, Yan Shi, and Tie Jun Cui

Abstract Electromagnetic metamaterials/metasurfaces have witnessed a spectacular development. The unthinkable manipulation ability of electromagnetic waves makes metamaterials/metasurfaces a promising solution in practical engineering applications. This chapter summarizes the content of this book about the electromagnetic metamaterials and metasurfaces from theory to applications.

Keywords Electromagnetic metamaterials · Electromagnetic metasurfaces · From theory to applications

12.1 Conclusion

The past several decades have witnessed a rapid growth of the metamaterials and metasurfaces from early metamaterial designs, such as left-hand material and photonic bandgap structure, to recent groundbreaking works in various fields, including cloak, antenna, electromagnetic interference, wireless power transfer, etc. As of today, metamaterials and metasurfaces have become novel paradigm in science and technology. There are countless examples of metamaterials/metasurface designs and applications ranging from radio frequencies to visible wavelengths that have been reported in the literatures. This book only covers a small portion of the cutting-edge contributions of the metamaterial/metasurfaces in theoretical, numerical, and experimental aspects.

In the fundamental part of the book, the theoretical model for metamaterials and metasurfaces including electrodynamics of the double negative metamaterial, generalized snell's law, digital coding metamaterials/ metasurfaces are reviewed. With the physical insights, analysis and design methods of the metamaterials and

L. Li (✉) · Y. Shi
School of Electronic Engineering, Xidian University, Xi'an 710071, Shaanxi, China
e-mail: lilong@mail.xidian.edu.cn

T. J. Cui
State Key Laboratory of Millimeter Waves, Southeast University, Nanjing 210096, China

L. Li et al. (eds.), *Electromagnetic Metamaterials and Metasurfaces: From Theory To Applications*, https://doi.org/10.1007/978-981-99-7914-1_12

metasurfaces are elaborated. The physics-based local resonant cavity cell model for electromagnetic bandgap structure, effective medium theory for metamaterial structure, and equivalent circuit-based model for the metasurface structures have been, respectively, developed. Moreover, a full-wave method based on the multilevel Green's function interpolation method has been given to highly efficiently simulate the periodic structures.

In the application part of the book, the metamaterials and metasurfaces are designed in several important research fields. The electromagnetic bandgap structures have surface-wave suppression and in-phase reflectivity characteristic. With the surface-wave suppression, the ultra-wideband ground bounce noise is suppressed by integrating the well-designed electromagnetic bandgap structures with the high-speed circuits. When applying the electromagnetic bandgap structures into the phased array, the coupling between the different elements can be reduced to overcome the scanning "blindness". In addition, the electromagnetic bandgap structures as the superstrate are used to improve the performance of the antenna, for example, dual-band, beam scanning, and circular polarization, etc.

Metamaterial/metasurface-based absorbers are one of the important applications. Different from the conventional absorbers composed of absorbing materials, the metamaterial/metasurface-based absorbers with very low profiles can achieve nearly perfect absorption performance. Graphene is a good candidate to design the absorbers due to its unique electrical, optical, and mechanical properties, including finite conductivity, flexible structure, and optical transparency. Several graphene-based absorbers including transparent shielding enclosure, quasi-TEM wave microstrip absorber, microwave frequency selective surface absorber, millimeter-wave wide-band absorber, and switchable THz absorber, are given to demonstrate the huge potential of the metamaterial/metasurface in the absorber designs.

Flexible manipulation of electromagnetic waves, including the amplitude, phase, polarization, frequency, orbital angular momentum (OAM), etc., is one of the unique applications of metasurfaces. When some tunable components, for example, PIN diode, varactor, field programmable gata array (FPGA) are applied to the metasurface, reconfigurable functions and even real-time manipulations become possible. Several examples including beam scanning by the metasurface-based reflectarray, reconfigurable metasurfaces imaging design, dynamical OAM vortex wave generators, reflection-transmission reconfigurable metasurface, and frequency-spatial-domain reconfigurable metasurfaces are given to show the advantages of the metasurfaces in the wave manipulation, thus pushing its rapid development in various fields including wireless communication, imaging, etc.

The demands for electromagnetic invisibility and radar cross section reduction have rapidly increased in the radar and communication system, and thus metamaterial/metasurface-based invisible design has been the subject of intensive investigations. The coordinate transformation method as an intuitive approach is discussed to systematically design the cloaks for perfect invisibility, including 2D/3D cloak, illusion cloak, complementary cloak, etc. In addition to the coordinate transformation, the scattering cancellation technique based on Mie series and characteristic mode theory can be used to design canonical cloaks and arbitrarily shaped cloaks, respectively. However, the theoretically perfect cloaks are composed of complicated materials including high anisotropy and inhomogeneity. By comparison, the metasurface-based designs for the radar cross section reduction are easier to fabricate. Several metasurface designs with the low radar cross section are given, and moreover, with the integration of the metasurfaces with the antenna, the antenna designs with low radar cross sections are demonstrated.

The 5G/6G communication technologies enable the progress of the internet of things, and further promote the development of wireless power transfer (WPT) and wireless energy harvesting (WEH). Metasurface-based designs for wireless power transfer and wireless energy harvesting have caused great interests. The metasurface-based wireless power transfer covers various application scenarios from magnetic coupling resonance for short-distance applications to near-field focusing for medium-distance applications, and (microwave or laser) radiation for long-distance (greater than 1 m) applications. Several cases including magnetic negative (MNG) metasurfaces, double negative metamaterials, and negative refractive index (NRI) metasurfaces are designed to improve the performance of the magnetic coupling resonance based wireless power transfer. For the near-field focusing scenario, metasurface designs for single-feed and single-focus, single-feed and dual-focus, dual-feed and single-focus, and dual-feed and dual-focus are, respectively, elaborated. Furthermore, a new MIMO-WPT system architecture is proposed based on electromagnetic metasurfaces. On the other hand, in wireless energy harvesting, with the state-of-the arts of the energy harvesting reviewed, two metasurface designs are given to demonstrate wide-angle and polarization-insensitive energy harvesting performance. More importantly, a novel method for integrating rectifying diodes with the metasurface has been developed to simplify the structure of the ambient energy harvester.

With the progress of artificial intelligence, the intelligence metamaterials have become an inevitable development trend. The digital coding metamaterial provides a feasible way for adjustment of the metamaterial operation states in real-time. The introduction of information in the digital world into the coding metamaterial brings the concept of information metamaterials. Far beyond the manipulation of EM waves, the information metamaterials and metasurfaces operate with digital signals and electromagnetic fields simultaneously for information processing, transmission, and even recognition. Combining with the artificial intelligence, the information metamaterials can evolve towards the intelligence metamaterials. Some recent advances from digital metamaterials to information metamaterials to intelligence metamaterials have been demonstrated. Accordingly, some important concepts including information theories, operational theorems, and transformation of systematic architectures of information metamaterials have been proposed.

In summary, electromagnetic metamaterials/metasurfaces have opened up the doors to an exciting frontier in the creation of novel fundamental theories and design methods for engineering applications. This book comprehensively introduces the state-of-the-art of metamaterials/metasurfaces from theory to applications. The theoretical side involves electrodynamics of left-handed medium, generalized

Snell's Law, digital coding metamaterials/metasurfaces, group theory of metamaterials, information metamaterials and metasurfaces, etc. On the application side, a broad range of design examples have been discussed, including metamaterial antennas, electromagnetic interference, frequency selective surfaces, wireless power transfer and energy harvesting, cloak and radar cross section reduction, orbital angular momentum, wireless communication, imaging, etc. The book can provide researchers, engineers, and graduate students a great number of new discoveries, results, information, and knowledge in the field of metamaterials and metasurfaces.